W9-CWR-972

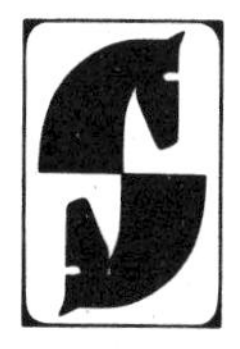

H. S. M. Coxeter

$48^{00}$

# The Geometric Vein

## The Coxeter Festschrift

Edited by
Chandler Davis   Branko Grünbaum
F. A. Sherk

With Contributions by
Patrice Assouad   C. M. Campbell   Jeffrey Cohen
H. S. M. Coxeter   Donald W. Crowe   Patrick Du Val
W. L. Edge   Erich W. Ellers   G. Ewald   L. Fejes Tóth
J. C. Fisher   David Ford   Cyril W. L. Garner
William J. Gilbert   J. M. Goethals   P. R. Goodey
Branko Grünbaum   N. I. Haritonova   Howard L. Hiller
S. G. Hoggar   Norman W. Johnson   William M. Kantor
I. N. Kashirina   Ignace I. Kolodner   Joseph Malkevitch
John McKay   P. McMullen   J. C. P. Miller   W. O. J. Moser
Stanley E. Payne   Jean J. Pedersen   J. F. Rigby
E. F. Robertson   C. A. Rogers   B. A. Rosenfeld   D. Ruoff
I. J. Schoenberg   J. J. Seidel   G. C. Shephard   J. Shilleto
J. Tits   W. T. Tutte   Harold N. Ward   Asia Weiss
J. B. Wilker   J. M. Wills   M. M. Woodcock   I. M. Yaglom

With 5 Color Plates,
6 Halftones, and 211 Line Illustrations

Springer-Verlag
New York   Heidelberg   Berlin

Chandler Davis
Department of Mathematics
University of Toronto
Toronto M5S 1A1
Canada

Branko Grünbaum
Department of Mathematics
University of Washington
Seattle, WA 98195
U.S.A

F. A. Sherk
Department of Mathematics
University of Toronto
Toronto M5S 1A1
Canada

AMS Subject Classifications (1980): 51–06, 52–06

Library of Congress Cataloging in Publication Data
Main entry under title:

The Geometric vein.

Bibliography: p.
1. Geometry—Addresses, essays, lectures.
2. Coxeter, H. S. M. (Harold Scott Macdonald), 1907– . I. Coxeter, H. S. M. (Harold Scott Macdonald), 1907– . II. Davis, Chandler. III. Grünbaum, Branko. IV. Sherk, F. A.
QA446.G46 516 81-9171
AACR2



Printed in the United States of America.

9 8 7 6 5 4 3 2 1

ISBN 0-387-90587-1 Springer-Verlag New York Heidelberg Berlin
ISBN 3-540-90587-1 Springer-Verlag Berlin Heidelberg New York

# Contents

## Part V: The Combinatorial Side

# Introduction

Geometry has been defined as that part of mathematics which makes appeal to the sense of sight; but this definition is thrown in doubt by the existence of great geometers who were blind or nearly so, such as Leonhard Euler. Sometimes it seems that geometric methods in analysis, so-called, consist in having recourse to notions outside those apparently relevant, so that geometry must be the joining of unlike strands; but then what shall we say of the importance of axiomatic programmes in geometry, where reference to notions outside a restricted repertory is banned? Whatever its definition, geometry clearly has been more than the sum of its results, more than the consequences of some few axiom sets. It has been a major current in mathematics, with a distinctive approach and a distinctive spirit.

A current, furthermore, which has not been constant. In the 1930s, after a period of pervasive prominence, it appeared to be in decline, even passé. These same years were those in which H.S.M. Coxeter was beginning his scientific work. Undeterred by the unfashionability of geometry, Coxeter pursued it with devotion and inspiration. By the 1950s he appeared to the broader mathematical world as a consummate practitioner of a peculiar, out-of-the-way art. Today there is no longer anything that out-of-the-way about it. Coxeter has contributed to, exemplified, we could almost say presided over an unanticipated and dramatic revival of geometry.

Coxeter's work, though faithful to the 19th-century traditions and style which it continues, has gained repeatedly from his openness to other sources: to the uses of mathematics in physics, and of symmetry in the arts; to the power of combinatorics, especially its introduction into group theory; to the progress in recent decades in extremal problems and inequalities, especially those involving convexity. Thus his work ties together much of the geometry of our times.

This was plain from the success of the Coxeter Symposium held at the University of Toronto, 21–25 May 1979. Close to a hundred mathematicians, from eight countries, gathered for five days of papers. The invited speakers (not all of whom were able to attend) included, beside Coxeter himself, the following: John Horton Conway, Erich Ellers, Gunter Ewald, L. Fejes Tóth, Branko Grünbaum, William Kantor, P. McMullen, C. A. Rogers, B. A. Rosenfeld, J.J. Seidel, G.C. Shephard, J. Tits, W.T. Tutte, and I.M. Yaglom. Rich fare indeed, enriched further by a roomful of geometrical models (assembled by Barry Monson) and an evening of films (assembled by Seymour Schuster).

The present volume, growing out of the Symposium, aims to document the broad range to which some root ideas have ramified. It will be noticed here as it was at the Symposium, how many of the papers refer to concerns and contributions of Coxeter directly, and most of the rest do so indirectly. Several of the papers have the character of surveys, especially those of Kantor, Fejes Tóth, and Tutte, and the first by Yaglom. Most of the papers, though, while tracing a relationship to a wider area, culminate in a new contribution. (It is characteristic of the subject that this takes pictorial form almost as often as definition-theorem-proof.) Despite the closeness of their subject matter, the articles do not refer to each other. We think it will be helpful, therefore, if we say a few words about the collection overall.

We have begun it with a listing of the scientific papers of H.S.M Coxeter. This list, though it is still being added to, surely deserves a place in the literature, and there could be no more appropriate place.

The papers which follow fall naturally into five divisions, and we have made this subdivision in presenting them here, though there are manifold interrelationships and overlaps between them. Two fields not recognized in this scheme, but recurring in the papers, are classical algebraic geometry and the 19th century's independent synthetic development of projective and other geometries.

The intuitive idea of symmetry, already extended beyond the reach of intuition by the ancients, underlies the extension of their "regular figures" to wider definitions, to higher dimensions, and to different geometries. This subject was surveyed in Coxeter's *Regular Polytopes* (1948), and has proliferated since. We hope some sense of its achievements and potentialities is conveyed by the papers in Part I of this book.

The symmetry classes of crystallography are expressed by discrete subgroups of the Euclidean group; various other sorts of symmetry types by discrete subgroups of other classical groups. The interplay between the regular figures and the transformations to which they are subjected is a familiar feature of this study. If the figures occupy the foreground in the papers of Part I, the transformation takes precedence in those of Part III, and group in Part IV.

Part II grows out of the study of regular figures in another way: the minimality and covering properties which they enjoy, and which account for so many of their applications, are sometimes shared by nonsymmetrical figures. (Even a tiling of the plane by equal squares illustrates this: it may, but need not, have translational symmetry in two directions.) Thus there is an area where the study of figures bound by restrictions of symmetry meets the extremal problem unbound by any such restriction. Several of the papers in both Parts I and II are properly in the area of overlap.

Finally, though combinatorial methods and style are seen throughout the volume, we have somewhat arbitrarily set aside Part V for the few most combinatorial of our collection. Their subjects, again, range from the nonsymmetric to the highly symmetric.

CHANDLER DAVIS
BRANKO GRÜNBAUM
F.A. SHERK

# H. S. M. Coxeter: Published Works

## A. Books or Chapters in Books

1. (with P. Du Val, H. T. Flather, and J. F. Petrie) *The* 59 *Icosahedra*, University of Toronto Studies (Math. Series, No. 6), 1938.

2. Polyhedra, in Rouse Ball's *Mathematical Recreations and Essays*, London, 1939. *MR* **8**, 440.

2a. Projective geometry, in *The Tree of Mathematics* (ed. Glenn James), Digest Press, Pacixma CA, 1957, 173–194.

3. *Non-Euclidean Geometry* (Mathematical Expositions, No. 2), Toronto, 1942. *MR* **4**, 50. (5th ed.) University of Toronto Press, 1965.

4. *Regular Polytopes*, London, 1948, New York, 1949. *MR* **10**, 261. (2nd ed.) Macmillan, New York, 1963. *MR* **27**, 1856. (3rd ed.), Dover, New York, 1973. *MR* **51**, 6554.

5. *The Real Projective Plane*, New York, 1949. *MR* **10**, 129. (2nd ed.) Camb. Univ. Press, 1955. *MR* **16**, 1143. *Reele Projektive Geometrie der Ebene*, (transl. of above), Oldenbourg, Munich, 1955. *MR* **17**, 183.

6. (with W. O. J. Moser) *Generators and Relations for Discrete Groups*, Springer-Verlag, Berlin, 1957. *MR* **19**, 527. (2nd ed.) 1965. *MR* **30**, 4818. (3rd ed.) 1972. *MR* **50**, 4229.

7. *Introduction to Geometry*, Wiley, New York, 1961. *MR* **23** A1251. (2nd ed.) 1969. *MR* **49**, 11369. *Unvergängliche Geometrie* (translation of above) Birkhäuser, Basel, 1963. (Japanese translation) Charles E. Tuttle Co., Tokyo, 1965. (Russian translation), Nauka, Moscow, 1966. *MR* **35**, 3516. *Wstep do Geometrii dawnej i nowej* (Polish translation), Panstwowe Wydawnictwo Naukowe, Warsaw, 1967. (Spanish translation), Limusa-Wiley

S.A., Mexico, 1971. (Hungarian translation), Mûszaki Könyvkiado, Budapest, 1973.

8. *Projective Geometry*, Blaisdell, New York, 1964. *MR* **30**, 2380. (2nd ed.) University of Toronto Press, 1973. *MR* **49**, 11377. *Projektivna Geometrija* (translation of above), Skolska knjiga, Zagreb, 1977.
9. The total length of the edges of a non-Euclidean polyhedron, in *Studies in Mathematical Analysis and Related Topics* (Essays in honour of George Pólya, ed. G. Szegö, C. Loewner et al.), Stanford Univ. Press, 1962, 62–69. *MR* **26**, 2944.
10. Geometry, in T. L. Saaty's *Lectures on Modern Mathematics*, Wiley, New York, 1965, 58–94. *MR* **31**, 2647.
11. Solids, geometric, in *Encyclopaedia Britannica*.
12. Geometry, Non-Euclidean, in *Encyclopaedia Britannica*.
13. Non-Euclidean geometry, in Collier's *Merit Students' Encyclopedia*.
14. Reflected light signals, in B. Hoffman's *Perspectives in Geometry and Relativity*, Indiana University Press, 1966, 58–70. *MR* **36**, 2378.
15. (with S. L. Greitzer) *Geometry Revisited*, Random House, New York, 1967. *Redécouvrons la Géométrie* (translation of above), Dunod, Paris, 1971.
16. *Twelve Geometric Essays*, Southern Illinois University Press, Carbondale, 1968. *MR* **46**, 9843.

16a. Non-Euclidean geometry, in *The Mathematical Sciences*, MIT Press, 1969, 52–59.

17. The mathematical implications of Escher's prints, in *The World of M. C. Escher* (ed. J. L. Locher), Abrams, New York, 1971, 49–52.
18. Virus macromolecules and geodesic domes, in *A Spectrum of Mathematics* (ed. J. C. Butcher), Auckland and Oxford University Presses, 1971, 98–107. *MR* **56**, 6547.
19. The inversive plane with four points on each circle, in *Studies in Pure Mathematics* (ed. L. Mirsky), Academic Press, London, 1971, 39–52. *MR* **43**, 6814.
20. Inversive geometry, in *The Teaching of Geometry at the Pre-College Level*, (ed. H. G. Steiner), Reidel, Dordrecht, 1971, 34–45.
21. Cayley diagrams and regular complex polygons, (in *A Survey of Combinatorial Theory*, (ed. J. N. Srivastava), North-Holland, 1973, 85–93. *MR* **51**, 789.
22. (with W. W. Rouse Ball) *Mathematical Recreations and Essays* (12th ed.), University of Toronto Press, 1974. *MR* **50**, 4229.
23. *Regular Complex Polytopes*, Cambridge University Press, 1974. *MR* **51**, 6555.
24. Polyhedral numbers, in *Boston Studies in the Philosophy of Science*, Vol. 15, (ed. R. S. Cohen, J. Stachel, and M. W. Wartofsky), Boston, Mass, 1974, 59–69.

25. Kepler and mathematics, in *Vistas in Astronomy*, Vol. 18, (ed. Arthur Beer), London, 1974.
26. Angels and Devils, in *The Mathematical Gardner* (ed. David Klarner), Wadsworth International, Belmont, CA, 1981, 197–209.
27. (with Roberto Frucht and David L. Powers) *Zero-Symmetric graphs: trivalent graphical regular representations of groups*, Academic Press, New York, 1981.
28. (with Roberto Frucht and D. L. Powers) *Zero-Symmetric Graphs*, Academic Press, 1981.

## B. Research Papers and Expository Articles

1. The pure Archimedean polytopes in six and seven dimensions, *Proc. Camb. Phil. Soc.* **24** (1928), 1–9.
2. The polytopes with regular-prismatic vertex figures I, *Phil. Trans. Royal Soc.* (A) **229** (1930), 329–425.
3. Groups whose fundamental regions are simplexes, *Journal London Math. Soc.* **6** (1931), 132–136.
4. The densities of the regular polytopes I, *Proc. Camb. Phil. Soc.* **27** (1931), 201–211.
5. The polytopes with regular-prismatic vertex figures II, *Proc. London Math. Soc.* (2) **34** (1932), 126–189.
6. The densities of the regular polytopes II, *Proc. Camb. Phil. Soc.* **28** (1932), 509–521.
7. The densities of the regular polytopes III, *ibid.* **29** (1933), 1–22.
8. Regular compound polytopes in more than four dimensions, *Journal of Math. and Phys.* **12** (1933), 334–345.
9. Discrete groups generated by reflections, *Annals of Math.* **35** (1934), 588–621.
10. On simple isomorphism between abstract groups, *Journal London Math. Soc.* **9** (1934), 211–212.
11. Abstract groups of the form $V_i^k = V_j^3 = (V_i V_j)^2 = 1$, *ibid*, 213–219.
12. (with J. A. Todd) On points with arbitrarily assigned mutual distances, *Proc. Camb. Phil. Soc.* **30** (1934), 1–3.
13. Finite groups generated by reflections, and their subgroups . . . , *Proc. Camb. Phil. Soc.* **30** (1935), 466–482.
14. The functions of Schläfli and Lobatschefsky, *Quarterly Journal of Math.* **6** (1935), 13–29.
15. (with P. S. Donchian) An $n$-dimensional extension of Pythagoras' theorem, *Math. Gazette* **19** (1935), 206.
16. The complete enumeration of finite groups $R_i^2 = (R_i R_j)^{k_{ij}} = 1$, *Journal London Math. Soc.* **10** (1935), 21–25.

17. Wythoff's construction for uniform polytopes, *Proc. London Math. Soc.* (2) **38** (1935), 327–339.

18. The representation of conformal space on a quadric, *Annals of Math.* **37** (1936), 416–426.

19. The groups determined by the relations $S^l = T^m = (S^{-1}T^{-1}ST)^p = 1$, *Duke Math. Journal* **2** (1936), 61–73.

20. An abstract definition for the alternating group . . . , *Journal London Math. Soc.* **11** (1936), 150–156.

21. (with J. A. Todd) Abstract definitions for the symmetry groups of the regular polytopes in terms of two generators I, *Proc. Camb. Phil. Soc.* **32** (1936), 194–200.

22. The abstract groups $R^m = S^m = (R^jS^j)^{p_j} = 1, \ldots,$ *Proc. London Math. Soc.* (2) **41** (1936), 278–301.

23. (with J. A. Todd) A practical method of enumerating cosets of a finite abstract group, *Proc. Edinburgh Math. Soc.* (2) **5** (1936), 26–34.

24. On Schläfli's generalization of Napier's pentagramma mirificum, *Bull. Calcutta Math. Soc.* **28** (1936), 123–144.

25. (with J. A. Todd) Abstract definitions for the symmetry groups of the regular polytopes in terms of two generators II, *Proc. Camb. Phil. Soc.*. **33** (1937), 315–324.

26. Regular skew polyhedra in three and four dimensions . . . , *Proc. London Math. Soc.* (2) **43** (1937), 33–62.

27. An easy method for constructing polyhedral group-pictures, *Amer. Math. Monthly* **45** (1938), 522–525.

28. The abstract groups $G^{m,n,p}$, *Trans. Amer. Math. Soc.* **45** (1939), 73–150.

29. The regular sponges, or skew polyhedra, *Scripta Mathematica* **6** (1939), 240–244.

30. Regular and semi-regular polytopes, *Math. Zeitschrift* **46** (1940), 380–407, *MR* **2**, 10.

31. A method for proving certain abstract groups to be infinite, *Bull. Amer. Math. Soc.* **46** (1940), 246–251, *MR* **1**, 258.

32. (with R. Brauer) A generalization of theorems of Schönhardt and Mehmke on polytopes, *Trans. Royal Soc. of Canada* **34** (3), (1940), 29–34, *MR* **2**, 125.

33. The polytope $2_{21}$ whose 27 vertices correspond to the lines on the general cubic surface, *Amer. Journal of Math.* **62** (1940), 457–486, *MR* **2**, 10.

34. The binary polyhedral groups, and other generalizations of the quaternion group, *Duke Math. Journal* **7** (1940), 367–379, *MR* **2**, 214.

35. The map-coloring of unorientable surfaces, *ibid.* **10** (1943), 293–304, *MR* **5**, 48.

36. A geometrical background for de Sitter's world, *Amer. Math. Monthly* **50** (1943), 217–227, *MR* **4**, 236.

37. Quaternions and reflections, *Amer. Math. Monthly* **53** (1946), 136–146, *MR* **7**, 387.
38. Integral Cayley numbers, *Duke Math. Journal* **13**, (1946), 561–578, *MR* **8**, 370.
39. The nine regular solids, *Proc. Canadian Math. Congress* **1** (1947), 252–264, *MR* **8**, 482.
40. The product of three reflections, *Quarterly of Applied Math.* **5** (1947), 217–222, *MR* **9**, 549.
41. A problem of collinear points, *Amer. Math. Monthly* **55** (1948), 26–28, 247, *MR* **9**, 458.
42. Configurations and maps, *Rep. Math. Colloq.* (2) **8** (1948), 18–38, *MR* **10**, 616.
43. Projective geometry, *Math. Magazine* **23** (1949), 79–97, *MR* **11**, 384.
44. Self-dual configurations and regular graphs, *Bull. Amer. Math. Soc.* **56** (1950), 413–455, *MR* **12**, 350.
45. (with A. J. Whitrow) World structure and non-Euclidean honeycombs, *Proc. R. S.* **A201** (1950), 417–437, *MR* **12**, 866.
46. Extreme forms, *Proc. Internat. Congress of Math.* (1950), 294–295.
47. Extreme forms, *Canad. J. Math.* **3** (1951), 391–441. *MR* **13**, 443.
48. The product of the generators of a finite group generated by reflections, *Duke Math. J.* **18** (1951), 765–782, *MR* **13**, 528.
49. Interlocked rings of spheres, *Scripta Math.* **18** (1952), 113–121, *MR* **14**, 494.
50. (with J. A. Todd) An extreme duodenary form, *Canad. J. Math.* **5** (1953), 384–392, *MR* **14**, 1066.
51. The golden section, phyllotaxis, and Wythoff's game, *Scripta Math.* **19** (1953), 135–143, *MR* **15**, 246.
52. (with M. S. Longuet-Higgins and J. C. P. Miller), Uniform polyhedra, *Phil. Trans. Royal Soc.* (A) **246** (1954), 401–450, *MR* **15**, 980.
53. Regular honeycombs in elliptic space, *Proc. London Math. Soc.* (3) **4** (1954), 471–501, *MR* **16**, 1145.
54. Six uniform polyhedra, *Scripta Math.* **20** (1954), 227.
55. An extension of Pascal's theorem, *Amer. Math. Monthly* **61** (1954), 723.
56. Arrangements of equal spheres in non-Euclidean spaces, *Acta Math. Acad. Sci. Hungaricae* **5** (1954), 263–276, *MR* **17**, 523.
57. The area of a hyperbolic sector, *Math. Gazette* **39** (1955), 318.
58. On Laves' graph of girth ten, *Canad. J. Math.* **7** (1955), 18–23, *MR* **16**, 739.
59. The affine plane, *Scripta Math.* **21** (1955), 5–14, *MR* **16**, 949.
60. Hyperbolic triangles, *Scripta Math.* **22** (1956), 5–13, *MR* **18**, 412.

61. Regular honeycombs in hyperbolic space, *Proc. Internat. Congress of Mathematicians* (1956) 155–169, *MR* **19**, 304.

62. The collineation groups of the finite affine and projective planes with four lines through each point, *Abh. Math. Sem. Univ. Hamburg* **20** (1956), 165–177, *MR* **18**, 378.

63. Groups generated by unitary reflections of period two, *Canad. J. Math.* **9** (1957), 263–272, *MR* **19**, 248.

64. Map-coloring problems, *Scripta Math.* **23** (1957), 11–25, *MR* **20**, 7277.

65. Crystal symmetry and its generalizations (Presidential Address), *Trans. Royal Society of Canada* (III) **51** (1957), 1–13.

66. Lebesgue's minimal problem, *Eureka* **21** (1958), 13.

67. The chords of the non-ruled quadric in $PG(3,3)$, *Canad. J. Math.* **10** (1958), 484–488, *MR* **21**, 841.

68. Twelve points in $PG(5,3)$ with 95040 self-transformations, *Proc. Royal Soc. London* A **247** (1958), 279–293, *MR* **22**, 1104.

69. On subgroups of the modular group, *J. de Math. Pures Appl.* **37** (1958), 317–319.

70. Close packing and froth, *Illinois J. Math.* **2** (1958), 746–758, *MR* **21**, 848.

71. Factor groups of the braid group, *Proc. 4th Canad. Math. Congress*, Toronto Univ. Press, (1959), 95–122.

72. The four-color map problem, 1840–1890, *Math. Teacher* **52** (1959), 283–289, *MR* **21**, 4414.

73. Symmetrical definitions for the binary polyhedral groups, *Proceedings of Symposia in Pure Mathematics, Amer. Math. Soc.* **1**, (1959), 64–87, *MR* **22**, 6850.

74. Polytopes over $GF(2)$ and their relevance for the cubic surface group, *Canad. J. Math.* **11** (1959), 646–650, *MR* **21**, 7476.

75. (with L. Few and C. A. Rogers) Covering space with equal spheres, *Mathematika* **6** (1959), 147–157, *MR* **23**, A2131.

76. On Wigner's problem of reflected light signals in de Sitter space, *Proceedings of the Royal Society of London* A **A261** (1961), 435–442, *MR* **22**, 13231.

77. Similarities and conformal transformations, *Annali di Mat. pura ed appl.* **53** (1961), 165–172, *MR* **26**, 648.

78. Music and mathematics, *Canadian Music Journal* **6** (1962), 13–24.

79. The problem of packing a number of equal nonoverlapping circles on a sphere, *Trans. New York Acad. Sci.* (II) **24** (1962), 320–331.

80. The classification of zonohedra by means of projective diagrams, *J. de Math. pures appl.* **41** (1962), 137–156, *MR* **25**, 4417.

81. The symmetry groups of the regular complex polygons, *Archiv der Math.* **13** (1962), 86–97, *MR* **26**, 2516.

82. The abstract group $G^{3,7,16}$, *Proc. Edinburgh Math. Soc.* (II) **13** (1962), 47–61 and 189, *MR* **26**, 190 and 6267.

83. Projective line geometry, *Mathematicae Notae Universidad Nacional del Litoral, Rosario, Argentina* **1** (1962), 197–216.

84. An upper bound for the number of equal nonoverlapping spheres that can touch another of the same size, *Proc. Symposia in Pure Mathematics* **7** (1963), 53–71, *MR* **29**, 1437.

85. (with L. Fejes Tóth) The total length of the edges of a non-Euclidean polyhedron with triangular faces, *Quarterly J. Math.* **14** (1963), 273–284, *MR* **28**, 1532.

86. (with B. L. Chilton) Polar zonohedra, *Amer. Math. Monthly* **70** (1963), 946–951.

87. (with S. L. Greitzer) L'hexagramme de Pascal, un essai pour reconstituer cette découverte, *Le Jeune Scientifique* **2** (1963), 70–72.

88. Regular compound tessellations of the hyperbolic plane, *Proc. Royal Soc.* **A278** (1964), 147–167.

89. Achievement in maths, *Varsity Graduate* (Spring 1966) 15–18.

90. The inversive plane and hyperbolic space, *Abh. Math. Sem. Univ. Hamburg* **29** (1966), 217–242, *MR* **33**, 7920.

91. Inversive distance, *Annali di Matematica* (4) **71** (1966), 73–83, *MR* **34**, 3418.

92. Finite groups generated by unitary reflections, *Abh. Math. Sem. Univ. Hamburg* **31** (1967), 125–135, *MR* **37**, 6358.

93. The Lorentz group and the group of homographies, *Proc. Internat. Conf. on the Theory of Groups*, (ed. L. G. Kovacs and B. H. Neumann), Gordon and Breach, New York (1967), 73–77.

94. Transformation groups from the geometric viewpoint, *Proceedings of the CUPM Geometry Conference* (1967), 1–72.

95. The Ontario K-13 geometry report, *Ontario Mathematics Gazette* **5.3** (1967), 12–16.

96. Music and mathematics. (reprinted from the *Canadian Music Journal*), *Mathematics Teacher* **61** (1968), 312–320.

97. The problem of Apollonius, *Amer. Math. Monthly*, **75** (1968), 5–15, *MR* **37**, 5767.

98. Mid-circles and loxodromes, *Math. Gazette* **52** (1968), 1–8.

99. Loxodromic sequences of tangent spheres, *Aequationes Mathematicae* **1** (1968), 104–121, *MR* **38**, 3765.

100. Affinities and their fixed points, *The New Zealand Mathematics Magazine* **6** (1969), 114–117.

101. Helices and concho-spirals, in *Nobel Symposium* **11**, ed. Arne Engström and Bror Strandberg, Almqvist and Wiksell, Stockholm, (1969), 29–34.

102. Affinely regular polygons, *Abhandlungen aus dem Mathematischen Seminar der Universität Hamburg*, **34**, Heft 1/2, (1969), Vandenhoeck & Ruprecht in Göttingen, *MR* **42**, 2349.

103. Products of shears in an affine Pappian plane, *Rendiconti di Matematica*, VI **3** (1970), 1–6, *MR* **42**, 2350.

104. Twisted honeycombs, (Regional Conference Series in Mathematics, No. 4), American Mathematical Society, Providence, R. I., (1970), *MR* **46**, 3639.

105. Inversive geometry, *Vinculum* **7** (1970), 72–76.

106. The mathematics of map coloring, *Leonardo* **4** (1971), 273–277.

107. Frieze patterns, *Acta Arithmetica* **18** (1971), 297–310, *MR* **44**, 3980.

108. Cyclic sequences and frieze patterns, *Vinculum* **8** (1971), 4–7.

109. An ancient tragedy, *Mathematical Gazette* **55** (1971), 312.

110. The role of intermediate convergents in Tait's explanation for phyllotaxis, *Journal of Algebra* **20** (1972), 168–175, *MR* **45**, 4255.

111. (with J. H. Conway and G. C. Shephard) The centre of a finitely generated group, *Tensor* **25** (1972), 405–418, **26** (1972), 477, *MR* **48**, 11326.

112. (with J. H. Conway) Triangulated polygons and frieze patterns, *Mathematical Gazette* **57** (1973), 87–94, 175–186, *MR* **57**, 1254, 1255.

113. The Dirac matrix group and other generalizations of the quaternion group, *Communications on Pure and Applied Mathematics* **26** (1973), 693–698, *MR* **48**, 11275.

114. The equianharmonic surface and the Hessian polyhedron, *Annali di Matematica* (4) **98** (1974), 77–92, *MR* **51**, 5605.

115. Desargues configurations and their collineation groups, *Math. Proc. Cambridge Philos. Soc.* **78** (1975), 227–246, *MR* **52**, 9070.

116. The space-time continuum, *Historia Mathematica* **2** (1975), 289–298.

117. The Pappus configuration and its groups, *K. Nederl. Akad. Wetensch, Amsterdam Verslag Afd. Natuurkunde* **85** (1976), 44–46, *MR* **52**, 14303.

118. The Erlangen program, *The Mathematical Intelligencer* **0** (1977), 22.

119. The Pappus configuration and its groups, *Pi Mu Epsilon J.* **6** (1977), 331–336, *MR* **56**, 5330.

120. (with G. C. Shephard) Regular 3-complexes with toroidal cells, *Journal of Combinatorial Theory* **22** (1977), 131–138, *MR* **55**, 11140.

121. The Pappus configuration and the self-inscribed octagon, *Proc. K. Nederl. Akad. Wetensch.* A **80** (1977), 256–300.

122. (with C. M. Campbell and E. F. Robertson) Some families of finite groups having two generators and two relations, *Proc. Royal Soc. London* **A357** (1977), 423–438, *MR* **57**, 463.

123. Gauss as a geometer, *Historia Mathematica* **4** (1977), 379–396.

124. Polytopes in the Netherlands, *Nieuw Archief voor Wiskunde* (3) **26** (1978), 116–141.

125. Review of *Three-Dimensional Nets and Polyhedra* by A. F. Wells, *Bull. Amer. Math. Soc.* **84** (1978), 466–470.
126. The amplitude of a Petrie polygon, *C. R. Math. Rep. Acad. Sci. Canada* **1** (1978), 9–12.
127. Parallel lines, *Canad. Math. Bull.* **21** (1978), 385–397.
128. On R. M. Foster's regular maps with large faces, *AMS Proc. Symp. Pure Math.* **34** (1979), 117–128.
129. (with R. W. Frucht) A new trivalent symmetrical graph with 110 vertices, *Annals New York Acad. Sci.* **319** (1979), 141–152.
130. The non-Euclidean symmetry of Escher's picture 'Circle Limit III,' *Leonardo* **12** (1979), 19–25.
131. The derivation of Schoenberg's star polytopes from Schoute's simplex nets, *C. R. Math. Rep. Acad. Sci. Canada* **1** (1979), 195.
132. (with Pieter Huybers) A new approach to the chiral Archimedean solids, *C. R. Math. Rep. Acad. Sci. Canada* **1** (1979), 269–274.
133. Angles and arcs in the hyperbolic plane, *Math. Chronicle* **9** (1980), 17–33.
134. Higher-dimensional analogues of the tetrahedrite crystal twin, *Match* **9** (1980), 67–72.
135. The edges and faces of a 4-dimensional polytope, *Congressus Numerantium* **28** (1980), 309–334.
136. A systematic notation for the Coxeter graph, *C. R. Math. Rep. Acad. Sci. Canada* **3** (1981), 329–332.

# Part I: Polytopes and Honeycombs

# Uniform Tilings with Hollow Tiles[1]

Branko Grünbaum*
J. C. P. Miller†
G. C. Shephard††

## 1. Introduction

Tilings of the plane in which each tile is a closed topological disk, or some other simple kind of set, have been extensively studied from many points of view; see [18]. Of special interest are tilings whose tiles are regular polygons; there are many such tilings and they occur frequently in practical applications. In particular, the three *regular* tilings have been known since ancient times and the *uniform* (or Archimedean) tilings have been known since Kepler's pioneering work in the seventeenth century [19].

The purpose of this paper is to present a generalization of these ideas that leads to a great variety of new and visually attractive tilings. The generalization is based on the concept of a "hollow tile", which can also be traced back to Kepler. He was, we believe, the first to define a polygon to be a circuit of edges and vertices—and *not* the boundary of a "patch" or "piece" of the plane. In this context the regular star-polygons arise naturally, and by a slight extension one is led to the consideration of the regular infinite polygons, namely the apeirogon and zigzags. It was at the end of the nineteenth century that Badoureau [2], [3] first explored the possibility of using these polygons in a tiling, and the only later work in this direction we have been able to trace is Miller's thesis of 1933 [21] and the paper of Coxeter et al. [7], which reproduced Miller's list of tilings with one addition.

[1] This material is based upon work supported by the National Science Foundation Grant No. MCS77-01629 A01.

* Department of Mathematics, University of Washington, Seattle, Washington 98195, U.S.A.

† Department of Pure Mathematics and Mathematical Statistics, University of Cambridge, Cambridge, England.

†† University of East Anglia, Norwich NR4 7TJ, England.

*We regret to announce that on April 24, 1981, while this chapter was in proof, Dr. J. C. P. Miller died from a heart attack.*

One difficulty in the treatment of tilings by hollow tiles is that many of the traditional definitions have to be recast. Even the usual definition of a tiling—a family of sets (tiles) that cover the plane without gaps or overlaps—is clearly inapplicable here. We must therefore begin by reformulating the definitions of many terms we shall use. For the most part we shall be concerned with uniform tilings (vertex-transitive tilings by regular polygons) and we shall give diagrams of all those whose existence is known. These represent a true generalization of the "traditional" uniform tilings in that the latter are included when suitably interpreted in terms of our new definitions. We must point out, however, that we can only conjecture that our enumeration is complete. We are essentially sure that this is so when no infinite polygons occur, though we cannot produce a full proof even in this case.

In the final section of the paper we give a short history of the subject as well as indicating generalizations, related problems, and areas for further investigation.

## 2. Regular Polygons

Following Poinsot's memoir of 1810 [22], which was to a certain extent anticipated by the work of Kepler in 1619 [19], Girard in 1626 [10], Meister in 1769 [20], and others, we define a *finite polygon* (or *n-gon*) $P$ in the plane to be a sequence of $n$ distinct points $V_1, V_2, \ldots, V_n$ (the *vertices* of $P$) together with $n$ line segments $[V_i, V_{i+1}]$, $i = 1, 2, \ldots, n-1$ and $[V_n, V_1]$ (the *edges* of $P$). Such an $n$-gon will be denoted by $[V_1, V_2, \ldots, V_n]$ and will be called a *hollow polygon* when we wish to emphasize the distinction between this and the more usual definition of a polygon as a plane topological disk with piecewise linear boundary. When we need to refer to the latter interpretation, we shall use the phrase "polygonal region".

In an analogous way we define an *infinite polygon* $[\ldots, V_{-2}, V_{-1}, V_0, V_1, V_2, \ldots]$ to be a doubly infinite sequence of vertices $V_i$ and of edges $[V_i, V_{i+1}]$ as $i$ runs through integer values, subject to a local finiteness condition: no circular disk in the plane meets more than a finite number of edges or vertices.

An edge $E$ of a (finite or infinite) hollow polygon is said to be *incident* with the two vertices at its endpoints—and with no other vertices or edges even if the latter lie on $E$, or have non-empty intersection with $E$.

A hollow triangle is just the set of vertices and edges of the boundary of a triangular region. In Figure 1, besides a triangle, we show three examples of quadrangles (that is, 4-gons) and eleven pentagons (5-gons). Under certain restrictions on the polygons considered (no point of the plane belongs to more than two edges, and no three vertices are collinear) there is a natural classification of polygons into "types", in which the polygons of Figure 1 are all of different types, and all types of polygons with at most five edges are represented (Girard [10]). (Additional information about this classification of polygons and its history may be found in [11]. The question posed in [11] concerning the

**Figure 1.** Representatives of the different types of polygons with at most five sides.

number of types of hexagons has recently been solved by D. Buset [5], who proved that there are precisely 73 different types.)

Examples of infinite polygons are shown in Figure 2(a), and in Figure 2(b) we have indicated two infinite sets of points and segments that are *not* infinite polygons according to the above definition.

Although our definition of a polygon may at first glance appear strange, it is in many respects a natural extension of the naive concept. If a polygon is finite and has no self-intersections, then by the Jordan curve theorem (or its much more easily proved polygonal version) it has a well-defined interior and this is a polygonal region—that is to say, a polygon in the traditional sense of the word.

Here we shall be interested mainly in regular polygons. A *flag* of a hollow polygon $P$ is a pair $(V, E)$ consisting of a vertex $V$ and an edge $E$ which is incident with $V$. A polygon $P$ is *regular* if the symmetries of $P$ act transitively on the flags of $P$. It is well known (see, for example, [6], [12]) and easily proved that all finite regular polygons in the plane can be obtained in the following manner: the *vertices* are $n$ equidistant points on a circle, and for some integer $d$ with $1 \leqslant d < n/2$ that is relatively prime to $n$, an *edge* connects two vertices whenever they are separated by $d-1$ other vertices on the circle; the polygon just described is usually denoted by the symbol $\{n/d\}$. Thus, the $n$ vertices and edges of a regular convex $n$-gon form the $n$-gon $\{n/1\}$; for simplicity, this symbol is usually abbreviated to $\{n\}$. In Figure 3 we show all the distinct finite regular $n$-gons that occur in the tilings we shall describe in this paper. As usual, we call the polygons $\{n\} = \{n/1\}$ *convex*, and the polygons $\{n/d\}$ with $d \geqslant 2$ *star-polygons*.

Each infinite regular polygon in the plane is either an *apeirogon* (denoted by $\{\infty\}$) consisting of congruent segments that together form a straight line, or else

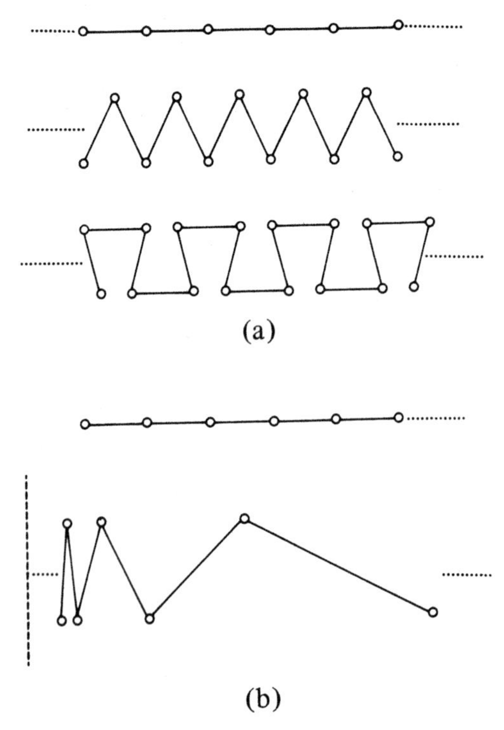


**Figure 2.** Examples of (*a*) infinite polygons, and (*b*) infinite sets of points and line segments which do not satisfy the definition of a polygon.

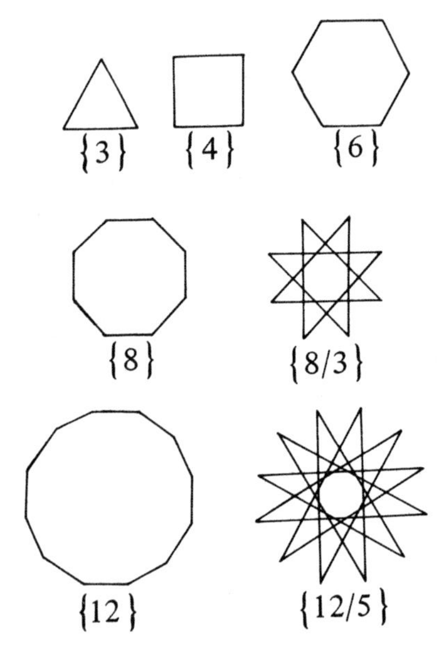


**Figure 3.** The finite regular polygons that occur in uniform tilings.

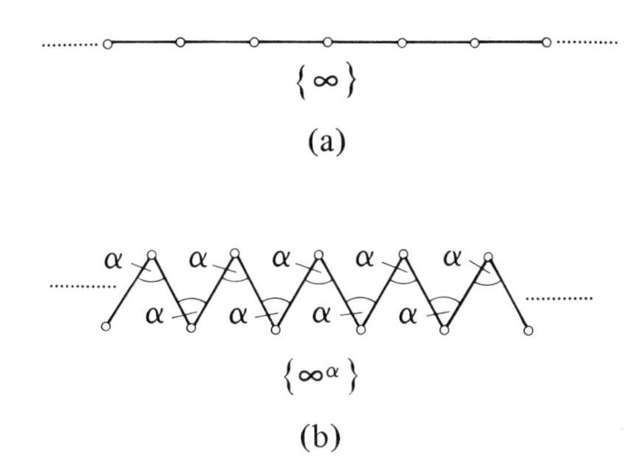


**Figure 4.** Infinite regular polygons: (*a*) the apeirogon $\{\infty\}$, and (*b*) the zigzag $\{\infty^{\alpha}\}$ where $0 < \alpha < \pi$. Here $\alpha$ is the angle between each two consecutive line segments (edges).

a *zigzag with angle* $\alpha$ (denoted by $\{\infty^{\alpha}\}$), where $0 < \alpha < \pi$ (see Figure 4). Clearly, the apeirogon $\{\infty\}$ may be considered as a special case of a zigzag $\{\infty^{\alpha}\}$ with angle $\alpha$ equal to $\pi$.

## 3. Tilings

Since the regular hollow polygons $\{n/d\}$, $\{\infty\}$, and $\{\infty^{\alpha}\}$ are not polygonal regions of the plane, the usual definition of a tiling is not applicable and we need to provide a replacement. This uses several new terms which we must explain first.

A family $\mathfrak{T}$ of polygons is called *edge-sharing* provided each edge of one of the polygons of $\mathfrak{T}$ is an edge of precisely one other polygon of $\mathfrak{T}$. We recall that in many investigations of tilings of the plane by convex polygonal regions only so-called "edge-to-edge" tilings are considered—and the property of being edge-sharing is the natural analogue for hollow tiles. For edge-sharing families $\mathfrak{T}$ each vertex or edge of a polygon in $\mathfrak{T}$ is also said to be a *vertex* or *edge* of $\mathfrak{T}$.

Throughout this paper, we shall restrict attention to *locally finite* families $\mathfrak{T}$, that is, families such that each circular disk meets only a finite number of polygons of $\mathfrak{T}$.

An edge-sharing family $\mathfrak{T}$ of polygons is called *connected* provided that, for each two polygons $P$ and $P^*$ in $\mathfrak{T}$, there is a finite family of polygons $P = P_0, P_1, P_2, \ldots, P_k = P^*$ in $\mathfrak{T}$ such that $P_{j-1}$ shares an edge with $P_j$ for $j = 1, 2, \ldots, k$. It is clear that the boundaries of the tiles in an edge-to-edge tiling of the plane by polygonal regions are hollow polygons that form a connected family in this sense.

If $\mathfrak{T}$ is an edge-sharing family of polygons and if $V$ is a vertex of $\mathfrak{T}$, the *vertex figure* $V(\mathfrak{T})$ of $\mathfrak{T}$ at $V$ is a family of points and segments obtained as follows. A point $A$ is a *vertex* of $V(\mathfrak{T})$ if and only if the segment $[V, A]$ is an edge of $\mathfrak{T}$. If $A, B$ are vertices of $V(\mathfrak{T})$, then the segment $[A, B]$ is an *edge* of $V(\mathfrak{T})$ if and only if $[V, A]$ and $[V, B]$ are edges of the same polygon $P$ in $\mathfrak{T}$; we shall say that

$[A, B]$ is the edge of $V(\mathfrak{T})$ that *corresponds* to $P$. It is clear that in an edge-sharing family the vertex figure at each vertex is either a (repeated) line segment, or a polygon, or a union of such segments and polygons. If a vertex figure is a single finite polygon we shall say that it is *unicursal*.

As examples we show in Figure 5 five edge-sharing, connected finite families of regular polygons, with all vertex figures unicursal. The vertex figures at some vertices in each family are also shown. In each we have marked an $n$ or $n/d$ near an edge that corresponds to a polygon $\{n\}$ or $\{n/d\}$, and we have also indicated the position of the vertex to which the vertex figure corresponds. It will be observed that the squares in Figure 5(c) have partly overlapping edges, but according to our definition this does not disqualify the family from being edge-sharing.

We are now ready for the new definition of a tiling:

A locally finite family $\mathfrak{T}$ of polygons in the plane is a *tiling* if and only if $\mathfrak{T}$ is edge-sharing, connected and each vertex figure is unicursal.

Hence all the families in Figure 5 are tilings, as are the ones in Figure 6; moreover, each edge-to-edge tiling by convex polygonal regions may be interpreted as a tiling in the new sense if each region is replaced by its bounding polygon.

The examples in Figures 5 and 6 show that, in contrast to the traditional situation, a tiling in the new sense can be finite, contained in a strip of finite width, or "occupy" only some other portion of the plane.

It is useful to consider some further properties of tilings by hollow tiles. To begin with, it is clear that each polygon can be assigned one of two possible orientations. We shall say that a tiling $\mathfrak{T}$ is *orientable* if the polygons of $\mathfrak{T}$ can be *coherently* oriented, that is to say, oriented so that every two polygons that share an edge induce opposite orientations on that edge. It is not hard to verify that all the tilings in Figure 5, and all the tilings in Figure 6 except the one designated $(3.6.-3.-6; 3.6.6.\infty.6.6)$, are orientable. It is also obvious that all edge-to-edge tilings by convex polygonal regions correspond to orientable tilings.

If $P$ is any oriented finite polygon and $A$ is any point of the plane not on any edge of $P$, we define the $P$-density of $A$ (or "winding number" of $P$ about $A$) as follows. Let $L$ be any ray with endpoint $A$ that contains no vertex of $P$. To each edge $E$ of $P$ we assign an *index*, equal to 0 if $E$ does not meet $L$, to $+1$ if $E$ crosses $L$ in an anticlockwise direction about $A$, and to $-1$ if $E$ crosses $L$ in a clockwise direction about $A$. The $P$-density of $A$ is the sum of indices of all edges of $P$. This $P$-density can be shown to depend only on $A, P$ and the orientation chosen for $P$, but not on $L$; moreover, any two points in the same connected component of the complement of $P$ in the plane have the same $P$-density.

Let $\mathfrak{T}$ be an orientable tiling by finite polygons, with a chosen coherent orientation for its polygons. If $A$ is a point not on any edge of $\mathfrak{T}$, a *$\mathfrak{T}$-density* of $A$ is the sum of the $P$-densities of $A$ for all the polygons $P$ in $\mathfrak{T}$. We shall say that $\mathfrak{T}$ has density $d$, where $d \geqslant 0$, if the coherent orientations of the polygons can be chosen so that the $\mathfrak{T}$-density of each point $A$, not on any edge of $\mathfrak{T}$, equals $d$. It

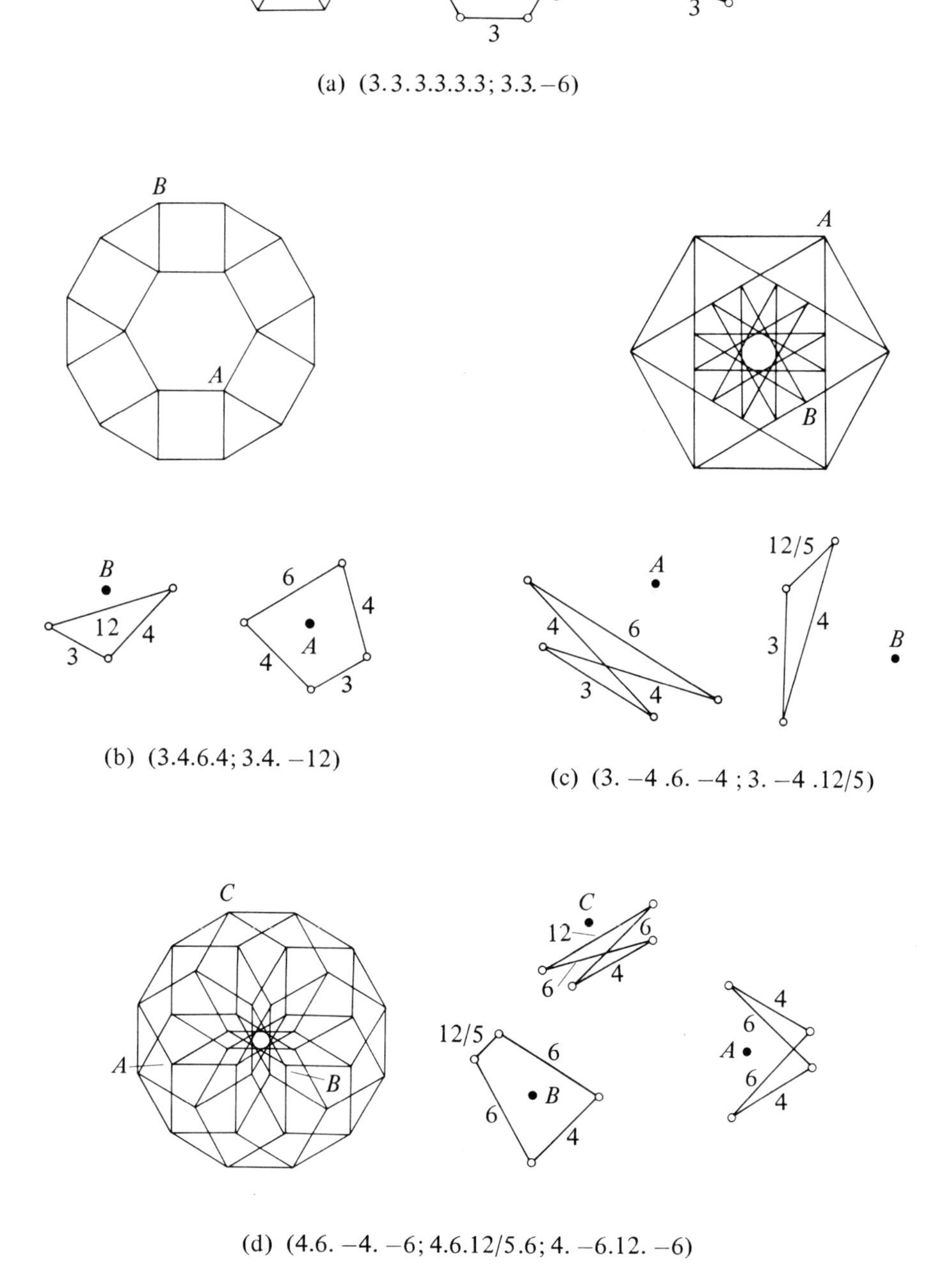


(a) (3.3.3.3.3.3; 3.3.−6)

(b) (3.4.6.4; 3.4. −12)

(c) (3. −4 .6. −4 ; 3. −4 .12/5)

(d) (4.6. −4. −6; 4.6.12/5.6; 4. −6.12. −6)

**Figure 5** (a–d). Five finite edge-sharing families of regular polygons (rosettes). (Continued on page 24.)

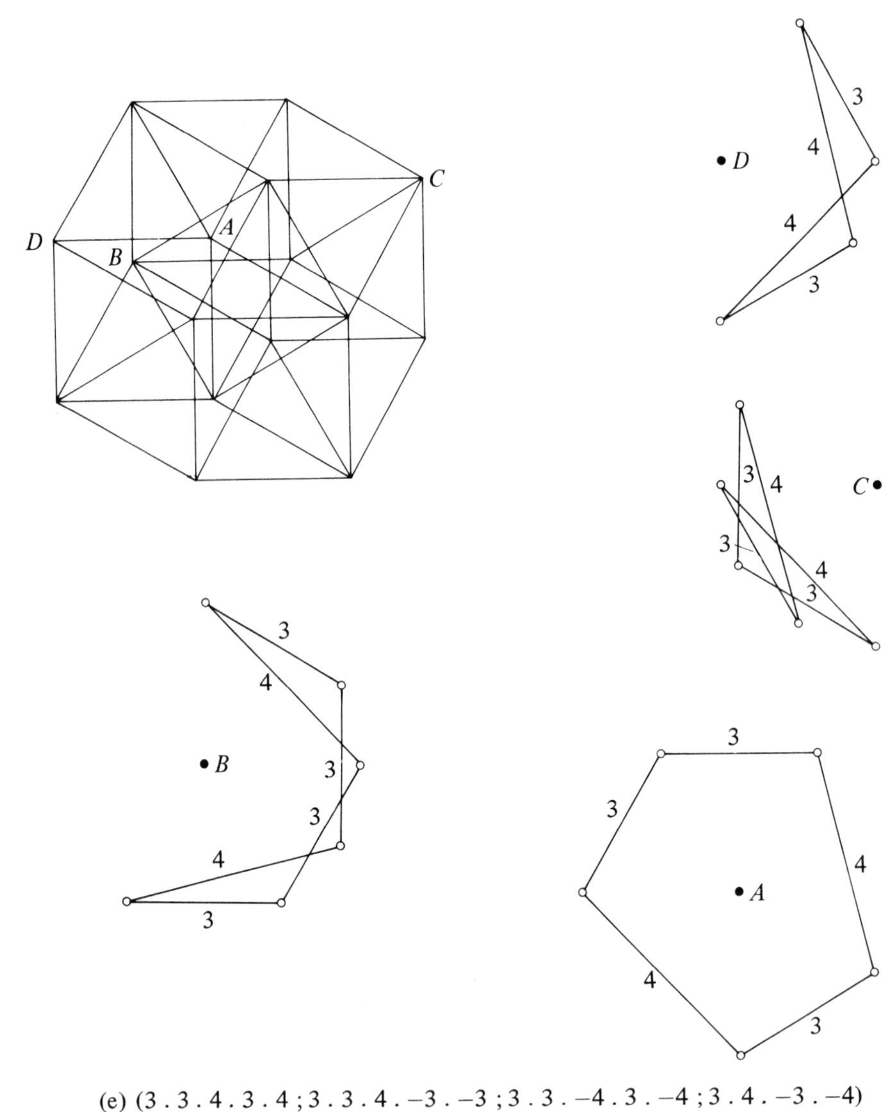


(e) (3 . 3 . 4 . 3 . 4 ; 3 . 3 . 4 . −3 . −3 ; 3 . 3 . −4 . 3 . −4 ; 3 . 4 . −3 . −4)

**Figure 5** (e). See legend on page 23.

follows easily that all the orientable tilings by finite polygons shown in Figures 5 and 6 have density 0, while the traditional tilings by polygonal regions correspond to orientable tilings by hollow tiles that have density 1.

## 4. Uniform Tilings

A tiling $\mathcal{T}$ is called *k-uniform* provided $\mathcal{T}$ consists of regular polygons and the vertices of $\mathcal{T}$ form $k$ equivalence classes under the symmetries of $\mathcal{T}$; if $k = 1$ we shall say that $\mathcal{T}$ is *uniform*, and if $k = 2$ we shall say that $\mathcal{T}$ is *biuniform*. It is easy

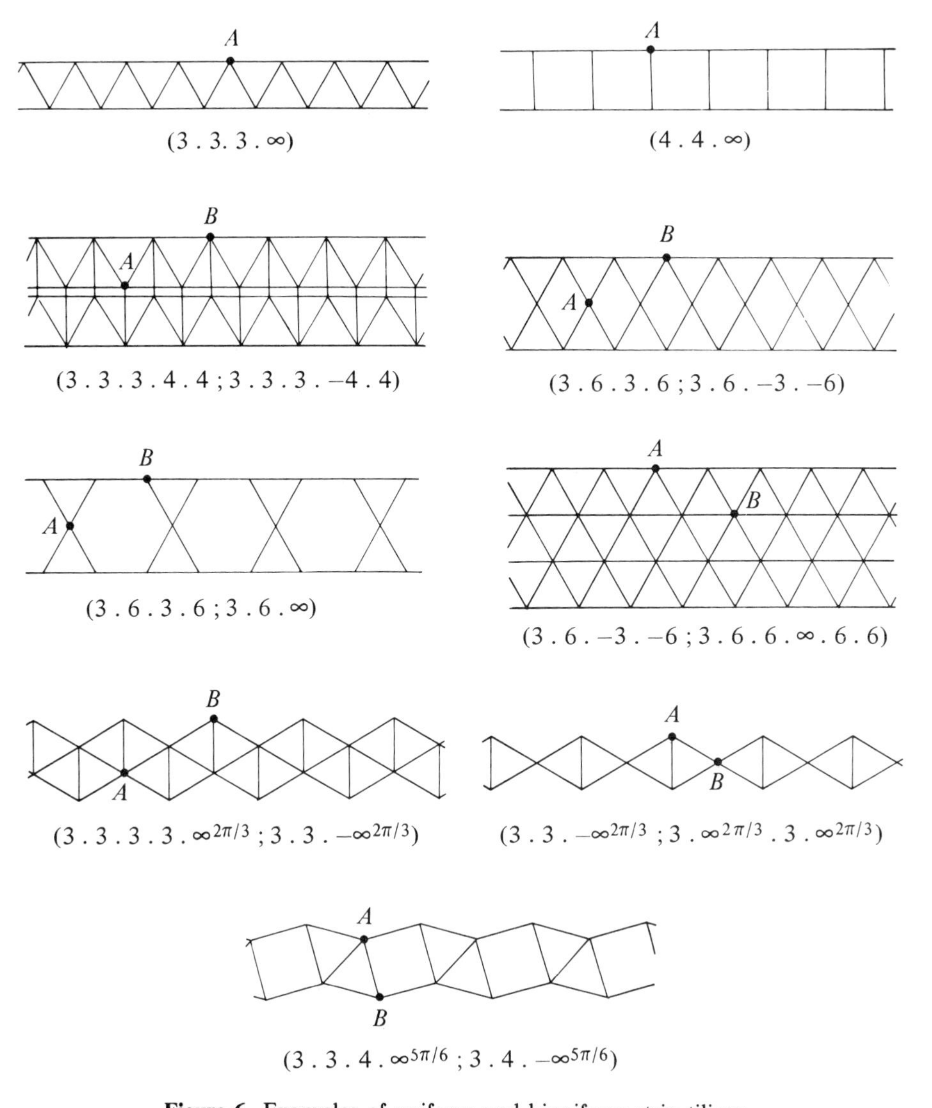


**Figure 6.** Examples of uniform and biuniform strip tilings.

to check that the traditional tilings usually called $k$-uniform or uniform (= Archimedean) remain so with the new definitions, as soon as we replace each convex polygonal "region" by its boundary. But the new definition is more general, as can be seen from the examples in Figures 5 and 6. In Tables 1, 2, and 3 we shall list all the uniform tilings known to us, but before doing that it is convenient to introduce some further notation. The key to this is the fact that each vertex in a uniform tiling is surrounded by polygons arranged in the same way.

Let $\{n_1\}, \{n_2\}, \ldots, \{n_k\} = \{n_0\}$ be a family of regular polygons with common vertex $V$, such that each edge incident with $V$ of one of the polygons is an edge of precisely one other; we assume the family and the notation chosen so that $n_{j-1}$ and $n_j$ share such an edge $E_j$ for $j = 1, 2, \ldots, k$. Then we associate with the family of polygons the vertex neighborhood symbol $\epsilon_1 n_1 . \epsilon_2 n_2 . \ \ldots \ . \epsilon_k n_k$ in which each $\epsilon_j$ is either $+1$ or $-1$. We choose $\epsilon_1$ arbitrarily, and the other $\epsilon_j$ are determined, step by step, as follows: for $j \geqslant 2$, choose $\epsilon_j$ so that $\epsilon_{j-1} . \epsilon_j$ is $+1$ or $-1$ depending on whether $E_{j-1}$ and $E_{j+1}$ are separated by the line that carries $E_j$, or not. When an apeirogon ($n_j = \infty$) is incident with $V$, then this rule fails since the line carrying $E_j$ contains the whole apeirogon. In this case it is convenient to imagine the apeirogon to be a zigzag $\{\infty^\alpha\}$ with $\alpha$ slightly less than $\pi$, and to stick to this same choice of $\alpha$ throughout the assignment of signs in the vertex neighborhood symbol. Of course the fact that we have a choice for $\alpha$ means that the sign of $\infty$ in the vertex neighborhood symbol is arbitrary, so we shall usually take it to be positive. In Figures 7 and 8 we show some examples of vertex neighborhood symbols. Clearly they are arbitrary within cyclic permuta-

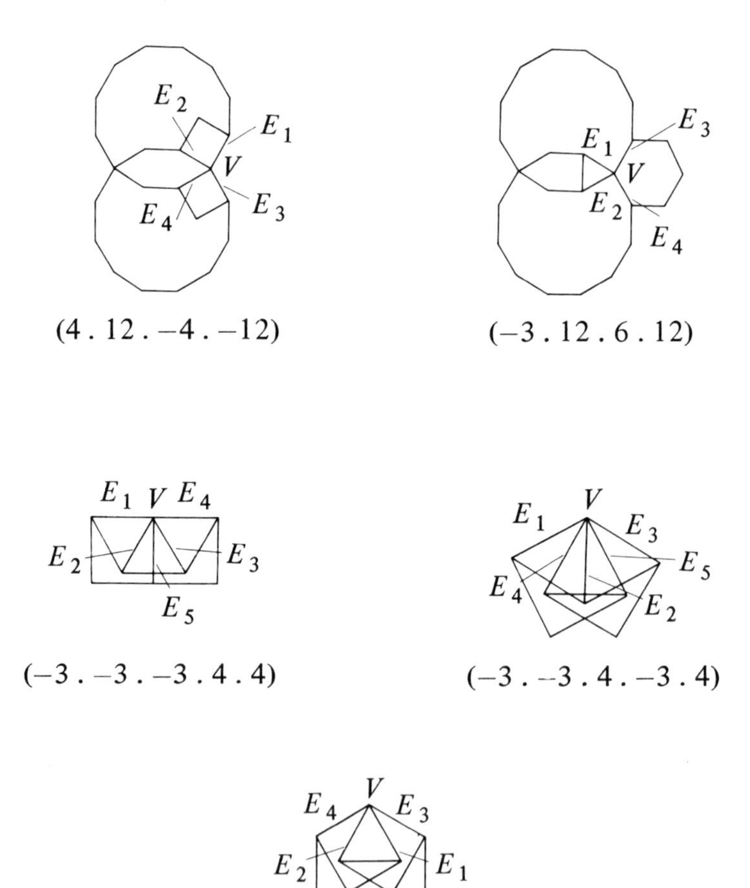


**Figure 7.** Examples for the assignment of symbols to families of convex polygons whose members have a common vertex.

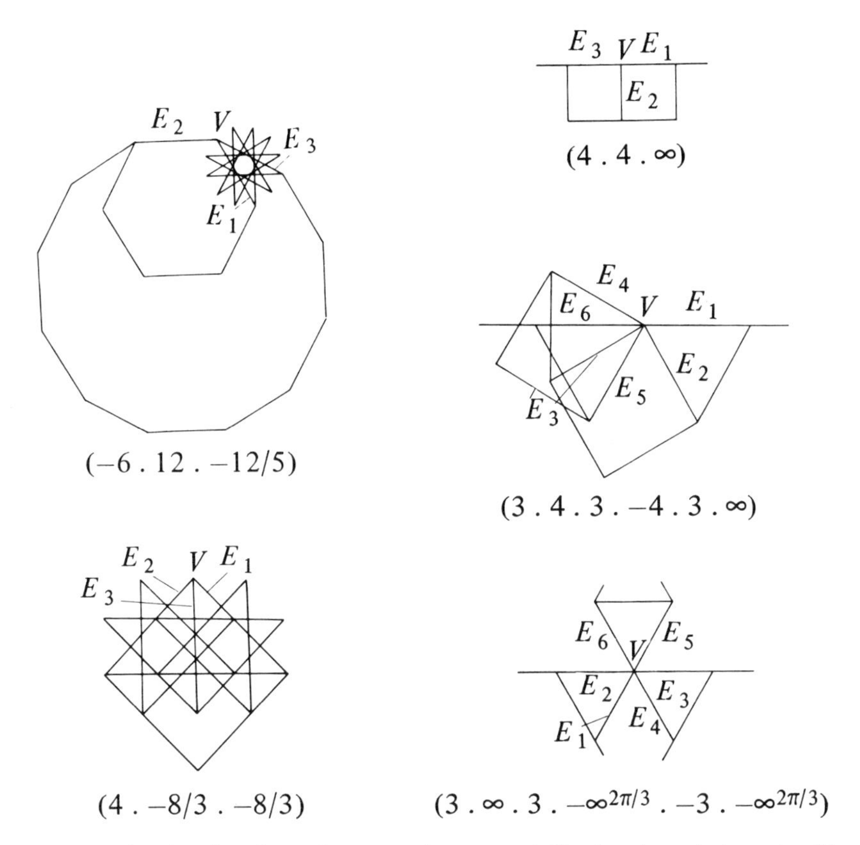


**Figure 8.** Examples showing the assignment of vertex neighborhood symbols to families of polygons which include star polygons or infinite polygons.

tions, reversal of order, or change of sign; we shall usually (but not always) select that which is lexicographically first among the various possibilities and includes a minimal number of negative signs.

The notation for uniform tilings is now easy to obtain: a tiling is denoted by the symbol (its *Schläfli symbol*) that is obtained by enclosing between parentheses the vertex neighborhood symbol assigned to the family of polygons incident with one (and hence with any) vertex of the tiling. Thus the traditional uniform tilings by polygonal regions lead to the same symbols as the uniform tilings (in our sense) obtained by considering the bounding polygons of the regions as hollow tiles. For $k$-uniform tilings the Schläfli symbol consists of parentheses enclosing the vertex neighborhood symbols, associated with the various transitivity classes of vertices, separated by semicolons.

It is not hard to verify that no finite tiling is uniform, and that the only uniform tilings contained in a convex, proper subset of the plane are the two tilings with Schläfli symbols (3.3.3.∞) and (4.4.∞) shown in Figure 6. The other tilings in Figures 5 and 6 are $k$-uniform for $k = 2$, 3, or 4, and have the indicated symbols.

The "new" uniform tilings described below differ from the traditional ones either because star-polygons or infinite polygons are admitted as tiles, or because the "interiors" of the polygons used are not disjoint. In many of the diagrams it is harder to pick out the tiles that form the tilings than in the traditional case. Therefore, in the following illustrations, we have indicated the polygons incident with one vertex in the tiling. This is especially necessary in those cases where the same set of edges and vertices corresponds to a number of different tilings, or if edges of different polygons partly overlap. The reader is invited to verify in each case that the tiling indeed exists and is uniform.

For convenience, we have distributed the tilings in three groups according to the nature of the polygons involved. The lists in Tables 1, 2, and 3 contain additional information about the tilings. We would have liked very much to present theorems asserting—as we believe to be true—that the list in each table is complete. But we can assert only the following:

**Theorem 1.** *There exist at least* 25 *types of uniform tilings in the plane in which all polygons are finite. They are listed in Table* 1 *and illustrated by the diagrams in Figure* 9.

Theorem 1 and Table 1 are based on the results of Badoureau [3], augmented by those of Coxeter et al. [7]; see Section 5. A proof of Theorem 1, which parallels the usual proof for the enumeration of Archimedean tilings (see for example [13]), can be constructed utilizing our knowledge of the size of angles that occur at vertices of the polygons under consideration. Thus we can draw up a long list of candidates for vertex figures—many of which can be eliminated on simple combinatorial or geometric grounds—and then each remaining one can be checked in detail as to the possibility of its realization in a tiling. The details are intricate and the checking long and tedious, but we strongly believe that our enumeration is complete.

Concerning tilings with finite polygons and apeirogons the situation is in principle similar, but our checking has not been so systematic. Therefore we can only conjecture that the enumeration in Table 2 is complete.

**Theorem 2.** *There exist at least* 28 *distinct uniform tilings with finite polygons and apeirogons. Except for the two uniform strip tilings* $(3.3.3.\infty)$ *and* $(4.4.\infty)$ *shown in Figure* 6, *they are listed in Table* 2 *and illustrated in the diagrams of Figure* 11.

In searching for uniform tilings using finite polygons and apeirogons, the following procedure is useful. For any such tiling $\mathfrak{T}$ of unit edge length, denote by $\mathfrak{N}$ the set of edges and vertices of $\mathfrak{T}$ (the *edge net* of $\mathfrak{T}$). Now find a vertex figure for $\mathfrak{N}$ (as defined above) using in the construction *all* the regular polygons of unit edge length in $\mathfrak{N}$, and not just those that are tiles of $\mathfrak{T}$. The result is a finite graph $G$ whose vertices lie on a unit circle centered at $V$, see Figure 14. The vertex figure of $\mathfrak{T}$ is a Hamiltonian circuit in $G$, that is to say, a circuit in $G$ which visits every vertex of $G$ exactly once. The various tilings with the same edge-net can then be determined by selecting from $G$, as potential vertex figures, all possible Hamiltonian circuits in $G$ and testing whether or not these circuits correspond to actual tilings. There seems to be no very easy way to carry out this last step; nevertheless the usefulness of the procedure is shown by the fact that several new tilings have been discovered by its means.

The completeness of our enumeration of uniform tilings is even less certain if zigzags are included. Here the analysis of possible vertex figures is difficult due to the fact that any length of line segment in a vertex figure can correspond to a zigzag of appropriately chosen angle. So again the most we can assert is the following:

**Theorem 3.** *The uniform tilings that contain zigzags form at least ten infinite families of tilings, of which eight depend on continuous parameters and two on discrete parameters, and* 23 *individual tilings. These are listed in Table* 3 *and illustrated by diagrams in Figure* 15.

Examination of the tables and diagrams reveals several facts that are remarkable in that they show the contrast between the uniform tilings by hollow tiles and the traditional tilings. In particular, we mention the following:

(i) While the eleven traditional uniform tilings have as their groups of symmetry only five of the 17 crystallographic planar groups (*cmm*, *p*4*g*, *p*4*m*, *p*6, *p*6*m*), and the symmetry groups of the 14 "new" tilings with finite polygons are also among these five, the symmetry groups of the uniform tilings with apeirogons or zigzags include also *p*2, *p*3, *p*4, *cm*, *pmg*, *pgg*, *p*3*m*1 and *p*31*m*.

(ii) In contrast to the traditional situation, the Schläfli symbol (or the vertex figure) of a tiling does not suffice in all cases to determine a tiling uniquely. Examples of distinct tilings with the same vertex figures appear in both Tables 2 and 3. It is also possible for every vertex of a biuniform tiling to have the same vertex figure, and for this vertex figure to coincide with that of a uniform tiling; see Figure 13. Hence, unlike the situation that occurs in the traditional case (see [13]) there exist non-uniform Archimedean tilings by hollow tiles.

## 5. Notes on the Past and Challenges for the Future

The concept of a hollow tile possibly originated many centuries ago when astrologers and other mystics used the pentagram as a magical symbol and drew it with five straight strokes of the pen. Illustrations by Leonardo da Vinci for Luca Pacioli's book *De Divina Proportione* (1509) show polyhedra bounded by hollow polygons, but the first mathematical treatment of the idea seems to be that of Johannes Kepler. In his book *Harmonices Mundi* published in 1619 [19], Kepler defined a pentagram $\{5/2\}$ in a very modern spirit as a regular hollow pentagon, and it seems likely that this led him to the discovery of the two regular star polyhedra that bear his name. However, there appears to be a certain lack of
(continued on page 48)

**Table 1.** Uniform Tilings with Finite Polygons[a]

| List number (1) | Edge net (2) | Schläfli symbol (3) | Symmetry group (4) | Orient-ability (5) | Density (6) | References: Badoureau (7) | References: Coxeter et al. (8) | References: Miller (9) | References: Kepler (10) |
|---|---|---|---|---|---|---|---|---|---|
| 1 | 19*a* | (3.3.3.4.4) | *cmm* | *O* | 1 | — | * | 105 | *M** |
| 2 | 19*c* | (3.3.3. − 4. − 4) | *cmm* | *O* | 1 | — | * | 106 | — |
| 3 | 35*b* | (3.3.4.3.4) | *p*4*g* | *O* | 1 | 66 | $\|\,244 = s\{{4 \atop 4}\}$ | 102 | *N* |
| 4 | 35*d* | (3.3. − 4.3. − 4) | *p*4*g* | *O* | 1 | — | $\|\,2\tfrac{4}{3}\tfrac{4}{3} = s'\{{4 \atop 4}\}$ | 103 | — |
| 5 | 38*a* | (4.8.8) | *p*4*m* | *O* | 1 | 65 | $24\,\|\,4 = t\{4,4\}$ | 79 | *V* |
| 6 | 38*b* | (4. − 8.8/3) | *p*4*m* | *O* | 1 | 152 | $2\tfrac{4}{3}4\,\| = t'\{{4 \atop 4}\}$ | 84 | — |
| 7 | 38*c* | (8.8/3. − 8. − 8/3) | *p*4*m* | *O* | 0 | 151 | $\tfrac{4}{3}4\tfrac{2}{\infty}\,\|$ | 83 | — |
| 8 | 38*d* | (− 4.8/3.8/3) | *p*4*m* | *O* | 1 | — | $24\,\|\,\tfrac{4}{3} = t'\{4,4\}$ | 80 | — |
| 9 | 41*a* | (4.4.4.4) (*R*) | *p*4*m* | *O* | 1 | — | $4\,\|\,24 = \{4,4\}$ | 71 | *E* |
| 10 | 42*a* | (3.3.3.3.6) (*E*) | *p*6 | *O* | 1 | — | $\|\,236 = s\{{3 \atop 6}\}$ | 104 | *L* |
| 11 | 46*a* | (4.6.12) | *p*6*m* | *O* | 1 | 63 | $236\,\| = t\{{3 \atop 6}\}$ | 96 | *Mm* |
| 12 | 46*c* | (6. − 12.12/5) | *p*6*m* | *O* | 2 | 158 | $3\tfrac{6}{5}6\,\|$ | 91 | — |
| 13 | 46*d* | (4. − 6.12/5) | *p*6*m* | *O* | 1 | 155 | $23\tfrac{6}{5} = t'\{{3 \atop 6}\}$ | 100 | — |

| 14 | $48Aa$ | $(3.4.6.4)$ | $p6m$ | $O$ | 1 | 64 | $36\,|\,2 = r\left\{{3 \atop 6}\right\}$ | 93 | $Ii$ |
|---|---|---|---|---|---|---|---|---|---|
| 15 | $48Aa$ | $(-3.12.6.12)$ | $p6m$ | $O$ | 2 | 64 | $\frac{3}{2}6\,|\,6$ | 94 | $Ii$ |
| 16 | $48Aa$ | $(4.12.-4.-12)$ | $p6m$ | $N$ | — | 64 | $26^{3/2}_{3}\,|$ | 95 | $Ii$ |
| 17 | $48Ab$ | $(3.-4.6.-4)$ | $p6m$ | $O$ | 1 | 156 | $\frac{3}{2}6\,|\,2 = r'\left\{{3 \atop 6}\right\}$ | 97 | — |
| 18 | $48Ab$ | $(3.12/5.-6.12/5)$ | $p6m$ | $O$ | 2 | 156 | $36\frac{6}{5}$ | 98 | — |
| 19 | $48Ab$ | $(4.12/5.-4.-12/5)$ | $p6m$ | $N$ | — | 156 | $2\frac{6}{5}\,{}^{3/2}_{3}\,|$ | 99 | — |
| 20 | $48Ba$ | $(3.12.12)$ | $p6m$ | $O$ | 1 | 61 | $23\,|\,6 = t\{6,3\}$ | 86 | $S$ |
| 21 | $48Bb$ | $(12.12/5.-12.-12/5)$ | $p6m$ | $N$ | — | 154 | $\frac{6}{5}6^{3}_{\infty}\,|$ | 90 | — |
| 22 | $48Bd$ | $(-3.12/5.12/5)$ | $p6m$ | $O$ | 1 | — | $23\,|\,\frac{6}{5} = t'\{6,3\}$ | 87 | — |
| 23 | $49a$ | $(3.6.3.6)$ | $p6m$ | $O$ | 1 | 62 | $2\,|\,36 = \left\{{3 \atop 6}\right\}$ | 76 | $P$ |
| 24 | $50a$ | $(6.6.6)\ (R)$ | $p6m$ | $O$ | 1 | — | $\{6,3\}$ | 73 | $F$ |
| 25 | $51a$ | $(3.3.3.3.3.3)\ (R)$ | $p6m$ | $O$ | 1 | — | $6\,|\,23 = \{3,6\}$ | 72 | $D$ |

[a] The list number in *column (1)* refers both to the diagram of the tiling in Figure 9 and to its vertex figure in Figure 10. The *edge-net symbol* in *column (2)* is composed of two parts: first an integer indicating the dot pattern type of the set of vertices, in the notation of [18, Section 5.3], then a letter $a, b, c, \ldots$ to distinguish the edge-nets with the same set of vertices. *Column (3)* gives the Schläfli symbol of the tiling; the tilings are ordered lexicographically by their edge-net symbols and then—for tilings with the same edge-net—by their Schläfli symbols. *Column (4)* shows the symmetry group of the tiling using the International Symbol for the crystallographic groups [18, Section 1.4]. In *column (5)* we indicate by $O$ or $N$ the orientability or non-orientability of the tiling, and in *column (6)* the density, if this exists. The last four columns give references to diagrams in the paper of Badoureau [3], symbols used by Coxeter et al. [7], and diagrams in the works of Miller [21] and Kepler [19]. An asterisk in *column (8)* means that the tiling is mentioned without a symbol being assigned, and one in *column (10)* means that the diagram contains an error. In *column (3)* an entry $(E)$ indicates that the tiling occurs in two enantiomorphic forms, and $(R)$ denotes a regular tiling.

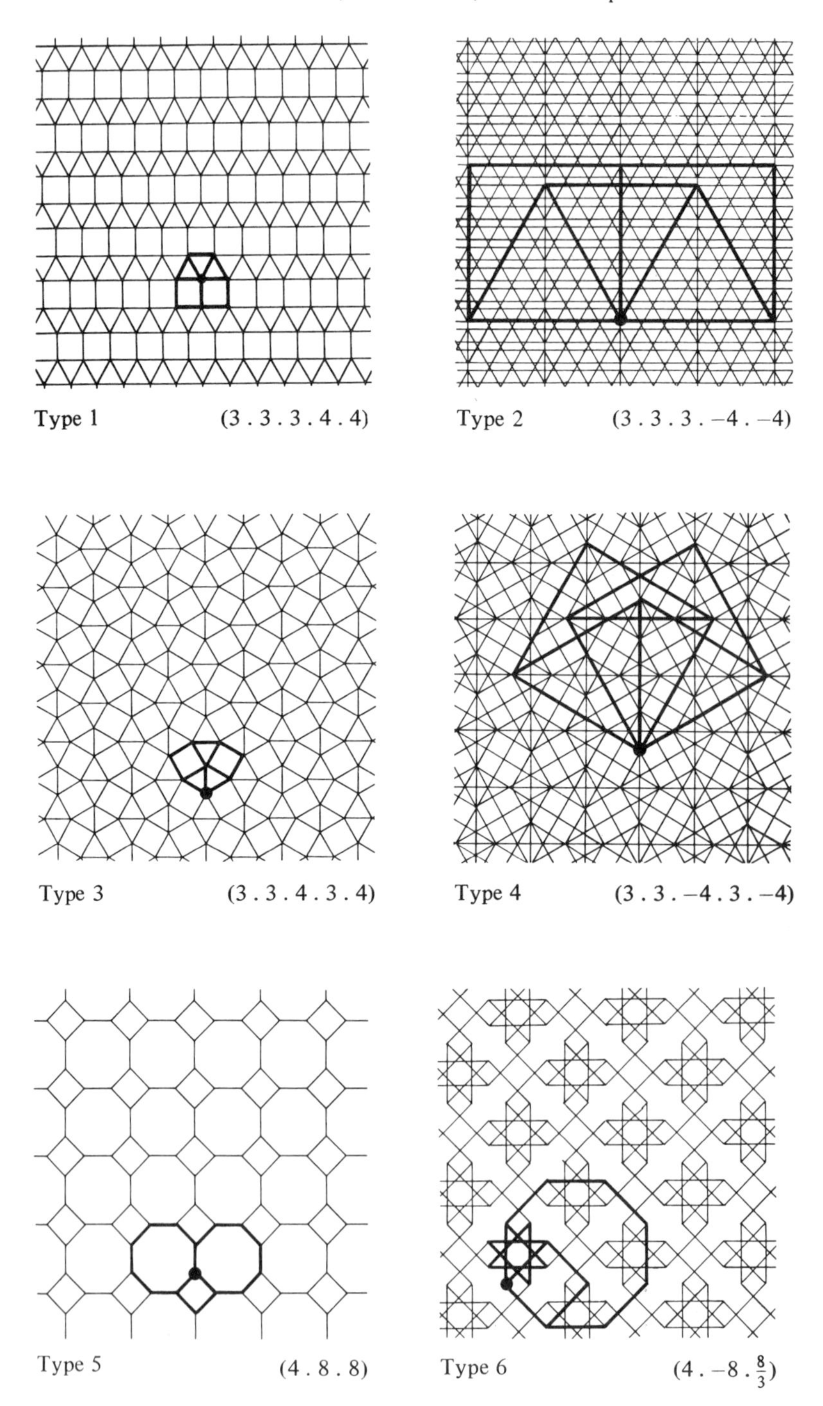


**Figure 9.** The 25 uniform tilings with finite polygons. For each tiling one vertex is emphasized by a solid dot, and all polygons which share that vertex are indicated by heavy lines. (Continued on pages 33–36.)

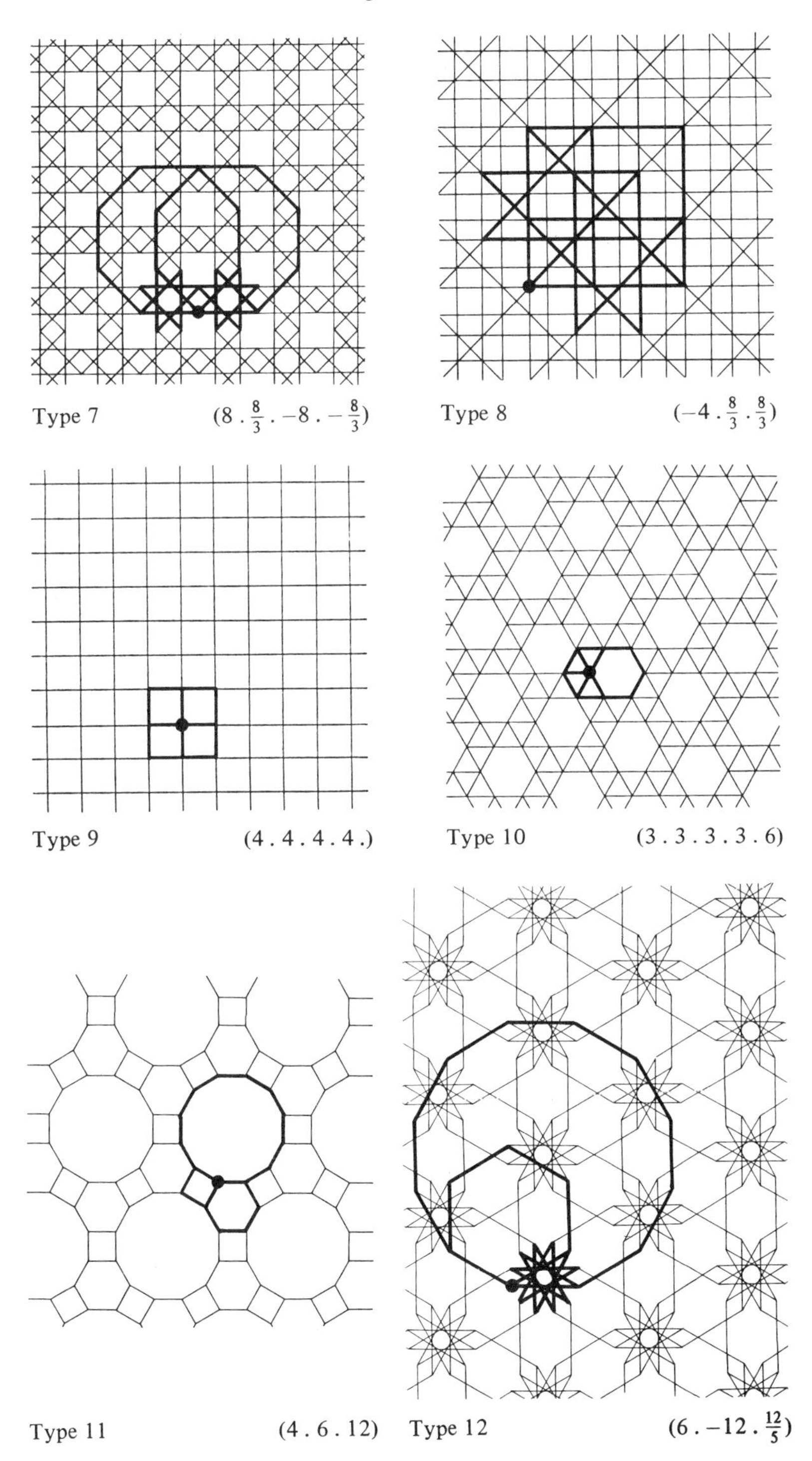


**Figure 9** (*continued*). See legend on page 32.

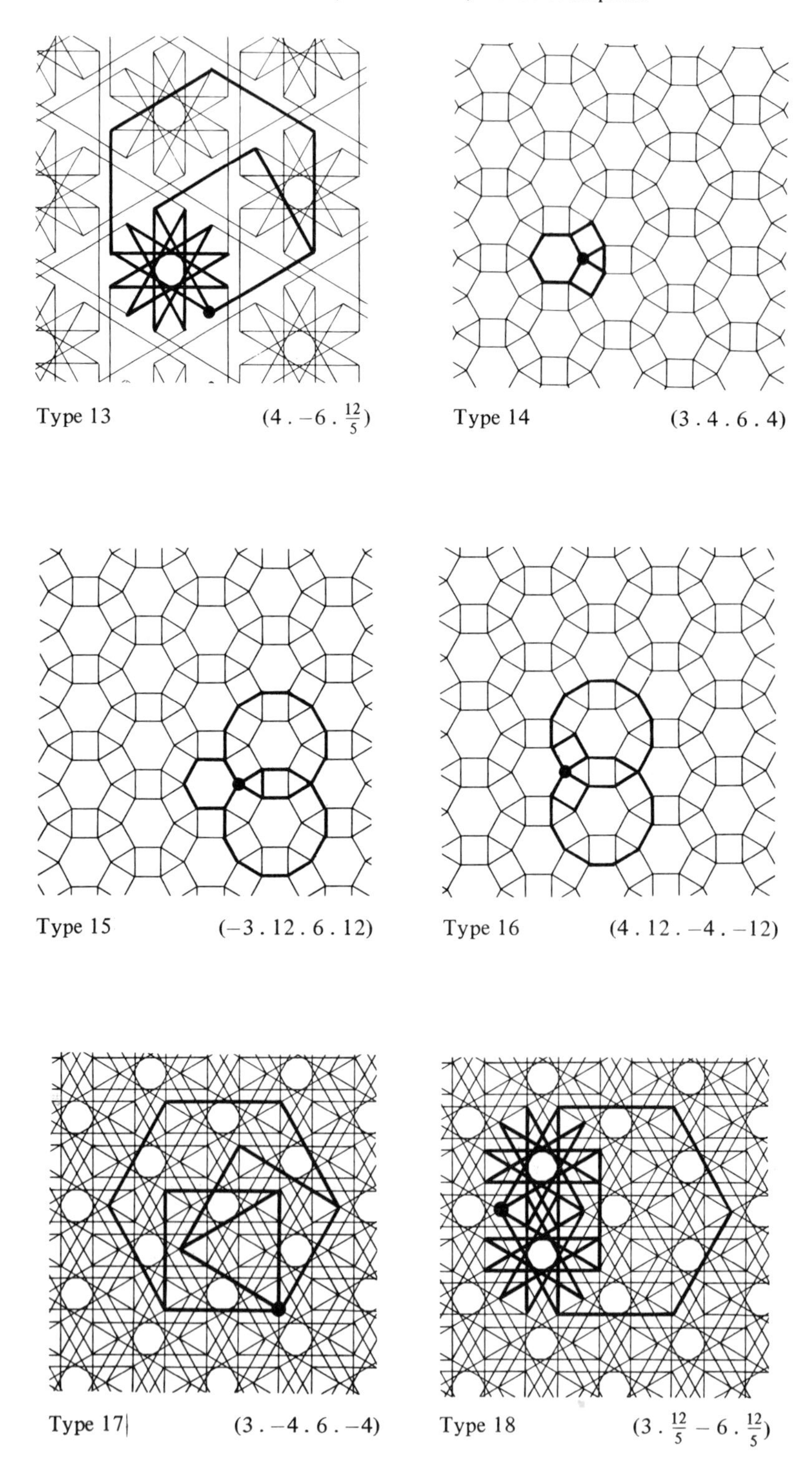


**Figure 9** (*continued*). See legend on page 32.

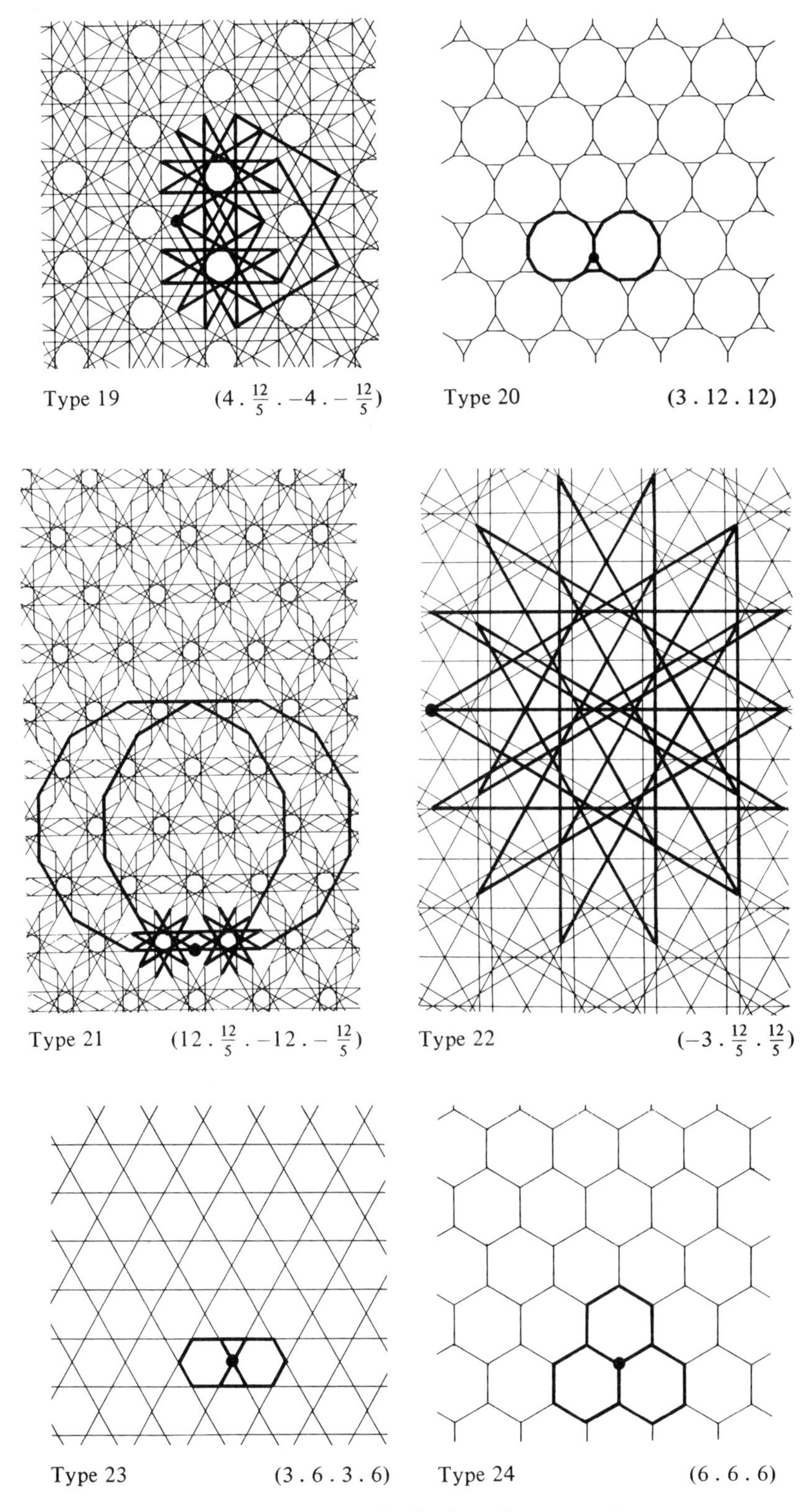


**Figure 9** (*continued*). See legend on page 32.

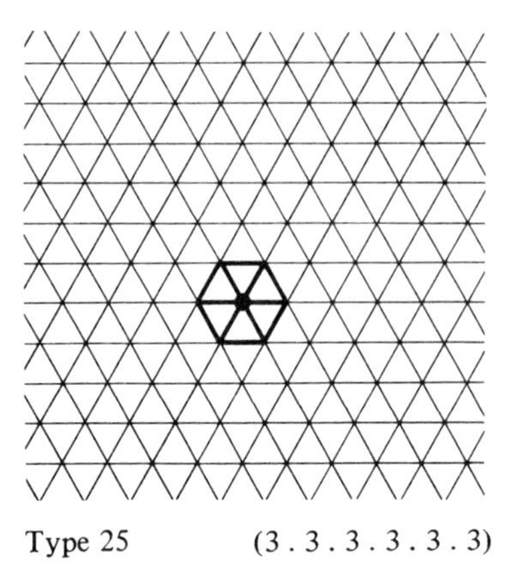


**Figure 9** (*continued*). See legend on page 32.

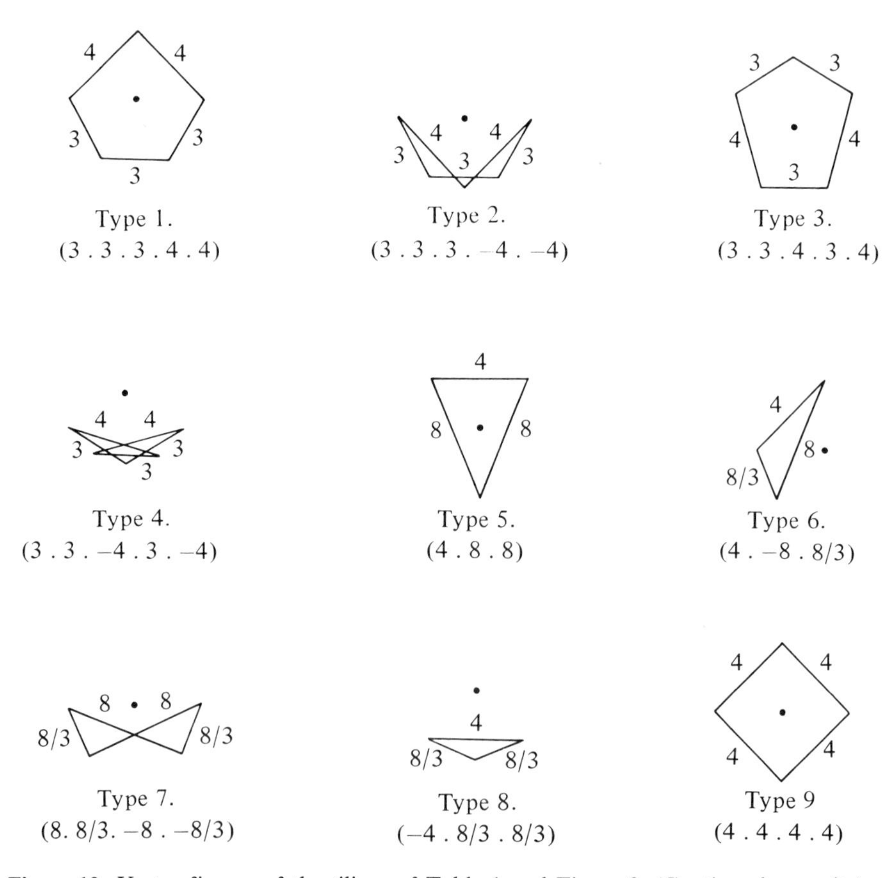


**Figure 10.** Vertex figures of the tilings of Table 1 and Figure 9. (Continued opposite).

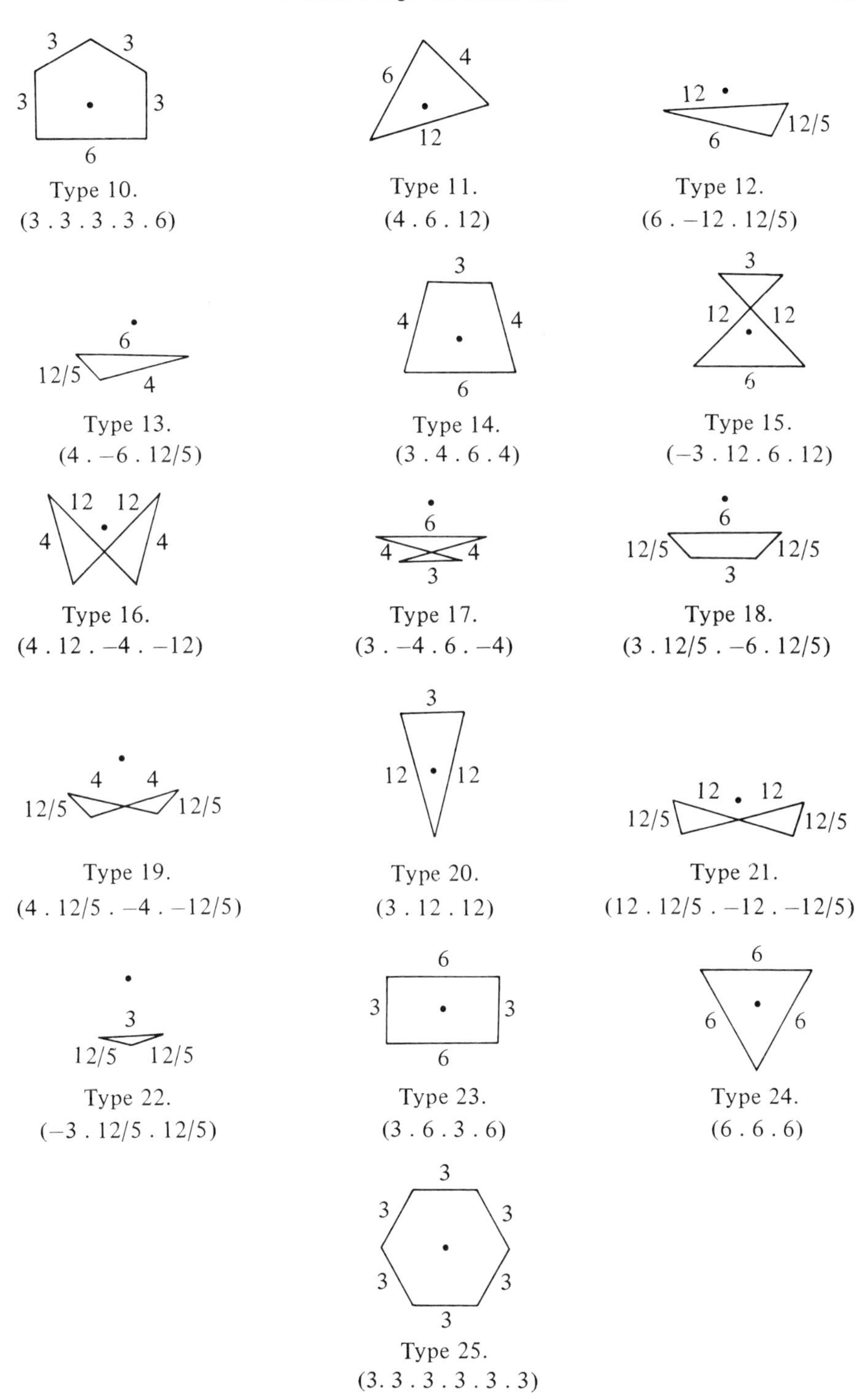


**Figure 10** (*continued*). See legend opposite.

**Table 2.** Uniform Tilings with Finite Polygons and Apeirogons[a]

| List number (1) | Edge net (2) | Schläfli symbol (3) | Symmetry group (4) | Orient-ability (5) | References | | | |
|---|---|---|---|---|---|---|---|---|
| | | | | | Density (6) | Badoureau (7) | Coxeter et al. (8) | Miller (9) |
| 1 | 19*b* | $(3.4.4.3.\infty.3.\infty)$ | *cmm* | *O* | 1 or 3 | — | — | — |
| 2 | 19*b* | $(3.-4.-4.3.\infty.3.\infty)$ | *cmm* | *O* | 1 | — | — | — |
| 3 | 25*c* | $(3.4.4.3.-4.-4.3.\infty)\ (E)$ | *p*3 | *N* | — | — | — | — |
| 4 | 35*c* | $(3.4.3.-4.3.\infty)\ (E)$ | *p*4 | *O* | 1 or 3 | — | $\mid \frac{4}{3} 4 \infty$ | — |
| 5 | 37*a* | $(8.8/3.\infty)$ | *p*4*m* | *O* | 1 or 3 | 153 | $\frac{4}{3} 4 \infty \mid$ | 85 |
| 6 | 38*c* | $(-4.8.\infty.8)$ | *p*4*m* | *O* | 3 or 5 | 151 | $\frac{4}{3} \infty \mid 4$ | 81 |
| 7 | 38*c* | $(4.8/3.\infty.8/3)$ | *p*4*m* | *O* | 1 or 3 | 151 | $4 \infty \mid \frac{4}{3}$ | 82 |
| 8 | 38*e* | $(4.8.8/3.-4.\infty)$ | *p*4*m* | *O* | 2 | — | — | — |
| 9 | 38*e* | $(-4.8.8/3.4.\infty)$ | *p*4*m* | *O* | 2 | — | — | — |
| 10 | 38*f* | $(4.8.-4.8.-4.\infty)_1$ | *p*4*m* | *N* | — | — | — | — |
| 11 | 38*f* | $(4.8.-4.8.-4.\infty)_2$ | *p*4*g* | *N* | — | — | — | — |
| 12 | 38*f* | $(4.8/3.4.8/3.-4.\infty)_1$ | *p*4*m* | *N* | — | — | — | — |
| 13 | 38*f* | $(4.8/3.4.8/3.-4.\infty)_2$ | *p*4*g* | *N* | — | — | — | — |

| | | | | | | | | |
|---|---|---|---|---|---|---|---|---|
| 14 | 41*a* | $(4.\infty.-4.\infty)$ | *p4m* | *O* | 1 | — | $\frac{4}{3}4 \mid \infty$ | 74 |
| 15 | 46*b* | $(12.12/5.\infty)$ | *p6m* | *O* | 3 or 6 | 157 | $\frac{6}{5}6\infty \mid$ | 92 |
| 16 | 48*Bb* | $(-6.12.\infty.12)$ | *p6m* | *O* | 9 or 12 | 154 | $\frac{6}{5}\infty \mid 6$ | 88 |
| 17 | 48*Bb* | $(6.12/5.\infty.12/5)$ | *p6m* | *O* | 3 or 6 | 154 | $6\infty \mid \frac{6}{5}$ | 89 |
| 18 | 48*Bc* | $(3.4.4.3.4.4.3.\infty)$ | *p6m* | *O* | 12 | — | — | — |
| 19 | 48*Bc* | $(3.-4.-4.3.-4.-4.3.\infty)$ | *p6m* | *O* | 3 | — | — | — |
| 20 | 48*Bc* | $(3.12.-6.12.3.\infty.3.\infty)$ | *p6m* | *O* | 12 or 15 | — | — | — |
| 21 | 48*Bc* | $(3.-12.6.-12.3.\infty.3.\infty)$ | *p6m* | *O* | 6 or 9 | — | — | — |
| 22 | 48*Bc* | $(3.12/5.6.12/5.3.\infty.3.\infty)$ | *p6m* | *O* | 6 or 9 | — | — | — |
| 23 | 48*Bc* | $(3.-12/5.-6.-12/5.3.\infty.3.\infty)$ | *p6m* | *O* | 0 or 3 | — | — | — |
| 24 | 49*a* | $(3.\infty.-3.\infty)$ | *p6m* | *O* | — | 62 | $\frac{3}{2}3 \mid \infty$ | 77 |
| 25 | 49*a* | $(6.\infty.-6.\infty)$ | *p6m* | *N* | — | 62 | $\frac{6}{5}6 \mid \infty$ | 78 |
| 26 | 51*a* | $(3.\infty.3.\infty.3.\infty)$ | *p3m1* | *O* | — | — | $\frac{3}{2} \mid 3\infty$ | 75 |

[a] All known tilings of this kind are listed except for the strip tilings $(3.3.3.\infty)$ and $(4.4.\infty)$ shown in Figure 6. The information given corresponds exactly to that of Table 1. Diagrams of the tilings are shown in Figure 11 and their vertex figures in Figure 12. The densities given in *column (6)* were obtained as explained in Section 3 with the following modification. If the tiling is orientable and the apeirogons can be paired so that the members of each pair are parallel but oriented in opposite directions, we interpret each such pair as a polygon $P$ with $P$-density equal to 1 or $-1$ for all points in the strip determined by $P$ (that is, lying between the two apeirogons of the pair). As the pairing can be carried out in more than one way, the $P$-density at each point of the plane is not uniquely determined and this may lead to different values of the density $d$. The values of $d$ given in *column (6)* were obtained, in each case, by pairing oppositely oriented parallel apeirogons nearest to each other.

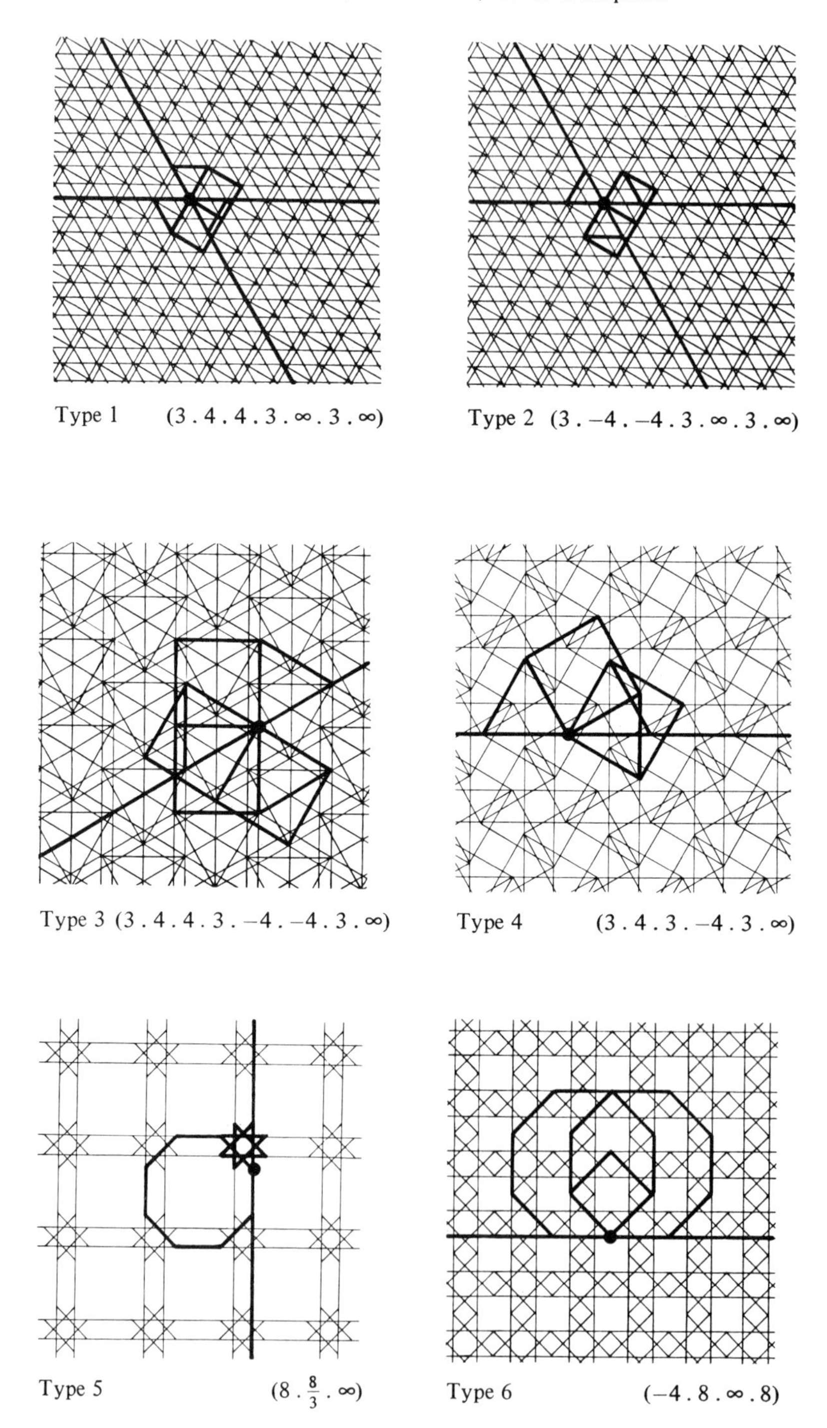


**Figure 11.** Uniform tilings with finite polygons and apeirogons. (Continued on pages 41–44.)

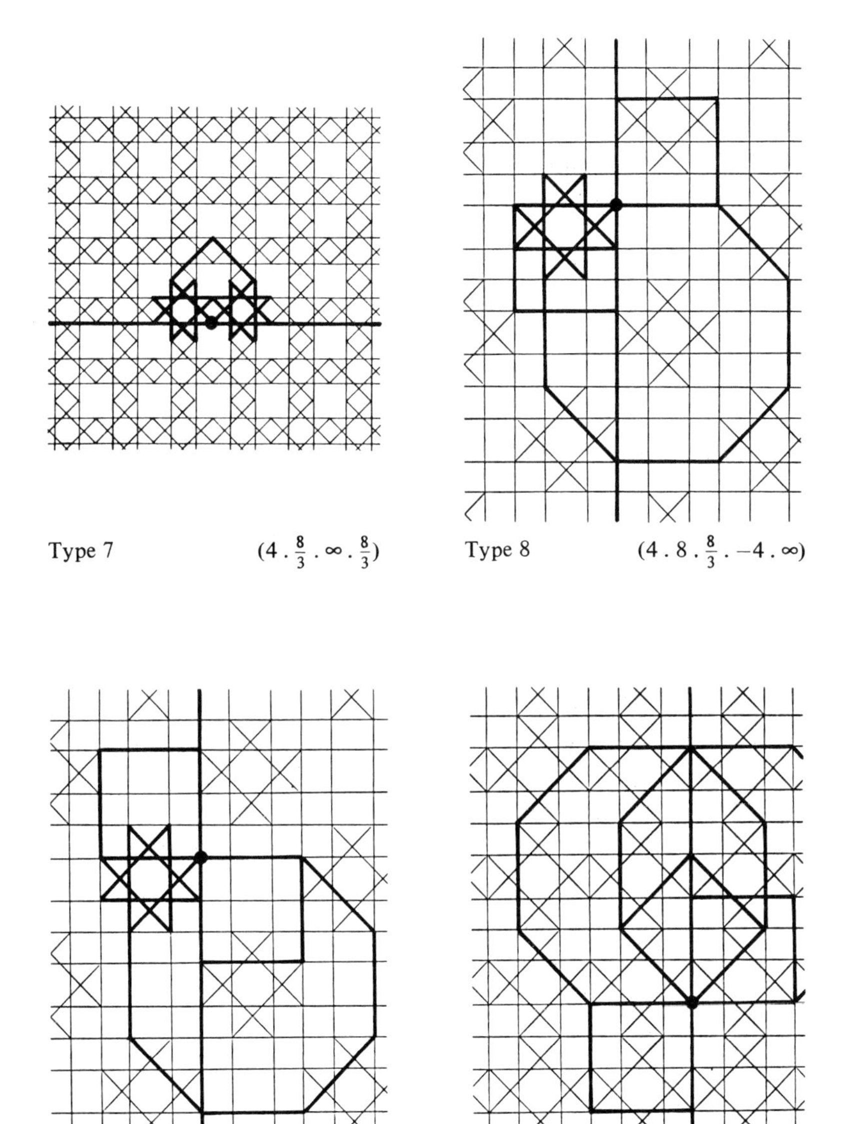


Type 9 $(-4 . 8 . \frac{8}{3} . 4 . \infty)$

Types 10, 11 $(4 . 8 . -4 . 8 . -4 . \infty)$

**Figure 11** (*continued*). See legend on page 40.

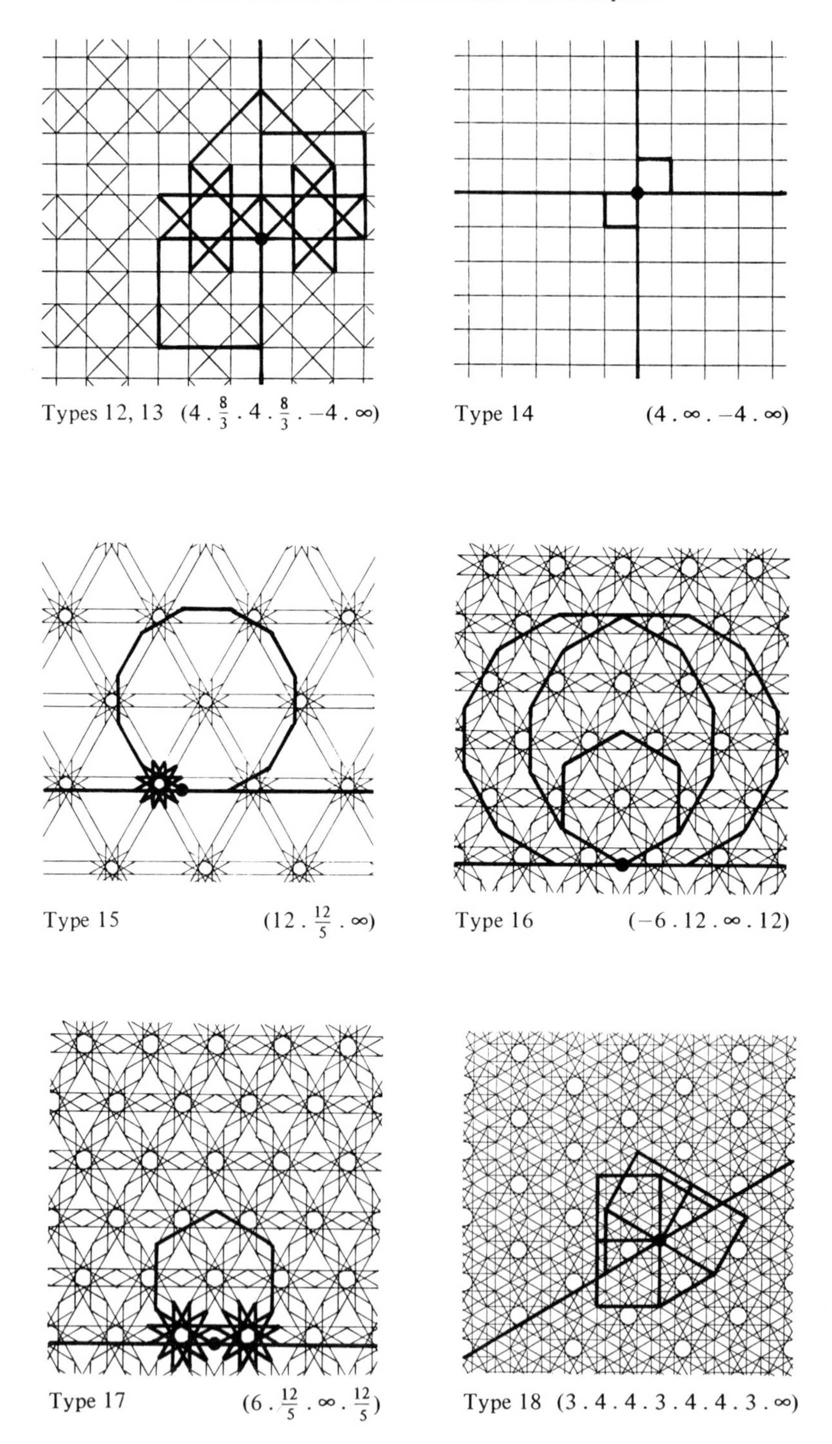


**Figure 11** (*continued*). See legend on page 40.

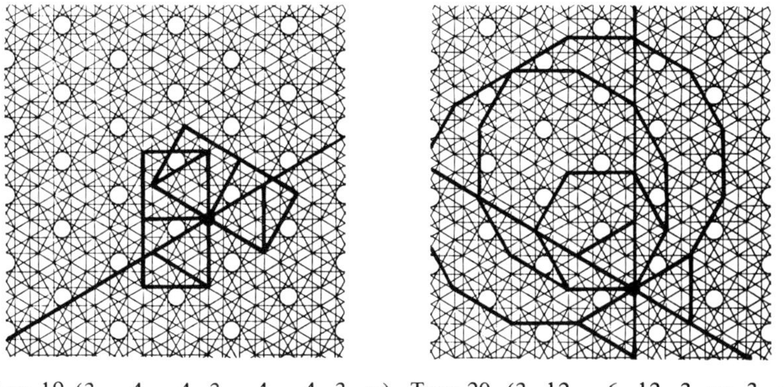

Type 19 (3 . −4 . −4 . 3 . −4 . −4 . 3 . ∞) Type 20 (3 . 12 . −6 . 12 . 3 . ∞ . 3 . ∞)

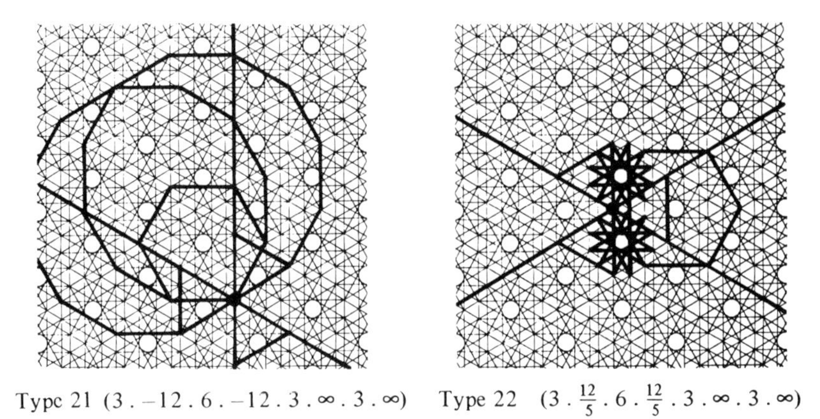

Type 21 (3 . −12 . 6 . −12 . 3 . ∞ . 3 . ∞) Type 22 $(3 . \frac{12}{5} . 6 . \frac{12}{5} . 3 . \infty . 3 . \infty)$

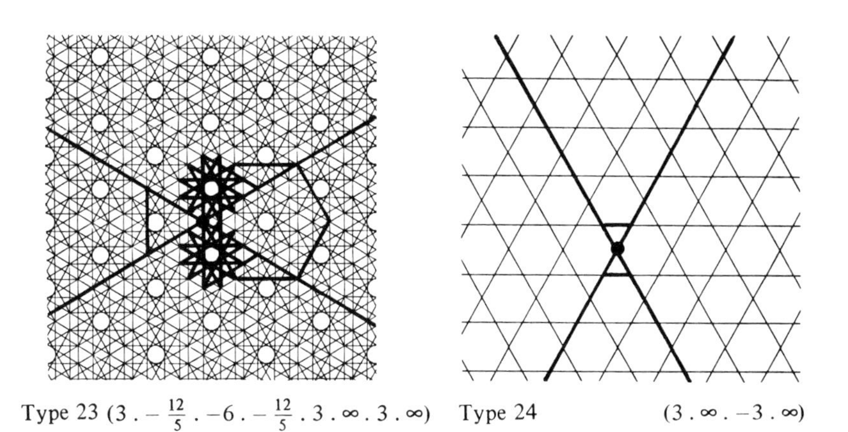

Type 23 $(3 . -\frac{12}{5} . -6 . -\frac{12}{5} . 3 . \infty . 3 . \infty)$ Type 24 (3 . ∞ . −3 . ∞)

**Figure 11** (*continued*). See legend on page 40.

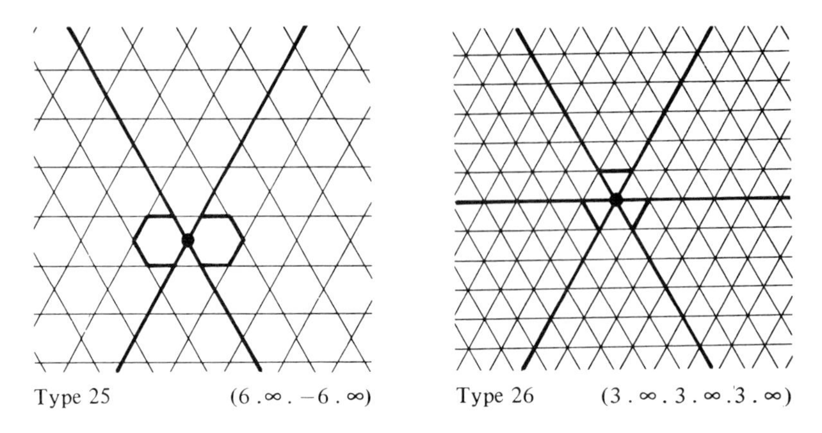


**Figure 11** (*continued*). See legend on page 40.

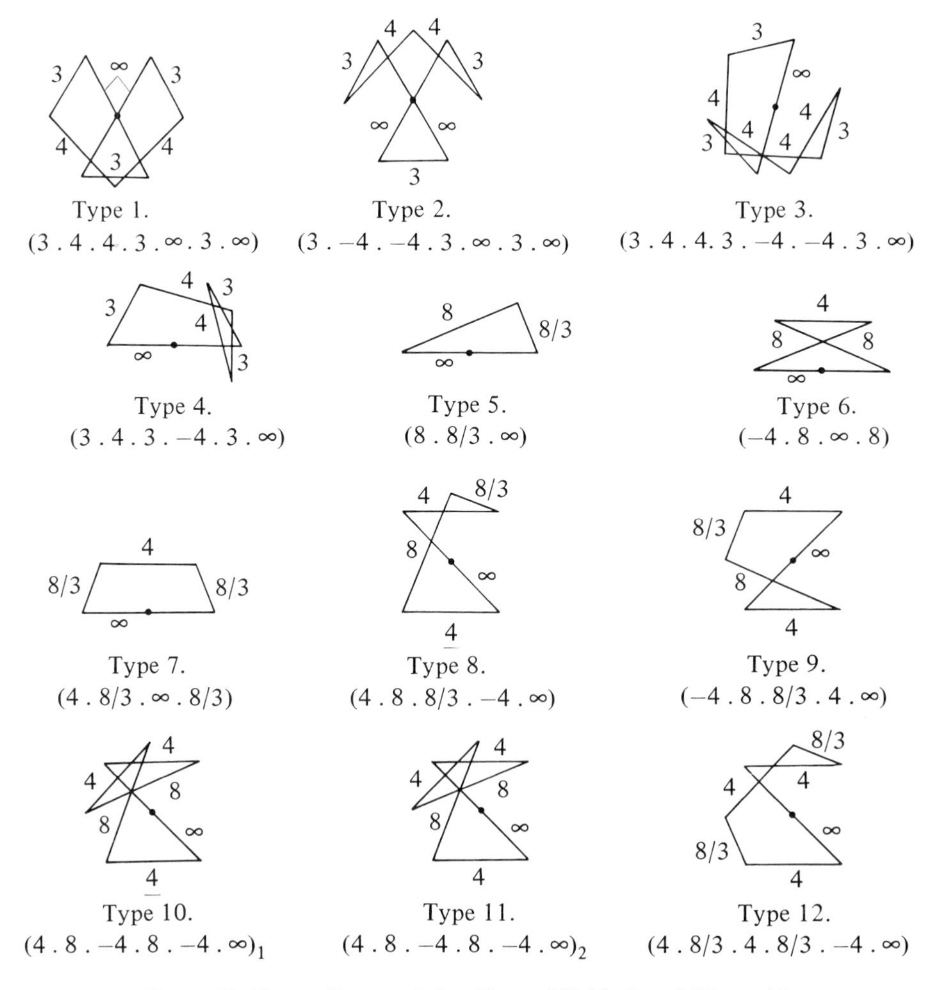


**Figure 12.** Vertex figures of the tilings of Table 2 and Figure 11.

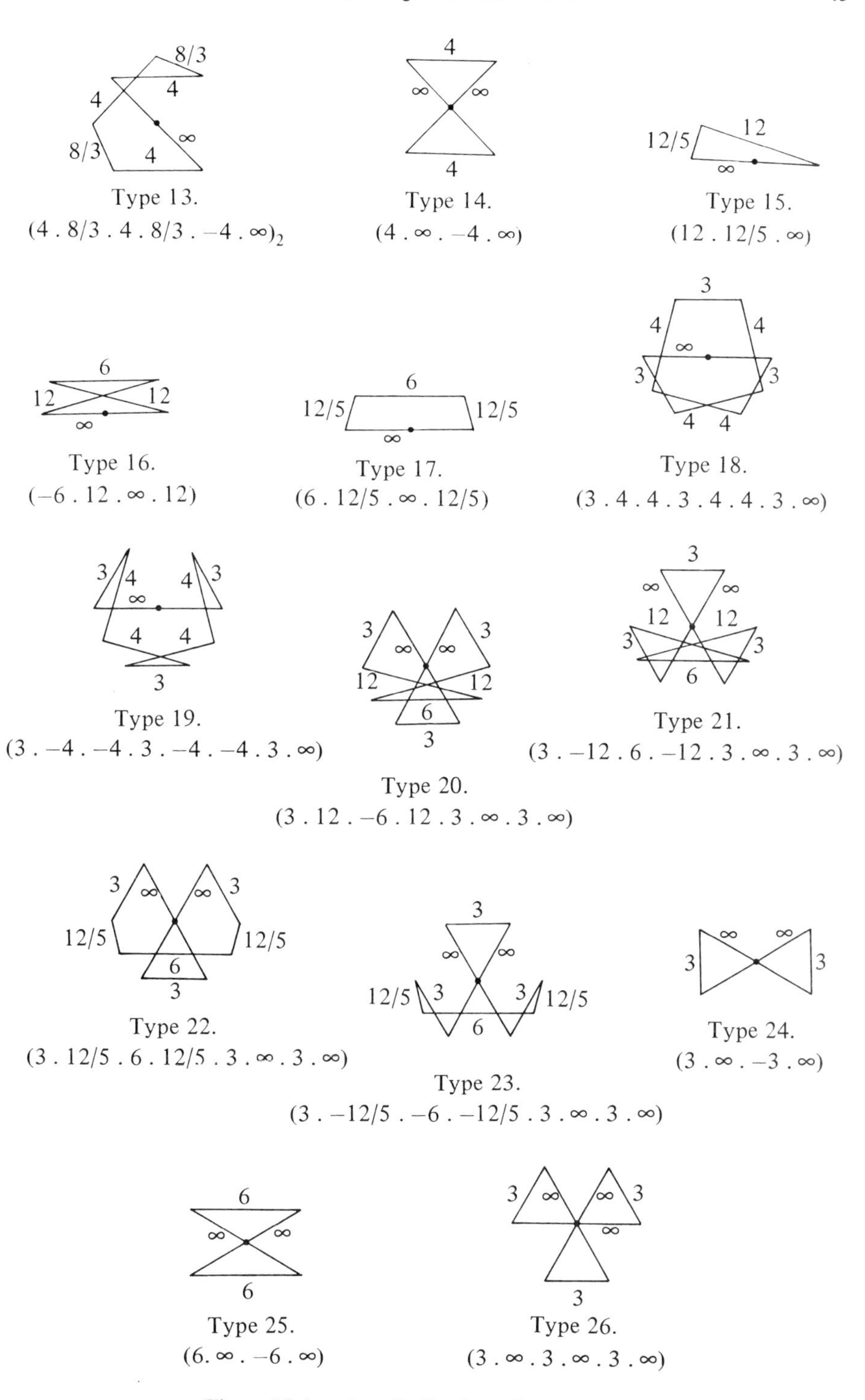


**Figure 12** (*continued*). See legend on page 44.

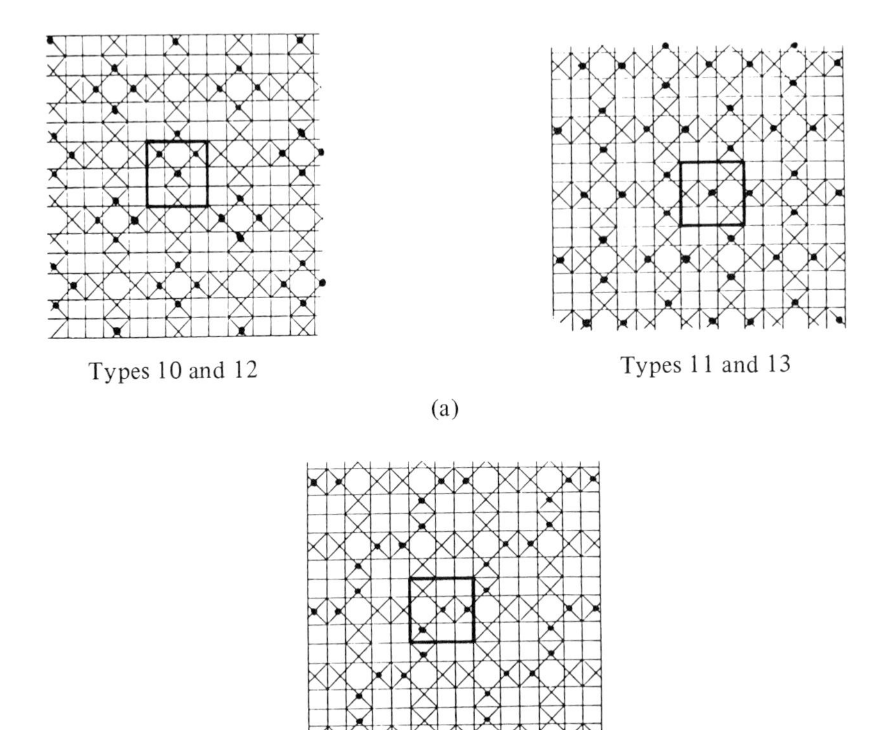


**Figure 13.** (*a*) The difference between tilings $(4.8.-4.8.-4.\infty)_j$ for $j = 1, 2$ (types 10 and 11 in Table 2) and between tilings $(4.8/3.4.8/3.-4.\infty)_j$ for $j = 1, 2$ (types 12 and 13 in Table 2). In the illustration the dots indicate the centers of the square tiles with horizontal and vertical edges, one of which is indicated by heavy lines. (*b*) Biuniform Archimedean tilings with the same Schläfli symbols and vertex figure as the uniform tilings shown in (*a*). Again the squares with horizontal and vertical edges are indicated by central dots.

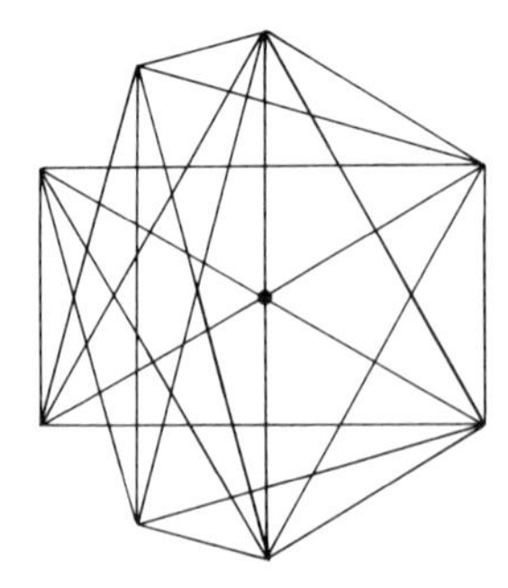

**Figure 14.** The vertex figure of the edge-net $48Bc$. Six Hamiltonian circuits in this graph are the vertex figures of uniform tilings (types 18 to 23 in Table 2 and Figure 11), see Figure 12.

**Table 3.** Uniform Tilings Which Include Zigzags.[a] (continued on page 48.)

| List number (1) | Edge net (2) | Schläfli symbol (3) | Symmetry group (4) | Orient-ability (5) |
|---|---|---|---|---|
| 1 | $7a$ | $(\infty^{\alpha}.\infty^{\beta}.\infty^{\gamma})$ for $\alpha+\beta+\gamma=2\pi$ $(R)(E)$ | $p2$ | $O$ |
| 2 | $7a$ | $(\infty^{\alpha}.\infty^{\beta}.-\infty^{\alpha+\beta})$ for $0<\alpha+\beta\leqslant\pi$ $(E)$ | $p2$ | $O$ |
| 3 | $9a$ | $(3.3.\infty^{\pi-\alpha}.-3.\infty^{\alpha+2\pi/3})$ for $0\leqslant\alpha\leqslant\pi/6$ | $pgg$ | $O$ |
| 4 | $9a$ | $(3.3.-\infty^{\pi-\alpha}.-3.\infty^{-\alpha+2\pi/3})$ for $0\leqslant\alpha<\pi/3$ | $pgg$ | $O$ |
| 5 | $16a$ | $(4.4.\infty^{\phi}.4.4.-\infty^{\phi})$ for $\phi=2\arctan(n/k)$, with $nk$ even and $(n,k)=1$ | $pmg$ | $O$ |
| 6 | $16a$ | $(4.4.\infty^{\phi}.-4.-4.-\infty^{\phi})$ for $\phi=2\arctan(n/k)$, with $nk$ even and $(n,k)=1$ | $pmg$ | $N$ |
| 7 | $19b$ | $(3.4.4.3.-\infty^{2\pi/3}.-3.-\infty^{2\pi/3})$ | $cmm$ | $O$ |
| 8 | $19b$ | $(3.-4.-4.3.-\infty^{2\pi/3}.-3.-\infty^{2\pi/3})$ | $cmm$ | $O$ |
| 9 | $19b$ | $(4.4.\infty^{\pi/3}.\infty.-\infty^{\pi/3})$ $(E)$ | $p2$ | $O$ |
| 10 | $19b$ | $(4.4.\infty^{2\pi/3}.\infty.-\infty^{2\pi/3})$ $(E)$ | $p2$ | $O$ |
| 11 | $20a$ | $(\infty.\infty^{\alpha}.\infty.-\infty^{\alpha})$ for $0<\alpha<\pi$ | $cmm$ | $O$ |
| 12 | $20a$ | $(\infty^{\alpha}.\infty^{\pi-\alpha}.\infty^{\alpha}.\infty^{\pi-\alpha})$ for $0<\alpha\leqslant\pi/2$ $(R)$ | $cmm$ | $O$ |
| 13 | $25a$ | $(3.\infty^{\alpha}.-3.-\infty^{\alpha})$ for $\pi/3<\alpha<\pi$ | $p31m$ | $O$ |
| 14 | $25b$ | $(4.4.\infty^{2\pi/3}.4.4.-\infty^{2\pi/3})$ | $p31m$ | $O$ |
| 15 | $25b$ | $(4.4.\infty^{\pi/3}.-4.-4.-\infty^{\pi/3})$ | $p31m$ | $O$ |
| 16 | $35a$ | $(4.\infty^{\alpha}.-4.-\infty^{\alpha})$ for $0<\alpha<\pi$, $\alpha\neq\pi/2$ | $p4g$ | $O$ |
| 17 | $38f$ | $(4.-8.\infty^{\pi/2}.\infty.-\infty^{\pi/2}.-8)_1$ | $cmm$ | $O$ |

[a] All known tilings of this kind are listed. The information given corresponds to that in Tables 1 and 2. Diagrams of the tilings are given in Figure 15 and their vertex figures in Figure 16. If a tiling belongs to a family which depends on a parameter, the permissible range of this parameter is shown in *column (3)*. In these cases the symmetry group indicated in *column (4)* refers to the "general" case; for special values of the parameter the tiling may have a larger symmetry group, or even be regular (this is indicated by $(R)$ in *column (3)*). Also, for particular values of the parameter there may occur "accidental" coincidences which destroy the unicursality of the vertex figures and hence the uniform character of the tiling. An $(E)$ in *column (3)* indicates that the tiling exists in two enantiomorphic forms.

**Table 3.** (*continued*)

| List number (1) | Edge net (2) | Schläfli symbol (3) | Symmetry group (4) | Orientability (5) |
|---|---|---|---|---|
| 18 | 38*f* | $(4.-8.\infty^{\pi/2}.\infty.-\infty^{\pi/2}.-8)_2\ (E)$ | *p*4 | *O* |
| 19 | 38*f* | $(4.8/3.\infty^{\pi/2}.\infty.-\infty^{\pi/2}.8/3)_1$ | *cmm* | *O* |
| 20 | 38*f* | $(4.8/3.\infty^{\pi/2}.\infty.-\infty^{\pi/2}.8/3)_2\ (E)$ | *p*4 | *O* |
| 21 | 48*Bc* | $(6.-12.\infty^{\pi/3}.\infty.-\infty^{\pi/3}.-12)\ (E)$ | *p*6 | *O* |
| 22 | 48*Bc* | $(6.-12.\infty^{2\pi/3}.\infty.-\infty^{2\pi/3}.-12)\ (E)$ | *p*6 | *O* |
| 23 | 48*Bc* | $(6.12/5.\infty^{\pi/3}.\infty.-\infty^{\pi/3}.12/5)\ (E)$ | *p*6 | *O* |
| 24 | 48*Bc* | $(6.12/5.\infty^{2\pi/3}.\infty.-\infty^{2\pi/3}.12/5)\ (E)$ | *p*6 | *O* |
| 25 | 51*a* | $(3.3.3.\infty^{2\pi/3}.-3.\infty^{2\pi/3})$ | *p*31*m* | *N* |
| 26 | 51*a* | $(3.\infty.3.-\infty^{2\pi/3}.-3.-\infty^{2\pi/3})$ | *cm* | *O* |
| 27 | 51*a* | $(3.\infty.-\infty^{2\pi/3}.\infty.-\infty^{2\pi/3}.\infty)$ | *p*31*m* | *N* |
| 28 | 51*a* | $(3.\infty^{2\pi/3}.\infty^{2\pi/3}.-3.-\infty^{2\pi/3}.-\infty^{2\pi/3})$ | *p*31*m* | *O* |
| 29 | 51*a* | $(\infty.\infty^{\pi/3}.\infty^{\pi/3}.\infty.-\infty^{\pi/3}.-\infty^{\pi/3})$ | *cmm* | *O* |
| 30 | 51*a* | $(\infty.\infty^{\pi/3}.-\infty^{2\pi/3}.\infty.\infty^{2\pi/3}.-\infty^{\pi/3})\ (E)$ | *p*2 | *O* |
| 31 | 51*a* | $(\infty.\infty^{2\pi/3}.\infty^{2\pi/3}.\infty.-\infty^{2\pi/3}.-\infty^{2\pi/3})$ | *cmm* | *O* |
| 32 | 51*a* | $(\infty^{\pi/3}.\infty^{\pi/3}.\infty^{\pi/3}.\infty^{\pi/3}.\infty^{\pi/3}.\infty^{\pi/3})\ (R)$ | *p*6*m* | *O* |
| 33 | 51*a* | $(\infty^{\pi/3}.-\infty^{2\pi/3}.-\infty^{2\pi/3}.\infty^{\pi/3}.-\infty^{2\pi/3}.-\infty^{2\pi/3})$ | *cmm* | *O* |

consistency in his work, for when he discusses tilings with star-polygons, he treats a pentagram as a non-convex planar region bounded by ten line segments. This is surprising, for Kepler was also the first to consider plane tilings as analogues of polyhedra. We wholeheartedly agree with this point of view and note that the only difference between tilings as defined here and polyhedra (as defined, for example, in [12]) is the requirement that all the polygons of a tiling lie in a plane. However, this close relationship was at times misunderstood—for example by Gergonne [9] who, in his account of regular and semiregular tilings and polyhedra composed of convex polygons, speaks of tilings of the sphere by congruent infinitely small squares (or equilateral triangles, or regular hexagons)!

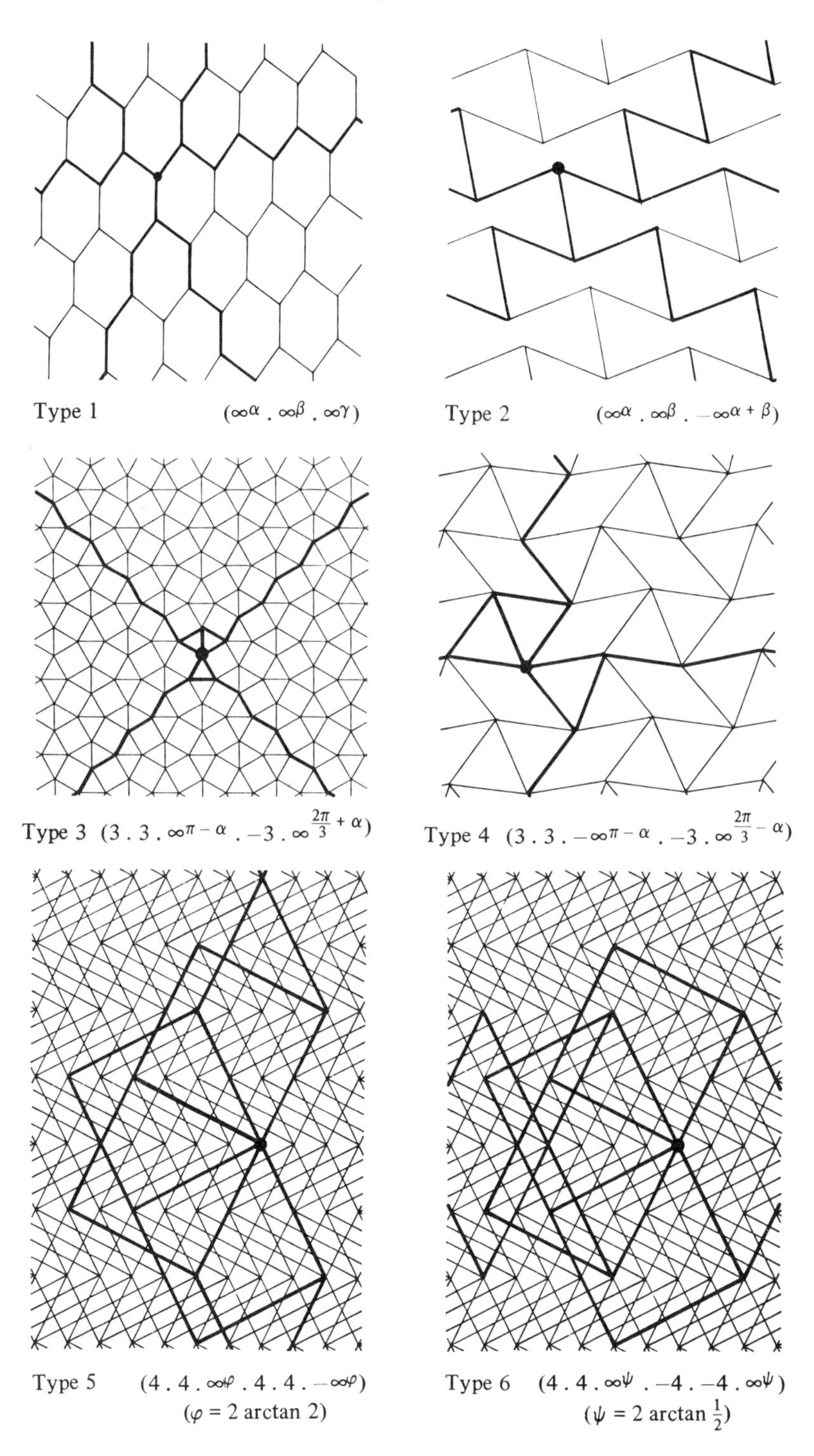


**Figure 15.** Uniform tilings which include zigzags. (Continued on pages 50–54.)

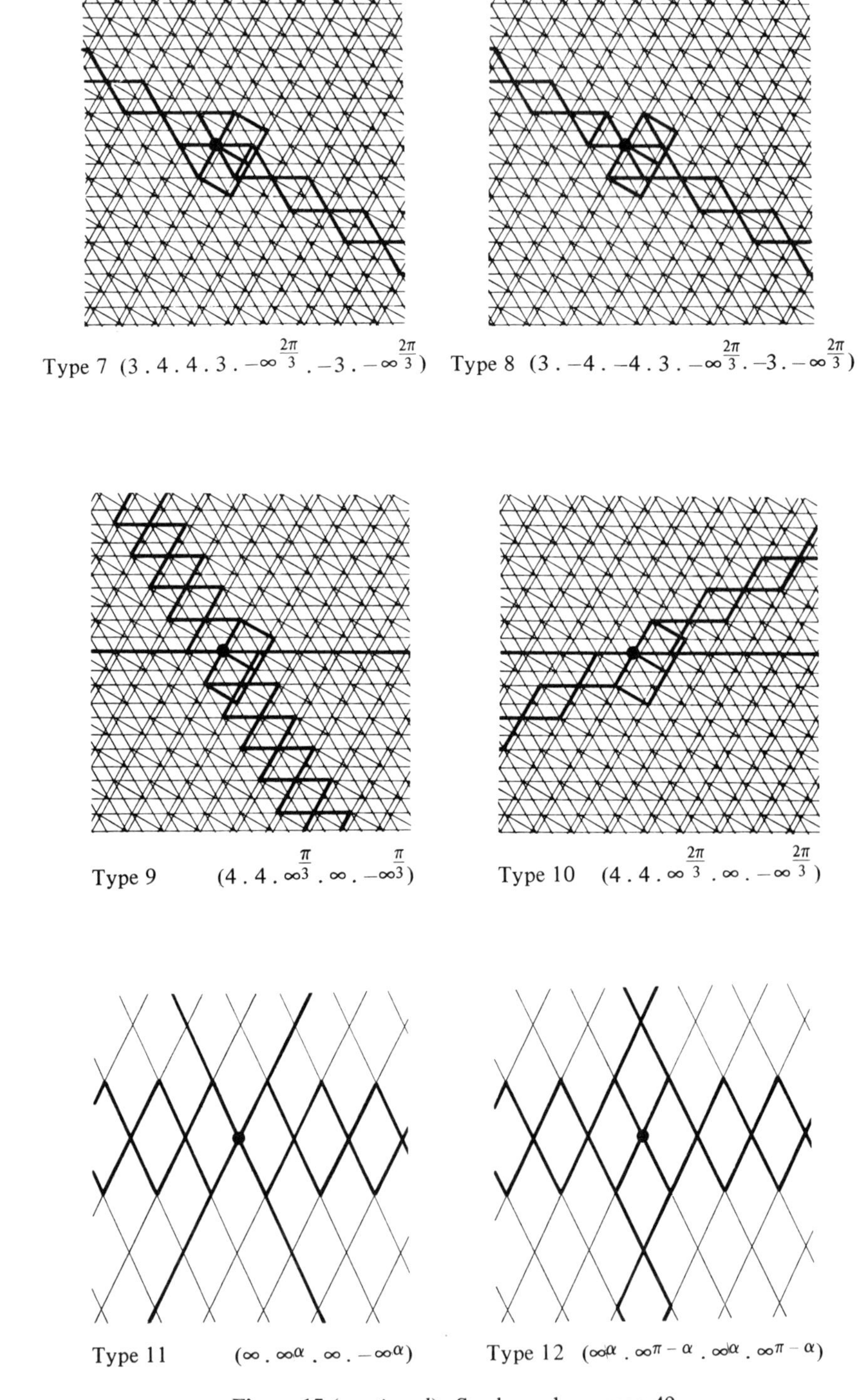


**Figure 15** (*continued*). See legend on page 49.

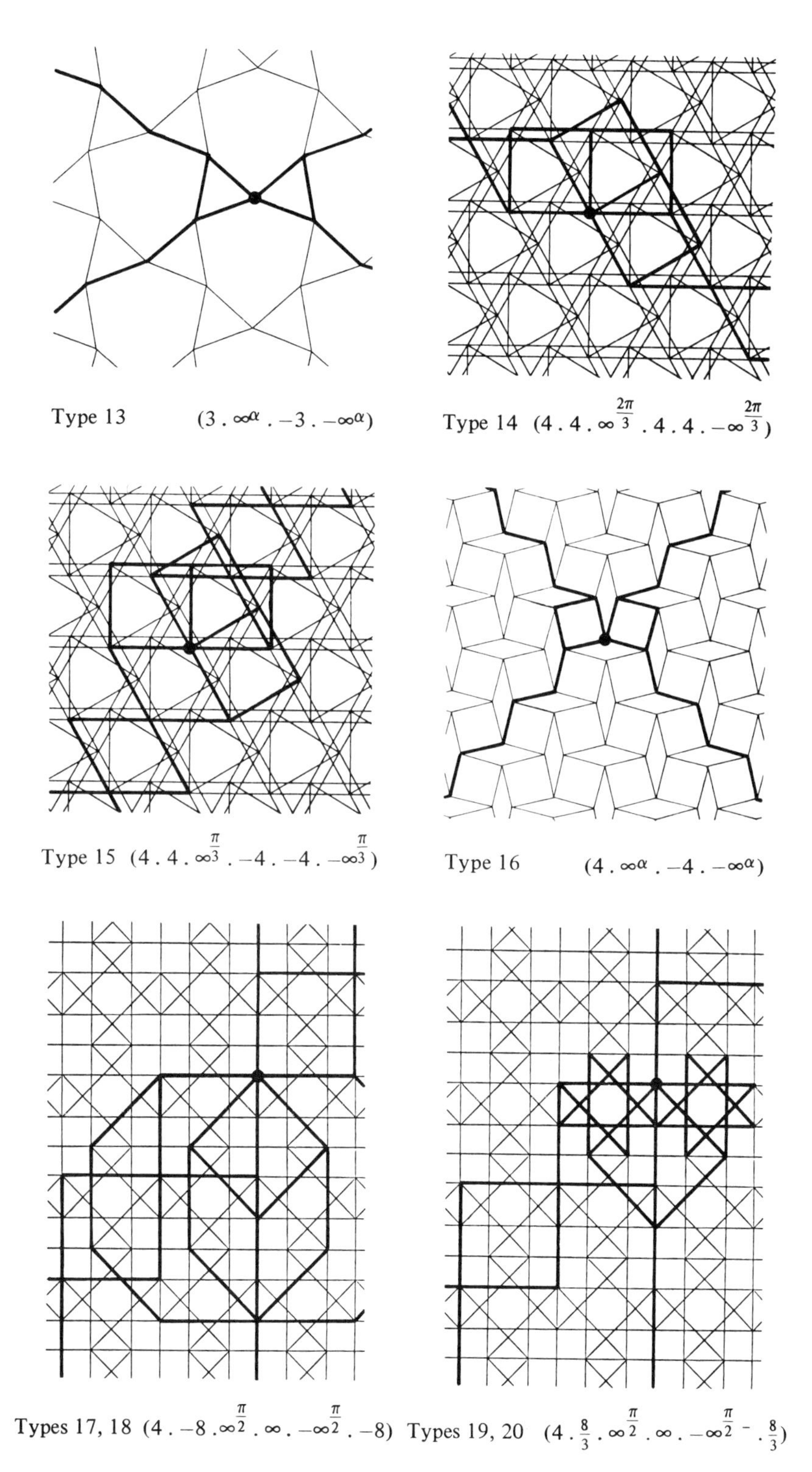


**Figure 15** (*continued*). See legend on page 49.

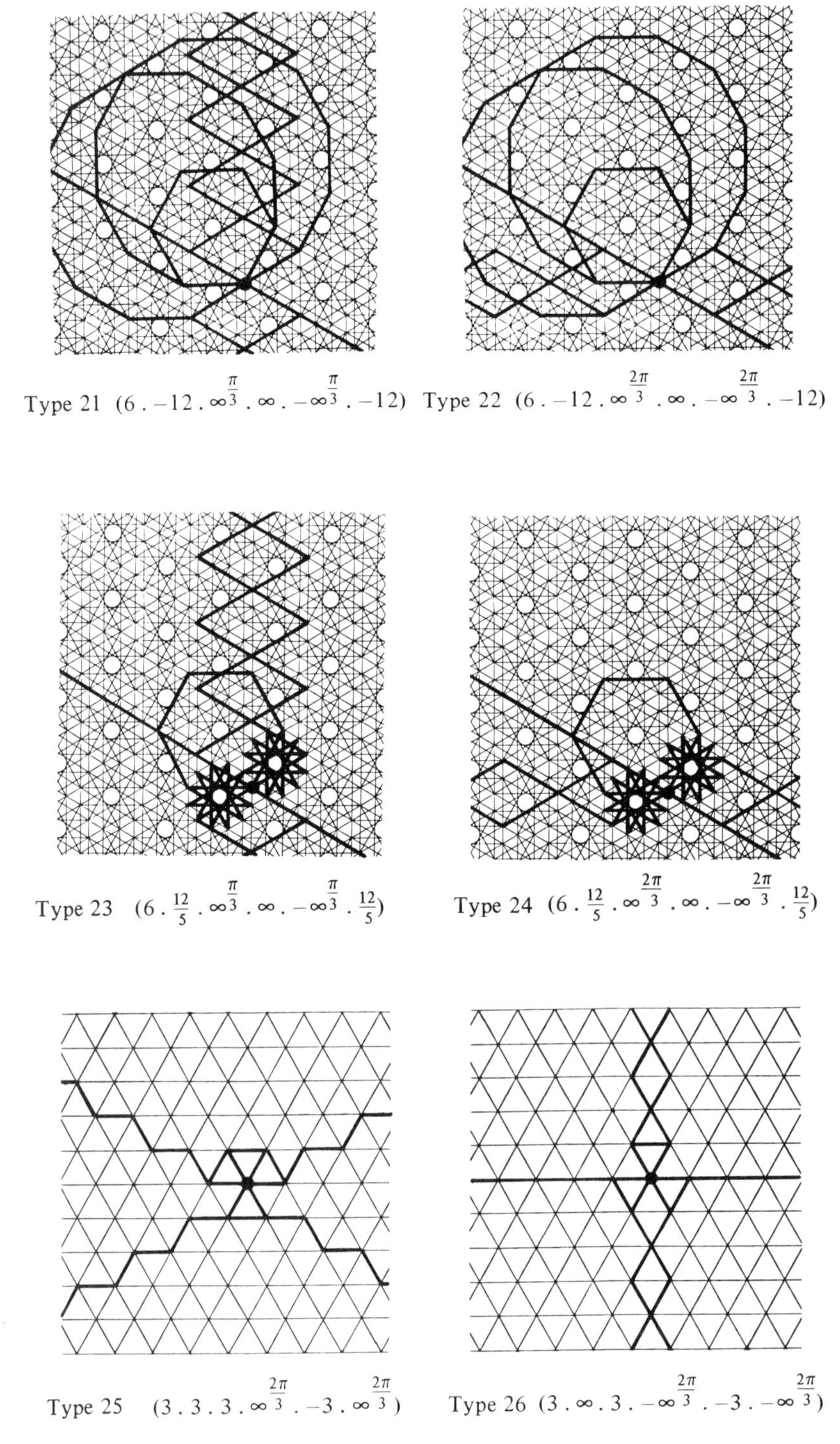

Type 21 $(6\,.\,-12\,.\,\infty^{\frac{\pi}{3}}\,.\,\infty\,.\,-\infty^{\frac{\pi}{3}}\,.\,-12)$ Type 22 $(6\,.\,-12\,.\,\infty^{\frac{2\pi}{3}}\,.\,\infty\,.\,-\infty^{\frac{2\pi}{3}}\,.\,-12)$

Type 23 $(6\,.\,\frac{12}{5}\,.\,\infty^{\frac{\pi}{3}}\,.\,\infty\,.\,-\infty^{\frac{\pi}{3}}\,.\,\frac{12}{5})$ Type 24 $(6\,.\,\frac{12}{5}\,.\,\infty^{\frac{2\pi}{3}}\,.\,\infty\,.\,-\infty^{\frac{2\pi}{3}}\,.\,\frac{12}{5})$

Type 25 $(3\,.\,3\,.\,3\,.\,\infty^{\frac{2\pi}{3}}\,.\,-3\,.\,\infty^{\frac{2\pi}{3}})$ Type 26 $(3\,.\,\infty\,.\,3\,.\,-\infty^{\frac{2\pi}{3}}\,.\,-3\,.\,-\infty^{\frac{2\pi}{3}})$

**Figure 15** (*continued*). See legend on page 49.

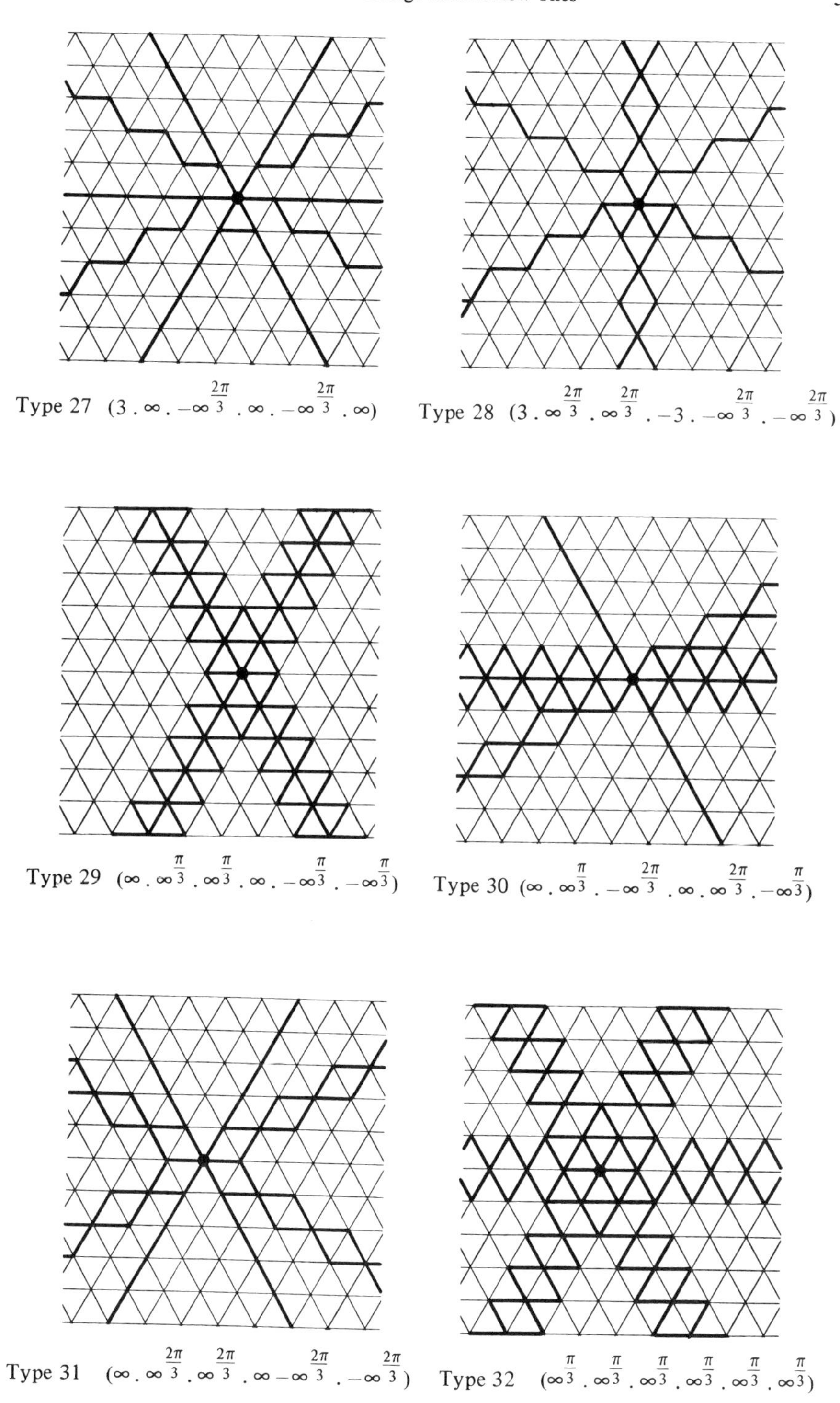

Type 27 $(3 . \infty . -\infty^{\frac{2\pi}{3}} . \infty . -\infty^{\frac{2\pi}{3}} . \infty)$

Type 28 $(3 . \infty^{\frac{2\pi}{3}} . \infty^{\frac{2\pi}{3}} . -3 . -\infty^{\frac{2\pi}{3}} . -\infty^{\frac{2\pi}{3}})$

Type 29 $(\infty . \infty^{\frac{\pi}{3}} . \infty^{\frac{\pi}{3}} . \infty . -\infty^{\frac{\pi}{3}} . -\infty^{\frac{\pi}{3}})$

Type 30 $(\infty . \infty^{\frac{\pi}{3}} . -\infty^{\frac{2\pi}{3}} . \infty . \infty^{\frac{2\pi}{3}} . -\infty^{\frac{\pi}{3}})$

Type 31 $(\infty . \infty^{\frac{2\pi}{3}} . \infty^{\frac{2\pi}{3}} . \infty - \infty^{\frac{2\pi}{3}} . -\infty^{\frac{2\pi}{3}})$

Type 32 $(\infty^{\frac{\pi}{3}} . \infty^{\frac{\pi}{3}} . \infty^{\frac{\pi}{3}} . \infty^{\frac{\pi}{3}} . \infty^{\frac{\pi}{3}} . \infty^{\frac{\pi}{3}})$

**Figure 15** (*continued*). See legend on page 49.

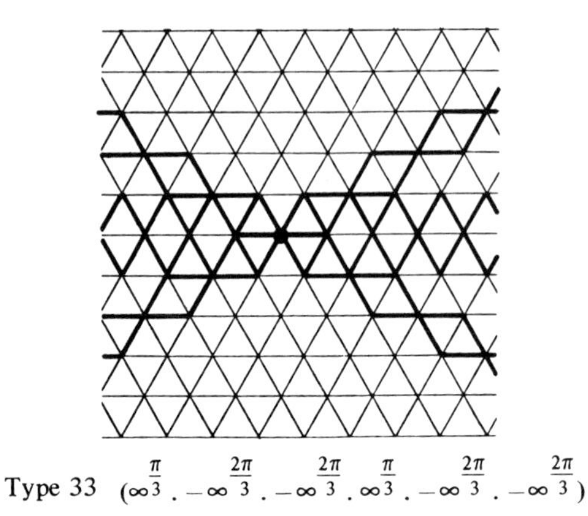

Type 33 $(\infty^{\pi/3} . -\infty^{2\pi/3} . -\infty^{2\pi/3} . \infty^{\pi/3} . -\infty^{2\pi/3} . -\infty^{2\pi/3})$

**Figure 15** (*continued*). See legend on page 49.

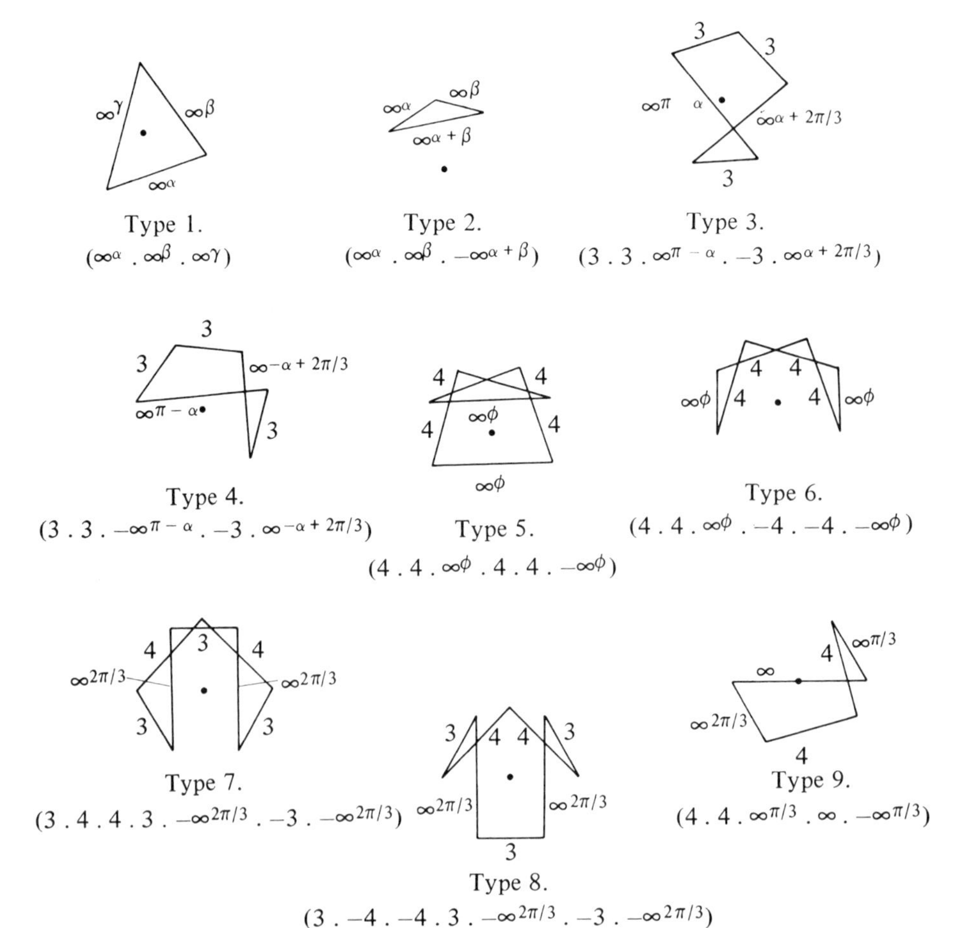


Type 1. $(\infty^{\alpha} . \infty^{\beta} . \infty^{\gamma})$

Type 2. $(\infty^{\alpha} . \infty^{\beta} . -\infty^{\alpha+\beta})$

Type 3. $(3 . 3 . \infty^{\pi-\alpha} . -3 . \infty^{\alpha+2\pi/3})$

Type 4. $(3 . 3 . -\infty^{\pi-\alpha} . -3 . \infty^{-\alpha+2\pi/3})$

Type 5. $(4 . 4 . \infty^{\phi} . 4 . 4 . -\infty^{\phi})$

Type 6. $(4 . 4 . \infty^{\phi} . -4 . -4 . -\infty^{\phi})$

Type 7. $(3 . 4 . 4 . 3 . -\infty^{2\pi/3} . -3 . -\infty^{2\pi/3})$

Type 8. $(3 . -4 . -4 . 3 . -\infty^{2\pi/3} . -3 . -\infty^{2\pi/3})$

Type 9. $(4 . 4 . \infty^{\pi/3} . \infty . -\infty^{\pi/3})$

**Figure 16.** Vertex figures of the tilings in Table 3 and Figure 15. (Continued on pages 55–57.)

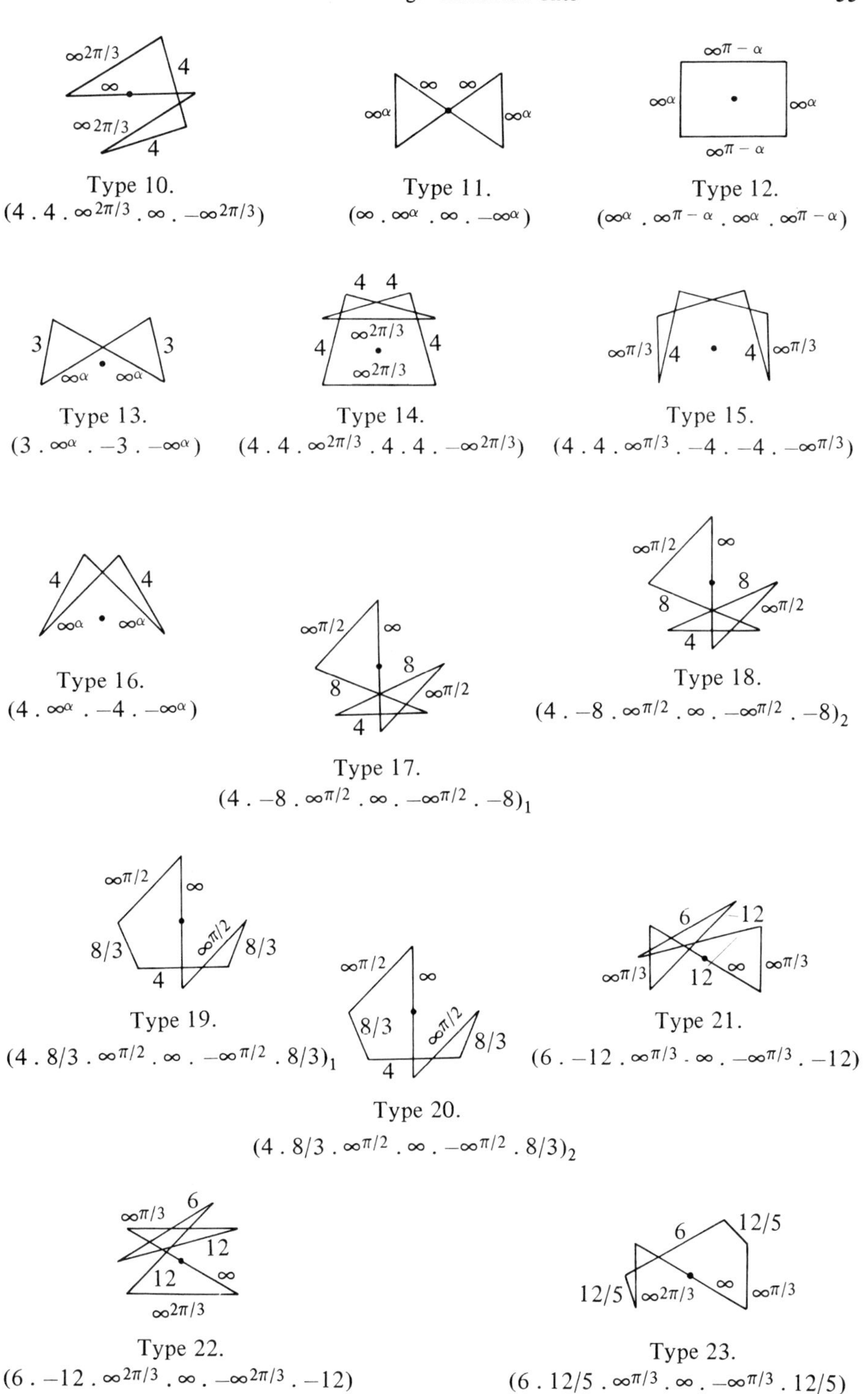


**Figure 16** (*continued*). See legend on page 54.

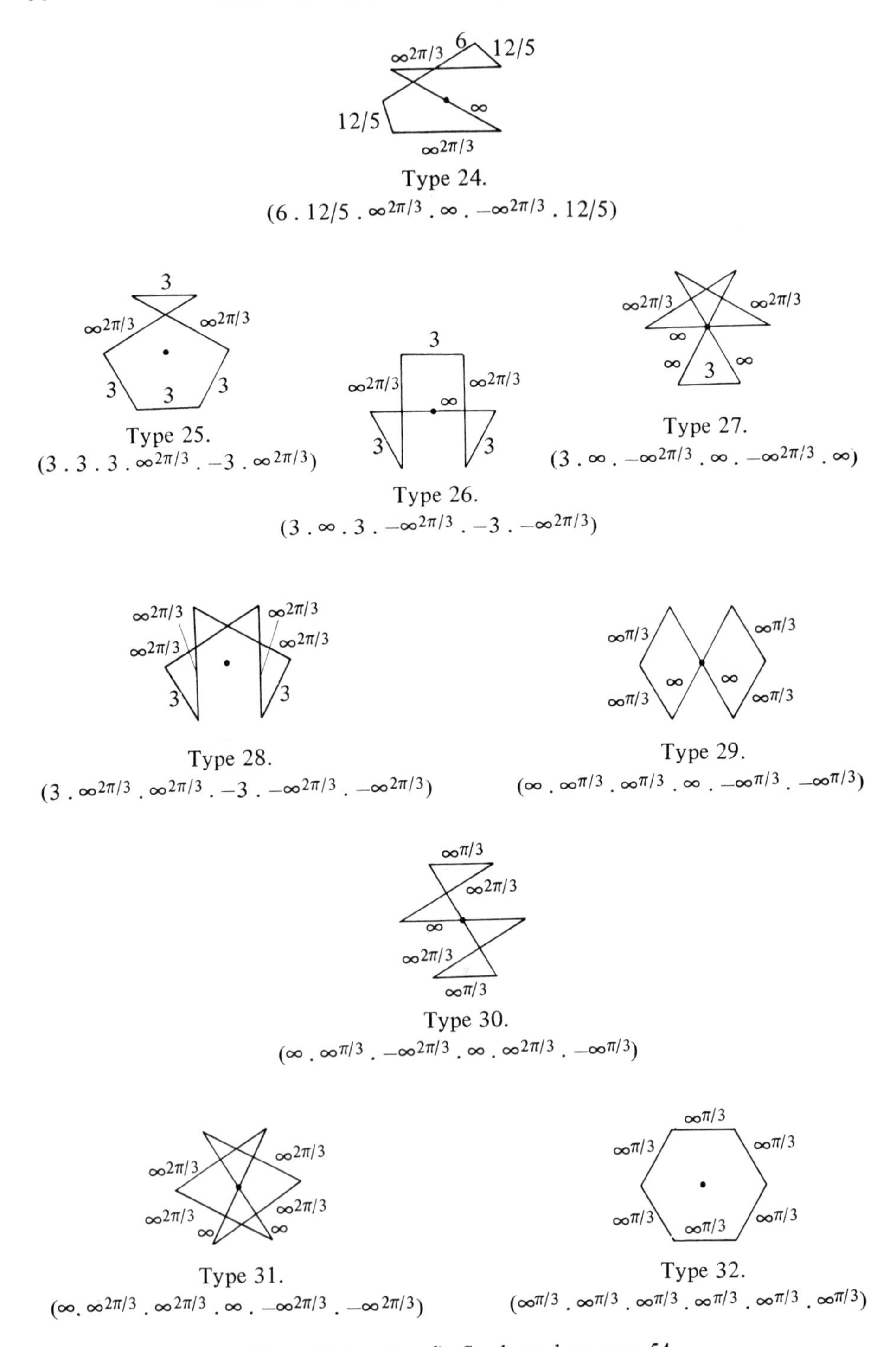


**Figure 16** (*continued*). See legend on page 54.

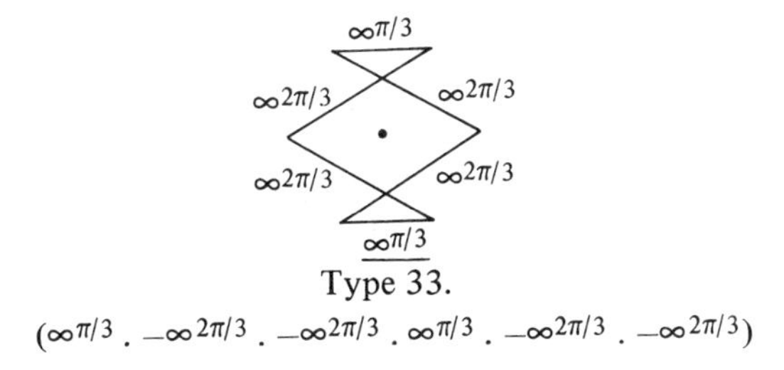


Type 33.

$(\infty^{\pi/3} . -\infty^{2\pi/3} . -\infty^{2\pi/3} . \infty^{\pi/3} . -\infty^{2\pi/3} . -\infty^{2\pi/3})$

**Figure 16** (*continued*). See legend on page 54.

The next significant contribution to the theory of uniform tilings was made by Badoureau [2], [3] just over a hundred years ago, in investigations dealing with uniform polyhedra as well. Badoureau was the first to admit the apeirogon and the regular star polygons $\{n/d\}$ as tiles. In many respects, his work was much ahead of his time and was treated accordingly. For example, Brückner [4] does not even mention Badoureau's work on tilings that include star-polygons or apeirogons, although he discusses uniform tilings and also Badoureau's results on polyhedra with star-polygons as faces. Ahrens [1] and others even misunderstood Badoureau's definition of a uniform tiling—which relies on the equivalence of vertices under symmetries of the tiling and not, as in most previous and many subsequent works, on the congruence of neighborhoods of vertices. Unfortunately, following a custom widespread in his times (but not restricted to them), Badoureau failed to give clear and unambiguous definitions. At any rate he did admit among the uniform tilings such previously unrecognized possibilities as $(4.4.\infty)$, $(4.\infty.-4.\infty)$, $(-3, 12.6.12)$, $(3.-4.6.-4)$, etc. (in our notation), so his ideas must have been very close to those described here. It is hardly surprising that Badoureau failed to discover many of the possible uniform tilings. However, these sins of omission are more than atoned for by the virtues of his novel ideas!

Badoureau's work on tilings with star polygons and apeirogons seems to have been completely ignored for about half a century, when an extension of his list appeared in the 1933 Ph.D. thesis of J. C. P. Miller [21]. Diagrams of 36 uniform tilings are given as Figures 71 to 106 in this thesis. At that time Coxeter and Miller thought of the polygons in the traditional sense as regions of the plane; for this reason they had some misgivings about the use of apeirogons and they did not admit zigzags as tiles.

Miller's list (together with one additional tiling discovered by M. S. Longuet-Higgins, namely $(3.4.3.-4.3.\infty)$) is reproduced without explanation as Table 8 in the paper by Coxeter et al. [7] in 1953. Although no claim for completeness was made, it will be seen that their list of 25 tilings with finite polygons coincides with ours (in Table 1 and Figure 9), a fact that strengthens our conviction that the enumeration is now complete.

Since 1953 several new tilings using apeirogons have been discovered, though all of these seem to have remained unpublished. Apart from those known to one or another of the present authors for various lengths of time, we must mention the discovery of four new tilings with apeirogons by Simon Norton about ten years ago. At about the same time J. H. Conway and M. J. T. Guy started a computer-assisted investigation of the completeness of the list of uniform tilings with finite polygons, though we have not been able to ascertain whether or not they settled the question conclusively.

We believe that the most important contribution of the present paper is the clarification of the notions involved, with precise definitions. This follows the analogous treatment of polyhedra in [12] which was, in turn, inspired by an account of regular polygons by Coxeter [6]. Further, this paper includes, we believe for the first time, an account of the uniform tilings with zigzags, though regular polyhedra with zigzag faces are mentioned in [12].

The consideration of tilings by hollow tiles, instead of following the more traditional approach, casts new light on old problems and leads to a great variety of possible directions for further investigations. We shall conclude by discussing a few of these that seem to us to be particularly attractive.

(i) Even in rather special cases, the investigation of $k$-uniform tilings is likely to produce esthetically pleasing results, besides posing mathematical challenges of various kinds. For example, by restricting attention to $k = 2$, to finite or strip tilings, to finite polygons or just convex polygons, different variants of the enumeration problems arise that appear to be tractable although not trivial to solve.

(ii) An "addition" can be defined for some pairs of tilings by superimposing them and deleting duplicate polygons. For example, if one copy of the 2-uniform finite tiling in Figure 5(b) is superimposed over another that has been rotated 30°, the biuniform finite tiling (3.4.−3.−4; 3.4.6.4) Figure 17a is obtained. In a similar way, from two copies of the tiling in Figure 5(c) we can obtain a biuniform tiling (3.4.−3.−4; 3.−4.6.−4) (Figure 17b). In [18] it has been conjectured that all finite tilings by regular polygons can be obtained by repeated "additions", starting from the five finite tilings (*rosettes*) shown in Figure 5.

(iii) If the connectedness assumption is deleted from the definition, new types of uniform tilings become possible. The question is of interest even for regular tilings. Restricting attention to tilings which are "edge-disjoint" (that is, in which no segment is contained in edges of more than two polygons) and have unicursal

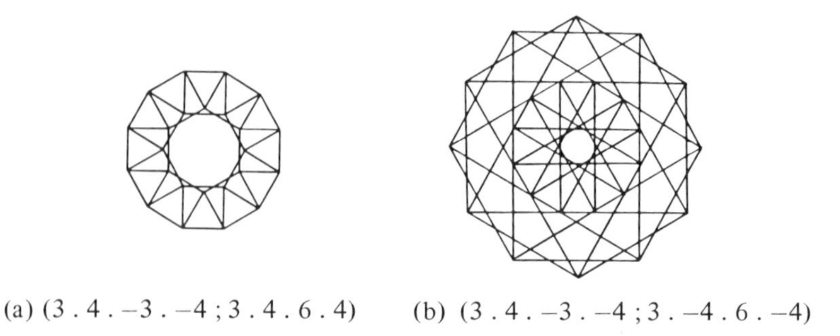

(a) (3 . 4 . −3 . −4 ; 3 . 4 . 6 . 4) (b) (3 . 4 . −3 . −4 ; 3 . −4 . 6 . −4)

**Figure 17** (a, b). Two biuniform finite tilings obtained by "addition" of tilings in Figure 5.

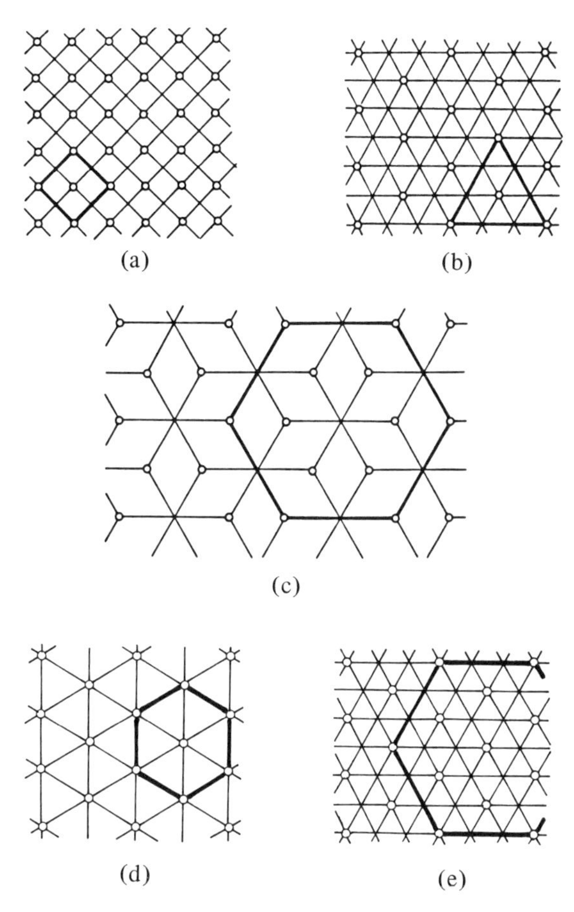


**Figure 18.** The known regular and edge-disjoint but not connected tilings by finite polygons. The vertices are indicated by small circles, and one polygon of each tiling is emphasized. The three unicursal tilings are shown in $(a, b, c)$. The first consists of two copies of $(4^4)$, the second of three copies of $(3^6)$, and the third of four copies of $(6^3)$. The tilings in $(d)$ and $(e)$ are not unicursal; the former consists of three copies of $(6^3)$, the latter of nine copies.

vertex figures, it is easy to see that the three tilings in Figure 18 (a, b, c) are regular. It has been conjectured in [18] that these are the only edge-disjoint but not connected regular tilings by finite polygons and with unicursal vertex figures. If the unicursality requirement is also dropped, other possibilities arise; the only known additional examples are the two shown in Figure 18(d, e).

(iv) There exist a great variety of tilings by regular polygons (in the traditional sense) in which the tiles are not edge-to-edge; see [13]. In a similar manner, many new possibilities arise if we delete from the definition of a tiling by hollow tiles the requirement that the tiles are edge-sharing, while retaining the condition that each edge of the tiling belongs to (that is, is an edge of, or part of an edge of) just two tiles. Examples of such tilings appear in Figure 19. For these a modification

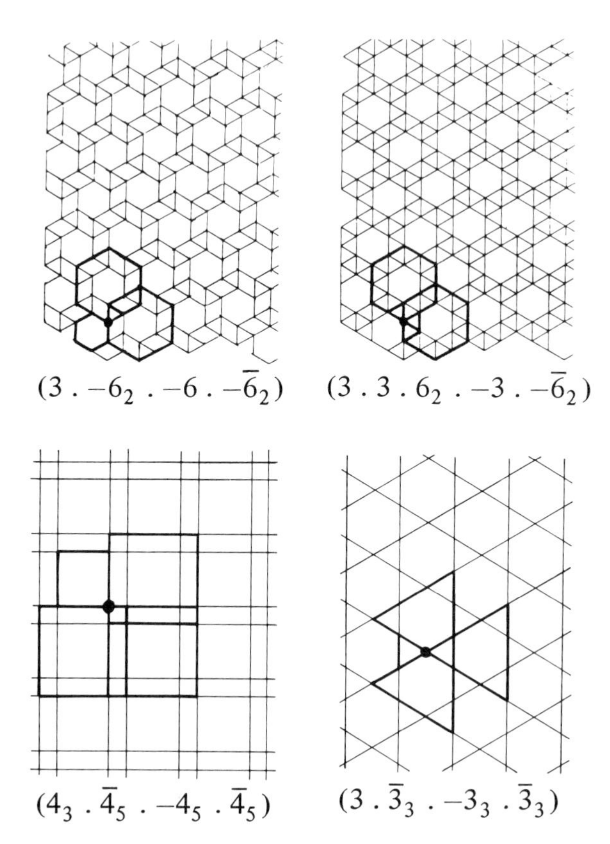


**Figure 19.** Four uniform but not edge-sharing tilings with regular convex tiles. In the Schläfli symbol a bar over a numeral indicates that the vertex of the tiling is *not* a vertex of the corresponding tile, and a subscript indicates the edge-length of the polygon if it is different from the unit.

of the Schläfli symbol is clearly necessary and a suggestion is made in the caption to the figure.

Some striking examples of finite 2-uniform and 3-uniform tilings which are not edge-sharing are shown in Figure 20. These were inspired by a "wreath" of Eberhart [8].

(v) Other possibilities arise if we drop the requirement that a tiling has unicursal vertex figures; in other words, if we allow more general types of vertex figures than a single hollow polygon. Two examples were shown in Figure 18(d, e); many other exist.

Another possibility is to allow, as a vertex figure, a "repeated" line segment. An example of this is the finite tiling or "plate" (mentioned in Miller's thesis [21]) consisting of just two superimposed $n$-gons (this is even regular!) and less trivial examples are the biuniform tilings, in which one of the two vertex figures is of this type, shown in Figure 21.

The method described after Theorem 2 for constructing a tiling from a given edge-net can be applied to construct tilings with a great variety of non-unicursal

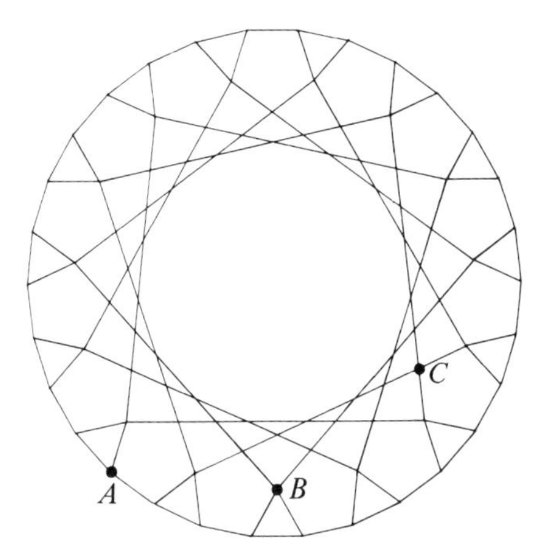


$(3 \,.\, 5 \,.\, -30 \,;\, 3 \,.\, 5 \,.\, (15/4)_\gamma \,.\, 5 \,;\, 5 \,.\, (\overline{15/4})_\gamma \,.\, 5_\delta \,.\, (\overline{15/4})_\gamma)$
with $\gamma = 2 \sin 48° \cos 36°/\sin 12° = 5.783 \ldots$
$\delta = 2 \cos 48° \sin 36°/\sin 12° = 3.783 \ldots$

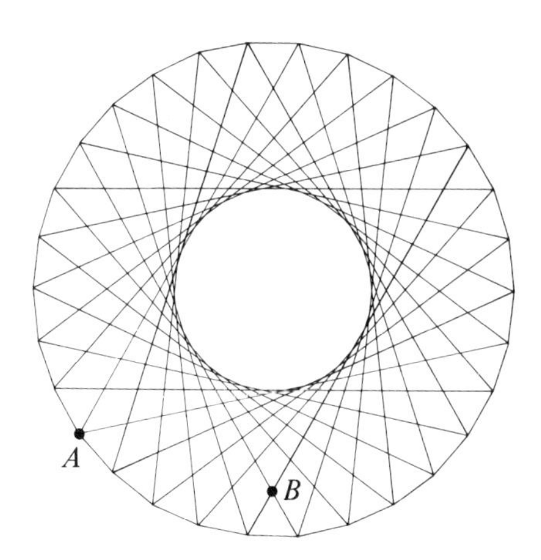


$(3 \,.\, -30 \,.\, 3 \,.\, (30/11)_\alpha \,;\, 3 \,.\, (\overline{30/11})_\alpha \,.\, -3_\beta \,.\, (\overline{30/11})_\alpha)$
with $\alpha = \sin 66°/\sin 6° = 8.740 \ldots$
$\beta = \sqrt{3} \cos 66°/\sin 6° = 6.740 \ldots$

**Figure 20.** Two finite tilings which are not edge-sharing. In each, one vertex of each transitivity class has been marked.

vertex figures. However, we know of no examples in which *new* edge-nets arise. An example of a connected "uniform tiling" with unconnected vertex figure can be obtained from No. 2 in Table 3 by choosing the values $\alpha = \beta = \pi/3$ for the parameters.

In another generalization, we might require that every edge of the tiling belongs to exactly *three* tiles. A simple example of such a tiling can be con-

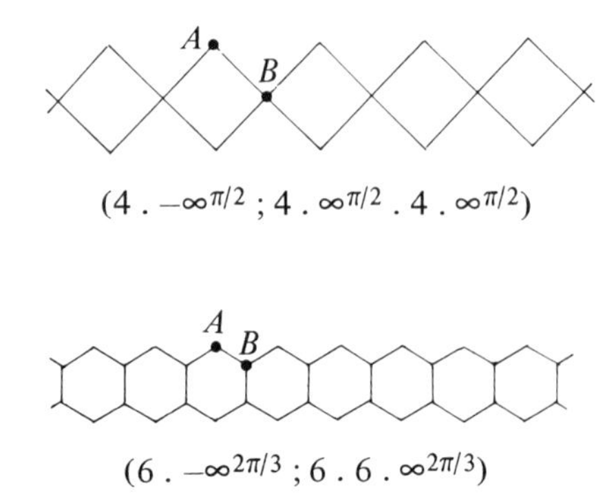


**Figure 21.** Two biuniform strip tilings in which some vertex figures are repeated segments.

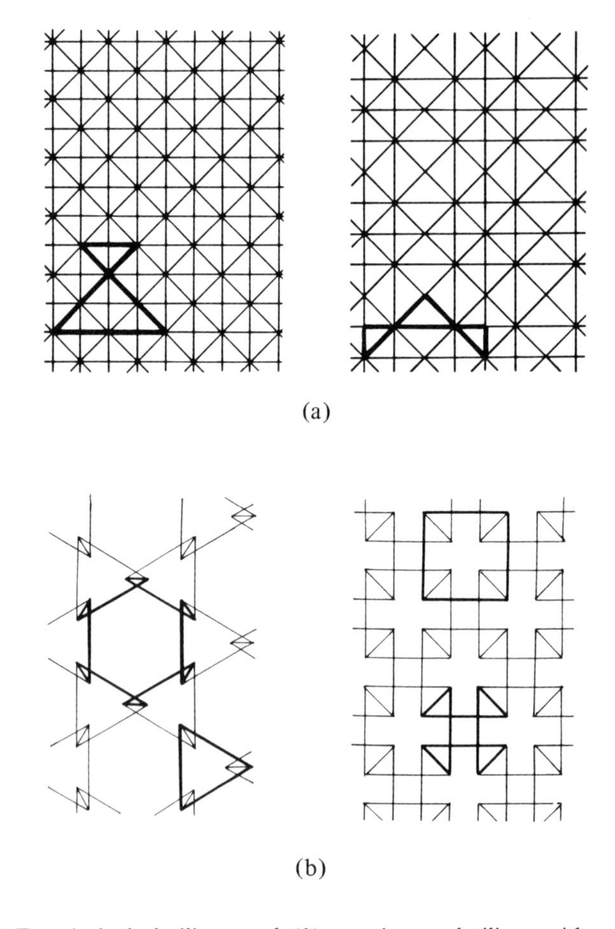


**Figure 22.** (*a*) Two isohedral tilings and (*b*) two isogonal tilings with self-intersecting (hollow) polygons. In (*b*) the polygons are "uniform." One polygon of each type is emphasized in each tiling.

structed with the edge-net $51a$ (that of (3.3.3.3.3.3) in Figure 9). The tiles, all of which have the same edge-length as the tiling, comprise all the hexagons and half the triangles in the figure. The triangles to be chosen are those which are translates of each other—in other words, are of one of the two aspects that occur. For another example we can add to the hollow polygons of the tiling (3.4.6.4) all the 12-gons of the same edge-net.

(vi) If the polygons making up the tiling are not restricted to be regular, many other interesting kinds of tilings can occur. We mention as examples the isohedral, isogonal, and isotoxal tilings—the analogues of which in the traditional setting have recently been determined in [14], [15], [16] and [17]. In Figure 22(a) we show two isohedral tilings—one by self-intersecting quadrangles, and the other by pentagons—and in Figure 22(b) we show two isogonal tilings. In the latter the tiles themselves may be called "uniform", since their symmetry groups are transitive on their vertices (though not, of course, on their flags). Infinite "uniform" polygons also exist; an example is the third polygon shown in Figure 2(a).

In general it would seem desirable to reinterpret, in the context of hollow tiles, all the types of tilings that are familiar in the traditional setting. New and interesting possibilities abound.

(vii) Finally we remark that all these ideas have natural analogues in three or more dimensions. Apart from the well-known (finite) polyhedra bounded by regular convex and star polygons there exist infinite (unbounded) uniform polyhedra. These can, of course, use both finite and infinite polygons as faces. Though examples of regular polyhedra with zigzag faces are known (see [12]), large numbers of uniform polyhedra of this kind probably exist and await discovery.

## References

[1] Ahrens, W., *Mathematische Unterhaltungen und Spiele*. Teubner, Leipzig 1901.

[2] Badoureau, A., Sur les figures isocèles. *C. R. Acad. Sci. Paris* **87** (1878), 823–825.

[3] Badoureau, A., Mémoire sur les figures isoscèles. *J. École Polytechn.* **30** (1881), 47–172.

[4] Brückner, M., *Vielecke und Vielflache*. Teubner, Leipzig 1900.

[5] Buset, D., (in preparation).

[6] Coxeter, H. S. M., *Regular Complex Polytopes*. Cambridge Univ. Press, New York 1974.

[7] Coxeter, H. S. M., Longuet-Higgins, M. S. and Miller, J. C. P., Uniform polyhedra. *Philos. Trans. Roy. Soc. London (A)* **246** (1953/54), 401–450.

[8] Eberhart, S., New and old problems. *Mathematical-physical correspondence*, No. **12** (1975), 4–8.

[9] Gergonne, J. D., Recherches sur les polyèdres, renfermant en particulier un commencement de solution du problème proposé à la page 256 du VII^e vol. des Annales. *Annales de Gergonne*, **9** (1819), 321–ff.

[10] Girard, A., *Table des sines, tangentes & secantes, selon le raid de 100000 parties*. Avec un traicte succint de la trigonometrie tant des triangles plan, que sphericques. Où sont plusiers operations nouvelles, non auparavant mises en lumière, très-utiles & necessaires, non seulement aux apprentifs; mais aussi aux plus doctes practiciens des mathematiques. Elzevier, La Haye 1626.

[11] Grünbaum, B., Polygons. In *The Geometry of Metric and Linear Spaces*, edited by L. M. Kelly. Lecture Notes in Mathematics No. 490. Springer-Verlag, Berlin–Heidelberg–New York 1975.

[12] Grünbaum, B., Regular polyhedra—old and new. *Aequationes Math.* **16** (1977), 1–20.

[13] Grünbaum, B. and Shephard, G. C., Tilings by regular polygons. *Mathematics Magazine*, **50** (1977), 227–247 and **51** (1978), 205–206.

[14] Grünbaum, B. and Shephard, G. C., The eighty-one types of isohedral tilings in the plane. *Math. Proc. Cambridge Philos. Soc.* **82** (1977), 177–196.

[15] Grünbaum, B. and Shephard, G. C., The ninety-one types of isogonal tilings in the plane. *Trans. Amer. Math. Soc.* **242** (1978), 335–353.

[16] Grünbaum, B. and Shephard, G. C., Isotoxal tilings. *Pacif. J. Math.* **76** (1978), 407–430.

[17] Grünbaum, B. and Shephard, G. C., Isohedral tilings of the plane by polygons. *Comment. Math. Helvet.* **53** (1978), 542–571.

[18] Grünbaum, B. and Shephard, G. C., *Tilings and Patterns*. Freeman and Co., San Francisco (1981).

[19] Kepler, J., *Harmonices Mundi*, Libri V. Lincii 1619. (Reprinted in: Johannes Kepler, *Gesammelte Werke*, edited by M. Caspar, Band VI. Beck, München 1940. German translation: *Welt-Harmonik*, M. Caspar, transl., Oldenbourg, München 1939.)

[20] Meister, A. L. F., Generalia de genesi figurarum planarum et independentibus earum affectionibus. *Novi Comm. Soc. Reg. Scient. Götting.* **1** (1769/70), 144–ff.

[21] Miller, J. C. P., On Stellar Constitution, on Statistical Geophysics, and on Uniform Polyhedra (Part 3: Regular and Archimedean Polyhedra), Ph.D. Thesis, 1933. (Copy deposited in Cambridge University Library.)

[22] Poinsot, L., Mémoire sur les polygones et les polyèdres. *J. École Polytechn.* **10** (1810), 16–48.

## *Note added in proof.*

(1) The fact that there exist precisely 73 "types" (in the sense discussed on page 19) of hexagons was independently established also by L. Togliani (see his "Morfologia degli esagoni piani", *Archimede* **30** (1978), 201–206).

(2) An annotated English translation of [19] appears in "Kepler's Star Polyhedra" by J. V. Field (*Vistas in Astronomy* **23** (1979), 109–141). The assertion (made in footnote 38 on page 118) to the effect that (3 . 3 . 4 . 3 . 4) occurs in two enantiomorphic forms is, naturally, erroneous.

# Spherical Tilings with Transitivity Properties[1]

Branko Grünbaum*
G. C. Shephard†

H. S. M. Coxeter's work on regular and uniform polytopes is, perhaps, his best-known contribution to geometry. By central projection one can relate each of these polytopes to a tiling on a sphere, and the symmetry properties of the polytopes then lead naturally to various transitivity properties of the corresponding tilings. The main purpose of this paper is to classify *all* tilings on the 2-sphere with these transitivity properties (and not just those obtained from three-dimensional polytopes). Our results are exhibited in Tables 3 and 4. Here we enumerate all "types" of tilings whose symmetry groups are transitive on the tiles (isohedral tilings), on the edges (isotoxal tilings), or on the vertices (isogonal tilings). The word "type" is used here in the sense of "homeomeric type" for details of which we refer the reader to recent literature on the subjects of patterns and plane tilings, especially [18] and [20].

## 1. Introduction

A *tiling* on the 2-sphere $S^2$ in $E^3$ is defined as a finite family $\mathcal{T} = \{T_1, \ldots, T_n\}$ of sets $T_i$ (the *tiles*) which cover $S^2$ without gaps or overlaps. More precisely $\bigcup_{i=1}^{n} T_i = S^2$ and $T_i \cap T_j$ has zero measure whenever $i \neq j$. It is convenient to place the following restrictions on the tilings under consideration:

**SN1.** *Each tile is a topological disk.*

[1]This material is based upon work supported by the National Science Foundation Grant No. MCS77-01629 A01.

*Department of Mathematics, University of Washington, Seattle, Washington 98195, U.S.A.
†University of East Anglia, Norwich NR4 7TJ, England.

**SN2.** *The intersection of any set of tiles of $\mathfrak{T}$ is a connected (possibly empty) set.*

With these restrictions each arc which is the intersection of two tiles is called an *edge* of the tiling $\mathfrak{T}$, and each point which is the intersection of three or more tiles is called a *vertex* of $\mathfrak{T}$.

**SN3.** *Each edge of $\mathfrak{T}$ has two endpoints which are vertices of $\mathfrak{T}$.*

Condition SN1 implies that the number of tiles in a tiling is at least two; conditions SN2 and SN3 ensure that every tile contains at least three edges on its boundary and that the *valence* of each vertex (that is, the number of edges whose endpoints coincide with the vertex) is at least three. These conditions also eliminate tiresome and uninteresting examples such as the tiling with two hemispherical tiles, one edge and no vertices, and the tiling with three tiles and three edges connecting two diametrically opposite vertices on $S^2$.

Any tiling $\mathfrak{T}$ which satisfies SN1, SN2, and SN3 will be called *normal*, and (except in the last section) we shall restrict attention to normal tilings. It is well known that for all normal tilings the Euler relation

$$v - e + t = 2 \tag{1}$$

holds. Here $v$, $e$, and $t$ represent the numbers of vertices, edges, and tiles in $\mathfrak{T}$.

A set $R \subset S^2$ is called *convex* if $R$ contains no diametrically opposite points of $S^2$ and if, for every two points $x, y \in R$, the minor arc of the great circle through $x$ and $y$ lies entirely in $R$. A tiling $\mathfrak{T}$ is called *convex* if every tile of $\mathfrak{T}$ is the closure of a convex set. It is easy to see that in such a tiling every tile is the intersection of a finite number of closed hemispheres—it is therefore either a hemisphere, or a *lune* bounded by two great semicircles, or a spherical $m$-gon (polygon) bounded by $m$ minor arcs of great circles with $m \geqslant 3$.

Familiar examples of convex tilings arise by radial projection of a convex polyhedron (3-polytope) $P$ from some interior point $z$ onto a 2-sphere centered at $z$. The tiles, edges, and vertices of $\mathfrak{T}$ are the images of the faces, edges, and vertices of $P$, and we shall say that $\mathfrak{T}$ is *obtained* from $P$ by radial projection. Any tiling $\mathfrak{T}$ that can be obtained in this way from a polyhedron $P$ will be called a *polyhedral tiling*, and such tilings will form the topic of Section 2 of this paper.

Associated with every tiling $\mathfrak{T}$ on $S^2$ is the group of all isometries which map $\mathfrak{T}$ into itself. This is known as the *symmetry group* of $\mathfrak{T}$ and is denoted by $S(\mathfrak{T})$. Clearly $S(\mathfrak{T})$ is a finite group, and every element of $S(\mathfrak{T})$ maps the center $z$ of $S^2$ onto itself. The possible groups of symmetries are well-known. Apart from the trivial group of order 1, all are listed in Table 1 which has been compiled from information in the book of Coxeter and Moser [1].

A tiling $\mathfrak{T}$ on $S^2$ is called *isohedral*[2] if $S(\mathfrak{T})$ is transitive on the tiles of $\mathfrak{T}$, is *isotoxal* if $S(\mathfrak{T})$ is transitive on the edges of $\mathfrak{T}$, and is *isogonal* if $S(\mathfrak{T})$ is transitive on the vertices of $\mathfrak{T}$. These are the transitivity properties referred to in the title of

[2]Such transitivity requirements are much stronger than the condition of being *monohedral*, that is, that all the tiles have the same shape, or the analogously defined conditions of being *monotoxal* or *monogonal*.

**Table 1.** The Discrete Groups of Isometries of $E^3$ That Leave One Point Fixed[a]

| Symbol (1) | Description (2) | Order (3) | Diagrams (4) |
|---|---|---|---|
| $[q]$ $q \geqslant 1$ | $q = 1$: one plane of reflection. $q \geqslant 2$: $q$ equally inclined planes of reflection passing through a $q$-fold axis of rotation. | $2q$ |  [1] [2] [6] |
| $[q]^+$ $q \geqslant 1$ | One $q$-fold axis of rotation | $q$ | 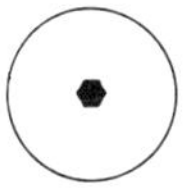 $[6]^+$ |
| $[2, q]$ $q \geqslant 2$ | $q$ equally inclined planes of reflection passing through a $q$-fold axis of rotation, and reflection in an equatorial plane. $q$ 2-fold axes of rotation. Central reflection if and only if $q$ is even. | $4q$ |  [2, 2] [2, 3] [2, 6] |
| $[2, q]^+$ $q \geqslant 2$ | One $q$-fold axis of rotation and $q$ 2-fold axes of rotation equally inclined in the equatorial (perpendicular) plane. $q = 2$: 3 mutually perpendicular 2-fold axes of rotation. | $2q$ | 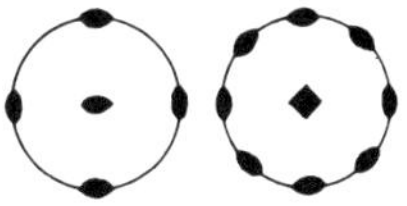 $[2, 2]^+$ $[2, 4]^+$ |
| $[2, q^+]$ $q \geqslant 2$ | One $q$-fold axis of rotation together with reflection in the equatorial plane. Central reflection if and only if $q$ is even. | $2q$ | 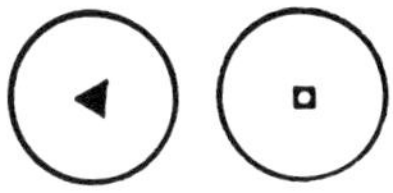 $[2, 3^+]$ $[2, 4^+]$ |

[a] *Column (1)* shows a symbol for the group in the notation of Coxeter and Moser [6]. The permissible values of the parameter $q$ are also indicated. *Columns (2) and (3)* give a brief description of the group and its order. *Column (4)* contains diagrammatic representation of the group. Each figure represents the "northern hemisphere", including the "equator", of a sphere $S^2$ on the surface of which elements of the group are marked as follows: A solid curve represents the intersection of $S^2$ with a plane of reflection. A small $r$-gon (including the 2-gon ⬬ ) represents the intersection of $S^2$ with an axis of $r$-fold rotation ($r \geqslant 2$). A dashed curve represents a plane of rotary reflection; the half arrowheads indicate the angle of rotation associated with the reflection.

**Table 1** (*continued*)

| Symbol (1) | Description (2) | Order (3) | Diagrams (4) |
|---|---|---|---|
| $[2^+, 2q]$ $q \geqslant 2$ | $q$ equally inclined planes of reflection passing through a $q$-fold axis of rotation, and $q$ 2-fold axes of rotation bisecting the angles between the planes of reflection. Central reflection if and only if $q$ is odd. | $4q$ |  $[2^+, 4]$  $[2^+, 6]$ 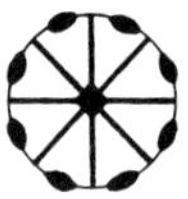 $[2^+, 8]$ |
| $[2^+, 2q^+]$ $q \geqslant 1$ | $q = 1$: central reflection only. $q \geqslant 2$: rotary reflection of order $2q$ together with a $q$-fold axis of rotation. Central reflection if and only if $q$ is odd. | $2q$ | 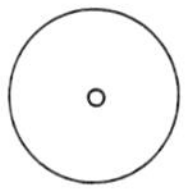 $[2^+, 2^+]$  $[2^+, 6^+]$  $[2^+, 8^+]$ |
| $[3, 3]$ | Symmetry group of regular tetrahedron. 4 3-fold and 3 2-fold axes of rotation. 6 planes of reflection. | 24 |  $[3, 3]$ |
| $[3, 3]^+$ | Rotational symmetries of regular tetrahedron. 4 3-fold and 3 2-fold axes of rotation. | 12 |  $[3, 3]^+$ |
| $[3, 4]$ | Symmetry group of cube. 3 4-fold, 4 3-fold, and 6 2-fold axes of rotation. 9 planes of reflection. | 48 |  $[3, 4]$ |
| $[3, 4]^+$ | Rotational symmetries of cube. 3 4-fold 4 3-fold and 6 2-fold axes of rotation. | 24 | 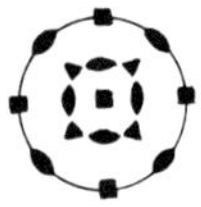 $[3, 4]^+$ |
| $[3^+, 4]$ | Reflections in 3 mutually perpendicular planes. 3 2-fold and 4 3-fold axes of rotation. | 24 |  $[3^+, 4]$ |

**Table 1** (*continued*)

| Symbol (1) | Description (2) | Order (3) | Diagrams (4) |
|---|---|---|---|
| [3, 5] | Symmetry group of regular dodecahedron. 6 5-fold, 10 3-fold, and 15 2-fold axes of rotation. 15 planes of reflection. | 120 | [3, 5] |
| $[3, 5]^+$ | Rotational symmetries of regular dodecahedron. 6 5-fold, 10 3-fold, and 15 2-fold axes of rotation. | 60 | $[3, 5]^+$ |

this paper. Our object is to determine every "type" of tiling with any of these three properties. In this endeavor it is clearly essential to define exactly what we mean by "type". The most natural definition is that by homeomerism, for details of which we refer the reader to [18] or [20]; this coincides with the classification by incidence symbols (see [19]). Moreover, incidence symbols can be used in an enumeration of the types. The method of doing this has already been described in detail for plane tilings with transitivity properties (see [15–17], and [19]), and exactly similar considerations apply to the spherical case. Here we shall describe a different approach to the enumeration problem which is feasible because the symmetry groups of spherical tilings are finite and their subgroups are easy to determine.

Besides carrying out the enumeration we shall also investigate, in Section 3, the different possible realizations of each tiling—here the details differ from the plane case and exhibit some new and interesting features.

Many related questions have been discussed in the literature, mostly in connection with polyhedra. More or less complete enumerations of isohedral or isogonal convex polyhedra have been given by several authors; see, in particular, [23], [12], and [1] (where references to the earlier literature can also be found). A related investigation of the topological character of the "space" of isogonal polyhedra (but with a coarser definition of "type") is given in [28]. The classification of "crystal forms" and "coordination polyhedra" according to [8], [25], and [9], overlaps the classification of isohedral and isogonal convex polyhedra by the isohedral or isogonal type of the associated polyhedral tilings. The classification of isohedral convex polyhedra in [14] is based on criteria different from the ones considered here. In [27] certain monohedral decompositions of the surfaces of some convex polyhedra are considered. Despite the use of the word "isogonal" the topic of [22] is much more restricted; see [4] for relations of this concept to other polyhedra.

Isogonal, isohedral, or monohedral tilings of the sphere, or at least their topological types, were investigated in [31–34]; a complete enumeration of monohedral tilings by spherical triangles is given in [7]. A number of authors have considered the topological types of isogonal or isotoxal tilings of the sphere in a graph-theoretic formulation, all except the last without realizing the connections to geometry; see [35], [36], [26] (in this paper the types (4.6.8) and (4.6.10) are missing), [13].

Illustrations of non-polyhedral tilings of the sphere are encountered much less frequently. An isogonal tiling of type SIG 28 (see below for an explanation of this notation), with regular spherical pentagons and equilateral spherical triangles, is shown in [2]; in [24] tilings of types SIG 6 and SIG 21 are similarly illustrated. Escher's "Sphere with fish" [11, Figure 112] is essentially a tiling of type SIH 50, "Polyhedron with flowers" [11, Figure 226] is isogonal of type SIG 28, and "Gravity" [11, Figures 177, 178] can be interpreted as being of type SIH 55; these tilings are reproduced in [10], and the first two also in [29] and [5]. The adaptations of Escher's tilings of the plane to polyhedral nets, presented in [30], have projections on the sphere that can be interpreted as being of types SIH 38, SIH 55, SIH 50, and SIH 43.

## 2. Polyhedral Tilings

If a convex polyhedron $P$ is radially projected from an interior point $z$ onto a sphere $S^2$ centered at $z$, then the resultant tiling will be denoted by $\mathfrak{T}(P,z)$. In this section we shall explore the properties of such tilings.

We begin by observing that although for every convex polyhedron $P$, and every choice of $z$, the tiling $\mathfrak{T}(P,z)$ is convex, not every convex tiling can arise in this way. A simple example is shown in Figure 1. Let $P$ be a prism whose base is an equilateral triangle (Figure 1(a)), and $z$ be the centroid of $P$. The tiling $\mathfrak{T}(P,z)$ is shown in Figure 1(b) where, for clarity, the vertices of $P$ and of $\mathfrak{T}(P,z)$ are lettered in a corresponding manner. Now suppose that we "twist" one of the triangular tiles, $ABC$, of $\mathfrak{T}$ relative to the other, $DEF$, to obtain the tiling $\mathfrak{T}'$ of Figure 1(d). From the fact that the corresponding twist of $P$ (see Figure 1(c)) necessarily introduces new edges which have no counterpart in $\mathfrak{T}'$, it is easily demonstrated that $\mathfrak{T}'$ cannot be obtained from any convex polyhedron.

The topological type of a tiling $\mathfrak{T}(P,z)$ is independent of $z$, but its symmetry group depends essentially on the choice of $z$. In fact, whatever the symmetry group of $P$, $z$ can always be chosen in such a way that the symmetry group of $\mathfrak{T}(P,z)$ is trivial. If $P$ has a "natural center" (which may be the centroid, incenter, circumcenter, etc.), then choosing $z$ to be this point "maximizes" the symmetry group of $\mathfrak{T}(P,z)$. In this case we shall write $\mathfrak{T}(P)$ for $\mathfrak{T}(P,z)$ and say that $\mathfrak{T}(P)$ is obtained from $P$ by *central projection*.

However, even in the case of central projection, the symmetry groups of $P$ and of $\mathfrak{T}(P)$ may differ. For example, let $P$ be a non-regular octahedron defined as the convex hull of three unequal mutually perpendicular lines whose centers coincide at $z$. Then the symmetry group of $P$ is [2, 2], but $\mathfrak{T}(P)$, which consists of

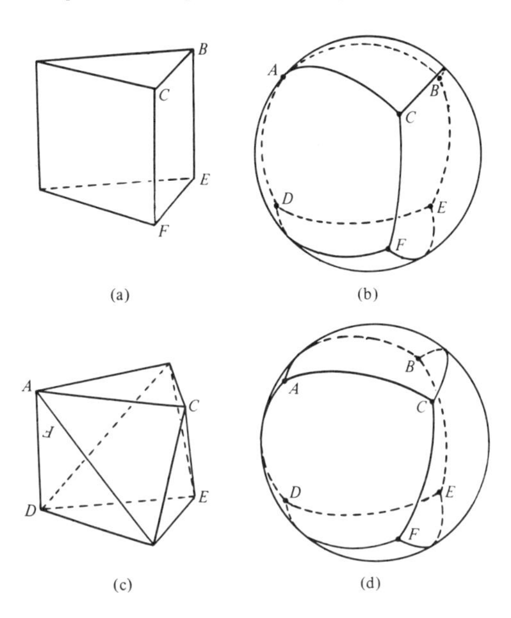


**Figure 1.** (b) A convex tiling obtained by central projection from the triangular prism shown in (a). In (d) one of the triangular tiles of (b) has been "twisted". The corresponding twist of the prism introduces three new edges, see (c). In this way it can be shown that the convex tiling in (d) cannot be obtained from any convex polyhedron by radial projection.

eight equal octants on the sphere, has symmetry group [3, 4] in the notation of Table 1 (see Figure 2(a)).

In spite of this, the symmetry properties of $P$ and of $\mathcal{T}(P)$ are closely related. For example, if $S(P)$ is transitive on the faces, edges, or vertices of $P$, then $\mathcal{T}(P)$ will be isohedral, isotoxal, or isogonal, respectively. Thus central projections of

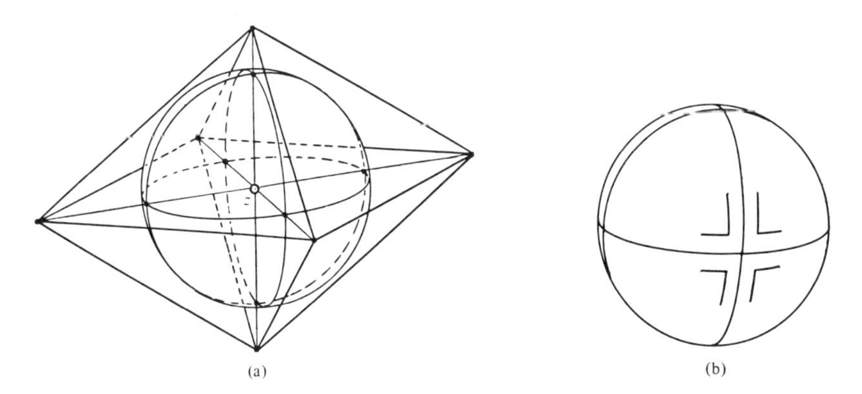


**Figure 2.** (a) A non-regular octahedron (the convex hull of three unequal mutually perpendicular segments whose centers coincide at $z$) and the tiling $\mathcal{T}$ obtained by central projection onto a sphere centered at $z$. The symmetry group of the polyhedron is [2, 2] of order 8, while that of the tiling is [3, 4] of order 48. (b) The tiling $\mathcal{T}$ in which each of the tiles has been marked with an L-shaped motif.

the five regular (Platonic) solids lead to tilings with all three transitivity properties.

In Table 2 we list all the tilings which are obtained by central projection from the regular, uniform, and dual uniform polyhedra. The common names of these polyhedra appear in column (8) of the table. For each polyhedron $P$ we give a symbol for the topological type of $\mathcal{T}(P)$ (which is the same as that of $\mathcal{T}(P, z)$ for all choices of $z$ in the interior of $P$) and also for the homeomeric type of the tilings $\mathcal{T}(P)$ obtained by central projection. The latter may be taken as referring to the complete list of homeomeric types which will be given in Table 3.

The explanation of the symbols for the topological types is as follows. If the tiling $\mathcal{T}$ is isohedral, then the sequence of valences $v_1, v_2, \ldots, v_r$ of the vertices round any tile of $\mathcal{T}$ must be the same for each tile. The topological type of the tiling $\mathcal{T}$ is denoted by $[v_1.v_2. \ldots .v_r]$, where among the various possible symbols we choose that which is lexicographically first; wherever possible we also use exponents to abbreviate in the obvious manner. In a similar way, the topological type of an isogonal tiling is denoted by $(t_1.t_2. \ldots .t_r)$, where each vertex is incident with (in cyclic order) tiles which have $t_1$ edges, $t_2$ edges, $\ldots$, $t_r$ edges. For an isotoxal tiling the corresponding symbol is $\langle t_1.t_2; v_1.v_2 \rangle$, which signifies that each edge joins vertices of valences $v_1$ and $v_2$ and is common to two tiles with $t_1$ and $t_2$ edges.

It is easily verified that each entry in Table 2 corresponds to a normal tiling. However, from our point of view, the following partial converse to this statement is particularly important.

**Theorem 1.** *Every normal tiling which is isohedral, isotoxal, or isogonal is of a topological type that is listed in columns* (2), (4), *or* (6) *of Table* 2. *Thus every such tiling is topologically equivalent to one obtained by central projection of a regular, uniform, or dual uniform polyhedron.*

For isohedral tilings, a proof of Theorem 1 appears in Heesch [21]. Alternative proofs can be based on the fact that a normal isohedral tiling of type $[v_1.v_2. \ldots .v_r]$ can exist only if

$$\sum_i \frac{v_i - 2}{v_i} < 2, \tag{2}$$

a normal isogonal tiling of type $(t_1.t_2. \ldots .t_r)$ can exist only if

$$\sum_i \frac{t_i - 2}{t_i} < 2, \tag{3}$$

and a normal isotoxal tiling of type $\langle t_1.t_2; v_1.v_2 \rangle$ can exist only if

$$\frac{1}{t_1} + \frac{1}{t_2} + \frac{1}{v_1} + \frac{1}{v_2} > 1, \tag{4}$$

where all $v_i$, $t_i$, and $r$ are at least 3. These conditions, which can be obtained from the Euler relation (1), are necessary but not sufficient for the existence of tilings of the corresponding topological types. Hence to determine all such types it is necessary to consider solutions of (2), (3), and (4) in integers greater than 2, and then eliminate by combinatorial arguments those possibilities that do not correspond to tilings. For example, by considering the tiles incident with a given tile $T$ as we go round $T$, we see that no isogonal tiling of topological type $(t_1.t_2.t_3)$ can exist when $t_1$, $t_2$, and $t_3$ are all unequal and any one of them is odd. In this way we arrive without difficulty at a full list of topological types, and Theorem 1 is established.

The topological universality of the tilings $\mathfrak{T}(P)$ asserted in Theorem 1 has a group-theoretical counterpart. For any tiling on $S^2$ denote by $S_h(\mathfrak{T})$ the group of combinatorial isomorphisms of $\mathfrak{T}$.

**Theorem 2.** *For each of the tilings $\mathfrak{T}(P)$ obtained from regular, uniform, or dual uniform polyhedra by central projection, the groups $S_h(\mathfrak{T})$ and $S(\mathfrak{T})$ are isomorphic.*

For a proof of Theorem 2 it is only necessary to check the existence of an isomorphism for each of the tilings $\mathfrak{T}(P)$ of Table 2. We illustrate the argument by one example only: let $P$ be a cube, so that $\mathfrak{T}(P)$ is of topological isohedral type $[3^4]$ (see Figure 5(a)). Every combinatorial isomorphism of $\mathfrak{T}(P)$ must permute the eight vertices of $\mathfrak{T}(P)$ and must do so in such a way that *adjacent* vertices in $\mathfrak{T}(P)$ (that is, those joined by an edge) are mapped into adjacent vertices. It is easily verified that any such combinatorial isomorphism is uniquely specified by the image of a vertex $V$ and of two adjacents of $V$. But exactly the same holds for the symmetries of $\mathfrak{T}(P)$, and hence the isomorphism can be established.

## 3. Isohedral, Isotoxal, and Isogonal Tilings

We now show how the results of the previous section can be used to enumerate all the homeomeric types of isohedral, isotoxal, and isogonal normal tilings on $S^2$.

We begin by considering isohedral tilings. Let $\mathfrak{T}$ be any isohedral tiling of given topological type $[v_1. \ldots .v_r]$. By Theorem 1 this is the same topological type as some tiling $\mathfrak{T}(P)$ obtained by central projection from a regular, uniform or dual-uniform polyhedron $P$, as listed in Table 2. Since $S(\mathfrak{T})$ is, by definition, transitive on the tiles of $\mathfrak{T}$, we may interpret $S(\mathfrak{T})$ as a tile-transitive group of combinatorial isomorphisms of $\mathfrak{T}$, and so as a subgroup of $S_h(\mathfrak{T}(P))$. However, Theorem 2 asserts that the latter group is isomorphic to $S(\mathfrak{T}(P))$, and so we deduce that $S(\mathfrak{T})$ is a subgroup of $S(\mathfrak{T}(P))$.

It follows that all isohedral tilings of the same topological type as $\mathfrak{T}(P)$ can be obtained by specifying a tile-transitive subgroup of $S(\mathfrak{T}(P))$—and all such types of tiling can be enumerated by listing the tile-transitive subgroups of $S(\mathfrak{T}(P))$. From the well-known geometrical properties of the polyhedra $P$ it is easy to determine all these subgroups, and a list of possibilities appears in Table 3. The table also contains the corresponding information for isotoxal tilings (edge-

**Table 2.** Topological and Homeomeric Types of Tilings on the 2-Sphere Obtained from the Regular, Uniform and Dual Uniform Polyhedra by Central Projection[a]

| Ref. No. | Topological and homeomeric type of $\mathcal{T}(P)$: *IH* | | *IT* | | *IG* | | Polyhedron P | Dual | Symmetry group | Numerical data: $v$ | $e$ | $t$ |
|---|---|---|---|---|---|---|---|---|---|---|---|---|
| (1) | (2) | (3) | (4) | (5) | (6) | (7) | (8) | (9) | (10) | (11) | (12) | (13) |
| P1 | $[3^3]$ | SIH 5 | $(3^2; 3^2)$ | SIT 2 | $(3^3)$ | SIG 5 | Tetrahedron | $s$ | [3, 3] | 4 | 6 | 4 |
| P2 | $[3^4]$ | SIH 37 | $(4^2; 3^2)$ | SIT 19 | $(4^3)$ | SIG 20 | Cube | } | [3, 4] | 8 | 12 | 6 |
| P3 | $[4^3]$ | SIH 20 | $(3^2; 4^2)$ | SIT 7 | $(3^4)$ | SIG 37 | Octahedron | | | 6 | 12 | 8 |
| P4 | $[3^5]$ | SIH 53 | $(5^2; 3^2)$ | SIT 26 | $(5^3)$ | SIG 27 | (Pentagonal) Dodecahedron | } | [3, 5] | 20 | 30 | 12 |
| P5 | $[5^3]$ | SIH 27 | $(3^2; 5^2)$ | SIT 9 | $(3^5)$ | SIG 53 | Icosahedron | | | 12 | 30 | 20 |
| P6 | $[3.6^2]$ | SIH 7 | | | | | Triakis tetrahedron | } | [3, 3] | 8 | 18 | 12 |
| P7 | | | | | $(3.6^2)$ | SIG 7 | Truncated tetrahedron | | | 12 | 18 | 8 |
| P8 | [3.4.3.4] | SIH 42 | $(4^2; 3.4)$ | SIT 22 | | | Rhombic dodecahedron | } | [3, 4] | 14 | 24 | 12 |
| P9 | | | $(3.4; 4^2)$ | SIT 12 | (3.4.3.4) | SIG 42 | Cuboctahedron | | | 12 | 24 | 14 |
| P10 | $[4.6^2]$ | SIH 23 | | | | | Tetrakis hexahedron | } | [3, 4] | 14 | 36 | 24 |
| P11 | | | | | $(4.6^2)$ | SIG 23 | Truncated octahedron | | | 24 | 36 | 14 |
| P12 | $[3.8^2]$ | SIH 10 | | | | | Triakis octahedron | } | [3, 4] | 14 | 36 | 24 |
| P13 | | | | | $(3.8^2)$ | SIG 10 | Truncated cube | | | 24 | 36 | 14 |
| P14 | $[3.4^3]$ | SIH 45 | | | | | Trapezoidal icositetrahedron | } | [3, 4] | 26 | 48 | 24 |
| P15 | | | | | $(3.4^3)$ | SIG 45 | Rhombicuboctahedron | | | 24 | 48 | 26 |
| P16 | [4.6.8] | SIH 24 | | | | | Hexakis octahedron | } | [3, 4] | 26 | 72 | 48 |
| P17 | | | | | (4.6.8) | SIG 24 | Truncated cuboctahedron | | | 48 | 72 | 26 |

| | | | | | | | | | | | | |
|---|---|---|---|---|---|---|---|---|---|---|---|---|
| P18 | $[3^4.4]$ | SIH 54 | | | | | Pentagonal icositetrahedron | } | $[3, 4]^+$ | 38 | 60 | 24 |
| P19 | | | | | $(3^4.4)$ | SIG 54 | Snub cube | | | 24 | 60 | 38 |
| P20 | [3.5.3.5] | SIH 49 | $(4^2; 3.5)$ | SIT 24 | | | Rhombic triacontahedron | } | [3, 5] | 32 | 60 | 30 |
| P21 | | | $(3.5; 4^2)$ | SIT 14 | (3.5.3.5) | SIG 49 | Icosidodecahedron | | | 30 | 60 | 32 |
| P22 | $[5.6^2]$ | SIH 29 | | | | | Pentakis dodecahedron | } | [3, 5] | 32 | 90 | 60 |
| P23 | | | | | $(5.6^2)$ | SIG 29 | Truncated icosahedron | | | 60 | 90 | 32 |
| P24 | $[3.10^2]$ | SIH 12 | | | | | Triakis icosahedron | } | [3, 5] | 32 | 90 | 60 |
| P25 | | | | | $(3.10^2)$ | SIG 12 | Truncated dodecahedron | | | 60 | 90 | 32 |
| P26 | [3.4.5.4] | SIH 47 | | | | | Trapezoidal hexacontahedron | } | [3, 5] | 62 | 120 | 60 |
| P27 | | | | | (3.4.5.4) | SIG 47 | Rhombicosidodecahedron | | | 60 | 120 | 62 |
| P28 | [4.6.10] | SIH 25 | | | | | Hexakis icosahedron | } | [3, 5] | 62 | 180 | 120 |
| P29 | | | | | (4.6.10) | SIG 25 | Truncated icosidodecahedron | | | 120 | 180 | 62 |
| P30 | $[3^4.5]$ | SIH 55 | | | | | Pentagonal hexacontahedron | } | $[3, 5]^+$ | 92 | 150 | 60 |
| P31 | | | | | $(3^4.5)$ | SIG 55 | Snub dodecahedron | | | 60 | 150 | 92 |
| P32($r$) | $[4^2.r]$ | SIH 60($r$) | | | | | $r$-gonal dipyramid | } | $[2, r]$ | $r + 2$ | $3r$ | $2r$ |
| P33($r$) | | | | | $(4^2.r)$ | SIG 60($r$) | $r$-gonal prism | | | $2r$ | $3r$ | $r + 2$ |
| P34($r$) | $[3^3.r]$ | SIH 63($r$) | | | | | $r$-gonal trapezohedron | } | $[2^+, 2r]$ | $2r + 2$ | $4r$ | $2r$ |
| P35($r$) | | | | | $(3^3.r)$ | SIG 63($r$) | $r$-gonal antiprism | | | $2r$ | $4r$ | $2r + 2$ |

[a] *Column (1)* gives a reference number to the polyhedron as shown in Figure 3. The prefixes SIH, SIT, and SIG mean "spherical isohedral", "spherical isotoxal", and "spherical isogonal" respectively. *Columns (2) to (7)* show the topological and homeomeric types of the tilings $\mathcal{T}(P)$ obtained by central projection; columns (2) and (3) show isohedral types, columns (4) and (5) show isotoxal types, and columns (6) and (7) show isogonal types. *Column (8)* gives the common name for the polyhedron $P$. *Column (9)* indicates the relationship between the polyhedra by duality, $s$ means self-dual, and dual pairs are bracketed. *Column (10)* gives the symmetry group of the polyhedron $P$ and also of the tiling $\mathcal{T}(P)$ in the notation of Table 1. *Columns (11) to (13)* indicate the numbers of vertices, edges, and faces of $P$. These are also the numbers of vertices, edges, and tiles of the tiling $\mathcal{T}(P)$.

**Table 3.** The Homeomeric Types of Isohedral, Isotoxal, and Isogonal Normal Tilings on the 2-Sphere[a]

| Homeomeric type (1) | Topological type (2) | Symmetry group (3) | Order (4) | Induced group (5) | Incidence symbol (6) |
|---|---|---|---|---|---|
| Isohedral Tilings | | | | | |
| SIH 1 | $[3^3]$ | $[2, 2]^+$ | 4 | $[1]^+$ | $[a^+b^+c^+; a^+b^+c^+]$ |
| SIH 2 | | $[2^+, 4^+]$ | 4 | $[1]^+$ | $[a^+b^+c^+; a^+c^-b^-]$ |
| SIH 3 | | $[2^+, 4]$ | 8 | $[1]$ | $[ab^+b^-; ab^+]$ |
| SIH 4 | | $[3, 3]^+$ | 12 | $[3]^+$ | $[a^+a^+a^+; a^+]$ |
| SIH 5 | | $[3, 3]$ | 24 | $[3]$ | $[aaa; a]$ |
| SIH 6 | $[3.6^2]$ | $[3, 3]^+$ | 12 | $[1]^+$ | $[a^+b^+c^+; a^+c^+b^+]$ |
| SIH 7 | | $[3, 3]$ | 24 | $[1]$ | $[ab^+b^-; ab^-]$ |
| SIH 8 | $[3.8^2]$ | $[3, 4]^+$ | 24 | $[1]^+$ | $[a^+b^+c^+; a^+c^+b^+]$ |
| SIH 9 | | $[3^+, 4]$ | 24 | $[1]^+$ | $[a^+b^+c^+; a^-c^+b^+]$ |
| SIH 10 | | $[3, 4]$ | 48 | $[1]$ | $[ab^+b^-; ab^-]$ |
| SIH 11 | $[3.10^2]$ | $[3, 5]^+$ | 60 | $[1]^+$ | $[a^+b^+c^+; a^+c^+b^+]$ |
| SIH 12 | | $[3, 5]$ | 120 | $[1]$ | $[ab^+b^-; ab^-]$ |
| SIH 13 | $[4^3]$ | $[2^+, 4]$ | 8 | $[1]^+$ | $[a^+b^+c^+; a^+b^-c^-]$ |
| SIH 14 | | $[2, 2]$ | 8 | $[1]^+$ | $[a^+b^+c^+; a^-b^-c^-]$ |
| SIH 15 | | $[2, 4^+]$ | 8 | $[1]^+$ | $[a^+b^+c^+; a^-c^+b^+]$ |
| SIH 16 | | $[2, 4]^+$ | 8 | $[1]^+$ | $[a^+b^+c^+; a^+c^+b^+]$ |
| SIH 17 | | $[3, 4]^+$ | 24 | $[3]^+$ | $[a^+a^+a^+; a^+]$ |
| SIH 18 | | $[3^+, 4]$ | 24 | $[3]^+$ | $[a^+a^+a^+; a^-]$ |
| SIH 19 | | $[2, 4]$ | 16 | $[1]$ | $[ab^+b^-; ab^-]$ |
| SIH 20 | | $[3, 4]$ | 48 | $[3]$ | $[aaa; a]$ |
| SIH 21 | $[4.6^2]$ | $[3, 4]^+$ | 24 | $[1]^+$ | $[a^+b^+c^+; a^+c^+b^+]$ |
| SIH 22 | | $[3, 3]$ | 24 | $[1]^+$ | $[a^+b^+c^+; a^-b^-c^-]$ |
| SIH 23 | | $[3, 4]$ | 48 | $[1]$ | $[ab^+b^-; ab^-]$ |
| SIH 24 | $[4.6.8]$ | $[3, 4]$ | 48 | $[1]^+$ | $[a^+b^+c^+; a^-b^-c^-]$ |
| SIH 25 | $[4.6.10]$ | $[3, 5]$ | 120 | $[1]^+$ | $[a^+b^+c^+; a^-b^-c^-]$ |

[a] *Column (1)* indicates the homeomeric type of the tiling. The prefixes SIH, SIT, and SIG stand for "spherical isohedral type", "spherical isotoxal type" and "spherical isogonal type" respectively. *Column (2)* contains symbols for the topological type of the tiling. The notation is explained in Section 2. *Columns (3), (4) and (5)* show the symmetry group of the tiling (in the notation of Table 1), the order of the group, and the induced group. The latter is the subgroup of the symmetry group that leaves one element (tile, edge, or vertex in the three cases) of the tiling fixed. *Column (6)* gives the incidence symbol of the tiling. This comprises (before the semicolon) a tile, edge, or vertex symbol and (after the semicolon) an adjacency symbol. These terms are explained fully in [15], [16], [17], [20], and especially [19]. *Column (7)* lists the edges, vertices and tiles by transitivity classes. The notation is the

| Transitivity classes (7) | | Realizations (8) | Cross references (9) | | | References (10) | | | | |
|---|---|---|---|---|---|---|---|---|---|---|
| Edges | Vertices | | | | | | | | | |
| $\alpha\beta\gamma$ | $\alpha\alpha\alpha$ | *N*, *C*, *P* | | SIG 1 | | H 17, | F 34, | Br 7, | Bu 28, | N 8.1 |
| $\alpha\beta\beta$ | $\alpha\alpha\alpha$ | *N* | | SIG 2 | | | | | | |
| $\alpha\beta\beta$ | $\alpha\alpha\alpha$ | *N*, *C*, *P* | | SIG 3 | | H 14, | | Br 4, | Bu 48, | N 8.2 |
| $\alpha\alpha\alpha$ | $\alpha\alpha\alpha$ | *N* | SIT 1 | SIG 4 | | | | | | |
| $\alpha\alpha\alpha$ | $\alpha\alpha\alpha$ | *C*, *P* | SIT 2 | SIG 5 | *P*1 | H 3, | | | Bu 138, | N 8.3 |
| $\alpha\beta\beta$ | $\alpha\alpha\beta$ | *N* | | | | | | | | |
| $\alpha\beta\beta$ | $\alpha\alpha\beta$ | *C*, *P* | | | *P*6 | H 9, | F 37, | Br 16, | Bu 143, | N 19 |
| $\alpha\beta\beta$ | $\alpha\alpha\beta$ | *N* | | | | | | | | |
| $\alpha\beta\beta$ | $\alpha\alpha\beta$ | *N* | | | | | | | | |
| $\alpha\beta\beta$ | $\alpha\alpha\beta$ | *C*, *P* | | | *P*12 | H 12, | F 38, | Br 10, | Bu 148, | N 26 |
| $\alpha\beta\beta$ | $\alpha\alpha\beta$ | *N* | | | | | | | | |
| $\alpha\beta\beta$ | $\alpha\alpha\beta$ | *C*, *P* | | | *P*24 | H 13, | F 38, | Br 22 | | |
| $\alpha\beta\gamma$ | $\alpha\alpha\beta$ | *N*, *C*, *P* | | | | H 18, | F 35, | Br 5, | Bu 51, | N 15.1 |
| $\alpha\beta\gamma$ | $\alpha\beta\gamma$ | *P* | | | | H 8a, | | Br 2, | Bu 30, | N 15.2 |
| $\alpha\beta\beta$ | $\alpha\alpha\beta$ | *N* | | | | | | | | |
| $\alpha\beta\beta$ | $\alpha\alpha\beta$ | *N* | | | | | | | | |
| $\alpha\alpha\alpha$ | $\alpha\alpha\alpha$ | *N* | SIT 4 | SIG 33 | | | | | | |
| $\alpha\alpha\alpha$ | $\alpha\alpha\alpha$ | – | SIT 5 | SIG 36 | | | | | | |
| $\alpha\beta\beta$ | $\alpha\alpha\beta$ | *P* | | | | H 8, | | Br 3, | Bu 49, | N 15.3 |
| $\alpha\alpha\alpha$ | $\alpha\alpha\alpha$ | *C*, *P* | SIT 7 | SIG 37 | *P*3 | H 4, | F 36, | | Bu 140, | N 15.4 |
| $\alpha\beta\beta$ | $\alpha\alpha\beta$ | *N* | | | | | | | | |
| $\alpha\beta\gamma$ | $\alpha\alpha\beta$ | *P* | | | | H 10a, | F 40b, | Br 17, | Bu 149, | N 27.1 |
| $\alpha\beta\beta$ | $\alpha\alpha\beta$ | *C*, *P* | | | *P*10 | H 10, | F 40a, | Br 8, | Bu 146, | N 27.2 |
| $\alpha\beta\gamma$ | $\alpha\beta\gamma$ | *C*, *P* | | | *P*16 | H 15, | F 41, | Br 11, | Bu 152, | N 30 |
| $\alpha\beta\gamma$ | $\alpha\beta\gamma$ | *C*, *P* | | | *P*28 | H 16, | F 42, | Br 23 | | |

same as in the literature on plane tilings mentioned above. *Column (8)* indicates all the realizations of each type of tiling in the notation explained in Section 3. All types can be realized by marked tilings. *Column (9)* gives cross references both to Table 2 and to other parts of Table 3. For example an isohedral tiling of type SIH 5 is also of types SIT 2 and SIG 5, besides appearing in the entry *P*1 of Table 2. *Column (10)* gives references to illustrations or description of polyhedra that yield the polyhedral tilings in question. H indicates diagrams in Hess [23], with Arabic numerals substituted for the Roman ones in the original; F indicates Fedorov [12], where each figure shows a dual pair of polyhedra; Br indicates Brückner [1, pp. 142–150]; and Bu indicates Buerger [3, Chapter 10]. N stands for the symbol in Niggli [25] and in Donnay, Hellner and Niggli [9].

**Table 3** (*continued*)

| Homeomeric type (1) | Topological type (2) | Symmetry group (3) | Order (4) | Induced group (5) | Incidence symbol (6) |
|---|---|---|---|---|---|
| Isohedral Tilings | | | | | |
| SIH 26 | $[5^3]$ | $[3, 5]^+$ | 60 | $[3]^+$ | $[a^+a^+a^+; a^+]$ |
| SIH 27 | | $[3, 5]$ | 120 | $[3]$ | $[aaa; a]$ |
| SIH 28 | $[5.6^2]$ | $[3.5]^+$ | 60 | $[1]^+$ | $[a^+b^+c^+; a^+c^+b^+]$ |
| SIH 29 | | $[3, 5]$ | 120 | $[1]$ | $[ab^+b^-; ab^-]$ |
| SIH 30 | $[3^4]$ | $[2, 3]^+$ | 6 | $[1]^+$ | $[a^+b^+c^+d^+; a^+b^+d^+c^+]$ |
| SIH 31 | | $[2^+, 6^+]$ | 6 | $[1]^+$ | $[a^+b^+c^+d^+; b^+a^+d^-c^-]$ |
| SIH 32 | | $[3, 3]^+$ | 12 | $[2]^+$ | $[a^+b^+a^+b^+; b^+a^+]$ |
| SIH 33 | | $[3, 4]^+$ | 24 | $[4]^+$ | $[a^+a^+a^+a^+; a^+]$ |
| SIH 34 | | $[2^+, 6]$ | 12 | $[1](l)$ | $[a^+b^+b^-a^-; a^-b^+]$ |
| SIH 35 | | $[3, 3]$ | 24 | $[2](l)$ | $[a^+a^-a^+a^-; a^-]$ |
| SIH 36 | | $[3^+, 4]$ | 24 | $[2](s)$ | $[abab; ba]$ |
| SIH 37 | | $[3, 4]$ | 48 | $[4]$ | $[aaaa; a]$ |
| SIH 38 | $[3.4.3.4]$ | $[3, 3]^+$ | 12 | $[1]^+$ | $[a^+b^+c^+d^+; b^+a^+d^+c^+]$ |
| SIH 39 | | $[3, 4]^+$ | 24 | $[2]^+$ | $[a^+b^+a^+b^+; b^+a^+]$ |
| SIH 40 | | $[3^+, 4]$ | 24 | $[1](l)$ | $[a^+b^+b^-a^-; b^+a^+]$ |
| SIH 41 | | $[3, 3]$ | 24 | $[1](l)$ | $[a^+b^+b^-a^-; a^-b^-]$ |
| SIH 42 | | $[3, 4]$ | 48 | $[2](l)$ | $[a^+a^-a^+a^-; a^-]$ |
| SIH 43 | $[3.4^3]$ | $[3, 4]^+$ | 24 | $[1]^+$ | $[a^+b^+c^+d^+; b^+a^+d^+c^+]$ |
| SIH 44 | | $[3^+, 4]$ | 24 | $[1]^+$ | $[a^+b^+c^+d^+; b^+a^+c^-d^-]$ |
| SIH 45 | | $[3, 4]$ | 48 | $[1](l)$ | $[a^+a^-b^+b^-; a^-b^-]$ |
| SIH 46 | $[3.4.5.4]$ | $[3, 5]^+$ | 60 | $[1]^+$ | $[a^+b^+c^+d^+; b^+a^+d^+c^+]$ |
| SIH 47 | | $[3, 5]$ | 120 | $[1](l)$ | $[a^+a^-b^+b^-; a^-b^-]$ |
| SIH 48 | $[3.5.3.5]$ | $[3, 5]^+$ | 60 | $[2]^+$ | $[a^+b^+a^+b^+; b^+a^+]$ |
| SIH 49 | | $[3, 5]$ | 120 | $[2](l)$ | $[a^+a^-a^+a^-; a^-]$ |
| SIH 50 | $[3^5]$ | $[3, 3]^+$ | 12 | $[1]^+$ | $[a^+b^+c^+d^+e^+; a^+c^+b^+e^+d^+]$ |
| SIH 51 | | $[3, 5]^+$ | 60 | $[5]^+$ | $[a^+a^+a^+a^+a^+; a^+]$ |
| SIH 52 | | $[3^+, 4]$ | 24 | $[1]$ | $[ab^+c^+c^-b^-; ac^+b^+]$ |
| SIH 53 | | $[3, 5]$ | 120 | $[5]$ | $[aaaaa; a]$ |
| SIH 54 | $[3^4.4]$ | $[3, 4]^+$ | 24 | $[1]^+$ | $[a^+b^+c^+d^+e^+; a^+c^+b^+e^+d^+]$ |
| SIH 55 | $[3^4.5]$ | $[3, 5]$ | 60 | $[1]^+$ | $[a^+b^+c^+d^+e^+; a^+c^+b^+e^+d^+]$ |
| SIH 56($r$) | $[4^2.r]$ | $[2, r]^+$ | $2r$ | $[1]^+$ | $[a^+b^+c^+; a^+c^+b^+]$ |
| SIH 57($r$) | $r \neq 4$ | $[2, r^+]$ | $2r$ | $[1]^+$ | $[a^+b^+c^+; a^-c^+b^+]$ |
| SIH 58($r$) | | $[2^+, r]$ $2\|r$ | $2r$ | $[1]^+$ | $[a^+b^+c^+; a^+b^-c^-]$ |
| SIH 59($r$) | | $[2, \frac{1}{2}r]$ $2\|r$ | $2r$ | $[1]^+$ | $[a^+b^+c^+; a^-b^-c^-]$ |
| SIH 60($r$) | | $[2, r]$ | $4r$ | $[1]$ | $[ab^+b^-; ab^-]$ |

| Transitivity classes (7) | | Realizations (8) | Cross references (9) | | | References (10) | | | | |
|---|---|---|---|---|---|---|---|---|---|---|
| Edges | Vertices | | | | | | | | | |
| $\alpha\alpha\alpha$ | $\alpha\alpha\alpha$ | *N* | SIT 8 | SIG 51 | | | | | | |
| $\alpha\alpha\alpha$ | $\alpha\alpha\alpha$ | *C*, *P* | SIT 9 | SIG 53 | *P*5 | H 5, | F 43 | | | |
| $\alpha\beta\beta$ | $\alpha\alpha\beta$ | *N* | | | | | | | | |
| $\alpha\beta\beta$ | $\alpha\alpha\beta$ | *C*, *P* | | | *P*22 | H 11, | F 44, | Br 20 | | |
| $\alpha\beta\gamma\gamma$ | $\alpha\alpha\alpha\beta$ | *N*, *C*, *P* | | | | H 25, | | Br 6, | Bu 83, | N 12.1 |
| $\alpha\alpha\beta\beta$ | $\alpha\beta\alpha\alpha$ | *N* | | | | | | | | |
| $\alpha\alpha\alpha\alpha$ | $\alpha\beta\alpha\beta$ | *N* | SIT 15 | | | | | | | |
| $\alpha\alpha\alpha\alpha$ | $\alpha\alpha\alpha\alpha$ | *N* | SIT 16 | SIG 17 | | | | | | |
| $\alpha\beta\beta\alpha$ | $\alpha\beta\beta\beta$ | *N*, *C*, *P* | | | | H 24, | | Br 3, | Bu 82, | N 12.2 |
| $\alpha\alpha\alpha\alpha$ | $\alpha\beta\alpha\beta$ | – | SIT 17 | | | | | | | |
| $\alpha\alpha\alpha\alpha$ | $\alpha\alpha\alpha\alpha$ | *N* | SIT 18 | SIG 18 | | | | | | |
| $\alpha\alpha\alpha\alpha$ | $\alpha\alpha\alpha\alpha$ | *C*, *P* | SIT 19 | SIG 20 | *P*2 | H 6, | F 46, | | Bu 139, | N 12.3 |
| $\alpha\alpha\beta\beta$ | $\alpha\beta\alpha\gamma$ | *N* | | | | | | | | |
| $\alpha\alpha\alpha\alpha$ | $\alpha\beta\alpha\beta$ | *N* | SIT 20 | | | | | | | |
| $\alpha\alpha\alpha\alpha$ | $\alpha\beta\alpha\beta$ | *N* | SIT 21 | | | | | | | |
| $\alpha\beta\beta\alpha$ | $\alpha\beta\gamma\beta$ | *P* | | | | H 19a, | F 48, | Br 18, | Bu 144, | N 21.1 |
| $\alpha\alpha\alpha\alpha$ | $\alpha\beta\alpha\beta$ | *C*, *P* | SIT 22 | | *P*8 | H 19, | F 47, | Br 9, | Bu 141, | N 21.2 |
| $\alpha\alpha\beta\beta$ | $\alpha\beta\alpha\gamma$ | *N* | | | | | | | | |
| $\alpha\alpha\beta\gamma$ | $\alpha\beta\alpha\gamma$ | *N*, *C*, *P* | | | | H 23, | F 51, | Br 15, | Bu 150, | N 28.1 |
| $\alpha\alpha\beta\beta$ | $\alpha\beta\alpha\gamma$ | *C*, *P* | | | *P*14 | H 21, | F 50, | Br 12, | Bu 147, | N 28.2 |
| $\alpha\alpha\beta\beta$ | $\alpha\beta\alpha\gamma$ | *N* | | | | | | | | |
| $\alpha\alpha\beta\beta$ | $\alpha\beta\alpha\gamma$ | *C*, *P* | | | *P*26 | H 22, | F 52, | Br 24 | | |
| $\alpha\alpha\alpha\alpha$ | $\alpha\beta\alpha\beta$ | *N* | SIT 23 | | | | | | | |
| $\alpha\alpha\alpha\alpha$ | $\alpha\beta\alpha\beta$ | *C*, *P* | SIT 24 | | *P*20 | H 20, | F 49, | Br 21 | | |
| $\alpha\beta\beta\gamma\gamma$ | $\alpha\alpha\beta\alpha\gamma$ | *N*, *C*, *P* | | | | H 28, | F 55, | Br 19, | Bu 145, | N 23.1 |
| $\alpha\alpha\alpha\alpha\alpha$ | $\alpha\alpha\alpha\alpha\alpha$ | *N* | SIT 25 | SIG 26 | | | | | | |
| $\alpha\beta\beta\beta\beta$ | $\alpha\alpha\beta\alpha\beta$ | *N*, *C*, *P* | | | | H 29, | F 54, | Br 14, | Bu 142, | N 23.2 |
| $\alpha\alpha\alpha\alpha\alpha$ | $\alpha\alpha\alpha\alpha\alpha$ | *C*, *P* | SIT 26 | SIG 27 | *P*4 | H 7, | F 53, | | | N 23.3 |
| $\alpha\alpha\beta\beta\gamma\gamma$ | $\alpha\alpha\beta\alpha\gamma$ | *N*, *C*, *P* | | | *P*18 | H 26, | F 56, | Br 13, | Bu 151, | N 29 |
| $\alpha\alpha\beta\beta\gamma\gamma$ | $\alpha\alpha\beta\alpha\gamma$ | *N*, *C*, *P* | | | *P*30 | H 27, | F 57, | Br 25 | | |
| $\alpha\beta\beta$ | $\alpha\alpha\beta$ | *N* | | | | | | | | |
| $\alpha\beta\beta$ | $\alpha\alpha\beta$ | *N* | | | | | | | | |
| $\alpha\beta\beta$ | $\alpha\alpha\beta$ | *N*, *C*, *P* | | | | H 18, | | Br 5, | Bu 85 | |
| $\alpha\beta\beta$ | $\alpha\beta\gamma$ | *P* | | | | H 8a, | | Br 2, | Bu 53, 91, 92 | |
| $\alpha\beta\beta$ | $\alpha\alpha\beta$ | *C*, *P* | | | *P*32(*r*) | H 8, | | Br 1, | Bu 87, 88 | |

**Table 3** (*continued*)

| Homeomeric type (1) | Topological type (2) | Symmetry group (3) | Order (4) | Induced group (5) | Incidence symbol (6) |
|---|---|---|---|---|---|
| Isohedral Tilings | | | | | |
| SIH 61($r$) | $[3^3.r]$ | $[2, r]^+$ | $2r$ | $[1]^+$ | $[a^+b^+c^+d^+; a^+b^+d^+c^+]$ |
| SIH 62($r$) | $r \neq 3$ | $[2^+, 2r^+]$ | $2r$ | $[1]^+$ | $[a^+b^+c^+d^+; b^-a^-d^+c^+]$ |
| SIH 63($r$) | | $[2^+, 2r]$ | $4r$ | $[1](l)$ | $[a^+a^-b^+b^-; a^+b^-]$ |
| Isotoxal Tilings | | | | | |
| SIT 1 | $\langle 3^2; 3^2 \rangle$ | $[3, 3]^+$ | 12 | $[2]^+$ | $\langle a^+a^+; a^+a^+a^+ \rangle$ |
| SIT 2 | | $[3, 3]$ | 24 | $[2]$ | $\langle aa; aaa \rangle$ |
| SIT 3 | $\langle 3^2; 4^2 \rangle$ | $[3, 3]^+$ | 12 | $[1]^+$ | $\langle a^+b^+; a^+a^+a^+, b^+b^+b^+ \rangle$ |
| SIT 4 | | $[3, 4]^+$ | 24 | $[2]^+$ | $\langle a^+a^+; a^+a^+a^+ \rangle$ |
| SIT 5 | | $[3^+, 4]$ | 24 | $[1](l)$ | $\langle a^+a^-; a^+a^+a^+ \rangle$ |
| SIT 6 | | $[3, 3]$ | 24 | $[1](p)$ | $\langle ab; aaa, bbb \rangle$ |
| SIT 7 | | $[3, 4]$ | 48 | $[2]$ | $\langle aa; aaa \rangle$ |
| SIT 8 | $\langle 3^2; 5^2 \rangle$ | $[3, 5]^+$ | 60 | $[2]^+$ | $\langle a^+a^+; a^+a^+a^+ \rangle$ |
| SIT 9 | | $[3, 5]$ | 120 | $[2]$ | $\langle aa; aaa \rangle$ |
| SIT 10 | $\langle 3.4; 4^2 \rangle$ | $[3, 4]^+$ | 24 | $[1]^+$ | $\langle a^+b^+; a^+a^+a^+, b^+b^+b^+b^+ \rangle$ |
| SIT 11 | | $[3^+, 4]$ | 24 | $[1]^+$ | $\langle a^+b^+; a^+a^+a^+, b^+b^-b^+b^- \rangle$ |
| SIT 12 | | $[3, 4]$ | 48 | $[1](p)$ | $\langle ab; aaa, bbbb \rangle$ |
| SIT 13 | $\langle 3.5; 4^2 \rangle$ | $[3, 5]^+$ | 60 | $[1]^+$ | $\langle a^+b^+; a^+a^+a^+, b^+b^+b^+b^+b^+ \rangle$ |
| SIT 14 | | $[3, 5]$ | 120 | $[1](p)$ | $\langle ab; aaa, bbbbb \rangle$ |
| SIT 15 | $\langle 4^2; 3^2 \rangle$ | $[3, 3]^+$ | 12 | $[1]^+$ | $\langle a^+b^+; a^+b^+a^+b^+ \rangle$ |
| SIT 16 | | $[3, 4]^+$ | 24 | $[2]^+$ | $\langle a^+a^+; a^+a^+a^+a^+ \rangle$ |
| SIT 17 | | $[3, 3]$ | 24 | $[1](l)$ | $\langle a^+a^-; a^+a^-a^+a^- \rangle$ |
| SIT 18 | | $[3^+, 4]$ | 24 | $[1](p)$ | $\langle ab; abab \rangle$ |
| SIT 19 | | $[3, 4]$ | 48 | $[2]$ | $\langle aa; aaaa \rangle$ |
| SIT 20 | $\langle 4^2; 3.4 \rangle$ | $[3, 4]^+$ | 24 | $[1]^+$ | $\langle a^+b^+; a^+b^+a^+b^+ \rangle$ |
| SIT 21 | | $[3^+, 4]$ | 24 | $[1]^+$ | $\langle a^+b^+; a^+b^+b^-a^- \rangle$ |
| SIT 22 | | $[3, 4]$ | 48 | $[1](l)$ | $\langle a^+a^-; a^+a^-a^+a^- \rangle$ |
| SIT 23 | $\langle 4^2; 3.5 \rangle$ | $[3, 5]^+$ | 60 | $[1]^+$ | $\langle a^+b^+; a^+b^+a^+b^+ \rangle$ |
| SIT 24 | | $[3, 5]$ | 120 | $[1](l)$ | $\langle a^+a^-; a^+a^-a^+a^- \rangle$ |
| SIT 25 | $\langle 5^2; 3^2 \rangle$ | $[3, 5]^+$ | 60 | $[2]^+$ | $\langle a^+a^+; a^+a^+a^+a^+a^+ \rangle$ |
| SIT 26 | | $[3, 5]$ | 120 | $[2]$ | $\langle aa; aaaaa \rangle$ |
| Isogonal Tilings | | | | | |
| SIG 1 | $(3^3)$ | $[2, 2]^+$ | 4 | $[1]^+$ | $(a^+b^+c^+; a^+b^+c^+)$ |
| SIG 2 | | $[2^+, 4^+]$ | 4 | $[1]^+$ | $(a^+b^+c^+; a^+c^-b^-)$ |
| SIG 3 | | $[2^+, 4]$ | 8 | $[1]$ | $(ab^+b^-; ab^+)$ |
| SIG 4 | | $[3, 3]^+$ | 12 | $[3]^+$ | $(a^+a^+a^+; a^+)$ |
| SIG 5 | | $[3, 3]$ | 24 | $[3]$ | $(aaa; a)$ |

| Transitivity classes (7) | | Realizations (8) | Cross references (9) | | | References (10) |
|---|---|---|---|---|---|---|
| Edges | Vertices | | | | | |
| $\alpha\beta\gamma\gamma$ | $\alpha\alpha\alpha\beta$ | $N, C, P$ | | | | H 25, F 45, Br 6 |
| $\alpha\alpha\beta\beta$ | $\alpha\alpha\alpha\beta$ | $N$ | | | | |
| $\alpha\alpha\beta\beta$ | $\alpha\alpha\alpha\beta$ | $N, C, P$ | | | $P34(r)$ | H 24, Br 3, Bu 50, 89 |
| Vertices | Tiles | | | | | |
| $\alpha\alpha$ | $TT$ | $N$ | SIH 4 | SIG 4 | | |
| $\alpha\alpha$ | $TT$ | $C, P$ | SIH 5 | SIG 5 | $P1$ | |
| $\alpha\alpha$ | $T_1T_2$ | $N$ | | SIG 32 | | |
| $\alpha\alpha$ | $TT$ | $N$ | SIH 17 | SIG 33 | | |
| $\alpha\alpha$ | $TT$ | – | SIH 18 | SIG 36 | | |
| $\alpha\alpha$ | $T_1T_2$ | $N$ | | SIG 35 | | |
| $\alpha\alpha$ | $TT$ | $C, P$ | SIH 20 | SIG 37 | $P3$ | |
| $\alpha\alpha$ | $TT$ | $N$ | SIH 26 | SIG 51 | | |
| $\alpha\alpha$ | $TT$ | $C, P$ | SIH 27 | SIG 53 | $P5$ | |
| $\alpha\alpha$ | $TQ$ | $N$ | | SIG 39 | | |
| $\alpha\alpha$ | $TQ$ | $N, C$ | | SIG 40 | | |
| $\alpha\alpha$ | $TQ$ | $N, C, P$ | | SIG 42 | $P9$ | |
| $\alpha\alpha$ | $TP$ | $N$ | | SIG 48 | | |
| $\alpha\alpha$ | $TP$ | $N, C, P$ | | SIG 49 | $P21$ | |
| $\alpha\beta$ | $QQ$ | $N$ | SIH 32 | | | |
| $\alpha\alpha$ | $QQ$ | $N$ | SIH 33 | SIG 17 | | |
| $\alpha\beta$ | $QQ$ | – | SIH 35 | | | |
| $\alpha\alpha$ | $QQ$ | $N$ | SIH 36 | SIG 18 | | |
| $\alpha\alpha$ | $QQ$ | $C, P$ | SIH 37 | SIG 20 | $P2$ | |
| $\alpha\beta$ | $QQ$ | $N$ | SIH 39 | | | |
| $\alpha\beta$ | $QQ$ | $N$ | SIH 40 | | | |
| $\alpha\beta$ | $QQ$ | $C, P$ | SIH 42 | | $P8$ | |
| $\alpha\beta$ | $QQ$ | $N$ | SIH 48 | | | |
| $\alpha\beta$ | $QQ$ | $C, P$ | SIH 49 | | $P20$ | |
| $\alpha\alpha$ | $PP$ | $N$ | SIH 51 | SIG 26 | | |
| $\alpha\alpha$ | $PP$ | $C, P$ | SIH 53 | SIG 27 | $P4$ | |
| Edges | Vertices | | | | | |
| $\alpha\beta\gamma$ | $TTT$ | $N, C, P$ | | SIH 1 | | |
| $\alpha\beta\beta$ | $TTT$ | $N$ | | SIH 2 | | |
| $\alpha\beta\beta$ | $TTT$ | $N, C, P$ | | SIH 3 | | |
| $\alpha\alpha\alpha$ | $TTT$ | $N$ | SIT 1 | SIH 4 | | |
| $\alpha\alpha\alpha$ | $TTT$ | $C, P$ | SIT 2 | SIH 5 | $P1$ | |

**Table 3** (*continued*)

| Homeomeric type (1) | Topological type (2) | Symmetry group (3) | Order (4) | Induced group (5) | Incidence symbol (6) |
|---|---|---|---|---|---|
| Isogonal Tilings | | | | | |
| SIG 6 | $(3.6^2)$ | $[3,3]^+$ | 12 | $[1]^+$ | $(a^+b^+c^+; a^+c^+b^+)$ |
| SIG 7 | | $[3,3]$ | 24 | $[1]$ | $(ab^+b^-; ab^-)$ |
| SIG 8 | $(3.8^2)$ | $[3,4]^+$ | 24 | $[1]^+$ | $(a^+b^+c^+; a^+c^+b^+)$ |
| SIG 9 | | $[3^+,4]$ | 24 | $[1]^+$ | $(a^+b^+c^+; a^-c^+b^+)$ |
| SIG 10 | | $[3,4]$ | 48 | $[1]$ | $(ab^+b^-; ab^-)$ |
| SIG 11 | $(3.10^2)$ | $[3,5]^+$ | 60 | $[1]^+$ | $(a^+b^+c^+; a^+c^+b^+)$ |
| SIG 12 | | $[3,5]$ | 120 | $[1]$ | $(ab^+b^-; ab^-)$ |
| SIG 13 | $(4^3)$ | $[2^+,4]$ | 8 | $[1]^+$ | $(a^+b^+c^+; a^+b^-c^-)$ |
| SIG 14 | | $[2,2]$ | 8 | $[1]^+$ | $(a^+b^+c^+; a^-b^-c^-)$ |
| SIG 15 | | $[2,4^+]$ | 8 | $[1]^+$ | $(a^+b^+c^+; a^-c^+b^+)$ |
| SIG 16 | | $[2,4]^+$ | 8 | $[1]^+$ | $(a^+b^+c^+; a^+c^+b^+)$ |
| SIG 17 | | $[3,4]^+$ | 24 | $[3]^+$ | $(a^+a^+a^+; a^+)$ |
| SIG 18 | | $[3^+,4]$ | 24 | $[3]^+$ | $(a^+a^+a^+; a^-)$ |
| SIG 19 | | $[2,4]$ | 16 | $[1]$ | $(ab^+b^-; ab^-)$ |
| SIG 20 | | $[3,4]$ | 48 | $[3]$ | $(aaa; a)$ |
| SIG 21 | $(4.6^2)$ | $[3,4]^+$ | 24 | $[1]^+$ | $(a^+b^+c^+; a^+c^+b^+)$ |
| SIG 22 | | $[3,3]$ | 24 | $[1]^+$ | $(a^+b^+c^+; a^-b^-c^-)$ |
| SIG 23 | | $[3,4]$ | 48 | $[1]$ | $(ab^+b^-; ab^-)$ |
| SIG 24 | $(4.6.8)$ | $[3,4]$ | 48 | $[1]^+$ | $(a^+b^+c^+; a^-b^-c^-)$ |
| SIG 25 | $(4.6.10)$ | $[3,5]$ | 120 | $[1]^+$ | $(a^+b^+c^+; a^-b^-c^-)$ |
| SIG 26 | $(5^3)$ | $[3,5]^+$ | 60 | $[3]^+$ | $(a^+a^+a^+; a^+)$ |
| SIG 27 | | $[3,5]$ | 120 | $[3]$ | $(a, a, a; a)$ |
| SIG 28 | $(5.6^2)$ | $[3,5]^+$ | 60 | $[1]^+$ | $(a^+b^+c^+; a^+c^+b^+)$ |
| SIG 29 | | $[3,5]$ | 120 | $[1]$ | $(ab^+b^-; ab^-)$ |
| SIG 30 | $(3^4)$ | $[2,3]^+$ | 6 | $[1]^+$ | $(a^+b^+c^+d^+; a^+b^+d^+c^+)$ |
| SIG 31 | | $[2^+,6^+]$ | 6 | $[1]^+$ | $(a^+b^+c^+d^+; b^+a^+d^-c^-)$ |
| SIG 32 | | $[3,3]^+$ | 12 | $[2]^+$ | $(a^+b^+a^+b^+; b^+a^+)$ |
| SIG 33 | | $[3,4]^+$ | 24 | $[4]^+$ | $(a^+a^+a^+a^+; a^+)$ |
| SIG 34 | | $[2^+,6]$ | 12 | $[1](l)$ | $(a^+b^+b^-a^-; a^-b^+)$ |
| SIG 35 | | $[3,3]$ | 24 | $[2](l)$ | $(a^+a^-a^+a^-; a^-)$ |
| SIG 36 | | $[3^+,4]$ | 24 | $[2](s)$ | $(abab; ba)$ |
| SIG 37 | | $[3,4]$ | 48 | $[4]$ | $(aaaa; a)$ |
| SIG 38 | $(3.4.3.4)$ | $[3,3]^+$ | 12 | $[1]^+$ | $(a^+b^+c^+d^+; b^+a^+d^+c^+)$ |
| SIG 39 | | $[3,4]^+$ | 24 | $[2]^+$ | $(a^+b^+a^+b^+; b^+a^+)$ |
| SIG 40 | | $[3^+,4]$ | 24 | $[1](l)$ | $(a^+b^+b^-a^-; b^+a^+)$ |
| SIG 41 | | $[3,3]$ | 24 | $[1](l)$ | $(a^+b^+b^-a^-; a^-b^-)$ |
| SIG 42 | | $[3,4]$ | 48 | $[2](l)$ | $(a^+a^-a^+a^-; a^-)$ |

| Transitivity classes (7) | | Realizations (8) | Cross references (9) | | | References (10) |
|---|---|---|---|---|---|---|
| Edges | Vertices | | | | | |
| $\alpha\beta\beta$ | $HHT$ | $N, C$ | | | | |
| $\alpha\beta\beta$ | $HHT$ | $N, C, P$ | | | $P7$ | |
| $\alpha\beta\beta$ | $OOT$ | $N, C$ | | | | |
| $\alpha\beta\beta$ | $OOT$ | $N, C$ | | | | |
| $\alpha\beta\beta$ | $OOT$ | $N, C, P$ | | | $P13$ | |
| $\alpha\beta\beta$ | $DDT$ | $N, C$ | | | | |
| $\alpha\beta\beta$ | $DDT$ | $N, C, P$ | | | $P25$ | |
| $\alpha\beta\gamma$ | $Q_1Q_2Q_3$ | $N, C, P$ | | | | |
| $\alpha\beta\gamma$ | $Q_1Q_2Q_3$ | $N, C, P$ | | | | |
| $\alpha\beta\beta$ | $Q_1Q_1Q_2$ | $N$ | | | | |
| $\alpha\beta\beta$ | $Q_1Q_1Q_2$ | $N, C$ | | | | |
| $\alpha\alpha\alpha$ | $QQQ$ | $N$ | SIT 16 | SIH 33 | | |
| $\alpha\alpha\alpha$ | $QQQ$ | $N$ | SIT 18 | SIH 36 | | |
| $\alpha\beta\beta$ | $Q_1Q_1Q_2$ | $N, C, P$ | | | | |
| $\alpha\alpha\alpha$ | $QQQ$ | $C, P$ | SIT 19 | SIH 37 | $P2$ | |
| $\alpha\beta\beta$ | $HHQ$ | $N, C$ | | | | |
| $\alpha\beta\gamma$ | $H_1H_2Q$ | $N, C, P$ | | | | |
| $\alpha\beta\beta$ | $HHQ$ | $N, C, P$ | | | $P11$ | |
| $\alpha\beta\gamma$ | $HOQ$ | $N, C, P$ | | | $P17$ | |
| $\alpha\beta\gamma$ | $DHQ$ | $N, C, P$ | | | $P28$ | |
| $\alpha\alpha\alpha$ | $TTT$ | $N$ | SIT 25 | SIH 51 | | |
| $\alpha\alpha\alpha$ | $TTT$ | $C, P$ | SIT 26 | SIH 53 | $P4$ | |
| $\alpha\beta\beta$ | $HHP$ | $N, C$ | | | | |
| $\alpha\beta\beta$ | $HHP$ | $N, C, P$ | | | $P23$ | |
| $\alpha\beta\gamma\gamma$ | $T_1T_1T_1T_2$ | $N, C, P$ | | | | |
| $\alpha\alpha\beta\beta$ | $T_1T_2T_1T_1$ | $N$ | | | | |
| $\alpha\alpha\alpha\alpha$ | $T_1T_2T_1T_2$ | $N$ | SIT 3 | | | |
| $\alpha\alpha\alpha\alpha$ | $TTTT$ | $N$ | SIT 4 | SIH 17 | | |
| $\alpha\alpha\beta\beta$ | $T_1T_2T_2T_2$ | $N, C, P$ | | | | |
| $\alpha\alpha\alpha\alpha$ | $T_1T_2T_1T_2$ | $N$ | SIT 6 | | | |
| $\alpha\alpha\alpha\alpha$ | $TTTT$ | – | SIT 5 | SIH 18 | | |
| $\alpha\alpha\alpha\alpha$ | $TTTT$ | $C, P$ | SIT 7 | SIH 20 | $P3$ | |
| $\alpha\alpha\beta\beta$ | $QT_1QT_2$ | $N, C$ | | | | |
| $\alpha\alpha\alpha\alpha$ | $QTQT$ | $N$ | SIT 10 | | | |
| $\alpha\alpha\alpha\alpha$ | $QTQT$ | $N, C$ | SIT 11 | | | |
| $\alpha\alpha\beta\beta$ | $QT_1QT_2$ | $N, C, P$ | | | | |
| $\alpha\alpha\alpha\alpha$ | $QTQT$ | $N, C, P$ | SIT 12 | | $P9$ | |

**Table 3** (*continued*)

| Homeomeric type (1) | Topological type (2) | Symmetry group (3) | Order (4) | Induced group (5) | Incidence symbol (6) |
|---|---|---|---|---|---|
| Isogonal Tilings | | | | | |
| SIG 43 | $(3.4^3)$ | $[3,4]^+$ | 24 | $[1]^+$ | $(a^+b^+c^+d^+; b^+a^+d^+c^+)$ |
| SIG 44 | | $[3^+,4]$ | 24 | $[1]^+$ | $(a^+b^+c^+d^+; b^+a^+c^-d^-)$ |
| SIG 45 | | $[3,4]$ | 48 | $[1](l)$ | $(a^+a^-b^+b^-; a^-b^-)$ |
| SIG 46 | $(3.4.5.4)$ | $[3,5]^+$ | 60 | $[1]^+$ | $(a^+b^+c^+d^+; b^+a^+d^+c^+)$ |
| SIG 47 | | $[3,5]$ | 120 | $[1](l)$ | $(a^+a^-b^+b^-; a^-b^-)$ |
| SIG 48 | $(3.5.3.5)$ | $[3,5]^+$ | 60 | $[2]^+$ | $(a^+b^+a^+b^+; b^+a^+)$ |
| SIG 49 | | $[3,5]$ | 120 | $[2](l)$ | $(a^+a^-a^+a^-; a^-)$ |
| SIG 50 | $(3^5)$ | $[3,3]^+$ | 12 | $[1]^+$ | $(a^+b^+c^+d^+e^+; a^+c^+b^+e^+d^+)$ |
| SIG 51 | | $[3,5]^+$ | 60 | $[5]^+$ | $(a^+a^+a^+a^+a^+; a^+)$ |
| SIG 52 | $(3^5)$ | $[3^+,4]$ | 24 | $[1]$ | $(ab^+c^+c^-b^-; ac^+b^+)$ |
| SIG 53 | | $[3,5]$ | 120 | $[5]$ | $(aaaaa; a)$ |
| SIG 54 | $(3^4.4)$ | $[3,4]^+$ | 24 | $[1]^+$ | $(a^+b^+c^+d^+e^+; a^+c^+b^+e^+d^+)$ |
| SIG 55 | $(3^4.5)$ | $[3,5]^+$ | 60 | $[1]^+$ | $(a^+b^+c^+d^+e^+; a^+c^+b^+e^+d^+)$ |
| SIG 56($r$) | $(4^2.r)$ | $[2,r]^+$ | $2r$ | $[1]^+$ | $(a^+b^+c^+; a^+c^+b^+)$ |
| SIG 57($r$) | $r \neq 4$ | $[2,r^+]$ | $2r$ | $[1]^+$ | $(a^+b^+c^+; a^-c^+b^+)$ |
| SIG 58($r$) | | $[2^+,r]$ $2\vert r$ | $2r$ | $[1]^+$ | $(a^+b^+c^+; a^+b^-c^-)$ |
| SIG 59($r$) | | $[2,\frac{1}{2}r]$ $2\vert r$ | $2r$ | $[1]^+$ | $(a^+b^+c^+; a^-b^-c^-)$ |
| SIG 60($r$) | | $[2,r]$ | $4r$ | $[1]$ | $(ab^+b^-; ab^-)$ |
| SIG 61($r$) | $(3^3.r)$ | $[2,r]^+$ | $2r$ | $[1]^+$ | $(a^+b^+c^+d^+; a^+b^+d^+c^+)$ |
| SIG 62($r$) | $r \neq 3$ | $[2^+,2r^+]$ | $2r$ | $[1]^+$ | $(a^+b^+c^+d^+; b^-a^-d^+c^+)$ |
| SIG 63($r$) | | $[2^+,2r]$ | $4r$ | $[1](l)$ | $(a^+a^-b^+b^-; a^+b^-)$ |

transitive subgroups) and isogonal tilings (vertex-transitive subgroups) for every topological type.

It is clear that different tile-transitive subgroups of $S(\mathfrak{T}(P))$ must correspond to different homeomeric types of isohedral tilings (see [18]) and conversely that every homeomeric type of tiling must correspond to some such subgroup. However, it remains to be shown that our listing of types is correct in the sense that the chosen subgroup can be *realized* as the symmetry group of a suitable tiling. As we shall see shortly, the concept of a tiling has to be generalized slightly if *every* subgroup is to be realized in this way. Exactly similar considerations apply to the isotoxal and isogonal tilings.

Our procedure is to examine each entry in Table 3 in turn and determine possible realizations of it. These are listed in column (8) of the table using the letters $C$, $N$, or $P$, whose meaning will now be explained.

The letter $C$ means that a type has a *convex realization*, that is, there exists a convex tiling with the stated symmetry group. Clearly all the tilings listed in Table 2 have convex realizations, but there are other possibilities as well. For

| Transitivity classes (7) | | Realizations (8) | Cross references (9) | | | References (10) |
|---|---|---|---|---|---|---|
| Edges | Vertices | | | | | |
| $\alpha\alpha\beta\beta$ | $Q_1TQ_1Q_2$ | $N$ | | | | |
| $\alpha\alpha\beta\gamma$ | $Q_1TQ_1Q_2$ | $N, C, P$ | | | | |
| $\alpha\alpha\beta\beta$ | $Q_1TQ_1Q_2$ | $N, C, P$ | | | $P15$ | |
| $\alpha\alpha\beta\beta$ | $QTQP$ | $N, C$ | | | | |
| $\alpha\alpha\beta\beta$ | $QTQP$ | $N, C, P$ | | | $P27$ | |
| $\alpha\alpha\alpha\alpha$ | $PTPT$ | $N$ | SIT 13 | | | |
| $\alpha\alpha\alpha\alpha$ | $PTPT$ | $N, C, P$ | SIT 14 | | $P21$ | |
| $\alpha\beta\beta\gamma\gamma$ | $T_1T_1T_2T_1T_3$ | $N, C, P$ | | | | |
| $\alpha\alpha\alpha\alpha\alpha$ | $TTTTT$ | $N$ | SIT 8 | SIH 26 | | |
| $\alpha\beta\beta\beta\beta$ | $T_1T_1T_2T_1T_2$ | $N, C, P$ | | | | |
| $\alpha\alpha\alpha\alpha\alpha$ | $TTTTT$ | $C, P$ | SIT 9 | SIH 27 | $P5$ | |
| $\alpha\beta\beta\gamma\gamma$ | $T_1T_1T_2T_1Q$ | $N, C, P$ | | | $P19$ | |
| $\alpha\beta\beta\gamma\gamma$ | $T_1T_1T_2T_1P$ | $N, C, P$ | | | $P31$ | |
| $\alpha\beta\beta$ | $QQR$ | $N, C$ | | | | |
| $\alpha\beta\beta$ | $QQR$ | $N$ | | | | |
| $\alpha\beta\gamma$ | $QQR$ | $N, C, P$ | | | | |
| $\alpha\beta\gamma$ | $Q_1Q_2R$ | $N, C, P$ | | | | |
| $\alpha\beta\beta$ | $QQR$ | $N, C, P$ | | | $P33(r)$ | |
| $\alpha\beta\gamma\gamma$ | $TTTR$ | $N, C, P$ | | | | |
| $\alpha\alpha\beta\beta$ | $TTTR$ | $N$ | | | | |
| $\alpha\alpha\beta\beta$ | $TTTR$ | $N, C, P$ | | | $P35(r)$ | |

example, the isogonal type SIG 56(3) has the convex realization shown in Figure 1(d), yet this type, as we have already remarked, cannot be obtained from a polyhedron by radial projection.

The letter $N$ means that a type has a *non-convex realization* that is, there exists a tiling by non-convex tiles with the stated symmetry group. Examples of non-convex realizations of isohedral, isotoxal, and isogonal tilings are shown in Figures 3(b), 4(b) and 6(b).

The properties of having convex and non-convex realizations are independent in the sense that types of tilings exist with both, either, or neither. For example the types SIH 37, SIT 19, and SIG 20 are all represented by the convex tiling of Figure 5(a) (the central projection of a cube), and none of these has non-convex realizations. The types SIH 9, SIT 20, and SIG 40, of which non-convex realizations are shown in Figures 3(b), 4(b), and 6(b), do not have convex realizations, whereas the type SIG 56(r) has both convex and non-convex realizations.

On the other hand a type such as SIH 14 has neither a convex nor a

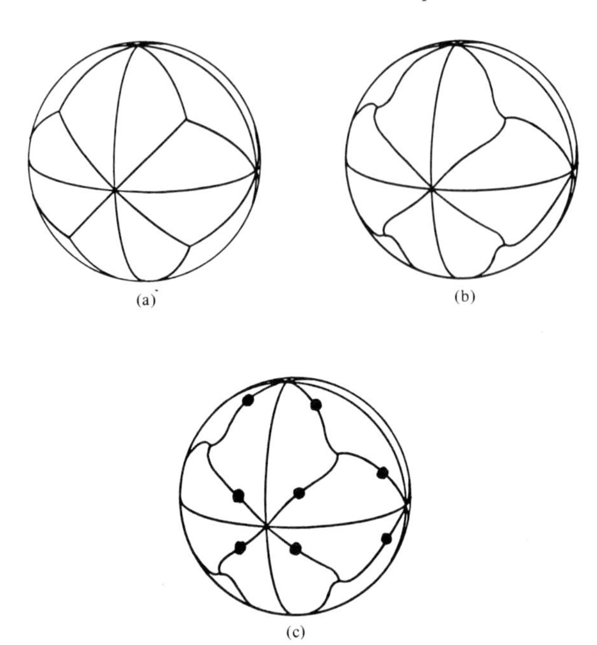


**Figure 3.** (a) The isohedral tiling $\mathcal{T}(P)$ obtained from the triakis octahedron $P$ by central projection. Its topological type is $[3.8^2]$, its homeomeric type is SIH 10, and its symmetry group is $[3,4]$. (b) An isohedral tiling with the same topological type $[3.8^2]$, corresponding to the tile-transitive subgroup $[3^+,4]$ of $[3,4]$. Its homeomeric type is SIH 9; a realization of this type by non-convex tiles is shown. (c) A non-normal isohedral tiling obtained from that of (b) by edge-division (inserting pseudovertices on some of the edges).

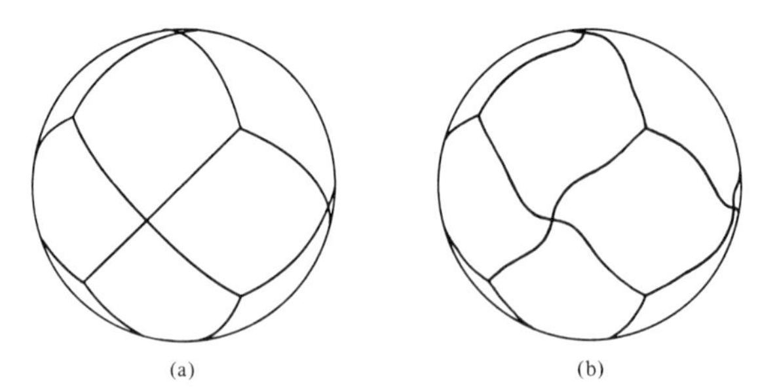


**Figure 4.** (a) The isotoxal tiling $\mathcal{T}(P)$ obtained from the rhombic dodecahedron $P$ by central projection. Its topological type is $\langle 4^2; 3.4\rangle$, its homeomeric type is SIT 22, and its symmetry group is $[3,4]$. (b) An isotoxal tiling of the same topological type $\langle 4^2; 3.4\rangle$ corresponding to the edge-transitive subgroup $[3,4]^+$ of $[3,4]$. The homeomeric type is SIT 20; a realization of this type by non-convex tiles is shown. In this case no non-normal isotoxal tilings can be derived by edge-division or edge-splitting.

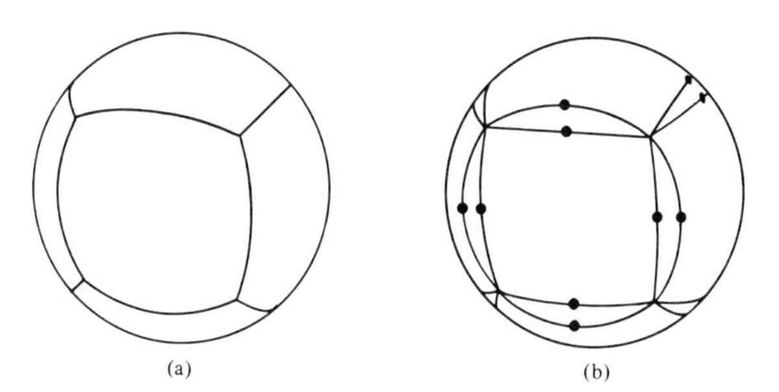


**Figure 5.** (a) The isotoxal tiling $\mathcal{T}(P)$ obtained from the cube $P$ by central projection. Its topological type is $\langle 4^2; 3^2 \rangle$, its homeomeric type is SIT 19, and its symmetry group is $[3,4]$. (b) A non-normal isotoxal tiling derived from that in (a) by first splitting each edge into a digon and then dividing each edge by a pseudovertex.

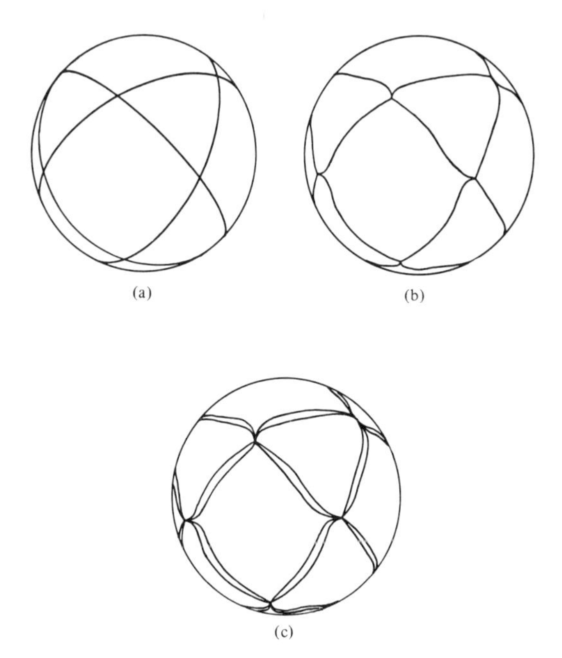


**Figure 6.** (a) The isogonal tiling $\mathcal{T}(P)$ obtained from the cuboctahedron $P$ by central projection. Its topological type is (3.4.3.4), its homeomeric type is SIG 42, and its symmetry group is $[3,4]$. (b) An isogonal tiling of the same topological type (3.4.3.4) corresponding to the vertex transitive subgroup $[3^+,4]$ of $[3,4]$. The homeomeric type is SIG 40; a realization by non-convex tiles is shown. (c) A non-normal isogonal tiling derived from the tiling in (b) by edge-splitting. Here, in order to maintain isogonality every edge must be split, but it may be split into arbitrarily many digons. In other cases, such as that of Figure 7, isogonality can be preserved by splitting a proper subset of the edges.

non-convex realization. To see this, let us suppose that a tiling $\mathcal{T}$ of the given type exists. The fact that the symmetry group of $\mathcal{T}$ is to be [2, 2], which contains reflections that carry each tile of $\mathcal{T}$ into each of its adjacent tiles, implies that the edges of $\mathcal{T}$ must be arcs of great circles. Hence each tile of $\mathcal{T}$ must be an octant of the sphere (see Figure 2(a)); its symmetry group is therefore [3, 4] and *not* the required [2, 2]. We deduce that no tiling (either convex or non-convex) can realize the type SIH 14.

There exists a systematic procedure for finding all convex and non-convex realizations of a tiling. This can be simply described as a set of rules for replacing the edges of a tiling $\mathcal{T}(P)$ by suitable curves. The method has been explained in detail in the context of plane tilings in [15–17], and [20]. As the procedure in the spherical case is completely analogous, we omit details here and refer the reader to the publications just mentioned.

From examination of Table 3 we see that 9 types (SIH 18, SIH 19, SIH 22, SIH 35, SIH 41, SIH 59(r), SIT 5, SIT 17, and SIG 36) have neither convex nor non-convex realizations, so strictly speaking these do not correspond to actual tilings (as defined in Section 1) at all. However, they *can* be realized if we are prepared to extend our definitions slightly. This can be done in two ways.

The first, which provides a universal method in the sense that all the types in Table 3 can be realized, is by the use of marked tilings. A *marked tiling* $\mathcal{T}^*$ on $S^2$ consists of a tiling $\mathcal{T}$ in the original sense, on each tile of which is a *mark* or *motif* which simply means a subset of the interior of the tile. The symmetry group of $\mathcal{T}^*$ is then defined as the group of isometries which not only map $\mathcal{T}$ onto itself but also map the mark on each tile onto the mark on the image tile. An example of a marked tiling is shown in Figure 2(b). Here we have chosen a small L-shape as the marking. It is easy to see that the symmetry group of $\mathcal{T}^*$ is the tile-transitive subgroup [2, 2] of the symmetry group [3, 4] of $\mathcal{T}$. We can therefore say that the marked isohedral tiling shown is of type SIH 14 (which, as we already know, admits no convex or non-convex realizations).

The fact that every isohedral type in Table 3 can be realized by a marked tiling is easy to see. We begin by choosing a tiling $\mathcal{T}(P)$ from Table 2 and specify a tile-transitive subgroup $G$ of $S(\mathcal{T}(P))$. Impose an arbitrary (unsymmetrical) marking $M'$ on one of the tiles of $\mathcal{T}(P)$ and take all the images of $M'$ under the operations of the group $G$. Notice that at least one copy of $M'$ will appear on every tile and that more than one copy of $M'$ may appear on the same (and therefore on every) tile. The latter possibility will occur if $G$ contains operations other than the identity which map a tile of $\mathcal{T}(P)$ onto itself. The union of the copies of $M'$ on each tile $T$ is taken to be the marking on $T$, and it is clear that in general the resulting marked tiling will have exactly the required symmetry group $G$. It is worth remarking that the homeomeric classification of marked tilings coincides with that of Table 3 so long as we only use marks which are closed topological disks (compare [9]). Similar considerations apply to isotoxal and isogonal tilings, though here the details are slightly more complicated.

In addition to the realizations of a tiling by convex, non-convex, or marked tilings, there is a fourth possibility which has no counterpart in the case of plane tilings. This is what we shall call a polyhedrally induced realization, and is indicated by the letter $P$ in column (8) of Table 3. Let $\mathcal{T}(P)$ be a tiling obtained

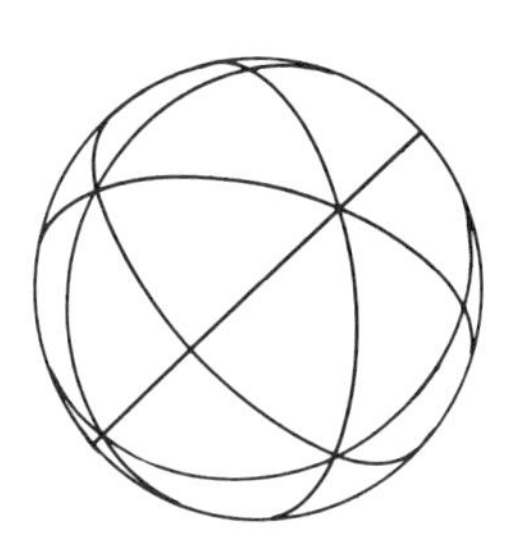
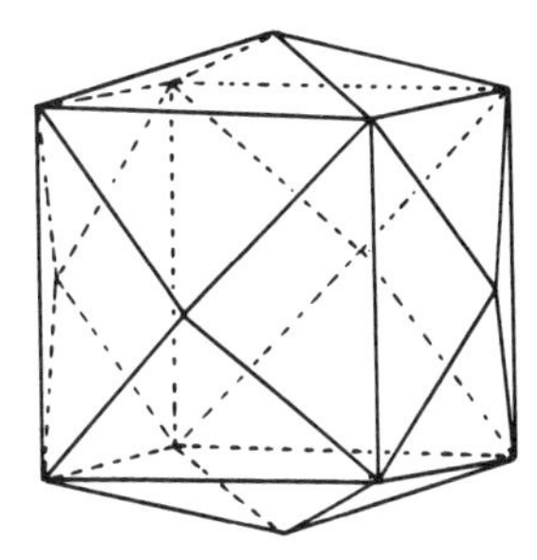

**Figure 7.** The tetrakis hexahedron and its central projection which is of topological type $[4.6^2]$.

by radial projection from a convex polyhedron $P$, and let $G$ be any given subgroup of $S(\mathfrak{T}(P))$, which is, for example, tile-transitive. If $G$ is isomorphic to $S(P)$ then we shall say that the isohedral type corresponding to the subgroup $G$ has a *polyhedrally induced realization*. Similar considerations apply to isotoxal and isogonal tilings. For an example we again refer to Figure 2. We have shown that isohedral type SIH 14 has no convex or non-convex realization; a realization by marked tiles is shown in Figure 2(b). However, this same type has a polyhedrally induced realization. We choose $P$ to be the (non-regular) octahedron shown in Figure 2(a) and notice that the symmetry group of $P$ is the required tile-transitive group $[2, 2]$.

All the tilings in Table 2 clearly have polyhedrally induced realizations—this is a consequence of Theorem 2—but as can be seen from Table 3, there are 30 additional types. Of these 25 have convex realizations and 5 (all isohedral) have neither convex nor non-convex realizations. There are no types with polyhedrally induced and non-convex realizations that do not also have convex realizations.

To illustrate the procedure described above we shall consider a particular example in detail, namely isohedral tilings of topological type $[4.6^2]$. The corresponding polyhedron is the tetrakis hexahedron denoted by $P10$ in Table 2. This and its central projection are shown in Figure 7. The symmetry group of the central projection is $[3, 4]$, of which there are three tile-transitive subgroups, namely $[3, 4]^+$, $[3, 3]$, and $[3, 4]$ itself. Marked tilings corresponding to these three groups are shown in Figure 8, and they represent the homeomeric types SIH 21,

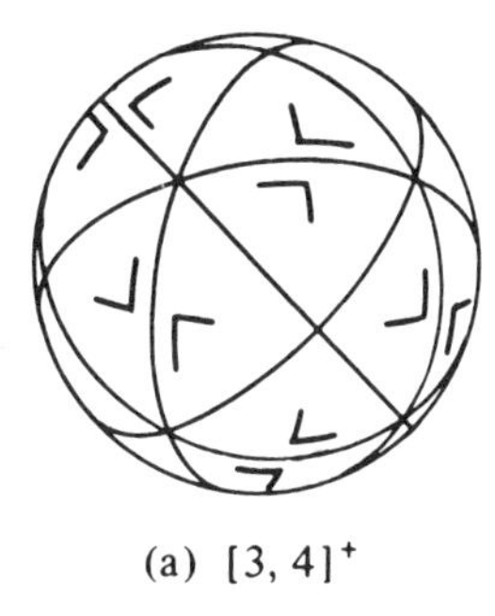
(a) $[3, 4]^+$

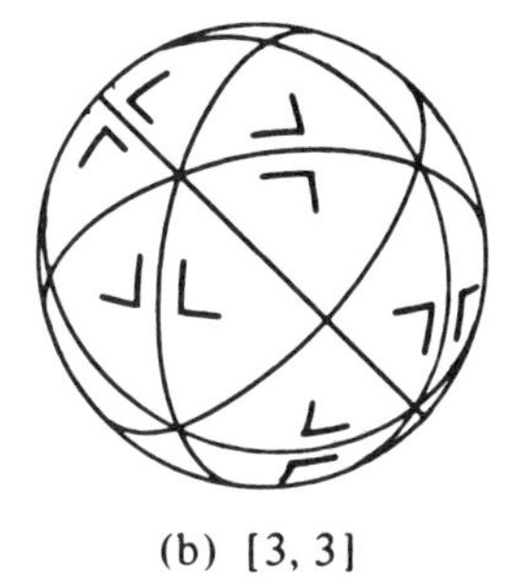
(b) $[3, 3]$

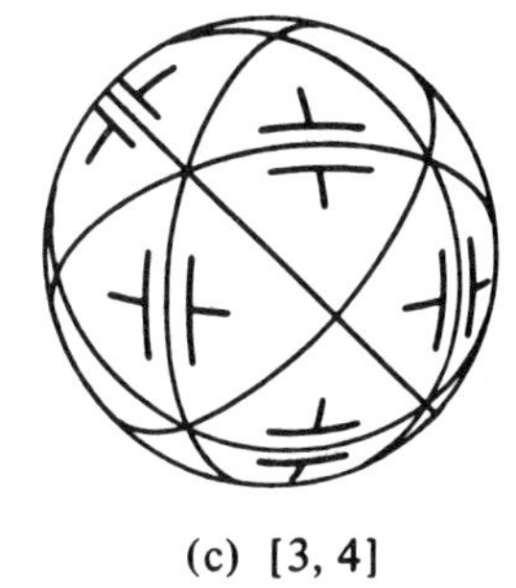
(c) $[3, 4]$

**Figure 8.** Markings of the tiling shown in Figure 7 corresponding to the tile-transitive groups $[3, 4]^+$, $[3, 3]$, and $[3, 4]$.

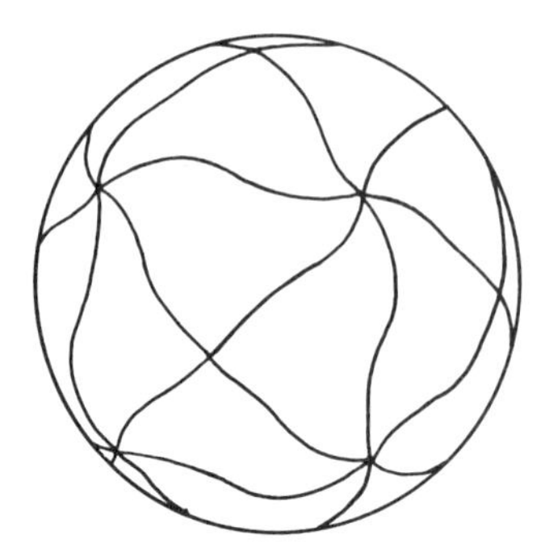

**Figure 9.** A non-convex realization of the isohedral type SIH 21. A realization of this same type by a marked tiling is shown in Figure 8(a).

SIH 22, and SIH 23 of Table 3. Figure 7 shows that SIH 23 has both a convex and a polyhedrally induced realization, and it is easy to see that no non-convex realization exists. In an obvious sense here the marks on the tiles are redundant. Type SIH 21 has no convex or polyhedrally induced realizations, but a non-convex realization is possible: see Figure 9. Type SIH 22 has no convex or non-convex realization, but a polyhedrally induced realization is possible; the corresponding polyhedron $P$ is shown in Figure 10. It may be described as a distorted triakis hexahedron—four of its vertices (arranged as the vertices of a regular tetrahedron) are pushed towards the center, and four others (arranged as the vertices of a dual tetrahedron) are pulled away from the center. If this is done in such a way that the combinatorial type is unchanged, then the symmetry group of $P$ is the required [3, 3], while the central projection on to $S^2$ is unaltered by the distortion.

In Figure 11 we show how, for each of these three isohedral types, the sides of the tiles can be oriented and labelled in accordance with the procedure described in [15] and [20, Chapter 6]. The corresponding incidence symbols are shown in column (6) of Table 3.

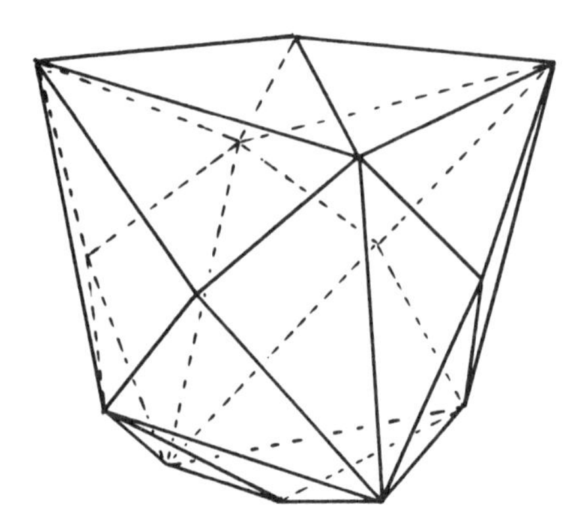

**Figure 10.** A distorted tetrakis hexahedron (see Figure 7) whose symmetry group is [3, 3]. This leads to a polyhedrally induced realization of the marked tiling of type SIH 22 shown in Figure 8(b).

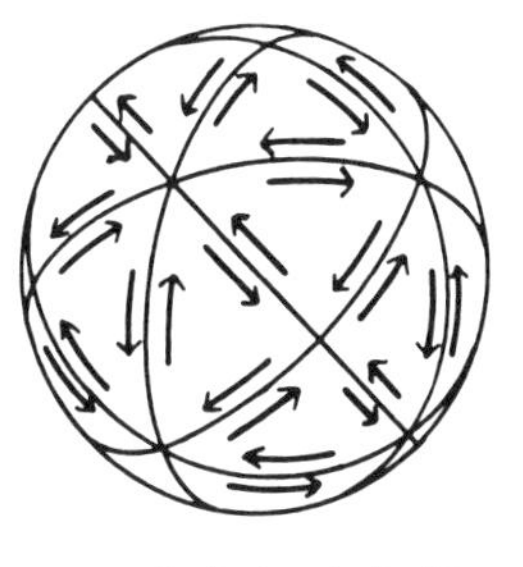

(a) $[a^+b^+c^+;a^+c^+b^+]$

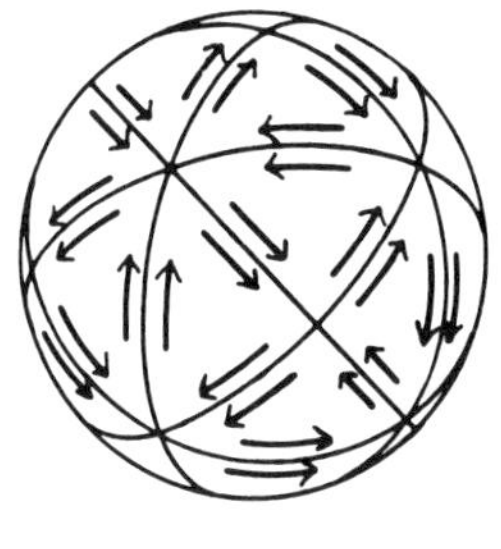

(b) $[a^+b^+c^+;a^-b^-c^-]$

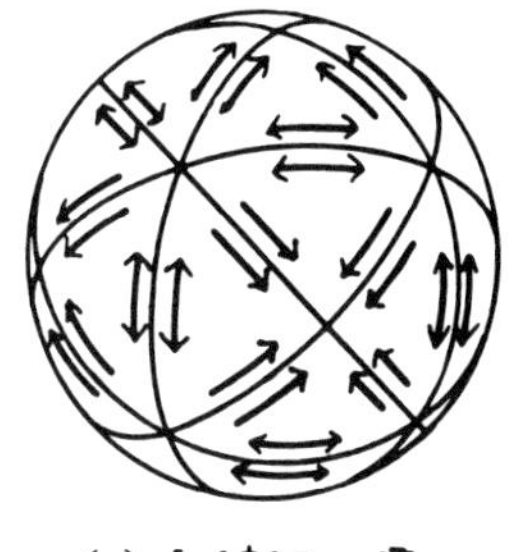

(c) $[ab^+b^-;a\overline{b}]$

**Figure 11.** Labelling of the sides of the tiles corresponding to the three isohedral types SIH 21, SIH 22, and SIH 23. Realizations of these three types by marked tilings are shown in Figure 8. The incidence symbols for the types are indicated below the diagrams.

The investigation of isotoxal and isogonal tilings can be carried out in an exactly similar manner. We summarize the results obtained, and exhibited in Table 3, in the following theorem.

**Theorem 3.** *There exist*

8 *infinite families and* 55 *other types of normal isohedral tilings,*
26 *types of normal isotoxal tilings, and*
8 *infinite families and* 55 *other types of normal isogonal tilings.*

*Each infinite family depends on a positive integer variable r, and the term "type" is to be understood in the sense of "homeomeric type". All can be realized by convex or non-convex tilings except for ten types, and of these five have polyhedrally induced realizations.*

## 4. Non-normal Tilings

The classification of normal isohedral, isotoxal and isogonal tilings given in Table 3 is complete. However, new possibilities (in fact an infinite number of new types) arise if we extend our investigation to non-normal tilings. In this section we shall briefly indicate some of the possibilities.

We shall, for convenience, restrict attention to tilings in which SN1 and SN3 hold, but condition SN2 may be violated. One problem that immediately arises is that if we allow the intersection of two tiles to be non-connected, then the definitions of vertices and edges given in Section 1 are no longer appropriate. From now on we shall use the word *vertex* to mean *any* point (specified as a vertex) which lies in the intersection of two or more distinct tiles, and *edge* to mean *any* (connected) arc which lies in the intersection of two distinct tiles, has distinct vertices as endpoints and contains no vertex in its relative interior. We shall also insist that the number of vertices and edges is finite, so that the boundary of each tile consists of a finite union of edges and vertices. For such tilings Euler's relation (1) continues to hold.

With these definitions, non-normal tilings can differ from normal tilings in two respects:

(1) There may exist *vertices of valence* 2 (that is, vertices which are the endpoints of precisely two edges). These are sometimes called *pseudovertices*.
(2) Tiles may occur whose boundaries are the union of just two edges of the tiling. We shall call these *digons* without implying that they are spherical polygons in the usual sense of the word.

From a given normal tiling $\mathcal{T}$ with transitivity properties it may be possible to obtain new types by the following two processes.

(1) *Edge-division.* This means that we insert (finitely many) pseudovertices on the edges of the original tiling.
(2) *Edge-splitting.* This means that an edge with endpoints $V_1$, $V_2$ is replaced by a digon, or a finite bunch of digons, whose boundaries each contain $V_1$ and $V_2$.

In Figure 3(c) we show a non-normal isohedral tiling obtained from the normal tiling of Figure 3(b) by edge division. In Figure 6(c) we show a non-normal isogonal tiling obtained from the normal tiling of Figure 6(b) by splitting each edge into a digon, and in Figure 12(b) a non-normal tiling obtained from the normal tiling of Figure 12(a) by splitting some of the edges into bunches of four digons. In Figure 5(b) we show a non-normal isotoxal tiling obtained from the normal tiling of Figure 5(a) by applying both edge-splitting and edge-division. It is worth noticing that for some tilings, such as the isotoxal tiling of Figure 4(b), no non-normal types can be produced by either of the above procedures. On the other hand it is sometimes possible to employ one operation, or even both, several times without destroying the transitivity properties.

Non-normal tilings produced by the above methods are not of much interest since their properties are simply related to those of the normal tilings from which they were derived. We shall say that they are *trivially related* to the normal tilings.

Of much more interest are the non-normal tilings with transitivity properties that are not trivially related to normal tilings. By an easy extension of the methods used in the proof of Theorem 1 (in which we allow $v_i$, $t_i$, and $r$ to take

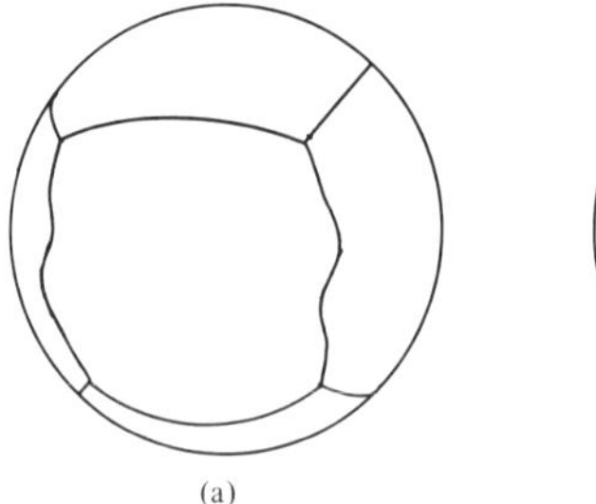
(a)

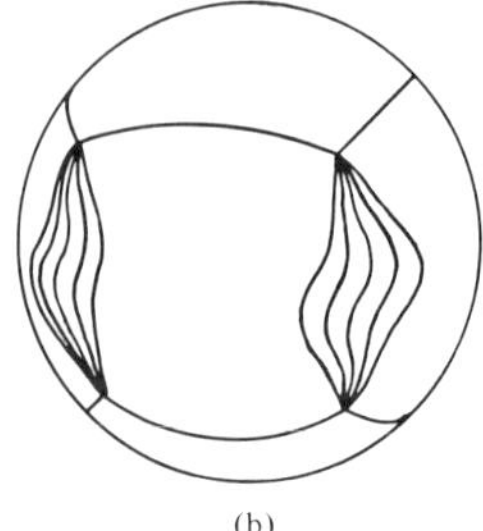
(b)

**Figure 12.** (a) A normal isogonal tiling of type SIG 16. (b) A non-normal isogonal tiling derived from that in (a) by splitting some of the edges into bunches of (four) digons.

the value 2), it can be shown that all such tilings must be of the following topological types:

$[2^r]$ or $[r^2]$ for isohedral tilings,
$\langle 2^2; r^2 \rangle$ or $\langle r^2; 2^2 \rangle$ for isotoxal tilings, and
$(2^r)$ or $(r^2)$ for isogonal tilings.

(These types may be thought of as resulting from the central projections of two "fictitious" polyhedra—one with just two $r$-gonal faces, and the other bounded by $r$ lunes.)

The investigation of the homeomeric types of tilings of these topological types now proceeds exactly as in the previous section. We determine all the tile-, edge-, and vertex-transitive subgroups of the corresponding symmetry groups, and then study their realizations. Table 4 displays information concerning these types, organized as in Table 3. The entries are what may be called "basic types" in the sense that many additional types can be obtained from these by edge-division and edge-splitting. Our classification is, in this sense, complete. More precisely we have the following.

**Theorem 4.** *All (normal and non-normal) homeomeric types of tilings on $S^2$ that are isohedral, isotoxal, or isogonal are listed in Tables 3 and 4 or are trivially related to these.*

To prevent any misunderstanding of the meaning of this result we remark that it may appear that certain types are missing from the tables. For example the isotoxal tiling with topological type $\langle 2^2; r^2 \rangle$ (where $2 \mid r$), symmetry group $[2, r]^+$ of order $2r$, and with incidence symbol $\langle a^+ b^+ ; a^+ a^+ , b^+ b^+ \rangle$ does not appear in the tables. This is because it is trivially related to the type SIT*3($\frac{1}{2}r$) by edge-splitting. Each edge of the latter is replaced by a single digon.

*Note added in proof.* Since the completion of the present paper a detailed account of the homeomeric classification in the plane has appeared (B. Grünbaum and G. C. Shephard, A hierarchy of classification methods for patterns. *Z. Kristallogr.* **154** (1981), 163–187). We also learned of several other works which are relevant to spherical tilings. An enumeration of isohedral tilings of the sphere in which adjacent polygons are related by reflection in the common edge is given by A. Kawaguchi (Polygons filling a sphere by reflexion or rabattement. *Tôhoku Math. J.* **28** (1928), 87–96). Beautiful illustrations of various tilings of the sphere by convex polygons are presented in M. J. Wenninger's book *Spherical Models* (Cambridge University Press, London 1979). Classifications of isohedral convex polyhedra in which the admitted symmetries are restricted by striations and other modifications of the faces have been studied by G. B. Bokiǐ (The number of physically distinct simple forms of crystals. (In Russian) *Trav. Labor. Crist. Acad. Sci. URSS* **2** (1940), 13–37) and by I. I. Šafranovskiǐ (The forms of crystals. (In Russian) *Trudy Inst. Kristallog. Akad. Nauk SSSR* **4** (1948), 13–166); these investigations deal also with unbounded polyhedra, which have no counterparts among spherical tilings.

**Table 4.** Homeomeric Types of Isohedral, Isotoxal and Isogonal Non-normal Tilings Not Trivially Related to Normal Tilings.[a] (Continued on pages 95–96.)

| Homeomeric type (1) | Topological type (2) | Symmetry group (3) | Order (4) | Induced group (5) | Incidence symbol (6) | Transitivity classes (7) | | Realizations (8) | Cross references (9) | |
|---|---|---|---|---|---|---|---|---|---|---|
| | | | | | | Edges | Vertices | | | |
| SIH*1($r$) | $[r^2]$ | $[2,\frac{1}{2}r]^+\ 2\mid r$ | $r$ | $[1]^+$ | $[a^+b^+;a^+b^+]$ | $\alpha\beta$ | $\alpha\alpha$ | $N$ | | SIG*13 |
| SIH*2($r$) | | $[2^+,\frac{1}{2}r]\ 4\mid r$ | $r$ | $[1]^+$ | $[a^+b^+;a^+b^-]$ | $\alpha\beta$ | $\alpha\alpha$ | $N$ | | SIG*19 |
| SIH*3($r$) | | $[\frac{1}{2}r]\ 2\mid r$ | $r$ | $[1]^+$ | $[a^+b^+;a^-b^-]$ | $\alpha\beta$ | $\alpha\beta$ | — | | |
| SIH*4($r$) | | $[r]^+$ | $r$ | $[1]^+$ | $[a^+b^+;b^+a^+]$ | $\alpha\alpha$ | $\alpha\beta$ | $N$ | SIT*1 | |
| SIH*5($r$) | | $[2^+,r^+]\ 2\mid r$ | $r$ | $[1]^+$ | $[a^+b^+;b^-a^-]$ | $\alpha\alpha$ | $\alpha\alpha$ | $N$ | SIT*2 | SIG*14 |
| SIH*6($r$) | | $[2,r]^+$ | $2r$ | $[2]^+$ | $[a^+a^+;a^+]$ | $\alpha\alpha$ | $\alpha\alpha$ | $N$ | SIT*3 | SIG*15 |
| SIH*7($r$) | | $[2^+,r]\ 2\mid r$ | $2r$ | $[2]^+$ | $[a^+a^+;a^-]$ | $\alpha\alpha$ | $\alpha\alpha$ | — | SIT*5 | SIG*18 |
| SIH*8($r$) | | $[2^+,r]\ 2\mid r$ | $2r$ | $[1](l)$ | $[a^+a^-;a^+]$ | $\alpha\alpha$ | $\alpha\alpha$ | $N$ | SIT*4 | SIG*17 |
| SIH*9($r$) | | $[r]$ | $2r$ | $[1](l)$ | $[a^+a^-;a^-]$ | $\alpha\alpha$ | $\alpha\beta$ | — | SIT*6 | |
| SIH*10($r$) | | $[2,r^+]$ | $2r$ | $[1](s)$ | $[ab;ba]$ | $\alpha\alpha$ | $\alpha\alpha$ | $N$ | SIT*7 | SIG*16 |
| SIH*11($r$) | | $[2,\frac{1}{2}r]\ 2\mid r$ | | $[1](s)$ | $[ab;ab]$ | $\alpha\beta$ | $\alpha\alpha$ | — | | SIG*20 |
| SIH*12($r$) | | $[2,r]$ | $4r$ | $[2]$ | $[aa;a]$ | $\alpha\alpha$ | $\alpha\alpha$ | $C$ | SIT*8 | SIG*21 |
| SIH*13($r$) | $[2^r]$ | | $r$ | $[r/2]^+$ | $[(a^+b^+)^{r/2};a^+b^+]$ | $(\alpha\beta)^{r/2}$ | $\alpha^r$ | $N$ | | SIG*1 |
| SIH*14($r$) | | $[2^+,r^+]\ 2\mid r$ | $r$ | $[r/2]^+$ | $[(a^+b^+)^{r/2};b^-a^-]$ | $\alpha^r$ | $\alpha^r$ | $N$ | SIT*10 | SIG*5 |
| SIH*15($r$) | | $[2,r]^+$ | $2r$ | $[r]^+$ | $[(a^+)^r;a^+]$ | $\alpha^r$ | $\alpha^r$ | $N$ | SIT*11 | SIG*6 |
| SIH*16($r$) | | $[2,r^+]$ | $2r$ | $[r]^+$ | $[(a^+)^r;a^-]$ | $\alpha^r$ | $\alpha^r$ | — | SIT*13 | SIG*10 |
| SIH*17($r$) | | $[2^+,r]\ 2\mid r$ | $2r$ | $[r/2](l)$ | $[(a^+a^-)^{r/2};a^+]$ | $\alpha^r$ | $\alpha^r$ | $N$ | SIT*12 | SIG*8 |
| SIH*18($r$) | | $[2^+,r]\ 2\mid r$ | $2r$ | $[r/2](s)$ | $[(ab)^{r/2};ba]$ | $\alpha^r$ | $\alpha^r$ | $N$ | SIT*14 | SIG*7 |
| SIH*19($r$) | | $[2^+,\frac{1}{2}r]\ 4\mid r$ | $r$ | $[r/4](s)$ | $[(a^+ba^-c)^{r/4};a^+cb]$ | $(\alpha\beta)^{r/2}$ | $\alpha^r$ | $N$ | | SIG*2 |
| SIH*20($r$) | | $[2,\frac{1}{2}r]\ 2\mid r$ | $2r$ | $[r/2](s)$ | $[(ab)^{r/2};ab]$ | $(\alpha\beta)^{r/2}$ | $\alpha^r$ | — | | SIG*11 |
| SIH*21($r$) | | $[2,r]$ | $4r$ | $[r](s)$ | $[a^r;a]$ | $\alpha^r$ | $\alpha^r$ | $C$ | SIT*16 | SIG*12 |

| | | | | | | Vertices | Tiles | | | |
|---|---|---|---|---|---|---|---|---|---|---|
| SIT*1($r$) | $\langle 2^2; r^2\rangle$ | $[r]^+$ | $r$ | $[1]^+$ | $\langle a^+b^+; a^+b^+\rangle$ | $\alpha\beta$ | $DD$ | $N$ | | SIH*4 |
| SIT*2($r$) | | $[2^+, r^+]\ 2\|r$ | $r$ | $[1]^+$ | $\langle a^+b^+; a^+b^-\rangle$ | $\alpha\alpha$ | $DD$ | $N$ | SIG*14 | SIH*5 |
| SIT*3($r$) | | $[2, r]^+$ | $2r$ | $[2]^+$ | $\langle a^+a^+; a^+a^+\rangle$ | $\alpha\alpha$ | $DD$ | $N$ | SIG*15 | SIH*6 |
| SIT*4($r$) | | $[2^+, r]\ 2\|r$ | $2r$ | $[2]^+$ | $\langle a^+a^+; a^+a^-\rangle$ | $\alpha\alpha$ | $DD$ | $N$ | SIG*17 | SIH*8 |
| SIT*5($r$) | | $[2^+, r]\ 2\|r$ | $2r$ | $[1](l)$ | $\langle a^+a^-; a^+a^+\rangle$ | $\alpha\alpha$ | $DD$ | — | SIG*18 | SIH*7 |
| SIT*6($r$) | | $[r]$ | $2r$ | $[1](l)$ | $\langle a^+a^-; a^+a^-\rangle$ | $\alpha\beta$ | $DD$ | — | | SIH*9 |
| SIT*7($r$) | | $[2, r^+]$ | $2r$ | $[1](p)$ | $\langle ab; ab\rangle$ | $\alpha\alpha$ | $DD$ | $N$ | SIG*16 | SIH*10 |
| SIT*8($r$) | | $[2, r]$ | $4r$ | $[2]$ | $\langle aa; aa\rangle$ | $\alpha\alpha$ | $DD$ | $C$ | SIG*21 | SIH*12 |
| SIT*9($r$) | $\langle r^2; 2^2\rangle$ | $[r]^+$ | $r$ | $[1]^+$ | $\langle a^+b^+; (a^+)^r(b^+)^r\rangle$ | $\alpha\alpha$ | $R_1R_2$ | $N$ | SIG*4 | |
| SIT*10($r$) | | $[2^+, r^+]\ 2\|r$ | $r$ | $[1]^+$ | $\langle a^+b^+; (a^+b^-)^{r/2}\rangle$ | $\alpha\alpha$ | $RR$ | $N$ | SIG*5 | SIH*14 |
| SIT*11($r$) | | $[2, r]^+$ | $2r$ | $[2]^+$ | $\langle a^+a^+; (a^+)^r\rangle$ | $\alpha\alpha$ | $RR$ | $N$ | SIG*6 | SIH*15 |
| SIT*12($r$) | | $[2^+, r]\ 2\|r$ | $2r$ | $[2]^+$ | $\langle a^+a^+; (a^+a^-)^{r/2}\rangle$ | $\alpha\alpha$ | $RR$ | $N$ | SIG*8 | SIH*17 |
| SIT*13($r$) | | $[2, r^+]$ | $2r$ | $[1](l)$ | $\langle a^+a^-; (a^+)^r\rangle$ | $\alpha\alpha$ | $RR$ | — | SIG*10 | SIH*16 |
| SIT*14($r$) | | $[2^+, r]\ 2\|r$ | $2r$ | $[1](p)$ | $\langle ab; (ab)^{r/2}\rangle$ | $\alpha\alpha$ | $RR$ | $N$ | SIG*7 | SIH*18 |
| SIT*15($r$) | | $[r]$ | $2r$ | $[1](p)$ | $\langle ab; a^r, b^r\rangle$ | $\alpha\alpha$ | $R_1R_2$ | $N$ | SIG*9 | |
| SIT*16($r$) | | $[2, r]$ | $4r$ | $[2]$ | $\langle aa; a^r\rangle$ | $\alpha\alpha$ | $RR$ | $C$ | SIG*12 | SIH*21 |
| | | | | | | Edges | Tiles | | | |
| SIG*1($r$) | $(r^2)$ | $[2, \frac{1}{2}r]^+\ 2\|r$ | $r$ | $[1]^+$ | $(a^+b^+; a^+b^+)$ | $\alpha\beta$ | $RR$ | $N$ | | SIH*13 |
| SIG*2($r$) | | $[2^+, \frac{1}{2}r]\ 4\|r$ | $r$ | $[1]^+$ | $(a^+b^+; a^+b^-)$ | $\alpha\beta$ | $RR$ | $N$ | | SIH*19 |
| SIG*3($r$) | | $[\frac{1}{2}r]\ 2\|r$ | $r$ | $[1]^+$ | $(a^+b^+; a^-b^-)$ | $\alpha\beta$ | $R_1R_2$ | $N$ | | |
| SIG*4($r$) | | $[r^+]$ | $r$ | $[1]^+$ | $(a^+b^+; b^+a^+)$ | $\alpha\alpha$ | $R_1R_2$ | $N$ | SIT*9 | |
| SIG*5($r$) | | $[2^+, r^+]\ 2\|r$ | $r$ | $[1]^+$ | $(a^+b^+; b^-a^-)$ | $\alpha\alpha$ | $RR$ | $N$ | SIT*10 | SIH*14 |
| SIG*6($r$) | | $[2, r]^+$ | $2r$ | $[2]^+$ | $(a^+a^+; a^+)$ | $\alpha\alpha$ | $RR$ | $N$ | SIT*11 | SIH*15 |
| SIG*7($r$) | | $[2^+, r]\ 2\|r$ | $2r$ | $[2]^+$ | $(a^+a^+; a^-)$ | $\alpha\alpha$ | $RR$ | $N$ | SIT*14 | SIH*18 |
| SIG*8($r$) | | $[2^+, r]\ 2\|r$ | $2r$ | $[1](l)$ | $(a^+a^-; a^+)$ | $\alpha\alpha$ | $RR$ | $N$ | SIT*12 | SIH*17 |
| SIG*9($r$) | | $[r]$ | $2r$ | $[1](l)$ | $(a^+a^-; a^-)$ | $\alpha\alpha$ | $R_1R_2$ | $N$ | SIT*15 | |
| SIG*10($r$) | | $[2, r^+]$ | $2r$ | $[1](s)$ | $(ab; ba)$ | $\alpha\alpha$ | $RR$ | — | SIT*13 | SIH*16 |
| SIG*11($r$) | | $[2, \frac{1}{2}r]\ 2\|r$ | $2r$ | $[1](s)$ | $(ab; ab)$ | $\alpha\beta$ | $RR$ | $C$ | | SIH*20 |
| SIG*12($r$) | | $[2, r]$ | $4r$ | $[2]$ | $(aa; a)$ | $\alpha\alpha$ | $RR$ | $C$ | SIT*16 | SIH*21 |

[a] The information in this table corresponds to that in Table 3. The reference number for each homeomeric type bears an asterisk to signify that the tilings are not normal.

**Table 4.** (*continued*)

| Homeomeric type (1) | Topological type (2) | Symmetry group (3) | Order (4) | Induced group (5) | Incidence symbol (6) | Transitivity classes (7) | | Realizations (8) | Cross references (9) | |
|---|---|---|---|---|---|---|---|---|---|---|
| | | | | | | Edges | Vertices | | | |
| SIG*13($r$) | ($2^r$) | $[2, \frac{1}{2}r]^+ \; 2\|r$ | $r$ | $[r/2]^+$ | $((a^+b^+)^{r/2}; a^+b^+)$ | $(\alpha\beta)^{r/2}$ | $D^r$ | $N$ | | SIH*1 |
| SIG*14($r$) | | $[2^+, r^+] \; 2\|r$ | $r$ | $[r/2]^+$ | $((a^+b^+)^{r/2}; b^-a^-)$ | $\alpha^r$ | $D^r$ | $N$ | SIT*2 | SIH*5 |
| SIG*15($r$) | | $[2, r]^+$ | $2r$ | $[r]^+$ | $((a^+)^r; a^+)$ | $\alpha^r$ | $D^r$ | $N$ | SIT*3 | SIH*6 |
| SIG*16($r$) | | $[2, r^+]$ | $2r$ | $[r]^+$ | $((a^+)^r a^-)$ | $\alpha^r$ | $D^r$ | $N$ | SIT*7 | SIH*10 |
| SIG*17($r$) | | $[2^+, r] \; 2\|r$ | $2r$ | $[r/2](l)$ | $((a^+a^-)^{r/2}; a^+)$ | $\alpha^r$ | $D^r$ | $N$ | SIT*4 | SIH*8 |
| SIG*18($r$) | | $[2^+, r] \; 2\|r$ | $2r$ | $[r/2](s)$ | $((ab)^{r/2}; ba)$ | $\alpha^r$ | $D^r$ | — | SIT*5 | SIH*7 |
| SIG*19($r$) | | $[2^+, \frac{1}{2}r] \; 4\|r$ | $r$ | $[r/4](s)$ | $((a^+ba^-c)^{r/4}; a^+cb)$ | $(\alpha\beta)^{r/2}$ | $D^r$ | $N$ | | SIH*2 |
| SIG*20($r$) | | $[2, \frac{1}{2}r] \; 2\|r$ | $2r$ | $[r/2](s)$ | $((ab)^{r/2}; ab)$ | $(\alpha\beta)^{r/2}$ | $D^r$ | — | | SIH*11 |
| SIG*21($r$) | | $[2, r]$ | $4r$ | $[r](s)$ | $(a^r; a)$ | $\alpha^r$ | $D^r$ | $C$ | SIT*8 | SIH*12 |

## References

[1] Brückner, M., *Vielecke und Vielflache*. Teubner, Leipzig 1900.

[2] Brun, V., Some theorems on the partitioning of the sphere, inspired by virus research. (Norwegian, with summary in English) *Nordisk Matem. Tidskrift* **20** (1972), 87–91, 120.

[3] Buerger, M. J., *Elementary Crystallography*. John Wiley and Sons, New York–London–Sydney 1963.

[4] Coxeter, H. S. M., Review of [22]. *Math. Reviews* **30** (1965), #3406.

[5] Coxeter, H. S. M., Angels and devils. In *The Mathematical Gardner*, edited by D. A. Klarner. Prindle, Weber and Schmidt, Boston 1981.

[6] Coxeter, H. S. M. and Moser, W. O. J., *Generators and Relations for Discrete Groups*. 3rd ed. Springer-Verlag, Berlin–Heidelberg–New York 1972.

[7] Davies, H. L., Packings of spherical triangles and tetrahedra. In *Proc. Colloq. on Convexity (Copenhagen 1965)*. Københavns Univ. Mat. Institut, Copenhagen 1967.

[8] Delone, B. N., On regular partitions of spaces. (In Russian) *Priroda* 1963, No. 2, pp. 60–63.

[9] Donnay, J. D. H., Hellner, E., and Niggli, A., Coordination polyhedra. *Z. Kristallogr.* **120** (1964), 364–374.

[10] Ernst, B., *The Magic Mirror of M. C. Escher*. Random House, New York, 1976.

[11] Escher, M. C., *The World of M. C. Escher*. Abrams, New York 1970.

[12] Fedorov, E. S., *Elements of the Theory of Figures*. (In Russian) Akad. Nauk, St. Peterburg 1885. Reprinted by Akad. Nauk SSSR, Moscow 1953.

[13] Fleischner, H. and Imrich, W., Transitive planar graphs. *Math. Slovaca* **29** (1979), 97–105.

[14] Galiulin, R. V., Holohedral variants of simple crystal forms. (In Russian) *Kristallografiya* **23** (1978), 1125–1132.

[15] Grünbaum, B. and Shephard, G. C., The eighty-one types of isohedral tilings in the plane. *Math. Proc. Cambridge Philos. Soc.* **82** (1977), 177–196.

[16] Grünbaum, B. and Shephard, G. C., The ninety-one types of isogonal tilings in the plane. *Trans. Amer. Math. Soc.* **242** (1978), 335–353.

[17] Grünbaum, B. and Shephard, G. C., Isotoxal tilings. *Pacific J. Math.* **78** (1978), 407–430.

[18] Grünbaum, B. and Shephard, G. C., The homeomeric classification of tilings. *C. R. Math. Reports, Acad. Sci.* **1** (1978), 57–60.

[19] Grünbaum, B. and Shephard, G. C., Incidence symbols and their applications. In *Proc. Sympos. on Relations between Combinatorics and Other Parts of Mathematics, Columbus, Ohio 1978. Proc. Sympos. Pure Math.* **34** (1979) 199–244.

[20] Grünbaum, B. and Shephard, G. C., *Tilings and Patterns*. Freeman, San Francisco (to appear).

[21] Heesch, H., Über Kugelteilung. *Comment. Math. Helv.* **6** (1933–34), 144–153.

[22] Heppes, A., Isogonale sphärische Netze. *Ann. Univ. Sci. Budapest. Eötvös Sect. Math.* **7** (1964), 41–48.

[23] Hess, E., *Einleitung in die Lehre von der Kugelteilung*. Teubner, Leipzig 1883.

[24] Killingbergtrø, H. G., Remarks on Brun's spherical tessellations. (Norwegian, with summary in English) *Nordisk. Matem. Tidskrift* **24** (1976), 53–55, 75.

[25] Niggli, A., Zur Topologie, Metrik und Symmetrie der einfachen Kristallformen. *Schweiz. Mineral. und Petrograph. Mitt.* **43** (1963), 49–58.

[26] Ozawa, T. and Akaike, S., Uniform plane graphs. *Mem. Fac. Engin. Kyoto Univ.* **39** (1977), 495–503.

[27] Pawley, G. S., Plane groups on polyhedra. *Acta Cryst.* **15** (1962), 49–53.

[28] Robertson, S. A. and Carter, S., On the Platonic and Archimedean solids. *J. London Math. Soc. (2)* **2** (1970), 125–132.

[29] Sakane, I., *Natural History of Games*. (In Japanese) Asahi-Shinbun, Tokyo 1977.

[30] Schattschneider, D. and Walker, W., *M. C. Escher Kaleidocycles*. Ballantine, New York 1977.

[31] Schlegel, V., Theorie der homogen zusammengesetzten Raumgebilde. *Nova Acta Leop. Carol.* **44** (1883), 343–459.

[32] Sommerville, D. M. Y., Semi-regular networks of the plane in absolute geometry. *Trans. Roy. Soc. Edin.* **41** (1905), 725–747.

[33] Sommerville, D. M. Y., Division of space by congruent triangles and tetrahedra. *Proc. Roy. Soc. Edin.* **43** (1922–23), 85–116.

[34] Sommerville, D. M. Y., Isohedral and isogonal generalizations of the regular polyhedra. *Proc. Roy. Soc. Edin.* **52** (1931–32), 251–263.

[35] Zelinka, B., Finite vertex-transitive planar graphs of the regularity degree four or five. *Mat. Čas.* **25** (1975), 271–280.

[36] Zelinka, B., Finite vertex-transitive planar graphs of the regularity degree three. *Časop. Pěstov. Mat.* **102** (1977), 1–9.

# Some Isonemal Fabrics on Polyhedral Surfaces

Jean J. Pedersen*

The motivation for the mathematics presented here should really be viewed as originating with the practitioners of the weaver's craft. The catalyst that resulted in this particular effort, however, was some recent work of Branko Grünbaum and G. C. Shephard [4, 5]. They have carefully analyzed certain geometric objects which represent an idealization of woven fabrics in the plane and their investigations lead, among other things, to remarkable theorems concerning the number and nature of the different kinds of what they call "isonemal"[1] fabrics in the plane. They have posed many open problems. The models described and pictured here (see Plates A–E, following page 120) were the result of my investigating one of their problems. The resulting models were a joy to discover and are truly beautiful to behold, but as so frequently happens in mathematics, as the existence of the answer to the original question was unveiled other similar questions seemed to spring forth. And herein lies the major difficulty involved with presenting such embryonic material. It is tempting (and, of course, desirable in the long run) to attack the problem with a great deal of mathematical rigor and preciseness (a) because it will certainly yield to that kind of discussion and (b) because there are beautiful and psychologically satisfying results. I will choose not to do that here because I believe that it is beneficial for the reader to observe first some of the natural beauty and surprise that is felt when viewing these models for the first time (unencumbered by technical detail). My second reason is that I wish, right now, to write an article—not a book. This brings us, in our intuitive approach to:

## 1. The Main Question

To put it informally, for the moment, we will investigate the question: What is the nature of fabrics woven on topological spheres where you use, for strands,

* Department of Mathematics, University of Santa Clara, Santa Clara, CA 95053 USA.

[1] "Isonemal" is a term derived from the Greek words ισοσ (the same) and νημα (a thread or yarn).

cylindrical-like rings instead of straight strips (which were used in the plane), and then require that the closures of the strands completely cover the "sphere" in some specified uniform and symmetric way? The notion is considered metrically, and the resulting fabrics are referred to as "woven polyhedra."

As you will see, the investigation leads quite naturally to many possibilities and variations. In consequence it is necessary to omit the details for many classes of woven polyhedra. Given the constraint of brevity, it seems reasonable to present in detail only the models that appear to have the most beautiful appearance on esthetic grounds. It is quite satisfying, logically, that these very symmetric and visually appealing models turn out to be woven polyhedra that might be considered analogous to the most symmetrical isonemal fabrics (in two dimensions) described by Grünbaum and Shephard. Other possibilities will be mentioned in order to give the reader some idea of the scope of the subject. On the basis of these ideas, interested readers will have ample opportunity to do some explorations of their own choosing.

## 2. Plan of Attack

The material is presented so that it outlines a sequence of thought closely resembling the actual development of the subject.

Section 3 is an intentionally terse account of the Grünbaum–Shephard theorem. It is presented (a) so that the reader will have easy access to that result and (b) so that it can be used, by making minor adjustments to its various constituent parts, and by adding appropriate additional requirements, in order to obtain a suitable result in three dimensions.

Section 4 outlines the reasoning, by way of analogy, that led to the discovery of the first "regular" isonemal coverings of polyhedral surfaces.

Section 5 contains specific instructions and diagrams that can be used for constructing the woven polyhedra.

Section 6 gives suggestions for research.

## 3. A Grünbaum–Shephard Theorem

**Definitions.** A *strand* (see Figure 1(a)) is a doubly infinite open strip of constant width, that is, the set of points of the plane which lie strictly between two parallel straight lines (think of it as a strip of paper having zero thickness). In diagrams it is sometimes useful to shade the strand, as in Figure 1(a), to indicate its direction —this helps to interpret the diagram when only small portions of a strand are visible as in Figure 2(b).

A *layer* is a collection of disjoint (parallel) strands such that each point of the plane either belongs to one of the strands or is on the boundary of two adjacent strands (see Figure 1(b)).

A *fabric* is, roughly speaking, two or more layers of connected strands in the same plane $E$ such that the strands of different layers are nonparallel and they

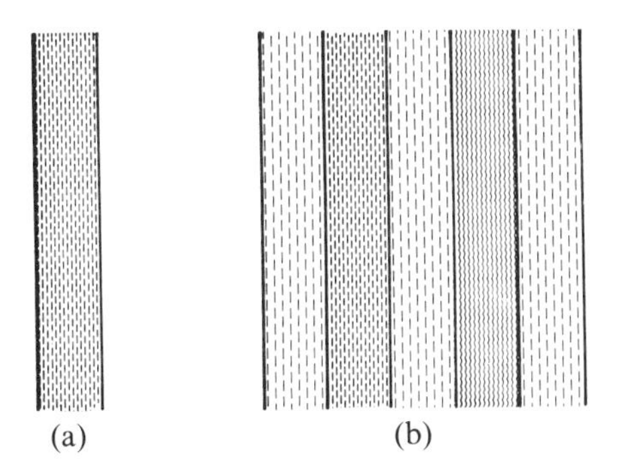


**Figure 1.** (a) single strand; (b) parallel strands.

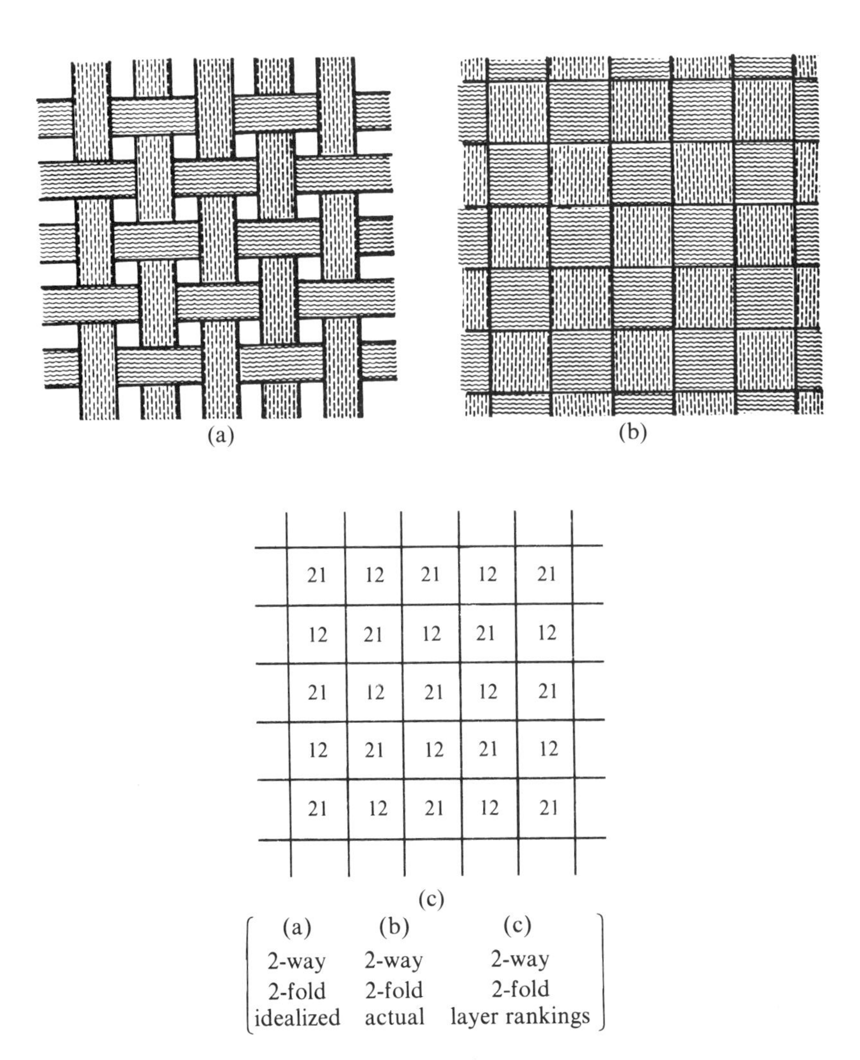


**Figure 2**

"weave" over and under each other in such a way that the fabric "hangs together." To be precise, *weaving* means that at any point $Q$ of $E$ which does not lie on the boundary of a strand, the two strands containing $Q$ have a stated *ranking*, and this ranking is the same for each point $Q$ contained in both strands. This ranking may be conveniently expressed by saying that one strand is *higher than* or *passes over* the other, in accordance with the obvious practical interpretation. Saying that the fabric *hangs together* means that it is impossible to partition the set of all strands into two nonempty subsets so that each strand of the first subset passes over every strand of the second subset.

If a fabric consists of $n$ layers it is called an *n-fold fabric*. Figures 2(a) and (b) represent the same 2-fold fabric (in (a) the strands have been "separated" for clarity—this diagram may be regarded as representing the "real" fabric corresponding to the "idealized" fabric of (b)). This is the most common and familiar of all fabrics, known variously as the *over-and-under*, *plain*, *calico*, or *tabby* weave.

A systematic way of graphically representing fabrics can be achieved by drawing all the straight lines that are boundaries of the various strands; this determines a tiling of the plane. The individual strands are then labeled according to the layer to which they belong, and their ranking in each tile is indicated by writing the labels in succession as required by the ranking (the top layer first, the second layer next, etc.). In Figure 2(c) this method is illustrated for the plain weave.

A fabric $F$ is said to be *k-way* provided the layers that form $F$ are parallel to $k$ directions. For example, Figure 2 illustrates a 2-way 2-fold fabric. If the strands in Figure 2 were all "doubled up," the resulting fabric would be 2-way and 4-fold, and if only the horizontal strands were doubled up, it would be a 2-way 3-fold fabric. A 3-way 3-fold fabric is indicated in Figure 3. It is often encountered in basketry and in the weaving of straws and reeds.

A *symmetry* of a fabric $F$ is any isometry of the plane of $F$ onto itself which—possibly in conjunction with the reversal of all rankings—maps each strand of $F$ onto a strand of $F$. All symmetries of $F$ clearly form a group under composition; it is denoted by $S(F)$ and called the *group of symmetries* of $F$. The subgroup of $S(F)$ that consists of those symmetries of $F$ that do not reverse the rankings of the strands is denoted by $S_0(F)$—those are the symmetries that preserve the *sides* of $F$ (so that rotations and translations of the fabric in its plane are permitted, but turning the fabric over is not allowed).

Finally, a fabric $F$ is called *isonemal* if its group of symmetries $S(F)$ acts transitively on its strands. The 2-way 3-fold variation of Figure 2, mentioned above, serves as an example of a fabric that is *not* isonemal.

**Theorem.**[2] *If $F$ is a $k$-way $n$-fold isonemal fabric, then the pair $(k, n)$ is one of the following six:* $(2,2)$, $(2,4)$, $(3,3)$, $(3,6)$, $(4,4)$, *or* $(6,6)$. *Conversely for each of these six pairs $(k, n)$ there exist infinitely many distinct $k$-way $n$-fold periodic isonemal fabrics.*

[2]The proof of this theorem appears in [5].

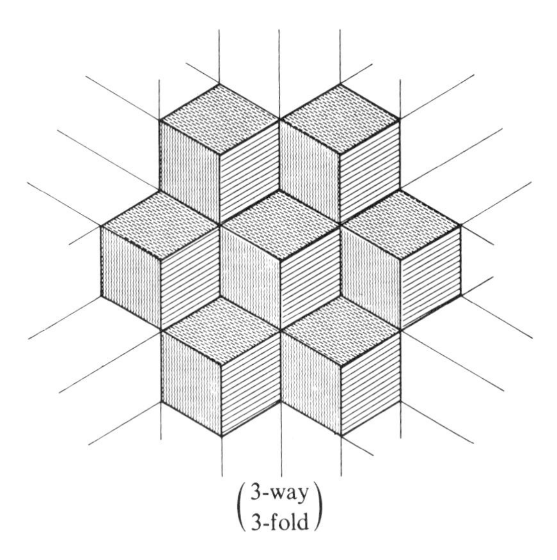


**Figure 3**

Examples of the first three have already been given. To obtain an example of a 3-way 6-fold isonemal fabric you can simply double up all the strands in the fabric illustrated by Figure 3. The 4-way 4-fold isonemal fabric is obtained by using two copies of a 2-way 2-fold isonemal fabric where each strand of that fabric "floats" at regular intervals over several other strands. Figure 4 shows what is known as a "sponge weave," and it is one of many fabrics which is suitable for this purpose. If you think of the black squares as being the visible portions of the vertical strands and the white squares as being the visible portions of the horizontal strands, then you can verify from the illustration that each strand in this fabric repeats the following sequence, with regard to the number of strands it goes over ($O$) and under ($U$) in succession:

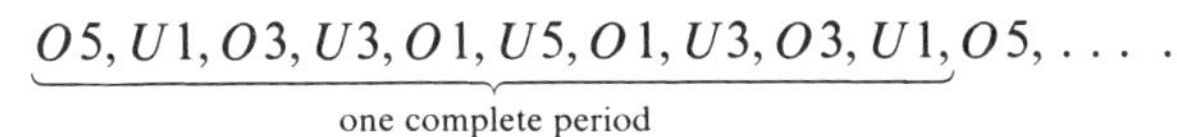

$$\underbrace{O5, U1, O3, U3, O1, U5, O1, U3, O3, U1,}_{\text{one complete period}} O5, \ldots .$$

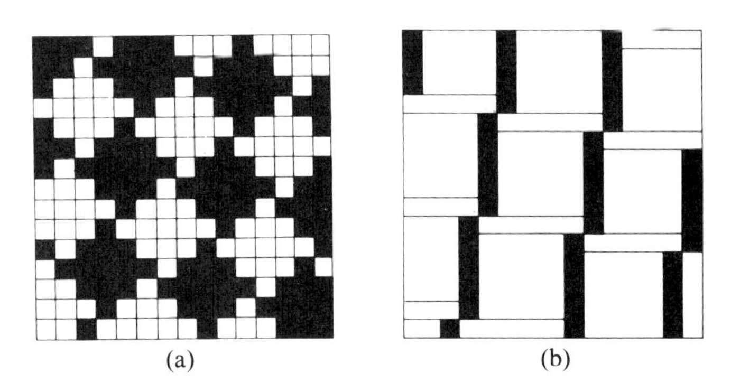


**Figure 4.** (a) A sponge weave. (b) The *longest* floating strands for the sponge weave shown in (a).

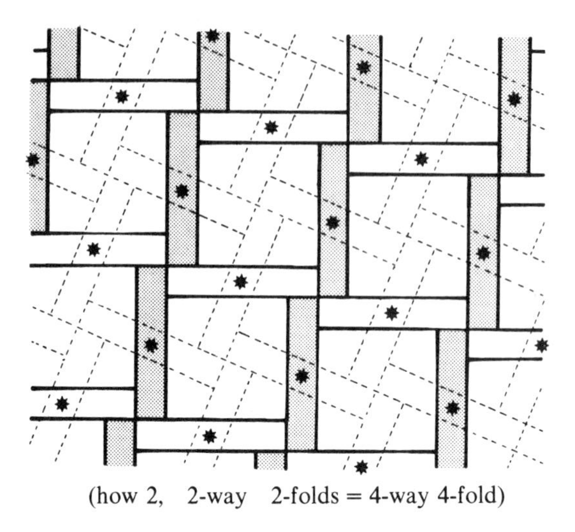

(how 2, 2-way 2-folds = 4-way 4-fold)

**Figure 5**

The "over 5" portions of the strands in this fabric constitute an array of "floating" parts. If you take two copies of this fabric, they can be oriented so that those floating portions can be interwoven producing a 4-way 4-fold isonemal fabric. Figure 5 shows only the floating strands of the two copies which are interwoven at the places marked by the stars.

The 6-way 6-fold fabric is realizable in an analogous way by interweaving the floating strands from two copies of a suitable 3-way 3-fold fabric (see [5]).

Some of the infinitely many distinct kinds of 2-way 2-fold isonemal fabrics are discussed in [4].

## 4. A Helpful Analogy

We return now to the question raised in Section 1, and we begin by making the appropriate changes in the definitions of Section 3, so that the weaving becomes a linkage of strands on the surface of a polyhedron. A *strand* (or ring) is isometric to the curved surface of a (short) cylinder. It may be scored to produce flat faces, such as squares, equilateral triangles, etc. We say that a $k$-way $n$-fold polyhedral fabric is a set of $k$ strands on a topological sphere (or polyhedron) such that every point on the surface not on the boundary of a strand belongs to exactly $n$ strands. And we require that the fabric "hang together" (in the same way as was defined for the fabrics in the plane). Thus for any point $Q$ on the surface of the polyhedron and not on the boundary of a strand, the strands containing $Q$ have a stated ranking—this will be denoted on the net (or netlike) diagrams exactly as was done in the plane. A symmetry of the polyhedral weaving is any isometry of the polyhedron onto itself which maps each strand of the polyhedron onto a strand of the polyhedron. (In the case of a plane fabric you can actually turn the fabric over—turning a polyhedron inside out is not completely equivalent, so we omit this concept.) If $P$ denotes the polyhedral weaving then all symmetries of $P$ form a group under composition; call it $S(P)$, the *group of symmetries* of $P$. The subgroup of $S(P)$ that consists of those symmetries of $P$ that do not reverse the rankings of the strands (i.e., the proper rotations in space) is denoted by $S_0(P)$—and these are the symmetries with which we are principally concerned, since they don't require the mental gymnastics of turning a polyhedron inside out. A polyhedral weaving is *isonemal* if its group of symmetries $S_0(P)$ acts transitively on its strands.

Now we wish to investigate the permissible values of $(k, n)$ in an isonemal polyhedral weaving. We hope, of course, to find a theorem resembling the Grünbaum–Shephard result of Section 3. But it is natural to look first at the most symmetric arrangements possible (many other arrangements exist, as will be pointed out in Section 6), so we observe that the theorem for isonemal fabrics includes especially symmetric fabrics which are related to symmetries of the polygons that form tesselations in the plane (see Figure 6).

By way of example, for the 2-way 2-fold isonemal fabric with strands crossing at right angles, we can think of all the strands as being perpendicular to one of the two axes of symmetry that join opposite sides of some square in the plane of the fabric. Alternately, we could view those strands as being perpendicular to one of the two axes of symmetry that join opposite vertices of some square in the plane of the fabric. Of course, because of the way the corresponding 2-way 4-fold fabrics are constructed, their strands will relate to the axes of symmetry for some square in identically the same way. The 4-way 4-fold isonemal fabric in Figure 5 has strands that can be partitioned into exactly four sets, so that the strands of each set are perpendicular to either one of the axes of symmetry joining opposite sides, or opposite vertices, of some parallelogram.

It is not difficult to see that the structure of the 3-way 3-fold (and 3-way 6-fold) isonemal fabrics is, in a similar way, related to the axes of symmetry for

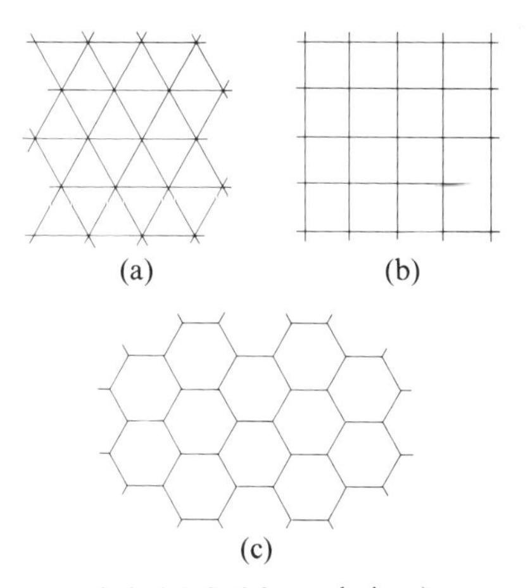


(3-6, 4-4 & 6-3 tesselations)

**Figure 6**

an equilateral triangle (or, equivalently, the sets of axes joining opposite faces or vertices for the regular hexagon).

Likewise, the 6-way 6-fold isonemal fabric is, of necessity, structured so that its strands can be partitioned into six nonempty sets, with the strands in each set being perpendicular to one of the six axes of symmetry of some hexagon which tiles the fabric's plane.

The relationship between strands of certain especially symmetric isonemal fabrics in the plane and the axes of symmetry for the polygons that form tesselations in the plane suggests that an analogous situation might exist for symmetric isonemal weavings on the surface of polyhedra. And since, in some sense, the Platonic solids are the three-dimensional analogs of the regular tesselations in the plane, it seems plausible that we could find isonemal weavings for polyhedra such that the rings "go around" each of the axes in the various sets of axes for the platonic solids.

To begin investigating this idea we first enumerate all of the various sets of axes of symmetry related to the Platonic solids. In a straightforward manner we determine that those axes of symmetry occur in sets of 3, 4, 6, 10, and 15 (see Figure 7). Consequently we now look for polyhedra that can be woven with 3, 4, 6, 10, or 15 identical rings.

We commence our search by carefully examining a cube as it rotates about an axis through the center of two opposing faces. If the axis of rotation is perpendicular to our line of vision, then what comes into view are successive square faces —four of them—forming the vertical faces of a square prism, whose bases are the two squares whose centers determined the axis of this rotation. There are three such structures. If these are identified with colors 1, 2, and 3 we see that a cube can be constructed from three identical strands, properly sized and scored so that each contains four squares. The net (stereographic projection) for such a cube, along with a typical pattern piece (including a "tab" used for overlap), and an illustration of the finished 3-way 2-fold isonemal cube are shown in Figure 8. This model, and all of the other 2-fold fabrics mentioned in this article, can be constructed without gluing the tab in place, since friction will hold the strips in place.

Now, what happens if we try the same technique for the dual of the cube, that is, the octahedron? You can see this for yourself if you rotate a regular octahedron about an axis through the center of each of two opposing faces and, with the axis of rotation perpendicular to your line of sight, look at the faces that come into view. What appears is a succession of triangular faces, six of them, forming an antiprism, whose bases are the two triangles whose centers lie on the axis of rotation. There are four of these baseless antiprisms, and consequently four closed rings, consisting of six equilateral triangles each, can be woven together in space to form a 4-way 3-fold isonemal octahedron. The net, a typical pattern piece, and illustrations of the finished model are shown in Figure 9.

In a similar way the 4-way 2-fold isonemal cube shown in Figure 10 may be obtained. (You begin by observing the surface of a cube as it rotates around an axis connecting opposite vertices.) An interesting variant of this cube is shown in Figure 11. The two models are isometric (the second is just a "squashed" version

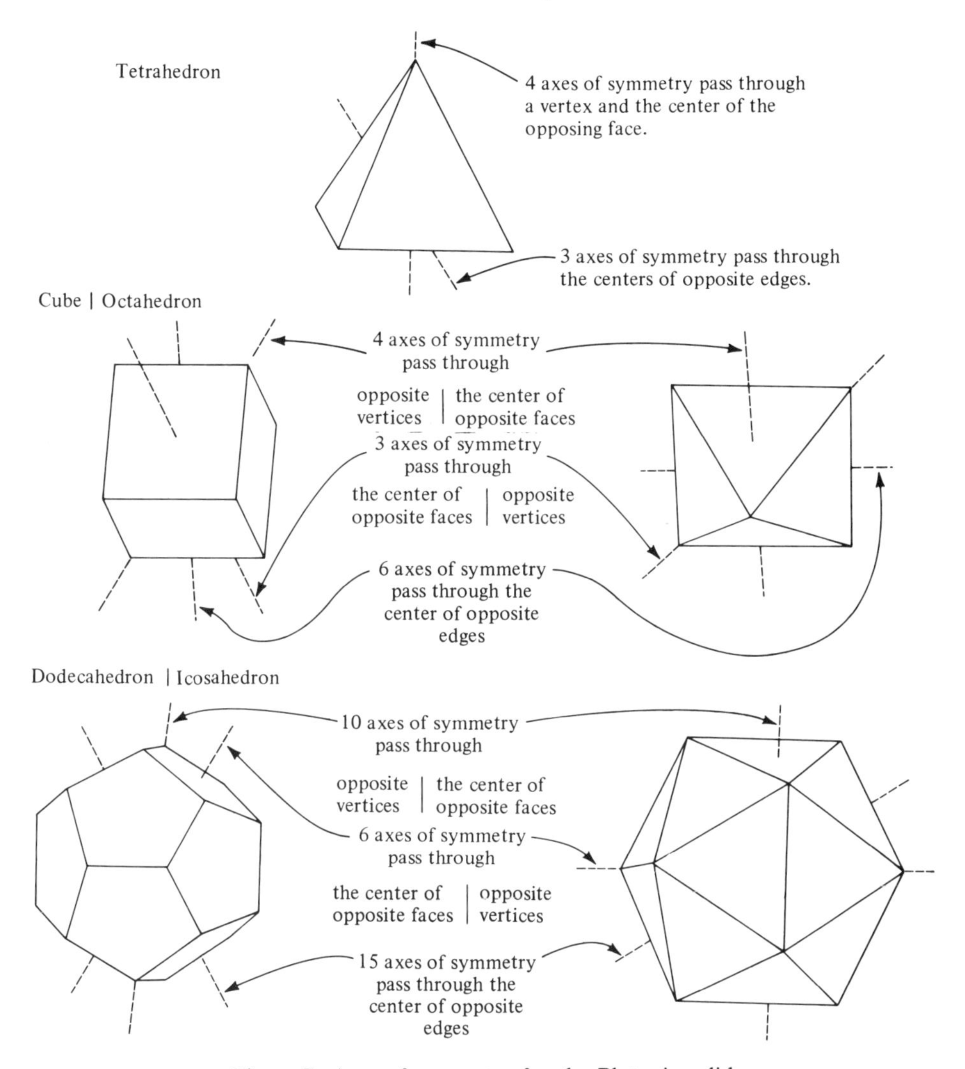


**Figure 7.** Axes of symmetry for the Platonic solids.

of the first). The model in Figure 11 is particularly useful, as we will see, from a conceptual point of view. Its surface may be viewed as an octahedron on which each triangular face has been replaced by a baseless pyramid whose three faces are all right isosceles triangles. It is unusual to draw net diagrams with joins along the center of faces, but this shows more clearly the relationship between the models in Figures 10 and 11.

As promising as this approach seems, without new insights, we would have reached an impasse at this point. Rotating the platonic solids about the axes of symmetry not yet discussed will not yield any new isonemal polyhedra. In all of the remaining cases there are prohibitive features. For example, if you rotate the icosahedron about an axis of symmetry through opposite vertices it is easy to

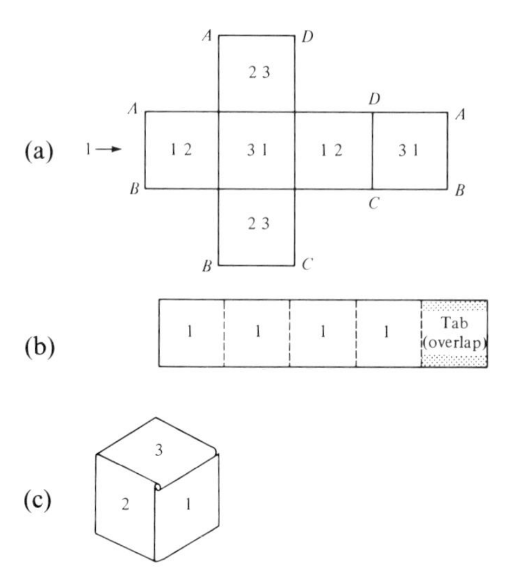


**Figure 8.** Cube. (a) Net diagram shows ranking of each strand that crosses that face. (b) A typical strand (make three and label them with 1's, 2's, and 3's, respectively). (c) The 2-way 2-fold isonemal cube produced by weaving together the three strands so that they have the ranking indicated in the net diagram in (a).

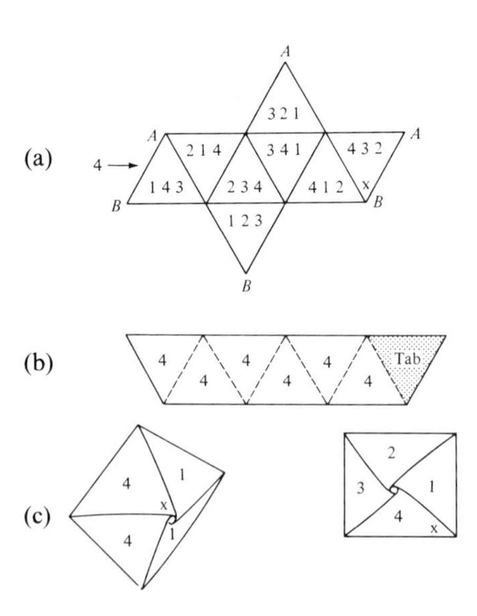


**Figure 9.** Octahedron. (a) Net diagram: Numbers indicate the ranking of strands crossing that face. (b) A typical strand (make four and label them with 1's, 2's, 3's, and 4's, respectively). (c) Two views of the 4-way 3-fold isonemal octahedron produced by weaving together the four strands so that they have the ranking indicated in the net diagram (a). The x provides orientation.

The edges of this net are joined along the centers of what become triangular faces on the finished model.

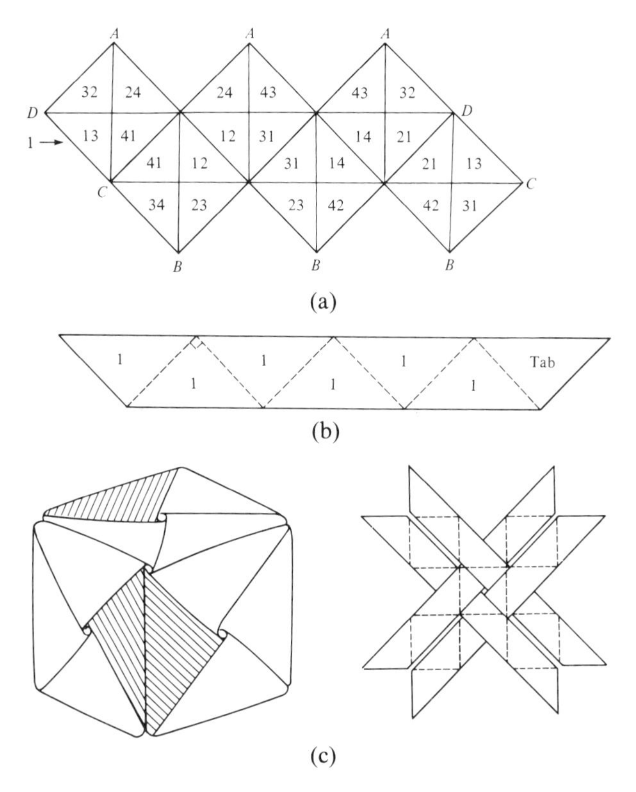


**Figure 10.** (a) Numbers indicate the rankings of the strands crossing each face. (b) A typical strand (make four and label them with 1's, 2's, 3's and 4's, respectively). (c) A 4-way 2-fold isonemal cube. Begin the construction by arranging the strips as shown on the right. The center square forms one face of the completed cube.

identify a ring of ten equilateral triangles circling that axis—but, when you try to construct the woven model, it is not possible to weave more than five such rings together. This is because the five rings cover so many edges that the sixth ring cannot get from the outside to the inside of the model. But without the sixth ring you cannot have a uniform covering of the 20 triangular faces (the 50 triangles contributed by the first five strips cannot be evenly distributed among the 20 faces).

Fortunately we do have a new insight, provided by the model in Figure 11. The idea, in its most restricted form, is to take a familiar polyhedron, composed of equilateral triangles, and create a new polyhedron by replacing all of the original faces with pyramids consisting of three right isosceles triangles. As we have already seen, the effect of this replacement on the octahedron results in a 4-way 2-fold isonemal polyhedron. A similar replacement of the icosahedron results in a 6-way 2-fold isonemal polyhedron; and if we do this on the

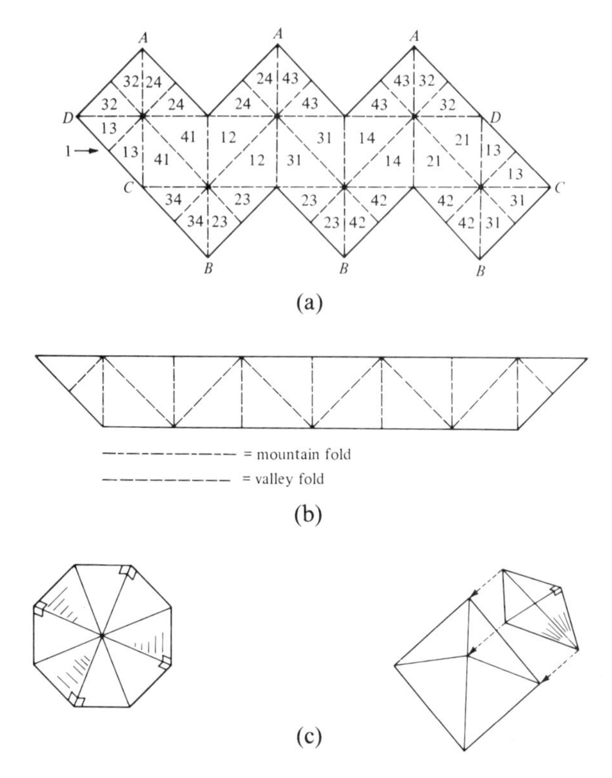


**Figure 11.** (a) The edges of this net are joined along the centers of what become triangular faces on the finished model. (b) A typical strand (make four, label them with 1's, 2's, 3's, and 4's, respectively). (c) Think of this model as an octahedron where each face is replaced by a pyramid, as shown at the right for one face.

tetrahedron, the result is the cube of Figure 8, with fold lines on its faces corresponding with the edges of the inscribed tetrahedron that generated the model.

What we need now is a way of describing these models. By way of example, we redo the model of Figure 11 as shown in Figure 12. To use this new netlike plan, interpret each triangle from the parent octahedron net as a pyramid having three right isosceles triangles for faces (only one right angle, which serves as a representative case, is marked). The numbers, as usual, denote the rankings for the $n$ (4 in this case) numbered rings. In general, a portion of a strand, or ring, for the particular weaving will be shown and the number of parts required for each ring will be indicated. Figure 13 shows the netlike plan for the 6-way 2-fold isonemal polyhedron which is an offspring of the icosahedron.

Although this technique is valuable, it becomes even more so if we are not so restrictive in its use. We could, for example, replace the triangular faces by pyramids consisting of three equilateral triangles (and obtain some 3-fold models), or we could make an analogous replacement of squares or pentagons by pyramids containing four or five equilateral triangles, respectively. If we now consider these variations in combination with each other, or with repeated use,

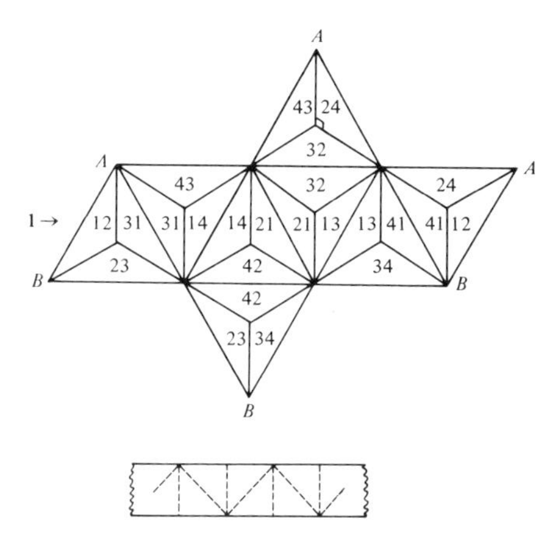


Figure 12. The 4-way 2-fold "octahedron" requires 4 strands. Each strand consists of 6 square sections plus a tab.

the results are astounding. For the moment we confine ourselves to the more symmetric results that satisfy our original search.

The regular tetrahedron, whose faces have been replaced by pyramids having three equilateral triangles for faces, results in a model which admits a 3-way 3-fold isonemal woven covering (see Figure 14).

Another example of this type can be obtained by replacing the faces of the octahedron with pyramids consisting of three equilateral triangles. The result is the well-known Stella Octangula. Its surface yields a 4-way 3-fold isonemal weaving as diagrammed in Figure 15. It is interesting to compare the model of Figure 15 with its mate in Figure 12. First notice that the rings on the 2-fold model are oriented so that the successive layer levels of any strip has a period of 2 (over, under, over, under, etc.); but the orientation of the rings on the 3-fold model is such that the successive layer levels of any strip has a period of 6 (over,

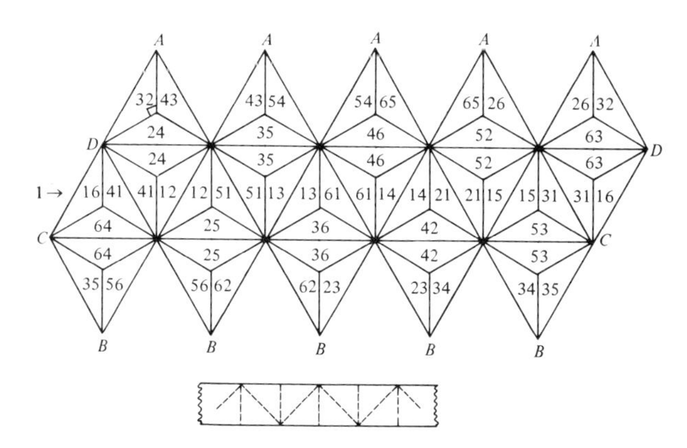


Figure 13. The 6-way 2-fold "icosahedron" requires 6 strands. Each strand consists of 10 square sections plus a tab.

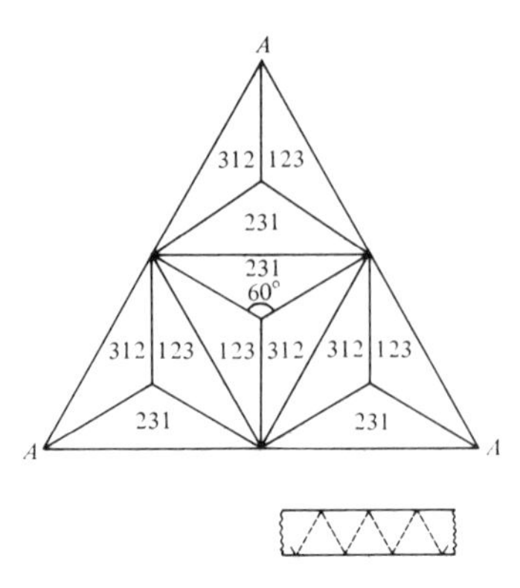


**Figure 14.** The 3-way 3-fold "tetrahedron" requires 3 strands. Each strand consists of 12 equilateral triangles plus a tab.

over, middle, under, under, middle, etc.). So what we have is two related models, constructed from different types of strands, that go over and under each other in very different ways; and yet the rings can be woven so that the color arrangement is the *same* on both models. To see that this is the case, observe that if you remove the middle number from the ranking numbers in the net of Figure 15 you obtain the ranking numbers for the "net" of Figure 12. Thus the coloring of both models is seen to be the same on the inside as well as on the outside! Plates D and E illustrate this phenomenon for the 6- and 10-way examples, respectively.

Perhaps what is even more important to observe is that there are exactly three symbols used in the three rankings within any equilateral triangle in the net of Figure 12; and if you think of separating the two symbols, in each of the three rankings, then all you need to do is insert the missing symbol between the two

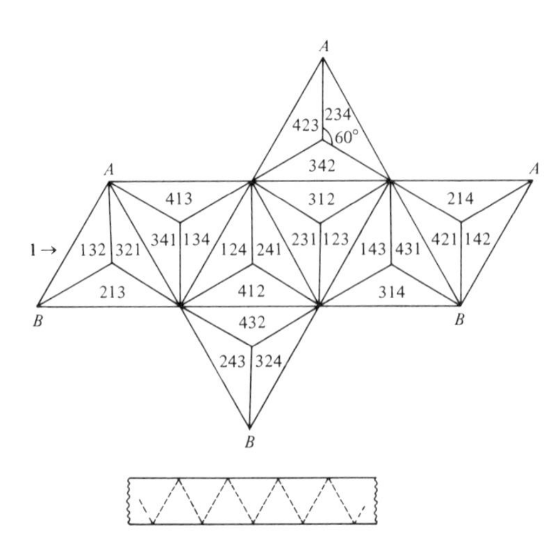


**Figure 15.** The 4-way 3-fold "octahedron" requires 4 strands. (The surface is the well-known Stella Octangula.) Each strand consists of 18 equilateral triangles plus a tab.

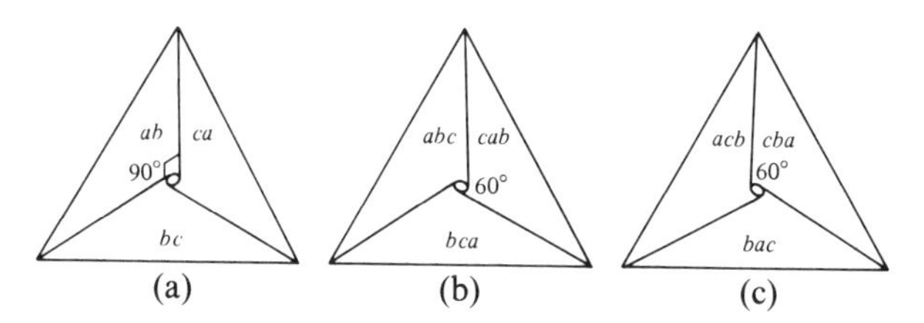


**Figure 16.** (a) A typical equilateral triangle (containing three right isosceles triangles) from a 2-fold net. The strands overlap around the center in a clockwise direction. (b) A typical equilateral triangle (containing three equilateral triangles) from a 3-fold net with the arrangement of the 3 surface colors the same as in (a) and with the strands overlapping in a clockwise direction. Note that the inside colors are *not* the same as in (a). (c) The same as in (b) except the strands overlap in a counterclockwise direction. This results in an arrangement where both the outside and inside colors are the same as in (a).

original symbols, to obtain the appropriate ranking for the 3-fold model. This observation is true in general, and it provides an efficient method for obtaining the plan for 3-fold models whenever the plan for a 2-fold model is known. Thus, given any 2-fold model, you can construct its 3-fold companion, with corresponding coloring, by starting with the same arrangement around some vertex, and if the strands on one (say the 2-fold) are overlapped in a clockwise direction, then the corresponding strands on the other (the 3-fold) must then be overlapped in a counterclockwise direction. To see why this is so study Figure 16, which illustrates the two possible directions for overlapping the strands $a$, $c$, $b$ so that they are arranged in that clockwise order on the surface.

We have already constructed $k$-way 2-fold isonemal woven coverings for polyhedra where $k = 3$, 4, and 6; and we now know how to obtain a 3-fold model from the structure of certain 2-fold models. A reasonable expectation (we hope) at this point is that we will now be able to find $k$-way isonemal polyhedra in the 2-fold versions for $k = 10$ and 15—and then use those results to find the 10-way 3-fold and 15-way 3-fold isonemal woven coverings. This (with some bonuses), as a matter of fact, does turn out to be possible. And if the reader armed with the ideas already given is (a) familiar with the platonic and archimedean solids and (b) willing to try the modifications mentioned above, then he or she can now discover the required models.

What will happen, however, if you begin this search, is that almost everything you try yields a woven model of some sort—and many of them are isonemal in the broadest sense. It is then the classification and appropriate description of these models that becomes the interesting question. We will discuss this later when we formalize the results. We first list in Table 1 some of the most symmetric of the 2-fold isonemal coverings of polyhedra.

Each of the models in this list is woven from strands composed of squares. If we assume the strands will cross every other strand exactly twice (and that they don't cross over themselves at all), then for any number $k$ the number of squares necessary for the total woven model is just the number of ways you can take $k$ things two at a time, multiplied first by 2 (because they cross twice) and then again by 2 (since every crossing involves two layers). But this total number of

**Table 1.** Some 2-Fold Isonemal Woven Coverings

| Ref. no. | Figure no. | $k$(-way) | Model description* | $S^{\dagger}$ | $s = 2S/k = 2(k-1)^{\ddagger}$ |
|---|---|---|---|---|---|
| 1 | 8 | 3 | Cube, or tetrahedron with all 4 △s built out by  (pyramid, 90°) | 6 | 4 |
| 2 | 12 | 4 | Octahedron with all 8 △s built out by 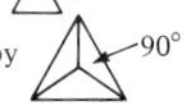 (pyramid, 90°) | 12 | 6 |
| 3 | 13, Plate D (left) | 6 | Icosahedron with all 20 △s built out by 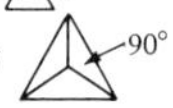 (pyramid, 90°) | 30 | 10 |
| 4 | Plate A | 10 | Dodecahedron with all 12 pentagons replaced by pyramids consisting of 5 equilateral triangles. Then all 60 △s built out by  (pyramid, 90°) | 90 | 18 |
| 5 | Plate B | 15 | Snub dodecahedron with all 12 pentagons replaced by pyramids consisting of 5 equilateral triangles. Then all 140 △s built out by  (pyramid, 90°) | 210 | 28 |

* "Built out" means "replaced by a pyramid consisting of 3 right triangles."

† Total number of squares constituting the model—including *both* layers.

‡ Number of square sections in each strand (not including the tab).

squares must then be divided equally among the $k$ strands; hence the number of squares, denoted $s$, required for such a model in each strand is

$$\frac{\binom{k}{2}(2)(2)}{k} = 2(k-1).$$

It is also useful to note that if the faces of the model's surface constitute a total of $S$ square sections (where two right isosceles triangles constitute one square), then the number of square sections required for each ring on such a $k$-way model (if it exists) must be $2S/k$. Determining the values for $S$ and $s$ in advance enables us to make more judicious choices as we search for appropriate $k$-way 2-fold models.

It would now be possible to construct an analogous table for "some 3-fold isonemal woven coverings." It is a fairly direct process in which you simply (a) change all of the 90° angles on the built out pyramids to 60° angles; (b) compute $T$, the total number of triangles constituting the model, including all three layers; and (c) compute $t$, the number of equilateral triangles in each strand. Except for the following statements about how to compute $T$ and $t$, the construction of this table will be left as an exercise for the reader.

Thus, for models on the list of 3-fold isonemal coverings we assume that the strands will cross every other strand exactly twice (and that they do not cross over themselves at all), and that two triangles from each crossing will be visible. Then for any number $k$ the number of triangles necessary for the total model is just the number of ways you can take $k$ things two at a time, multiplied by 2 (because they cross twice) and then by 6 (since every crossing involves two visible triangles, and there are three layers). But these triangles must then be divided equally among the $k$ strands; hence the number of triangles, $t$, required in each strand is

$$\frac{\binom{k}{2}(2)(6)}{k} = 6(k-1).$$

If the model's surface has a total of $T$ visible equilateral triangles, then the number of triangular sections required for each strand on such a $k$-way model (if it exists) must be $3T/k$. As in the case of the 2-fold models, determining the values for $T$ and $t$ in advance enables us to make more judicious choices as we search for appropriate $k$-way 3-fold models.

You may recall that we obtained a 4-way 3-fold model that does not seem to fit into this second table (see Figure 9). This is because the crossings on that woven model are not all visible, and this may happen on 3-fold models whenever certain special conditions are realized. Thus, if you don't require all crossings be visible, then you can focus your attention on the *edge* where the crossing takes place and determine that you need a model with $\binom{k}{2}(2)$ edges (denoted $E$). But since all faces are equilateral triangles, $3F = 2E = 2k(k-1)$ (where $F$ is the number of faces) and $F = 2k(k-1)/3$. And, because every face is covered three times, there are a total of $3(2)k(k-1)/3 = 2k(k-1)$ triangles on the woven model. Now these triangles must be divided equally among the $k$ strands, so each strand would have $2(k-1)$ equilateral triangles. If we recall that the weaving sequence on our models, involving equilateral triangles, had a period of six (there are other possibilities, but they are all multiples of six), then we see that only when $2(k-1) \equiv 0 \pmod 6$ will we be able to construct such a *minimal* $k$-way 3-fold isonemal polyhedron. For the values of $k$ that concern us, only $k = 4$ and $k = 10$ satisfy this last requirement. The model for $k = 4$ is the octahedron of Figure 9. A model for $k = 10$ may be obtained by replacing each pentagon on the dodecahedron with pyramids consisting of five equilateral triangles. If you weave together the 10 strands on the surface of this model with the pyramids pointing "in," it is very sturdy (and pretty), as seen in the photograph of Plate C.

It would seem at this point that we have completed the set of required polyhedra. However, if you look at the symmetries of the woven models and compare them with the symmetries of the models that gave rise to the $k$, you realize that the set is deficient. For example, the value of $k = 3$ comes from axes of both the tetrahedron and the cube, yet we only have a model with the symmetry of the cube. We get around this difficulty by considering the woven model with the score lines on its strands that outline one of the two inscribed tetrahedra. This marked model would then have, of course, just the symmetries of the tetrahedron. You should notice that the rotation of the cube about the four axes joining opposite vertices always leaves this inscribed tetrahedron invariant, and the same is true for the three 180° rotations through opposite faces of the cube. The twelve rotations of the cube that are lost by taking into account this marking are the six 180° rotations through opposite edges and the six 90° rotations (three clockwise and three counterclockwise) through opposite faces.

Similarly on the 4-way 2-fold model we can outline, on the edges of the strips, those portions that would coincide with one of the inscribed tetrahedra and see that this marked woven model now has only the symmetries of the inscribed tetrahedron. Of course, the same can be done with the 4-way 3-fold model.

Six is the only other number occurring in two different symmetry sets in Figure 7. You might expect that outlining the edges of the strips that correspond with one of the five inscribed cubes of the dodecahedron would produce a model having six strips and octahedral symmetry. This is not so, because the orientation of the inscribed cubes (five of them) within the dodecahedron is such that rotation of the dodecahedron about any axis joining the centers of opposite faces does not leave any of the five cubes invariant. What does work, however, is to mark on the strands the fold lines that correspond with the edges of the icosahedron that would lie on the faces of a cube circumscribed about the icosahedron. These "marked" 6-way models (both 2- and 3-fold) will then have just octahedral symmetry.

What we now know is that each of the possible $k$-way regular isonemal weavings does exist for 2-fold and 3-fold models; furthermore, in some cases (4-way 3-fold and 6-way 3-fold) more than one model is possible. In fact, as seen in Figure 17, there is yet another 6-way 2-fold regular isonemal polyhedron.[3] We give no explanation for this figure, but trust that it will give you a new idea to explore. Perhaps you would like to search for others. The next section gives some practical hints for carrying out woven constructions. Unless you are better at mental weaving than most of us, you will probably find it necessary to construct at least a few of these in order to get a feel for how to draw the nets of models you devise.

## 5. Suggestions for Constructing Woven Polyhedra

The references [6] through [12] give many specific and detailed instructions for constructing various types of woven polyhedra. Although some of the models referred to in this paper are mentioned explicitly, many of the polyhedra in those articles are *not* isonemal weavings—some have holes in them, and some do not cover the polyhedron uniformly. Nevertheless they may be useful in terms of giving you ideas for possible arrangements of strips on polyhedra. The article on

[3] This net was sent to me by Geoffrey Shephard, during the course of our correspondence about these matters.

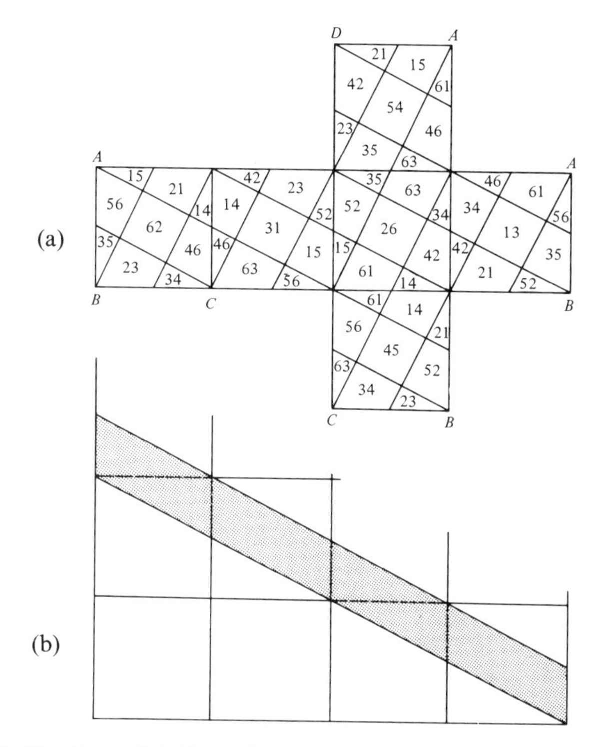


**Figure 17.** The 6-way 2-fold regular isonemal weaving with octahedral symmetry in (a) is formed with six strands like the shaded portion of (b).

collapsoids [10] may be of special interest because all of the models discussed there can be used as surfaces for nonregular isonemal weavings (because strands, in pairs, cross over each other more than two times). They provide, in fact, an infinite class of suitable models for weaving $2m$ strands where $m \geqslant 3$.

### 5.1. For 2-Fold Models

First prepare colored strands from paper of a reasonable weight (about the weight of the paper used for file folders). Next, overlap the strands to correspond with some part of the net diagram (fixing them in place, if necessary, with transparent tape on what will be the inside of the finished model). You can then proceed to "weave" the strands together, remembering that every strand must go alternately over and under the succession of strands it crosses all the way around the model. One of the annoyances, for beginners, is nearing the completion of the model and realizing that you have oriented the strands so that the overlap takes place on the top layer. If this happens you can hold the ends of the strands together with a paper clip, glue them together, or cut off one section and attach it to the other end of the strand so that the last section will tuck in neatly. After building a few models you realize that this possibility materializes quite frequently, and one way you can cope with the problem is to start with strands longer than you will need, so that when you finish the model you can cut off the unnecessary parts at the point that will allow the last part to tuck in.

To give step-by-step instructions for the actual weaving of these models would be long and tedious (and probably not productive). What you need to realize is that this is mostly a matter of eye–hand coordination—and most of us haven't had much experience with these matters, in this form. So try it, be patient, and be careful; but if it seems hopeless, don't despair. You can *cheat*! And, no one will know from the looks of the finished model that you didn't actually weave it together. To achieve this harmless deception you first construct the base model in the usual way (see [2] or [13] for details); then write the ranking numbers on its faces. Next take the colored strands that are required to weave this model and cut them into sections like

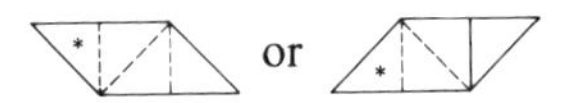


Check the ranking numbers on your model to see which type of sections you need, if your paper isn't reversible. Then, following the ranking scheme, you can glue down the triangle marked with an asterisk so that that section goes around the model in the right direction according to the ranking scheme. Once these pieces are all glued in place they can be interwoven so that the center square is on the first layer and the triangle on the other end tucks in.

### 5.2. For 3-Fold Models

Here the deception becomes almost imperative because of the complicated way the strands are interwoven. So construct the base model and record the ranking numbers on its faces. Then cut the strands into sections like

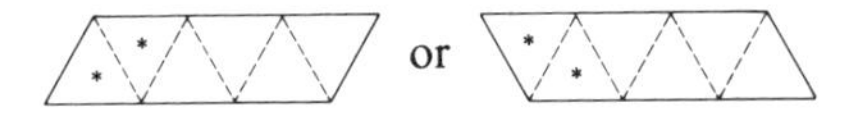


Again, check your model to see which type of sections you need if your paper isn't reversible. Then take a section and glue down the two triangles marked with an asterisk, placing them where the ranking numbers indicate that this strand is on the third layer. Check the surrounding rankings to make certain the section "goes in the right direction."

As on the 2-fold models, once all the sections are glued in place, you can "weave" them together so that each strand goes under, under, middle, over, over, middle, etc. And since you have already glued down the two "under, under" triangles, you know the next triangle is sandwiched between two layers and then goes on the top layer for two triangles. Finally, the last triangle is tucked in between two strands. This is in fact how the 3-fold models in the photographs of Plates B, C, D, and E were constructed.

An alternate to all of this is just to construct the base model and color the faces as though it were woven. Even more effective is to color the faces with concentric holes in their centers showing the colors of the successive layers. This might be viewed as a constructive type of existence proof.

## 6. Research Questions

The following is an incomplete list of the questions that remain to be investigated.

1. If it is not required that every strand cross over every other strand in exactly two places, you can obtain an infinite number of isonemal woven coverings by weaving them on the surface of what are called collapsoids [10]. An open question is: How many other kinds of models of this type exist, and is there a reasonable way to classify them?

2. If we allow a strand to cross over itself, then the triangular dipyramid and pentagonal dipyramids (when constructed from equilateral triangles) serve as examples of surfaces that require only *one* strand to produce an isonemal weaving (the strand on the completed model forms a knot in space). There are other such models but, the only ones I have found have the same symmetry group as one of the two already mentioned. The question is: Do there exist models that can be woven from just one strand that have symmetry other than the dihedral symmetry of an equilateral triangle or a regular pentagon?[4]

3. The only odd values for $k$ that I have found for isonemal weavings are 1, 3, and 15 (of which 3 and 15 are regular). Is an isonemal polyhedral weaving possible for any other odd value of $k$?

4. It was not mentioned specifically in this article, but the strands of isonemal fabrics can cross each other at any angle, not just the 90°, 45°, 120°, or 60° mentioned here. For all of the isonemal coverings of polyhedra that we have seen, the regions formed by the crossing of the strands are either squares, or rhombs whose smallest interior angle is 60°. These quadrilaterals are usually scored and folded along a diagonal so as to produce two adjacent triangular surfaces on the finished model. An open question is: Do other possibilities exist, or can it be proven that these are the only possibilities?

5. The model illustrated in Figure 17 seems like a much more genuine example of a 6-way 2-fold model that possesses octahedral symmetry (that is, it is less contrived than our example obtained by adding marks on a version originally possessing icosahedral symmetry). Does there exist a corresponding 6-way 3-fold model with octahedral symmetry?

6. If we allow the strands on the polyhedral weaving to go over and under each other in a more complicated way, what new isonemal weavings will result? For example, the cube of Figure 8 could be woven with each strand going over two sections and then under two sections, as could be done in Figure 18 as well.

7. The symmetry requirements we imposed for regularity guarantee that for every strand there is a symmetry which leaves the strand fixed but interchanges its edges (this is easily seen on all of the 2-fold models mentioned). Now suppose that every strand (on the 2-fold models) is split down its center into two strands.

[4]Since preparing the original manuscript for this article (two years ago) I have discovered that there are an infinite number of these models, all of which have dihedral symmetry. I don't know, however, whether or not any such models exist with other kinds of symmetry.

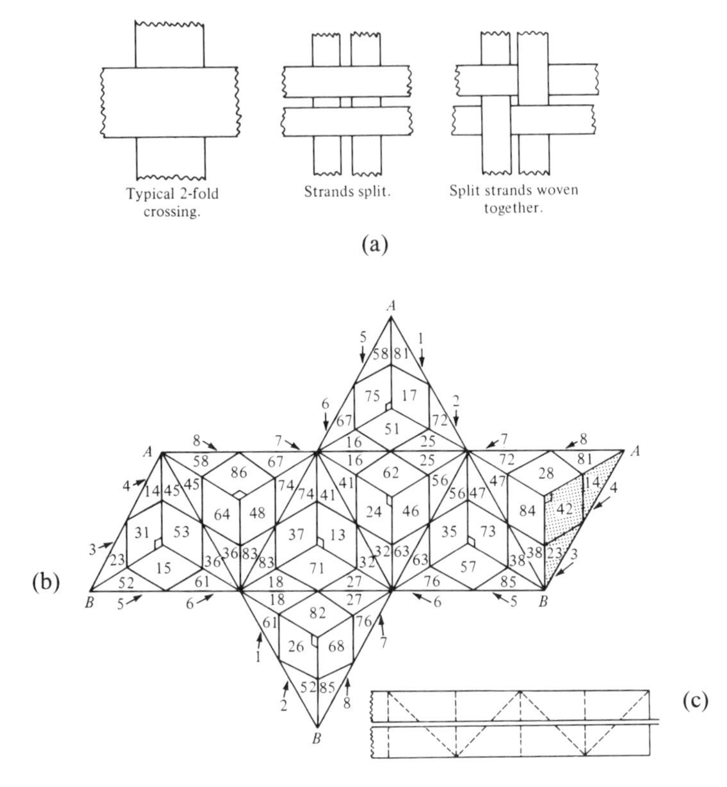


**Figure 18.** (a) Splitting and weaving. (b) 8 strands, 2-fold (a 4-way, 2-fold woven polyhedron). Note: cross-hatched face is flat. (c) Stretch of typical strand, as constructed from original strand.

The fabric will no longer be regular, because each strand now fails to cross over its "parallel" partner, but if we split the strands and proceed to weave them at every crossing as shown in Figure 18(a), then some very attractive models result. If we use this variation on the infinite class of polyhedra mentioned in question 10 of this section, it may be possible to weave "twills" or "satins" on the cube. As a simple example of this variation, the "octahedron" of Figure 12 is shown with its strands split and rewoven in Figure 18(b).

8. Weavings resulting from division and reweaving the strands on models involving equilateral triangles could also be investigated.

9. Another set of models can be made as follows: Suppose the polyhedron has a right-angled pyramid on each face of some "base" model—that is, it is made of strips like this:

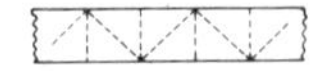

Each strip, of necessity, goes around a total of *two* faces of each pyramid. In our examples the strand always goes across 1 face and 1 face, but suppose the strand

# Plates

Plate A

Plate B

Plate C

Plate D

Plate E

goes instead around $\frac{1}{2}$ face, 1 face, $\frac{1}{2}$ face, thus:

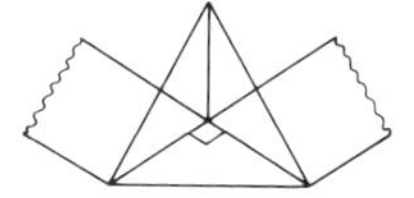

The strand would then look like this:

The "octahedron" of Figure 12, so modified, would have 6 strips and be 2-fold. But it would be a 3-way 2-fold model. This idea, of course, could be investigated on all the models constructed of right-angled pyramids. Can a similar thing be done on the models involving equilateral triangles?

10. An infinite number of 2-fold isonemal cubes can be woven. The diagonal cube of Figure 10 is the first member of the set. The cube in Figure 17 is the second member of the set. The third member of the set has four (self-overlapping) strands. The fourth, fifth, and sixth members of this set have 3, 4, and 6 strands respectively. Of course each member of this set must have 3, 4, or 6 strands (because they all have octahedral symmetry), but is there any easy rule to decide the number of strands?[5]

11. Consider the polyhedron that results from the triangulation of the net for the cuboctahedron in Figure 19. If every triangle of that "net" is an equilateral triangle, then this model is composed of sixty such triangles. One strand containing thirty triangles goes around this model in two distinct ways, and it is *impossible* to weave this surface with the strands going in the usual 6-cycle pattern. The question is: Can you weave this model in any reasonably symmetric way? When do models consisting of equilateral triangles admit either a regular or nonregular isonemal weaving?

12. It may be useful to arrange the subject by some method of classification. One approach might be to classify the models according to the number of transitivity classes, $t$, of their vertices. Thus, for example, in this paper we have

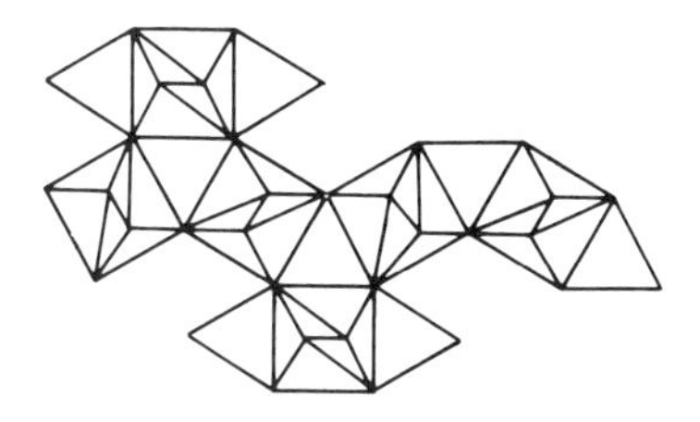

**Figure 19**

[5]The answer turns out to be, *yes*.

for

$t = 1$, the models in Figures 8 and 9;
$t = 2$, the models in Figures 10, 11, 13, 14, 15, 17 and Plates C and D;
$t = 3$, the models in Plates A and E;
$t = 4$, the model in Plate B.

It might be worthwhile to try to enumerate all cases with $t \leqslant 2$ or $t \leqslant 3$.

## References

[1] Ball, W. W. Rouse (revised by Coxeter, H. S. M.) *Mathematical Recreations and Essays*, 11th ed. Macmillan 1939.

[2] Cundy, H. Martyn and Rollett, A. P., *Mathematical Models*, 2nd ed. Oxford University Press 1973.

[3] Coxeter, H. S. M., *Regular Polytopes*. Methuen 1948.

[4] Grünbaum, Branko and Shephard, G. C., Satins and twills—an introduction to the geometry of fabrics. *Mathematics Magazine* Vol. 53 No. 3, (May, 1980) 139–166.

[5] Grünbaum, Branko and Shephard, G. C., *Isonemal Fabrics* (to appear).

[6] Pedersen, Jean J., Asymptotic euclidean type constructions without euclidean tools. *The Fibonacci Quarterly*, **9**, No. 2 (Apr. 1971), 199–216.

[7] Pedersen, Jean J., Some whimsical geometry. *The Mathematics Teacher*, **LXV**, (Oct. 1972), No. 6, 513–521.

[8] Pedersen, Jean J., Plaited platonic puzzles. *The Two Year College Mathematics Journal*, **4**, No. 3 (Fall 1973), 22–37.

[9] Pedersen, Jean J., Platonic solids from strips and clips. *The Australian Mathematics Teacher*, **30**, No. 4 (Aug. 1974), 130–133.

[10] Pedersen, Jean J., Collapsoids, *The Mathematical Gazette*, **59** (1975), 81–94.

[11] Pedersen, Jean J., Braided rotating rings. *The Mathematical Gazette*, **62** (1978), 15–18.

[12] Pedersen, Jean J., Visualizing parallel divisions of space. *The Mathematical Gazette*, **62** (1978), 250–262.

[13] Wenninger, Magnus J., *Polyhedron Models*. Cambridge University Press 1971.

# Convex Bodies which Tile Space

P. McMullen*

## 1. Introduction

We say that the convex body (compact convex set with nonempty interior) $K$ *tiles* $d$-dimensional Euclidean space $E^d$ (*by translation*) if there is some family $T$ of translation vectors, such that (i) $\mathcal{K} = \{K + t \mid t \in T\}$ covers $E^d$, and (ii) if $t_i \in T$ $(i = 1, 2)$ with $t_1 \neq t_2$, then $K + t_1$ and $K + t_2$ have disjoint interiors; that is, $\mathcal{K}$ is simultaneously a covering and packing of $E^d$. We call $\mathcal{K}$ a *tiling* of $E^d$ (*by translation*), and call $K$ and its translates in $\mathcal{K}$ *tiles*. A particularly important case is when $T$ is a lattice (discrete additive subgroup of $E^d$), when we call $\mathcal{K}$ a *lattice tiling*.

We shall take as our starting point the investigation of lattice tilings in $E^3$ by the Russian crystallographer E. S. Fedorov in 1885. In [4] appeared the first classification of the five combinatorial types of tile (or parallelohedron) in $E^3$. Subsequently, B. N. Delaunay (Delone) [3] classified the 51 types of lattice tile in $E^4$, but apart from some more special results (which we shall mention below), little further progress has been made in this particular direction.

One possible reason for this lack of progress may have been the absence of a general characterization of tiles (among all convex bodies). (But see also the *Note added in proof* preceding the references.) We shall describe such a characterization here. The fact that this characterization applies to general tiles has a number of interesting consequences, and we suggest more which might also follow.

## 2. Necessary Conditions on a Tile

Let $K$ tile $E^d$; we do not assume here that the corresponding set $T$ of translations is a lattice. Then $K$ satisfies a number of conditions.

*University College, London, England.

**I.** *K is a polytope.*

Here, as elsewhere in this article, we shall merely sketch proofs. The details may be found in [9], as will more extensive references to background material.

There is no loss in generality in assuming that $o \in T$. For $I$, then, we need only note that the boundary of $K$ is a finite union of sets $K \cap (K + t)$, with $o \neq t \in T$.

**II.** *K is centrally symmetric.*

For, $K$ is a finite union of centrally symmetric sets of the form $K \cap (-K - t)$ $(t \in T)$, which have disjoint interiors, and so, by a result of Minkowski [10], $K$ is centrally symmetric.

**III.** *Each facet F of K is centrally symmetric.*

There is again no loss of generality in assuming that $o$ is the centre of $K$, so that the facet ($(d-1)$-dimensional face) of $K$ opposite $F$ is $-F$. Then $F$ is a finite union of centrally symmetric sets of the form $F \cap (-F + t)$ $(t \in T)$, whose interiors relative to the affine hull of $F$ are disjoint. By [10] again, $F$ is centrally symmetric.

If $G$ is a subfacet ($(d-2)$-dimensional face) of $K$, then $G$ lies in two facets of $K$, say $F$ and $F'$. Since $F'$ is centrally symmetric, it has a subfacet $G'$ opposite to $G$, which is the intersection of $F'$ with another facet $F''$, say. Carrying on in this way, we find a *belt* of facets, $F, F', F'', \ldots, F^{(k)} = F$, say, such that each $F^{(i-1)} \cap F^{(i)}$ is a translate of $G$ or of $-G$. (Belts are the same as zones if $d = 3$, but the term zone, which is naturally restricted to *zonotopes* or vector sums of line segments, has a different meaning if $d > 3$.) We then have:

**IV.** *Each belt of K contains* 4 *or* 6 *facets.*

Let the belt of $K$ containing the subfacet $G$ have $m$ pairs of opposite facets, and let $g \in G$ lie in no $j$-face of any tile $K + t$ with $j < d - 2$. The sum of the dihedral angles of $K$ at its subfacets parallel to $G$ is $2(m-1)\pi$. Suppose $m \geqslant 4$. Since the sum of the dihedral angles at two non-opposite subfacets of the belt is greater than $(m-1)\pi - (m-2)\pi = \pi$, $g$ cannot lie in the relative interior of any facet of a tile. So $g$ lies in subfacets alone. But, similarly, the sum of the dihedral angles at three non-opposite subfacets is greater than $(m-1)\pi - (m-3)\pi = 2\pi$, so three tiles cannot fit around $g$. Thus $m \leqslant 3$, as was claimed.

Conditions I, II, and III are due to Minkowski [10], albeit in the context of lattice tilings in $E^3$. But his proofs extend without change of language to arbitrary tilings in $E^d$. The origin of condition IV is more obscure. While it was clear to earlier investigators that the condition was important, it appears to have been Coxeter [2] who first explicitly suggested that it might be a crucial condition, at least for zonotopes.

## 3. The Sufficiency of the Conditions

The central result of our discussion is:

**Theorem 1.** *Conditions I, II, III, and IV are necessary and sufficient for a convex body $K$ to tile $E^d$.*

We have already demonstrated that the conditions are necessary. To show that they are sufficient, we must first describe a suitable candidate for a tiling of $E^d$ by $K$. If $F$ is a facet of $K$, then the opposite facet $-F$ is a translate of $F$, and so there is a translation vector $t_F$ carrying $-F$ into $F$. Thus we have $(K + t_F) \cap K = F = -F + t_F$. If we do this for each facet of $K$, by iteration of the process, we obtain the family $\mathcal{K} = \{K + t \mid t \in T\}$, where $T = \{\sum_F n_F t_F \mid n_F \in Z\}$. This is our candidate.

It is not obvious that $\mathcal{K}$ is either a covering or a packing of $E^d$, but if it is, and so is a tiling of $E^d$, it is a special kind, called a *face-to-face tiling*, since the intersection of two tiles $K + t_1$ and $K + t_2$ will be empty or a common face of each tile. Naturally, such a tiling is a lattice tiling.

In sketching a proof of the theorem, we introduce some useful terminology and notation. Let $g \in bdK$. We say that $g'$ is *equivalent* to $g$, written $g' \sim g$, if there is a sequence $g = g_0, g_1, \ldots, g_m = g'$ in $K$ such that, for $i = 1, \ldots, m$, $g_i$ is obtained from $g_{i-1}$ by successive reflexions in the centre of a facet $F$ of $K$ (to which it belongs) and in the centre of $K$ itself; in other words, $g_i = g_{i-1} + t_F$. Equivalent faces of $K$ are related in the same way. If $G$ is a face of $K$, we define $\mathcal{K}_G$ to be the subfamily of tiles $K'$ in $\mathcal{K}$ such that there is a sequence $K = K_0, K_1, \ldots, K_m = K'$ in $\mathcal{K}$, with, for $i = 1, \ldots, m$, $K_{i-1}$ and $K_i$ meeting in a common facet of each tile which contains $G$. If $g \in \operatorname{relint} G$ is any point, we note that $\mathcal{K}_G = \{K + (g' - g) \mid g' \sim g\}$.

We further say that $\mathcal{K}_G$ *surrounds* the face $G$ of $K$ if $\operatorname{relint} G \subseteq \operatorname{int}(\bigcup \mathcal{K}_G)$, and that $\mathcal{K}_G$ *fits around* $G$ if, for each $K', K'' \in \mathcal{K}_G$, $K'$ and $K''$ meet in a common face of each. If we write $\mathcal{K}_\varnothing = \mathcal{K}$, the analogues of these concepts for $\mathcal{K}$ are the defining properties for a covering or packing.

It is straightforward to prove that $\mathcal{K}_G$ surrounds $G$, under the given conditions I–IV (in fact, IV is not needed at this stage at all), and the idea of the proof extends to show that $\mathcal{K}$ covers $E^d$. We use induction on $d - r$, where $r = \dim G$, and show that $\mathcal{K}_G$ covers a small $(d - r - 1)$-sphere $S$ centred at a point of $\operatorname{relint} G$ and orthogonal to $\operatorname{aff} G$. An important role is played by the sets $C_{F,G} = \bigcup\{\bigcup \mathcal{K}_{F'} \mid G \subset F' \subseteq F\}$, which are neighbourhoods of the faces $F$ of $K$ which contain $G$ in the sphere $S$.

To show that $\mathcal{K}_G$ fits around $G$, we need only show that the intersections $K' \cap S$ $(K' \in \mathcal{K}_G)$ fit together in $S$. If $\dim G = r \geqslant d - 2$, this is ensured by the initial conditions (here IV enters). If $r \leqslant d - 3$, we again use an induction argument on $d - r$. If we suppose that $\mathcal{K}_G$ does not fit around $G$, then two tiles $K'$ and $K''$ in $\mathcal{K}_G$ overlap without coinciding. There is a sequence $K' = K_0$, $K_1, \ldots, K_m = K''$ in $\mathcal{K}_G$ such that, for $i = 1, \ldots, m$, $K'$ and $K''$ meet in a common facet which contains $G$. We then join a fixed point $x \in \operatorname{int} K' \cap \operatorname{int} K''$

to itself by a loop lying in $S$, which successively passes from $\operatorname{int} K_{i-1}$ to $\operatorname{int} K_i$ through $\operatorname{relint}(K_{i-1} \cap K_i)$. We then contract the loop across $S$ to $x$, using the inductive hypothesis that $\mathcal{K}_F$ fits around $F$ if $\dim F > r$, and the fact that, for $r \leqslant d-3$, a $(d-r-1)$-sphere is simply connected. This eventually shows that $K'$ and $K''$ after all coincide. An exactly analogous argument proves that $\mathcal{K}$ is a packing, although the argument needs some modification to account for the fact that $\mathcal{K}$ is infinite.

## 4. Some Further Consequences

The first thing we may note follows directly from the argument of the last section.

**Theorem 2.** *If $K$ tiles $E^d$ by translation, then $K$ admits a face-to-face, and hence a lattice, tiling.*

This, of course, impinges on the 18th problem of Hilbert [7]. It is noteworthy that if we relax "convex" to "star-shaped," then Stein [12] has shown that there are (nonconvex) star-shaped polyhedral sets in $E^d$ (for $d \geqslant 5$), even centrally symmetric (if $d \geqslant 10$), which tile space by translation, but do not admit any lattice tiling.

We say that *$K$ tiles $E^d$ by homothety* if there is some closed interval $[\alpha, \beta]$ of positive real numbers such that some family of homothetic copies $K' = \lambda K + t$ of $K$, with ratio of homothety $\lambda \in [\alpha, \beta]$, tiles $E^d$. Now Groemer [5] showed that a homothety tile $K$ satisfies conditions I, II, and III, and it is clear that IV depends only on the shapes of the tiles, and not on their relative sizes. Thus we have:

**Theorem 3.** *A homothety tile is also a translation tile.*

Groemer [6] showed that, for $d \leqslant 4$, if $K$ admits a *proper* tiling by homothety, in which not all the tiles are actually translates, then $K$ is a prism over a $(d-1)$-dimensional translation tile. In fact, this result holds generally:

**Theorem 4.** *A proper homothety tile is a prism.*

## 5. Voronoĭ Polytopes

If $L$ is a lattice in $E^d$, its *Voronoĭ polytope* (or *Dirichlet region*) $V$ is the set of points no further from $o$ than from any other point of $L$. More conveniently, we can define an affine version of this, as follows. If $\varphi$ is a positive definite quadratic form, the Voronoĭ polytope $V(\varphi, L)$ is defined by $V(\varphi, L) = \{x \in E^d \mid \varphi(x)$

$\leqslant \varphi(x - t)$ for all $t \in L \setminus \{o\}\}$. Implicitly, the following was proposed by Voronoĭ [13]:

**Conjecture 1.** *Every tile is (a translate of) a Voronoĭ polytope.*

We say a tile is *r-primitive* if, in its face-to-face tiling, each $r$-face belongs to exactly $d - r + 1$ tiles. 0-primitive is usually called *primitive*. Voronoĭ [13] showed that primitive tiles are Voronoi polytopes, Žitomirskiĭ [14] extended the result to $(d - 2)$-primitive tiles (that is, every belt has six facets), and Delaunay [3] proved the general result for $d \leqslant 4$. (Incidentally, Voronoĭ showed that there are three types of primitive tile in $E^4$, and Baranovskiĭ and Ryškov [1] enumerated the 221 types of primitive tile in $E^5$.)

Since it is clear that a limit (in the Hausdorff metric) of a sequence of Voronoĭ polytopes is again a Voronoĭ polytope, one possible approach would be to show:

**Conjecture 2.** *Every tile is a limit of primitive tiles.*

All that one really needs to establish Conjecture 1 is to prove that every tile is a limit of $(d - 2)$-primitive tiles. Maybe the characterization of tiles in Theorem 1 would help to show this.

## 6. Zonotopes

Zonotopes are particular types of convex bodies satisfying conditions I, II, and III. Zonotopes which tile space were investigated by Shephard [11], who proposed that certain conditions were equivalent to the tiling property. (One of these is IV, which, as we earlier remarked, was put forward by Coxeter [2].) Shephard verified the equivalence for $d \leqslant 4$, and the general result was established by McMullen [8].

Something close to Conjecture 1 has been proved for zonotopes. We say that two zonotopes are *equivalent* if, up to affinity, one can be obtained from the other by varying the lengths of its component line segments. Then McMullen [8] showed:

**Theorem 5.** *A zonotope which tiles $E^d$ is equivalent to a Voronoĭ polytope.*

In a rather different direction from what we have been discussing hitherto, we have the following. A $d$-zonotope $Z$ which is the sum of $n$ line segments (which we allow to be parallel, or even of zero length) is affinely equivalent to the image of a regular $n$-cube under orthogonal projection. Any $(n - d)$-zonotope $\overline{Z}$ which is affinely equivalent to the image of the same cube under orthogonal projection onto the orthogonal complementary subspace is said to be associated with $Z$. Among the conditions of Shephard [11], again proved by him for $d \leqslant 4$ and by

McMullen [8] generally, was:

**Theorem 6.** *If the zonotope $Z$ tiles $E^d$, then its associated zonotope $\bar{Z}$ tiles $E^{n-d}$.*

*Note added in proof (August 1981):* We recently learned that the main result, Theorem 1, and its immediate corollary, Theorem 2, were proved earlier by B. A. Venkov [17] in 1954. His method is much the same as ours. A. D. Aleksandrov [15] has also proved a far reaching generalization, to tilings by congruent copies of a finite set of polytopes in spaces of constant curvature; the proof above carries over easily. (See also [16] for more historical detail.)

## References

[1] Baranovskiĭ, E. P. and Ryškov, S. S., Primitive five-dimensional parallelohedra. *Dokl. Akad. Nauk SSSR* **212** (1973), 532–535 = *Soviet Math. Dokl.* **14** (1973), 1391–1395 (1974).

[2] Coxeter, H. S. M., The classification of zonohedra by means of projective diagrams. *J. Math. Pures Appl.* **41** (1962), 137–156.

[3] Delaunay (Delone), B. N., Sur la partition regulière de l'espace à 4 dimensions, I, II. *Izvestia Akad. Nauk SSSR, Ser. VII* (1929), 79–110, 147–164.

[4] Fedorov, E. S., *Elements of the Study of Figures* (in Russian). St. Petersburg 1885 (Leningrad 1953).

[5] Groemer, H., Ueber Zerlegungen des Euklidischen Raumes. *Math. Z.* **79** (1962), 364–375.

[6] Groemer, H., Ueber die Zerlegungen des Raumes in homothetische konvexe Körper. *Monatsh. Math.* **68** (1964), 21–32.

[7] Hilbert, D., Problèmes futurs des mathématiques. In *Proc. II Internat. Congr. Math. 1900*. Paris, 1902.

[8] McMullen, P., Space tiling zonotopes. *Mathematika* **22** (1975), 202–211.

[9] McMullen, P., Convex bodies which tile space by translation *Mathematika* **27** (1980), 113–121.

[10] Minkowski, H., Allgemeine Lehrsätze über konvexen Polyeder. *Nachr. K. Akad. Wiss. Göttingen, Math.-Phys. Kl.* **ii** (1897), 198–219.

[11] Shephard, G. C., Space filling zonotopes. *Mathematika* **21** (1974), 261–269.

[12] Stein, S. K., A symmetric star body that tiles but not as a lattice. *Proc. Amer. Math. Soc.* **36** (1972), 543–548.

[13] Voronoi, G. F., Nouvelles applications des paramètres continus à la théorie des formes quadratiques. Deuxième Mémoire: Recherche sur les parallèloèdres primitifs. *J. Reine Angew. Math.* **134** (1908), 198–287; **136** (1909), 67–181.

[14] Žitomirskiĭ, O. K., Verschärfung eines Satzes von Woronoi. *Ž. Leingr. fiz.-mat. Obšč.* **2** (1929), 131–151.

[15] Aleksandrov, A. D., On filling of space by polytopes (in Russian). *Vestnik Leningrad. Univ. (Ser. Mat. Fiz. Him.)* **9** (1954), 33–43.

[16] McMullen, P., Convex bodies which tile space by translation: Acknowledgment of priority. *Mathematika* **28** (1981).

[17] Venkov, B. A., On a class of euclidean polytopes (in Russian). *Vestnik Leningrad. Univ. (Ser. Mat. Fiz. Him.)* **9** (1954), 11–31.

# Geometry of Radix Representations

William J. Gilbert*

## 1. Introduction

The aim of this paper is to illuminate the connection between the geometry and the arithmetic of the radix representations of the complex numbers and other algebraic number fields. We indicate how these representations yield a variety of naturally defined fractal curves and surfaces of higher dimensions.

As is well known, the natural numbers can all be represented using any integer $b$, larger than one, as base, with the digits $0, 1, 2, \ldots, b-1$. All the integers, both positive and negative, can be represented without signs by means of the negative integral base $b$, less than minus one, using the natural numbers $0, 1, 2, \ldots, |b|-1$ as digits [6, §4.1]. Each Gaussian integer may be uniquely represented in binary form as $\sum_{k=0}^{r} a_k(-1+i)^k$, where each $a_k = 0$ or 1, [1, §4.3; 6, §4.1]. We will unify and generalize such representations.

## 2. Algebraic Number Fields

We now describe more precisely what we mean by a radix representation in an algebraic number field. Let $\rho$ be an algebraic integer whose minimum polynomial is $x^n + p_{n-1}x^{n-1} + \cdots + p_1x + p_0$; *let*

$$N = |\mathrm{Norm}(\rho)| = |(-1)^n p_0|.$$

We will try to represent elements of the algebraic number field $\mathbb{Q}(\rho)$ using the radix $\rho$ and natural numbers as digits. We restrict ourselves here to only considering digits which are natural numbers, as this appears to be the obvious

*Department of Pure Mathematics, University of Waterloo, Waterloo, Ontario, Canada N2L 3G1.

generalization of the familiar number systems and it is more convenient for doing arithmetical calculations. However, usually for geometric reasons, it is sometimes necessary to use nonintegral digits. In such cases, the results obtained may be slightly different. The largest set of algebraic numbers we could expect to represent, without using negative powers of the radix, is the ring $\mathbb{Z}[\rho]$. Note that this ring may not be the whole ring of algebraic integers $\mathbb{A} \cap \mathbb{Q}(\rho)$ in the number field. We say that $\rho$ is the base (or radix) of a *full radix representation* of $\mathbb{Z}[\rho]$ if each element $z$ of $\mathbb{Z}[\rho]$ can be written in the form $z = \sum_{k=0}^{r} a_k \rho^k$, where the digits $a_k$ are natural numbers such that $0 \leqslant a_k < N$. We denote this representation by $z = (a_r a_{r-1} \ldots a_1 a_0)_\rho$.

The reason that the norm yields the correct number of digits is due to the following observation.

**Lemma .** *Let $c$ and $d$ be two integers in $\mathbb{Z}$. Then $c \equiv d \pmod{\rho}$ in $\mathbb{Z}[\rho]$ if and only if $c \equiv d \pmod{N}$ in $\mathbb{Z}$.*

*Proof.* Suppose $c \equiv d \pmod{\rho}$ in $\mathbb{Z}[\rho]$. Then there exist rational integers $q_i$ such that

$$\begin{aligned} c - d &= \rho\left(q_n \rho^{n-1} + \cdots + q_2 \rho + q_1\right) = q_n \rho^n + \cdots + q_2 \rho^2 + q_1 \rho \\ &= -q_n\left(p_{n-1}\rho^{n-1} + \cdots + p_1 \rho + p_0\right) + q_{n-1}\rho^{n-1} + \cdots + q_2 \rho^2 + q_1 \rho \\ &= (q_{n-1} - q_n p_{n-1})\rho^{n-1} + \cdots + (q_1 - q_n p_1)\rho - q_n p_0. \end{aligned}$$

Since $1, \rho, \rho^2, \ldots, \rho^{n-1}$ are linearly independent over $\mathbb{Q}$, it follows that $c - d = -q_n p_0$. As $N = |p_0|$, we have $c \equiv d \pmod{N}$.

Now $N = \pm p_0 = \mp \rho(\rho^{n-1} + p_{n-1}\rho^{n-2} + \cdots + p_1)$, so that $N$ is divisible by $\rho$ in $\mathbb{Z}[\rho]$ and the converse implication follows. □

This lemma implies that the quotient ring $\mathbb{Z}[\rho]/(\rho)$ is isomorphic to $\mathbb{Z}_N$ and that $0, 1, 2, \ldots, N-1$ form a complete set of representatives of the congruence classes modulo $\rho$ in $\mathbb{Z}[\rho]$. Clearly, the digits of any radix representation of $\mathbb{Z}[\rho]$ must form a complete set of representatives of these classes.

If an element of $\mathbb{Z}[\rho]$ can be represented using the base $\rho$ and digits $0, 1, 2, \ldots, N-1$, the representation is unique. It does not matter whether $\rho$ yields a full or only a partial radix representation. The proof of the uniqueness uses the above lemma and is the same as for ordinary decimals.

Katai and Szabo [5] show that, for each positive integer $m$, the Gaussian integers can be represented by the radix $-m + i$ (and $-m - i$) using the digits $0, 1, 2, \ldots, m^2$. In particular, the complex numbers can be written as "decimals" in base $-3 + i$; for example, $(241)_{-3+i} = 2(-3+i)^2 + 4(-3+i) + 1 = 5 - 8i$. The bases mentioned above are the only ones that will represent all the Gaussian integers in the required form. Knuth (see [6, §4.1]) has defined a "quater-imaginary" number system for the complex numbers based on the radix $2i$, which has norm 4. All the elements of $\mathbb{Z}[2i]$, that is, Gaussian integers with even imaginary parts, can be uniquely represented in this system. Gaussian integers with odd imaginary parts can be represented if we allow expansions to one radix place; for example $(31.2)_{2i} = 3(2i) + 1 + 2(2i)^{-1} = 1 + 5i$.

For the complex quadratic fields $\mathbb{Q}(\sqrt{-m})$, where $-m \equiv 2, 3 \pmod 4$ and $-m \neq -1$, one good base is provided by $\sqrt{-m}$ itself. Given any integer $a + b\sqrt{-m}$ in the field, first write the rational integers $a = (a_r \ldots a_1 a_0)_{-m}$ and $b = (b_s \ldots b_1 b_0)_{-m}$ in base $-m$. It then follows that

$$a + b\sqrt{-m} = (b_t a_t b_{t-1} a_{t-1} \ldots b_1 a_1 b_0 a_0)_{\sqrt{-m}},$$

where $t = \max(r, s)$.

Given an arbitrary number field, it is not always possible to find a base for its integers. For example, in the biquadratic field $\mathbb{Q}(\sqrt{7}, \sqrt{10})$ there is no integer $\alpha$ such that $\mathbb{Z}[\alpha] = \mathbb{A} \cap \mathbb{Q}(\sqrt{7}, \sqrt{10})$ (see [8, p. 46]). However, it may still be possible to represent the integers in the field by allowing radix expansions using negative powers of the base.

The usual arithmetic operations of addition and multiplication can be performed using these radix representations in much the same way as ordinary arithmetic base $N$. The only difference is in the carry digits. For example, the root $\rho$ of the cubic $P(x) = x^3 + x^2 + x + 2$ is a base for $\mathbb{Z}[\rho]$. Since $\rho$ is also a root of $(x-1)P(x)$, we have $\rho^4 + \rho = 2$ and so $2 = (10010)_\rho$. Hence, whenever we have an overflow of 2 in any one column when doing an arithmetical operation, we have to carry 1001 to the next four higher columns.

## 3. Geometry of Representations

The elements of $\mathbb{Q}(\rho)$ can be pictured as points in $\mathbb{Q}^n$ using coordinates $1, \rho, \rho^2, \ldots, \rho^{n-1}$. However, if $\mathbb{Q}(\rho) = \mathbb{Q}(i)$ it is often more useful to use the Argand diagram instead. In $\mathbb{Q}^n$, the points of $\mathbb{Z}[\rho]$ correspond to the integer lattice points. The radix representations in base $\rho$ map injectively to the lattice points. If $\rho$ is a base for a full representation of $\mathbb{Z}[\rho]$, then all the lattice points will be covered; if not, the image will be some infinite subset.

These images can be viewed as $n$-dimensional jigsaw puzzles whose $r$th piece consists of the union of unit $n$-dimensional cubes centered at the points whose base $\rho$ representation is of length $r$. The $(r+1)$st piece is formed from $N-1$ copies of the first $r$ pieces translated in $\mathbb{Q}^n$ along the directions of $\rho^r$, $2\rho^r, \ldots, (N-2)\rho^r$, and $(N-1)\rho^r$. For example, in the jigsaw in the Argand diagram in Figure 1 derived from the base $1 - i$, each piece is twice the size of the previous piece. Each little square corresponds to one Gaussian integer with the origin at the center black square. Since the jigsaw only fills up half the Argand diagram, $1 - i$ only provides a partial radix representation of the Gaussian integers. However, exactly the same pieces put together in Figure 2 using base $-1 + i$ fill the entire plane; this demonstrates the fact that $-1 + i$ is a base for all the Gaussian integers.

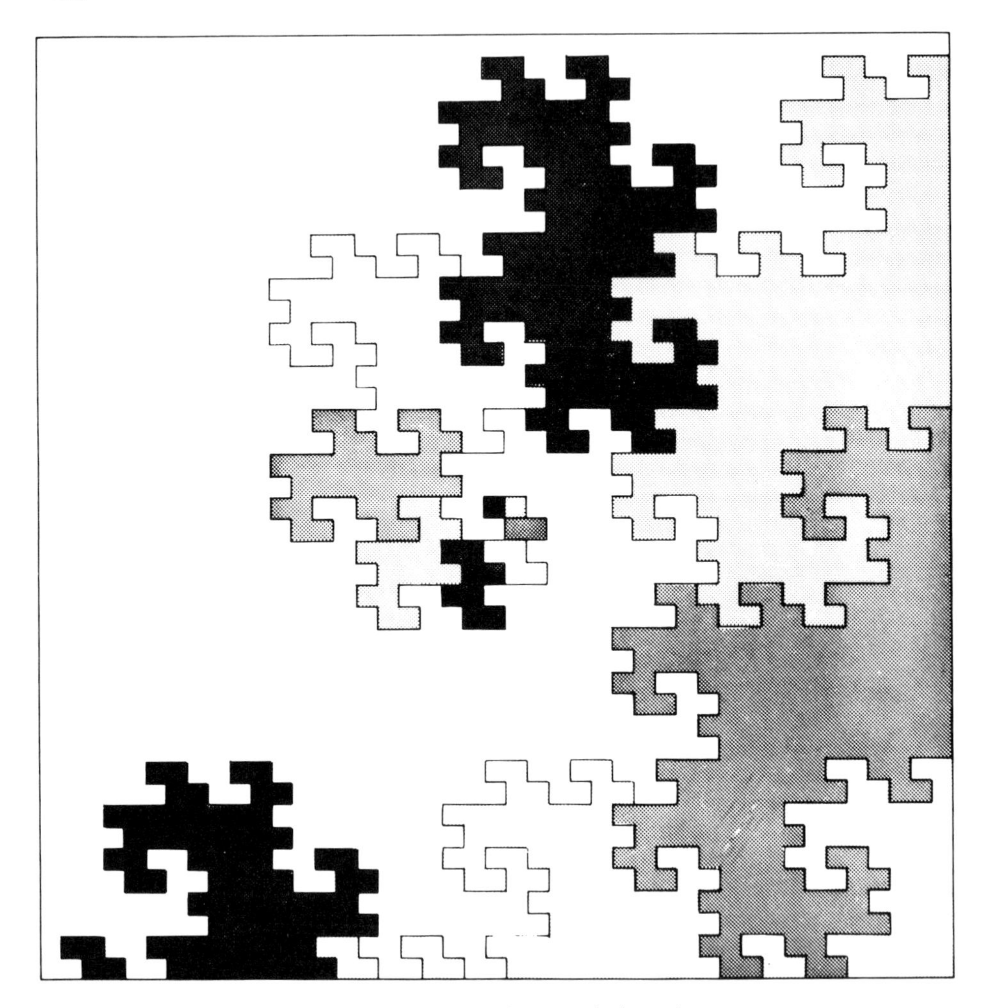

**Figure 1.** The Gaussian integers in base $1 - i$.

In Figure 3, the elements of $\mathbb{Z}[\omega]$ are represented in the base $-2 - \omega$, where $\omega$ is a complex cube root of unity. This base is a root of $x^2 + 3x + 3$, so it has norm 3. Each element of $\mathbb{Z}[\omega]$ is pictured as a unit hexagon in the Argand diagram with the origin being the black one. The figure shows the radix representation up to six places, and if continued it would fill the plane, since $-2 - \omega$ is a good base for $\mathbb{Z}[\omega]$.

Figures 4 and 5 show three-dimensional models derived from the cubic fields generated by the polynomials $x^3 + x^2 + x - 2$ and $x^3 + x^2 + x + 2$ respectively. The former only yields a partial representation, while the latter, if extended, would fill the whole of $\mathbb{Z}^3$ and so provide a full radix representation.

C. Davis and D. Knuth [2] use bases $1 + i$ and $1 + 2\omega$ in their investigation of the dragon and ter-dragon curves in the Argand diagram.

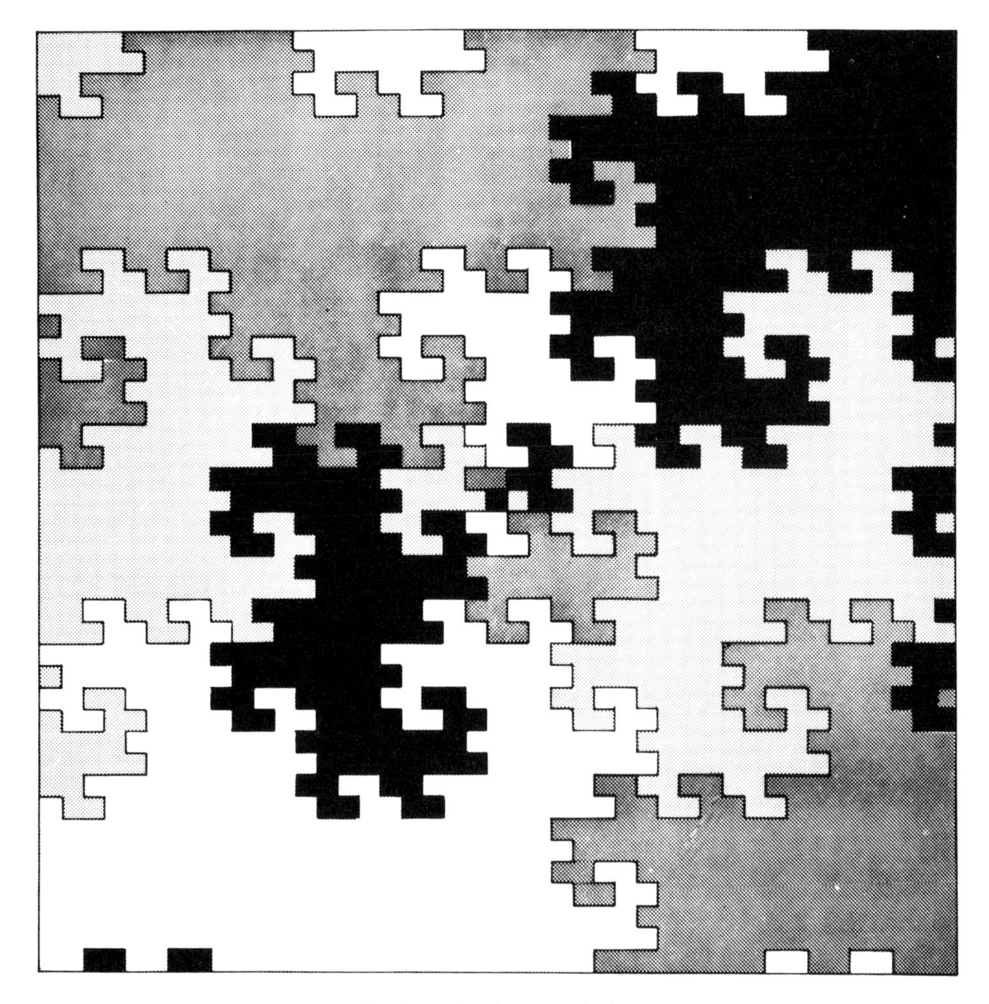

**Figure 2.** The Gaussian integers in base $-1 + i$.

## 4. Fractal Curves and Surfaces

It is natural to extend a radix representation to an infinite expansion using negative powers of the base. We say that an element of $\mathbb{Q}(\rho)$ can be written in base $\rho$ if it has an expansion of the form $\sum_{k=-\infty}^{r} a_k \rho^k$ where $0 \leqslant a_k < N$ for all $k$; we denote this expansion by $(a_r a_{r-1} \ldots a_0 \cdot a_{-1} a_{-2} \ldots)_\rho$. Terminating expansions correspond to elements of $\mathbb{Q}(\rho)$ whose denominators are some power of the norm.

From a geometric point of view, it is tempting to try to complete the representations of $\mathbb{Q}^n$ to representations of $\mathbb{R}^n$. However, if $1, \rho, \rho^2, \ldots, \rho^{n-1}$ are linearly dependent over $\mathbb{R}$, different points of $\mathbb{R}^n$ would correspond to the same number. Therefore, besides the rational numbers, the only fields whose represen-

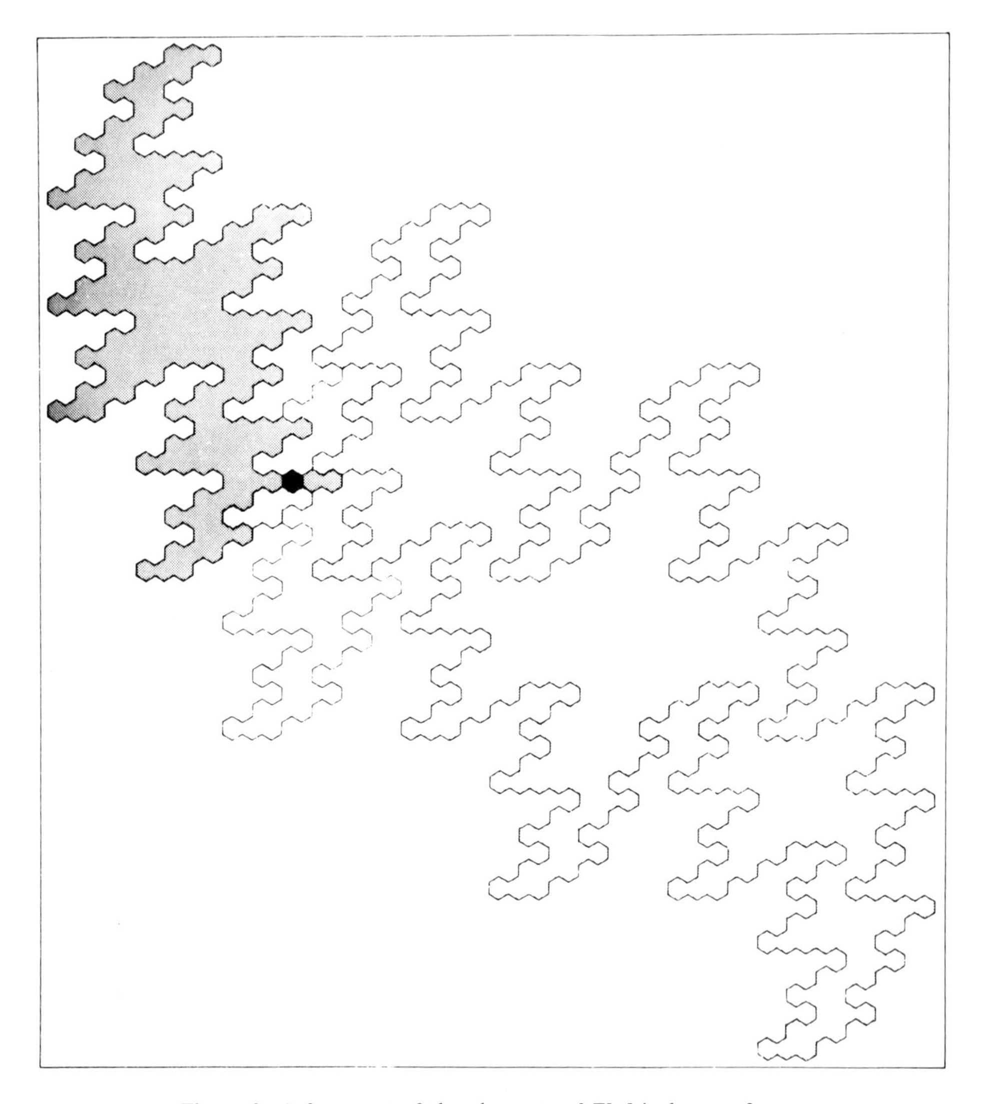

**Figure 3.** A fragment of the elements of $\mathbb{Z}[\omega]$ in base $-2-\omega$.

tations we could complete are the complex quadratic fields; these fields can all be represented in the Argand diagram.

We find that these complex quadratic fields yield some fascinating geometry by examining the regions of the Argand diagram corresponding to radix expansions of a given form. The regions whose points have expansions of the form $(a_r \ldots a_0 \cdot a_{-1} \ldots)_\rho$, for some fixed power $r$, have boundaries that are naturally defined fractal curves. Figure 6 shows all the complex numbers that are representable in base $1-i$ using expansions of any length. This region is in fact two space-filling dragon curves joined tail to tail. Mandelbrot [7, p. 313] has calculated the fractal (i.e. Hausdorff) dimension of the dragon's "skin," and it is approximately 1.5236.

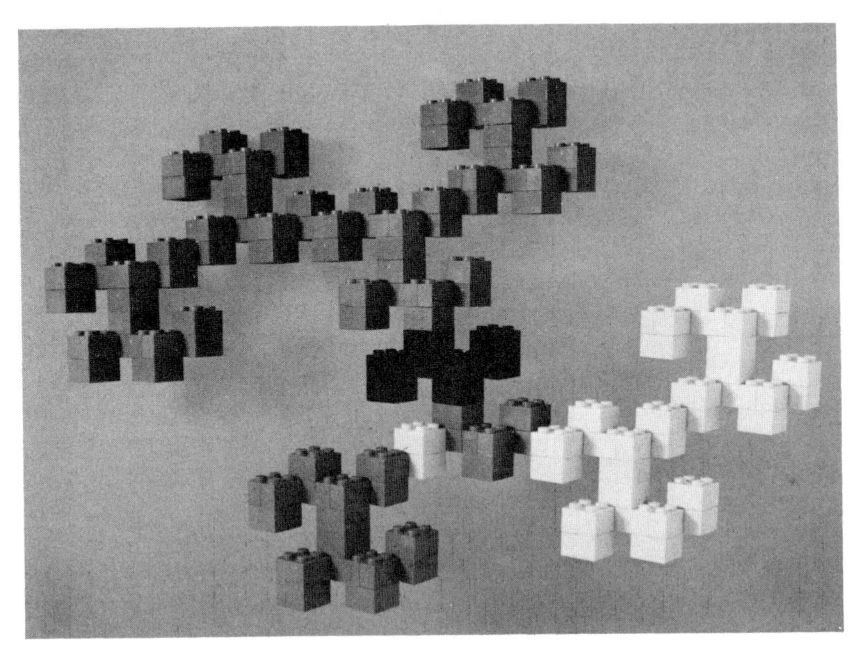

**Figure 4.** The elements of $\mathbb{Z}[\rho]$ in base $\rho$ where $\rho^3 + \rho^2 + \rho - 2 = 0$.

Figure 7 is a close-up of the Argand diagram in which each region consists of numbers having a fixed integer part in base $-1 + i$. (The axes in this figure are at 45° to the edges.) The boundaries have the same fractal dimension as that of Figure 6. Points on the boundary of two regions will have two representations in base $-1 + i$; each have different integral parts. Since the Argand diagram is two-dimensional, there must be some points that lie on the boundary of three regions, and they have three different representations; for example, $(2 + i)/5 = (0.\overline{011})_{-1+i} = (1.\overline{110})_{-i+1} = (1110.\overline{101})_{-1+i}$, where the bars over the digits indicate that they are to be repeated indefinitely.

For each base $-m + i$ of the complex numbers, we can show [3] that the fractal dimension of the boundary of the resulting regions is

$$(\log \lambda_m)/\log\sqrt{m^2+1}\ ,$$

where $\lambda_m$ is the positive root of $\lambda^3 - (2m-1)\lambda^2 - (m-1)^2\lambda - (m^2+1)$.

For an arbitrary number field $\mathbb{Q}(\rho)$, the boundary of the resulting regions in $\mathbb{Q}^n$ may not contain as many points as we desire, because $\mathbb{Q}^n$ is not complete. However, we can still define the fractal dimension of the edge of a region $S$ in $\mathbb{Q}^n$ as follows. Let $\epsilon > 0$ and let $E_\epsilon$ be the set of points within $\epsilon$ of the edge of $S$, that is, points whose $\epsilon$-neighborhood contains points of $S$ and points not in $S$. For each positive number $d$, cover $E$ by balls of radius $\sigma_i < 2\epsilon$ and take the following infimum over all such coverings:

$$m_d^\epsilon = \inf \sum_i \sigma_i^d .$$

**Figure 5.** A fragment of the elements of $\mathbb{Z}[\rho]$ in base $\rho$ where $\rho^3 + \rho^2 + \rho + 2 = 0$.

Now let $m_d = \sup_{\epsilon > 0} m_d^\epsilon$; this number is proportional to the $d$-dimensional measure of the edge. The edge is said to have *fractal dimension D* if

$$m_d = \begin{cases} \infty & \text{for all } d < D, \\ 0 & \text{for all } d > D. \end{cases}$$

This fractal dimension is a metric invariant and hence will remain unchanged under a linear transformation. Therefore, whether we represent a complex quadratic field by points in $\mathbb{Q}^2$ or by points in the Argand diagram, we will obtain the same dimension.

Some bases $\rho$ which only yield partial radix representations of $\mathbb{Z}[\rho]$ may not give any infinite convergent radix expansions. For example, all infinite radix expansions using the base of Figure 4 diverge because one of the roots of the minimum polynomial, $x^3 + x^2 + x - 2$, has modulus smaller than one. Therefore fractal surfaces cannot be constructed from this base. On the other hand, the periodic radix expansions using the base $\rho$ of Figure 5 do converge to points of

**Figure 6.** All the complex numbers representable in base $1 - i$.

$\mathbb{Q}(\rho)$, and this base will yield a fractal surface of dimension between two and three.

## 5. Problems

These radix representations suggest many interesting problems, both geometric and arithmetic. We mention three here.

Firstly, which algebraic integers yield full radix representations? For the quadratic fields we can show that a root of the irreducible polynomial $x^2 + cx + d$ gives a full radix representation if and only if $d \geqslant 2$ and $-1 \leqslant c \leqslant d$. A root of the linear polynomial $x + d$ yields a complete representation if and only $d \geqslant 2$.

Secondly, find an algorithm for dividing a number in base $\rho$ by a rational

**Figure 7.** Complex numbers with given integer parts in base $-1 + i$. (Thanks are due to John Beatty, who programmed this at Lawrence Livermore Lab.)

integer. In [4] we give such an algorithm in the case of the negative integral bases.

Thirdly, calculate the fractal dimensions of the edges of the regions derived from the representation whose base is the root of a given polynomial. In occasional cases, such as $\rho = \sqrt{-m}$, this dimension will be integral, but it seems that most bases yield fractal curves or surfaces of nonintegral dimension.

## References


[1] Akushskiǐ, I. Ia., Amerbaev, V. M., and Pak, I. T., *Osnovy Mashinnoǐ Arifmetiki Kompleksnykh Chisel.* Nauka, Alma-Ata, Kazakhstan SSR 1970.

[2] Davis, Chandler and Knuth, Donald E., Number representations and dragon curves—I, II. *J. Recreational Math.* **3** (1970), 66–81, 133–149.

[3] Gilbert, William J., The fractal dimension of snowflake spirals, *Notices Amer. Math. Soc.* **25** (1978), A–641.

[4] Gilbert, William J. and Green, R. James, Negative based number systems, *Math. Mag.* **52** (1979), 240–244.

[5] Katai, I. and Szabo, J., Canonical number systems for complex integers, *Acta Sci. Math. (Szeged)* **37** (1975), 255–260.

[6] Knuth, Donald E., *The Art of Computer Programming, Vol. 2, Seminumerical Algorithms.* Addison-Wesley, Reading, Mass. 1969.

[7] Mandelbrot, Benoit B., *Fractals; Form, Chance and Dimension*. Freeman, San Francisco 1977.

[8] Marcus, Daniel A., *Number Fields*. Springer-Verlag, New York 1977.

# Embeddability of Regular Polytopes and Honeycombs in Hypercubes[1]

Patrice Assouad*

## 1. Notation

(1) Let $X$ be a set with at least two points, and $d$ a symmetric nonnegative function on $X \times X$ vanishing on the diagonal. Then:

(a) $d$ is said to be $h$-embeddable if there is an integer $n$ and an application $f$ of $X$ into $\mathbb{Z}^n$ (considered as the set of vertices of the $n$-cubic regular honeycomb in $\mathbb{R}^N$ and equipped with the corresponding graphmetric $d_n$) such that

$$\forall\, x, y \in X, \qquad d(x, y) = d_n(f(x), f(y))$$

(the least value of such integers $n$ is called the $h$-rank of $d$);

(b) $d$ is said to be $L^1$-embeddable if there is a measurable space $(\Omega, \mathcal{Q})$, a nonnegative measure $\mu$ on it and an application $f$ of $X$ into $L^1(\Omega, \mathcal{Q}, \mu)$ such that

$$\forall\, x, y \in X, \qquad d(x, y) = \|f(x) - f(y)\|_{L^1(\Omega, \mathcal{Q}, \mu)}\,.$$

(2) We will consider polytopes or tesselations as metric spaces $(X, d)$ in the following way: $X$ is the set of vertices and $d$ the graphmetric of the graph of vertices and edges.

## 2. Embeddability into $L^1$

We begin with the following lemma (an equivalent form of a result of [2]):

**Lemma 1.** *Let $X$ be a set and $d$ a semimetric on $X$. Then $d$ is $L^1$-embeddable if and only if there is a measurable space $(\Omega, \mathcal{Q})$, a nonnegative measure $\mu$ on it, and an*

[1] Supported by CNRS and Université de Paris XI (Orsay).

* CNRS and Université de Paris XI (Orsay). Present address: Université de Paris Sud, Centre d'Orsay, Mathématiques, bâtiment 425, 91405 Orsay Cedex, France.

*application* $x \to A(x)$ *from* $X$ *into* $\mathcal{Q}$ *such that*

$$\forall\, x, y \in X, \qquad d(x, y) = \mu(A(x) \triangle A(y))$$

*(where* $\triangle$ *denotes the symmetric difference). In the same way, d is h-embeddable if and only if there is a set* $\Omega$ *and an application* $x \to A(x)$ *from* $X$ *into* $2^{\Omega}$ *such that*

$$\forall\, x, y \in X, \qquad d(x, y) = |A(x) \triangle A(y)|.$$

*Proof*. We give the proof in the second case (for the first it is quite the same). We have only to prove the result when $(X,d)$ is the $n$-cubic regular honeycomb $(\mathbb{Z}^n, d_n)$. We take $\Omega = \{1, \ldots, n\} \times \mathbb{Z}$, and for each $x = (x_1, \ldots, x_n)$ belonging to $\mathbb{Z}^r$, we set $A(x) = \{(i, j) \in \Omega \mid x_i \leqslant j\}$. Then we have

$$\forall\, x, y \in \mathbb{Z}^n, \qquad d_n(x, y) = |A(x) \triangle A(y)|. \quad \square$$

We note that if $X$ is finite and $d$ is $h$-embeddable, we can assume that $\Omega$ is finite. Thus for a finite metric space, embeddability in some space $\mathbb{Z}^n$ (considered as the cubic regular honeycomb) is the same as embeddability in some hypercube (by hypercube, we mean its graph; the least value of the dimension of such a hypercube is called the $h$-content of $d$). So it seems natural to study the $h$-embeddability of the regular polytopes and regular honeycombs (considered as graphs and thus as metric spaces). This will be done in the next section.

We recall that an $L^1$-embeddable semimetric $d$ is necessarily $(2n+1)$-polygonal (cf. [5]) for each positive integer $n$, i.e.,

$$\forall\, x_1, \ldots, x_n, y_1, \ldots, y_{n+1} \in X,$$

$$\sum_{i=1}^{n} \sum_{j=1}^{n} d(x_i, x_j) + \sum_{k=1}^{n+1} \sum_{l=1}^{n+1} d(y_k, y_l) \leqslant 2 \sum_{i=1}^{n} \sum_{k=1}^{n+1} d(x_i, y_k).$$

This enables us to get easily some elementary non-$L^1$-embeddable metrics:

**Lemma 2.** *Set* $X_5 = \{1,2,3,4,5\}$ *and* $X_7 = \{1,2,3,4,5,6,7\}$. *We define* $F_1 = \{(1,3), (1,4), (1,5), (2,3), (2,4), (2,5)\}$, $F_2 = F_1 \cup \{(1,2)\}$, $F_3 = F_1 \cup \{(3,4)\}$, $F_4 = F_1 \setminus \{(1,3)\}$, *and* $F' = \{(2,3), (3,4), (3,5), (4,5), (4,6), (5,6), (6,7)\} \cup \{(1,i) \mid i = 2, \ldots, 7\}$. *Let* $d_1$, $d_2$, $d_3$, $d_4$ *be the metrics on* $X_5$ *defined as follows*: $d_i$ *is the infinum of* 2 *and of the pathmetric of the graph* $(X_5, F_i)$ *for each* $i = 1, \ldots, 4$ *(infinum is needed only for* $i = 4$*). Let* $d'$ *be the pathmetric of the graph* $(X_7, F')$, *Then* $d_1$, $d_2$, $d_3$, *and* $d_4$ *are not* 5*-polygonal*; $d'$ *is not* 7*-polygonal. In particular* $d_1$, $d_2$, $d_3$, $d_4$, *and* $d'$ *are not* $L^1$*-embeddable.*

(Note that $d_1$ is an extremal metric on $X_5$).

Finally we mention the following result.

**Lemma 3.** *An h-embeddable semimetric is* $L^1$*-embeddable. Conversely, let* $X$ *be a finite set and* $d$ *an integer-valued* $L^1$*-embeddable semimetric on it; then*

$$\eta(d) \stackrel{\text{def}}{=} \inf\{\lambda \in \mathbb{Q} \setminus \{0\} \mid \lambda d \quad \text{is } h\text{-embeddable}\}$$

*is finite* ($\eta(d)$ *is called the scale of* $d$).

For a systematic survey on $L^1$-embeddability, see [1].

# 3. Graphs of Regular Honeycombs and Regular Convex Polytopes

Each polytope and honeycomb is considered here as the graph of its vertices and edges. In each case $d$ will denote the graphmetric, and the $L^1$-embeddability of $d$ will be studied (giving evaluation of $\eta(d)$).

We examine first the case of polygons. In this case the graph is $C_n$ (a cycle with $n$ edges) for some $n$, and we have:

**Lemma 4** [7]. *For each integer $n \geqslant 3$, $C_n$ is $L^1$-embeddable. (We can add that more precisely $d$ ($2d$) is h-embeddable when n is even (odd).)*

*Proof.*

(a) Suppose first $n$ is even. We fix a vertex $a$ of $C_n$ and take $\Omega$ the set of pairs of antipodal edges of $C_n$. For each vertex $x$ of $C_n$, we take $A(x)$ the set of elements of $\Omega$ having an edge between $a$ and $x$ (i.e. on the shortest path joining them).

(b) Suppose now $n$ is odd. We take $\Omega$ the set of vertices of $C_n$, and for each vertex $x$ of $C_n$, we take $A(x) = \{y \mid d(x,y) \leqslant n/2\}$. □

Let us consider the case of regular polyhedra.

We remark first that there are non-$L^1$-embeddable polyhedra (when considered as graphs): consider for example the cube truncated on the supporting plane of the star of a vertex; then it contains as metric subspace the metric space $(X_5, d_4)$ of Lemma 2 and thus is non-$L^1$-embeddable.

It is shown in [7] that the five platonic solids are $L^1$-embeddable. We leave the tetrahedron, cube, and octahedron to be studied with simplexes, hypercubes, and crosspolytopes. For the two other polyhedra we give the result of [7] in our notations:

**Lemma 5** [7]. *Denote by d the graphmetric of the dodecahedron or of the icosahedron. Then in both cases $4d$ is h-embeddable.*

*Proof.* We take for $\Omega$ the set of vertices, and for each vertex $x$, $A(x) = \{y \mid d(x,y) \leqslant 1\}$ for the icosahedron and $A(x) = \{y \mid d(x,y) \leqslant 2\}$ for the dodecahedron. □

We consider now the three convex regular polytopes $\{3,4,3\}$, $\{3,3,5\}$, and $\{5,3,3\}$. Then we have:

**Lemma 6.** *The polytopes $\{3,4,3\}$ and $\{3,3,5\}$ are non-$L^1$-embeddable.*

*Proof.*

(1) We recall first that the graph of the cross polytope in $\mathbb{R}^4$ is $H_1 = (Y_4, E_1)$ defined as follows: $Y_4 = \{1_+, 2_+, 3_+, 4_+\} \cup \{1_-, 2_-, 3_-, 4_-\}$ is the set of vertices; each pair belongs to $E_1$ except $(1_+, 1_-)$, $(2_+, 2_-)$, $(3_+, 3_-)$, $(4_+, 4_-)$. Thus

using the Cesaro construction as given in [4, p. 148], the graph of $\{3,4,3\}$ can be defined in the following way:

its vertices are the elements of $E_1$;
two vertices (i.e. two edges of $(Y_4, E_1)$) are linked if they belong to the same triangle of $H_1$.

Then the metric subspace $\{(1_+,2_+), (1_+,3_+), (1_+,4_+), (1_+,4_-), (2_+,3_+)\}$ is exactly the metric space $(X_5, d_2)$ of Lemma 2.

(2) We give now the graph of the polytope $\{3,4,3\}$ in its dual presentation $H_2 = (Z, E_2)$. The set of vertices is $Z = Y_4 \cup Z_0$, where $Z_0$ is the algebra of all subsets of $\{1,2,3,4\}$ and $Y_4$ is as above; the elements of $E_2$ (i.e. the edges) are listed as follows:

$(A,B)$ such that $A, B \in Z_0$, $|A \triangle B| = 1$ (i.e. edges of the 4-cube),
$(i_+, A)$ (respectively $(i_-, A)$) such that $i \in A \in Z_0$ (respectively $i \notin A \in Z_0$).

We will give an orientation to each edge in the following way:

if $|A|$ is odd, $(i_+, A)$, $(i_-, A)$, and $(A,B)$ are oriented respectively from $i_+$ to $A$, from $i_-$ to $A$, and from $A$ to $B$;
if $|A|$ is even, orientation is reversed.

We note that the 3-facets of $\{3,4,3\}$ (in this presentation) correspond exactly to the elements of $E_1$ (the vertices in the first presentation); we will denote for example by $f((i_+, j_+)) = \{i_+, j_+\} \cup \{A \mid A \ni i, A \ni j\}$ the 3-facet corresponding to the element $(i_+, j_+)$ of $E_1$.

Thus using Gosset's construction as given in [4, p. 153], the graph of $\{3,3,5\}$ can be defined in the following way: its vertices are the elements of $E_1 \cup E_2$, and its edges can be listed as follows:

(i) $(u,v)$ for $u, v$ elements of $E_2$ included in the same triangle of $H_2$,
(ii) $(u,v)$ for $u, v$ elements of $E_2$ pointing out (for the above orientation) from the same vertex of $H_2$,
(iii) $(u,v)$ for $u \in E_1$, $v \in E_2$, and the edge $v$ (of $H_2$) included in the 3-facet $f(u)$ (of $H_2$).

Then the metric subspace $\{(1_+,2_+), (2_+,\{1,2\}), (1_+,\{1,2\}), (1_+,\{1,2,3\}), (1_+,\{1,2,4\}), (1_+,\{1,2,3,4\}), (2_+,\{1,2,3,4\})\}$ is exactly the metric space $(X_5, d')$ of Lemma 2. (Note that the first vertex is in $E_1$, and the other vertices are in $E_2$ and belong to the same icosahedral cell of $s\{3,4,3\}$ (see below).) □

We note also that deleting the edges (iii) (respectively the edges (ii) and (iii)), one will obtain the polytope

$$s\{3,4,3\} \qquad \left(\text{respectively} \quad \left\{\begin{matrix} 3 \\ 4,3 \end{matrix}\right\}\right)$$

(see [4, pp. 151–152]).

We have not yet studied embeddability of $\{5,3,3\}$.

We examine now the three infinite series of regular convex polytopes:

**Proposition 1.** *Simplexes, hypercubes, and cross polytopes in* $\mathbb{R}^n$ *have an* $L^1$-*embeddable graphmetric for each* $n$. *Precisely, for simplexes and* $n \geqslant 2$ *we have* $\eta(d) = 2$, *for hypercubes (obviously) we have* $\eta(d) = 1$, *and for cross polytopes in* $\mathbb{R}^n$, $\eta(d)$ *has the following evaluation*:

$$\frac{n}{2} \leqslant \eta(d) = 2\inf\{N \mid C(N) \geqslant n\} < n,$$

*where* $C(N)$ *is the largest cardinality of a subset of the* $4N$-*cube in which distinct points are at mutual distance* $2N$ (*cross polytopes are considered in* [7] *only for* $n = 3$).

*Proof.*

(1) Hypercubes are obviously $h$-embeddable. For the simplex in $\mathbb{R}^n$ the graph is $K_{n+1}$ (the vertices are $0, 1, 2, \ldots, n$ and each pair is an edge). Then we take $\Omega = \{0, 1, 2, \ldots, n\}$, $A(x) = \{x\}$ for each vertex $x$. Thus $2d$ is $h$-embeddable. Moreover for $n \geqslant 2$, $d$ is not $h$-embeddable (since triangles have an odd perimeter).

(2) The graph of the cross polytope in $\mathbb{R}^n$ is $(Y_n, E)$ defined as follows:

$$Y_n = \{1_+, 2_+, \ldots, n_+\} \cup \{1_-, 2_-, \ldots, n_-\}.$$

Each pair belongs to $E$ except $(1_+, 1_-)$, $(2_+, 2_-)$, $\ldots$, $(n_+, n_-)$. Since triangles must have an even perimeter, we have only to consider the $h$-embeddability of $2Nd$ ($N$ integer).

By Lemma 1, $2Nd$ is $h$-embeddable if and only if there is a finite set $\Omega$ and subsets $A(x)$ (for each vertex $x \in Y_n$) satisfying $\forall\ x, y \in Y_n$, $d(x, y) = |A(x) \triangle A(y)|$. This condition means exactly (setting $B_i = A(i_+)$ for each $i = 1, \ldots, n$ and $B = A(n_+) \cup A(n_-)$) that there is a set $B$ with $|B| = 4N$ and $n$ subsets $B_1, \ldots, B_n$ of $B$ at mutual distance $2N$. Thus we must have $C(N) \geqslant n$. The converse is obvious. We note that the equality $C(N) = 4N$ occurs exactly when there exists an Hadamard matrix of rank $4N$, in particular when $N$ is a power of 2. This gives the evaluation of $\eta(d)$ (the use of Hadamard matrices comes from [3]). □

Finally we consider the case of regular honeycombs.

**Proposition 2.** *All regular honeycombs except* $\{3,4,3,3\}$ *are* $L^1$-*embeddable. Precisely, if* $d$ *denotes in each case the graph metric, we have*:

(*a*) *for the hexagonal honeycomb in* $\mathbb{R}^2$ *and for the cubic honeycomb in* $\mathbb{R}^n$ (*for each* $n$), $d$ *is* $h$-*embeddable*;

(*b*) *for the triangular honeycomb in* $\mathbb{R}^2$ *and for the honeycomb* $\{3,3,4,3\}$ *in* $\mathbb{R}^4$, $2d$ *is* $h$-*embeddable*.

*Proof.*

(a) Naturally each cubic honeycomb is $h$-embeddable (it is the definition). We remark also that the hexagonal honeycomb (considered as metric space) is exactly $\{x \in \mathbb{Z}^3 \mid x_1 + x_2 + x_3 \in \{0, 1\}\}$ (where $\mathbb{Z}^3$ is the set of vertices of the cubic honeycomb in $\mathbb{R}^3$ with the corresponding metric).

(b) We note that $\{3, 3, 4, 3\}$ (resp. the triangular honeycomb) with the metric $2d$ is exactly the metric subspace of the cubic honeycomb in $\mathbb{R}^4$ (resp. of the hexagonal honeycomb) defined by the condition $x_1 + x_2 + x_3 + x_4$ even (resp. $x_1 + x_2 + x_3$ even).

The cells of the regular honeycomb $\{3, 4, 3, 3\}$ are polytopes $\{3, 4, 3\}$. Thus $\{3, 4, 3, 3\}$ admits $\{3, 4, 3\}$ as metric subspace and therefore is not $L^1$-embeddable (by Lemma 6). □

## 4. Some Results in the Hyperbolic Plane

We recall first a characterization due to D. Ž. Djoković [6] of the $h$-embeddable graphs:

**Theorem** [6]. *A finite connected graph has an h-embeddable graph metric if and only if:*

(1) *it is bipartite;*
(2) *for each edge $(a, b)$, the set $G(a, b)$ of the vertices closer to $a$ than to $b$ is metrically closed (i.e. for each $x, y \in G(a, b)$, each shortest path joining them is in $G(a, b)$).*

Using these criteria, we can prove:

**Proposition 3.** *Let $n, p \in \mathbb{N}$ with $1/2n + 1/p < \frac{1}{2}$. Denote by $d_{n,p}$ the graph metric of the regular honeycomb of the hyperbolic plane $\{2n, p\}$ (symbol of Schläffli). Then $d_{n,p}$ is $L^1$-embeddable. More precisely, $d_{n,p}$ is h-embeddable when restricted to any finite subset.*

*Proof.* To see that $d_{n,p}$ is $L^1$-embeddable, it is sufficient to see that $d_{n,p}$ is $L^1$-embeddable when restricted to any finite subset (for this character of finiteness, see for example [2]). Thus we have only to verify conditions (1) and (2) of the above theorem for the infinite graph $\{2n, p\}$. The verification is easy. □

*Remark.* The same method would work for the cubic and hexagonal honeycombs in $\mathbb{R}^2$ (with symbols $\{4, 4\}$ and $\{6, 3\}$).

Finally we give a last example:

**Lemma 7.** *The hyperbolic regular honeycombs of symbol* $\{\infty, p\}$ *(with* $p \geqslant 2$*) are* $L^1$*-embeddable.*

*Proof.* They are trees, and as remarked in [7], finite trees are $h$-embeddable. □

## References


[1] Assouad, P. and Deza, M., Embeddings of metric spaces in combinatorics and analysis (to appear).

[2] Assouad, P., Plongements isométriques dans $L^1$: aspect analytique, Séminaire d'Initiation à l'Analyse 1979–1980 (Paris 6).

[3] Blake, I. F. and Gilchrist, J. H., Addresses for graphs. *IEEE Transactions on Information Theory* **IT-19** (No. 5, 1973), 683–688.

[4] Coxeter, H. S. M., *Regular Polytopes*, 2^d ed. MacMillan, New York 1963.

[5] Deza (Tylkin), M., On Hamming geometry of unitary cubes (Russian). *Doklady Akad. Nauk SSSR* **134** (No. 5, 1960), 1037–1040.

[6] Djoković, D. Ž., Distance preserving subgraphs of hypercubes. *J. Comb. Th. Ser. B* **14** (1973), 263–267.

[7] Kelly, J. B., Hypermetric spaces. In *The Geometry of Metric and Linear Spaces*. Lecture Notes in Math. 490. Springer 1975.

# The Derivation of Schoenberg's Star-Polytopes from Schoute's Simplex Nets

H. S. M. Coxeter*

## 1. Introduction

On a square billiard table with corners $(\pm 1, \pm 1)$, the path of a ball is easily seen to be periodic if and only if it begins with a line

$$Xx + Yy = N,$$

where $X$ and $Y$ are integers and $|N| < |X| + |Y|$ [16, p. 82]. Ignoring a trivial case, we shall assume $XY \neq 0$. We lose no generality by taking these integers to be positive and relatively prime. After any number of bounces, the path is still of the form

$$\pm Xx \pm Yy = N \pm 2k,$$

where $k$ is an integer. Among these paths for various values of $k$, those that come closest to the origin are of the form

$$\pm Xx \pm Yy = N',$$

where $0 \leqslant N' \leqslant 1$. The distance of such a path from the origin is

$$N'/\sqrt{X^2 + Y^2}\,.$$

Schoenberg [18, p. 8] was looking for the values of $X$, $Y$, $N$ which will maximize this distance. For this purpose we must have

$$N' = X = Y = 1,$$

so that $N$ is an odd integer. Since $|N| < |X| + |Y| = 2$, this implies $N = \pm 1$. The paths

$$\pm x \pm y = 1$$

form a square whose vertices are the midpoints of the edges of the billiard table.

*Department of Mathematics, University of Toronto, Toronto (M5S 1A1) Ontario, Canada.

Analogously, in a kaleidoscope whose mirrors are the bounding hyperplanes

$$x_\nu = \pm 1 \qquad (\nu = 1, 2, \ldots, n)$$

of an $n$-cube $\gamma_n$, consider an $(n-1)$-dimensional pencil of light rays in the hyperplane

$$\sum X_\nu x_\nu = N,$$

where the $X_\nu$ are positive integers with no common divisor greater than 1, and $|N| < \sum X_\nu$. The mirror $x_\mu = 1$ will reflect this hyperplane so as to yield

$$X_\mu(2 - x_\mu) + \sum_{\nu \neq \mu} X_\nu x_\nu = N,$$

and any number of such reflections will produce

$$\sum \pm X_\nu x_\nu = N \pm 2k,$$

where $k$ is an integer. Among these hyperplanes for various values of $k$, those nearest to the origin are of the form

$$\sum \pm X_\nu x_\nu = N',$$

where $0 \leqslant N' \leqslant 1$. The distance of such a hyperplane from the origin, namely

$$N'/\sqrt{\sum X_\nu^2}\,,$$

attains its greatest possible value when

$$N' = X_1 = X_2 = \cdots = X_n = 1,$$

so that $N$ is an odd integer. Since each reflection reverses the sign of one coordinate and changes by one unit the $k$ in the equation

$$\sum \pm x_\nu = N \pm 2k,$$

the number of minus signs on the left has the same parity as $k$. Thus, if we begin with

$$\sum x_\nu = 1,$$

all the hyperplanes are given by

$$\sum \epsilon_\nu x_\nu = (1 \pm 4m) \prod \epsilon_\nu,$$

where $\epsilon_\nu = \pm 1$ and $m = 0, 1, \ldots,$ the possible values of $m$ being limited by the requirement that

$$|1 \pm 4m| \leqslant n.$$

Since such a hyperplane is unchanged when we reverse the signs on both sides of the equation, the list can be simplified to

$$\sum \epsilon_\nu x_\nu = 1, 3, 5, \ldots, n-1 \qquad (\epsilon_\nu = \pm 1)$$

when $n$ is even, and to

$$\sum \epsilon_\nu x_\nu = 1, -3, 5, -7, \ldots, \pm n, \qquad \prod \epsilon_\nu = 1$$

when $n$ is odd.

The figure formed by all these hyperplanes is simply a square when $n = 2$ and a tetrahedron when $n = 3$ [16, p. 87]. When $n > 3$, the facets intersect one another internally, like the sides of a pentagram, so we shall call the figure *Schoenberg's star-polytope* [19, p. 55]. His symbol for it is $\tilde{\Pi}_n^{n-1}$.

## 2. Schoute's Simplex Nets

A natural variant of the ordinary toy kaleidoscope consists of three rectangular mirrors joined together so as to form the side faces of a tall prism based on an equilateral triangle. Any object placed inside has a theoretically unlimited number of images. In terms of Cartesian coordinates $x_1, x_2, x_3$, we may specify this prism by the inequalities

$$x_1 \geqslant x_2, \qquad x_2 \geqslant x_3, \qquad x_3 \geqslant x_1 - 1,$$

so that the images of the point $(0, 0, 0)$ on the edge $x_1 = x_2 = x_3$ are all the points whose coordinates are integers having sum zero; and the images of the point $(\frac{1}{2}, 0, -\frac{1}{2})$ on the third mirror, midway between the first and second, are all the points whose coordinates, with sum zero, consist of one integer and two halves of odd integers. Since all the mirrors are perpendicular to the plane $x_1 + x_2 + x_3 = 0$, we may work entirely in this plane and regard the cross-section of the prism as a 2-dimensional kaleidoscope.

The lattice points that lie in this plane are the vertices of the regular tessellation $\{3, 6\}$ of equilateral triangles, and the other points just mentioned are the vertices of the quasiregular tessellation $\left\{ {3 \atop 6} \right\}$ of triangles and hexagons [9, p. 60].

This familiar kaleidoscope is the case $n = 3$ of the $(n - 1)$-dimensional kaleidoscope formed by the simplex

$$x_1 \geqslant x_2, \qquad x_2 \geqslant x_3, \ldots, \qquad x_{n-1} \geqslant x_n, \qquad x_n \geqslant x_1 - 1$$

in the hyperplane $x_1 + \cdots + x_n = 0$ of Cartesian $n$-space [2, p. 162]. The $n$ mirrors are conveniently represented by the vertices of an $n$-gon. The sides of the $n$-gon indicate that adjacent mirrors are inclined at $\pi/3$, so that the reflections in them, say R and S, satisfy RSR = SRS, and it is understood that nonadjacent mirrors (represented by vertices not directly joined) are at right angles, so that the reflections in them commute: RT = TR. These $n$ reflections generate an infinite discrete group for which the simplex serves as a fundamental region. The $n + \binom{n}{2}$ relations

$$\mathrm{R}_\nu^2 = 1, \quad \mathrm{R}_1\mathrm{R}_2\mathrm{R}_1 = \mathrm{R}_2\mathrm{R}_1\mathrm{R}_2, \ldots, \quad \mathrm{R}_n\mathrm{R}_1\mathrm{R}_n = \mathrm{R}_1\mathrm{R}_n\mathrm{R}_1, \quad \mathrm{R}_1\mathrm{R}_3 = \mathrm{R}_3\mathrm{R}_1, \ldots$$

provide a presentation [2, p. 145; 9, p. 188].

When $n = 3$, this is the group **p3m1** with the presentation

$$\mathrm{R}_\nu^2 = 1, \quad \mathrm{R}_1\mathrm{R}_2\mathrm{R}_1 = \mathrm{R}_2\mathrm{R}_1\mathrm{R}_2, \quad \mathrm{R}_2\mathrm{R}_3\mathrm{R}_2 = \mathrm{R}_3\mathrm{R}_2\mathrm{R}_3, \quad \mathrm{R}_3\mathrm{R}_1\mathrm{R}_3 = \mathrm{R}_1\mathrm{R}_3\mathrm{R}_1$$

[17, p. 444].

By placing a ring round one vertex of the $n$-gon we obtain the "graphical symbol" for the $(n - 1)$-dimensional honeycomb whose vertices are the images of one vertex of the simplex, namely the vertex opposite to the indicated mirror.

More generally, by ringing two or more vertices of the $n$-gon we symbolize the honeycomb whose vertices are the images of a point which is equidistant from the indicated mirrors while lying on the intersection of the remaining mirrors.

When $n = 3$, this notation yields

$$\triangle = \{3, 6\}, \quad \triangle = \left\{\begin{matrix}3\\6\end{matrix}\right\}, \quad \triangle = \{6, 3\}.$$

The case $n = 4$ has been described elsewhere [5, pp. 402, 403; 14].

Returning to $n$ dimensions, where the $n$ mirrors are

$$x_1 = x_2, \quad x_2 = x_3, \ldots, \quad x_{n-1} = x_n, \quad x_n = x_1 - 1,$$

the $n$-gon with one ring symbolizes the honeycomb $\alpha_{n-1}h$ whose vertices are the images of the origin, namely the lattice points with $x_1 + \cdots + x_n = 0$ [1, p. 366]. Its cells, symbolized by the same graph minus each of its $n - 1$ "unringed" vertices in turn, are the regular simplex $\alpha_{n-1}$ and each of its principal truncations $t_\mu\alpha_{n-1}$ [10, pp. 18, 164; 11, p. 127]. Such a polytope $t_\mu\alpha_{n-1}$ has $\binom{n}{\mu+1}$ vertices, one for each $\alpha_\mu$ of $\alpha_{n-1}$ ($\mu = 0, 1, \ldots, n-2$), with the natural convention that $t_\mu\alpha_{n-1}$ is simply $\alpha_{n-1}$ itself when $\mu = 0$ or $n - 2$. Each vertex of the honeycomb $\alpha_{n-1}h$ belongs to $\binom{n}{\mu+1}$ such cells [1, p. 367]. Thus the honeycomb contains (in a natural sense), for each vertex, one $t_\mu\alpha_{n-1}$ (and of course also one $t_{n-2-\mu}\alpha_{n-1}$, which is just like it). In particular, $\alpha_2h$ is $\{3,6\}$, as we have seen, and $\alpha_3h$ is the familiar honeycomb of tetrahedra and octahedra whose vertices form the "face-centered cubic lattice."

When $r$ vertices of the $n$-gon are adorned with rings, suppose these are the $a$th, $b$th, ..., $k$th, and $n$th (in their natural order round the $n$-gon, beginning just after one of the rings, so that $a < b < \cdots < k < n$). To avoid fractions, let us apply a dilatation so that the $n$th mirror is not $x_n = x_1 - 1$ but $x_n = x_1 - r$, and the sum of the $n$ coordinates is not zero but $a + b + \cdots + k$. Then the chosen point, whose images we seek, being equidistant from the hyperplanes

$$x_a = x_{a+1}, \quad x_b = x_{b+1}, \ldots, \quad x_n = x_1 - r$$

and lying on the remaining mirrors, has coordinates consisting of

$$a\ (r-1)\text{'s}, \quad b - a\ (r-2)\text{'s}, \quad c - b\ (r-3)\text{'s}, \ldots, \quad n - k\ 0\text{'s}. \tag{2.1}$$

Applying the reflections, we see that the vertices of the honeycomb have for coordinates all the sets of $n$ integers that satisfy the equation

$$x_1 + x_2 + \cdots + x_n = a + b + \cdots + k,$$

while their residues modulo $r$ are some permutation of the numbers (2.1) [20, p. 43].

For instance, when there are just two rings, on adjacent vertices of the $n$-gon (say the first and $n$th), the coordinates are all the sets of $n$ integers, one odd and $n - 1$ even, with sum 1. For lack of a better name, let us call this honeycomb $\alpha_{n-1}h^2$. Its cells are $\alpha_{n-1}$ and each of its *intermediate* truncations $t_{\mu-1,\mu}\alpha_{n-1}$ [3, p. 20; 7, p. 70]. Such a polytope $t_{\mu-1,\mu}\alpha_{n-1}$ has $n\binom{n-1}{\mu}$ vertices ($\mu = 0, 1, \ldots, n-1$) [8, pp. 43–44] if we make the conventions

$$t_{-1,0}\alpha_{n-1} = \alpha_{n-1} = t_{n-2,n-1}\alpha_{n-1}.$$

Since each vertex of $\alpha_{n-1}h^2$ belongs to $\binom{n-1}{\mu}$ such cells, the honeycomb contains, for every $n$ vertices, one $t_{\mu-1,\mu}\alpha_{n-1}$ (and of course also one $t_{n-2-\mu,n-1-\mu}\alpha_{n-1}$, which is just like it). In particular, $\alpha_3h^2$ is the honeycomb of tetrahedra and truncated tetrahedra which Coxeter and Wells [14, p. 469] call $q\delta_4$.

When *every* vertex of the $n$-gon is ringed, we have the Hinton–Schoute honeycomb [8, pp. 48, 73] for which the coordinates are all sets of $n$ integers, mutually incongruent modulo $n$, with a constant sum. Analogy suggests that this honeycomb might be denoted by $\alpha_{n-1}h^n$, so that (for instance)

$$\alpha_2h = \{3,6\}, \qquad \alpha_2h^2 = \left\{\begin{smallmatrix}3\\6\end{smallmatrix}\right\}, \qquad \alpha_2h^3 = \{6,3\}.$$

The cells of $\alpha_{n-1}h^n$ are all alike, namely $t_{0,1,\ldots,n-2}\alpha_{n-1}$, and each vertex belongs to $n$ of them. Thus $\alpha_3h^4$ is the well-known honeycomb $t_{1,2}\delta_4$ of truncated octahedra [5, p. 402], and $\alpha_4h^5$ is the principal topic of a fascinating old book by Hinton [15].

## 3. Finite Complexes

Each of these honeycombs yields a finite topological complex when we restrict the coordinates to their residues modulo $2r$ and thus identify all pairs of points related by translations that add $2r$ to one coordinate while subtracting $2r$ from another.

In this manner, $\alpha_2h$, $\alpha_2h^2$, $\alpha_2h^3$ yield (at first) three maps on a torus:

$$\{3,6\}_{2,0}, \quad \left\{\begin{matrix}3\\6\end{matrix}\right\}_{2,0}, \quad \{6,3\}_{2,2},$$

in the notation of Coxeter and Moser [13, p. 107]. However, in the first case, four of the eight triangles coincide with the remaining four, so that the torus collapses to a tetrahedron. (See Figure 1, where $(-1,2,-1)$, for instance, has been abbreviated to $\bar{1}2\bar{1}$.)

The second map, which has 8 triangles and 4 hexagons, was denoted earlier by $\{3/6\}_4$ [4, p. 132, Figure 13]; we shall return to it in Section 5.

The third map, with 12 hexagons [8, p. 131, Figure 20] is the Levi graph for the projective configuration $12_3$.

## 4. Analogs of the Tetrahedron of König and Szücs

In general, when we restrict the $n$ coordinates to their residues modulo 2, $\alpha_{n-1}h$ yields a complex having $2^{n-1}$ vertices. Their coordinates consist of $2\mu$ ones and $n-2\mu$ zeros for $\mu = 0, 1, \ldots, [n/2]$. In $\alpha_{n-1}h$ itself, the *vertex figure*, formed by

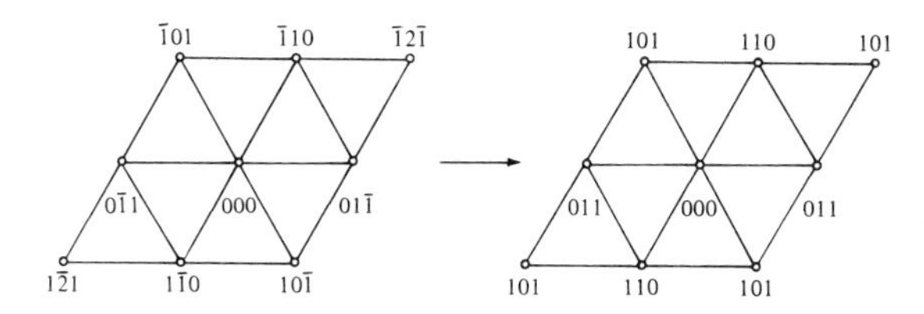


**Figure 1.** How $\alpha_2h$ (mod 2) yields a tetrahedron.

the farther ends of all the edges at one vertex, namely the $n(n-1)$ permutations of

$$(1, 0^{n-2}, -1),$$

is the "expanded simplex" $e\alpha_{n-1}$:

[1, p. 366; 8, p. 50], whose cells or *facets* consist of $\binom{n}{\nu}$ "prisms" $\alpha_{\nu-1} \times \alpha_{n-\nu-1}$ for $\nu = 1, 2, \ldots, n-1$ (with the natural convention $\alpha_0 \times \alpha_{n-2} = \alpha_{n-2}$). This "prism," or *Cartesian product* $\alpha_{\nu-1} \times \alpha_{n-\nu-1}$, is the vertex figure of the cell $t_{\nu-1}\alpha_{n-1}$ of $\alpha_{n-1}h$ [1, p. 359, where square brackets were used instead of the multiplication sign]. When we restrict the coordinates to their residues modulo 2, the $n(n-1)$ vertices of $e\alpha_{n-1}$ become fused in pairs of opposites, as when we pass from spherical space to elliptic space [6; 8, pp. 114–115]. This modified vertex figure, whose vertices are the $\binom{n}{2}$ permutations of

$$(1^2, 0^{n-2}),$$

may accordingly be denoted by $e\alpha_{n-1}/2$. For the moment, it is a topological complex of prisms $\alpha_{\nu-1} \times \alpha_{n-\nu-1}$; but we may use the same symbol $e\alpha_{n-1}/2$ to describe the nonconvex $(n-1)$-dimensional polytope which arises when we interpret the coordinates 1 and 0 as ordinary integers instead of the classes "odd" and "even." (When $n > 3$, this is a *star*-polytope, somewhat analogous to Poinsot's *great dodecahedron* $\{5, \frac{5}{2}\}$ [10, p. 12] which has the same vertices and edges as the icosahedron $\{3, 5\}$.) Its typical $(n-2)$-dimensional facet $\alpha_{\nu-1} \times \alpha_{n-\nu-1}$ has the $\nu(n-\nu)$ vertices

$$(1, 0^{\nu-1}; 1, 0^{n-\nu-1}),$$

where the semicolon separates the sets of coordinates that are permuted among themselves. Thus there are $\binom{n}{\nu}$ such facets for $\nu = 1, 2, \ldots, [(n-1)/2]$, and $\frac{1}{2}\binom{n}{n/2}$ "central" facets if $n$ is even. The facet $\alpha_{\nu-1} \times \alpha_{n-\nu-1}$ has its own typical $(n-3)$-dimensional facet $\alpha_{\nu-2} \times \alpha_{n-\nu-1}$, whose vertices are

$$(1, 0^{\nu-2}; 0; 1, 0^{n-\nu-1});$$

this is its interface with another facet of $e\alpha_{n-1}/2$, namely the $\alpha_{\nu-2} \times \alpha_{n-\nu}$ whose vertices are

$$(1, 0^{\nu-2}; 1, 0^{n-\nu}).$$

We have already remarked that the $(n-1)$-dimensional honeycomb $\alpha_{n-1}h$, whose vertices have $n$ integral coordinates with sum zero, yields a finite complex of $t_{\nu-1}\alpha_{n-1}$'s ($\nu = 1, 2, \ldots$) when we restrict these coordinates to their residues modulo 2. This complex, like its vertex figure $e\alpha_{n-1}/2$, is isomorphic to a nonconvex polytope in which the odd and even coordinates are replaced by ordinary ones and zeros. Since the coordinates still have an even sum, they are precisely

$$(1^{2\mu}, 0^{n-2\mu}) \qquad (\mu = 0, 1, \ldots, [n/2]).$$

Accordingly, this $n$-dimensional star-polytope has the same $2^{n-1}$ vertices as the *half-measure polytope* h$\gamma_n$ [1, p. 363; 9, p. 155], whose facets consist of $2^{n-1}$ $\alpha_{n-1}$'s and $2n$ h$\gamma_{n-1}$'s. Since the star-polytope has $e\alpha_{n-1}/2$ for its vertex figure, let us call it

$$[e\alpha_{n-1}/2]^{+1}$$

[1, p. 368]. Its facets at one vertex have for their vertex figures the facets of $e\alpha_{n-1}/2$, which are, as we have seen, $\binom{n}{\nu}$ $\alpha_{\nu-1} \times \alpha_{n-\nu-1}$'s for $\nu = 1, 2, \ldots, [(n-1)/2]$ and $\frac{1}{2}\binom{n}{\nu}$ $\alpha_{\nu-1} \times \alpha_{\nu-1}$'s when $n$ is even and $\nu = n/2$. Since $t_{\nu-1}\alpha_{n-1}$ (whose vertex figure is $\alpha_{\nu-1} \times \alpha_{n-\nu-1}$) has $\binom{n}{\nu}$ vertices, while $[e\alpha_{n-1}/2]^{+1}$ has $2^{n-1}$, the facets of this $n$-dimensional star-polytope consist of $2^{n-1}$ $t_{\nu-1}\alpha_{n-1}$'s for $\nu = 1, 2, \ldots, [(n-1)/2]$ and, if $n$ is even (and $\nu = n/2$), $2^{n-2}$ $t_{\nu-1}\alpha_{n-1}$'s lying in hyperplanes through the center. In other words, the facets of $[e\alpha_{n-1}/2]^{+1}$ consist of $2^{n-2}$ $t_{\nu-1}\alpha_{n-1}$'s for $\nu = 1, 2, \ldots, n-1$.

Since the coordinates of the vertices are any even number of ones, and the rest zeros, the vertices of a facet $t_{\nu-1}\alpha_{n-1}$ are given by the permutations of

$$(1^{\nu}, 0^{n-\nu}) \qquad (\nu = 1, 2, \ldots, n-1)$$

with a certain number of the coordinates subtracted from 1. This "certain number" is arbitrary except that it must have the same parity as $\nu$ (to ensure that the sum of all the coordinates remains even). However, this symmetrical description counts every facet twice. For instance, the permutations of $(1^{\nu}, 0^{n-\nu})$ with the *first* $\mu$ coordinates subtracted from 1 are the same as the permutations of $(1^{n-\nu}, 0^{\nu})$ with the *last* $n-\mu$ coordinates subtracted from 1. If $n$ is odd, we can avoid this duplication by taking $\nu = 1, 2, \ldots, [(n-1)/2]$.

Trivially, when $n = 2$ or 3, $[e\alpha_{n-1}/2]^{+1}$ is the *same* as h$\gamma_n$, namely a line segment when $n = 2$, and a tetrahedron when $n = 3$ [9, p. 156]. But $e\alpha_3$ is the cuboctahedron [1, p. 368] and $e\alpha_3/2$ is the *tetratrihedron*, denoted elsewhere by $\frac{3}{2}3\,|\,2$ or $\mathrm{r}'\{\frac{3}{3}\}$ [12, pp. 415, 435, 440 (Figure 36)], whose $4 + 3$ faces consist of alternate faces of the regular octahedron along with the three "equatorial" squares. Therefore $[e\alpha_3/2]^{+1}$ is the analogous 4-dimensional star-polytope whose $8 + 4$ facets consist of alternate facets of the regular "cross polytope" h$\gamma_4 = \beta_4$ along with the four equatorial octahedra.

In 4 dimensions, the convex polytope $t_1\alpha_4$ and the star-polytope $e\alpha_4/2$ have the same ten vertices

$$12, 13, \ldots, 45,$$

meaning $(1,1,0,0,0)$, $(1,0,1,0,0)$, $\ldots$, $(0,0,0,1,1)$. Figure 2 represents both these polytopes, as they have not only the same vertices but also the same edges. The facets of $t_1\alpha_4$ consist of 5 tetrahedra such as 12 13 14 15, and 5 octahedra such as 23 24 25 34 35 45. The facets of $e\alpha_4/2$ are the same 5 tetrahedra along with 10 triangular prisms. One of these prisms joins the triangle 14 24 34 to 15 25 35; another joins 13 23 34 to 15 25 45.

This 4-dimensional star-polytope $e\alpha_4/2$ is the vertex figure of the 5-dimensional star-polytope $[e\alpha_4/2]^{+1}$ whose $1 + 10 + 5 = 16$ vertices have for coordinates the permutations of

$$(0,0,0,0,0),\ (1,1,0,0,0),\ (1,1,1,1,0).$$

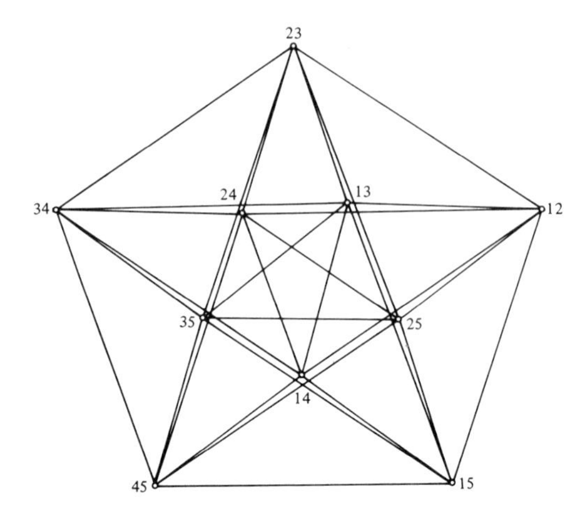


**Figure 2.** $t_1\alpha_4$ or $e\alpha_4/2$.

Since 5 is an odd number, we can describe the facets as 16 $\alpha_4$'s and 16 $t_1\alpha_4$'s. A typical $\alpha_4$ is

$$(1,0,0,0,0)$$

permuted, with an *odd* number of the coordinates subtracted from 1; for instance, if these are the last three coordinates we have

$$(1,0,1,1,1)\ (0,1,1,1,1)\ (0,0,0,1,1)\ (0,0,1,0,1)\ (0,0,1,1,0).$$

A typical $t_1\alpha_4$ is

$$(1,1,0,0,0)$$

permuted, with an *even* number of the coordinates subtracted from 1.

We have found, for the $2^{n-1}$ vertices of $[e\alpha_{n-1}/2]^{+1}$, coordinates consisting of all the permutations of $2\mu$ ones and $n-2\mu$ zeros, for $\mu = 0, 1, \ldots, [n/2]$. The center of the polytope is $(\frac{1}{2}, \frac{1}{2}, \ldots, \frac{1}{2})$. Sometimes it is more convenient to have the center at the origin. Accordingly, let us replace each coordinate $x$ by $1-2x$ (so that 0 becomes 1 while 1 becomes $-1$). The coordinates for the $2^{n-1}$ vertices may now be described more simply as

$$(\pm 1, \pm 1, \ldots, \pm 1)$$

with any *even* number of minus signs [9, p. 158].

When we use this coordinate system, the vertices of a typical facet $t_{\nu-1}\alpha_{n-1}$ are given by the permutations of $\nu$ minus-ones and $n-\nu$ ones, with a certain number of the coordinates reversed in sign, this number having the same parity as $\nu$. Thus the facet lies in the hyperplane

$$\sum \epsilon_\mu x_\mu = n - 2\nu,$$

where $\epsilon_\mu = \pm 1$ and $\prod \epsilon_\mu = (-1)^\nu$. If $n$ is even, the "central" facets $t_{(n/2)-1}\alpha_{n-1}$ lie in hyperplanes

$$\sum \epsilon_\mu x_\mu = 0, \qquad \prod \epsilon_\mu = (-1)^{n/2}$$

passing through the center $(0, 0, \ldots, 0)$. But if $n$ is odd, the "most nearly central" facets lie in hyperplanes

$$\sum \epsilon_\mu x_\mu = 1, \qquad \prod \epsilon_\mu = (-1)^{(n-1)/2}$$

which are tangent to the $(n-1)$-sphere

$$\sum x_\mu^2 = 1/n.$$

In other words, *if $n$ is odd, all the hyperplanes that contain facets of $[e\alpha_{n-1}/2]^{+1}$ are outside the open ball of radius $n^{-1/2}$.*

Looking back at Section 1, we thus see that, when $n$ is odd, Schoenberg's star-polytope $\tilde{\Pi}_n^{n-1}$ is $[e\alpha_{n-1}/2]^{+1}$.

## 5. The Octatetrahedron and Its Analogs

Turning now to $\alpha_{n-1}h^2$, symbolized by an $n$-gon with two consecutive vertices ringed, we have seen that the vertices have coordinates consisting of $n$ integers, one odd and $n-1$ even, with sum 1. A typical cell $t_{\nu-1,\nu}\alpha_{n-1}$ has the $n\binom{n-1}{\nu}$ vertices

$$(2^\nu, 1, 0^{n-\nu-1})$$

translated by a vector whose components are $n$ even integers with sum $-2\nu$ (to restore the sum of all the coordinates to 1). In other words, the typical cell is determined by the basic equation $x_1 + \cdots + x_n = 1$ along with the inequalities

$$2a_\mu \leqslant x_\mu \leqslant 2(a_\mu + 1) \qquad (\mu = 1, \ldots, n),$$

where $a_1, \ldots, a_n$ are fixed integers. In fact, the whole honeycomb is cut out from the $(n-1)$-space $x_1 + \cdots + x_n = 1$ by the $n$ families of parallel hyperplanes

$$x_\mu = 2a_\mu \qquad (\mu = 1, \ldots, n).$$

Reducing the $n$ coordinates to their residues modulo 4, we obtain the "net" for a finite $n$-dimensional star-polytope which we shall denote by

$$\alpha_{n-1}h^2 \text{ (mod 4)}.$$

The process of folding the net out of the hyperplane $\sum x_\nu = 1$ may be described in terms of a ray beginning inside the cell $(1, 0^{n-1})$ (permuted). Whenever the ray penetrates one of the hyperplanes $x_\mu = 2a_\mu$, the cell that it enters is "reflected" so that, for each vertex of that cell, the sign of $x_\mu$ is reversed. (In Figure 3 we have

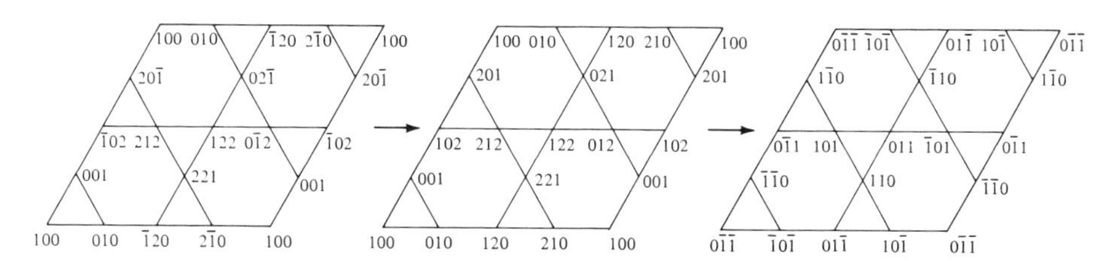


**Figure 3.** Folding up $\alpha_2 h^2$ (mod 4) from net to solid.

$n = 3$.) Accordingly, the

$$n\sum\binom{n-1}{\nu} = 2^{n-1}n$$

vertices, whose coordinates consist of $\nu$ twos, one $(-1)^\nu$, and $n - \nu - 1$ zeros (mod 4), are transformed into the permutations of

$$(2^\nu, 1, 0^{n-\nu-1}) \qquad (\nu = 0, 1, \ldots, n-1). \tag{5.1}$$

To put these coordinates into a more agreeable form, we may subtract 1 from each so as to obtain the permutations of

$$(0, \pm 1, \ldots, \pm 1). \tag{5.2}$$

Since these $2^{n-1}n$ points are the midpoints of the edges of the $n$-cube $\gamma_n$ [1, p. 360], the star-polytope $\alpha_{n-1}h^2$ (mod 4) has the same vertices (and edges) as the convex polytope $t_1\gamma_n$. But it has different facets. In fact, since the infinite honeycomb $\alpha_{n-1}h^2$ has one cell $t_{\nu-1,\nu}\alpha_{n-1}$ for every $n$ vertices ($\nu = 0, 1, \ldots, n-1$), while the star-polytope has $2^{n-1}n$ vertices, the latter has $2^{n-1}$ facets $t_{\nu-1,\nu}\alpha_{n-1}$ for each value of $\nu$.

For instance, when $n = 3$ we have the *cuboctahedron* $t_1\gamma_3$, whose 12 vertices are the midpoints of the edges of the cube $\gamma_3$. The star-polyhedron $\alpha_2h^2$ (mod 4) has the same 12 vertices, the same 24 edges, and the same 8 triangular faces; but the cuboctahedron's 6 square faces are replaced by its 4 "equatorial" hexagons, as in Figure 4. Thus $\alpha_2h^2$ (mod 4) is the *octatetrahedron*

$$\tfrac{3}{2}3\,|\,3$$

[16, p. 88 (Figures 7 and 8); 12, pp. 417, 435, 440 (Figure 37)].

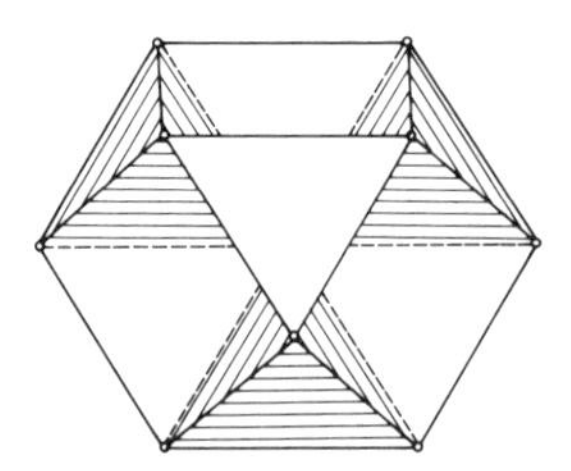

**Figure 4.** The octatetrahedron $\alpha_2h^2$ (mod 4).

As we saw in Section 2, the honeycomb $\alpha_3h^2$ is not only the second member of the sequence $\alpha_nh^2$ ($n = 2, 3, \ldots$)

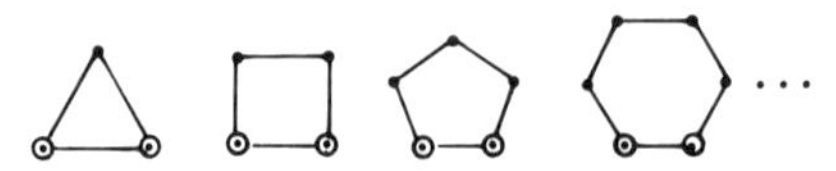

but also the first member of the otherwise distinct sequence $q\delta_n$ ($n = 4, 5, \ldots$)

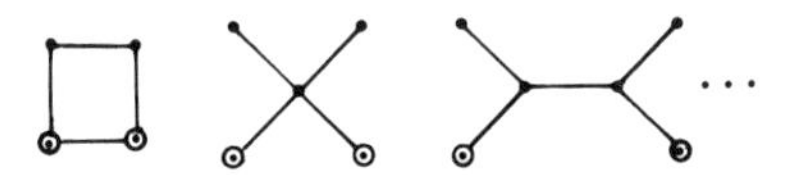

In a single formula,

$$\alpha_3 h^2 = q\delta_4.$$

However, this does *not* imply that we should use $q\delta_4$ (mod 4) as an alternative name for $\alpha_3 h^2$ (mod 4), because, although the vertices of $q\delta_4$ may be described as the permutations of

$$(0,0,0),\ (0,1,1),\ (0,2,2),\ (0,3,3),\ (1,2,3) \pmod 4$$

[14, p. 468], there is no longer a direct transition, like what happened in Figure 3, to the 4-dimensional coordinates $(0, \pm 1, \pm 1, \pm 1)$.

As a consequence of the transition from (5.1) to (5.2), the vertices of a typical facet $t_{\nu-1,\nu}\alpha_{n-1}$ of $\alpha_{n-1}h^2$ (mod 4) are now given by the permutations of $\nu$ ones, 1 zero, and $n - \nu - 1$ minus-ones. The $2^n$ possible reversals of sign (of any number of the coordinates) yield the $2^n$ such facets

$$t_{\nu-1,\nu}\alpha_{n-1} = t_{n-\nu-2,n-\nu-1}\alpha_{n-1}.$$

Since the first facet lies in the hyperplane $\sum x_\mu = 2\nu + 1 - n$, the general facet of this kind lies in

$$\sum \epsilon_\mu x_\mu = n - 2\nu - 1, \qquad \epsilon_\mu = \pm 1.$$

(See Figures 5, 6, 7, 8 for the case $n = 4$.)

If $n$ is odd, the central facets, given by $\nu = (n-1)/2$, lie in hyerplanes $\sum \epsilon_\mu x_\mu = 0$ passing through the center $(0, 0, \ldots, 0)$. But if $n$ is even, the "most nearly central" facets lie in the hyperplanes

$$\sum \epsilon_\mu x_\mu = 1,$$

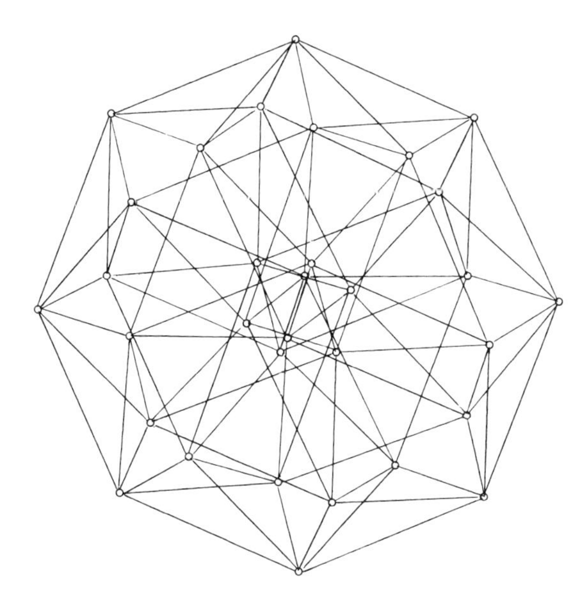

**Figure 5.** The 4-dimensional star-polytope $\alpha_3 h^2$ (mod 4), or the convex polytope $t_1\gamma_4$.

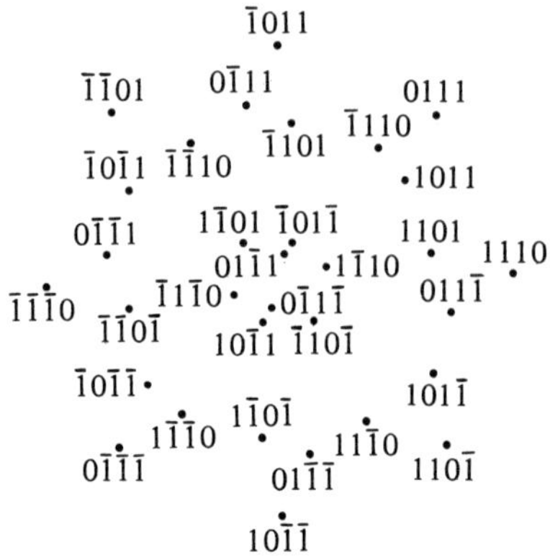

$\bar{1}011$

$\bar{1}\bar{1}01$ $0\bar{1}11$ $0111$

$\bar{1}101$ $\bar{1}110$

$\bar{1}0\bar{1}1$ $\bar{1}\bar{1}10$ $1011$

$0\bar{1}\bar{1}1$ $1\bar{1}01$ $\bar{1}01\bar{1}$ $1101$

$01\bar{1}1$ $1\bar{1}10$ $1110$

$\bar{1}1\bar{1}0$ $\bar{1}1\bar{1}0$ $0\bar{1}1\bar{1}$ $011\bar{1}$

$\bar{1}\bar{1}0\bar{1}$ $10\bar{1}1$ $\bar{1}10\bar{1}$

$\bar{1}0\bar{1}\bar{1}$ $101\bar{1}$

$1\bar{1}\bar{1}0$ $1\bar{1}0\bar{1}$ $11\bar{1}0$

$0\bar{1}\bar{1}\bar{1}$ $01\bar{1}\bar{1}$ $110\bar{1}$

$10\bar{1}\bar{1}$

**Figure 6.** The 32 vertices of $t_1\gamma_4$ or of $\alpha_3 h^2$ (mod 4).

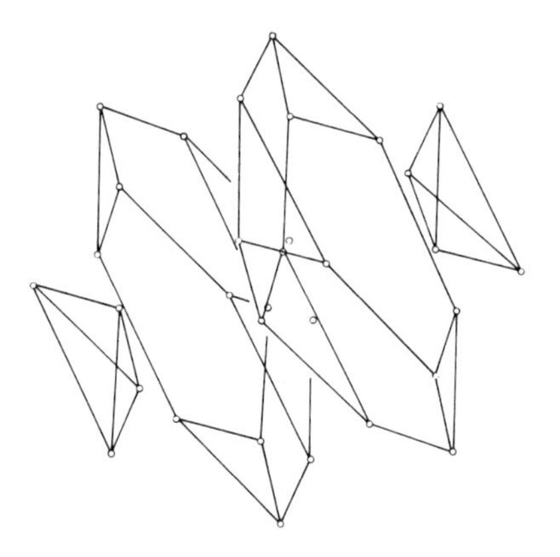

**Figure 7.** The tetrahedra $\sum x = \pm 3$ and the truncated tetrahedra $\sum x = \pm 1$, facets of $\alpha_3 h^2$ (mod 4).

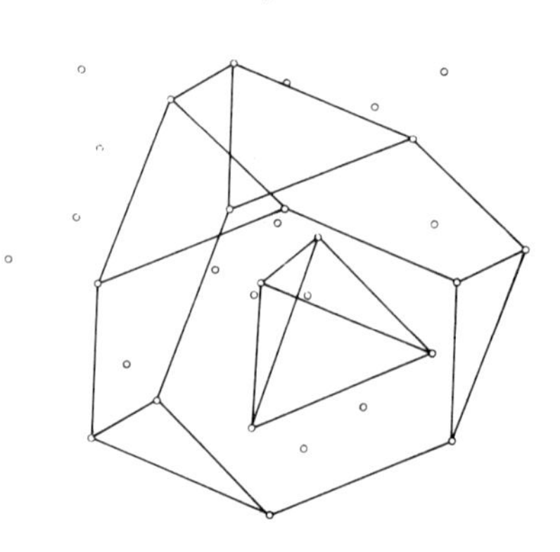

**Figure 8.** The tetrahedron $x_1 - x_2 + x_3 - x_4 = 3$ "inside" the truncated tetrahedron $x_1 - x_2 + x_3 - x_4 = 1$.

which are tangent to the $(n-1)$-sphere

$$\sum x_\mu^2 = 1/n.$$

In other words, *if n is even, all the hyperplanes that contain facets of* $\alpha_{n-1}h^2$ (mod 4) *are outside the open ball of radius* $n^{-1/2}$.

Looking back at Section 1, we thus see that, when $n$ is even, Schoenberg's star-polytope $\tilde{\Pi}_n^{n-1}$ is $\alpha_{n-1}h^2$ (mod 4).

## 6. A Twisted Antiprism

To see more clearly how the facets of $\alpha_{n-1}h^2$ (mod 4) are arranged, it is possibly helpful to examine its vertex figure. It has long been known [1, p. 352; 3, p. 21] that the vertex figure of $t_{\mu-1,\mu}\alpha_{n-1}$ is the generalized pyramid

$$\left(\alpha_{\mu-1} \underset{\sqrt{3}}{\text{———}} \alpha_{n-\mu-2}\right)$$

formed by joining all the vertices of a simplex $\alpha_{\mu-1}$ to all the vertices of a simplex $\alpha_{n-\mu-2}$ by lines of length $\sqrt{3}$ (vertex figures of hexagons). The vertex figure of $\alpha_{n-1}h^2$, having such pyramids for its facets, is thus seen to be the generalized antiprism

$$\left(\alpha_{n-2} \underset{\sqrt{3}}{=\!=\!=} \alpha_{n-2}\right)$$

[1, p. 366].

Now, the vertex figure of $t_1\gamma_n$ is the generalized prism $\alpha_{n-2} \times \beta_1$, whose 2-dimensional side faces are rectangles $\alpha_1 \times \beta_1$, vertex figures of cuboctahedra $t_1\gamma_3$ [1, p. 360]. The vertex figure of $\alpha_{n-1}h^2$ (mod 4), having the same $2(n-1)$ vertices as this tall prism, is evidently the "twisted antiprism" whose side edges (of length $\sqrt{3}$) are the diagonals of these rectangles. In the ordinary antiprism, the two bases $\alpha_{n-2}$ are *oppositely* oriented, so that the vertices and facets of the "upper" one are "above" the facets and vertices of the "lower" one. In the twisted antiprism the two bases are *similarly* oriented (as of course they are in the prism), but each vertex of either $\alpha_{n-2}$ is joined to the noncorresponding vertices of the other.

## 7. Wythoff's Construction

By referring to the octatetrahedron $\alpha_2h^2$ (mod 4) as $\frac{3}{2}3\,|\,3$, we were in effect placing rings round two vertices of a graphical symbol consisting of an equilateral triangle with one side marked $\frac{3}{2}$. In other words, we were regarding the vertices of this star-polyhedron as the images of a point suitably placed on one of the three mirrors of a kaleidoscope whose dihedral angles are $\pi/3$, $\pi/3$, $2\pi/3$. The solid angle formed by these three mirrors may most simply be described by the inequalities

$$x_1 \geqslant x_2 \geqslant x_3 \geqslant -x_1$$

for 3 ordinary Cartesian coordinates $x_1, x_2, x_3$. On a sphere with its center at the origin, this solid angle cuts out a spherical triangle with angles $\pi/3$, $\pi/3$, $2\pi/3$. The bisector of the obtuse angle decomposes this triangle into two triangles with angles $\pi/3$, $\pi/3$, $\pi/2$; in symbols,

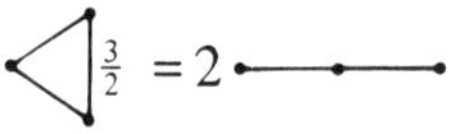

$$\triangleleft\tfrac{3}{2} = 2\,\bullet\!\!-\!\!\!-\!\!\bullet\!\!-\!\!\!-\!\!\bullet$$

In fact, two mirrors inclined at $2\pi/3$ have the same effect as if one of them were replaced by the bisecting plane of that angle; thus the new kaleidoscope yields the same group $[3,3] \simeq S_4$ as the familiar tetrahedral kaleidoscope [10, p. 16].

A point on the mirror $x_1 = x_2$, equidistant from $x_2 = x_3$ and $x_3 = -x_1$, is conveniently taken to be

$$(1, 1, 0).$$

The last two mirrors reflect this to the neighboring vertices $(1, 0, 1)$ and $(0, 1, -1)$, in agreement with Figure 3. Similarly, the point $(1, 1, 1)$, on the line of intersection of $x_1 = x_2$ and $x_2 = x_3$, is reflected by the third mirror to $(-1, 1, -1)$, which is another one of the four vertices of the regular tetrahedron $\alpha_3 = [e\alpha_2/2]^{+1}$.

These simple examples suggest that, for all values of $n$, the passage from $\alpha_{n-1}h$ to $[e\alpha_{n-1}/2]^{+1}$, and from $\alpha_{n-1}h^2$ to $\alpha_{n-1}h^2$ (mod 4), may be achieved by *affixing the mark* $\frac{3}{2}$ *to one edge of the representative n-gon* (with rings on one or two vertices, respectively), so as to change one of the dihedral angles from $\pi/3$ to $2\pi/3$. To prove that this is indeed the correct procedure, let us replace the "prismatic" kaleidoscope

$$x_1 \geqslant x_2 \geqslant \cdots \geqslant x_{n-1} \geqslant x_n \geqslant x_1 - 1$$

(whose mirrors are all perpendicular to $\sum x = 0$) by the "pyramidal" kaleidoscope

$$x_1 \geqslant x_2 \geqslant \cdots \geqslant x_{n-1} \geqslant x_n \geqslant -x_1$$

(whose mirrors all pass through the origin).

By placing a ring round the $n$th vertex of the $n$-gon, we indicate that a typical vertex of the polytope will be a point, such as $(1, 1, \ldots, 1, 1)$, which lies on the line of intersection of the first $n-1$ mirrors. The $n$th mirror, $x_n = -x_1$ (or $x_1 + x_n = 0$), reflects this to the neighboring vertex $(-1, 1, \ldots, 1, -1)$, the other mirrors permute these coordinates, the $n$th mirror takes

$$(1, -1, -1, 1, \ldots, 1, 1) \quad \text{to} \quad (-1, -1, -1, 1, \ldots, 1\ -1),$$

and so on. Eventually we obtain all possible arrangements involving $-1$ an even number of times, in agreement with the known coordinates for the vertices of $[e\alpha_{n-1}/2]^{+1}$.

Similarly, by placing rings round the first vertex of the $n$-gon as well as the $n$th, we indicate a point, such as $(0, 1, \ldots, 1, 1)$, lying on all the mirrors except the first and $n$th. The first mirror yields $(1, 0, 1, \ldots, 1)$ and the $n$th replaces the final 1 by $-1$; thus ultimately we obtain all the permutations of $(0, \pm 1, \pm 1, \ldots, \pm 1)$, in agreement with the known coordinates for the vertices of $\alpha_{n-1}h^2$ (mod 4).

## 8. The Decomposition of a Spherical Simplex

It is easy to verify that reflections in the $n$ hyperplanes

$$x_1 = x_2, \quad x_2 = x_3, \ldots, \quad x_{n-1} = x_n, \quad x_1 + x_n = 0 \tag{8.1}$$

generate the same group as reflections in

$$x_1 = x_2, \quad x_2 = x_3, \ldots, \quad x_{n-1} = x_n, \quad x_{n-1} + x_n = 0 \tag{8.2}$$

[9, Nos. 200, 297], namely $[3^{n-3,1,1}]$, of order $2^{n-1}n!$, the symmetry group of $h\gamma_n$. (This is also the symmetry group of $[e\alpha_n/2]^{+1}$, whereas that of $\alpha_{n-1}h^2$ (mod 4) is $[3^{n-2},4]$, the complete group of $\gamma_n$, which includes the reflection in $x_n = 0$.) In the kaleidoscope (8.1), the first and last mirrors form a dihedral angle $2\pi/3$. Its bisecting hyperplane $x_2 + x_n = 0$ yields two smaller kaleidoscopes which each exhibit another such obtuse angle. Repeated bisection, using hyperplanes such as $x_\nu + x_n = 0$ for $\nu = 3, 4, \ldots, n-1$, ultimately yields $2^{n-2}$ copies of the basic kaleidoscope (8.2).

Without any appeal to coordinates, Barry Monson has obtained the same result as follows. By repeated application of decompositions such as

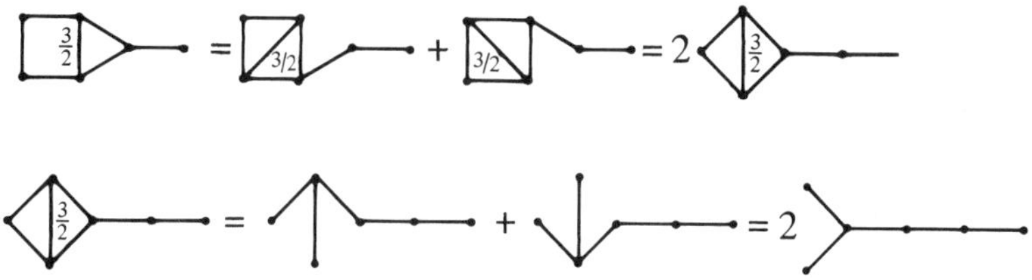

(compare [8, pp. 209, 211; 9, pp. 111, 281, 297]) he deduced

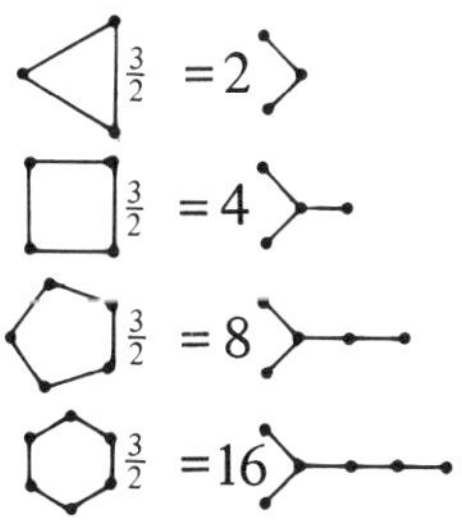

$= 2B_3$

$= 4B_4$

$= 8B_5$

$= 16B_6$

and so on. Thus the simplex (on an $(n-1)$-sphere) represented by an $n$-gon with one edge marked $\frac{3}{2}$ decomposes into $2^{n-2}$ copies of the simplex $B_n$ which is represented by the Y-shaped graph.

## 9. Conclusion

As we saw at the end of Sections 4 and 5, Schoenberg's star-polytope $\tilde{\Pi}_n^{n-1}$ is $[e\alpha_{n-1}/2]^{+1}$ when $n$ is odd, and $\alpha_{n-1}h^2$ (mod 4) when $n$ is even. The graphical symbol consists of an $n$-gon with one edge marked $\frac{3}{2}$ and one or two vertices ringed, namely one vertex when $n$ is odd, two adjacent vertices when $n$ is even.

In 3 dimensions, $[e\alpha_2/2]^{+1} =$

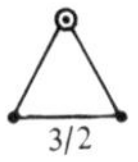

3/2

In 4 dimensions, $\alpha_3 h^2 \pmod 4 =$

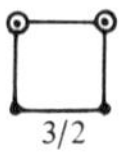

3/2

In 5 dimensions, $[e\alpha_4/2]^{+1} =$

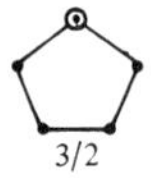

3/2

In 6 dimensions, $\alpha_5 h^2 \pmod 4 =$

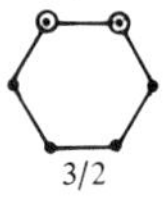

3/2

And so on!

## References


[1] Coxeter, H. S. M., The polytopes with regular-prismatic vertex figures. *Philos. Trans. Royal Soc. A* **229** (1930), 329–425.

[2] Coxeter, H. S. M., The polytopes with regular-prismatic vertex figures (Part 2). *Proc. London Math. Soc. (2)* **34** (1931), 126–189.

[3] Coxeter, H. S. M., The densities of the regular polytopes (Part 3). *Proc. Camb. Philos. Soc.* **29** (1933), 1–22.

[4] Coxeter, H. S. M., The abstract groups $G^{m,n,p}$. *Trans. Amer. Math. Soc.* **45** (1939), 73–150.

[5] Coxeter, H. S. M., Regular and semi-regular polytopes (Part 1). *Math. Z.* **46** (1940), 380–407.

[6] Coxeter, H. S. M., Regular honeycombs in elliptic space. *Proc. London Math. Soc. (3)* **4** (1954), 471–501.

[7] Coxeter, H. S. M., Symmetrical definitions for the binary polyhedral groups. *Proc. Symposia in Pure Mathematics (Amer. Math. Soc.)* **1** (1959), 64–87.

[8] Coxeter, H. S. M., *Twelve Geometric Essays*. Southern Illinois University Press, Carbondale 1968.

[9] Coxeter, H. S. M., *Regular Polytopes* (3rd ed.). Dover, New York 1973.

[10] Coxeter, H. S. M., *Regular Complex Polytopes*. Cambridge University Press 1974.

[11] Coxeter, H. S. M., Polytopes in the Netherlands. *Nieuw Archiev voor Wiskunde (3)* **26** (1978), 116–141.

[12] Coxeter, H. S. M., Longuet-Higgins, M. S., and Miller, J. C. P., Uniform polyhedra. *Philos. Trans. Royal Soc. A* **246** (1954), 401–450.

[13] Coxeter, H. S. M. and Moser, W. O. J., *Generators and Relations for Discrete Groups* (4th ed.). Springer-Verlag, Berlin 1980.

[14] Coxeter, H. S. M., Review of Three-Dimensional Nets and Polyhedra by A. F. Wells. *Bull. Amer. Math. Soc.* **84** (1978), 466–470.

[15] Hinton, C. H., *The Fourth Dimension*. London 1906.

[16] König, D. and Szücs, A., Mouvement d'un point abandonné à l'intérieur d'un cube. *Rend. Circ. Mat. di Palermo* **36** (1913), 79–90.

[17] Schattschneider, Doris, The plane symmetry groups. *Amer. Math. Monthly* **85** (1978), 439–450.

[18] Schoenberg, I. J., On the motion of a billiard ball in two dimensions. *Delta* **5** (1975), 1–18.

[19] Schoenberg, I. J., Extremum problems for the motions of a billiard ball III: The multi-dimensional case of König and Szücs. *Studia Scientiarum Mathematicarum Hungarica* **13** (1978), 53–78.

[20] Schoute, P. H., Analytical treatment of the polytopes regularly derived from the regular polytopes I. *Verh. K. Akad van Wetensch. te Amsterdam (eerste sectie)*, 11.3 (1911).

# The Harmonic Analysis of Skew Polygons as a Source of Outdoor Sculptures[1]

I. J. Schoenberg*

## 1. Introduction

The previous paper [4] on the subject of the finite Fourier series (f.F.s.) dealt with some known and some new applications to problems of elementary geometry. In the present second paper we apply it to a beautiful theorem of Jesse Douglas [3] on skew pentagons in space. It is shown here that Douglas's theorem amounts to the graphical harmonic analysis of skew pentagons and that it is also the source of striking outdoor sculptures. This last opinion is shared by two great art experts, Allan and Marjorie McNab, whom I wish to thank for their encouragement.

The case of a pentagon is discussed in Sections 2 and 3. Again with possible sculptures in mind, we present in Sections 4 and 5 the harmonic analysis of a skew heptagon.

The theorem mentioned above is as follows. (See Figure 1.)

**Theorem 1** (J. Douglas). *Let*

$$\Pi = (z_0, z_1, z_2, z_3, z_4) \qquad (z_{\nu+5} = z_\nu) \tag{1.1}$$

*be a skew closed pentagon in* $\mathbb{R}^3$, *viewed as a vector space. Let*

$$z'_\nu = \tfrac{1}{2}(z_{\nu+2} + z_{\nu-2}) \qquad (\nu = 0, 1, 2, 3, 4) \tag{1.2}$$

*be the midpoint of the side* $[z_{\nu-2}, z_{\nu+2}]$ *which is opposite to the vertex* $z_\nu$.

[1]Sponsored by U. S. Army Research Office, P. O. Box 12211, Research Triangle Park, North Carolina 27709, under Contract No. DAAG29-75-C-0024.

*Mathematics Research Center, University of Wisconsin, Madison, Wisconsin.

*For each $\nu$ determine, on the line joining $z_\nu$ to $z'_\nu$, the points $f_\nu^1$, $f_\nu^2$ such that*

$$f_\nu^1 - z'_\nu = \frac{1}{\sqrt{5}}(z'_\nu - z_\nu), \qquad f_\nu^2 - z'_\nu = -\frac{1}{\sqrt{5}}(z'_\nu - z_\nu). \tag{1.3}$$

*Then*

$$\Pi^1 = (f_0^1, f_1^1, f_2^1, f_3^1, f_4^1) \tag{1.4}$$

*is a plane and affine regular pentagon, and*

$$\Pi^2 = (f_0^2, f_1^2, f_2^2, f_3^2, f_4^2) \tag{1.5}$$

*is a plane and affine regular star-shaped pentagon.*

By an *affine regular* (*star-shaped*) *pentagon* we mean an affine image of a regular (star-shaped) pentagon.

Theorem 1 was easy to verify, but was not easily discovered. In several papers [1–3], Douglas thoroughly explores these problems. He uses the classical eigenvalue properties of cyclic (or circulant) square matrices. Theorem 1 is stated as an example of general results in [1, p. 125], and is also proved directly in [3], with a short ad hoc proof which does not seem to be particularly transparent. The author's contributions go in two different directions.

1. The natural foundation of Douglas's theory seems to be the finite Fourier series. To be sure, the f.F.s. is essentially equivalent to the properties of cyclic matrices used by Douglas. However, it is shown in Section 2 that if we invert the f.F.s. for a pentagon, not in its usual complex form, but in its so-called real form, we are inevitably led to Douglas's Theorem 1. From this point of view Douglas's idea easily generalizes to the harmonic analysis of skew heptagons in $\mathbb{R}^3$ (Theorem 2 of Section 4).

2. The author constructed out of 20 thin wooden sticks a 3-dimensional model, well over two feet in size, illustrating Theorem 1. The appearance of the plane affine regular pentagons $\Pi^1$ and $\Pi^2$ was expected, but enjoyable just the same, especially as they lie in two different planes. For contrast, the sides of the pentagons $\Pi$, $\Pi^1$, $\Pi^2$ were painted in three different colors. The *shape* of the entire structure, i.e. ignoring rigid motions, depends on 9 real parameters. This diversity and total lack of symmetry allows for artistic effects and makes the presence of the affine regular pentagons more striking: order out of chaos. Made of metal bars and of a more heroic size, it would provide a striking outdoor sculpture. Our Figure 1 shows the case when the pentagon $\Pi$, having the vertices $z_0, z_1, z_2, z_3, z_4$, is in a plane. This, however, gives only a faint idea of the aspect of a 3-dimensional structure.

We also constructed a 3-dimensional illustration of Theorem 2 out of 63 thin wooden sticks. Based on a skew heptagon $\Pi$, it shows the three affine regular heptagons $\Pi^1$, $\Pi^2$, $\Pi^3$, painted in three contrasting colors. This model is yet to be shown to the art experts for their comments on its suitability as an outdoor sculpture. Our Figure 3 shows an example when the heptagon $\Pi = (z_0, z_1, \ldots, z_6)$ is in the plane.

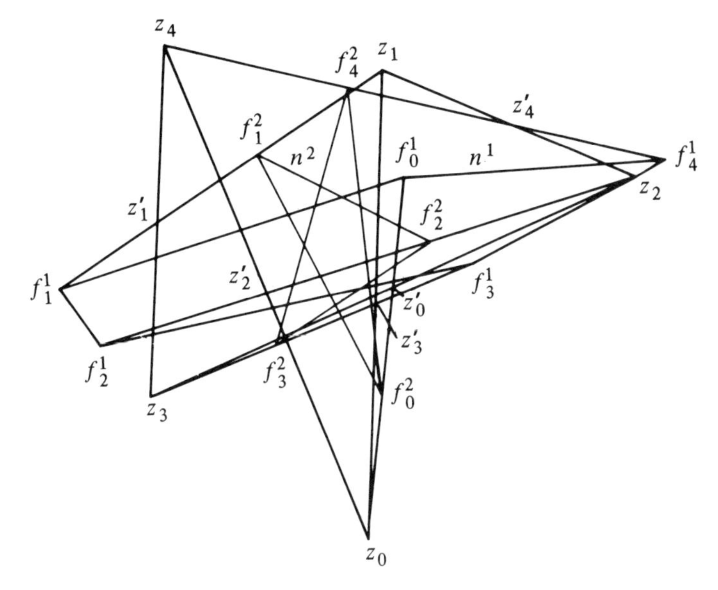


**Figure 1**

## 2. A Proof of Theorem 1 for Pentagons $\Pi$ in the Complex Plane

If $\Pi \subset \mathbb{C}$, we can consider all symbols $z_\nu$, $z'_\nu$, $f_\nu^1$, $f_\nu^2$, of Theorem 1, as complex numbers. With $\omega_\nu = \exp(2\pi i\nu/5)$, the f.F.s. of the $z_\nu$ is the expansion

$$z_\nu = \zeta_0 + \zeta_1\omega_\nu + \zeta_2\omega_\nu^2 + \zeta_3\omega_\nu^3 + \zeta_4\omega_\nu^4 \qquad (\nu = 0, \ldots, 4), \tag{2.1}$$

where the f.F.s. coefficients $\zeta_\nu$ are given by the inverse formulae

$$\zeta_\nu = \tfrac{1}{5}\left(z_0 + z_1\bar{\omega}_\nu + z_2\bar{\omega}_\nu^2 + z_3\bar{\omega}_\nu^3 + z_4\bar{\omega}_\nu^4\right). \tag{2.2}$$

Both formulae extend the definitions of $(z_\nu)$ and $(\zeta_\nu)$ to periodic sequences of period 5. Since $\zeta_3 = \zeta_{-2}$, $\zeta_4 = \zeta_{-1}$, we may rewrite (2.1) as

$$z_\nu = \zeta_0 + \left(\zeta_1\omega_\nu + \zeta_{-1}\omega_\nu^{-1}\right) + \left(\zeta_2\omega_\nu^2 + \zeta_{-2}\omega_\nu^{-2}\right), \tag{2.3}$$

which is the so-called *real* f.F.s. of the $(z_\nu)$. Writing

$$\tilde{f}_\nu^1 = \zeta_1\omega_\nu + \zeta_{-1}\omega_\nu^{-1}, \qquad \tilde{f}_\nu^2 = \zeta_2\omega_\nu^2 + \zeta_{-2}\omega_\nu^{-2}, \tag{2.4}$$

we obtain the final form of the f.F.s. as

$$z_\nu = \zeta_0 + \tilde{f}_\nu^1 + \tilde{f}_\nu^2. \tag{2.5}$$

By (2.2) $\zeta_0$ is the centroid of the $z_\nu$. Selecting this centroid as the origin 0 of the complex plane, (2.5) simplifies to

$$z_\nu = \tilde{f}_\nu^1 + \tilde{f}_\nu^2 \qquad (\nu = 0, \ldots, 4). \tag{2.6}$$

Introducing the two new pentagons

$$\tilde{\Pi}^1 = (\tilde{f}_\nu^1) \quad \text{and} \quad \Pi^2 = (\tilde{f}_\nu^2), \tag{2.7}$$

we may represent the pentagon $\Pi = (z_\nu)$ in the form

$$\Pi = \tilde{\Pi}^1 + \tilde{\Pi}^2. \tag{2.8}$$

The simple nature of the pentagons (2.7) is shown by the following statements:

$$\tilde{\Pi}^1 \text{ is an affine regular pentagon} \tag{2.9}$$

$$\tilde{\Pi}^2 \text{ is an affine regular star-shaped pentagon} \tag{2.10}$$

A proof is immediate: Setting in the first relation (2.4) $\tilde{f}_\nu^1 = x_\nu + iy_\nu$, $\zeta_1 = a + b$, $\zeta_{-1} = c + d$, we find that

$$x_\nu = (a + c)\cos\frac{2\pi\nu}{5} + (-b + d)\sin\frac{2\pi\nu}{5}$$

$$y_\nu = (b + d)\cos\frac{2\pi\nu}{5} + (a - c)\sin\frac{2\pi\nu}{5},$$

and (2.9) is established. Replacing in the right sides $\nu$ by $2\nu$, we obtain (2.10).

So far we have only made general remarks on the f.F.s. of 5 terms which readily extend to the series for $k$ terms. To obtain Theorem 1 we want to invert the real f.F.s. (2.6), i.e. find the individual terms $\tilde{f}_\nu^1$ and $\tilde{f}_\nu^2$. This is where Douglas's idea comes in. From (2.3), with $\zeta_0 = 0$, and writing $\omega = \omega_1$, we obtain

$$z_{\nu+2} = (\zeta_1\omega_\nu\omega^2 + \zeta_{-1}\omega_\nu^{-1}\omega^{-2}) + (\zeta_2\omega_\nu^2\omega^{-1} + \zeta_{-2}\omega_\nu^{-2}\omega),$$

$$z_{\nu-2} = (\zeta_1\omega_\nu\omega^{-2} + \zeta_{-1}\omega_\nu^{-1}\omega^2) + (\zeta_2\omega_\nu^2\omega + \zeta_{-2}\omega_\nu^{-2}\omega^{-1}),$$

and therefore

$$\begin{aligned} z_\nu' &= \tfrac{1}{2}(z_{\nu+2} + z_{\nu-2}) \\ &= \tfrac{1}{2}(\omega^2 + \omega^{-2})(\zeta_1\omega_\nu + \zeta_{-1}\omega_\nu^{-1}) + \tfrac{1}{2}(\omega + \omega^{-1})(\zeta_2\omega_\nu^2 + \zeta_{-2}\omega_\nu^{-2}). \end{aligned}$$

But then, by (2.4), we have

$$z_\nu' = \tilde{f}_\nu^1 \cos\frac{4\pi}{5} + \tilde{f}_\nu^2\cos\frac{2\pi}{5}. \tag{2.11}$$

Since $\cos(4\pi/5) = -\cos(\pi/5)$, all that we have to do now is invert the system of equations

$$\begin{aligned} z_\nu &= \tilde{f}_\nu^1 + \tilde{f}_\nu^2 \\ z_\nu' &= -\tilde{f}_\nu^1\cos\frac{\pi}{5} + \tilde{f}_\nu^2\cos\frac{2\pi}{5}. \end{aligned} \tag{2.12}$$

Since $\cos(\pi/5) = (1 + \sqrt{5})/2$, $\cos(2\pi/5) = (-1 + \sqrt{5})/4$, we readily find the solution of (2.12) to be given by

$$\begin{aligned} \tilde{f}_\nu^1 &= \left(-\frac{1}{\sqrt{5}}z_\nu + \left(1 + \frac{1}{\sqrt{5}}\right)z_\nu'\right)\cdot\frac{1-\sqrt{5}}{2} \\ \tilde{f}^2 &= \left(\frac{1}{\sqrt{5}}z + \left(1 - \frac{1}{\sqrt{5}}\right)z_\nu'\right)\cdot\frac{1+\sqrt{5}}{2}. \end{aligned} \tag{2.13}$$

Introducing the new points

$$f_\nu^1 = -\frac{1}{\sqrt{5}} z_\nu + \left(1 + \frac{1}{\sqrt{5}}\right) z'_\nu,$$
$$f_\nu^2 = \frac{1}{\sqrt{5}} z_\nu + \left(1 - \frac{1}{\sqrt{5}}\right) z'_\nu, \tag{2.14}$$

we obtain the f.F.s. (2.6) in the form

$$z_\nu = \frac{1 - \sqrt{5}}{2} f_\nu^1 + \frac{1 + \sqrt{5}}{2} f_\nu^2. \tag{2.15}$$

Let us now establish Theorem 1 for the case where $\Pi \subset \mathbb{C}$. From the first relation (2.14) we find that

$$f_\nu^1 - z'_\nu = \frac{1}{\sqrt{5}} (z'_\nu - z_\nu), \tag{2.16}$$

while the second relation (2.14) shows that

$$f_\nu^2 - z'_\nu = -\frac{1}{\sqrt{5}} (z'_\nu - z_\nu). \tag{2.17}$$

(2.16), (2.17) are identical with the relations (1.3) that we wished to establish.

Why are the polygons $\Pi^1$ and $\Pi^2$, defined by (1.4) and (1.5), affine regular? From (2.13) and (2.14) we find that

$$f_\nu^1 = \frac{\tilde{f}_\nu^1}{\frac{1}{2}(1 - \sqrt{5})}, \qquad f_\nu^2 = \frac{\tilde{f}_\nu^2}{\frac{1}{2}(1 + \sqrt{5})} \tag{2.18}$$

while we know by (2.7), (2.9), (2.10) that the polygons $\tilde{\Pi}^1$ and $\tilde{\Pi}^2$ are affine regular. A proof of Theorem 1, for the case where $\Pi \subset \mathbb{C}$, follows from the relations (2.18).

## 3. A Proof of Theorem 1 if $\Pi \subset \mathbb{R}^3$

We point out first that the definition of the pentagons (1.4) and (1.5), by the relations (1.2) and (1.3), remains valid in any real vector space, in particular for $\mathbb{R}^3$. The only statements still in doubt are (2.9) and (2.10).

Let

$$F = (\Pi, \Pi^1, \Pi^2) \tag{3.1}$$

denote the space figure obtained by (1.2) and (1.3), and let

$$F_{xy} = (\Pi_{xy}, \Pi^1_{xy}, \Pi^2_{xy}), \qquad F_{xz} = (\Pi_{xz}, \Pi^1_{xz}, \Pi^2_{xz}) \tag{3.2}$$

be its orthogonal projections onto the coordinate planes $xOy$ and $xOz$, respectively. Since the construction of $F$ is affine invariant, it is clear that we can apply to the plane figures (3.2) the results of the last section; in particular

*the pentagons $\Pi^1_{xy}$ and $\Pi^1_{xz}$ are affine regular.* (3.3)

We now appeal to the following most elementary

**Lemma 1.** *If the space pentagon*

$$\Pi^1 = (x_\nu, y_\nu, z_\nu) \qquad (\nu = 0, 1, 2, 3, 4) \tag{3.4}$$

*has plane projections*

$$\Pi^1_{xy} = (x_\nu, y_\nu), \qquad \Pi^1_{xz} = (x_\nu, z_\nu) \tag{3.5}$$

*which are affine regular pentagons, then* $\Pi^1$ *itself is a plane pentagon which is affine regular.*

*Proof.* The affine regular pentagons (3.5) admit representations of the form

$$\begin{aligned} x_\nu &= a\cos\frac{2\pi\nu}{5} + b\sin\frac{2\pi\nu}{5} \qquad & x_\nu &= a'\cos\frac{2\pi\nu}{5} + b'\sin\frac{2\pi\nu}{5}, \\ y_\nu &= c\cos\frac{2\pi\nu}{5} + d\sin\frac{2\pi\nu}{5} \qquad & z_\nu &= e\cos\frac{2\pi\nu}{5} + f\sin\frac{2\pi\nu}{5}. \end{aligned} \tag{3.6}$$

On comparing the first two equations of (3.6) we conclude that we must have $a = a'$, $b = b'$, and so

$$\begin{aligned} x_\nu &= a\cos\frac{2\pi\nu}{5} + b\sin\frac{2\pi\nu}{5}, \\ y_\nu &= c\cos\frac{2\pi\nu}{5} + d\sin\frac{2\pi\nu}{5}, \\ z_\nu &= e\cos\frac{2\pi\nu}{5} + f\sin\frac{2\pi\nu}{5}. \end{aligned} \tag{3.7}$$

It follows that $\Pi^1$ is an affine regular pentagon in the plane defined by the oblique coordinate system of the two vectors $u = (a, c, e)$ and $v = (b, d, f)$. This completes our proof of Theorem 1. □

*Remarks.* 1. The two pentagons $\Pi^1$ and $\Pi^2$ of Theorem 1 lie in different planes, but have as common center the centroid 0 of the vertices of $\Pi$. The problem of choosing $\Pi$ so as to maximize the artistic effect of the entire structure is not mathematical and is, of course, hopeless.

2. Douglas's fortunate idea is to construct the pentagons $\Pi^1$ and $\Pi^2$, and not the pentagons

$$\tilde{\Pi}^1 = \frac{1-\sqrt{5}}{2}\Pi^1, \qquad \tilde{\Pi}^2 = \frac{1+\sqrt{5}}{2}\Pi^2 \tag{3.8}$$

which provide the final harmonic analysis

$$\Pi = \tilde{\Pi}^1 + \tilde{\Pi}^2 \tag{3.9}$$

according to (2.8). This idea simplifies the final construction considerably, because finding the pentagons (3.8) themselves would require two homothetic images with center 0, a cumbersome complication.

## 4. The Graphical Harmonic Analysis of a Skew Heptagon

Our application of the f.F.s. to Douglas's theorem readily suggests the way to generalize his result to closed skew polygons having $k$ vertices. Having in mind further outdoor sculptures, we restrict our discussion to the case when $k = 7$; hence

$$\Pi = (z_0, z_1, \ldots, z_6) \tag{4.1}$$

is a heptagon. We have omitted the case when $k = 6$ for the reason that regular star-shaped hexagons are not particularly interesting. We commence our discussion by assuming that

$$\Pi \subset \mathbb{C}, \tag{4.2}$$

when the $z_\nu$ are complex numbers. Their f.F.s. and its inverse formulae are

$$z_\nu = \sum_{\alpha=0}^{6} \zeta_\alpha \omega_\nu^\alpha, \quad \zeta_\nu = \frac{1}{7} \sum_{\alpha=0}^{6} z_\alpha \bar{\omega}_\nu^\alpha \qquad (\nu = 0, \ldots, 6) \tag{4.3}$$

where $\omega_\nu = \exp(2\pi\nu/7)$. Again we assume that $z_0 + z_1 + \cdots + z_6 = 0$, hence $\zeta_0 = 0$, and folding the f.F.s., as in (2.3), we obtain

$$z_\nu = (\zeta_1\omega_\nu + \zeta_{-1}\omega_\nu^{-1}) + (\zeta_2\omega_\nu^2 + \zeta_{-2}\omega_\nu^{-2}) + (\zeta_3\omega_\nu^3 + \zeta_{-3}\omega^{-3}). \tag{4.4}$$

The midpoint of the side of $\Pi$ that is opposite to the vertex $z_\nu$ is

$$z'_\nu = \tfrac{1}{2}(z_{\nu+3} + z_{\nu-3}). \tag{4.5}$$

However, now we also need the further midpoint

$$z''_\nu = \tfrac{1}{2}(z_{\nu+2} + z_{\nu-2}). \tag{4.6}$$

From (4.4), and writing $\omega_1 = \omega$, we obtain

$$z_{\nu\pm3} = (\zeta_1\omega_\nu\omega^{\pm3} + \zeta_{-1}\omega_\nu^{-1}\omega^{\mp3}) + (\zeta_2\omega_\nu^2\omega^{\mp1} + \zeta_{-2}\omega_\nu^{-2}\omega^{\pm1}) + (\zeta_3\omega_\nu^3\omega^{\pm2} + \zeta_{-3}\omega_\nu^{-3}\omega^{\mp2}),$$

whence

$$z'_\nu = \frac{\omega^3 + \omega^{-3}}{2}\tilde{f}^1 + \frac{\omega + \omega^{-1}}{2}\tilde{f}^2 + \frac{\omega^2 + \omega^{-2}}{2}\tilde{f}^3, \tag{4.7}$$

if we write

$$\tilde{f}_\nu^j = \zeta_j\omega_\nu^j + \zeta_{-j}\omega_\nu^{-j} \qquad (j = 1, 2, 3, \quad \nu = 0, \ldots, 6). \tag{4.8}$$

Likewise we obtain from (4.4) that

$$z_{\nu\pm2} = (\zeta_1\omega_\nu\omega^{\pm2} + \zeta_{-1}\omega_\nu^{-1}\omega^{\mp2}) + (\zeta_2\omega_\nu^2\omega^{\mp3} + \zeta_{-2}\omega_\nu^{-2}\omega^{\pm3}) + (\zeta_3\omega_\nu^3\omega^{\mp1} + \zeta_{-3}\omega_\nu^{-3}\omega^{\pm1}),$$

whence

$$z''_\nu = \frac{\omega^2 + \omega^{-2}}{2}\tilde{f}_\nu^1 + \frac{\omega^3 + \omega^{-3}}{2}\tilde{f}_\nu^2 + \frac{\omega + \omega^{-1}}{2}\tilde{f}_\nu^3. \tag{4.9}$$

By (4.4) and (4.8) we see that the real f.F.s. of $\Pi$ is

$$z_\nu = \tilde{f}_\nu^1 + \tilde{f}_\nu^2 + \tilde{f}_\nu^3. \tag{4.10}$$

As in the case of pentagons, the analog of Douglas's theorem will arise if we invert the $3 \times 3$ system of equations (4.10), (4.7), (4.9). Writing

$$\Omega_j = \tfrac{1}{2}(\omega^j + \omega^{-j}) = \cos\frac{2\pi j}{7} \qquad (j = 1,2,3), \tag{4.11}$$

we are to solve the system

$$\begin{aligned} z_\nu &= \tilde{f}_\nu^1 + \tilde{f}_\nu^2 + \tilde{f}_\nu^3 \\ z'_\nu &= \Omega_3 \tilde{f}_\nu^1 + \Omega_1 \tilde{f}_\nu^2 + \Omega_2 \tilde{f}_\nu^3 \\ z''_\nu &= \Omega_2 \tilde{f}_\nu^1 + \Omega_3 \tilde{f}_\nu^2 + \Omega_1 \tilde{f}_\nu^3. \end{aligned} \tag{4.12}$$

In terms of the inverse matrix

$$\begin{Vmatrix} A_1 & B_1 & C_1 \\ A_2 & B_2 & C_2 \\ A_3 & B_3 & C_3 \end{Vmatrix} = \begin{Vmatrix} 1 & 1 & 1 \\ \Omega_3 & \Omega_1 & \Omega_2 \\ \Omega_2 & \Omega_3 & \Omega_1 \end{Vmatrix}^{-1} \tag{4.13}$$

the solutions are

$$\tilde{f}_\nu^j = A_j z_\nu + B_j z'_\nu + C_j z''_\nu \qquad (j = 1,2,3). \tag{4.14}$$

By (4.8) it is clear that the three heptagons

$$\tilde{\Pi}^j = (\tilde{f}_0^j, \tilde{f}_1^j, \tilde{f}_2^j, \tilde{f}_3^j, \tilde{f}_4^j, \tilde{f}_5^j, \tilde{f}_6^j) \qquad (j = 1,2,3), \tag{4.15}$$

are affine images of the three regular heptagons

$$(1,\omega,\omega^2,\omega^3,\omega^4,\omega^5,\omega^6), \quad (1,\omega^2,\omega^4,\omega^6,\omega,\omega^3,\omega^5) \quad (1,\omega^3,\omega^6,\omega^2,\omega^5,\omega,\omega^4), \tag{4.16}$$

respectively. In terms of the heptagons (4.15) we may write (4.10) as

$$\Pi = \tilde{\Pi}^1 + \tilde{\Pi}^2 + \tilde{\Pi}^3. \tag{4.17}$$

However, the heptagons (4.15) are *not* the ones that we wish to construct. Rather, following Douglas's lead, we introduce the weights

$$\alpha_j = \frac{A_j}{s_j}, \quad \beta_j = \frac{B_j}{s_j}, \quad \gamma_j = \frac{C_j}{s_j}, \qquad \text{where } s_j = A_j + B_j + C_j, \tag{4.18}$$

and want to construct the heptagons

$$\Pi^j = (f_0^j, f_1^j, f_2^j, f_3^j, f_4^j, f_5^j, f_6^j) \qquad (j = 1,2,3), \tag{4.19}$$

having vertices given by

$$f_\nu^j = \alpha_j z_\nu + \beta_j z'_\nu + \gamma_j z''_\nu \qquad (j = 1,2,3). \tag{4.20}$$

We state our results as

**Theorem 2.** *Let*

$$\Pi = (z_0, z_1, \ldots, z_6) \tag{4.21}$$

*be a skew heptagon in* $\mathbb{R}^3$, *and let*

$$z'_\nu = \tfrac{1}{2}(z_{\nu+3} + z_{\nu-3}), \qquad z''_\nu = \tfrac{1}{2}(z_{\nu+2} + z_{\nu-2}) \tag{4.22}$$

*be the midpoints of appropriate sides and chords of* $\Pi$. *By* (4.11), (4.13), *and* (4.18) *we define the three sets of numerical weights*

$$\alpha_j, \beta_j, \gamma_j, \quad \alpha_j + \beta_j + \gamma_j = 1 \qquad (j = 1, 2, 3). \tag{4.23}$$

*In each of the seven triangles*

$$T_\nu = (z_\nu, z'_\nu, z''_\nu) \qquad (\nu = 0, \ldots, 6) \tag{4.24}$$

*we define the three points*

$$f_\nu^1, f_\nu^2, f_\nu^3 \tag{4.25}$$

*as the centroids of* $T_\nu$ *with the three sets of weights* (4.23), *respectively. Equivalently,* (4.25) *are defined by the equation* (4.20). *Then the three heptagons*

$$\Pi^j = (f_0^j, f_1^j, f_2^j, f_3^j, f_4^j, f_5^j, f_6^j) \qquad (j = 1, 2, 3), \tag{4.26}$$

*are plane heptagons and they are affine images of the regular heptagons* (4.16), *respectively.*

Our Theorem 2 is, of course, fully established if we assume that $\Pi \subset \mathbb{C}$. That it remains true if $\Pi \subset \mathbb{R}^3$ follows from reasonings similar to those used in extending Theorem 1 from $\mathbb{R}^2$ to $\mathbb{R}^3$, in particular from the lemma: *If a heptagon* $\Pi$ *in* $\mathbb{R}^3$ *has two affine regular plane projections, then* $\Pi$ *itself is plane and affine regular.*

## 5. The Construction of a Space Model Illustrating Theorem 2

By this we mean the construction of the figure

$$F = (\Pi, \Pi^1, \Pi^2, \Pi^3), \tag{5.1}$$

where $\Pi$, $\Pi^1$, $\Pi^2$, $\Pi^3$, are the heptagons of Theorem 2. This could be done graphically on a sheet of paper by the methods of descriptive geometry. However, we have in mind a 3-dimensional structure made out of thin (wooden) sticks.

For this purpose we need the numerical values of the weights (4.18). With sufficient accuracy for any physical construction, these are as follows:

$$\begin{Vmatrix} \alpha_1 & \beta_1 & \gamma_1 \\ \alpha_2 & \beta_2 & \gamma_2 \\ \alpha_3 & \beta_3 & \gamma_3 \end{Vmatrix} = \begin{Vmatrix} -0.08627 & 0.69859 & 0.38768 \\ 0.78485 & 1.08626 & -0.87111 \\ 0.30141 & 0.21515 & 0.48344 \end{Vmatrix}, \tag{5.2}$$

$$s_1 = -1.24697, \qquad s_2 = 0.44504, \qquad s_3 = 1.80193. \tag{5.3}$$

The construction of the 14 points $z'_\nu$ and $z''_\nu$ by the formulae (4.22) presents no difficulties. These also determine the 7 triangles (4.24).

In the plane of each $T_\nu$ we are now to construct the centroids (4.25) for the three sets of weights (4.23). Here we use the following lemma, which is too elementary to require a proof (the reader is asked to supply a diagram).

**Lemma 2.** *Let*

$$T = (z, z', z'') \tag{5.4}$$

*be a triangle, and let*

$$f = \alpha z + \beta z' + \gamma z'' \tag{5.5}$$

*be its centroid for the weights* $\alpha$, $\beta$, $\gamma$, *with* $\alpha + \beta + \gamma = 1$.

*If $h$ denotes the intersection of the line joining $z$ to $z'$, with the line joining $z''$ to $f$, then the relations*

$$h - z' = \rho(z' - z), \qquad f - h = \sigma(h - z'') \tag{5.6}$$

*hold, where*

$$\rho = \frac{\alpha}{\alpha + \beta}, \qquad \sigma = -\gamma. \tag{5.7}$$

We apply Lemma 2 to each $T_\nu$ with the sets of weights (5.2). We drop the subscript $\nu$ and show in Figure 2 the location of the centroids $f^1$, $f^2$, $f^3$ in the plane of the triangle $T = (z, z', z'')$. Using Lemma 2 and the numerical values (5.2), we obtain the relations

$$\begin{aligned} h^1 - z' &= \rho_1(z' - z), \qquad & f^1 - h^1 &= \sigma_1(h^1 - z''), \\ h^2 - z' &= \rho_2(z' - z), \qquad & f^2 - h^2 &= \sigma_2(h^2 - z''), \\ h^3 - z' &= \rho_3(z' - z), \qquad & f^3 - h^3 &= \sigma_3(h^3 - z''). \end{aligned} \tag{5.8}$$

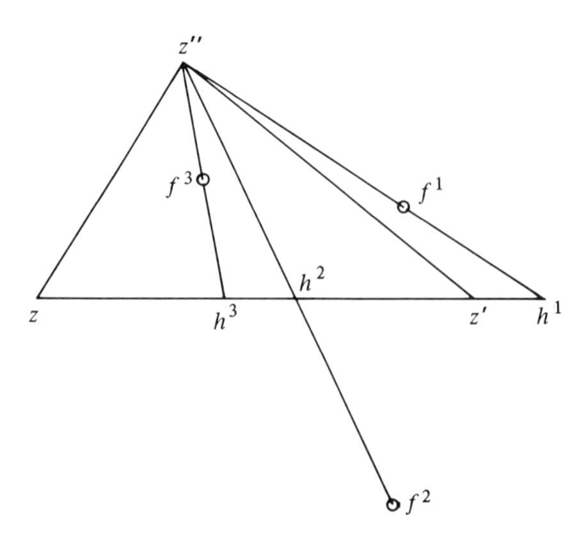


**Figure 2**

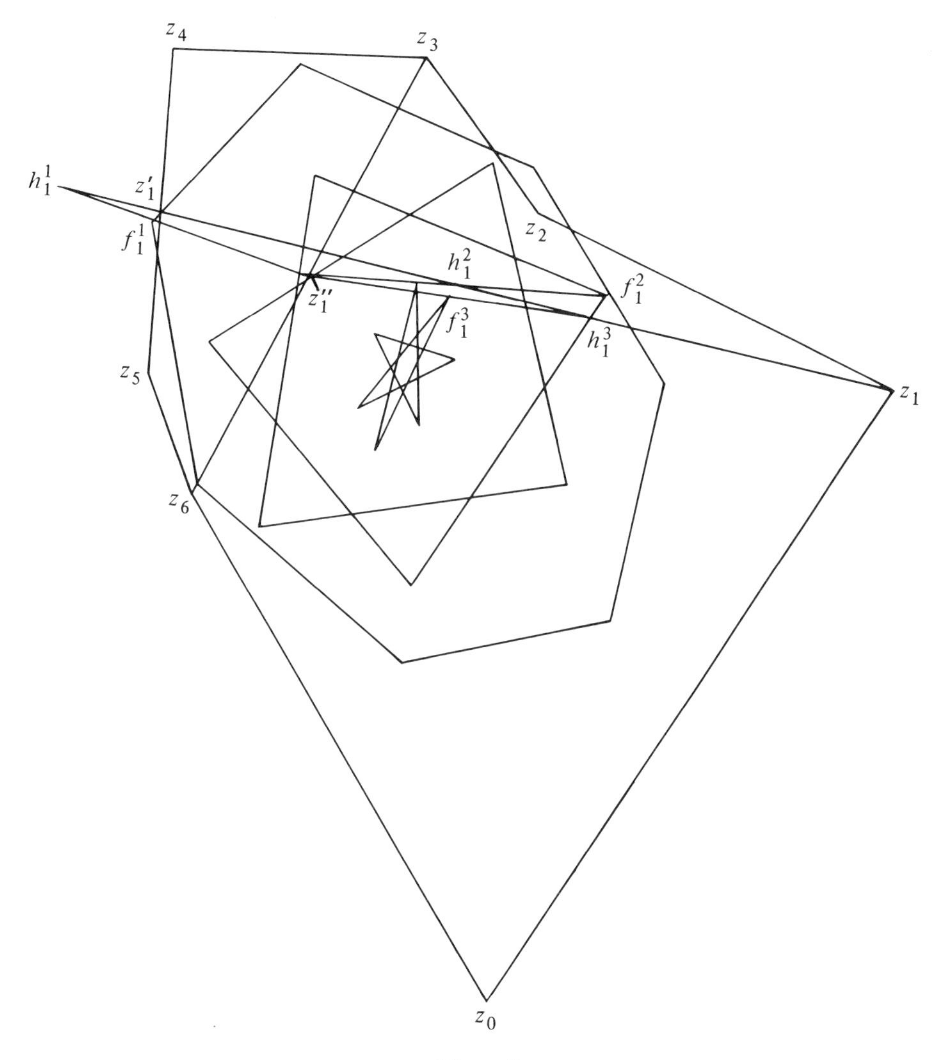


**Figure 3**

The numerical values of the ratios $\rho_j$ and $\sigma_j$, given by (5.7) and (5.2), are

$$\begin{aligned} \rho_1 &= 0.14089, & \sigma_1 &= -0.38768, \\ \rho_2 &= -0.41946, & \sigma_2 &= 0.87111, \\ \rho_3 &= -0.58350, & \sigma_3 &= -0.48344. \end{aligned} \tag{5.9}$$

The locations of the points $h^j$ and $f^j$ in Figure 2 are drawn to scale. For any other triangle $T_\nu = (z_\nu, z_\nu', z_\nu'')$ the corresponding diagram is the image of Figure 2 by the affine transformation mapping $T$ onto $T_\nu$.

Our Figure 3 shows a 2-dimensional illustration of Theorem 2. It shows the three affine regular heptagons $\Pi^1$, $\Pi^2$, and $\Pi^3$. In order to simplify the drawing it shows only the construction of the three vertices

$$f_1^1, f_1^2, f_1^3,$$

corresponding to the triangle $T_1 = (z_1, z_1', z_1'')$.

## References


[1] Douglas, Jesse, Geometry of polygons in the complex plane. *J. of Math. and Phys.* **19** (1940), 93–130.

[2] Douglas, Jesse, On linear polygon transformations. *Bull. Amer. Math. Soc.* **46** (1940), 551–560.

[3] Douglas, Jesse, A theorem on skew pentagons. *Scripta Math.* **25** (1960), 5–9.

[4] Schoenberg, I. J., The finite Fourier series and elementary geometry, *Amer. Math. Monthly* **57** (1950), 390–404.

# The Geometry of African Art
# III. The Smoking Pipes of Begho[1]

Donald W. Crowe*

## 1. Introduction

It is not generally known among archeologists that there is a universal, cross-cultural classification scheme for the repeated patterns occurring in such diverse media as textiles, pottery, basketry, wall decoration, and the art of M. C. Escher. In this paper we introduce this scheme by means of a "flowchart" which reduces the analysis of any particular pattern to a sequence of simple questions (mostly answered "yes" or "no"). We then apply it to the analysis of the decorated pipes excavated from the K2 site of the Kramo quarter of Begho (Ghana) in January–March, 1979, under the direction of Professor Merrick Posnansky. In the sequel we refer to this site as Begho K2.

The analysis uses some geometrical ideas which, although very simple, are unfamiliar to nonmathematicians. Section 2 is devoted to these ideas. Sections 3 and 4 describe the flowcharts for the 7 one-dimensional and 17 two-dimensional patterns. Section 5 describes the Begho pipes, and tabulates the pattern types appearing on them.

Attention is called to the two recent publications by Dorothy Washburn, and by B. Zaslow and A. E. Dittert, listed in the references. They present much of our geometric information, in different ways. The main geometric contribution of the present paper is the introduction and use of the flow charts to expedite the analysis of a given pattern.

[1]The author is indebted to Ebenezer Quarcoopome and Doris Volkhardt for the drawings in Figures 3 and 16, respectively; to Elizabeth Vaughan and Dorothy Washburn for helpful comments and encouragement; and to Merrick Posnansky and the University Research Expeditions Program for making possible his participation in the Begho K2 dig.

*University of Wisconsin, Madison, Wisconsin, U.S.A.

## 2. Geometric Prerequisites

The designs to be analyzed will always be thought of as lying in a plane. Many designs of particular interest, e.g. those on pottery bowls, are not actually in a plane, but can be thought of as "unrolled" with very little distortion. For the purpose of our analysis it is always imagined that this has been done.

A brief explanation of the special kinds of design which we call repeated pattern is in order. A *repeated pattern* in the plane may repeat in only one direction (like a wallpaper border along the top edge of a wall, or a narrow band around a pottery bowl); or in more than one direction (like the usual wallpaper patterns which cover an entire wall, or the hexagons on a tortoise shell). The former are called *one-dimensional patterns*; the latter are *two-dimensional patterns*. Another way of describing this difference is to say that a one-dimensional pattern can be slid along itself, in exactly one direction, in such a way that in its resulting position it cannot be noticed to have shifted. We say that such a pattern *admits a translation* in exactly one direction. One-dimensional patterns have often been called "strips" or "bands." On the other hand, a two-dimensional pattern is one which admits translations in more than one direction. Two-dimensional patterns have sometimes been called "allover" patterns. We reserve the phrase "repeated pattern" or simply "pattern" for such designs as admit a translation in at least one direction. Other designs, which admit no translation at all (although they may possibly admit rotations), will be called *finite designs*. In this paper we are concerned only with repeated patterns, not finite designs. Figures 1, 2, and 3 illustrate these concepts with designs from Begho pipes.

Note that, especially in fragmentary archeological specimens, the above distinctions are somewhat arbitrary. For example, the juxtaposition of several rows of circles, as in Figure 3(b), is a two-dimensional pattern. But if the pipe had been broken so that only a single row,  had been preserved, it would be called a one-dimensional pattern. In the extreme case where no more than a single circle remained, it would be called a finite design: .

It is a geometrical fact that there are only four possible isometries ("rigid motions") of a plane onto itself. These are *reflection* (in a line), *translation*, *rotation* (about a point), and *glide-reflection* (a reflection in a line, followed by a translation in the direction prescribed by the line). For this reason it is not surprising that any one-dimensional pattern admits one of only seven different combinations of these motions. That is, when classified according to their admissible rigid motions, there are only seven one-dimensional patterns. Similarly, there are only seventeen two-dimensional patterns.

Prototypes of each of the seven one-dimensional patterns, which are designated (1), (2), . . . , (7), are illustrated in Figure 4.

Although a few examples are given in [6, Figures 38, 47], the seventeen two-dimensional patterns had apparently not been reproduced in their entirety in archeological literature before the pioneering study by Dorothy Washburn, *A*

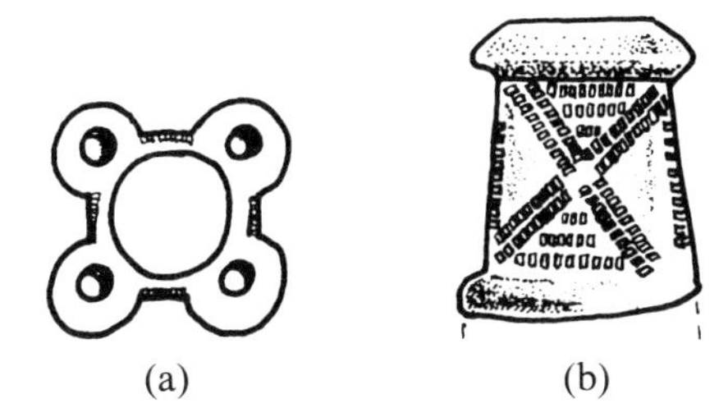

(a) (b)

**Figure 1**(a, b). Examples of finite designs.

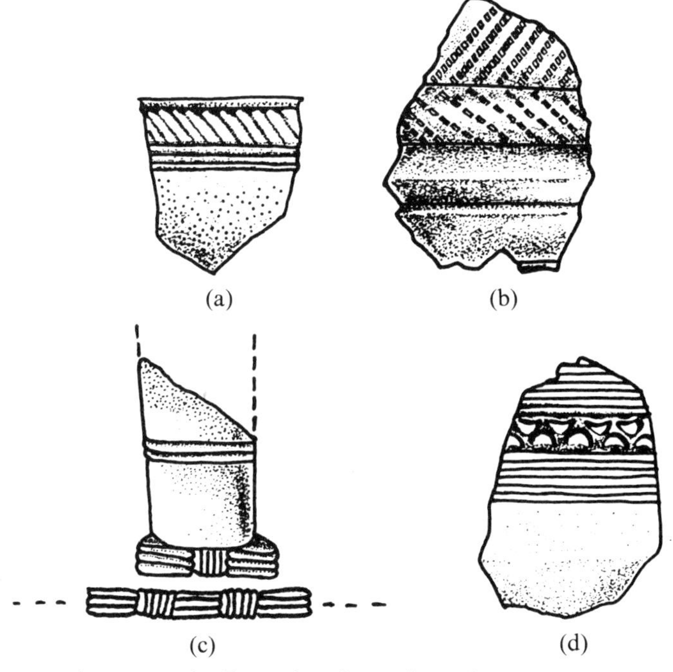

(a) (b)

(c) (d)

**Figure 2**(a–d). Examples of one-dimensional patterns.

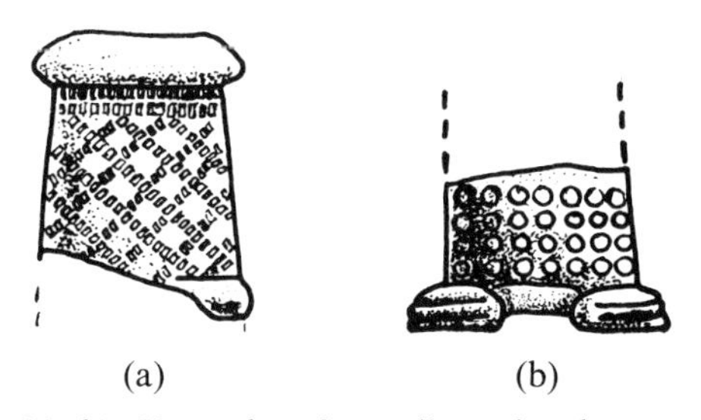

(a) (b)

**Figure 3**(a, b). Examples of two-dimensional patterns.

*Symmetry Analysis of Upper Gila Ceramic Design* (1977). Indeed, Washburn goes further and considers the broader class of two-colored ("counter-changed") designs. We give the seventeen patterns in Figure 5, essentially copied from A. Speiser, *Theorie der Gruppen von endlicher Ordnung*, Birkhauser, Basel 1956. Another convenient illustration of the seventeen prototypes can be found in

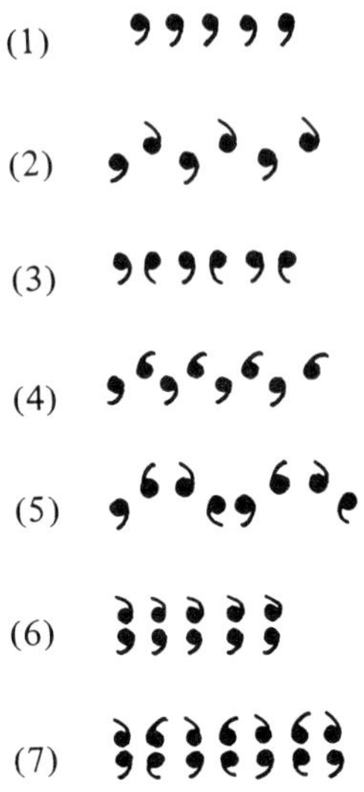

(1)

(2)

(3)

(4)

(5)

(6)

(7)

**Figure 4.** The seven one-dimensional patterns.

Chart 4 of [4]. The illustrations there are taken primarily from patterns in Chinese lattices. Apparently there is no single natural source where all seventeen can be found. We use the same notation as in Chart 4 (different from Speiser's), which is the standard modification of the notation used in the *International Tables of X-ray Crystallography* (N. F. M. Henry and K. Lonsdale, vol. 1, Kynoch Press, Birmingham 1952).

## 3. Flowchart for One-Dimensional Patterns

The flowchart of Figure 6 is a simple tool for classifying one-dimensional patterns. Before looking at the flowchart it is necessary to know what the questions in it mean. For the following explanations the reader should refer back to the seven patterns shown in Figure 4, *not* to the flowchart itself.

"Is there a vertical reflection?" means "Does the pattern admit a reflection in a ("vertical") line perpendicular to the length of the pattern?" For patterns (3), (5), (7) the answer is "yes." For patterns (1), (2), (4), (6) the answer is "no."

"Is there a horizontal reflection?" means "Does the pattern admit a reflection in a ("horizontal") line through the length of the pattern?" For patterns (6), (7) the answer is "yes." For patterns (1), (2), (3), (4), (5), the answer is "no."

"Is there a horizontal reflection or a glide-reflection?" means "Does the pattern admit either a reflection or a glide-reflection (or possibly both) in a horizontal line?" For patterns (2), (5), (6), (7), the answer is "yes." For patterns (1), (3), (4) the answer is "no."

"Is there a half turn?" means "Does the pattern admit a rotation (in its plane) by 180° (i.e. by a half turn)?" For patterns (4), (5), (7) the answer is "yes." For patterns (1), (2), (3), (6) the answer is "no."

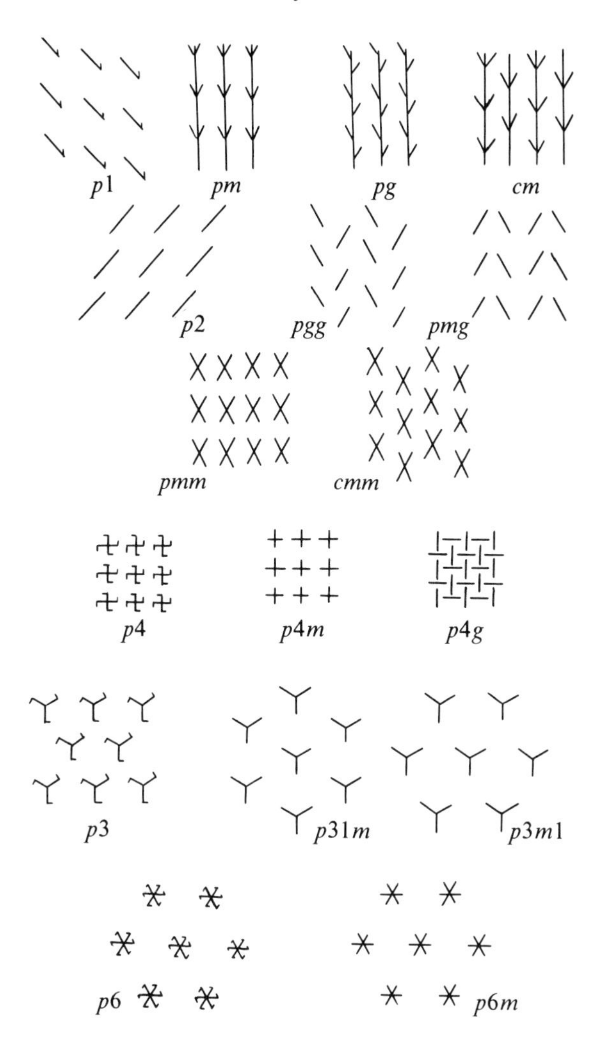


**Figure 5.** The seventeen two-dimensional patterns.

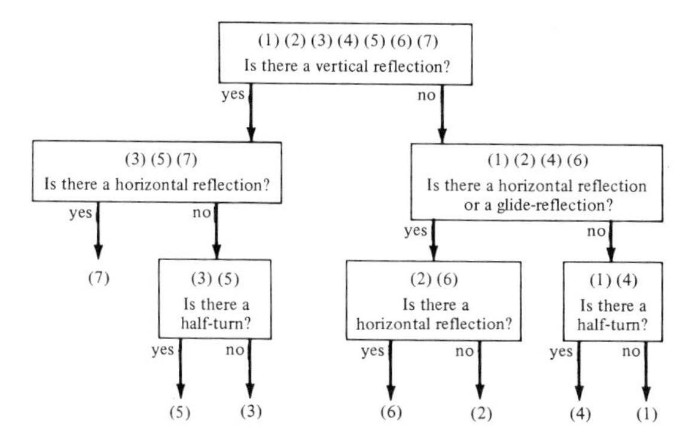


**Figure 6.** Flowchart for the seven one-dimensional patterns.

**Illustrative Example.** We apply the flowchart to the analysis of a schematic version of the pattern in Figure 2(d):

Question 1: Is there a vertical reflection? The answer is "yes," in any of the vertical lines shown in Figure 7. Hence the pattern is one of (3), (5), (7).

**Figure 7**

Question 2: Is there a horizontal reflection? The answer is "no," because a reflection in the center line of Figure 8 changes the pattern. Hence the pattern is one of (3), (5).

**Figure 8**

Question 3: Is there a half turn? The answer is "yes," about any of the points marked in Figure 9. Hence the pattern is of type (5).

**Figure 9**

## 4. Flowchart for Two-Dimensional Designs

The flowchart of Figure 11 will classify two-dimensional patterns in the same way that Figure 6 classifies the one-dimensional patterns. The first question, "What is the smallest rotation?" separates the 17 patterns into 5 classes. The reason for this is that the only admissible rotations for a two-dimensional pattern are "none," 180°, 120°, 90°, 60° (corresponding to no rotation and rotation by $\frac{1}{2}$, $\frac{1}{3}$, $\frac{1}{4}$, $\frac{1}{6}$ turns respectively). Typical patterns having each of these types of rotation are shown in Figure 10.

Since there is not necessarily any way to distinguish "vertical" and "horizontal" from any other direction, the question "Is there a reflection?" asks whether there is any line at all in which the pattern admits a reflection. For the remaining questions a "bifold center" is a point about which the pattern admits a 180° rotation (but not a 90° rotation), and a "3-fold center" is a point about which the pattern admits a 120° rotation (but not a 60° rotation).

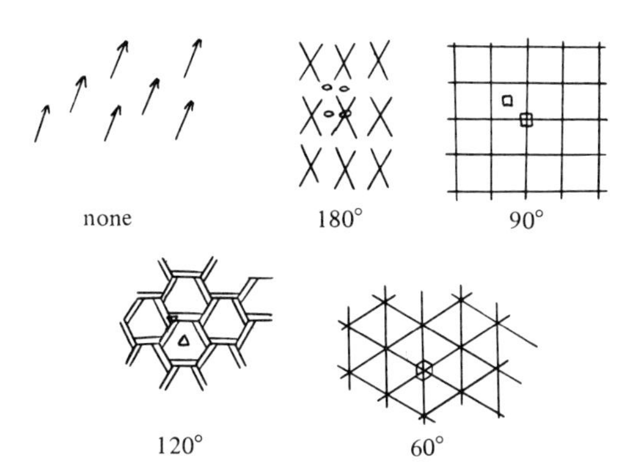


**Figure 10.** Illustrations of the five types of admissible rotations for two-dimensional patterns.

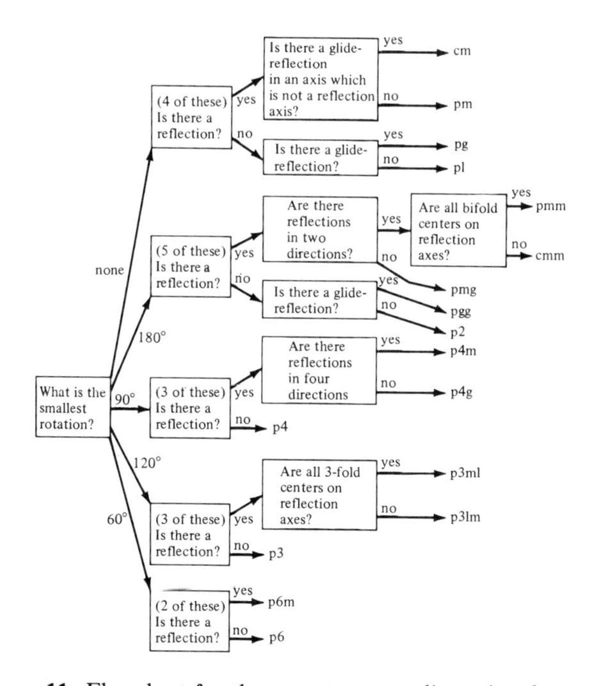


**Figure 11.** Flowchart for the seventeen two-dimensional patterns.

**Illustrative Example.** We apply the flowchart to the analysis of a schematic version of Figure 3(a):

Note that we have interpreted the pattern as not having bilateral symmetry. (See the next example for the analysis when bilateral symmetry is assumed.) We also interpret the lines of the pattern as not meeting at 90° angles.

Question 1: What is the smallest rotation? The smallest rotations are by 180°, about points indicated in Figure 12.

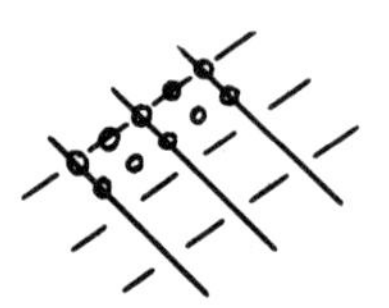

**Figure 12**

Question 2: Is there a reflection? The answer is "no." (If the lines in the pattern met at 90° angles the answer would be "yes.")
Question 3: Is there a glide-reflection? The answer is "no." Hence the pattern is of type $p2$.

Note that if the pattern of Figure 3(a) is interpreted as having bilateral symmetry, as in Figure 13, the answer to Question 1 is still 180°. The remaining questions and answers follow.

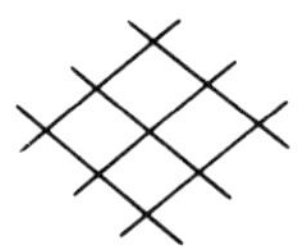

**Figure 13**

Question 2: Is there a reflection? The answer is "yes," in the dashed lines of Figure 14.

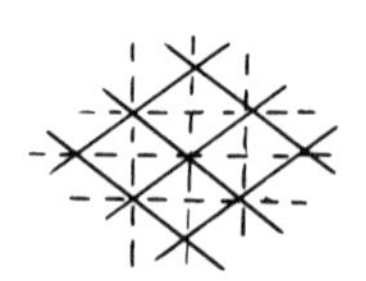

**Figure 14**

Question 3: Are there reflections in two directions? The answer is "yes." (The dashed lines in Figure 15 are in two directions.)
Question 4: Are all bifold centers on reflection axes? Some of the bifold centers are marked on Figure 15. The answer is "no." (The centers lying on sides

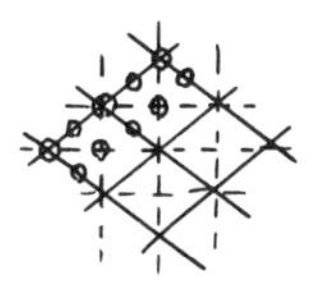

**Figure 15**

of the parallelograms are not on reflection axes.) The pattern is consequently of type *cmm*.

## 5. The Begho K2 Pipes

The tobacco pipes from earlier excavations at Begho have been discussed at length by I. K. Afeku in his 1976 University of Ghana honours dissertation [1]. At that time some 450 pipe fragments had been examined by him. The excavations at the Kramo quarter of Begho in 1979 were especially productive of pipe fragments, some 638 having been recorded as small finds. Many of these fragments either were undecorated or contained unusably small portions of design. However, 230 one-dimensional patterns and 49 two-dimensional patterns could be classified. (This excludes 100 designs consisting of two or three parallel lines, and 23 consisting of many parallel lines or "squiggly" lines.) The total number of decorated fragments is, however, somewhat smaller than this total of 279 because many fragments had two or more classifiable patterns.

Some of these pipe fragments are shown in Figure 16. Parts (a) and (b) show a stem, an intact base, and the lower part of a bowl which has been broken off. The incised decoration around the bottom or middle of many bowls weakens them, and they tend to break along this line of decoration. (For this reason some designs classified as one-dimensional may actually be parts of two-dimensional designs.) Figures 16(c), (d) and (e) show particularly elaborate bowl fragments. However, many fragments are undecorated, except for a uniform red slip. Figures 16(f) and (g) are photographs of two bowl fragments, slightly more than life size. (Figures 16(e) and (f) are different views of the same fragment.)

Afeku classified pipes into four main categories (aside from a fifth consisting of imported European kaolin pipes) according to types of base. Only two of these four types were found at the Begho K2 site, and of the 638 fragments only 146 had recognizable bases (58 of "ring" type, and 88 of "foliate" type). By using the 279 recognizable patterns we are able to classify more of the fragments than are classifiable according to types of base. Some typical patterns of each type are shown in Figure 17.

Table 1 gives the number of patterns of each of the seven one-dimensional types. The predominance of type (7) indicates a preference for symmetry on the part of the artists, since type (7) is the type which admits all possible strip symmetries (translation, horizontal reflection, vertical reflection, 180° rotation, and glide reflection). Each of the other types fails to admit at least one of the possible strip symmetries. In other cultures there is a preference for less symme-

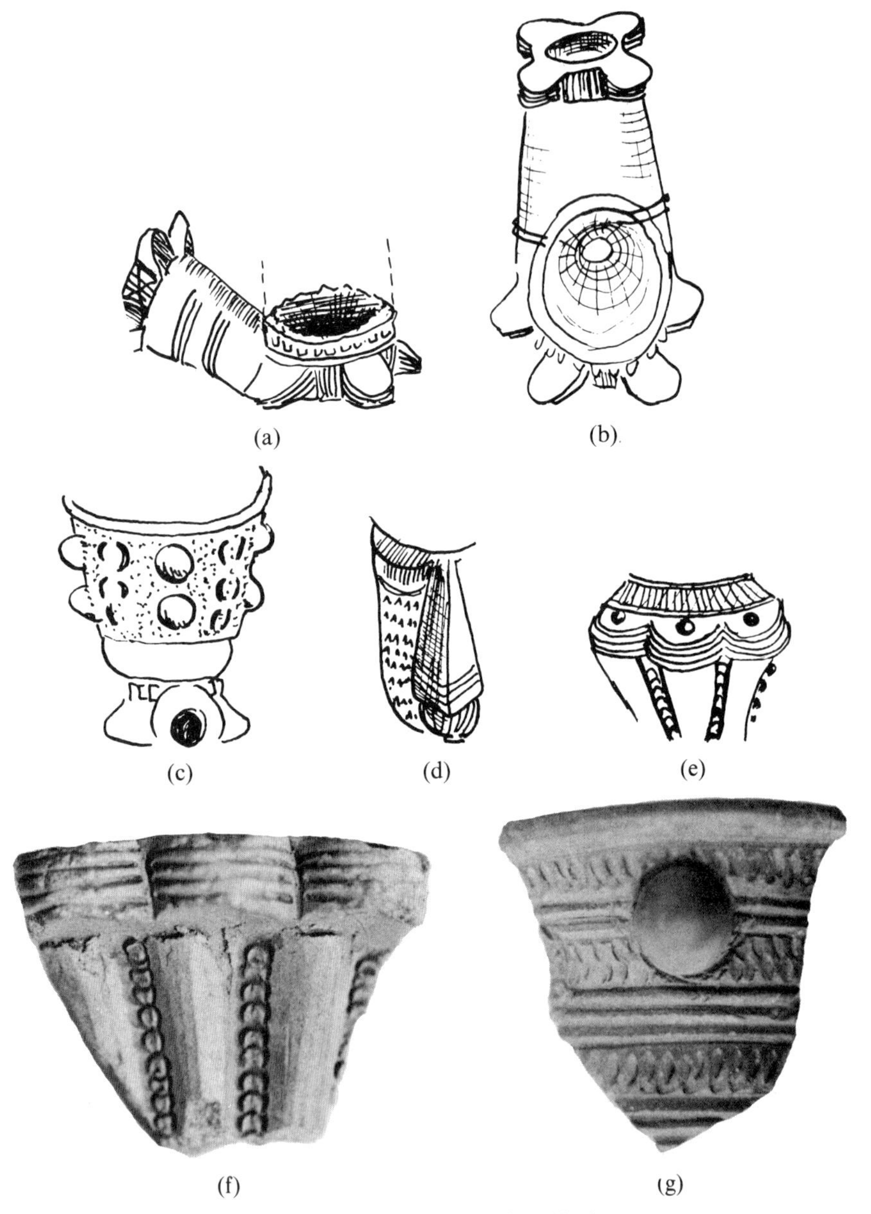


**Figure 16**(a–g). Some Begho K2 pipes.

try. A particularly well-documented example of this [5] is the "Early Rio Grande glaze-paint ware" of the U.S. southwest, where some 70% of strip patterns are of type (4).

Of the seventeen possible two-dimensional patterns, only seven were found on Begho K2 pipes. In fact two of the seven (*pgg* and *p6m*) are represented by only single examples. Of these the *pgg* example is essentially due to the unusual structure of the bowl on which it occurs, rather than to the design proper, while

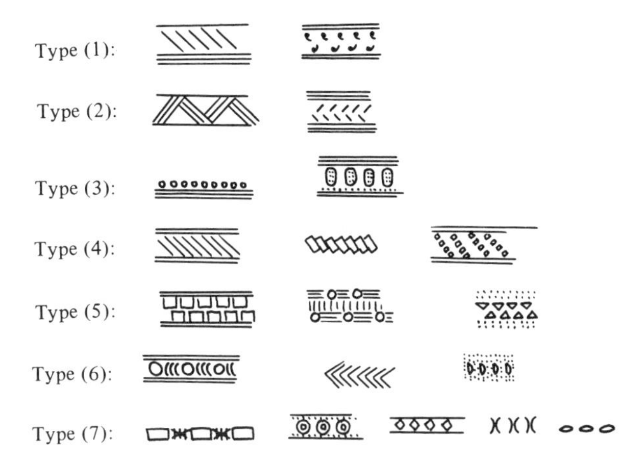


**Figure 17.** Some typical Begho K2 pipe patterns of each of the seven one-dimensional types.

**Table 1**

| Type | Number of Patterns | % of Total |
|---|---|---|
| (1) | 4 | 2 |
| (2) | 2 | 1 |
| (3) | 22 | 10 |
| (4) | 19 | 8 |
| (5) | 9 | 4 |
| (6) | 9 | 4 |
| (7) | 165 | 72 |
| Totals | 230 | 101% |

the *p*6*m* example is on such a small fragment as to be somewhat ambiguous. The great majority of the patterns are of types *p*4*m* (55%) and *pmm* (25%). This is not surprising, since each of these is readily obtained as parallel strips of suitable versions of the predominant type (7) one-dimensional pattern.

This last fact leads to the following ambiguity in the two-dimensional patterns which does not occur in the one-dimensional. A common pattern consists of parallel rows of "comb-stamping", i.e. rows of tiny rectangles. In the 18 examples of this we have read the rectangles as squares, and hence classified these 18 as of type *p*4*m*. However, because they are so minute, these "squares" might be read by another observer as (nonsquare) rectangles. In that case, the pattern would have been classified as type *pmm*. Indeed, if that were done, the total numbers for *p*4*m* and *pmm* would be more or less interchanged.

The two-dimensional results are given in Table 2.

At the Begho K2 site, the location of pipe fragments was recorded as Level I (humus), II, III, or IV, (and occasionally V, VI), with IV generally representing the lowest level. It might be hoped that some orderly transition from preference

**Table 2**

| Type | Number of Patterns | % of Total |
|---|---|---|
| *p*2 | 3 | 6 |
| *pm* | 2 | 4 |
| *pmm* | 12 | 25 |
| *pgg* | 1 | 2 |
| *cmm* | 3 | 6 |
| *p*4*m* | 27 | 55 |
| *p*6*m* | 1 | 2 |
| Totals | 49 | 100% |

for one pattern to preference for another could be seen, corresponding to the time sequence supposedly represented by the different levels. Table 3 gives the seven pattern types, and the percentage of each at the four levels. No significant correspondence is apparent to the present writer.

**Table 3.** Percentage of Strip Types According to Level[a]

| Strip type | Level I | Level II | Level III | Level IV |
|---|---|---|---|---|
| (1) | — | 1 | 7 | 5 |
| (2) | — | 2 | — | — |
| (3) | 4 | 12 | 13 | 9 |
| (4) | 12 | 7 | 7 | 5 |
| (5) | 4 | 4 | 13 | — |
| (6) | 8 | 4 | — | — |
| (7) | 73 | 69 | 60 | 82 |
| Total nos. of pipes | 26 | 102 | 15 | 22 |

[a] Neglecting surface finds and Levels V, VI.

## References

[1] Afeku, I. K., A study of smoking pipes from Begho. B.A. Honours Dissertation, University of Ghana, Legon, Apr. 1976.

[2] Crowe, D. W., The geometry of African art I. Bakuba art. *Journal of Geometry* **1** (1971), 169–182.

[3] Crowe, D. W., The geometry of African art II. A catalog of Benin patterns. *Historia Mathematika* **2** (1975), 253–271.

[4] Schattschneider, Doris, The plane symmetry groups: their recognition and notation. *Amer. Math. Monthly* **85** (1978), 439–450.

[5] Shepard, A. O., The symmetry of abstract design with special reference to ceramic decoration. In Carnegie Inst. Wash. Publ. 574, Contrib. 47, 1948.

[6] Shepard, A. O., *Ceramics for the Archeologist*. Carnegie Inst. Wash. Publ. 609, seventh printing, Washington, D.C. 1971.

[7] Speiser, A., *Theorie der Gruppen von endlicher Ordnung*. Birkhäuser, Basel 1956.

[8] Washburn, Dorothy, A symmetry analysis of Upper Gila ceramic design. Papers of the Peabody Museum, Harvard Univ., No. 78, 1977.

[9] Zaslow, B., and Dittert, A. E., Pattern mathematics and archeology. Arizona State University Anthropological Research Papers No. 2, 1977.

# Crystallography and Cremona Transformations

Patrick Du Val*

The note that follows is based essentially on some investigations which I undertook in about 1930 [4, 5], in response to Coxeter's earliest researches [1] on the pure Archimedean polytopes $(PA)_n$ in $n$ dimensions, later fitted into his more general notation [2] as $(n-4)_{2,1}$ $(3 \leqslant n \leqslant 9)$. It had been remarked that the 27 vertices of $(PA)_6$ correspond in an invariant manner to the 27 lines on a general cubic surface; and in the discussions that followed amongst the group of students that surrounded H. F. Baker, it soon emerged that there was a similar correspondence between $(PA)_n$ $(n = 3, 4, 5)$ and the lines on the del Pezzo surface of order $9 - n$, between $(PA)_7$ and the bitangents of a general plane quartic curve, and between $(PA)_8$ and the tritangent planes of a certain twisted sextic curve. The theory I propose now to outline provides a systematic explanation of all these correspondences, as well as others that were remarked later.

We consider, in $d$-dimensional complex projective space $S_d$, a set of $r \geqslant d + 1$ points $(P) = P_1, \ldots, P_r$ in general position (if we like we can think of them as a generic set, all of whose coordinates are independent transcendents over the complex number field), and denote a primal of order $n_0$ with an $n_i$-ple point in $P_i$ $(i = 1, \ldots, r)$ by the column vector of integers $\mathbf{n} = (n_0, n_1, \ldots, n_r)^T$. The virtual complete linear system (whether effective or not, i.e. whether it has any actual members or not) of all primals of order $n_0$ with $n_i$-ple base points in $P_i$ $(i = 1, \ldots, r)$ may be denoted by $|\mathbf{n}|$. Such a system, i.e. one defined by assigned order and base multiplicities in some or all of $P_1, \ldots, P_r$, will be said to be based on $(P)$.

A cremona transformation whose homaloidal system (the system transformed into that of all primes of $S_d$) is based on $(P)$ will itself be said to be based on $(P)$. In the plane $(d = 2)$ it is known that every cremona transformation is the resultant of a finite sequence of elementary transformations, whose homaloidal

*10 Gainsborough Close, Cambridge CB4 1SY, England.

nets consist of conics through three base points. For $d \geqslant 3$ there is no such general theorem; but we define an elementary transformation based on a given simplex, by taking this as simplex of reference for a homogeneous coordinate system $(\xi_0, \ldots, \xi_d)$, when the transformation is

$$(\xi_0 : \ldots : \xi_d) \rightarrow (1/\xi_0 : \ldots : 1/\xi_d).$$

The homaloidal system is of order $d$, with $(d-1)$-ple base points at the vertices of the simplex; it also passes $(d-2)$-ply through the edges, $(d-3)$-ply through the planc faces, . . . , and simply through the $(d-2)$-dimensional elements of the simplex; but these base elements are necessary consequences of the $(d-1)$-ple base points at the vertices, and do not need to be specified. The characteristic curves, which are transformed into the lines of $S_d$, are normal rational $d$-ic curves, passing through the vertices of the simplex. We define further a punctual transformation in $S_d$ to be one which is the resultant of a finite sequence of elementary transformations. This has the distinctive property that its homaloidal system is completely specified by its order and multiplicities in a finite set of base points, any base elements of higher dimension being necessary consequences of these.

A punctual transformation with $s$ base points destroys these $s$ points (i.e. they have no images); their neighborhoods are mapped birationally on $s$ primals; and conversely, there are $s$ primals which are mapped onto the neighborhoods of $s$ new points (the base points of the inverse transformation, which is likewise punctual) created by the transformation. In particular, the elementary transformation maps each vertex of the base simplex onto the opposite prime face, by a $(d-1)$-dimensional elementary transformation, so that a line in the neighborhood of a vertex is mapped on, and is the map of, a normal rational $(d-1)$-ic curve through the remaining vertices. If the transformation is based on $(P)$, we are thus presented with a new set $(P')$ of $r$ points, $s$ of which are the new points created by the transformation, and the remaining $r-s$ are the transforms of those of the points $(P)$ which, not being base points of the transformation, have not been destroyed by it. We can denote these $r$ points by $P_1', \ldots, P_r'$ in any order. We can then follow the transformation up by another punctual transformation based on $(P')$, and this by a third based on the new set $(P'')$ produced by the second, and so on. Regarding the object on which we are operating, rather formally, as a general set of $r$ points in $S_d$, together with the aggregate of virtual linear systems based on the set, we see that after each operation we have precisely the same object before us again, and that the aggregate of all punctual cremona transformations based on the $r$ points forms a group, which we denote by $G(r,d)$, and which is the object of our study. We note that owing to the arbitrariness of the ordering of each new set as it arises, $G(r,d)$ includes all permutations of the points $(P)$, or (perhaps better) of the ordinal indices $1, \ldots, r$.

For the elementary transformation with base simplex $P_1, \ldots, P_{d+1}$ we naturally take $P_i'$ to be $P_i$ $(i = 1, \ldots, d+1)$, and to be the transform of the undestroyed point $P_i$ $(i = d+2, \ldots, r)$. If $F$ is a general member of the system $|\mathbf{n}|$, let $F'$ be its transform, with order-and-multiplicity vector $\mathbf{n}'$. As the intersections of $F'$ with a line are the images of those of $F$ with a $d$-ic curve through $P_1, \ldots, P_{d+1}$, other than these points, which are destroyed, we have $n_0' = dn_0 - \sum_{i=1}^{d+1} n_i$; as the intersections of $F'$ with a line in the neighborhood of $P_1$ are the images of those of $F$ with a $(d-1)$-ic curve through $P_2, \ldots, P_{d+1}$, apart from these points, we have $n_1' = (d-1)n_0 - \sum_{i=2}^{d+1} n_i$, with similar expressions for $n_2', \ldots, n_{d+1}'$; and for $i = d+2, \ldots, r$, as the cremona mapping is regular at $P_i$, we have $n_i' = n_i$. This means that the virtual linear system $|\mathbf{n}|$ is transformed into the virtual linear system $|\mathbf{n}'| = |\mathbf{E}\mathbf{n}|$, where

$$\mathbf{E} = \left[\begin{array}{cccccc|c} d & -1 & -1 & -1 & \cdots & -1 & \\ d-1 & 0 & -1 & -1 & \cdots & -1 & \\ d-1 & -1 & 0 & -1 & \cdots & -1 & \\ d-1 & -1 & -1 & 0 & \cdots & -1 & \mathbf{0} \\ \cdots & \cdots & \cdots & \cdots & \cdots & \cdots & \\ d-1 & -1 & -1 & -1 & \cdots & 0 & \\ & & \mathbf{0}^T & & & & \mathbf{I}' \end{array}\right] \quad \begin{array}{l} \left.\right\} d+1 \text{ rows} \\ \\ r-d-1 \text{ rows} \end{array} \tag{1}$$

(with the first column followed by $d+1$ columns and then $r-d-1$ columns)

$\mathbf{I}'$ denoting the $(r-d-1) \times (r-d-1)$ identity matrix, and $\mathbf{0}$ a block of $(d+2) \times (r-d-1)$ zeros. This matrix can also be written

$$\mathbf{E} = \mathbf{I} + \mathbf{e}\mathbf{e}^T\mathbf{J}, \tag{2}$$

where $\mathbf{I}$ is the $(r+1) \times (r+1)$ identity matrix, $\mathbf{e}$ is a column of $d+2$ 1's followed by $r-d-1$ zeros, and $\mathbf{J} = \operatorname{diag}(d-1, -1, -1, \ldots, -1)$. Similarly, the elementary transformation with any base simplex chosen from $(P)$ transforms the system $|\mathbf{n}|$ into $|\mathbf{E}'\mathbf{n}| = |(\mathbf{I} + \mathbf{e}'\mathbf{e}'^T\mathbf{J})\mathbf{n}|$, where $e_0' = 1$, and for $i = 1, \ldots, r$, $e_i' = 1$ or 0 according as $P_i$ is a vertex of the base simplex or not. As every transformation in $G(r,d)$ is the resultant of a finite sequence of elementary transformations based on $(P)$, it transforms the system $|\mathbf{n}|$ into $|\mathbf{H}\mathbf{n}|$, where $\mathbf{H}$ is the product of the matrices $\mathbf{E}, \mathbf{E}', \ldots$ corresponding to the elementary factors; and the group of $(r+1) \times (r+1)$ matrices of integers generated by the matrices $\mathbf{E}, \mathbf{E}', \ldots$ corresponding to all the elementary transformations based on $(P)$ is naturally isomorphic with $G(r,d)$.

We verify at once that $\mathbf{e}^T\mathbf{J}\mathbf{e} = -2$; and from this it follows that

$$\begin{aligned} \mathbf{E}^T\mathbf{J}\mathbf{E} &= (\mathbf{I} + \mathbf{J}\mathbf{e}\mathbf{e}^T)\mathbf{J}(\mathbf{I} + \mathbf{e}\mathbf{e}^T\mathbf{J}) \\ &= \mathbf{J} + 2\mathbf{J}\mathbf{e}\mathbf{e}^T\mathbf{J} + \mathbf{J}\mathbf{e}(\mathbf{e}^T\mathbf{J}\mathbf{e})\mathbf{e}^T\mathbf{J} = \mathbf{J}, \end{aligned} \tag{3}$$

since the second and third terms of the penultimate member cancel. Similarly, $\mathbf{E}'^T\mathbf{J}\mathbf{E}' = \mathbf{J}$ for any other elementary transformation based on $(P)$. This means that the quadratic form $\mathbf{x}^T\mathbf{J}\mathbf{x}$ and the bilinear form $\mathbf{x}^T\mathbf{J}\mathbf{y}$, for any column vectors $\mathbf{x}$, $\mathbf{y}$ of real numbers, are invariant under the linear transformation $\mathbf{x} \longrightarrow \mathbf{E}\mathbf{x}$ (or $\mathbf{x} \longrightarrow \mathbf{E}'\mathbf{x}$) corresponding to any elementary transformation based on $(P)$, and hence also under the transformation $\mathbf{x} \longrightarrow \mathbf{H}\mathbf{x}$ corresponding to any element of $G(r,d)$. The invariance of this bilinear form is obvious geometrically for $d = 2$, as in this case $\mathbf{n}_1^T\mathbf{J}\mathbf{n}_2$ is the number of intersections, not absorbed in the base points, of general curves of the systems $|\mathbf{n}_1|$, $|\mathbf{n}_2|$; for $d > 2$, it is the number of intersections, not absorbed in the base points, of general primals of the two systems, with a general surface of an invariant system of equivalence, of order $d-1$ and with simple base points in all the points $(P)$.

Accordingly, taking the order-and-multiplicity vectors $\mathbf{n}$ in a real affine space $A_{r+1}$, and interpreting the components of a vector as the affine coordinates of a point of $A_{r+1}$, we can use the matrix $\mathbf{J}$ to define a scalar product $\mathbf{x} \cdot \mathbf{y} = \mathbf{x}^T\mathbf{J}\mathbf{y}$, and so impose metrical properties on $A_{r+1}$. This is not of course a Euclidean metric, but is of the kind familiar in relativity theory, with one dimension of time and $r$ dimensions of space. In terms of this metric, all the transformations of $G(r,d)$ are orthogonal; and as they leave invariant the lattice of integer points, this means that $G(r,d)$ acts on $A_{r+1}$ as a crystallographic group.

In terms of this metric, $\mathbf{x} \longrightarrow \mathbf{E}\mathbf{x}$ is the reflection in the mirror prime $\mathbf{e} \cdot \mathbf{x} = 0$; for $\mathbf{e}$, and hence also every vector parallel to $\mathbf{e}$, is reversed in sign, whereas every vector perpendicular to $\mathbf{e}$ is unchanged:

$$\begin{gathered} \mathbf{E}\mathbf{e} = \mathbf{e} + \mathbf{e}\mathbf{e}^T\mathbf{J}\mathbf{e} = \mathbf{e} - 2\mathbf{e} = -\mathbf{e}, \\ \mathbf{e}^T\mathbf{J}\mathbf{x} = 0 \quad \Rightarrow \quad \mathbf{E}\mathbf{x} = \mathbf{x} + \mathbf{e}\mathbf{e}^T\mathbf{J}\mathbf{x} = \mathbf{x}. \end{gathered} \tag{4}$$

Similarly, for any other elementary transformation based on $(P)$, $\mathbf{x} \longrightarrow \mathbf{E}'\mathbf{x}$ is the reflection in the mirror prime $\mathbf{e}' \cdot \mathbf{x} = 0$. Thus $G(r,d)$ is a crystallographic group generated by reflections.

There is one vector which is perpendicular to all the vectors $\mathbf{e}, \mathbf{e}', \ldots$ corresponding to the elementary transformations based on $(P)$, and which is consequently unchanged by any operation of $G(r,d)$, namely

$$\mathbf{k} = (d+1, d-1, d-1, \ldots, d-1)^T, \tag{5}$$

representing the linear system (effective for sufficiently low values of $r$) of $(d+1)$-ic primals with $(d-1)$-ple base points in all the points $(P)$. The invariance of this vector is obvious geometrically, since $|-\mathbf{k}|$ is the canonical system on the $d$-dimensional algebraic variety obtained by dilating all the points $(P)$ on $S_d$; i.e., on a general member of the system $|\mathbf{n}|$, its canonical system is traced by the system $|\mathbf{n} - \mathbf{k}|$.

All operations of $G(r,d)$ consequently leave invariant in $A_{r+1}$, not only all the concentric $r$-spheres $\mathbf{x} \cdot \mathbf{x} = \nu$ (which, of course, in the relativity space, are not convex quadrics, but hyperboloids, of two sheets or one according as $\nu$ is positive or negative), but also all the parallel primes $A_r$: $\mathbf{k} \cdot \mathbf{x} = \sigma$; and hence also in each $A_r$ $\mathbf{k} \cdot \mathbf{x} = \sigma$, they leave invariant all the concentric $(r$-$1)$-spheres traced on it by the $r$-spheres $\mathbf{x} \cdot \mathbf{x} = \nu$. As the mirror primes of the reflections generating $G(r,d)$ all contain the line through the origin of $A_{r+1}$ parallel to $\mathbf{k}$, they cut all the $A_r$'s $\mathbf{k} \cdot \mathbf{x} = \sigma$ in the same configuration of $A_{r-1}$'s, and $G(r,d)$ operates on each of these $A_r$'s as a crystallographic point group, generated by the reflections in these $A_{r-1}$'s; the fixed point being the intersection of $\mathbf{k} \cdot \mathbf{x} = \sigma$ with the line through the origin of $A_{r+1}$ parallel to $\mathbf{k}$, which is the point $\sigma\mathbf{k}/(\mathbf{k} \cdot \mathbf{k})$; and this is the center of the $(r-1)$-spheres $\mathbf{k} \cdot \mathbf{x} = \sigma$, $\mathbf{x} \cdot \mathbf{x} = \nu$.

As $\mathbf{k} \cdot \mathbf{k} = (d-1)[(d+1)^2 - r(d-1)]$,

$$\mathbf{k} \cdot \mathbf{k} \gtreqless 0 \quad \text{according as} \quad r \lesseqgtr \frac{(d+1)^2}{d-1} = d + 3 + \frac{4}{d-1}\,. \tag{6}$$

If $\mathbf{k} \cdot \mathbf{k} > 0$, i.e. if $r < (d+1)^2/(d-1)$, the geometry in each $A_r$ perpendicular to $\mathbf{k}$ is Euclidean, with the trivial awkwardness that the unit of length is pure imaginary, so that $\mathbf{x} \cdot \mathbf{x}$, the square of the length of a real vector $\mathbf{x}$ in $A_r$, is a negative number; in particular the square of the radius of the sphere $\mathbf{k} \cdot \mathbf{x} = \sigma$, $\mathbf{x} \cdot \mathbf{x} = \nu$, namely $\nu - \sigma^2/(\mathbf{k} \cdot \mathbf{k})$, must be negative for the sphere to have a real sheet. In this case, $G(r,d)$ is an ordinary crystallographic point group generated by reflections in Euclidean space, and is of course finite; we shall shortly identify it with one of the groups generated by reflections listed by Coxeter [3].

There are three cases in which $\mathbf{k} \cdot \mathbf{k} = 0$, namely $d = 2$, $r = 9$; $d = 3$, $r = 8$; and $d = 5$, $r = 9$ (the only cases in which $d - 1$ is a divisor of 4). In these cases the metric in any prime perpendicular to $\mathbf{k}$ (which contains vectors parallel to $\mathbf{k}$) is degenerate, every vector parallel to $\mathbf{k}$ being of zero length, and any two whose difference is parallel to $\mathbf{k}$ being of the same length. The $(d-1)$-spheres $\mathbf{k} \cdot \mathbf{x} = \sigma$, $\mathbf{x} \cdot \mathbf{x} = \nu$ are paraboloids, with axes parallel to $\mathbf{k}$, and their common center, the fixed point of $G(r,d)$ as a point group in $A_r$, is at infinity in the direction $\mathbf{k}$. The geometry on any of these $(d-1)$-spheres is that of the horosphere in hyperbolic geometry; the sphere projects isometrically, parallel to $\mathbf{k}$, onto any prime $A_{r-1}$ in $A_r$ not parallel to $\mathbf{k}$; and in this $A_{r-1}$ the geometry is Euclidean. In particular the prime faces of any polytope inscribed in the sphere project isometrically into $(r-1)$-dimensional polytopes forming a tessellation in $A_{r-1}$. Also, if two points in $A_r$ are interchanged by reflection in a prime $M_r$ in $A_{r+1}$, their projections on $A_{r-1}$ are interchanged by reflection in the prime $M_{r-2}$ of $A_{r-1}$, which is the intersection of $A_{r-1}$ with $M_r$; the angle between two mirror primes $M_r$, $M_r'$ in $A_{r+1}$ is equal to that between their traces $M_{r-2}$, $M_{r-2}'$, on $A_{r-2}$; and this angle may be zero, which means in $A_{r+1}$ that the intersection of $M_r$, $M_r'$ touches the isotropic cone $\mathbf{x} \cdot \mathbf{x} = 0$ along its generator parallel to $\mathbf{k}$, and in $A_{r-1}$ that $M_{r-2}$, $M_{r-2}'$ are parallel, so that the resultant of the reflections in them is a translation. Thus $G(r,d)$ operates on $A_{r-1}$ as a crystallographic space group (not a point group) in $(r-1)$-dimensional Euclidean space.

Finally, if $\mathbf{k} \cdot \mathbf{k} < 0$, i.e., if $r > (d+1)^2/(d-1)$, the metric in each prime perpendicular to $\mathbf{k}$ is the relativity metric, with one dimension of time and $r - 1$ dimensions of space; and $G(r,d)$ operates on each of these as a crystallographic point group in the $r$ dimensional relativity space. Such a group is infinite, and will in general contain, as well as ordinary rotations, what we may call hyperbolic rotations, i.e. essentially Lorenz transformations, reducible by a change of coordinates to

$$\begin{pmatrix} x_0 \\ x_1 \end{pmatrix} \longrightarrow \begin{pmatrix} \cosh u & \sinh u \\ \sinh u & \cosh u \end{pmatrix} \begin{pmatrix} x_0 \\ x_1 \end{pmatrix},$$

leaving the other coordinates unchanged. This will arise as the resultant of reflections in two primes $\mathbf{a} \cdot \mathbf{x} = 0$, $\mathbf{b} \cdot \mathbf{x} = 0$, where $\mathbf{a} \cdot \mathbf{a} = \mathbf{b} \cdot \mathbf{b} = -2$, and $\mathbf{a} \cdot \mathbf{b} = -2\cosh(u/2)$.

$G(r,d)$ being generated by all the elementary transformations based on $(P)$, is generated by any one of these, say that with the base simplex $P_1 \ldots P_{d+1}$, which is the reflection in the prime $\mathbf{e} \cdot \mathbf{x} = 0$, and all permutations of the coordinates $x_1, \ldots, x_r$, i.e. by the reflection in $\mathbf{e} \cdot \mathbf{x} = 0$ and the transpositions of consecutive coordinates $(x_1, x_2)$, $(x_2, x_3)$, $\ldots$, $(x_{r-1}, x_r)$. The transposition $(x_i, x_j)$ is the reflection in the prime $\mathbf{t}_{i,j} \cdot \mathbf{x} = 0$, where $\mathbf{t}_{i,j} \cdot \mathbf{x} = x_i - x_j$, so that $\mathbf{t}_{i,j}$ is the vector whose components are all zero, except those with indices $i$, $j$, which are $-1$, $1$ respectively. We note that $\mathbf{t}_{i,j} \cdot \mathbf{t}_{i,j} = -2$, and that $\mathbf{E}_{i,j} = I + \mathbf{t}_{i,j}\mathbf{t}_{i,j}^T\mathbf{J}$ is the matrix effecting the transposition $(x_i, x_j)$. Thus $G(r,d)$ is generated by the reflections in the primes

$$\mathbf{e} \cdot \mathbf{x} = 0, \quad \mathbf{t}_{1,2} \cdot \mathbf{x} = 0, \quad \mathbf{t}_{2,3} \cdot \mathbf{x} = 0, \ldots, \quad \mathbf{t}_{r-1,r} \cdot \mathbf{x} = 0. \tag{7}$$

Now if two vectors $\mathbf{a}$, $\mathbf{b}$ are both of length $\sqrt{-2}$, i.e. if $\mathbf{a} \cdot \mathbf{a} = \mathbf{b} \cdot \mathbf{b} = -2$, the angle $\theta$ between them is given by $-2\cos\theta = \mathbf{a} \cdot \mathbf{b}$; in particular, if $\mathbf{a} \cdot \mathbf{b} = 1$, then $\theta = 2\pi/3$. But it is easily verified that $\mathbf{t}_{i,j} \cdot \mathbf{t}_{j,k} = 1$, whereas $\mathbf{t}_{i,j} \cdot \mathbf{t}_{k,l} = 0$ if $i, j, k, l$ are all different; and that $\mathbf{e} \cdot \mathbf{t}_{d+1,d+2} = 1$, whereas $\mathbf{e} \cdot \mathbf{t}_{i,j} = 0$ if $i$, $j$ are both $\leqslant d+1$, or both $\geqslant d+2$. Thus the mirror primes (7) are represented by the Coxeter graph

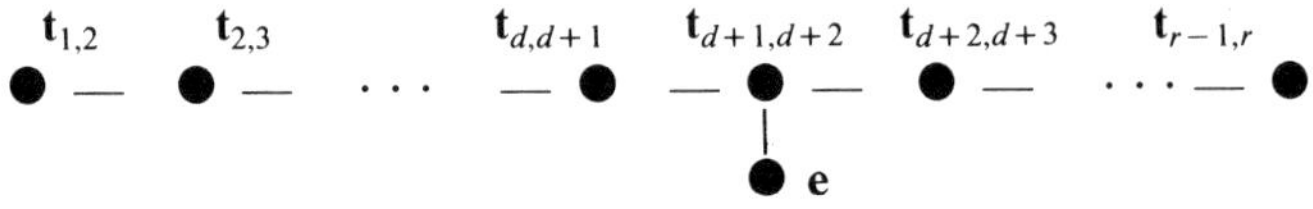

$$\begin{array}{ccccccccccccc} \mathbf{t}_{1,2} & & \mathbf{t}_{2,3} & & & & \mathbf{t}_{d,d+1} & & \mathbf{t}_{d+1,d+2} & & \mathbf{t}_{d+2,d+3} & & \mathbf{t}_{r-1,r} \\ \bullet & - & \bullet & - & \cdots & - & \bullet & - & \bullet & - & \bullet & - \cdots - & \bullet \\ & & & & & & & & | & & & & \\ & & & & & & & & \bullet\ \mathbf{e} & & & & \end{array} \tag{8}$$

where the dots represent primes perpendicular to the vectors named, inclined at $\pi/2$ or $2\pi/3$ according as the dots are unlinked or linked. This graph represents the generating reflections for Coxeter's group $[3^{d,d',1}]$, where $d' = r - d - 2$. The groups $[3^{n,p,q}]$ in Euclidean space, corresponding to graphs consisting of three concurrent chains, of $n$, $p$, and $q$ links, have been enumerated classically by Coxeter [3], and correspond to all values of $n$, $p$, $q$ satisfying

$$\frac{1}{n+1} + \frac{1}{p+1} + \frac{1}{q+1} \geqslant 1,$$

with equality for a crystallographic space group in $n + p + q$ dimensions, and strict inequality for a crystallographic point group in $n + p + q + 1$ dimensions. It can in fact be shown that $[3^{n,p,q}]$ exists for all integer values of $n$, $p$, $q$, and is a crystallographic point group in Euclidean space of $n + p + q + 1$ dimensions, a crystallographic space group in Euclidean space of $n + p + q$ dimensions, or an (infinite) crystallographic point group in relativity space of $n + p + q + 1$ dimensions, according as

$$\frac{1}{n+1} + \frac{1}{p+1} + \frac{1}{q+1} > 1, = 1, \text{ or } < 1; \tag{9}$$

and for $(n, p, q) = (d, r - d - 2, 1)$, this criterion reduces precisely to (6).

For $r < d + 1$ there are no cremona transformations based on $r$ points in $S_d$; and for $r = d + 1$, the graph (8) falls apart into

$$\begin{array}{cccccccc} \mathbf{t}_{1,2} & & \mathbf{t}_{2,3} & & & & \mathbf{t}_{d,d+1} & \quad \mathbf{e} \\ \bullet & - & \bullet & - & \cdots & - & \bullet & \quad \bullet \end{array}$$

corresponding to the fact that in this case there is only one proper cremona transformation, the elementary transformation based on $P_1, \ldots, P_{d+1}$; and $G(d+1,d)$ is the direct product of the group of order 2 generated by this elementary transformation, with the group of all permutations on the indices $1, \ldots, d+1$. Using Coxeter's notation [3], the point and space groups in Euclidean space for $d \leqslant 6$ are the following:

$$\begin{array}{lccccccc} & r=3 & r=4 & r=5 & r=6 & r=7 & r=8 & r=9 \\ d=2: & A_2\times A_1 & A_4 & B_5 & E_6 & E_7 & E_8 & T_9 \\ d=3: & & A_3\times A_1 & A_5 & B_6 & E_7 & T_8 & \\ d=4: & & & A_4\times A_1 & A_6 & B_7 & E_8 & \\ d=5: & & & & A_5\times A_1 & A_7 & B_8 & T_9 \\ d=6: & & & & & A_6\times A_1 & A_8 & B_9 \end{array} \qquad (10)$$

For all $d \geqslant 6$, the only groups in Euclidean space are $G(d+1,d) = A_d \times A_1$, $G(d+2,d) = A_{d+2}$, and $G(d+3,d) = B_{d+3}$.

Still taking $d' = r - d - 2$, it is clear that the graph (8) represents a generating set of reflections for $G(r,d')$ as well as one for $G(r,d)$, by taking $\mathbf{e} = 0$ to be the mirror prime corresponding to the elementary transformation in $S_{d'}$ based on $P_{d+2}, \ldots, P_r$, i.e.

$$e = \Big(1, \overbrace{0, \ldots, 0}^{d+1}, \overbrace{1, \ldots, 1}^{d'+1}\Big)^T \quad \text{instead of} \quad \Big(1, \overbrace{1, \ldots, 1}^{d+1}, \overbrace{0, \ldots, 0}^{d+1}\Big)^T.$$

This establishes an isomorphism between $G(r,d)$ and $G(r,d')$ $(d + d' = r - 2)$ in which every permutation of the indices $1, \ldots, r$ is self-corresponding, and the elementary transformation based on any selection of $d+1$ of the points $(P)$ in $S_d$ corresponds to that in $S_{d'}$ based on the points with the remaining $d'+1$ indices. Of the groups in Euclidean space tabulated in (10), this gives $G(7,2) = G(7,3) = E_7$, $G(8,2) = G(8,4) = E_8$, $G(9,2) = G(9,5) = T_9$; but of course similar isomorphisms hold also between groups in relativity space, $G(9,3) = G(9,4)$, $G(10,2) = G(10,6)$, $G(10,3) = G(10,5)$, and so on. Similarly, $G(2d+2,d)$ has an automorphism, in which every permutation of the indices $1, \ldots, 2d+2$ is self-corresponding, and the elementary transformation based on any set of $d+1$ of the points $(P)$ corresponds to that based on the remaining $d+1$ points. This may be an inner automorphism; for $G(6,2)$ for instance it is the inner automorphism induced by the operation $\mathbf{x} \longrightarrow (\mathbf{I} + \mathbf{c}\mathbf{c}^T\mathbf{J})\mathbf{x}$ of $G(6,2)$, where $\mathbf{c} = (2,1,1,1,1,1,1)^T$; this is the cremona transformation by quintics with double base points in $P_1, \ldots, P_6$, mapping the neighborhood of each base point on the conic through the other five. For $G(8,3)$ on the other hand the automorphism in question is an outer one; for it is easily seen that there is no homaloidal system in $S_3$, except $(1,0,\ldots,0)^T$, with eight base points all of the same multiplicity, and hence no operation of $G(8,3)$ except identity that commutes with all permutations of $x_1, \ldots, x_8$.

By a trajectory of any transformation group, finite or infinite, we mean a set of points which consists of the transforms, under all operations of the group, of any one point of the set. Any trajectory of $G(r,d)$ in $A_{r+1}$ is inscribed in one of the ($r$-1)-spheres $\mathbf{k} \cdot \mathbf{x} = \sigma$, $\mathbf{x} \cdot \mathbf{x} = \nu$, owing to the invariance of the forms $\mathbf{k} \cdot \mathbf{x}$, $\mathbf{x} \cdot \mathbf{x}$ under all operations of the group, and consists of the vertices of a polytope in the $A_r$ $\mathbf{k} \cdot \mathbf{x} = \sigma$. In the finite case $\mathbf{k} \cdot \mathbf{k} > 0$, this is obvious, the polytope, as a region of $A_r$, being the convex hull of the trajectory, the smallest closed convex region to which no point of the trajectory is exterior. But in all cases, even in relativity space, the polytope is clearly identifiable, its prime faces being those $A_{r-1}$'s in $A_r$ that are spanned by the points of the trajectory that lie in them, and are such that all points of the trajectory that are not in the $A_{r-1}$ are on the same side of it in $A_r$; and the elements of lower dimension of the polytope are defined in the same way from the prime faces, by induction downwards.

In particular, for any positive integers $n$, $p$, $q$, the trajectory of $[3^{n,p,q}]$ generated by a point lying in all the primes represented in the graph except that represented by the end dot of the chain of $n$ links, is classically (as in [3]) the vertices of the polytope $n_{p,q}$; this has two families of faces, namely $n_{p-1,q}$ and $n_{p,q-1}$, on each of which the group is transitive; and its vertex figure (the base of the pyramid of prime faces at any vertex) is $(n-1)_{p,q}$. The three polytopes $n_{p,q}$, $p_{q,n}$, $q_{n,p}$ are what is called semireciprocal, which means that the centers of the faces of each are the vertices of both the others; in fact, the centers of the faces $n_{p-1,q}$ of $n_{p,q}$ are the vertices of $p_{q,n}$, and those of the faces $n_{p,q-1}$ are the vertices of $q_{n,p}$.

Now the order-and-multiplicity vector representing the neighborhood of the base point $P_i$ is that denoting order 0 and multiplicity $-1$ at $P_i$ and 0 at all the other base points $(P)$; this can be seen by verifying that the transformation $\mathbf{x} \longrightarrow \mathbf{E}\mathbf{x}$ interchanges $(0, -1, 0, \ldots, 0)^T$ with $(1, 0, 1, \ldots, 1, 0, \ldots, 0)^T$ representing the prime $P_2 \ldots P_{d+1}$. But the lattice point $(0, 0, \ldots, 0, -1)^T$ lies in all the primes represented in the graph (8) except $\mathbf{t}_{r-1,r} \cdot \mathbf{x} = 0$, $x_{r-1} = x_r$; thus the trajectory of this point consists of the vertices of the polytope $d'_{d,1}$ $(d' = r - d - 2)$. In this context it is convenient, in order to extend the notation downwards to the cases $r = d + 2, d + 1$, to denote the vertex figure of $1_{d,1}$ by $0_{d,1}$, and that of $0_{d,1}$ by $(-1)_{d,1}$; these polytopes are a truncated simplex in $A_{d+2}$ and a simplicial prism, the product of a simplex in $A_d$ with a line segment. The coordinate vectors of the vertices of $d'_{d,1}$ are accordingly the order-and-multiplicity vectors of all primals of $S_d$ that can be transformed by any operation of $G(r, d)$ into the neighborhood of any of the points $(P)$; these are called the exceptional primals based on $(P)$. $d'_{d,1}$ is inscribed in the $(d-1)$-sphere $\mathbf{k} \cdot \mathbf{x} = d - 1$, $\mathbf{x} \cdot \mathbf{x} = -1$. (In particular, for $d = 2$, the exceptional curves are all the rational curves of grade $-1$ based on $(P)$; and these are all the rational curves that are uniquely determined by their order and multiplicities in $(P)$.)

Again, the point $(1, 0, \ldots, 0)^T$, representing the complete system of primes in $S_d$ without base points, lies in all the primes $\mathbf{t}_{i,i+1} \cdot \mathbf{x} = 0$, but not in $\mathbf{e} \cdot \mathbf{x} = 0$; thus the trajectory of this point consists of the vertices of the polytope $1_{d,d'}$; and the coordinate vectors of these vertices are the order-and-multiplicity vectors of all linear systems based on $(P)$ that can be transformed by any operation of $G(r, d)$ into the primes of space, i.e. of all the homaloidal systems based on $(P)$. $1_{d,d'}$ is inscribed in the $(d-1)$-sphere $\mathbf{k} \cdot \mathbf{x} = d^2 - 1$, $\mathbf{x} \cdot \mathbf{x} = d - 1$. The prime faces $d'_{d,0}$ of $d'_{d,1}$ are simplexes, since the removal of the single link to $\mathbf{e}$ from the graph (8) reduces it to the single chain $A_{r-1}$; and one of these simplexes has the vertices $(0,-1,0,\ldots,0)^T,\ldots,(0,0,\ldots,0,-1)^T$ corresponding to the neighborhoods of the points $(P)$, and the vertex of $1_{d,d'}$ corresponding to this is $(1,0,\ldots,0)^T$; thus the vertices of each simplificial face of $d'_{d,1}$ represent the $d+1$ exceptional primals which are transformed into the neighborhoods of the points $(P)$ by the cremona transformation whose homaloidal system is represented by the corresponding vertex of $1_{d,d'}$.

The third polytope $d_{d',1}$ has not much geometrical significance except for $d=2$, since a generating lattice point for this trajectory is $(1,d-1,0,\ldots,0)^T$, which for $d>2$ does not represent any effective linear system. For $d=2$ however, $(1,1,0,\ldots,0)^T$ represents the pencil of lines through $P_1$ in $S_2$, so that the vertices of $d_{d',1}=2_{r-4,1}$ represent all pencils of rational curves based on $(P)$ in $S_2$. The faces $d'_{d-1,1}=(r-4)_{1,1}$ of $d'_{d,1}=(r-4)_{2,1}$ are cross polytopes (the analog of the octahedron) in $A_{r-1}$, with $r-1$ pairs of diametrically opposite vertices on perpendicular diameters; and these pairs represent pairs of exceptional curves which together form a reducible curve in the pencil represented by the corresponding vertex of $2_{r-4,1}$.

A number of the finite groups $G(r,d)$ here considered contain the central symmetry, which interchanges diametrically opposite points in $A_r$; notable amongst these are $B_6$, $E_7$, $E_8$, i.e. $G(6,3)$, $G(7,2)$, $G(7,3)$, $G(8,2)$, and $G(8,4)$. This operation of $G(r,d)$ is in $S_d$ an involutory cremona transformation, generating an invariant subgroup of order 2, i.e. coinciding with its transform by any element of $G(r,d)$. As the center of $1_{d,d'}$, in the $A_r$ $(\mathbf{k}\cdot\mathbf{x}=d^2-1$, is the point $\alpha\mathbf{k}$, where $\alpha=(d^2-1)/(\mathbf{k}\cdot\mathbf{k})=(d+1)/[(d+1)^2-r(d-1)]$, the point that is interchanged with $(1,0,\ldots,0)^T$ is $2\alpha\mathbf{k}-(1,0,\ldots,0)^T$, namely the following:

$$\begin{array}{lll} \text{for } (r,d)= & (6,3), & (7,4,4,4,4,4,4)^T,\\ & (7,2), & (8,3,3,3,3,3,3,3)^T,\\ & (7,3), & (15,8,8,8,8,8,8,8)^T,\\ & (8,2), & (17,6,6,6,6,6,6,6,6)^T,\\ & (8,4), & (49,30,30,30,30,30,30,30,30)^T. \end{array} \tag{11}$$

These are accordingly the order-and-multiplicity vectors for the homaloidal systems of the cremona transformations in question. The first four of these interchange the pairs of points in well-known and classical involutions, namely those of Geiser based on six points in $S_3$, of Geiser based on seven points in the plane, of Kantor based on seven points in $S_3$, and of Bertini based on eight points in the plane. The fifth involution, based on eight points in $S_4$, has not, so far as I am aware, yet been studied.

We return now to the correspondences mentioned in our opening paragraph, which were the starting point of the whole study. The del Pezzo surface of order $n$ in $S_n$ is the projective model of the system of cubics with $9-n$ simple base points in the plane $(3\leqslant n\leqslant 9)$, i.e. of the system $|\mathbf{k}|$ for $d=2$, $r\leqslant 6$. The lines on the surface are the images of the exceptional curves based on $(P)$, which are the only effective irreducible curves satisfying $\mathbf{k} \cdot \mathbf{n} = 1$. They thus correspond to the vertices of $(r-4)_{2,1} = (5-n)_{2,1}$, for $n = 3, 4, 5, 6$. Moreover, if lines $l$, $l'$ on the surface correspond to exceptional curves (and vertices of the polytope) represented by vectors $\mathbf{n}$, $\mathbf{n}'$ as $\mathbf{n} \cdot \mathbf{n} = \mathbf{n}' \cdot \mathbf{n}' = -1$, then $(\mathbf{n} - \mathbf{n}') \cdot (\mathbf{n} - \mathbf{n}') = -2(1 + \mathbf{n} \cdot \mathbf{n}')$; thus according as the lines $l$, $l'$ are skew or intersect, the distance between the corresponding vertices is $\sqrt{2}\, i$ or $2i$, in terms of the rather artificial unit of length introduced with our scalar product $\mathbf{x} \cdot \mathbf{y} = \mathbf{x}^T J \mathbf{y}$, or $1, \sqrt{2}$ in terms of a more natural unit which is the edge length of the polytope.

Rather similarly, the projective model of the quadrics with $r \leqslant 5$ simple base points in $S_3$, i.e. the system $|\frac{1}{2}\mathbf{k}|$ for $d = 3$, is a three-dimensional variety of order $8 - r$, whose prime sections are del Pezzo surfaces, and on which are a finite number of planes, images of the exceptional surfaces based on the $r$ points, and which thus correspond invariantly to the vertices of $(r-5)_{3,1}$ for $r = 4, 5$. Vertices separated by distance $a$ (the natural unit introduced above), i.e. joined by an edge of the polytope, correspond to planes meeting only in a node of the variety (represented by a line joining two base points in $S_3$) or not at all, whereas vertices separated by distance $\sqrt{2}\, a$ correspond to planes meeting in a line.

The next members of these two sequences are the two involutions of Geiser; for $(r, d) = (7, 2)$, the system $|\mathbf{k}|$ of cubics through $(P) = P_1, \ldots, P_7$ in the plane is compounded with the involution of pairs of points which with $(P)$ make up nine associated points; its projective model is the Geiser double plane, branching on a general quartic curve. Each of the 28 pairs of diametrically opposite vertices, separated by distance $\sqrt{3}\, a$, corresponds to a pair of exceptional curves based on $(P)$, which together form a reducible curve of $|\mathbf{k}|$, and meet in two points outside $(P)$; and the corresponding lines on the double plane coincide in a bitangent of the branch curve, their two common points being the points of contact. Similarly, for $(r, d) = (6, 3)$, the system $|\frac{1}{2}\mathbf{k}|$ is compounded with the involution of pairs of points which with $(P)$ make up eight associated points; the projective model of the system is a double $S_3$, branching on a Kummer surface. Each of the 16 pairs of opposite vertices of $1_{3,1}$ corresponds to a pair of exceptional surfaces, whose images on the double $S_3$ are coincident planes, touching the branch surface and intersecting each other along a conic. The 16 nodes of the Kummer surface correspond to the 15 lines joining the base points by pairs, and the twisted cubic through all of them.

Turning now to $(r, d) = (8, 2)$, the system $|\mathbf{k}|$ is a pencil, with the ninth associated base point; but $|2\mathbf{k}| = (6, 2, \ldots, 2)^T$ is of grade 4 and freedom 3 and is compounded with the Bertini involution. Its projective model is a double quadric cone in $S_3$, whose generators are the images of the cubics $|\mathbf{k}|$, with an isolated branch point at the vertex (image of the ninth associated point) and branching also on a sextic curve, which is a general cubic section of the cone. Each of the 120 pairs of opposite vertices of $4_{2,1}$ (separated by distance $2a$) corresponds to a pair of exceptional curves, whose images on the double cone are coincident conics, forming together the section of the double cone by a tritangent plane of the branch curve, the points of contact being the images of the three common points (outside of $(P)$) of the two exceptional curves.

For $(r, d) = (7, 3)$, the system $|\frac{1}{2}\mathbf{k}|$ is the net of quadrics with the eighth associated base point; but $|\mathbf{k}| = (4, 2, \ldots, 2)^T$ is compounded with the Kantor involution. The projective model of the system is a double three-dimensional cone projecting a Veronese surface from a point in $S_6$. It has an isolated branch point at the vertex of the cone, and branches also on a surface of order 12, which is a cubic section of the cone—not a general cubic section, however, as it has 28 nodes, corresponding invariantly to the pairs of opposite vertices of $3_{2,1}$, and mapped in $S_3$ by the 21 lines joining the base points by pairs, and 7 twisted cubics through all but one of them. Each of the 63 pairs of opposite vertices of $2_{3,1}$ corresponds to a pair of exceptional surfaces, whose images on the double cone are coincident Veronese surfaces, forming together a prime section of the double cone, and touching the branch surface along an elliptic sextic curve, through 12 of the 28 nodes, which is the image of the curve of intersection of the two exceptional surfaces in $S_3$.

We have now looked at the involutions whose pairs are interchanged by the cremona transformations corresponding to four of the homaloidal systems (11). The fifth, based on eight points in $S_4$, is I believe quite unknown. We note however that each of the 1080 pairs of opposite vertices of $2_{4,1}$ corresponds to a pair of exceptional primals, which together form a reducible primal of the system $|2\mathbf{k}| = (10, 6, \ldots, 6)^T$. It is thus at least a plausible conjecture that the system $|2\mathbf{k}|$ in $S_4$ is compounded with this involution; that the projective model of the system is a four-dimensional variety $V_4$, doubled and branching on a $V_3$; and that the images on this of each of the 1080 pairs of exceptional primals coincide in the section of $V_4$ by a prime, which touches the branch locus $V_3$ at all points of a surface, which is the image of the intersection of the two exceptional primals in $S_4$. But the detailed study of this figure in the algebraic geometry of $S_4$ seems to present substantial difficulties.

## References


[1] Coxeter, H. S. M., The pure archimedian polytopes in six and seven dimensions. *Proc. Cambridge Phil. Soc.* **24** (1928), 1–9.

[2] Coxeter, H. S. M., The polytopes with regular-prismatic vertex figures. *Phil. Trans. Royal Soc. London (A)* **229** (1930), 329–425.

[3] Coxeter, H. S. M., Chapter 11 in *Regular Polytopes*. Methuen, London 1928; 2nd ed. Macmillan, New York 1963; 3rd ed. Dover, New York, 1973.

[4] Du Val, P., On the directrices of a set of points in a plane. *Proc. London Math. Soc. (2)* **35** (1932), 23–74.

[5] Du Val, P., On the Kantor group of a set of points in a plane. *Proc. London Math. Soc. (2)* **42** (1936), 18–51.

[6] Du Val, P., Application des idées cristallographiques a l'étude des groupes de transformations crémoniennes. In *3^e Colloque de Géometrie Algébrique*. Centre Belge de Recherches Mathématiques, Bruxelles 1959. (This is not referred to in the text above; I include it as being the only other publication, of my own or, so far as I know, of anybody, dealing with the present topic.)

# Cubature Formulae, Polytopes, and Spherical Designs

J. M. Goethals*
J. J. Seidel†

## 1. Introduction

The construction of a cubature formula of strength $t$ for the unit sphere $\Omega_d$ in $\mathbb{R}^d$ amounts to finding finite sets $X_1, \ldots, X_N \subset \Omega_d$ and coefficients $a_1, \ldots, a_N \in \mathbb{R}$ such that

$$|\Omega_d|^{-1} \int_{\Omega_d} f(\xi)\, d\omega(\xi) = \sum_{i=1}^{N} a_i |X_i|^{-1} \sum_{x \in X_i} f(x), \tag{1.1}$$

for all functions $f$ represented on $\Omega_d$ by polynomials of degree $\leqslant t$; cf. [16], [15], [11]. Sobolev [14, 15] introduced group theory into the construction of cubature formulae by considering orbits $X_i$ under a finite subgroup $G$ of the orthogonal group $O(d)$. Thus spherical polytopes and root systems (cf. Coxeter [3]) enter the discussion. There are further relations to Coxeter's work, since the obstruction to higher strength for a cubature formula is caused essentially by the existence of certain invariants. For finite groups generated by reflections, the theory of exponents and invariants goes back to Coxeter [4].

A spherical $t$-design $X \subset \Omega_d$ may be defined by (cf. [5], [7])

$$|\Omega_d|^{-1} \int_{\Omega_d} p(\xi)\, d\omega(\xi) = |X|^{-1} \sum_{x \in X} p(x), \tag{1.2}$$

for all polynomials $p$ of degree $\leqslant t$. Thus a spherical $t$-design provides a cubature formula of strength $t$ with one set, and the use of several spherical $t$-designs $X_i$ may provide cubature formulae (1.1) of strength $> t$. The condition (1.2) is equivalent to

$$\operatorname*{ave}_X f := |X|^{-1} \sum_{x \in X} f(x) = 0, \tag{1.3}$$

* Philips Research Laboratory, Brussels, Belgium.
† Technological University Eindhoven, the Netherlands.

for all harmonic homogeneous polynomials $f$ of degree $\leqslant t$. In the group case, when $X$ is a $G$-orbit, this says that there are no $G$-invariant harmonic polynomials on $\Omega_d$ of degrees $1, 2, \ldots, t$. In the general case an analogous statement holds. To that end we adapted the notion of invariant to the nongroup situation. This reflects the strategy of the present paper. In each of Sections 2, 3, 4 we start with the general situation, and later specialize to the group case. In the final Sections 5, 6, 7 specific examples are considered. The main contents are as follows.

In Section 2 consideration of the harmonic components reduces the construction of cubature formulae to the solution of a system of linear equations. In the group case this system may have much smaller size, depending on the coefficients of the harmonic Molien series for the group. The dihedral group of order 12 illustrates this. In Section 3 spherical $t$-designs $X$ are introduced in terms of tensors and in terms of the special polynomials

$$s_k(X,\xi) := |X|^{-1} \sum_{x \in X} (x,\xi)^k - |\Omega_d|^{-1} \int_{\Omega_d} (\eta,\xi)^k \, d\omega(\eta). \tag{1.4}$$

These definitions are equivalent to (1.2) and to (1.3). In Section 4 distance-invariant spherical $t$-designs $X$ are characterized by the absence of harmonic invariants for $X$ of degrees $1, \ldots, t$, where an invariant for $X$ is a function on $\Omega_d$ which takes nonzero constant values on $X$. Furthermore, for the case of finite reflection groups, the theory of exponents and invariants, and Flatto's results [6] on the actual invariants are reviewed. The theory is applied to the regular polytopes in $\mathbb{R}^3$ and $\mathbb{R}^4$ in Section 5, to the root systems in Section 6, and to the Leech lattice in Section 7. In each example harmonic invariants of the type (1.4) are "killed" either by taking suitable linear combinations of orbits, or by taking the orbit of a zero of the harmonic invariant. Thus we obtain spherical 15-designs in $\mathbb{R}^{24}$, 11-designs in $\mathbb{R}^8$, 19-designs in $\mathbb{R}^4$, 9-designs in $\mathbb{R}^3$ (and an improvement of the football).

## 2. Cubature Formulae

Let $\Omega_d$, with measure $\omega(\xi)$, denote the unit sphere in real Euclidean space $\mathbb{R}^d$. A *cubature formula of strength t* for $\Omega_d$ consists of a finite set of points $x_1, \ldots, x_N \in \Omega_d$ and coefficients $a_1, \ldots, a_N \in \mathbb{R}$ such that

$$|\Omega_d|^{-1} \int_{\Omega_d} f(\xi) \, d\omega(\xi) = \sum_{i=1}^{N} a_i f(x_i) \tag{2.1}$$

for all $f \in \mathrm{Pol}(t)$. Here $\mathrm{Pol}(t)$ denotes the linear space of all functions in $d$ variables which, restricted to $\Omega_d$, are represented by polynomials of degree $\leqslant t$. Let $\mathrm{Harm}(k)$ denote the linear subspace of $\mathrm{Pol}(k)$ consisting of all harmonic homogeneous polynomials of degree $k$, that is, all homogeneous polynomials of degree $k$ which satisfy Laplace's equation. With respect to the inner product

$$\langle f, g \rangle = |\Omega_d|^{-1} \int_{\Omega_d} f(\xi) g(\xi) \, d\omega(\xi)$$

we have the orthogonal decomposition

$$\mathrm{Pol}(t) = \mathrm{Harm}(t) \perp \mathrm{Harm}(t-1) \perp \ldots \perp \mathrm{Harm}(0)$$

(cf. [9], [7]). The dimensions of these spaces are

$$Q_K := \dim \mathrm{Harm}(k) = \binom{d+k-1}{d-1} - \binom{d+k-3}{d-1},$$

$$R_k := \dim \mathrm{Pol}(k) = \binom{d+k-1}{d-1} + \binom{d+k-2}{d-1}.$$

In terms of the harmonic components, the condition (2.1) reads

$$f_0 = \sum_{i=1}^{N} a_i \sum_{k=0}^{t} f_k(x_i) \tag{2.2}$$

for all $f_k \in \mathrm{Harm}(k)$, for $k = 0, 1, \ldots, t$. Let $f_{k,1}, \ldots, f_{k,Q_k}$ denote any orthonormal basis for Harm($k$). Then (2.2) is equivalent to

$$1 = \sum_{i=1}^{N} a_i, \qquad 0 = \sum_{i=1}^{N} a_i f_{k,j}(x_i) \tag{2.3}$$

for $j = 1, \ldots, Q_k$, for $k = 1, \ldots, t$. These are

$$1 + \sum_{k=1}^{t} Q_k = R_t$$

linear constraints on $a_1, \ldots, a_N$ with the coefficient matrix

$$H = [H_0 H_1 \ldots H_t], \quad \text{where } H_k = [f_{k,j}(x_i)].$$

There exist points $x_1, \ldots, x_N \in \Omega_d$ such that rank $H = R_t$; cf. [9, Theorem 3]. Therefore the problem of constructing cubature formulae of arbitrary strength $t$ is solved in principle. However, finding adequate points $x_1, \ldots, x_N$ and solving (2.3) may be a time-consuming affair.

Sobolev [14] introduced group theory into the construction of cubature formulae. Let $G$ denote a finite subgroup of the orthogonal group $O(d)$ in $\mathbb{R}^d$ and let

$$X_i = \{ g x_i \mid g \in G \}$$

be the orbit of any $x_i \in \Omega_d$. Any $g \in G$ acts on a function $f$ on the sphere by

$$f^g(\xi) := f(g^{-1}\xi), \qquad \xi \in \Omega_d.$$

The function $f$ is *G-invariant* whenever $f = f^g$ for all $g \in G$. For any function $f$ its *average*

$$\bar{f} = \operatorname*{ave}_G f := |G|^{-1} \sum_{g \in G} f^g$$

is $G$-invariant, and every $G$-invariant function is obtained in this way. Let $\mathrm{Harm}^G(k)$ denote the subspace of the $G$-invariant functions of Harm($k$). In Section 4 we shall see that

$$q_k := \dim \mathrm{Harm}^G(k)$$

may be calculated from the harmonic Molien–Poincaré series

$$|G|^{-1} \sum_{g \in G} \frac{1-\lambda^2}{\det(1-\lambda g)} = \sum_{k=0}^{\infty} q_k \lambda^k.$$

Now consider cubature formulae of strength $t$ for $\Omega_d$ consisting of finite subsets $X_1, \ldots, X_M$ of $\Omega_d$ and coefficients $c_1, \ldots, c_M \in \mathbb{R}$ such that

$$|\Omega_d|^{-1} \int_{\Omega_d} f(\xi)\, d\omega(\xi) = \sum_{i=1}^{M} c_i |X_i|^{-1} \sum_{x \in X_i} f(x) \tag{2.4}$$

for all $f \in \mathrm{Pol}(t)$. Taking for each $X_i$ the orbit under $G$ of some $x_i \in \Omega_d$, we have

$$|X_i|^{-1} \sum_{x \in X_i} f(x) = |G|^{-1} \sum_{g \in G} f^g(x_i) = \bar{f}(x_i).$$

The equations (2.3) reduce to

$$1 = \sum_{i=1}^{M} c_i, \qquad 0 = \sum_{i=1}^{M} c_i \bar{f}_k(x_i) \tag{2.5}$$

for all $\bar{f}_k \in \mathrm{Harm}^G(k)$, for $k = 1, \ldots, t$. These are

$$r_t := \sum_{k=0}^{t} q_k$$

linear constraints on $c_1, \ldots, c_M$. Thus cubature formulae of strength $t$ are constructed provided adequate orbits $X_i = \{ gx_i \,|\, g \in G \}$ are found to solve (2.5). Clearly the use of the appropriate groups may save a considerable amount of work. We illustrate the method by the following trivial example, leaving further examples for later sections.

**(2.6) Example.** We consider the case $d = 2$ (so $\Omega_2$ is the unit circle) and the group $G = W(G_2)$, the Weyl group of the hexagon, that is, the dihedral group of order 12. The harmonic Molien series reads

$$(1-\lambda^6)^{-1} = 1 + \lambda^6 + \lambda^{12} + \cdots,$$

and the $G$-invariant harmonic functions are

$$\cos 6k\theta \quad \text{for } k = 1, 2, \ldots .$$

Any starting point $x \in \Omega_2$ yields an orbit of 12 points (the vertices of two regular hexagons), providing a cubature formula of strength 5. If the starting point satisfies $\cos 6\theta = 0$, say $\theta = \pi/12$, then the orbit consists of the vertices of a regular 12-gon, providing a cubature formula of strength 11; we have killed the invariant $\cos 6\theta$.

In order to illustrate in our example the equations (2.5) with various coefficients $c_i$, we first calculate the principal determinants of the following matrix, with $x = \cos 6\theta_1$, $y = \cos 6\theta_2$, $z = \cos 6\theta_3$:

$$\left[\begin{array}{cc|c|c} 1 & 1 & 1 & 1 \\ 0 & x & y & z \\ \hline 0 & 2x^2-1 & 2y^2-1 & 2z^2-1 \\ \hline 0 & 4x^3-3x & 4y^3-3y & 4z^3-3z \end{array}\right].$$

They are $\det = x$, $\det = (y-x)(2xy+1)$,

$$\det = 4(x-y)(y-z)(z-x)(2xyz+x+y+z),$$

respectively. If $\det = 0$, then provided the denominators are nonzero, the coefficients are

$$c_1 = 1; \qquad c_1 = \frac{y}{y-x}, \quad c_2 = \frac{-x}{y-x};$$

$$c_1 = \frac{yz(z^2-y^2)}{\Delta}, c_2 = \frac{xz(x^2-z^2)}{\Delta}, c_3 = \frac{xy(y^2-x^2)}{\Delta},$$

$$\Delta = (x-y)(y-z)(z-x)(x+y+z);$$

respectively. Thus cubature formulae of strength 17 are obtained from two orbits whose starting points have $\theta_1$ and $\theta_2$ with

$$\cos 6\theta_1 \neq \cos 6\theta_2, \qquad 1 + 2\cos 6\theta_1 \cos 6\theta_2 = 0,$$

and the coefficients are

$$c_1 = \frac{2\cos^2 6\theta_2}{1+2\cos^2 6\theta_2}, \qquad c_2 = \frac{1}{1+2\cos^2 6\theta_2}.$$

In the special case

$$\cos 6\theta_1 = \tfrac{1}{2}\sqrt{2} = -\cos 6\theta_2, \qquad c_1 = c_2 = \tfrac{1}{2},$$

we have strength 23; indeed, in this case the two orbits together form the vertices of a regular 24-gon. Cubature formulae of strength 23 are also obtained from 3 orbits whose $\theta_1$, $\theta_2$, $\theta_3$ satisfy

$$2\cos 6\theta_1 \cos 6\theta_2 \cos 6\theta_3 + \cos 6\theta_1 + \cos 6\theta_2 + \cos 6\theta_3 = 0;$$

$\cos 6\theta_1$, $\cos 6\theta_2$, $\cos 6\theta_3$ distinct with sum $\neq 0$.

## 3. Spherical Designs

For any nonempty finite subset $X$ of the unit sphere $\Omega_d$, and for $k = 0, 1, 2, \ldots,$ we define the *symmetric k-tensor* $S^k(X)$, and the *special polynomial* $s_k(X, \xi)$ of degree $k$ as follows.

**(3.1) Definition.**

$$S^k(X) := |X|^{-1} \sum_{x \in X} \otimes^k x - |\Omega_d|^{-1} \int_{\Omega_d} \otimes^k \eta \, d\omega(\eta),$$

$$s_k(X, \xi) := \left(S^k(X), \otimes^k \xi\right), \qquad \xi \in \Omega_d.$$

We recall that, for any $\xi \in \mathbb{R}^d$ with orthogonal coordinates $\xi_1, \ldots, \xi_d$, the components of $\otimes^k \xi$ are the monomials in $\xi_1, \ldots, \xi_d$ of total degree $k$, and that the (trace) inner product of $\otimes^k \xi$ and $\otimes^k \eta$ equals

$$\left(\otimes^k \xi, \otimes^k \eta\right) = (\xi, \eta)^k.$$

The definitions (3.1) are justified by the following lemmas, which have straightforward proofs; cf. also [7, Theorem 3.1].

**(3.2) Lemma.**

$$\left(S^k(X), S^k(X)\right) = \begin{cases} |X|^{-2} \sum\limits_{x,y \in X} (x,y)^k & \text{for odd } k, \\ |X|^{-2} \sum\limits_{x,y \in X} (x,y)^k - \dfrac{1 \cdot 3 \cdot \ldots \cdot (k-1)}{d(d+2) \ldots (d+k-2)} & \text{for even } k. \end{cases}$$

**(3.3) Lemma.**

$$s_k(X,\xi) = \begin{cases} |X|^{-1} \sum\limits_{x \in X} (x,\xi)^k & \text{for odd } k, \\ |X|^{-1} \sum\limits_{x \in X} (x,\xi)^k - \dfrac{1 \cdot 3 \cdot \ldots \cdot (k-1)(\xi,\xi)^{k/2}}{d(d+2) \ldots (d+k-2)} & \text{for even } k. \end{cases}$$

**(3.4) Lemma.** *If* $s_k(X, y) = 0$ *for all* $y \in X$, *then* $s_k(X,\xi) = 0$ *for all* $\xi \in \Omega_d$.

**(3.5) Lemma.** *If* $s_k(X,X) := |X|^{-1}\sum_{x \in X} s_k(X,x) = 0$, *then* $s_k(X,\xi) = 0$ *for all* $\xi \in \Omega_d$.

**(3.6) Lemma.** $\int_{\Omega_d} s_k(X,\xi)\, d\omega(\xi) = 0$.

**(3.7) Lemma.** $\Delta s_k(X,\xi) = k(k-1)s_{k-2}(X,\xi)$, *where* $\Delta$ *is Laplace's operator.*

Since the inner product of a tensor with itself is nonnegative, Lemma (3.2) yields inequalities in terms of the inner products of the vectors of $X$; these inequalities are due to Sidelnikov [12]. Our interest will be in sets $X$ for which equality holds; cf. [5], [7].

**(3.8) Definition.** A finite nonempty set $X \subset \Omega_d$ is a *spherical t-design* whenever $S^k(X) = 0$ for $k = 1, 2, \ldots, t$.

The following equivalencies are immediate; cf. [7, Theorem 4.4].

**(3.9) Theorem.** *For a finite nonempty set* $X \subset \Omega_d$ *the following conditions are equivalent*:

(i) $X$ *is a spherical t-design*,
(ii) $s_k(X,X) = 0$ *for* $k = 1, 2, \ldots, t$,
(iii) $\sum_{x \in X} f(x) = 0$ *for all* $f \in \mathrm{Harm}(k)$, *for* $k = 1, 2, \ldots, t$,
(iv) $|X|^{-1}\sum_{x \in X} p(x) = |\Omega_d|^{-1}\int_{\Omega_d} p(\xi)\, d\omega(\xi)$ *for all* $p \in \mathrm{Pol}(t)$.

The equivalence of (i) and (iv) provides the link with the previous section:

**(3.10) Theorem.** *A spherical $t$-design yields a cubature formula of strength $t$.*

**(3.11) Example.** *The* 120 *vertices of the regular polytope* $\{3,3,5\}$ *in* $\mathbb{R}^4$ *form a spherical* 11*-design* [5, *Example* 8.6], *and hence yield a cubature formula for* $\Omega_4$ *of strength* 11; *cf. Salihov* [10] *and Section* 5.

Next we consider point sets $X$ which are point-orbits on $\Omega_d$ under a finite subgroup $G$ of the orthogonal group $O(d)$. The following theorem relates spherical designs and $G$-invariant polynomials.

**(3.12) Theorem.** *For a finite subgroup $G$ of $O(d)$ the following conditions are equivalent*:

(i) *every $G$-orbit is a spherical $t$-design,*
(ii) *there are no $G$-invariant harmonic polynomials of degrees* $1, 2, \ldots, t$.

*Proof.* By Theorem (3.9) every $G$-orbit $X_0 = \{ gx_0 \mid g \in G \}$ is a spherical $t$-design iff for all $x_0 \in \Omega_d$, for $k = 1, \ldots, t$, and for all $f \in \mathrm{Harm}(k)$

$$0 = |X_0|^{-1} \sum_{x \in X_0} f(x) = |G|^{-1} \sum_{g \in G} f^g(x_0) = \bar{f}(x_0).$$

□

(3.13) *Remark.* Further results are the following (cf. [7, Theorem 6.10]). For even $t$, every $G$-orbit is a spherical $t$-design if and only if, for $k = 1, 2, \ldots, \lfloor \frac{1}{2} t \rfloor$, the representation $\rho_k$ of $G$ on $\mathrm{Harm}(k)$ is real irreducible. For odd $t$, every $G$-orbit is a spherical $t$-design if, for $k = 1, 2, \ldots, \lfloor \frac{1}{2} t \rfloor$, the representation $\rho_k$ of $G$ on $\mathrm{Harm}(k)$ is real irreducible and has no common constitutents with $\rho_{\lfloor \frac{1}{2} t \rfloor + 1}$.

The implications of Theorem (3.12) for the second part of Section 2 are obvious. Clearly, suitable linear combinations of spherical $t$-designs may serve to kill invariants, and to push up the strength of cubature formulae. We shall meet specific examples in later sections.

Another way to obtain cubature formulae of strength $> t$ on the basis of Theorem (3.12) applies for instance to the case when $\dim \mathrm{Harm}^G(t) = 1$. If $x_0$ is a zero of the representing $G$-invariant polynomial, then the orbit generated by $x_0$ provides a spherical design of strength $> t$. The existence of such $x_0$ follows from the next Lemma; cf. [6].

**(3.14) Lemma.** *Any harmonic polynomial has a zero on* $\Omega_d$.

*Proof.* The integral over $\Omega_d$ of any harmonic polynomial $f$ equals zero. Therefore, if $f$ takes positive values on $\Omega_d$, then $f$ also takes negative values, and hence the value 0, on $\Omega_d$.

□

## 4. Invariants

It seems contradictory to discuss invariants in a nongroup situation. Yet, certain analogies lead to the following notions. We restrict to nonempty finite sets $X \subset \Omega_d$.

**(4.1) Definition.** A function $f$ defined on $\Omega_d$ is an *invariant* of the set $X \subset \Omega_d$ whenever it takes nonzero constant values on $X$.

Candidates for invariants of a set $X$ are its special polynomials $s_k(X, \cdot)$, introduced in Section 3, and its *valencies* $v_\alpha$, for $\alpha \in A := \{(x, y) \mid x, y \in X\}$, defined by

$$v_\alpha(\xi) := |\{x \in X \mid (x, \xi) = \alpha\}|, \qquad \xi \in \Omega_d.$$

**(4.2) Definition.** $X$ is *distance-invariant* whenever, for all $\alpha \in A$, the valency $v_\alpha$ is an invariant of $X$.

**(4.3) Lemma.** *If $X$ is distance-invariant, then for any $k$, any nonzero $s_k(X, \xi)$ is an invariant of $X$.*

*Proof.* The hypothesis implies that $s_k(X, y)$ is constant for any $y \in X$, since

$$\sum_{x \in X} (x, \xi)^k = \sum_{\alpha \in A} \alpha^k v_\alpha(\xi).$$

The statement then follows from Lemma (3.4).

□

**(4.4) Lemma.** *If $s_k(X, \xi)$ is constant on $X$ for $k = 1, \ldots, |A| - 1$, then $X$ is distance-invariant.*

*Proof.* For the unknowns $v_\alpha(y)$, $\alpha \in A$, $y \in X$, the hypothesis yields $|A|$ linear equations

$$\sum_{\alpha \in A} \alpha^k v_\alpha(y) = c_k, \qquad k = 0, 1, \ldots, |A| - 1,$$

with nonzero Vandermonde determinant. Hence the $v_\alpha(y)$ are uniquely determined and are independent of $y \in X$.

□

**(4.5) Theorem.** *A distance-invariant set $X$ is a spherical $t$-design iff it has no harmonic invariants of degrees $1, 2, \ldots, t$.*

*Proof.* If $X$ is a spherical $t$-design, then $\sum_{x \in X} f(x) = 0$ for all $f \in \mathrm{Harm}(k)$, for $k = 1, 2, \ldots, t$. Hence $f$ is not an invariant of $X$. Conversely, suppose that the distance-invariant set $X$ has no harmonic invariants of degrees $1, 2, \ldots, t$. The function $s_k(X, \xi)$ is zero for $k = 0$, is harmonic for $k = 1$, 2, and hence is zero for

$k = 0, 1, \ldots, t$ by application of Lemmas (4.3) and (3.7). By Theorem (3.9) this implies that $X$ is a spherical $t$-design.

□

In the group case there is an extensive theory of invariants. The dimensions $q_i$ of the spaces $\mathrm{Harm}^G(i)$ of the $G$-invariant harmonic polynomials of degree $i$ follow from the *harmonic Molien–Poincaré series*:

**(4.6) Theorem**

$$\sum_{i=0}^{\infty} q_i \lambda^i = |G|^{-1} \sum_{g \in G} \frac{1 - \lambda^2}{\det(1 - \lambda g)} .$$

*Proof*. The dimensions $h_i$ of the spaces $V_i^G$ of the $G$-invariant functions represented on $\Omega_d$ by homogeneous polynomials of degree $i$ are the coefficients in the Molien–Poincaré series [1, 13]

$$\sum_{i=0}^{\infty} h_i \lambda^i = |G|^{-1} \sum_{g \in G} \frac{1}{\det(1 - \lambda g)} .$$

The Laplace operator $\Delta$ is a $G$-invariant operator, and $\mathrm{Harm}^G(i)$ is the kernel of $\Delta : V_i^G \to V_{i-2}^G$. Hence $q_i = h_i - h_{i-2}$, and the harmonic formula follows.

□

We now restrict to the case of real finite reflection groups. For the following results we refer to Carter [1] and to Flatto [6]. Let $G$ denote an irreducible finite group generated by reflections in $\mathbb{R}^d$. The ring $R^G$ of the $G$-invariant polynomials has the following characteristic property. $R^G$ has an algebraic basis consisting of $d$ homogeneous polynomials, called *basic invariants*, of degrees $1 + m_i$, $i = 1, \ldots, d$. The *exponents* $1 = m_1 \leqslant m_2 \leqslant \ldots \leqslant m_d$ are the logarithms, to the base $\exp 2\pi i/h$, of the eigenvalues of the Coxeter–Killing transformation, and

$$m_i + m_{d+1-i} = h = \frac{2r}{d}, \qquad \sum_{i=1}^{d} m_i = r,$$

where $h$ is the period of that transformation and $r$ is the total number of reflections. All this was initiated by Coxeter [4], who also classified the finite reflection groups in terms of the root systems $\Phi$, with Weyl groups $W(\Phi)$, and calculated the exponents $m_i$, as shown in Table 1. The exponents also serve to calculate $q_i = \dim \mathrm{Harm}^G(i)$, since in the case of finite reflection groups the harmonic Molien–Poincaré series reads

$$\sum_{i=0}^{\infty} q_i \lambda^i = \prod_{i=2}^{d} (1 - \lambda^{1+m_i})^{-1}.$$

Once the dimensions and the degrees of $\mathrm{Harm}^G(i)$ are known, it remains to determine the invariant harmonic polynomials themselves. To that end, Flatto [6] proves the following theorem.

**Table 1**

| $\Phi$ | $r = \frac{1}{2}\|\Phi\|$ | $\|W(\Phi)\|$ | $h$ | $m_i$ |
|---|---|---|---|---|
| $A_d$ $(d \geqslant 1)$ | $\frac{1}{2}d(d+1)$ | $(d+1)!$ | $d+1$ | $1, 2, \ldots, d$ |
| $B_d$ $(d \geqslant 2)$ | $d^2$ | $2^d d!$ | $2d$ | $1, 3, \ldots, 2d-1$ |
| $D_d$ $(d \geqslant 4)$ | $d(d-1)$ | $2^{d-1}d!$ | $2d-2$ | $1, 3, \ldots, 2d-3, d-1$ |
| $H_2^p$ $(p \geqslant 5)$ | $p$ | $2p$ | $P$ | $1, p-1$ |
| $G_2$ | 6 | 12 | 6 | $1, 5$ |
| $I_3$ | 15 | 120 | 10 | $1, 5, 9$ |
| $F_4$ | 24 | $2^7 \cdot 3^2$ | 12 | $1, 5, 7, 11$ |
| $I_4$ | 60 | $120^2$ | 30 | $1, 11, 19, 29$ |
| $E_6$ | 36 | $2^7 \cdot 3^4 \cdot 5$ | 12 | $1, 4, 5, 7, 8, 11$ |
| $E_7$ | 63 | $2^{10} \cdot 3^4 \cdot 5 \cdot 7$ | 18 | $1, 5, 7, 9, 11, 13, 17$ |
| $E_8$ | 120 | $2^{14} \cdot 3^5 \cdot 5^2 \cdot 7$ | 30 | $1, 7, 11, 13, 17, 19, 23, 29$ |

**(4.7) Theorem.** *Let $P_k(\xi, \eta) := |G|^{-1}\sum_{g \in G}(\xi, g\eta)^k$, for a finite reflection group $G \neq D_{2d}$. Then*

$$\det\left[\frac{\partial(P_{1+m_1}, \ldots, P_{1+m_d})}{\partial(\xi_1, \ldots, \xi_d)}\right] = \prod_{i=1}^{d} J_i(\eta) \prod_{j=1}^{r} L_j(\xi),$$

*where $L_j(\xi) = 0$ denote the reflecting hyperplanes, and $J_1(\eta), \ldots, J_d(\eta)$ are the unique (up to constants) basic invariants satisfying $J_1(\eta) = (\eta, \eta)$, $J_k(\partial/\partial\eta)J_l = 0$ for $1 \leqslant k < l \leqslant d$.*

Clearly, the unique basic invariants $J_2, \ldots, J_d$ are harmonic. Theorem (4.7) implies that, provided $y \in \Omega_d$ satisfies $\prod_{i=1}^{d} J_i(y) \neq 0$, a set of basic invariants is given by the polynomials

$$P_{1+m_1}(\xi, y), \ldots, P_{1+m_d}(\xi, y).$$

It follows that, provided $Y$ is the orbit of $y \in \Omega_d$ satisfying $\prod_{i=1}^{d} J_i(y) \neq 0$, a set of basic invariants is also given by the special polynomials

$$(\xi, \xi), s_{1+m_2}(Y, \xi), \ldots, s_{1+m_d}(Y, \xi).$$

Indeed, up to a constant the Flatto polynomials and the special polynomials have the same Jacobian determinant.

## 5. The Regular Polytopes in $\mathbb{R}^3$ and $\mathbb{R}^4$

The full symmetry groups of the tetrahedron, the octahedron and the icosahedron in $\mathbb{R}^3$ (the binary polyhedral groups) are the Weyl groups of the root systems $A_3$, $B_3$, $I_3$, of the order 24, 48, 120, respectively. Their harmonic Molien series are

$$\frac{1}{(1-\lambda^3)(1-\lambda^4)}, \quad \frac{1}{(1-\lambda^4)(1-\lambda^6)}, \quad \frac{1}{(1-\lambda^6)(1-\lambda^{10})}.$$

We first illustrate the previous sections by $W(A_3)$, taking the set $X = \{x_0, x_1, x_2, x_3\}$ of the vertices of the tetrahedron to be $X = 3^{-1/2}\{(1,1,1), (1,-1,-1), (-1,1,-1), (-1,-1,1)\}$. The orbit $Y$ of the point $y_0 = 2^{-1/2}(0,1,1)$ consists of the centers of the edges of the tetrahedron and is the root system $A_3$. The orbit $Z$ of the point $z_0 = 5^{-1/4}(0, \tau^{1/2}, \tau^{-1/2})$ consists of the vertices of two icosahedra. The harmonic Molien series implies that $\mathrm{Harm}^W(k)$ has dimension $1,1,0$ for $k = 3,4,5$, respectively. The unique basic invariants $J_1(\xi) = (\xi,\xi)$, $J_2(\xi)$ of degree 3, and $J_3(\xi)$ of degree 4 are obtained from the orbit $X$ as follows, by use of Lemma (3.7):

$$J_2(\xi) = s_3(X,\xi) = |X|^{-1} \sum_{x\in X} (x,\xi)^3 = 2\sqrt{3}\,\xi_1\xi_2\xi_3,$$

$$J_3(\xi) = s_4(X,\xi) = |X|^{-1} \sum_{x\in X} (x,\xi)^4 - \tfrac{1}{5}(\xi,\xi)^2$$

$$= -\tfrac{2}{9}\left(\xi_1^4 + \xi_2^4 + \xi_3^4\right) + \tfrac{2}{15}\left(\xi_1^2 + \xi_2^2 + \xi_3^2\right)^2.$$

It follows that $J_2(\xi)$ vanishes on $Y$ and on $Z$, and that $J_3(\xi)$ vanishes on $Z$ but not on $Y$. This implies that $X$, $Y$, $Z$ are spherical designs of strength $2,3,5$, respectively. This also illustrates that the harmonic invariants $J_2$ and $J_3$ cannot be obtained from $Y$ and from $Z$, since

$$s_3(Y,\xi) = s_3(Z,\xi) = s_4(Z,\xi) = 0.$$

Leaving the octahedral group to the reader, we now turn to the icosahedral group $W(I_3)$ of order 120, taking the 12 vertices of the icosahedron as follows:

$$X = 5^{-1/4}\left\{(\pm\tau^{1/2}, \pm\tau^{-1/2}, 0), (\pm\tau^{-1/2}, 0, \pm\tau^{1/2}), (0, \pm\tau^{1/2}, \pm\tau^{-1/2})\right\}.$$

From the harmonic Molien series we infer that the basic invariants are $J_1(\xi) = (\xi,\xi)$, and the unique harmonic polynomials $J_2(\xi)$ of degree 6 and $J_3(\xi)$ of degree 10. This implies that every orbit of $W(I_3)$ on $\Omega_d$ is a spherical 5-design. Furthermore, the orbit of any zero of $J_2(\xi)$, which exists by Lemma (3.14), provides a spherical 9-design.

Thus we may "improve" the polytope used in the football game. The set of the 60 vertices of the current football (with its regular 5-gons and 6-gons) is the orbit under $W(I_3)$ of the point

$$\lambda(0, \tau^{1/2}, \tau^{-1/2}) + \mu(\tau^{-1/2}, 0, \tau^{1/2}), \qquad \lambda + \mu = 1,$$

with $\lambda = \frac{2}{3}$; it approximates the sphere by agreement in their moments of degrees $\leqslant 5$. For the proposed football we take $\lambda \approx 0.642$, thus killing $J_2$ and approximating the sphere by agreement in their moments of degrees $\leqslant 9$.

The invariants $J_2(\xi)$ and $J_3(\xi)$ are obtained from the orbit $X$ as follows:

$$J_2(\xi) = s_6(X,\xi) = |X|^{-1} \sum_{x\in X} (x,\xi)^6 - \tfrac{1}{7}(\xi,\xi)^3,$$

$$J_3(\xi) = 17 s_{10}(X,\xi) - 42 s_6(X,\xi)(\xi,\xi)^2$$

$$= |X|^{-1} \sum_{x\in X} \left(17(x,\xi)^{10} - 42(x,\xi)^6(\xi,\xi)^2\right) + \tfrac{49}{11}(\xi,\xi)^5.$$

Indeed, the polynomials are nonzero, $s_6(X,\xi)$ is harmonic, $s_{10}(X,\xi)$ is not, but the given linear combination is harmonic. By a computer search it turns out that $J_2(\xi)$ and $J_3(\xi)$ have no common zeros. However, suitable linear combinations of suitable orbits yield cubature formulae of strength 11, 15, etc. For instance, for the icosahedron $X$, the dodecahedron $Y$, and the icosidodecahedron $Z$ (cf. [3]), straightforward calculations show that

$$\begin{bmatrix} s_6(X,X) & s_6(X,Y) & s_6(X,Z) \\ s_6(Y,X) & s_6(Y,Y) & s_6(Y,Z) \\ s_6(Z,X) & s_6(Z,Y) & s_6(Z,Z) \end{bmatrix} = \frac{16}{21} \begin{bmatrix} \frac{1}{25} & -\frac{1}{45} & -\frac{1}{80} \\ -\frac{1}{45} & \frac{1}{81} & \frac{1}{144} \\ -\frac{1}{80} & \frac{1}{144} & \frac{1}{256} \end{bmatrix},$$

where

$$s_6(X_i,X_j) := |X_i|^{-1}|X_j|^{-1} \sum_{x\in X_i} \sum_{y \in X_j} s_6(x,y).$$

Hence the equations (2.5) for $X$ and $Y$ reduce to

$$1 = c_1 + c_2, \quad \tfrac{1}{5}c_1 - \tfrac{1}{9}c_2 = 0; \qquad c_1 = \tfrac{5}{14}, \quad c_2 = \tfrac{9}{14}.$$

Thus from the triacontahedron, which is the union of an icosahedron and a dodecahedron, a cubature formula of strength 9 is obtained by weighing the 32 vertices alternately by the numbers 25 and 27. Strength 11 is achieved by involving $Z$, and strength 15 by also involving the rhombicosidodecahedron on 60 vertices.

Finally, we consider the Weyl group $W(I_4)$, which is isomorphic to $A_5 \times A_5$. The $120 = 96 + 8 + 16$ vertices of the 600-cell $\{3,3,5\}$ are represented by the even permutations of

$$\left(0, \pm\tfrac{1}{2}, \pm\tfrac{1}{2}\tau, \pm\tfrac{1}{2}\tau^{-1}\right), \quad \text{and} \quad (\pm 1,0,0,0), \quad \left(\pm\tfrac{1}{2}, \pm\tfrac{1}{2}, \pm\tfrac{1}{2}, \pm\tfrac{1}{2}\right).$$

From the harmonic Molien series we obtain the degrees 2, 12, 20, 30 of the basic invariants; $\dim \operatorname{Harm}^W(k) = 1$ for $k = 12, 20, 24, 30, 32, 36, 40$, and $= 0$ for the remaining $k < 40$. By Lemma (3.14) the basic invariant $J_2(\xi) = s_{12}(I_4,\xi)$ has zeros. The orbits of these zeros are spherical 19-designs the smallest of which consists of 1440 points. Another cubature formula of strength 19 is obtained from a suitable linear combination of the 120 vertices of $I_4$ and the 600 centers of the faces, the 600-cell and the 120-cell; cf. [10].

## 6. The Root Systems

For a root system $\Phi$ with Weyl group $W(\Phi)$, we denote by $L(\Phi)$ the $\mathbb{Z}$-lattice it generates, and by $L^{\perp}(\Phi)$ the dual lattice of $L(\Phi)$. Every layer in these lattices consists of a union of orbits of $W(\Phi)$, and hence provides candidates for spherical designs and cubature formulae. We shall illustrate our methods by a few examples.

The lattices generated by the root systems $D_4$ and $F_4$ are related by $L(F_4) = \mathrm{L}^{\perp}(D_4)$. With respect to an orthonormal basis $\{e_1,e_2,e_3,e_4\}$ for $\mathbb{R}^4$, the root system $D_4$ consists of the 24 vectors $\pm e_i \pm e_j$, and forms an orbit for the group

$W(F_4)$. The root system $F_4$ consists of two disjoint orbits, one of which is the above, and the other is

$$X = \{\pm e_i\} \cup \{\pm \tfrac{1}{2}e_1 \pm \tfrac{1}{2}e_2 \pm \tfrac{1}{2}e_3 \pm \tfrac{1}{2}e_4\}.$$

By Theorem (4.5), each of these orbits is a spherical 5-design, since the harmonic Molien series for $W(F_4)$ reads $((1-\lambda^6)(1-\lambda^8)(1-\lambda^{12}))^{-1}$. The basic harmonic invariant of degree 6 for $W(F_4)$ is the polynomial

$$J_2(x) = 16S_3(x) - 20S_1(x)S_2(x) + 5S_1^3(x),$$

where $S_k(x) = x_1^{2k} + x_2^{2k} + x_3^{2k} + x_4^{2k}$ (cf. [17]). We observe that

$$J_2(x) = \begin{cases} 1 & \text{for } x = e_i, \\ -1 & \text{for } x = \dfrac{1}{\sqrt{2}}(e_i \pm e_j). \end{cases}$$

Hence by projecting $F_4$ on the sphere we obtain a spherical 7-design, since then $\sum J_2(x) = 0$. Thus the 48 vectors of $X \cup (1/\sqrt{2})D_4$ provide a cubature formula of strength 7.

The Weyl groups $W(E_6)$, $W(E_7)$, $W(E_8)$ are the groups of symmetries of certain polytopes in $\mathbb{R}^6$, $\mathbb{R}^7$, $\mathbb{R}^8$, known as Gosset's polytopes and described in [3]. The first few terms in the harmonic Molien series for these groups are as follows:

$$\begin{aligned} &1 + \lambda^5 + \lambda^6 + \lambda^8 + \cdots && \text{for } W(E_6), \\ &1 + \lambda^6 + \lambda^8 + \lambda^{10} + \cdots && \text{for } W(E_7), \\ &1 + \lambda^8 + \lambda^{12} + \lambda^{14} + \cdots && \text{for } W(E_8). \end{aligned}$$

Hence, by Theorem (4.5), any orbit has strength at least equal to 4, 5, 7, respectively. In order to construct cubature formulae of higher strength we proceed as follows.

For $W(E_6)$, we consider the orbits $X$ and $Y$ consisting of the projection on the sphere of the vectors at minimum distance from the origin in $L^{\perp}(E_6)$ and in $L(E_6)$, respectively. The set $X$ consists of 27 vectors; it is distance-invariant with the following distribution of inner products: $(1)^1(-\tfrac{1}{2})^{10}(\tfrac{1}{4})^{16}$. The set $Y$ consists of the 72 roots of $E_6$ with the inner products $(\pm 1)^1(0)^{30}(\pm\tfrac{1}{2})^{20}$. Any root $y \in Y$ has inner products 0 with 15, $\sqrt{\tfrac{3}{8}}$ with 12, and $-\sqrt{\tfrac{3}{8}}$ with 12 elements of $X$. From these data it follows that

$$s_5(X,X) = \tfrac{5}{192}, \qquad s_5(X,Y) = s_5(Y,Y) = 0,$$

$$s_6(X,X) = \tfrac{3}{256}, \qquad s_6(X,Y) = -\tfrac{1}{128}, \qquad s_6(Y,Y) = \tfrac{1}{192}.$$

Hence the first harmonic invariants $J_2$ of degree 5, and $J_3$ of degree 6, may be obtained from $X$ by defining

$$J_3(\xi) = s_5(X,\xi), \qquad J_3(\xi) = s_6(X,\xi).$$

Since $J_2(y) = 0$ for every root $y$, the root system $Y = E_6$ is a spherical 5-design. A cubature formula of strength 7 is obtained by taking the 3 orbits $X$, $-X$, and $Y$ with the coefficients $\tfrac{1}{5}$, $\tfrac{1}{5}$, $\tfrac{3}{5}$, respectively. Indeed, the invariants $J_2$ and $J_3$ are killed, and there are no harmonic invariants of degree 7.

Leaving the case of $E_7$ to the reader, we now turn to the Weyl group of $E_8$, for which the first harmonic invariants $J_2$ and $J_3$ have degrees 8 and 12, respectively. Simple calculations show that, for the orbit $X$ consisting of the 240 roots, we have $s_8(X,X) = \frac{6}{2560}$. Hence we may define $J_2(\xi) = s_8(X,\xi)$. Another orbit $Y$ consists of the $9 \times 240$ points at the second minimum distance from the origin in the lattice $L(E_8)$. For this orbit, we have

$$J_2(y) = s_8(X,y) = -\tfrac{1}{2560} \qquad \text{for all} \quad y \in Y.$$

Thus a cubature formula of strength 11 is obtained by taking the orbits $X$ and $Y$ with the coefficients $\frac{1}{7}$, $\frac{6}{7}$, respectively. Another way of obtaining a formula of strength 11 consists in finding a zero $\xi$ for the invariant $J_2$, and taking its orbit under the group $W(E_8)$. Such a zero can be obtained as follows. Let $r,s \in X$ denote any two mutually orthogonal roots, and let $\xi := r\cos\varphi + s\sin\varphi$. Then, by simple calculations, we obtain

$$s_8(X,\xi) = \tfrac{1}{2560}\big(6\cos^4(2\varphi) - \sin^4(2\varphi)\big).$$

Thus $\xi$ is a zero of $J_2$ for $\tan^4(2\varphi) = 6$. Notice that for $\varphi = 0, \pi/4$, we obtain an element $\xi$ in $X$ and $Y$, respectively.

## 7. The Leech Lattice

The Conway group .0 is the group of all orthogonal transformations in $\mathbb{R}^{24}$ which preserve the Leech lattice. For a description of the group and the lattice we refer to [2]. Huffman and Sloane [8] have recently obtained the harmonic Molien series for this group; its first few terms are as follows:

$$1 + \lambda^{12} + \lambda^{16} + \lambda^{18} + \lambda^{20} + \lambda^{22} + 3\lambda^{24} + \dots .$$

It should be noticed that this group is not of the Coxeter type; hence its Molien series cannot be given the simple form $\Pi(1 - \lambda^{1+m_i})^{-1}$. However, as the first few terms in the above expansion show, it has no harmonic invariant of degree $1,2,\dots,11$. Hence, by Theorem (4.5), every orbit is a spherical 11-design. In particular, the projection on the sphere of any layer of the Leech lattice is a spherical 11-design. Here we shall obtain a cubature formula of strength 15 by combining the two orbits consisting of the projection on the unit sphere of the first two layers. We give in Table 2 a description of the vectors in these two layers which, following Conway [2], we denote by $\Lambda_2$ and $\Lambda_3$. In Table 3 we give the distribution of the inner products $(x,\xi)$, where for each $\alpha$ we denote by $v_\alpha(\xi)$ the cardinality of the set $\{x \in \Lambda_2 \,|\, (x,\xi) = \alpha\}$.
Let $X$ and $Y$ denote the projection on the unit sphere of $\Lambda_2$ and $\Lambda_3$, respectively. Then, from the data in Table 3, we obtain

$$s_{12}(X,X) = |X|^{-1}\frac{23 \times 25 \times 9}{17 \times 2^{13}},$$

$$s_{12}(X,Y) = |X|^{-1}\frac{-23 \times 25}{9 \times 17 \times 2^{11}}.$$

**Table 2.** The vectors in $\Lambda_2$ and $\Lambda_3$

| Layer | Shape of vectors | Number of vectors |
|---|---|---|
| $\sqrt{8}\,\Lambda_2$ | $(\pm 4)^2(0)^{22}$ | $48 \times 23$ |
| | $(\pm 2)^8(0)^{16}$ | $48 \times 2024$ |
| | $(\mp 3)^1(\pm 1)^{23}$ | $48 \times 2048$ |
| $\sqrt{8}\,\Lambda_3$ | $(\mp 3)^3(\pm 1)^{21}$ | $2^{12} \times 2024$ |
| | $(\pm 5)^1(\pm 1)^{23}$ | $2^{12} \times 24$ |
| | $(\pm 2)^{12}(0)^{12}$ | $2^{12} \times 1288$ |
| | $(\pm 2)^8(\pm 4)^1(0)^{15}$ | $2^{12} \times 759$ |

**Table 3.** The distribution of the inner products

| | $\alpha$ | $v_\alpha(\xi)$ |
|---|---|---|
| $\xi \in \Lambda_2$ | $\pm 4$ | 1 |
| | $\pm 2$ | $23 \times 200$ |
| | $\pm 1$ | $23 \times 2048$ |
| | 0 | $23 \times 4050$ |
| $\xi \in \Lambda_3$ | $\pm 3$ | $23 \times 24$ |
| | $\pm 2$ | $23 \times 486$ |
| | $\pm 1$ | $24 \times 2025$ |
| | 0 | $23 \times 3300$ |

Hence the function defined by

$$s_{12}(X,\xi) = |X|^{-1} \sum_{x \in X} (x,\xi)^{12} - \frac{1 \cdot 3 \cdot 5 \cdot \ldots \cdot 11}{24 \cdot 26 \cdot 28 \cdot \ldots \cdot 34} (\xi,\xi)^6$$

is the harmonic invariant of the group .0. Since there are no other harmonic invariants of degree $\leqslant 15$, we can obtain a cubature formula of strength 15 by killing the above invariant. This can be done by taking the orbits $X$ and $Y$ with the coefficients $\frac{4}{85}$ and $\frac{81}{85}$, respectively. Furthermore, the orbit of any zero of $s_{12}(X,\xi)$ yields a spherical 15-design.

## References


[1] Carter, R. W., *Simple Groups of Lie Type*. Wiley 1972.

[2] Conway, J. H., A group of order 8, 315, 553, 613, 086, 720, 000. *Bull. London Math. Soc.* **1** (1969), 79–88.

[3] Coxeter, H. S. M., *Regular polytopes*, 3rd ed. Dover 1973.

[4] Coxeter, H. S. M., The product of the generators of a finite group generated by reflections. *Duke Math. J.* **18** (1951), 765–782.

[5] Delsarte, P., Goethals, J. M., and Seidel, J. J., Spherical codes and designs. *Geometriae Dedicata* **6** (1977), 363–388.

[6] Flatto, L., Invariants of finite reflection groups. *L'Enseignement mathém.* **24** (1978), 237–292.

[7] Goethals, J. M. and Seidel, J. J., *Spherical Designs.* In *Proc. Sympos. Pure Math.* **34**, edited by D. K. Ray Chaudhuri, Amer. Math. Soc. 1979.

[8] Huffman, W. C. and Sloane, N. J. A., Most primitive groups have messy invariants. *Advances in Math.* **32** (1979), 118–127.

[9] Müller, C., *Spherical Harmonics*. Lecture Notes Math. **17**. Springer-Verlag 1966.

[10] Salihov, G. N., Cubature formulas for a hypersphere that are invariant with respect to the group of the regular 600-face. *Dokl. Akad. Nauk SSSR* **223** (1975), 1075–1078; English translation *Soviet Math. Dokl.* **16** (1975), 1046–1049.

[11] Salihov, G. N., On the theory of cubature formulas for multidimensional spheres. Avtoreferat (Russian), Acad. Sci. USSR, Novosibirsk 1978; Dutch translation memo. 1978-09, Techn. Univ. Eindhoven.

[12] Sidelnikov, V. M., New bounds for the density of sphere packings in an $n$-dimensional Euclidean space. *Mat. Sbornik* **95** (1974); English translation *Math. USSR Sbornik* **24** (1974), 147–157.

[13] Sloane, N. J. A., Error-correcting codes and invariant theory: new applications of a nineteenth-century technique. *Amer. Math. Monthly* **84** (1977), 82–107.

[14] Sobolev, S. L., Cubature formulas on the sphere invariant under finite groups of rotations, *Dokl. Akad. Nauk SSSR* **146** (1962), 310–313; English translation *Soviet Math. Dokl.* **3** (1962), 1307–1310.

[15] Sobolev, S. L., *Introduction to the Theory of Cubature Formulas* (Russian). Nauka 1974.

[16] Stroud, A. H., *Approximate Calculation of Multiple Integrals*. Prentice-Hall 1971.

[17] van Asch, A. G., Modular forms and root systems, Thesis, Univ. Utrecht, 1975.

# Two Quaternionic 4-Polytopes

S. G. Hoggar*

## 1. What Is a Quaternionic Polytope?

One property of a (convex) polytope in $\mathbb{R}^n$ is that the vertex set defines the actual subdivision into edges, triangles, etc. The cells (dimension $n-1$) are the intersections of the convex hull of the vertices with its bounding hyperplanes. The cells intersect in $(n-2)$-dimensional elements, and so on. All these are finite. But for a polytope in $\mathbb{C}^n$ convexity is not available; there is some latitude as to the various elements (now subspaces), subject to suitable conditions on their incidences. For example the fractional polytope $\frac{1}{3}\gamma_3^3$ and generalized cross polytope $\beta_3^3$ [10] agree as to vertices and "edges," but the first has 18 "triangles" whereas the second has 27.

McMullen's definition of complex polytope [4] extends naturally to the quaternions, with the convention that multiplication of vectors by scalars, in quaternionic $n$-space $\mathbb{H}^n$ is on the left.

## 2. The Symmetry Group

A. M. Cohen [1] has classified the quaternionic reflection groups. Those in dimension $>2$ have reflections only of order 2. As we are using left scalar multiplication, the operation of reflection in a hyperplane in $\mathbb{H}^n$ with normal vector $m$ sends an arbitrary point $x$ to $x-(2(x\cdot m)/\|m\|^2)m$, where $x\cdot m$ is the inner product $(x_1,\ldots,x_n)\cdot(m_1,\ldots,m_n)=\sum_r x_r\overline{m}_r$.

We concentrate on a reflection group in $\mathbb{H}^4$ denoted $W(S_1)$ by Cohen [1], and here abbreviated to $G$. We find $G$ is generated by just four reflections and so is

*Department of Mathematics, University of Glasgow, University Gardens, Glasgow G12 8QW, Scotland.

convenient for application of Wythoff's construction [3,4]. We obtain two distinct polytopes, with properties possessed by no real or complex polytope.

It is useful to start with the group $W(S_3)$ which contains $G$ as a subgroup. Their orders are $2^{13} \cdot 3^4 \cdot 5$ and $2^8 \cdot 3^3$ respectively. The larger group is generated by the 180 reflections with normal vectors got by permuting coordinates in

$$(p, 1, 0, 0), \quad (1, 0, 0, 0), \quad (1, p, q, r),$$

where $p$, $q$, $r$ lie in the quaternion group $Q = \{\pm 1, \pm i, \pm j, \pm k\}$, subject to $pqr = \pm 1$. These partition into 45 tetrads of mutually perpendicular vectors. Each reflection permutes the tetrads among themselves, the same permutation of 45 tetrads being obtained from any two reflections corresponding to the same tetrad. Every tetrad can be reflected bijectively onto every other, though every reflection maps *some* tetrad nonbijectively. Henceforth we use the same symbol for a reflection, its fixed hyperplane, and a specified vector normal to that plane.

The following notation simplifies working with these reflections. The tetrads fall into three types:

*Type* 1: The single tetrad $A = \{A_1, A_2, A_3, A_4\}$, where

$$A_1 = (1,0,0,0), \ldots, \qquad A_4 = (0,0,0,1).$$

*Type* 2: Observe that the 8 vectors $(1, \pm 1, \pm 1, \pm 1)$ form two of the tetrads (we refer to such as *dual* tetrads):

$$B = \{B_1(1,1,1,1), B_2(1,1,-1,-1), B_3(1,-1,1,-1), B_4(1,-1,-1,1)\},$$
$$b = \{b_1(1,-1,-1,-1), b_2(1,-1,1,1), b_3(1,1,-1,1), b_4(1,1,1,-1)\}.$$

We are using capital letters to denote vectors with an even number of minus signs in their coordinates and lowercase to denote an odd number. The operation of taking the dual vector or tetrad is indicated by a prime, e.g. $b_3' = B_3$. The remaining type 2 tetrads are indicated below (some use superscripts):

$$\begin{array}{lllll} C(1,i,j,k) & F(1,i,k,j) & Q^2(1,1,i,i) & R^2(1,1,j,j) & S^2(1,1,k,k) \\ D(1,j,k,i) & G(1,j,i,k) & Q^3(1,i,1,i) & R^3(1,j,1,j) & S^3(1,k,1,k) \\ E(1,k,i,j) & H(1,k,j,i) & Q^4(1,i,i,1) & R^4(1,j,j,1) & S^4(1,k,k,1) \end{array}$$

**Example.** $Q_4^3 = (1, -i, -1, i)$.

If $\alpha$, $\beta$, $X_r$ are reflections of type 2, then

$$\alpha^{A_1} = \alpha', (\alpha^\beta)' = (\alpha')^{\beta'}, \qquad X_r^{x_r} = A_r = x_r^{X_r}. \tag{1}$$

One reason for the subscript notation is the following action of the $A_r$ on the reflection vectors in type-2 tetrads:

$$\begin{array}{ll|ll} & A_1 & \text{dualize} & \\ A_1 & A_2 & \text{subscript perm.} & 12.34 \\ A_1 & A_3 & & 13.24 \\ A_1 & A_4 & & 14.23 \end{array} \tag{2}$$

These three permutations abbreviate naturally to the numbers 2, 3, 4 respectively.

*Type* 3:

$$L^2 = \{L_1^2(1,i,0,0), L_2^2(1,-i,0,0), L_3^2(0,0,1,-i), L_4^2(0,0,1,i)\},$$
$$L^3 = \{L_1^3(1,0,i,0), L_2^3(1,0,-i,0), L_3^3(0,1,0,-i), L_4^3(0,1,0,i)\},$$
$$L^4 = \{L_1^4(1,0,0,i), L_2^4(1,0,0,-i), L_3^4(0,1,-i,0), L_4^4(0,1,i,0)\}.$$

Similarly for $M^r$, $N^r$, $P^r$ corresponding to $i$ replaced by $j$, $k$, 1 respectively ($r = 2$, 3, 4).

Out of many subgroups isomorphic to $G$ by conjugation in $W(S_3)$ we use the one (due also to John Conway) whose 36 reflections are the tetrads

$$A, B, b, C, c, D, d, E, e.$$

## 3. Multiplication Table for Reflections in $G$

The multiplication table is given in Table 1. For tetrads $X$, $Y$, $Z$, if the $X$, $Y$ entry of the table is $Z_n$ then $\alpha^\beta$ is in $Z$ for every $\alpha$ in $X$, $\beta$ in $Y$; the relationships between the subscripts are shown in matrix number $n$. We use $\alpha^\beta$ to mean conjugation $\beta^{-1}\alpha\beta$ (but $\beta^{-1} = \beta$ of course).

**Example.** $B_2^{C_3} = e_1$.

The rows corresponding to $b$, $c$, $d$, $e$ are omitted, for by (1), we have for example $b_3^{D_4} = (B_3^{d_4})' = E_2' = e_2$.

**Table 1**

| | $B$ | $b$ | $C$ | $c$ | $D$ | $d$ | $E$ | $e$ |
|---|---|---|---|---|---|---|---|---|
| $B$ | $B$<br>1 | $A$<br>2 | $e$<br>3 | $D$<br>4 | $c$<br>3 | $E$<br>4 | $d$<br>3 | $C$<br>4 |
| $C$ | $e$<br>4 | $d$<br>3 | $C$<br>1 | $A$<br>2 | $E$<br>3 | $b$<br>4 | $D$<br>4 | $B$<br>3 |
| $D$ | $c$<br>4 | $e$<br>3 | $E$<br>4 | $B$<br>3 | $D$<br>1 | $A$<br>2 | $C$<br>3 | $b$<br>4 |
| $E$ | $d$<br>4 | $c$<br>3 | $D$<br>3 | $b$<br>4 | $C$<br>4 | $B$<br>3 | $E$<br>1 | $A$<br>2 |

| | B | | | | b | | | | C | | | | c | | | |
|---|---|---|---|---|---|---|---|---|---|---|---|---|---|---|---|---|
| | 1 | 1 | 1 | 1 | 1 | 2 | 3 | 4 | 1 | 3 | 4 | 2 | 1 | 4 | 2 | 3 |
| B | 2 | 2 | 2 | 2 | 2 | 2 | 4 | 3 | 4 | 2 | 1 | 3 | 3 | 2 | 4 | 1 |
| | 3 | 3 | 3 | 3 | 3 | 4 | 3 | 2 | 2 | 4 | 3 | 1 | 4 | 1 | 3 | 2 |
| | 4 | 4 | 4 | 4 | 4 | 3 | 2 | 4 | 3 | 1 | 2 | 4 | 2 | 3 | 1 | 4 |
| Matrix no. → | 1 | | | | 2 | | | | 3 | | | | 4 | | | |

The row and column corresponding to $A$ are omitted because of (2) and the fact that $A_r^\alpha = \alpha^{A_r}$ for any reflection $\alpha$ of type 2. Indeed, the product of any two reflections $\alpha$, $\beta$ in distinct tetrads of $G$ has period 3. Hence $\alpha^\beta = \beta^\alpha$.

**Example.** $A_2^{e_3} = e_3^{A_2} = E_4$, since $A_2$ dualizes and does subscript perm. 12.34.

## 4. The Generating Reflections

The 36 reflections of $G$, and hence $G$ itself, are generated by just 4 reflections $A_1$, $B_1$, $A_2$, $C_1$, as may be verified directly from the multiplication table. Some of the relations satisfied by these four are shown in the following Coxeter diagram:

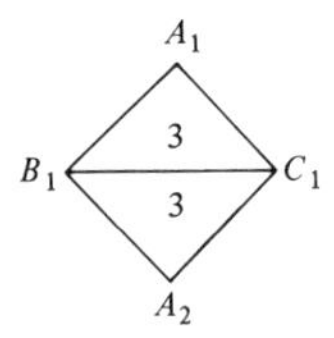


This means: the product of two generators has period 3 if they are joined by an (unmarked) edge and 2 if not; the products $A_1B_1C_1B_1$ and $A_2B_1C_1B_1$ have period 3. However, this is not a diagram for the group $G$, since Coxeter's work [5, p. 257] shows the above relations on their own define an infinite group. On the other hand, any subset of 3 quaternionic reflection vectors may be scaled so that their inner products lie in the field of reals extended by a square root of $-1$ (an observation due to Simon Norton). Thus they generate a complex reflection group which, if finite, can be identified by its Coxeter diagram from the classification of Shephard, Todd, and Coxeter [6, 11; see also 2]. In particular the triples $A_1$, $B_1$, $C_1$ and $B_1$, $A_2$, $C_1$ each have diagram

and so generate $G(3, 3, 3)$. The subgroup

$A_1 \quad C_1 \quad A_2$

of $G$ is even real, namely the reflection group of the regular tetrahedron. Certainly all subgroups generated by 2 reflections are real.

$A_1 \quad C_1$

is the dihedral group of order 6.

# 5. Quaternionic Polytope No. 1

We use Wythoff's construction and the diagram

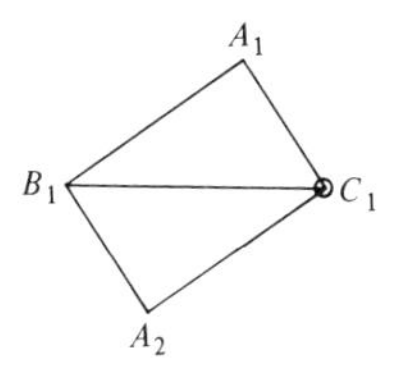


The set of all elements of some type $x$ (vertex, edge, . . . , cell) is the orbit under the group $G$ of an initial element of type $x$. Every such diagram has the connected graph containing the ringed vertex $A_2$ after deletion of certain vertices and their incident edges in the diagram.

### 5.1 Vertices

The ring on $A_2$ means our initial point $T_0$ lies on the hyperplanes in $\mathbb{H}^4$ corresponding to $A_1$, $B_1$, $C_1$. The 128 vertices we obtain can be described as follows. Let

$$T_0 = (0, w, \overline{w}, 1),$$
$$U_0 = (w, 0, 1, \overline{w}),$$
$$V_0 = (\overline{w}, 1, 0, w),$$
$$W_0 = (1, \overline{w}, w, 0) \qquad \left[w = \tfrac{1}{2}(-1 + i + j + k)\right].$$

Multiply these on the *right* by 1, $i$, $j$, $k$ (N.B. we use *left* vector spaces), and perform all sign changes of coordinates. Similarly to the tetrads of $G$, capitals denote an even number of minus signs and lower case letters an odd.

The pattern of signs on the 3 nonzero coordinates is indicated thus:

$$x_0: ---, \quad x_1: -++, \quad x_2: +-+, \quad x_3: ++-; \qquad X_r = -x_r \quad \text{(dual)}.$$

**Example.** $t_1 = (0, -w, \overline{w}, 1)$, $U_2 i = (-w, 0, i, -\overline{w})i$.

### 5.2 Edges

The initial edge is

$$\underset{A_2}{\odot}$$

spanned by $T_0$ and $T_0^{A_2} = t_1$. The set of points and edges defines a 27-regular graph on 128 points, with 1728 edges.

### 5.3 Triangles

The diagram dictates we take the (disjoint) orbits of two initial triangles

$A_2$ — $B_1$ and $A_2$ — $C_1$

but the symmetry of the incidences enables us to consider all triangles as one element type. There are (for instance) 54 triangles at each vertex of the polytope and 4 incident with each edge: 2304 triangles in all.

### 5.4 Cells

(1) 576 regular tetrahedra, starting from

$A_2$ — $B_1$ — $A_1$ and $A_2$ — $C_1$ — $A_1$

(2) 128 complex fractional polytopes $\frac{1}{3}\gamma_3^3$ [10] starting from

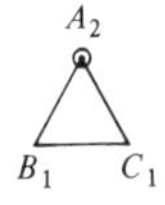


Every triangle is in exactly one tetrahedron and one fractional polytope. The 9 vertices of the initial $\frac{1}{3}\gamma_3^3$ can be grouped into columns as follows so that the 27 edges are defined by the pairs of points from distinct columns (its graph is of course the complete tripartite $K_{3,3,3}$):

$$\begin{matrix} T_0 & t_1 & U_1 \\ W_0 i & w_2 i & U_3 i \\ v_0 k & V_2 k & u_2 k \end{matrix}$$

### 5.5 Incidences

Table 2 displays $N_{xy}$, the number of $y$-elements incident with each $x$-element—and $N_{xx}$, the number of $x$-elements. We have $N_{xy} = |G_x|/|G_x \cap G_y|$, $N_{xx} = |G|/|G_x|$, where $G_x$ is the subgroup of $G$ fixing the initial $x$-element. A check is $N_x N_{xy} = N_y N_{yx}$. We have $|G| = 2^8 \cdot 3^3 = 6912$.

**Table 2**

| $x$ \ $y$ | Vertex | Edge | Triangle | Tetra. | $\frac{1}{3}\gamma_3^3$ | Symbol | $G_x$ | $\lvert G_x \rvert$ |
|---|---|---|---|---|---|---|---|---|
| Vertex | 128 | 27 | 54 | 18 | 9 | | $B_1$ △ $C_1$ ($A_1$) | 54 |
| Edge | 2 | 1728 | 4 | 4 | 2 | $A_2$ | $A_1$ $A_2$ | 4 |
| Triangle | 3 | 3 | 2304 | 1 | 1 | $A_2$ — $B_1$ $A_2$ — $C_1$ | | 6 |
| Tetra. | 4 | 6 | 4 | 576 | — | $A_2$ — $B_1$ — $A_1$ $A_2$ — $C_1$ — $A_1$ | | 24 |
| $\frac{1}{3}\gamma_3^3$ | 9 | 27 | 18 | — | 128 | $B_1$ △ $C_1$ ($A_2$) | | 54 |

### 5.6 Vertex Figure of No. 1

This is the polytope "surrounding a vertex." We obtain its 27 vertices as the orbit of $T_0^{A_2} = t_1$ under the group $G_{T_0} = \langle A_1, B_1, C_1 \rangle$. These vertices lie by threes in 9 diametral planes, as shown in the columns below:

$$\begin{array}{cccccc} t_1 & t_2 & t_3 & u_2 i & U_3 i & \dots \\ w_0 & u_0 & v_0 & v_3 i & V_2 i & \dots \\ V_1 & W_1 & U_1 & w_2 i & W_3 i & \dots \end{array}$$

(the last two columns with $i$ replaced by $j$, $k$).

There may be many ways of assigning edges and cells so as to give a polytope with symmetry group $G_{T_0}$. It will be essentially complex of course, since this group is generated by 3 reflections only. Its initial point $t_1$ lies on mirror $A_1$ but not on $B_1$ or $C_1$, so it may be described as a truncation

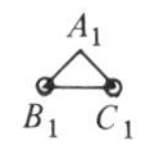


of $\frac{1}{3}\gamma_3^3$, *provided* the hexagon generated from $t_1$ by reflections $B_1$, $C_1$ is regular. A check shows this is so. In effect, we have verified that Coxeter's rule for the vertex figure of a real uniform polytope applies here also (see Coxeter, *Twelve Geometric Essays*, p. 50).

## 6. No. 1 as Projective Configuration

The 128 vertices lie on 64 diameters $\{x_r, X_r\}$. We define suitable diametral planes; we investigate incidences between diameters, diametral planes, and the 36 hyperplanes of symmetry given in 9 tetrads.

### 6.1 Diameters and Planes of Symmetry

We note the symmetry group $G$ is transitive on diameters and on the tetrads. Tetrad A partitions the diameters into $4 \times 16$, hence so does every tetrad. On the other hand, each diameter lies in 9 planes of symmetry, one from each tetrad, since $T_0$ is in precisely the hyperplanes $A_1$, $B_1$, $b_1$, $C_1$, $c_1$, $D_1$, $d_1$, $E_1$, $e_1$.

### 6.2 Diameters and Diametral Planes

Seeking planes through the origin containing more than two diameters, we observe $T_0 - wu_2 = \bar{w}W_2$, $T_0 - \bar{w}u_2 = wv_3$, so there is a plane $\langle T_0, U_2, V_3, W_2 \rangle$. We define our diametral planes to be the 48 planes in the orbit of this one under $G$. The action of $G_{T_0} = \langle A_1, B_1, C_1 \rangle$ shows each diameter lies in 3 diametral planes. The 4 diameters in each diametral plane can be read off from the following list:

$$\begin{array}{ll} \langle T_0 \ U_2 \ V_3 \ W_2 \rangle & \langle T_2 \ U_0 \ V_3 \ W_1 \rangle \\ \langle T_0 \ U_3 \ V_2 \ W_3 \rangle & \langle T_2 \ U_1 \ V_2 \ W_0 \rangle \\ \langle T_1 \ U_2 \ V_1 \ W_0 \rangle & \langle T_3 \ U_0 \ V_1 \ W_3 \rangle \\ \langle T_1 \ U_3 \ V_0 \ W_1 \rangle & \langle T_3 \ U_1 \ V_0 \ W_2 \rangle \end{array}$$

and $\{X_r X_r i X_r j X_r k\}$, where $X = T, U, V, W$ with $r = 0, 1, 2, 3$. Thus there are 16 of this type. The first two columns are repeated with right multiplication by $i, j, k$, giving 32 of this type. Hence the total is 48.

### 6.3 Diametral Planes and Planes of Symmetry

The hyperplanes of symmetry containing $U_2$ are $A_2$, $B_4$, $b_3$, $C_2$, $c_1$, $D_2$, $d_1$, $E_2$, $e_1$; hence the diametral plane $\langle T_0, U_2 \rangle$ lines in $c_1$, $d_1$, $e_1$ and each one lies in exactly 3 planes of symmetry, from distinct tetrads. Conversely, each plane of symmetry contains 4 diametral planes, since $A_1$ contains precisely the diametral planes $\langle T_r, T_r i \rangle$ for $r = 0, 1, 2, 3$.

Finally, here is the incidence matrix of No. 1 (projective):

| | Diameter | Diam. plane | Plane of symmetry |
|---|---|---|---|
| Diameter | 64 | 3 | 9 |
| Diametral plane | 4 | 48 | 3 |
| Plane of symmetry | 16 | 4 | 36 |

**Table 3.** Action of the Symmetry Group on the Vertices of No. 1

| $x$ | $x^{B_1}$ | $x^{C_1}$ | $(xi)^{C_1}$ | $(xj)^{C_1}$ | $(xk)^{C_1}$ | $x^{A_1}$ | $x^{A_2}$ |
|---|---|---|---|---|---|---|---|
| $t_0$ | $t_0$ | $t_0$ | $u_2$ | $v_3$ | $w_2$ | $t_0$ | $T_1$ |
| $t_1$ | $U_1$ | $U_3 i$ | $t_1 i$ | $W_1 i$ | $v_0 i$ | $t_1$ | $T_0$ |
| $t_2$ | $V_1$ | $V_2 j$ | $w_0 j$ | $t_2 j$ | $U_1 j$ | $t_2$ | $T_3$ |
| $t_3$ | $W_1$ | $W_3 k$ | $V_1 k$ | $u_0 k$ | $t_3 k$ | $t_3$ | $T_2$ |
| $u_0$ | $u_0$ | $W_3 j$ | $V_1 j$ | $u_0 j$ | $t_3 j$ | $U_1$ | $u_0$ |
| $u_1$ | $T_1$ | $v_2 k$ | $W_0 k$ | $T_2 k$ | $u_1 k$ | $U_0$ | $u_1$ |
| $u_2$ | $V_2$ | $t_0 i$ | $u_2 i$ | $v_3 i$ | $w_2 i$ | $U_3$ | $u_2$ |
| $u_3$ | $W_2$ | $u_3$ | $T_1$ | $w_1$ | $V_0$ | $U_2$ | $u_3$ |
| $v_0$ | $v_0$ | $U_3 k$ | $t_1 k$ | $W_1 k$ | $v_0 k$ | $V_1$ | $V_2$ |
| $v_1$ | $T_2$ | $w_3 i$ | $v_1 i$ | $U_0 i$ | $T_3 i$ | $V_0$ | $V_3$ |
| $v_2$ | $U_2$ | $v_2$ | $W_0$ | $T_2$ | $u_1$ | $V_3$ | $V_0$ |
| $v_3$ | $W_3$ | $t_0 j$ | $u_2 j$ | $v_3 j$ | $w_2 j$ | $V_2$ | $V_1$ |
| $w_0$ | $w_0$ | $V_2 i$ | $w_0 i$ | $t_2 i$ | $U_1 i$ | $W_1$ | $W_2$ |
| $w_1$ | $T_3$ | $u_2 j$ | $T_1 j$ | $w_1 j$ | $V_0 j$ | $W_0$ | $W_3$ |
| $w_2$ | $U_3$ | $t_0 k$ | $u_2 k$ | $v_3 k$ | $w_2 k$ | $W_3$ | $W_0$ |
| $w_3$ | $V_3$ | $w_3$ | $v_1$ | $U_0$ | $T_3$ | $W_2$ | $W_1$ |

The action of the symmetry group is shown in Table 3. Here $x^\alpha$ means the image of vertex $x$ under reflection $\alpha$.

## 7. Quaternionic Polytope No. 2

With diagram as shown here, the initial point may be taken as $P_3^2 = (0, 0, 1, -1)$, lying on mirrors $A_1$, $B_1$, $A_2$. Let $a = \frac{1}{2}(1 + i)$. Recall $w = \frac{1}{2}(-1 + i + j + k)$, $Q = \{\pm 1, \pm i, \pm j, \pm k\}$.

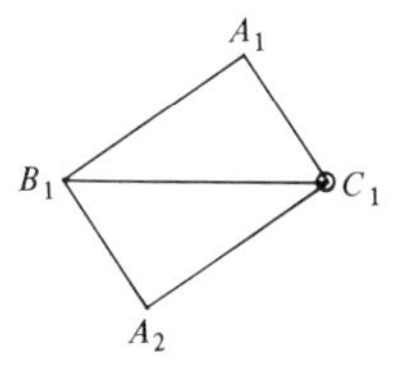


### 7.1 Vertices

With the reflection vectors of $W(S_3)$ in the standard form given earlier, the 288 vertices we obtain may be written

| | |
|---|---|
| $Q \cdot a \cdot F, Q \cdot a \cdot f$ and their images under cyclic perm. $(ijk)$: | $8 \cdot 8 \cdot 3$ |
| $Q \cdot P^2, Q \cdot w \cdot P^3, Q \cdot \overline{w} \cdot P^4$: | $8 \cdot 4 \cdot 3$ |
| | 288 |

### 7.2 Edges

Initial edge $\overset{\odot}{C_1}$, spanned by $P_3^2$ and $(P_3^2)^{C_1} = jaF_1$. The set of vertices and edges defines a 24-regular graph on 288 vertices, with 3456 edges.

### 7.3 Triangles

There are 3 types (i.e. orbits), of which two may be combined. We then have two types, say $\Delta_a$, $\Delta_b$.

### 7.4 Cells

As in No. 1, we have fractional polytopes $\frac{1}{3}\gamma_3^3$, here 256 of them, but rather than tetrahedra as the other type, we have 288 octahedra. Every type-$\Delta_a$ triangle lies in exactly one octahedron and one fractional polytope $\frac{1}{3}\gamma_3^3$, whereas one of type-$\Delta_b$ is in no octahedron but in two fractional polytopes.

### 7.5 Incidences

The incidences are shown in Table 4.

**Table 4**

| $x \setminus y$ | Vertex | Edge | $\Delta_a$ | $\Delta_b$ | Octa | $\frac{1}{3}\gamma_3^3$ | Symbol | $G_x$ | $\lvert G_x \rvert$ |
|---|---|---|---|---|---|---|---|---|---|
| Vertex | 288 | 24 | 24 | 14 | 4 | 8 | | $A_2$ $B_1$ $A_1$ | 24 |
| Edge | 2 | 3456 | 2 | 1 | 1 | 2 | $C_1$ | $C_1$ | 2 |
| $\Delta_a$ | 3 | 3 | 2304 | — | 1 | 1 | $C_1$ $A_1$, $C_1$ $A_2$ | | 6 |
| $\Delta_b$ | 3 | 3 | — | 1152 | 0 | 2 | $C_1$ $B_1$ | | 6 |
| Octa | 6 | 12 | 8 | 0 | 288 | — | $A_1$ $C_1$ $A_2$ | | 24 |
| $\frac{1}{3}\gamma_3^3$ | 9 | 27 | 9 | 9 | — | 256 | $A_2$ $B_1$ $C_1$, $A_1$ $B_1$ $C_1$ | | 54 |

### 7.6 Vertex Figure

This has 24 vertices, no two from any one diameter, which lie by threes in 8 diametral planes as follows. One plane contains $jaF_1$ (the initial point of the vertex figure) and its two images under the cyclic perm. $(ijk)$. Similarly for $-jaf_1$, $-jaF_2$, $jaf_2$, $-kaF_3$, $kaf_3$, $kaF_4$, $-kaf_4$. The initial point is on none of the 3 generating mirrors, and the vertex figure has the structure of a truncated octahedron

$A_1$ $B_1$ $A_2$

(though constructed via the proper subgroup •—•—• of the symmetry group of the octahedron).

## 8. No. 2 as Projective Configuration

The 288 vertices lie by eights (corresponding to the elements of $Q$) in 36 diameters. The polytope is 8-symmetric [4]. There are the same 36 (hyper)planes of symmetry as for No. 1.

### 8.1 Diameters and Planes of Symmetry

The initial point $P_3^2$ is in planes of symmetry $A_1$, $A_2$, $B_1$, $b_1$, $B_2$, $b_2$ only, so every diameter is in exactly 6 planes of symmetry, two from each of three tetrads. It follows every plane of symmetry contains 6 diameters.

### 8.2 Diameters and Diametral Planes

$\langle P_3^2, P_2^3, P_2^4 \rangle$ is our initial diametral plane. The symmetry-group action then generates a total of 48: 16 of type $\langle P_l^2, P_m^3, P_n^4 \rangle$, 16 of type $\langle F_q, G_r, H_s \rangle$, and 16 got by dualizing the latter. Each diameter lies in 4 diametral planes.

### 8.3 Diametral Planes and Planes of Symmetry

The planes of symmetry containing $P_2^3$ are $A_2$, $A_4$, $B_1$, $B_3$, $b_4$, $b_2$, so that diametral plane $\langle P_3^2, P_2^3 \rangle$ lies in just $A_2$, $B_1$, $b_2$. Also $A_1$ contains diametral planes $\langle P_3^2, P_4^3, P_4^4 \rangle$, $\langle P_3^2, P_3^3, P_3^4 \rangle$, $\langle P_4^2, P_3^3, P_4^4 \rangle$, $\langle P_4^2, P_4^3, P_3^4 \rangle$. It follows each diametral plane lies in 3 planes of symmetry, from distinct tetrads; each plane of symmetry contains 4 diametral planes in such a way that each diameter involved is contained in exactly 2 of these 4 (a total of 6 diameters).

The incidence matrix of No. 2 (projective) is

| | Diameter | Diam. Plane | Plane of Symmetry |
|---|---|---|---|
| Diameter | 36 | 4 | 6 |
| Diametral plane | 3 | 48 | 3 |
| Plane of Symmetry | 6 | 4 | 36 |

## 9. Concluding Remarks

1. From earlier work of the author [9], the 64 diameters of No. 1 form a set of lines through the origin in $\mathbb{H}^4$ which is of greatest possible size, given that the mutual angles have squared cosines $\frac{1}{3}$, $\frac{1}{9}$. The 36 diameters of No. 2, or equally the 36 relection vectors of the symmetry group $G$, constitute such a greatest-sized set with squared cosines 0, $\frac{1}{4}$. Complexifying the diameters of No. 1 yields a greatest-sized set (by [8]) of 64 lines through the origin in $C^8$ such that the squared cosines all equal $\frac{1}{9}$.

2. It would be interesting to know if the incidence matrices of the polytopes as projective configurations correspond to known geometrical configurations, or have special properties. Or, indeed, whether they can be realized in *complex* projective spaces.

3. Although the two polytopes have each vertex surrounded alike because the symmetry group $G$ is transitive on vertices, they are certainly not regular by McMullen's definition [4], since $G$ is not transitive on flags. This in turn is because our "diagrams" contain triangles. In fact, Norton's observation shows every such diagram must contain at least two triangles if it comes from a truly quaternionic reflection group.

4. It remains (for example) to verify whether (a) the complex and real forms are, as seems likely, truncations of complex and real cross-polytopes, and (b) there is any connection with extreme forms (cf. [7]).

## References

[1] Cohen, A. M., Finite quaternionic reflection groups. Memorandum 229. *J. Algebra* **64** (1980), 293–324.

[2] Cohen, A. M., Finite complex reflection groups. *Ann. Scient. Ec. Norm. Sup. (4)* **9** (1976), 379–436.

[3] Coxeter, H. S. M., *Regular Polytopes*. Dover, New York, 1973.

[4] Coxeter, H. S. M., *Regular Complex Polytopes*. C.U.P. 1974.

[5] Coxeter, H. S. M., Groups generated by unitary reflections of period two. *Canad. J. Math.* **9** (1957), 243–272.

[6] Coxeter, H. S. M., Finite groups generated by unitary reflections. *Abh. a.d. Math. Sem. d. Univ. Hamburg* **31** (1967), 125–135.

[7] Coxeter, H. S. M., and Todd, J. A., An extreme duodenary form. *Canad. J. Math.* **5** (1953), 384–392.

[8] Delsarte, P., Goethals, J. M., and Seidel, J. J., Bounds for systems of lines, and Jacobi polynomials. *Philips Research Reports* **30** (1975), 91–105.

[9] Hoggar, S. G., Bounds for quaternionic line systems and reflection groups. *Mathematica Scandinavica* **43** (1978), 241–249.

[10] Shephard, G. C., Unitary groups generated by reflections. *Canad. J. Math.* **5** (1953), 364–383.

[11] Shephard, G. C., and Todd, J. A., Finite unitary reflection groups. *Canad. J. Math.* **6** (1954), 274–304.

# Span-Symmetric Generalized Quadrangles

Stanley E. Payne*

## I. Introduction

A *generalized quadrangle* (GQ) *of order* $(s, t)$ is a point-line incidence geometry $\mathcal{S} = (\mathcal{P}, \mathcal{L}, I)$ with pointset $\mathcal{P}$, lineset $\mathcal{L}$, and incidence relation $I$ satisfying the following:

(1) Two points are incident with at most one line in common.
(2) If $x \in \mathcal{P}$, $L \in \mathcal{L}$, and $x \not\!I L$ (i.e. $x$ is not incident with $L$), there is a unique pair $(y, M) \in \mathcal{P} \times \mathcal{L}$ for which $x\, I\, M\, I\, y\, I\, L$.
(3) Each point (respectively, line) is incident with $1 + t$ lines (respectively, $1 + s$ points).

If $L_1$ and $L_2$ are lines of $\mathcal{S}$, $L_1 \sim L_2$ indicates that $L_1$ and $L_2$ are incident with a point in common (including the case $L_1 = L_2$). For distinct lines $L_1$, $L_2$, the *trace* of the pair $(L_1, L_2)$ is defined by $\operatorname{tr}(L_1, L_2) = \{M \in \mathcal{L} \mid L_1 \sim M$ and $L_2 \sim M\}$. The *span* of $(L_1, L_2)$ is defined by $\operatorname{sp}(L_1, L_2) = \{L \in \mathcal{L} \mid L \sim M$ for all $M \in \operatorname{tr}(L_1, L_2)\}$. Hence $\{L_1, L_2\} \subseteq \operatorname{sp}(L_1, L_2)$. The *closure* of $(L_1, L_2)$ is defined by $\operatorname{cl}(L_1, L_2) = \{M \in \mathcal{L} \mid M \sim L$ for some $L \in \operatorname{sp}(L_1, L_2)\}$. The *star* of a line $L$ is $\operatorname{st}(L) = \{M \in \mathcal{L} \mid L \sim M\}$. Hence it follows that $\operatorname{tr}(L_1, L_2) = \operatorname{st}(L_1) \cap \operatorname{st}(L_2)$, and $\operatorname{sp}(L_1, L_2) = \cap \{\operatorname{st}(M) \mid M \in \operatorname{tr}(L_1, L_2)\}$.

It is clear from the axioms for a GQ $\mathcal{S}$ of order $(s, t)$ that there is a point-line duality for which the dual $\mathcal{S}$ is a GQ of order $(t, s)$. Moreover, each of the definitions and notations for lines has a dual for points. We always assume these dual definitions to be given, and use a result in either its original form or its dual form without further mention.

Now let $\mathcal{S} = (\mathcal{P}, \mathcal{L}, I)$ be a generalized quadrangle of order $(s, t)$ with a fixed pair $(L_1, L_2)$ of nonconcurrent lines. Put $\operatorname{tr}(L_0, L_1) = \{M_0, M_1, \ldots, M_s\}$ and

*Miami University, Oxford, Ohio 45056.

$\mathrm{sp}(L_0, L_1) = \{L_0, L_1, \ldots, L_p\}$. A *symmetry about* $L_0$ is a collineation of $\mathcal{S}$ that fixes each line concurrent with $L_0$. It is well known that the group of symmetries about $L_0$ acts semiregularly on the lines of $\mathrm{sp}(L_0, L_1)\backslash\{L_0\}$. Moreover, by a result of Thas ([12]; cf. also [8]), $p \leqslant t^2/s$. If $p = t^2/s$ and if the full group of symmetries about $L_0$ has order $t^2/s$, we say that $L_0$ is an *axis of symmetry*. If $L_0$ is an axis of symmetry and if some $L_j$, $1 \leqslant j \leqslant p$, admits a nonidentity symmetry about it, then clearly each line of $\mathrm{sp}(L_0, L_1)$ is an axis of symmetry, in which case $\mathcal{S}$ is said to be span-symmetric with base span $\mathrm{sp}(L_0, L_1)$.

In this note we propose and begin to investigate the following:

**Problem.** Determine all span-symmetric generalized quadrangles.

Our main result takes care of the case $s = t$.

**Theorem.** *A span-symmetric GQ of order $(s,s)$ is isomorphic to the classical example of a nonsingular quadric $Q(4,s)$ in projective space $PG(4,s)$.*

The proof uses two main techniques: the coordinatization scheme developed in [5] and [9], and a characterization of span-symmetric GQ of order $(s,s)$ as a kind of group coset geometry.

Let $G$ be a group of order $s^3 - s$, $s \geqslant 2$, having a collection $\mathcal{T}$ of subgroups, $\mathcal{T} = \{S_0, \ldots, S_s\}$, where each of the subgroups $S_i$ has order $s$. We say $\mathcal{T}$ is a *4-gonal basis for $G$* provided the following three conditions are satisfied:

(4) $\mathcal{T}$ is a complete conjugacy class in $G$.
(5) $S_i \cap N_G(S_j) = \{e\}$ if $i \neq j$, $0 \leqslant i, j \leqslant s$.
(6) $S_i S_j \cap S_k = \{e\}$ for $i, j, k$ distinct.

If $\mathcal{S}$ is span-symmetric with base span $\mathrm{sp}(L_0, L_1)$, and if $S_i$ is the group of symmetries about $L_i$, then $\mathcal{T} = \{S_0, \ldots, S_s\}$ is a 4-gonal basis for the group $G = \langle S_i \mid 0 \leqslant i \leqslant s\rangle$. Conversely, given $G$ and $\mathcal{T}$, the generalized quadrangle $\mathcal{S}$ can be recaptured as a kind of coset geometry $(G, \mathcal{T})$. The details are worked out in Section 3.

In Section 4 the coordinatization scheme is recalled in detail and used to finish the proof of the Theorem. In Section 5 a brief look is taken at the classical case, where $G$ is isomorphic to $\mathrm{SL}(2,s)$.

Before considering the special case $s = t$, we make a few observations in a wider context.

## II. The Substructure Fixed by a Collineation

Let $\mathcal{S} = (\mathcal{P}, \mathcal{L}, I)$ be a GQ of order $(s,t)$, $s \geqslant 2$, $t \geqslant 2$. Let $L_0$ and $L_1$ be fixed nonconcurrent lines of $\mathcal{S}$. Put $\mathrm{tr}(L_0, L_1) = \{M_0, \ldots, M_s\}$, $\mathrm{sp}(L_0, L_1) = \{L_0, L_1, \ldots, L_p\}$. By a lemma of Thas [12], $p \leqslant t^2/s$. Moreover, if $p = t^2/s$, then $t \mid s$ and each line $M$ outside $\mathrm{cl}(L_0, L_1)$ meets $1 + s/t$ lines of $\mathrm{tr}(L_0, L_1)$. (An alternate treatment of this lemma appears in [8].)

For the remainder of this section we suppose $p = t^2/s$. If $s = t$, there are no lines outside $\mathrm{cl}(L_0, L_1)$. If $s \neq t$, then $s > t$ and $1 + s/t \geqslant 3$.

If $\theta$ is any collineation of $\mathcal{S}$, put $\mathcal{P}_\theta = \{x \in \mathcal{P} \mid x^\theta = x\}$, $\mathcal{L}_\theta = \{M \in \mathcal{L} \mid M^\theta = M\}$, and let $I_\theta$ be the restriction of $I$ to points and lines of $\mathcal{P}_\theta$ and $\mathcal{L}_\theta$, respectively. Then $\mathcal{S}_\theta = (\mathcal{P}_\theta, \mathcal{L}_\theta, I_\theta)$ is the *fixed substructure* of $\theta$. It is known (see [6] or [14]) that $\mathcal{S}_\theta$ must be one of the following types:

(i) $\mathcal{S}_\theta$ is a set of pairwise noncollinear points (partial ovoid) or a set of pairwise nonconcurrent lines (partial spread).
(ii) There is a point $x \in \mathcal{P}_\theta$ for which $\mathcal{P}_\theta \subseteq \mathrm{st}(x)$ and each line of $\mathcal{L}_\theta$ is incident with $x$; or there is a line $L \in \mathcal{L}_\theta$ for which $\mathcal{L}_\theta \subseteq \mathrm{st}(L)$ and each point of $\mathcal{P}_\theta$ is incident with $L$.
(iii) $\mathcal{S}_\theta$ is a grid or a dual grid (i.e. complete bipartite graph).
(iv) $\mathcal{S}_\theta$ is a subquadrangle of order $(s', t')$, $2 \leqslant s' \leqslant s$, $2 \leqslant t' \leqslant t$.

**II.1.** *Let $\theta$ be a collineation fixing each line of* $\mathrm{tr}(L_0, L_1)$, *and let* $\mathcal{S}_\theta = (\mathcal{P}_\theta, \mathcal{L}_\theta, I_\theta)$ *be the fixed substructure of $\theta$.*

(i) *If $\theta$ fixes some point $x$ not on any $M_i \in \mathrm{tr}(L_0, L_1)$, then* $\theta = \mathrm{id}$.
(ii) *If $\theta$ fixes some line $M \notin \mathrm{cl}(L_0, L_1)$ (so $s > t$) and $\theta \neq \mathrm{id}$, then $\mathcal{S}_\theta$ is a subquadrangle of order $(s/t, t)$ whose points all lie on the $M_i$'s but not on any of the $L_j$'s.*
(iii) *If $\theta$ fixes some line $L \in \mathrm{sp}(L_0, L_1)$, then $\mathcal{S}_\theta$ is a grid "lying in"* $\mathrm{sp}(L_0, L_1)$ *or a substructure of* $\mathrm{st}(L)$.

*Proof*. (i): Let $x^\theta = x$ with $x$ not on any $M_i$. Considering the possibilities listed prior to II.1, it follows readily that $\mathcal{S}_\theta$ must be a subquadrangle of order $(s', t')$, $2 \leqslant s' \leqslant s$, $2 \leqslant t' \leqslant t$. But each line through $x$ either meets some $M_j$ at a point on some $L_i$ and is fixed (since both $x$ and $M_j$ are), or meets $1 + s/t$ lines of $\mathrm{tr}(L_0, L_1)$ and is fixed. Hence $t = t'$. And a line through $x$ meeting $L_0$, for example, is fixed, since each $M_i$ is fixed. Then $L_0$ must be fixed along with all points of $L_0$, i.e. $s' = s$ and $\theta = \mathrm{id}$.

(ii): Suppose $\theta$ fixes some line $M \notin \mathrm{cl}(L_0, L_1)$, $\theta \neq \mathrm{id}$, so $s > t$. The $1 + s/t \geqslant 3$ points of $M$ on lines $M_i$ are fixed; and each line through such a point is fixed, since it either is an $M_i$ or meets $1 + s/t$ of the $M_i$'s. It follows that $\mathcal{S}_\theta$ is a subquadrangle of order $(s', t)$, where $s' \geqslant s/t$. By the theorem of Payne [4], either $s = s'$ (which is impossible, since $\theta \neq \mathrm{id}$), or $s \geqslant s't' = s't$. Hence $s' = s/t$, and $\theta$ fixes no point not on some $M_i$. If some line $L \in \mathrm{sp}(L_0, L_1)$ were fixed, as it would be if any point of $L$ were fixed, then $s'$ would equal $s$, an impossibility. The proof of (ii) is complete.

(iii): Suppose $\theta$ fixes some line $L \in \mathrm{sp}(L_0, L_1)$. If $\theta \neq \mathrm{id}$, by parts (i) and (ii) we know that $\theta$ fixes no point off the $M_i$'s and no line outside $\mathrm{cl}(L_0, L_1)$. If $\theta$ fixes a second line $L'$ of $\mathrm{sp}(L_0, L_1)$, then checking the possibilities preceding II.1, we see that $\mathcal{S}_\theta$ must be a grid (lying in $\mathrm{sp}(L_0, L_1)$) or a subquadrangle of order $(s', t')$. In the latter case, since each point of $L$ is fixed, $s' = s$, forcing all points of $M_0$ to be fixed. If $s = t$ this is O.K.: $\mathcal{S}_\theta$ is the grid composed of $\mathrm{sp}(L_0, L_1) \cup \mathrm{tr}(L_0, L_1)$ and their points of intersection. If $s > t$, $\mathcal{S}_\theta$ cannot be a subquadrangle of order $(s, t')$, since $\theta$ can fix no point of $M_0$ not on some line of $\mathrm{sp}(L_0, L_1)$. If $L$ is the unique line of $\mathrm{sp}(L_0, L_1)$ fixed by $\theta$, $\mathcal{S}_\theta$ is a substructure "contained in" $\mathrm{st}(L)$. □

We now reconsider the subquadrangle $\mathcal{S}_\theta$ of order $(s/t, t)$ in part (ii) of II.1. A line of $\mathcal{S}$ is called *external*, *tangent*, or *internal* according as it is incident with no point of $\mathcal{S}_\theta$, a unique point of $\mathcal{S}_\theta$, or more than one—and hence $1 + s/t$—points of $\mathcal{S}_\theta$. A point of $\mathcal{S}$ is called *external*, *tangent*, or *internal* according as it is on no line of $\mathcal{S}_\theta$, a unique line of $\mathcal{S}_\theta$, or more than one—and hence $1 + t$—lines of $\mathcal{S}_\theta$. By a result of Thas [11], since $t = t'$, each external line is concurrent with exactly $1 + ts' = 1 + s$ lines of $\mathcal{S}_\theta$. For lines of $\text{sp}(L_0, L_1)$, this is clear. If $L$ is a line tangent to $\mathcal{S}_\theta$ at a fixed point $y \in \mathcal{P}_\theta$, then $t' = t$ implies $L \in \mathcal{L}_\theta$, i.e., $L$ is not tangent. *So each line is external or internal.* Let $L$ be an external line. $L$ must be concurrent with exactly $1 + s$ fixed lines. If two of these fixed lines meet at a point $x$ of $L$, then $x$ is fixed and $L$ must be internal. Hence each point of $L$ lies on a unique fixed line but must not be fixed itself.

If $x$ is any point not on any $M_i$, then $x$ is on a line $M$ meeting $L_0$. $M$ must be external, so $x$ must be a tangent point. It follows that *each point is tangent or internal*. If $L$ is external, $L \not\sim L^\theta$. If $x$ is tangent, $x^\theta \sim x$.

Let $S_i$ be the group of symmetries about $L_i$, $0 \leqslant i \leqslant t^2/s$.

**II.2.** *Let $\theta = \sigma_i\sigma_j$, where $\sigma_i \neq \text{id} \neq \sigma_j$, $\sigma_i \in S_i$, $\sigma_j \in S_j$, $i \neq j$. If $x$ is a point not on any $M_i$, then $x^\theta \not\sim x$. Hence $\theta$ fixes no line outside* $\text{sp}(L_0, L_1) \cup \text{tr}(L_0, L_1)$, *and $\theta$ is not a symmetry. In particular, $S_iS_j \cap S_k = \{e\}$ if $i, j, k$ are distinct.*

*Proof.* $x$ and $x^{\sigma_i}$ lie on a line meeting $L_i$, $x \neq x^{\sigma_i}$. $x^{\sigma_i}$ and $x^\theta$ lie on a line meeting $L_j$, $x^{\sigma_i} \neq x^\theta$. Then $x \not\sim x^\theta$, since $\mathcal{S}$ has no triangles. □

## III. The Case $s = t$

Let $\mathcal{S} = (\mathcal{P}, \mathcal{L}, I)$ be a GQ of order $(s, s)$. A pair $(L, M)$ of distinct lines is *regular* provided $|\text{sp}(L_0, L_1)| = 1 + s$, and a line $L$ is *regular* provided $(L, M)$ is regular for all lines $M$ such that $L \not\sim M$. The notion of regularity (along with various generalizations) has played an important role in the study of GQ (e.g., see [5], [9], [13]). A pair $(L, M)$ is regular if and only if some pair of lines in $\text{sp}(L, M)$ or some pair of lines in $\text{tr}(L, M)$ is regular. This observation has the following consequence.

**III.1.** *Let $L_0$, $L_1$ be fixed, nonconcurrent lines. If each line of* $\text{sp}(L_0, L_1)$ *is regular, then each line of* $\text{tr}(L_0, L_1)$ *is regular.*

*Proof.* Let $\text{sp}(L_0, L_1) = \{L_0, L_1, \ldots, L_s\}$, with each $L_i$ regular, and put $\text{tr}(L_0, L_1) = \{M_0, M_1, \ldots, M_s\}$. Let $M$ be any line not concurrent with $M_i$ for some $i$, $0 \leqslant i \leqslant s$. Clearly $M$ must be incident with a point $x = L_j \cap M_k$ for some $j$ and $k$. Since $L_j$ is regular, it follows that the pair $(M_i, M)$ is regular, and hence $M_i$ must be regular. □

Put $S_i$ equal to the group of symmetries about $L_i$, $G = \langle S_i \mid 0 \leqslant i \leqslant s \rangle$, and $T_i = N_G(S_i)$. Then $|S_i| = s$ by hypothesis, and by II.1 $G$ acts semiregularly on the set $\Omega$ of $s^3 - s$ points not incident with any $M_j$. Let $x, y \in \Omega$, with $x \sim y$. If $xy$ is a line meeting $L_i$, some symmetry about $L_i$ moves $x$ to $y$. If $x, y \in \Omega$ with $x \not\sim y$, let $M$ be any line through $x$, and $L$ the line through $y$ meeting $M$ at a point $z$. If $z \in \Omega$, clearly some element of $G$ moves $x$ to $y$. Otherwise we may suppose each point of $\text{tr}(x, y)$ lies on some $M_j$. In that case it is easy to find points $z, w \in \Omega$ such that $x \sim z \sim w \sim y$. So $G$ contains an element moving $x$ to $y$. Hence $G$ is transitive on $\Omega$ and therefore regular on $\Omega$. In particular, $|G| = s^3 - s$.

Each $S_i$ is transitive on $\text{sp}(L_0, L_1) - \{L_i\}$, and $G$ leaves $\text{sp}(L_0, L_1)$ invariant and fixes each line of $\text{tr}(L_0, L_1)$. It follows that $\mathcal{S} = \{S_0, \ldots, S_s\}$ is a complete conjugacy class. It is easy to check that $T_i = N_G(S_i)$ is the stabilizer in $G$ of $L_i$, so that $|T_i| = |G|/(s + 1) = s(s - 1)$. As $S_i$ is transitive (acting by conjugation) on $\{S_j \,|\, 0 \leqslant j \leqslant s, j \neq i\}$, $T_i$ is also, so that $T_i$ is transitive on $\{T_j \,|\, 0 \leqslant j \leqslant s, j \neq i\}$. Thus the stabilizer of $T_j$ in $T_i$ has order $|T_i|/s$, implying $|T_i \cap T_j| = s - 1$. Let $x, y$ be on $L_i$, $L_j$, respectively, $x \not\sim y$. Then $|\text{tr}(x, y) \cap \Omega| = s - 1$, and $T_i \cap T_j$ acts regularly on $\text{tr}(x, y) \cap \Omega$. $S_i S_j \cap S_k = \{\text{id}\}$ by II.2. We have more than proved the following:

**III.2.** *$\mathcal{T}$ is a* 4-*gonal basis for $G$ as defined in Section* I.

The main purpose of this section is to show that in fact a span-symmetric GQ of order $(s, s)$ is equivalent to a group $G$ with a 4-gonal basis $\mathcal{T}$.

Let $x_0$ be a fixed point of $\Omega$. For each point $y \in \Omega$ there is a unique element $g \in G$ for which $x_0^g = y$. In this way each point of $\Omega$ is identified with a unique element of $G$. Let $N_i$ be the line through $x_0$ meeting $L_i$. Points of $N_i$ in $\Omega$ correspond to elements of $S_i$. Let $z_i$ be the point of $L_i$ on $N_i$, $0 \leqslant i \leqslant s$. For $i \neq j$, $T_i \cap T_j$ acts regularly on the points of $\text{tr}(z_i, z_j) \cap \Omega$. It follows that $S_i(T_i \cap T_j) = (T_i \cap T_j)S_i = T_i$ acts regularly on the points of $\text{st}(z_i) \cap \Omega$, so that the elements of a given coset $tS_i = S_i t$ of $S_i$ in $T_i$ correspond to the points of a fixed line through $z_i$. Hence we may identify $T_i$ with $z_i$. Now suppose that lines of $\text{tr}(L_0, L_1)$ are labeled so that $T_i = z_i$ is a point of $M_i$. Let $g$ be a collineation in $G$ mapping $x_0$ to a point collinear with $L_j \cap M_i$ (keep in mind that $G$ fixes $M_i$). Then each point of $x_0^{T_i g}$ is collinear with $L_j \cap M_i$, so identify $L_j \cap M_i$ with $T_i g$. In this way the points of $M_i$ are identified with the right cosets of $T_i$ in $G$, and a line through $T_i g$ (not in $\text{tr}(L_0, L_1) \cup \text{sp}(L_0, L_1)$) is a coset of $S_i$ contained in $T_i g$. Hence the points of $L_i$ consist of one coset of $T_j$ for each $j = 0, 1, \ldots, s$. If $T_i \cap T_j g$ contains a point $y = x_0^h$, then $S_i h$ is a line joining $T_i$ and $y$ and $S_j h$ is a line joining $T_j g$ and $y$. Hence if $T_j g$ is a point of $L_i$ (and hence collinear with $T_i$), it must be that $T_i \cap T_j g = \emptyset$. We show later that for each $j, j \neq i$, $T_i$ is disjoint from a unique right coset of $T_j$, so that the points of $L_i$ are uniquely determined as $T_i$ and the unique right coset of $T_j$ disjoint from $T_i$ for $j = 0, \ldots, s, j \neq i$.

Now let $G$ be an abstract group of order $s^3 - s$ with 4-gonal basis $\mathcal{T} = \{S_0, \ldots, S_s\}$. Put $T_i = N_G(S_i)$. Clearly $s + 1 = [G : T_i]$, so $|T_i| = s(s - 1)$. By (5), no two elements of $S_i$ are in the same coset of $T_j$ $(i \neq j)$, so $S_i$ acts regularly (by conjugation) on $\mathcal{T} \backslash \{S_i\}$, and hence $T_i$ acts transitively on $\mathcal{T} \backslash \{S_i\}$. Since any inner automorphism of $G$ moving $S_j$ to $S_k$ also moves $T_j$ to $T_k$, $T_i$ also acts transitively on $\{T_0, \ldots, T_s\} \backslash \{T_i\}$, and $\{T_0, \ldots, T_s\}$ is a complete conjugacy class in $G$. As the number of conjugates of $T_i$ in $G$ is $1 + s = [G : N_G(T_i)]$, and $1 + s = [G : T_i]$, it follows that $T_i = N_G(T_i)$. As $T_i$ acts transitively on $\mathfrak{T}\backslash\{S_i\}$, the subgroup of $T_i$ fixing $S_j$ has order $|T_i|/s = s - 1$, i.e. $|T_i \cap T_j| = s - 1$. The following result is now clear.

**III.3.** *For $i \neq j$, $|T_i \cap T_j| = s - 1$ and $T_i$ is a semidirect product of $S_i$ and $T_i \cap T_j$.*

**III.4.** *Let $T_ig_i$ and $T_jg_j$ be arbitrary cosets of $T_i$ and $T_j$, $i \neq j$. Then $T_ig_i \cap T_jg_j = \emptyset$ iff $g_jg_i^{-1}$ sends $T_j$ to $T_i$ under conjugation. Moreover, if $T_ig_i \cap T_jg_j \neq \emptyset$, then $|T_ig_i \cap T_jg_j| = s - 1$.*

*Proof.* If $x \in T_i \cap T_jg$, a standard argument shows that $T_i \cap T_jg = \{tx \mid t \in T_i \cap T_j\}$, so $|T_i \cap T_jg| = s - 1$. Since $|T_i| = s(s-1)$, $T_i$ meets $s$ cosets of $T_j$ and is disjoint from the one remaining. Two elements $x, y \in G$ send $T_j$ to the same $T_k$ iff $x$ and $y$ belong to the same right coset of $N_G(T_j) = T_j$; hence iff $xy^{-1} \in T_j$. Suppose $g$ maps $T_j$ to $T_i$: $T_i = g^{-1}T_jg$, $i \neq j$. Then $g \notin T_j$, so $\emptyset = g^{-1}T_j \cap T_j$, implying $\emptyset = g^{-1}T_jg \cap T_jg = T_i \cap T_jg$. Hence $T_i \cap T_jg = \emptyset$ for all $g$ in that coset of $T_j$ mapping $T_j$ to $T_i$. Translating by $g_i$, we have $T_ig_i \cap T_jgg_i = \emptyset$ iff $(gg_i)g_i^{-1} = g$ maps $T_j$ to $T_i$. □

**III.5.** *Let $i, j, k$ be distinct, and $T_ig_i$, $T_jg_j$, $T_kg_k$ be any three cosets of $T_i$, $T_j$, $T_k$. If $T_ig_i \cap T_jg_j = \emptyset$ and $T_ig_i \cap T_kg_k = \emptyset$, then $T_jg_j \cap T_kg_k = \emptyset$.*

*Proof.* If $T_ig_i \cap T_jg_j = \emptyset$ and $T_kg_k \cap T_ig_i = \emptyset$, then $g_jg_i^{-1}$ maps $T_j$ to $T_i$ and $g_ig_k^{-1}$ maps $T_i$ to $T_k$. Hence $(g_jg_i^{-1})(g_ig_k^{-1}) = g_jg_k^{-1}$ maps $T_j$ to $T_k$, implying $T_jg_j \cap T_kg_k = \emptyset$. □

The next step is to construct a generalized quadrangle $\mathcal{S}(G, \mathfrak{T}) = (\mathcal{P}_{\mathfrak{T}}, \mathcal{L}_{\mathfrak{T}}, I_{\mathfrak{T}})$ of order $(s, s)$ from the pair $(G, \mathfrak{T})$.

$\mathcal{P}_{\mathfrak{T}}$: there are two kinds of points:

(a) elements of $G$ ($s^3 - s$ of these),
(b) right cosets of the $T_i$'s ($(s+1)^2$ of these).

$\mathcal{L}_{\mathfrak{T}}$: There are three kinds of lines:

(i) right cosets of $S_i$, $0 \leqslant i \leqslant s$ ($(s+1)(s^2-1)$ of these),
(ii) sets $M_i = \{T_ig \mid g \in G\}$, $0 \leqslant i \leqslant s$ ($s + 1$ of these),
(iii) sets $L_i = \{T_jg \mid T_ig \cap T_j = \emptyset, 0 \leqslant j \leqslant s, j \neq i\} \cup \{T_i\}$ ($1 + s$ of these).

$I_{\mathfrak{T}}$: A line $S_ig$ of type (i) is incident with the $s$ points of type (a) contained in it, together with that point $T_ig$ of type (b) containing it. The lines of types (ii) and (iii) are already described as sets of those points with which they are to be incident. By III.5 two cosets of distinct $T_i$'s are collinear (on a line of type (iii)) iff they are disjoint.

Clearly each line of $\mathcal{S}(G, \mathfrak{T})$ is incident with $1 + s$ points. A point of type (a) is on $1 + s$ lines, as it lies in a unique coset of each $S_i$, $0 \leqslant i \leqslant s$. A point of (b) is incident with $s - 1$ lines of type (i), one line of type (ii), and one line of type (iii).

A routine check shows that two points never lie on two lines in common, basically because of (5). Since there are $1 + s + s^2 + s^3$ points and also that many lines, the resulting geometry wlll be a GQ of order $(s,s)$ provided no triangles occur. Condition (6) guarantees that no triangle occurs with the vertices of type (a). The lines $L_0, \ldots, L_s$ of type (iii) are all disjoint, and the lines $M_0, \ldots, M_s$ of type (ii) are all disjoint, with each $L_i$ meeting each $M_j$. Clearly no triangle occurs with one vertex of type (a) and opposite side a line of type (ii). Suppose some point $x$ of type (a) is a vertex of a triangle whose opposite side is a line of type (iii). We may assume with no loss in generality that one vertex is $T_i$ and the remaining vertex is $T_jg$, where $T_i = g^{-1}T_jg \neq T_j$. Then $x$ belongs to some coset of $S_i$ in $T_i$, i.e. $x \in T_i$, and $x$ belongs to some coset of $S_j$ in $T_jg$, say $x = tg$ with $t \in T_j$. Hence $x \in T_i \cap T_jg$, contradicting the assumption that $T_i$ and $T_jg$ are on a line of type (iii), i.e. $T_i \cap T_jg = \emptyset$. The only remaining case would be a triangle with two points $x$ and $y$ of type (a) and one vertex $T_ig$ of type (b). But here the line through $x$ and $T_ig$ is $S_ix \subseteq T_ig$ and the line through $y$ and $T_ig$ is $S_iy \subseteq T_ig$. But a line through $x$ and $y$ has the form $S_jx = S_jy$ for some $j$. Hence $xy^{-1} \in S_j \cap T_i$, implying either $x = y$ (yielding no triangle) or $j = i$. But if $j = i$, then $x$, $y$, $T_ig$ are all on the line $S_ix = S_iy$, yielding no triangle. It follows that $\mathcal{S}(G, \mathcal{T})$ is a GQ of order $(s,s)$.

In this construction $G$ acts on $\mathcal{S}(G, \mathcal{T})$ by right multiplication, so that $S_i$ is the full group of symmetries about $L_i$, $0 \leqslant i \leqslant s$, and $T_i$ is the stabilizer of $L_i$ in $G$. This can be seen as follows.

For $x \in G$, let $x^\backslash$ denote the collineation determined by right multiplication by $x$. Then $T_jg \cap T_i = \emptyset$ iff $T_jgx \cap T_ix = \emptyset$, and $T_ix = T_i$ iff $x \in T_i$. Hence $x$ fixes $L_i$ (and each point of $L_i$) iff $x \in T_i$. Let $L$ be some line of type (i) meeting $L_i$ at, say, $T_jg$ for some $j \neq i$, where $g^{-1}T_jg = T_i$ (implying $g^{-1}S_jg = S_i$). Then $L$ is some coset of $S_j$ contained in $T_jg$, say $L = S_jt_jg$, where $t_j \in T_j$ and $L^x = L$ iff $S_jt_jgx = S_jt_jg$ iff $g^{-1}(t_j^{-1}S_jt_j)gx = g^{-1}(t_j^{-1}S_jt)g$ iff $(g^{-1}S_jg)x = g^{-1}S_jg$ iff $S_ix = S_i$ iff $x \in S_i$. Hence $S_i$ is the set of all $g \in G$ for which $g^\backslash$ fixes each line of $\mathcal{S}(G, T)$ meeting $L_i$.

Starting with a span-symmetric GQ $\mathcal{S}$ of order $(s,s)$, with base span sp$(L_0, L_1)$, deriving the 4-gonal basis $\mathcal{T}$ of the group $G$ generated by symmetries about lines in sp$(L_0, L_1)$, and then constructing the GQ $\mathcal{S}(G, \mathcal{T})$ ensures that $\mathcal{S}$ and $\mathcal{S}(G, \mathcal{T})$ are isomorphic. Howcvcr, we leave the details to the reader. This completes the proof of the following major result.

**III.6.** *A span-symmetric generalized quadrangle of order $(s,s)$ with given base span* sp$(L_0, L_1)$ *is canonically equivalent to a group $G$ of order $s^3 - s$ with a 4-gonal basis $\mathcal{T}$.*

Any automorphism of $G$ leaving $\mathcal{T}$ invariant must induce a collineation $\mathcal{S}(G, \mathcal{T})$. In particular, for each $g \in G$, conjugation by $g$ yields a collineation $\hat{g}$ of $\mathcal{S}(G, \mathcal{T})$. But conjugation by $g$ followed by right multiplication by $g^{-1}$ yields a collineation ${}^/g$ given by left multiplication by $g^{-1}$. Then $g \rightarrow {}^/g$ is a representation of $G$ as a group of collineations of $\mathcal{S}(G, \mathcal{T})$ in which $S_i$ is a full group of symmetries about $M_i$, and $T_i$ is the stabilizer of $M_i$. This is easily checked, so that we have the following.

**III.7.** *If* $\mathcal{S}$ *is a span-symmetric GQ of order* $(s,s)$ *with base span* $\mathrm{sp}(L_0, L_1)$, *then each line of* $\mathrm{tr}(L_0, L_1)$ *is also an axis of symmetry.*

# IV. Coordinates for GQ with Axes of Symmetry

Let $\mathcal{S} = (\mathcal{P}, \mathcal{L}, I)$ be a GQ of order $(s,s)$ having two concurrent lines $L_0$ and $M_0$, each of which is an axis of symmetry. From [9] we know that $\mathcal{S}$ has a coordinatization as follows. First, there is a projective plane $\pi_\infty$ based at $L_0$ and coordinated by a planar ternary ring $(R, F)$ consisting of a set $R$ of size $s$ and a ternary operation $F: R^3 \to R$ satisfying the following axioms (here 0 and 1 denote distinguished elements of $R$ called "zero" and "one," respectively):

(7) $F(a,0,c) = F(0,b,c) = c$, for all $a,b,c \in R$.
(8) $F(1,a,0) = F(a,1,0) = a$, for all $a \in R$.
(9) Given $a,b,c,d \in \mathrm{R}$ with $a \neq c$, there is a unique $x \in R$ for which $F(x,a,b) = F(x,c,d)$.
(10) Given $a,b,c \in R$, there is a unique $x \in R$ for which $F(a,b,x) = c$.
(11) Given $a,b,c,d \in R$ with $a \neq c$, there is a unique ordered pair $(x, y) \in R^2$ for which $F(a,x,y) = b$ and $F(c,x,y) = d$.

These are the standard axioms given by Hall [3].

Let $H_0$ denote the hypothesized group of symmetries about $L_0$, so $|H_0| = s$. $H_0$ is written additively, but is not assumed to be Abelian. Then there is a 4-gonal function $U_0: R^3 \to H_0$ satisfying conditions (12)–(14):

(12) $U_0(0,0,m) = U_0(a,b,0) = 0$ for all $a,b,m \in R$ where "0" denotes the "zero" of $R$ or the additive identity of $H_0$, whichever is appropriate.
(13) Condition (7) of [9].
(14) Condition (8) of [9].

(These last two are messy conditions that will not play a role in our computations.)

The GQ $\mathcal{S}$ may then be described as follows. Points are of the form $(\infty)$, $(a)$, $(m, g)$, $(a,b, g)$, for arbitrary $a,b,m \in R$, $g \in H_0$. Lines are of the form $[\infty]$, $[m]$, $[a,k]$, $[m, g,k]$, for arbitrary $a,m,k \in R$, $g \in H_0$. Here $L_0$ may be identified with $[\infty]$ and $M_0$ may be identified with $[0]$. Incidence is described as follows:

(15) $(\infty)$ is on $[\infty]$ and on $[m]$, $m \in R$.
$(a)$ is on $[\infty]$ and on $[a,k]$, $a,k \in R$.
$(m, g)$ is on $[m]$ and on $[m, g,k]$, $m,k \in R$, $g \in H_0$.
$(a, F(a,m,k), U_0(a, F(a,m,k), m) + g)$ is on $[a, F(a,m,k)]$ and on $[m, g,k]$, $a,m,k \in R$, $g \in H_0$.

These incidences are summarized in the "incidence diagram" in Figure 1.

So far this coordinatization uses only the assumption that $L_0 = [\infty]$ is an axis of symmetry. By IV.1 of [9], the assumption that $M_0 = [0]$ is also an axis of symmetry yields the following:

(16) $U_0(a,k,m) = U_0(0,k,m) + U_0(a,0,m)$, $a,k,m \in R$.
(17) $F(t,m,F(a,m,k)) = F(F(t,1,a),m,k)$, $a,t,m,k \in R$.

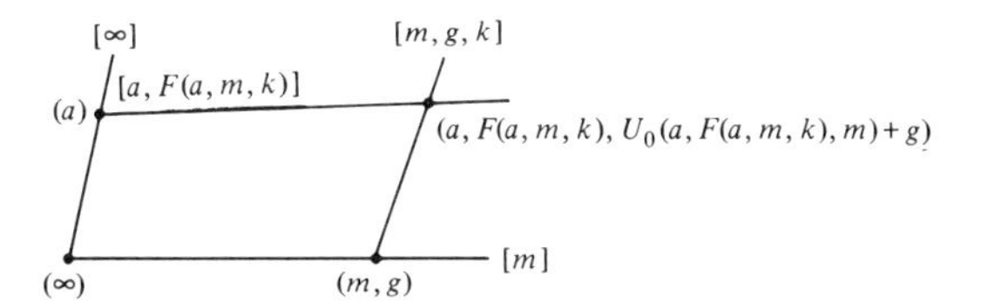


**Figure 1**

(18) $U_0(t,0,m) + U_0(a,0,m) = U_0(F(t,1,a),0,m)$, $a,t,m \in R$.
(19) For fixed $m \in R$, $m \neq 0$, the map $a \to U_0(a,0,m)$ is a group isomorphism from $(R, +)$ onto $H_0$.

Here the addition "+" on $R$ is defined as usual by

(20) $a + b = F(a,1,b)$, $a, b \in R$.

There is also a standard multiplication " $\circ$ " defined on $R$ by

(21) $a \circ m = F(a,m,0)$ for $a, m \in R$.

The projective plane $\pi_\infty$ based at $[\infty]$ is coordinated by $(R,F)$ as follows. Lines of $\pi_\infty$ are $[\infty], [m], [a,b]$, $a,b \in R$. Points of $\pi_\infty$ are $(\infty)$, $(a)$, and $((a,b))$, $a,b \in R$. Incidence is given as follows (cf. [5] and [9]):

(22) $(\infty)$ is on $[\infty]$ and on $[m]$, $m \in R$.
$(a)$ is on $[\infty]$ and on $[a,b]$, $a,b \in R$.
$((m,k))$ is on $[m]$ and on $[a,b]$ provided that $b = F(a,m,k)$ for $a,b,m,k \in R$.

In this coordinatization $\pi_\infty$ is the dual of the plane coordinatized by $(R,F)$ in the standard manner as in [2] or [3], under a duality in which by Theorem 20.5.4 of Hall [3] we know the following.

**IV.1.** $\pi_\infty$ *is* (0)–[0] *transitive* (*i.e. has all homologies with center* (0) *and axis* [0]) *if and only if both*

(i) $F(a,m,k) = (a \circ m) + k$ (i.e. *F is "linear"*) *and*
(ii) $(R^*, \circ)$ *is a group, where* $R^* = R \setminus \{0\}$.

We now suppose that $\mathcal{S}$ is span-symmetric with $L_0$ and $M_0$ being two of the hypothesized axes of symmetry. In setting up the coordinates in [5] there was a certain amount of arbitrariness. We may assume that any particular line meeting [0] at some point other than $(\infty)$ is one of the hypothesized axes of symmetry. So we suppose that $[0,0,0]$ meeting [0] at $(0,0)$ is an axis of symmetry. Then we have that each line of $\text{sp}([\infty], [0,0,0]) \cup \text{tr}([\infty], [0,0,0])$ is an axis of symmetry. An easy check shows that

(23) $\text{sp}([\infty], [0,0,0]) = \{[\infty]\} \cup \{[0, g, 0] \mid g \in H_0\}$,
(24) $\text{tr}([\infty], [0,0,0]) = \{[0]\} \cup \{[a,0] \mid a \in R\}$.

Let $G$ be the group generated by symmetries about lines of $\text{tr}([\infty],[0,0,0])$. In particular, by results of Section 3 we know that the stabilizers in $G$ of [0] and $[0,0,0]$ intersect in a group of order $s-1$ that fixes each point of [0], fixes (0), and acts regularly on the lines $[m]$, $m \neq 0$. These collineations induce homologies on $\pi_\infty$ with center (0) and axis [0], so that $\pi_\infty$ is indeed (0)–[0] transitive.

**IV.2.** *Assuming that each line of* tr([∞], [0, 0, 0]) *is an axis of symmetry, we have the following*:

(i) $F(a, m, k) = a \circ m + k$
(ii) $(R^*, \circ)$ *is a group, where* $R^* = R \setminus \{0\}$.

By (17) the right distributive law holds: $t \circ m + a \circ m = (t + a) \circ m$. Hence $(R, F)$ is a quasifield, implying that $(R, +)$ is elementary Abelian (cf. [2, p. 221]), so that $H_0$ is elementary Abelian. Since $[0, 0, 0]$ is an axis of symmetry, it must be regular. In particular, for each $k \in R$, $k \neq 0$, the pair $([0, 0, 0], [0, k])$ must be regular. Making free use of (16), (18), and IV.2, for each $m, x \in R$, $m \neq 0 \neq x$, the incidences shown in Figure 2 must hold. (Note: $(-a) \circ m = -(a \circ m)$, so $-a \circ m$ is unambiguous.)

By the regularity of $([0, 0, 0], [0, x \circ m])$, it must be that $[m, U_0(x, 0, m), x \circ m]$ meets $[m, -U_0(x, x \circ m, m), 0]$. It readily follows that the point of intersection cannot be of the type $(a, b, c)$, since then $b = F(a, m, x \circ m) = a \circ m + x \circ m$ and $b = F(a, m, 0) = a \circ m$, implying $x \circ m = 0$, an impossibility since $x \neq 0 \neq m$. Hence the point of intersection must be $(m, U_0(x, 0, m)) = (m, -U_0(x, x \circ m, m))$.

(25) $U_0(x, 0, m) = -U_0(x, x \circ m, m)$ if $x, m \in R$, $x \neq 0 \neq m$.

Of course (25) holds trivially if $x = 0$ or $m = 0$. Hence for any $x, m \in R$, $U_0(0, x \circ m, m) = U_0(0, x \circ m, m) + U_0(x, 0, m) - U_0(x, 0, m) = U_0(x, x \circ m, m) - U_0(x, 0, m) = -U_0(x, 0, m) - U_0(x, 0, m) = U_0(-2x, 0, m)$.

(26) $U_0(0, x \circ m, m) = U_0(-2x, 0, m)$ for all $x, m \in R$.

Using (16), (19), and (26) it is easy to see the following:

(27) *The map* $(a, b) \rightarrow U_0(a, b, m)$ *from* $R^2$ *to* $H_0 = (R, +)$ *is additive for each* $m \in R$.

By Section V of [9] we now have the following.

**IV.3.** $\mathcal{S}$ *is a translation GQ with base point* (∞), *so that each line through* (∞) *is regular.*

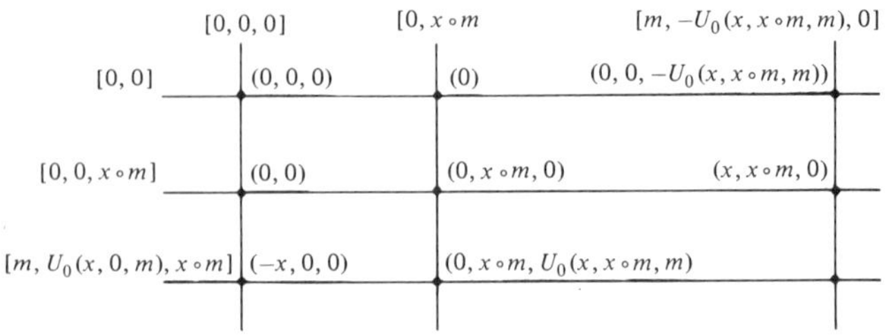

$[0, 0, 0]$ $[0, x \circ m$ $[m, -U_0(x, x \circ m, m), 0]$
$[0, 0]$ $(0, 0, 0)$ $(0)$ $(0, 0, -U_0(x, x \circ m, m))$
$[0, 0, x \circ m]$ $(0, 0)$ $(0, x \circ m, 0)$ $(x, x \circ m, 0)$
$[m, U_0(x, 0, m), x \circ m]$ $(-x, 0, 0)$ $(0, x \circ m, U_0(x, x \circ m, m)$

**Figure 2**

The above result must hold for each point on any line of sp([∞], [0,0,0]), since each such point may be taken as (∞) in some coordinatization of $\mathbb{S}$. This means that each line of $\mathbb{S}$ is regular, so that $\mathbb{S}$ is well known to be dual to $W_3(s)$, i.e. is isomorphic to the quadric $Q(4,s)$ (cf. V.1 and V.2 of [13]). This completes the proof of our main Theorem.

**IV.4.** *A span-symmetric GQ of order $(s,s)$ is isomorphic to $Q(4,s)$.*

## V. The Classical Case

The points and lines of a nondegenerate quadric surface $Q(4,s)$ in $PG(4,s)$, together with the natural incidence relation inherited from $PG(4,s)$, form a GQ of order $(s,s)$ in which all lines are regular and any span of nonconcurrent lines serves as a base span for viewing $Q(4,s)$ as span-symmetric. $Q(4,s)$ is the dual of the geometry $W_3(s)$ defined as follows (cf. [1] and [10]). Points of $W_3(s)$ are the points of $PG(3,s)$; the lines of $W_3(s)$ are the absolute lines of a null polarity of $PG(3,s)$; and incidence is just that of $PG(3,s)$ restricted to points and lines of $W_3(s)$. For example, if

$$C = \begin{pmatrix} C_0 & 0 \\ 0 & C_0 \end{pmatrix}, \quad \text{where } C_0 = \begin{pmatrix} 0 & 1 \\ -1 & 0 \end{pmatrix},$$

and if $\bar{x} = (x_0, x_1, x_2, x_3)$, $\bar{y} = (y_0, y_1, y_2, y_3)$ are homogeneous representations of points of $\bar{x}$ and $\bar{y}$ of $PG(3,s)$, then choosing the line of $PG(3,s)$ through $\bar{x}$ and $\bar{y}$ as a line of $W_3(s)$ iff $\bar{x}C\bar{y}^T = 0$ yields a model of $W_3(s)$. As a GQ of order $(s,s)$, $W_3(s)$ is completely characterized by the fact that all points are regular (see [13] for references to characterizations of the classical GQ). When $s$ is a power of 2, all lines of $W_3(s)$ are regular and $W_3(s)$ is self-dual. When $s$ is an odd prime power, each line of $W_3(s)$ is *antiregular*, i.e., if $L$ and $M$ are any two nonconcurrent lines of $W_3(s)$, then $L$ and $M$ are the only lines that meet more than two lines of $\operatorname{tr}(L,M)$. Consequently $\operatorname{sp}(L,M) = \{L, M\}$.

If $\bar{x} = (1,0,0,0)$ and $\bar{y} = (0,1,0,0)$, then in the specific model of $W_3(s)$ given above, $\operatorname{sp}(\bar{x}, \bar{y})$ consists of all linear combinations of $\bar{x}$ and $\bar{y}$, and the group $G$ generated by symmetries about points of $\operatorname{sp}(\bar{x}, \bar{y})$ consists of linear collineations of the form $\bar{x} \to \bar{x}A^T$, where

$$A = \begin{pmatrix} A_0 & 0 \\ 0 & I \end{pmatrix} \quad \text{and} \quad \det A_0 = 1.$$

(Here $A_0$ is $2 \times 2$.) Hence $G$ is isomorphic to $SL(2,s)$. Of course, it is well known that $|SL(2,s)| = s^3 - s$ and that $SL(2,s)$ possesses a conjugacy class $\mathfrak{T}$ satisfying the axioms for a 4-gonal basis of $G$. The main result of this paper guarantees that *any group $G$ with a 4-gonal basis must in fact be isomorphic to $SL(2,s)$.*

In general, the groups $SL(2,s)$ have received so much attention that no further comment here is needed. However, we note that the unique GQ of order $(4,4)$ (cf. [7]) has a concrete description as a span-symmetric GQ $\mathbb{S} = \mathbb{S}(G, \mathfrak{T})$ where $G = SL(2,4) \cong A_5$, the alternating group on $1, \ldots, 5$. Let $S_i$ be the Klein

4-group on the symbols $\{1,2,3,4,5\}\backslash\{i\}$, $1 \leqslant i \leqslant 5$. For example, $S_1 = \{e$, (23)(45), (24)(45), (25)(34)$\}$. Then $T_i = N_G(S_i)$ is the alternating group on the symbols $\{1,2,3,4,5\}\backslash\{i\}$. It follows that $\mathfrak{T} = \{S_1, \ldots, S_5\}$ is a 4-gonal basis for $A_5$.

## VI. Acknowledgments

The author would like to thank his colleagues C. E. Ealy, Jr. and C. S. Holmes for helpful conversations during the preparation of this work, as well as W. M. Kantor[1] for helpful criticisms of an early manuscript.

[1] *Note added in proof:* W. M. Kantor has recently shown that there is no GQ of order $(s, t)$ with $s < t < s^2$ having a pair of nonconcurrent lines about each of which the group of symmetries has order $s$.

## References

[1] Benson, C. T., On the structure of generalized quadrangles. *J. Algebra* **15** (1970), 443–454.

[2] Dembowski, P., *Finite Geometries*. Springer-Verlag, Berlin 1968.

[3] Hall, M., Jr., *The Theory of Groups*. MacMillan, New York 1959.

[4] Payne, S. E., A restriction on the parameters of a subquadrangle. *Bull. Amer. Math. Soc.* **79** (1973), 747–748.

[5] Payne, S. E., Generalized quadrangles of even order. *J. Algebra* **31** (1974), 367–391.

[6] Payne, S. E., Skew translation generalized quadrangles. In *Congressus Numerantium XIV, Proc. 6th S. E. Conf. Comb., Graph Theory, Comp.*, 1975.

[7] Payne, S. E., Generalized quadrangles of order 4, I and II. *J. Comb. Theory* **22** (1977), 267–279, 280–288.

[8] Payne, S. E., An inequality for generalized quadrangles. *Proc. Amer. Math. Soc.* **71** (1978), 147–152.

[9] Payne, S. E. and Thas, J. A., Generalized quadrangles with symmetry. *Simon Stevin* **49** (1976), 3–32, 81–103.

[10] Thas, J. A., Ovoidal translation planes. *Archiv der Mathematik* **XXIII** (1972), 110–112.

[11] Thas, J. A., 4-gonal subconfigurations of a given 4-gonal configuration. *Rend. Accad. Naz. Lincei* **53** (1972), 520–530.

[12] Thas, J. A., On generalized quadrangles with parameters $s = q^2$ and $t = q^3$. *Geometria Dedicata* **5** (1976), 485–496.

[13] Thas, J. A. and Payne, S. E., Classical finite generalized quadrangles: a combinatorial study. *Ars. Combinatoria* **2** (1976), 57–110.

[14] Walker, M., On the structure of finite collineation groups containing symmetries of generalized quadrangles. *Inventiones Mathematicae* **40** (1977), 245–265.

# On Coxeter's Loxodromic Sequences of Tangent Spheres

Asia Weiss*

## 1. Introduction

A loxodromic sequence of tangent spheres in $n$-space is an infinite sequence of $(n-1)$-spheres having the property that every $n+2$ consecutive members are mutually tangent. When considering mutually tangent spheres we'll always suppose they have distinct points of contact. Given any ordered set of $n+2$ mutually tangent $(n-1)$-spheres, we can invert into $n$ congruent $(n-1)$-spheres sandwiched between two parallel hyperplanes, and hence (since the centres of these $n$ are the vertices of a regular simplex) they are all inversively equivalent. Furthermore, any ordered set of $n+1$ mutually tangent $(n-1)$-spheres $\{C_0, C_1, \ldots, C_n\}$ can be completed to a set of $n+2$ spheres in exactly two ways. Hence the spheres belong to just one sequence

$$\ldots, C_{-1}, C_0, C_1, C_2, \ldots \tag{1}$$

with the property that every $n+2$ consecutive members are mutually tangent.

## 2. Inversive Geometry

We introduce coordinates for points and balls in inversive $n$-space. Let $\Sigma$ be the unit $n$-sphere in $\mathbb{R}^{n+1}$ centered at the origin, and let $\Pi$ be an $n$-flat through the origin. We use stereographic projection to establish a 1-1 correspondence between $\Pi$ and $\Sigma - \{\text{north pole}\}$, and hence with any point $X$ in $\Pi$ we can associate an $(n+2)$-tuple $(x, 1)$ where $x$ is the coordinate of the stereographic projection of $X$. We allow $X$ and $\lambda X$ for $\lambda > 0$ to name the same point. Similarly if $C$ is a ball (we allow half spaces as a special case) in $\Pi$, we can associate an

*Department of Mathematics, University of Toronto, Toronto M5S 1A1, Ontario, Canada.

$(n+2)$-vector $(\csc\theta \cdot c, \cot\theta)$ with it, where $c$ is the centre and $\theta$ the angular radius of the sphere on $\Sigma$ whose stereographic projection is $C$.

Between two $(n+2)$-vectors we define the indefinite bilinear form

$$\begin{aligned} A * B &= (a_1, \ldots, a_{n+2}) * (b_1, \ldots, b_{n+2}) \\ &= a_1 b_1 + a_2 b_2 + \cdots + a_{n+1} b_{n+1} - a_{n+2} b_{n+2}. \end{aligned}$$

A vector $A$ represents a point if and only if $A * A = 0$ and $a_{n+2} > 0$. A condition for a point $X$ to lie on the sphere bounding a ball $C$ is $X * C = 0$. Let $C$ and $D$ be two balls. The following properties of the multiplication $*$ will be often used:

| | |
|---|---|
| If $C$ and $D$ are tangent with nested interiors (in particular $C = D$), | then $C * D = 1$. |
| If $C$ and $D$ are tangent with disjoint interiors, | then $C * D = -1$. |
| If $C$ and $D$ have intersecting boundaries with angle $\theta$ between them, | then $C * D = \cos\theta$. |
| If $C$ and $D$ have disjoint boundaries with inversive distance $\delta$ and nested interiors, | then $C * D = \cosh\delta$. |
| If $C$ and $D$ have disjoint boundaries with inversive distance $\delta$ and disjoint interiors, | then $C * D = -\cosh\delta$. |

A cluster is an ordered set of $n+2$ mutually externally tangent balls. The names of the balls of a cluster form a basis for our coordinate space. The $n$-dimensional Möbius group $\mathfrak{M}_n$ is isomorphic to the (n + 2)-dimensional linear group $\mathfrak{L}_{n+2}$ (group of linear transformations which preserve the bilinear form $A * B$ and sign of $x_{n+2}$ on the cone $X * X = 0$). The group $\mathfrak{M}_n$ is sharply transitive on clusters. More details about $\mathfrak{M}_n$ and the inverse coordinates are available in [2].

## 3. Relation Between Spheres in the Sequence

In dimension 2, Coxeter [1] showed that the whole sequence (1) is mapped onto itself by a dilative rotation whose coefficient of dilatation is the real root $\lambda > 1$ of the equation

$$\lambda^4 - 2(\lambda^3 + \lambda^2 + \lambda) + 1 = 0. \tag{2}$$

It was also observed that if $\theta$ is the angle of rotation, then $e^{\pm\theta i}$ are the complex roots of the equation (2), but the reason for this was not known. Furthermore it was shown that points of contact of consecutive pairs of spheres lie on a

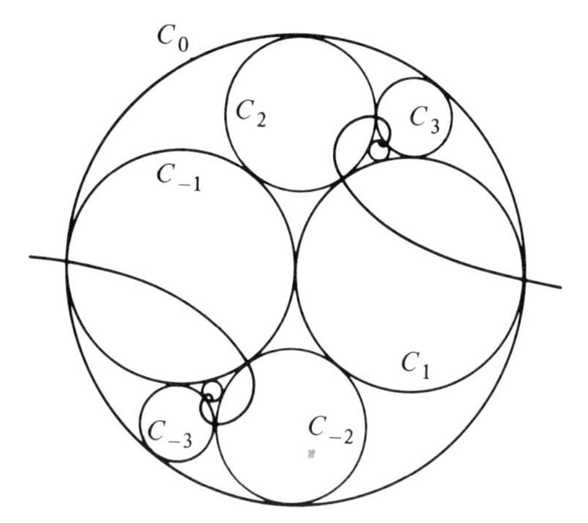


**Figure 1**

loxodrome, which being the inverse of an equiangular spiral has two poles (as in Figures 1 and 2).

Furthermore Coxeter proved that for every $n > 1$, just one loxodromic sequence exists with the radii in geometric progression, whose ratio is a unique real root $\lambda > 1$ of the following polynomial:

$$\lambda^{n+2} - \frac{2}{n-1}(\lambda^{n+1} + \lambda^{n} + \cdots + \lambda) + 1. \tag{3}$$

It will be shown that given any loxodromic sequence of tangent spheres there is a Möbius transformation mapping the sequence onto itself. In appropriate coordinates this is a linear transformation whose characteristic polynomial is (3). This transformation is a dilative rotation which is sense-preserving in even dimensions, while in odd dimensions it has a negative coefficient of dilatation and thus is sense-reversing. By considering the eigenvalues of this transformation we get

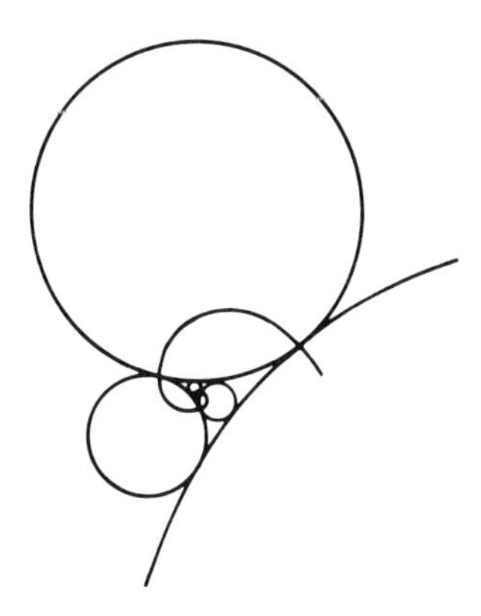

**Figure 2**

more information about the configuration of these spheres and their points of contact.

Let $\{C_0, C_1, \ldots, C_{n+1}\}$ be a set of $n+2$ mutually tangent spheres with distinct points of contact. To each sphere $C_i$ in the sequence (1) correspond two balls with disjoint interiors. One of them must have interior which does not intersect any of the spheres in the set. We give that ball the same name as the name of the corresponding sphere, so that $\{C_0, C_1, \ldots, C_{n+1}\}$ is a cluster. Then since $\{C_0, C_1, \ldots, C_{n+1}\}$ form a basis, we can write

$$C_{n+2} = a_0 C_0 + a_1 C_1 + \cdots + a_{n+1} C_{n+1}. \tag{4}$$

Hence

$$\begin{aligned}
-1 &= -a_0 + a_1 - a_2 - \cdots - a_n - a_{n+1},\\
-1 &= -a_0 - a_1 + a_2 - \cdots - a_n - a_{n+1},\\
&\ \vdots\\
-1 &= -a_0 - a_1 - a_2 - \cdots - a_n + a_{n+1},
\end{aligned}$$

where the relations are obtained from (4) by multiplying by $C_1, C_2, \ldots, C_{n+1}$ respectively. Clearly $a_1 = a_2 = \cdots = a_{n+1} = (1 - a_0)/(n-1)$, and hence

$$C_{n+2} - a_0 C_0 = \frac{1 - a_0}{n-1}(C_1 + \cdots + C_{n+1}).$$

Multiplying this relation by $C_0$ and by $C_{n+2}$, we get

$$\begin{aligned}
C_{n+2} * C_0 - a_0 &= \frac{1 - a_0}{n-1}(n+1)(-1),\\
1 - a_0 C_{n+2} * C_0 &= \frac{1 - a_0}{n-1}(n+1)(-1);
\end{aligned}$$

consequently $a_0 = -1$ and we get

$$C_{n+2} - \frac{2}{n-1}(C_{n+1} + C_n + \cdots + C_1) + C_0 = 0. \tag{5}$$

It can be easily seen that relation (5) remains valid when all the indices are increased or decreased by any integer, so that (5) provides a recursion formula for computing the coordinates for any sphere in the sequence (in terms of the basis $\{C_0, C_1, \ldots, C_{n+1}\}$).

We define the bend of a sphere to be the reciprocal of its radius, with a minus sign in appropriate cases. It is shown in [2] that $C * E$ is the bend of the sphere $C$ when $E = (0, \ldots, 0, -1, -1)$. It is easy to see that the vector relation (5) gives rise to the Euclidean relation

$$\epsilon_{n+2} - \frac{2}{n-1}(\epsilon_{n+1} + \epsilon_n + \cdots + \epsilon_1) + \epsilon_0 = 0 \tag{6}$$

where $\epsilon_i$ is the bend of the ball $C_i$. Hence (6) is a recursion formula for computing the radii of the spheres in the sequence provided we know the radii of $C_0, C_1, \ldots, C_{n+1}$. From (5) it also follows that $\cosh \delta_{n+2} = (3n+1)/(n-1)$, where $\delta_{n+2}$ is the inversive distance between $C_\nu$ and $C_{\nu+n+2}$. In general, for any

$\nu \in Z$, using (5) we can calculate the inversive distance $\delta_\nu$ between spheres $C_\mu$ and $C_{\mu+\nu}$.

## 4. A Transformation Mapping the Sequence into Itself

Since $\mathfrak{M}_n$ is sharply transitive on clusters, there is exactly one transformation $M$ in $\mathfrak{M}_n$ mapping $C_0, C_1, \ldots, C_{n+1}$ to $C_1, C_2, \ldots, C_{n+2}$ successively, and since $C_{n+2}$ is given by the relation (5) we have

$$M = \begin{bmatrix} 0 & 1 & 0 & \cdots & 0 \\ 0 & 0 & 1 & \cdots & 0 \\ \vdots & \vdots & \vdots & & \vdots \\ 0 & 0 & 0 & \cdots & 1 \\ -1 & \frac{2}{n-1} & \frac{2}{n-1} & \cdots & \frac{2}{n-1} \end{bmatrix}. \tag{7}$$

An inductive argument would show that this transformation maps the whole sequence onto itself, i.e., $C_i M = C_{i+1}$ for all $i \in Z$. $M$ can be expressed as the product of two Möbius involutions, one leaving $C_0$ fixed and interchanging $C_i$ with $C_{-i}$ [i.e., $(C_0)(C_1C_{-1})(C_2C_{-2}) \ldots$], and the other interchanging $C_i$ with $C_{-i+1}$ [i.e., $(C_0C_1)(C_{-1}C_2)(C_{-2}C_3) \ldots$].

The characteristic polynomial of the matrix (7) is the polynomial (3). Multiplying its characteristic equation by $\lambda - 1$ and rewriting it (see [1, p. 119]), we get

$$\frac{\lambda^{n+2} - 1}{\lambda^{n+2} + 1} = n\frac{\lambda - 1}{\lambda + 1}. \tag{8}$$

The substitution $\lambda = e^{2\delta}$ yields

$$\tanh(n+2)\delta = n \tanh \delta. \tag{9}$$

There is exactly one positive root of (9). Hence there are two (reciprocal) real roots $e^{\pm 2\delta}$ of (3), where $\delta$ is given by (9). If $n$ is odd, $-1$ is a real root of (3). The substitution $\lambda = e^{2i\theta}$ yields

$$\tan(n+2)\theta = n \tan \theta. \tag{10}$$

If $n$ is even, there are exactly $n$ complex roots $e^{\pm 2i\theta_1}, \ldots, e^{\pm 2i\theta_{n/2}}$ of (3), where the $\theta_j$'s are given by Equation (10). These together with the two real roots $e^{\pm 2\delta}$ give all the roots of (3). If $n$ is odd, there are exactly $n-1$ complex roots $e^{\pm 2i\theta_1}, \ldots, e^{\pm 2i\theta_{(n-1)/2}}$ of (3), where the $\theta_j$'s are given again by (10). These together with the roots $e^{\pm 2\delta}$ and $-1$ give $n+2$ roots of (3).

Now let $A$ be a characteristic vector of $N \in \mathfrak{M}_n$ with eigenvalue $\lambda$. Then $A * A = AN * AN = \lambda^2 A * A$, and hence characteristic vectors with eigenvalue different from $\pm 1$ represent fixed points of $N$. If $\lambda = \pm 1$ and $A * A \neq 0$, then $A$ represents a ball, and if $B$ is a point on the sphere bounding the ball $A$, then $0 = B * A = BN * AN = \pm BN * A$, i.e., the sphere is fixed under $N$. Furthermore, if $B$ is a fixed point and $A$ is a ball with the fixed sphere, then

$B * A = BN * AN = \pm \lambda B * A$, and if $\lambda \neq \pm 1$, the fixed point lies on the fixed sphere.

For an eigenvalue $-1$ of the transformation $M$ (given by (7)) in odd dimensions we compute the eigenvector

$$C = C_0 - \frac{n+1}{n-1} C_1 + C_2 - \frac{n+1}{n-1} C_3 + C_4 - \cdots - \frac{n+1}{n-1} C_n + C_{n+1}.$$

Furthermore we compute $C * C = 2n(n^2 + 2n - 1)/(n-1)^2 \neq 0$, and hence we see that $C$ represents a ball with a fixed bounding sphere.

We conclude that in even dimensions we have two fixed points, and in odd dimensions there is a fixed sphere containing two fixed points. The only Möbius transformations with these properties are (see [3]) "dilative rotations" in even dimensions and "dilative rotatory reflections" in odd dimensions (the quotation marks denote the inversive counterpart of the corresponding Euclidean transformations). If we allow dilatations with negative coefficients, then a dilative rotatory reflection with a coefficient of dilatation $e^{2\delta}$ and angles of rotation $2\theta_1, \ldots, 2\theta_{n/2}$ can be thought of as a dilative rotation with a coefficient of dilatation $-e^{2\delta}$ and angles of rotation $2\theta_1 + \pi, \ldots, 2\theta_{n/2} + \pi$. Following [3] we can denote by $[\circledcirc, \delta, \theta_1, \ldots, \theta_{n/2}]_n$ a "dilative rotation" (in even dimension $n$) whose coefficient of dilatation is $e^{2\delta}$ and angles of rotation are $2\theta_1, \ldots, 2\theta_{n/2}$, and by $[\ominus, \delta, \theta_1, \ldots, \theta_{n/2}]_{n+1}$ a "dilative rotatory reflection" (in odd dimension $n+1$) with the same coefficient and angles. The matrix representations for $[\circledcirc, \delta, \theta_1, \ldots, \theta_{n/2}]_n$ and $[\ominus, \delta, \theta_1, \ldots, \theta_{n/2}]_{n+1}$ are

$$\begin{bmatrix}
\cosh 2\delta & -\sinh 2\delta & 0 & 0 & \cdots & 0 & 0 \\
-\sinh 2\delta & \cosh 2\delta & 0 & 0 & \cdots & 0 & 0 \\
0 & 0 & \cos 2\theta_1 & \sin 2\theta_1 & \cdots & 0 & 0 \\
0 & 0 & -\sin 2\theta_1 & \cos 2\theta_1 & \cdots & 0 & 0 \\
\vdots & \vdots & \vdots & \vdots & & \vdots & \vdots \\
0 & 0 & 0 & 0 & \cdots & \cos 2\theta_{n/2} & \sin 2\theta_{n/2} \\
0 & 0 & 0 & 0 & \cdots & -\sin 2\theta_{n/2} & \cos 2\theta_{n/2}
\end{bmatrix} \tag{11}$$

and

$$\begin{bmatrix}
-1 & 0 & 0 & 0 & 0 & \cdots & 0 & 0 \\
0 & \cosh 2\delta & -\sinh 2\delta & 0 & 0 & \cdots & 0 & 0 \\
0 & -\sinh 2\delta & \cosh 2\delta & 0 & 0 & \cdots & 0 & 0 \\
0 & 0 & 0 & \cos 2\theta_1 & \sin 2\theta_1 & \cdots & 0 & 0 \\
0 & 0 & 0 & -\sin 2\theta_1 & \cos 2\theta_1 & \cdots & 0 & 0 \\
\vdots & \vdots & \vdots & \vdots & \vdots & & \vdots & \vdots \\
0 & 0 & 0 & 0 & 0 & \cdots & \cos 2\theta_{n/2} & \sin 2\theta_{n/2} \\
0 & 0 & 0 & 0 & 0 & \cdots & -\sin 2\theta_{n/2} & \cos 2\theta_{n/2}
\end{bmatrix} \tag{12}$$

respectively. The matrix $M$ is conjugate to (11) or (12) depending on the dimension, so that they have the same eigenvalues. Hence the coefficient of dilatation and angles of rotation of $M$ are given by the solutions of Equation (3).

## 5. Points of Contact

Let $X_k^\alpha$ be a point of contact of $C_k$ and $C_{k+\alpha}$; then $X_k^\alpha = C_k + C_{k+\alpha}$. To prove this observe that if $A$ and $B$ are two (externally) tangent balls with different boundaries, then $(A + B) * (A + B) = A * A + 2A * B + B * B = 1 - 2 + 1 = 0$. If we denote by $\theta_A$ and $\theta_B$ the angular radii of $A$ and $B$ respectively, then $\theta_A + \theta_B < \pi$, $\theta_A < \pi - \theta_B$, and consequently $\cot\theta_A > -\cot\theta_B$. This implies that the last component of $A + B$ is positive. Therefore $A + B$ is a point, and since $(A + B) * A = (A + B) * B = 0$, we see that it is a point of contact of $A$ and $B$. Now clearly $X_k^\alpha M = X_{k+1}^\alpha$ for all $k \in Z$ and all $\alpha \in \{1, \dots, n+1\}$, since $M$ is linear.

In odd dimensions $C_k * C = C_k M * CM = C_{k+1} * (-C)$, we have $X_k^1 * C = (C_k + C_{k+1}) * C = 0$, i.e., points of contact of consecutive spheres all lie on the fixed sphere $C$. Let $\psi$ be the angle between $C$ and $C_k$; then $C_{k+1} * C = C_k M * (-C)M = -C_k * C = -\cos\psi$, and $2\pi - \psi$ is the angle between $C_{k+1}$ and $C$. Hence all spheres in the sequence enclose the same angle with the fixed sphere, and their centres are alternately on opposite sides of the sphere. Inversion in some point on the fixed sphere $C$ (in Figure 3 it is a fixed point of $M$) maps $C$ to a hyperplane, and $C$ intersects the sequence as is shown in Figures 3 and 4 for the case $n = 3$ (balls which intersect $C$ in shaded disks all have their centers on the same side of $C$, and balls which intersect $C$ in nonshaded disks all have

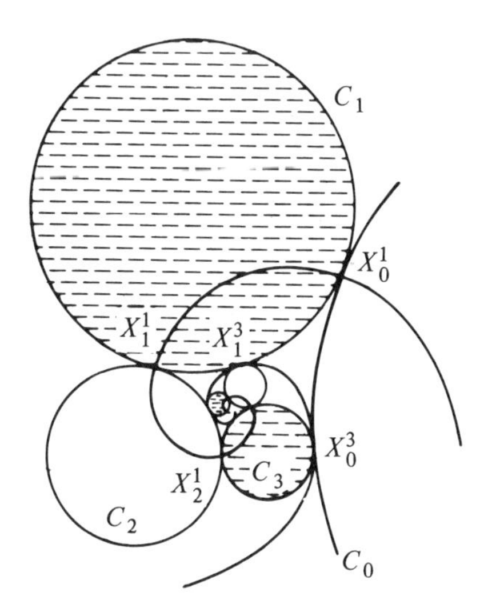


**Figure 3**

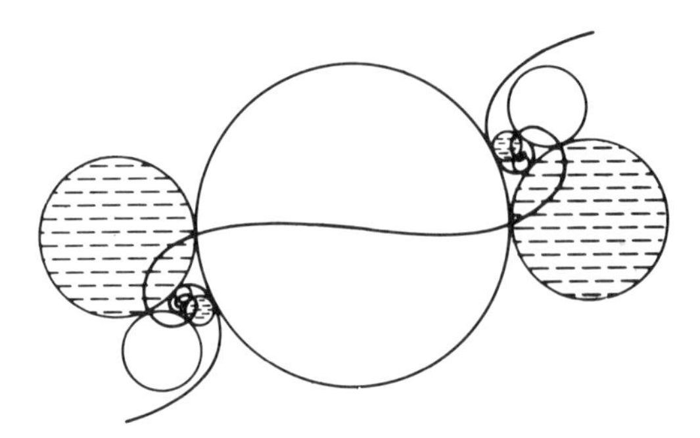

**Figure 4**

centres on the other side of $C$). Since

$$\ldots = -C_{-1} * C = C_0 * C = -C_1 * C = C_2 * C = \cdots$$
$$= C_{2j} * C = -C_{2j+1} * C = \ldots,$$

we see that points $X_k^\alpha = C_k + C_{k+\alpha}$ for all $k$ and all odd $\alpha \in \{1, \ldots, n+1\}$ lie on $C$. There are altogether $(n+1)/2$ disjoint ordered sets of points $\{X_k^\alpha\}_{k \in Z}$, $\alpha \in \{1, 3, 5, \ldots, n\}$, all of which lie on the sphere $C$.

Combining pairs of real coordinates $x$ and $y$ into complex coordinates $z = x + iy$, we can write the dilative rotation $[\circledcirc, \delta, \theta_1, \ldots, \theta_m]_{2m}$ as

$$(z_1, z_2, \ldots, z_m) \longrightarrow e^{2\delta}(z_1 e^{2i\theta_1}, z_2 e^{2i\theta_2}, \ldots, z_m e^{2i\theta_m}).$$

If a point $A$ has coordinate representation $(a_1, \ldots, a_m)$, then the orbit of the point $A$ will lie on the curve

$$\gamma(t) = e^{2\delta t}(a_1 e^{2i\theta_1 t}, \ldots, a_m e^{2i\theta_m t}). \tag{13}$$

In dimension 2 this curve is a loxodrome. For each $\alpha \in \{1, \ldots, n+1\}$ the set of points $\{X_k^\alpha\}_{k \in Z}$ lie on the curve (13) (for a proper choice of starting point). In odd dimensions a dilative rotatory reflection restricts on a fixed sphere $C$ to a dilative rotation, and hence for odd $\alpha \in \{1, \ldots, n+1\}$ the set of points $\{X_k^\alpha\}_{k \in Z}$ lies on the curve (13), which is contained in $C$. The fixed points of $M$ are the accumulation points of the curves. Figures 3 and 4 show the case $n = 3$, for which curves containing points of contact are loxodromes.

## References


[1] Coxeter, H. S. M., Loxodromic sequences of tangent spheres. *Aequationes Math.* **1** (1968), 104–121.

[2] Wilker, J. B., Inversive geometry (this volume, pp. 379–442).

[3] Wilker, J. B., Möbius transformations in dimension $n$ (to appear).

# Part II: Extremal Problems

# Elementary Geometry, Then and Now†

I. M. Yaglom*

## 1. Elementary Geometry of the 19th Century

What is elementary geometry, and when did it originate? The first of these questions—the content of elementary geometry—is not at all simple, and a clear-cut answer is not possible. The most natural answer for present purposes would be the following: "Elementary geometry is the collection of those geometric concepts and theorems taken up in secondary school, together with immediate consequences of these theorems." However, in spite of the seeming simplicity of this answer, it raises at once a host of objections. The appeal to the word "geometric" in the definition is in itself hard to interpret, since the question "what is geometry?" also admits no clear-cut answer (on that, more below); but in any case, the rapid rate of change in school curricula in all countries of the world, currently seeming to reach its maximum, would oblige us if we adopted that definition to accept the existence of indefinitely many elementary geometries. The concept would have to change not merely from country to country, but for each given country also from year to year if not even from school to school. In addition, such a definition clearly refers only to the content of the *school subject* "elementary geometry," while we are here asking about the content of the corresponding science—or, since the word "science" here may seem pompous, about the corresponding *direction of scientific thought*.

However, the difficulty of defining the notion of "elementary geometry" does not at all take away our right to use the term. Thus in the first half of this century much discussion surrounded consideration of the term "geometry." The first general definition of geometry, given in 1872 by the outstanding German mathematician Felix Klein (1849–1925) in his "Erlanger program," proved not

†Translated by Chandler Davis.
* 1-ĭ Goncharnyĭ per. 7, apt. 17, Moscow 109172, USSR.

applicable to the whole range of subdivisions of geometry—in particular to those which at that period attracted the most attention from mathematicians and physicists. But no substitute for it could be found. In this connection the eminent American geometer Oswald Veblen (1880–1960) proposed in 1932 that geometry be confined by definition to *that part of mathematics which a sufficient number of people of acknowledged competence in the matter thought it appropriate so to designate*, guided both by their inclinations and intuitive feelings, and by tradition. This "definition" is frankly ironic; yet for many years it stood as the only one generally accepted by scholars, and scientific articles and studies were devoted to defending it (less to analysis of Veblen's "definition," of course, than to demonstrating the impossibility of any other). We propose to follow this example, calling *elementary geometry* that portion of geometry which is recognized by a sufficiently large number of experts and connoisseurs as meriting the title.

With this understanding, it is clear that elementary geometry is the study of a multitude of properties of triangles and polygons, circles and systems of circles—quite nontrivial and in part entirely unexpected properties, set forth in specialized treatises on the subject (for instance, [1]), and well known only to a small number of specialists in the field (among whom, by the way, the author of these lines does not presume to include himself). The specialists are few, just as are the serious specialists in any sufficiently extensive and far advanced domain of knowledge: say, in postage-stamp collecting or algebraic $K$-theory.

Let us illustrate this for a not too large group of theorems fairly characteristic of "classical elementary geometry"—or, since the adjective "classical" here refers not to musty antiquity but to a relatively recent past on the scale of human history, of "elementary geometry of the 19th century." Consider an arbitrary quadrilateral $\Delta$, not a trapezoid, whose sides are the four lines $a_1$, $a_2$, $a_3$, $a_4$. (Or we may simply mean by $\Delta$ a quadruple of lines $a_i$, no two parallel, and no three passing through any single point.) Taken three at a time, these lines form four triangles $T_1, T_2, T_3, T_4$. Then the points of intersection of altitudes (orthocenters) of our triangles $T_i$ lie on a line $s$ (sometimes called the *Steiner line* of the quadrilateral $\Delta$ after the famous Swiss Jacob Steiner (1796–1863)); the midpoints of the diagonals of $\Delta$ and the midpoint of the segment joining the points of intersection of its pairs of opposite sides, lie on another line $g$ (this was discovered by the great Karl Friedrich Gauss (1777–1855), in whose honor $g$ is called the *Gauss line* of $\Delta$); here always $s \perp g$. Further, the circles circumscribed about the triangles $T_i$ intersect in a single point $C$ (the letter referring to the Englishman William Kingdon Clifford (1845–1879), in whose honor it would be appropriate to call $C$ the *Clifford point* of $\Delta$); the feet of the perpendiculars dropped from $C$ on the sides of $\Delta$ lie on a line $w$, which might be called the *Wallis line* of the quadrilateral after one of Newton's predecessors, the Englishman John Wallis (1616–1703). Also the nine-point circles of the $T_i$, which pass through the midpoints of the sides of these triangles, intersect in a point $E$, which we may call the *Euler point* of $\Delta$ after another Swiss, the renowned Leonhard Euler (1707–1783).

Consider now a pentagon $\Pi$ with sides $a_1$, $a_2$, $a_3$, $a_4$, $a_5$. The five quadruples of lines $(a_1, a_2, a_3, a_4), \ldots, (a_2, a_3, a_4, a_5)$ describe five quadrilaterals $\Delta_5$, $\Delta_4$, $\Delta_3$, $\Delta_2$, $\Delta_1$. The Gauss lines $g_5, \ldots, g_1$ of our quadrilaterals intersect in a point $G$ (the *Gauss point* of $\Pi$); their Clifford points $C_5, \ldots, C_1$ lie on a circle $c$ (the *Clifford circle* of $\Pi$); in case the pentagon $\Pi$ is inscribed in a circle one can also define the concepts of *Euler point* and *Wallis line* of $\Pi$ (see for example [2, Chapter II, Section 8]); and this array of theorems may be much extended (see [3, Chapter 5]).

Having given this answer to the question of the content of elementary geometry, we may pass to the second of the questions posed, on the date of its origin. To aficionados of elementary geometry the answer to this question is well known, but others may find it a bit surprising: the science of triangles and circles —elementary geometry—was founded in the 19th century. What? no earlier? not in ancient Greece?—I hear the doubting questions of the reader not too well informed on the history of mathematics—not by the great Euclid and Archimedes, but by some unknowns or other living no more than a hundred years ago? Yes, is my reply, even less than a hundred years ago; for the central body of elementary geometric theorems known today were discovered in the last third of the 19th century and (to a lesser extent) the first decade of the 20th.[1]

The point is that the giants of ancient Greek mathematics (and maybe this is exactly why they deserve to be called giants) seem not to have included anyone seriously concerned with elementary geometry. The great Euclid (around 300 B.C.) was the author of the first textbook of (elementary) geometry that has come down to us (and what a remarkable textbook it is!—which may have something to do with the disappearance of the texts which preceded it). However, Euclid's personal interests, and apparently also his personal contributions, seem to have lain in other areas (possibly in the study of numbers rather than figures: think of the famous Euclidean proof of the infinitude of primes). So limited was Euclid's knowledge of the theory of triangles that he did not even know the elementary theorem on the point of intersection of the altitudes, which Albert Einstein so prized for its nontriviality and beauty. The mighty Archimedes (3rd century B.C) was one of the founders of (theoretical or mathematical) mechanics, and one of the progenitors of modern "mathematical analysis" (calculus); but to the triangle, and the points and circles associated with it, he gave little attention. Apollonius of Perga, the younger contemporary of Archimedes, was deeply versed in all possible properties of conic sections (ellipse, parabola, and hyperbola)—but not of triangles and circles. Finally, the last of the great ancient Greek mathematicians, Diophantus of Alexandria (most likely 3rd century A.D.), was interested only in arithmetic and number theory, not in geometry.

Thus in the domain of elementary geometry, as the term is traditionally understood, the knowledge accumulated in ancient Greece was not especially profound; nor was any great progress made there in subsequent centuries, right up to the 19th. In the 19th century, on the other hand, especially the second half, through the work of a multitude of investigators, an appreciable portion of whom were secondary-school mathematics teachers,[2] a number of striking and unex-

[1] Note in this connection that whereas Gauss, Steiner, and Clifford (all mathematicians of the 19th century) really knew the theorems associated with their name, the designations "Wallis line" (Wallis was a 17th-century mathematician) and "Euler point" (Euler lived in the 18th century) are rather a matter of convention, for the corresponding theorems were not known to these authors (Wallis and Euler knew only simpler assertions related to those we have stated).

[2] Among them may be mentioned especially the short-lived Karl Wilhelm Feuerbach (1800–1834) (whose brother Ludwig became famous as a philosopher). To us today, K. W. Feuerbach appears as *the* classical representative of this movement. But the greatest scientists of the 19th century, like J. Steiner or even K. F. Gauss, were not at all disdainful of elementary-geometric research. (By the way, Steiner belongs to the intersection $M \cap T$, where $M$ is the set of outstanding mathematicians and $T$ the set of school teachers.)

pected theorems were discovered, an idea of which is given by those set forth above. These theorems were collected in many textbooks of elementary geometry (like the books [1] and [4]) or more narrowly of geometry of the triangle (cf. [5]) or "geometry of the circle" (see [6]), most of which appeared at the end of the 19th or the first third of the 20th century.

The following fact may serve to substantiate this account. At the turn of the 20th century F. Klein conceived the grandiose project of publishing an *Enzyklopädie der mathematische Wissenschaften*, which he envisaged as encompassing the whole accumulation to that time of knowledge of pure and applied mathematics. Klein set a lot of activity in motion on the project; he succeeded in enlisting a broad collective of leading scholars from many countries, and in getting out a work of many volumes, which now takes up more than a shelf in many a major library. To be sure, this project was never brought to a conclusion (it grew clearer and clearer that with the passage of time the quantity of material "not yet" included was not diminishing but increasing, for the growth of the "Encyclopedia" was being far outstripped by the progress of science), and now it has long been hopelessly out of date. To prepare the article on elementary geometry for this publication, Klein assigned the German teacher Max Simon, who enjoyed the reputation of being the strongest expert in this area. Subsequently, however, Klein decided against including a section in the Encyclopedia on elementary geometry, rightly considering that this area of knowledge, having more pedagogical significance than scientific, was out of place in a strictly scientific work. As a result, Simon's survey, which aspired to encyclopedic fullness of coverage of all that was known on elementary geometry at the beginning of the 20th century, had to be published as a separate book; this work [7] is still much prized by specialists and lovers of elementary geometry. In the foreword to his book M. Simon saw fit to emphasize that it dealt with the development of elementary geometry in a single century, the 19th—that it had become clear in the course of preparing the book that a complete survey of all that had been done in elementary geometry essentially coincided with what had been done in the last century.

Thus the 19th is the "golden age" of classical elementary geometry. The flowering of the study of triangles, circles, and their relationships did extend into the beginning of the 20th century, involving some of the prominent mathematicians of that time (for example, Henri Léon Lebesgue (1875–1941), who brought out a book of geometric constructions with circles and lines, and who had curious results on the so-called theorem of F. Morley on the trisectrices of a triangle—on which see [4]). But by about the end of the first quarter of this century one notes a definite falling off of interest in this area. To be sure, broad treatises appear as before on elementary geometry (as on the "geometry of the tetrahedron"), and journals are published devoted entirely or primarily to it.[3] Still

[3] Perhaps the publication of this type enjoying the greatest reputation was the Belgian journal *Mathesis*, appearing from 1881 on. This journal maintains its existence to this day, but the general falling off of interest in the subject it champions has taken its toll on the journal, and today few mathematicians and teachers have even heard of its existence. [Incomparably greater popularity is enjoyed at present by another journal, *Nico*, also published in Belgium and also directed primarily to teachers, which is in every way the exact opposite of *Mathesis* (the name *Nico* comes by abbreviation from the name Nicolas Bourbaki).]

it becomes noticeable that general interest in this part of geometry is lessening. Witness the almost complete disappearance of publications on this subject in serious mathematical journals, and of talks on the subject by eminent scientists at major conferences and congresses—both of which in the 19th century were almost the norm. And this relative decline of elementary geometry was not at all related to the exhaustion of the subject matter, since new theorems on elementary geometry—frequently still quite striking and unexpected—continued to be discovered; clearly some other, deeper circumstances must be involved.

In order to identify some of the causes both of the flowering and of the subsequent eclipse of classical elementary geometry, we will have to turn to some general laws of scientific development which are sometimes hard to formulate but are easy to observe and in principle fully explainable. It should be noted first of all that the keen interest in the study of triangles, quadrilaterals, and circles which we see throughout the 19th century was by no means an isolated phenomenon; it was intimately related with the flourishing in this period of so-called *synthetic geometry*, i.e., geometry based not on analytic devices involving the use of one or another system of coordinates, but on sequential deductive inference from axioms.[4] Synthetic geometry in this period was not studied just as an end in itself: it stimulated a number of important general mathematical ideas. At the core of this preoccupation was the concept of the non-uniqueness of the traditional (or "school") geometry of Euclid, of the existence of an abundance of in some sense equally deserving geometrical disciplines, such as the *hyperbolic geometry of Lobačevskiĭ and Bolyai* or *projective geometry*. They prepared the ground for serious general syntheses such as Klein's "Erlanger program" mentioned above. All of this facilitated also the serious posing of the question of the logical nature of geometry (or even of all mathematics, since in the 19th century the subject of the foundations of mathematics was analyzed almost exclusively for geometry), giving rise to several systems of axioms for geometry which were elaborated by a number of investigators (foremost among them Italians and Germans: Giuseppe Peano, Mario Pieri, Moritz Pasch, David Hilbert) at the turn of the 20th century; this played a very large role in the development of 20th-century mathematics.

An especially prominent place in the development of synthetic geometry in the 19th century was occupied by *projective geometry*. I would go so far as to assert that not only did projective geometry in a well-known sense grow out of elementary geometry (this approach to projective geometry is emphasized in the book [9] addressed to beginners), but also 19th-century elementary geometry was in a significant sense produced by projective geometry—a circumstance which Felix Klein liked to point out, and which stands out especially when one analyzes the elementary-geometric work of eminent leaders of 19th-century mathematics like K. F. Gauss or J. Steiner. So, for example, all of the "geometry of the

[4]Typical of the preferences of that time was the flat prohibition against solving a construction problem by an algebraic method—a prohibition which teachers, preserving attitudes typical of the early 20th century, often took as so self-evident that they didn't even express it. (I myself recall the time when a construction problem solved algebraically was often regarded as not solved, to the annoyance of pupils.) (For the relation between "geometric" solution of construction problems by such and such a choice of instruments prescribed in advance, and the axiomatic method in geometry, see, for instance, the book [8].)

triangle," with all the properties of the "special points" and "special circles" associated with a triangle, can be fitted neatly into the program of investigating projective properties of pentagons—the first polygons which have distinctive projective properties (quadrilaterals, being all "projectively equivalent," can't have individual projective properties). The transition from projective 5-gons to Euclidean 3-gons requires only the identification of two of the five vertices of the 5-gon with the "cyclic (ideal) points" whose fixing in the projective plane converts projective geometry to Euclidean. (See, for example, the classical book [10]; note that the conic section, which is the primary object of investigation in projective geometry, goes into a circle if it is required that it pass through the cyclic points.)

But then in the first half of the 20th century came a very palpable (though perhaps temporary) decline in synthetic geometry. "The Moor has done his work, the Moor may leave": those general ideas and understandings referred to above, which had grown out of synthetic geometry, were now established, and synthetic geometry was no longer required. It is well known that the history of science exhibits ebbs and flows; if the 19th century was the golden age of geometry, then our times are distinguished by the preeminence of algebra, by the distinctive "algebraization" of all branches of mathematics reflected in the acceptance of Nicolas Bourbaki's mathematical structures, converting even geometry virtually into a part of algebra. In this situation it is not surprising that projective geometry, for instance, while retaining its position as an important part of the school geometry course (see, e.g., the books [11] and [12]), has in the strictly scientific domain undergone inconspicuously such a transformation that today algebraic questions play if anything a bigger role than geometric (see for instance the old but still popular text [13]). Now remembering also that the general "algebraization" of mathematics, putting algebraic structures as much as possible in the foreground, squarely posed the question of revision of school geometry courses, which many mathematicians and educators proposed to base on the (essentially algebraic!) concept of *vector space*—we see that there has been significant erosion even at the core of the one possible "application" of classical elementary geometry, its use in the teaching of mathematics in secondary school.

A good illustration of the algebraization of geometry is provided by the popular axiomatic approach to geometry by Friedrich Bachmann [14], which gives priority to purely algebraic concepts (groups generated by their involutory elements). Another clear-cut example is the recent set of axioms of Walter Prenowitz (see [15]), specially suited to the analysis of ideas related to the notion of convexity: it permits the introduction of a novel "multiplication" of points, whereby the product $A \cdot B$ (or $AB$) of points $A$ and $B$ is to be thought of as the *segment* with endpoints $A$ and $B$; this multiplication is commutative, associative, idempotent, and distributive over set-theoretic addition. One might also cite the booklet on polygons [16], so typical of current trends: if its title and the majority of its results make it a work on elementary geometry, yet the tools and methods used identify it rather as a book on general algebra (theory of lattices). I point out also that I was able to fill the book [2] with elementary-geometric material under cover of the purely algebraic nature of the techniques used (the theory of hypercomplex numbers or diverse algebras having elements of the form $x + Iy$, where $x, y \in \mathbb{R}$ and $I^2 = -1$, 0 or

+1); but for that, the beautiful purely geometric constructions appearing in [2] would today interest nobody.

Especially surprising also to mathematicians of my generation is the transformation undergone before our eyes by *topology*. Whereas in our student years it was regarded as a part of geometry and worked extensively with intuitive visualization, modern (algebraic) topology by its tools and methods belongs not to geometry but to algebra. This revolution has driven out of topology several investigators of a more geometric turn of mind, even producing in isolated cases serious emotion trauma.[5]

Interesting in this connection is the position of Jean Dieudonné, one of the authoritative French mathematicians and a leader in the founding of the Bourbaki group. In accordance with the general orientation of this group, Dieudonné is flatly opposed to retaining in school mathematics teaching any trace whatever of classical elementary geometry, i.e., the "geometry of the triangle" and all its relatives and subdivisions. This despite apparently incontrovertible connection between the high level of French mathematics and the traditions of instruction at the French lycées, where students were trained in the solution of subtle and quite complicated elementary-geometric problems (see for example the classic schoolbooks [17] and [18], the second of which is by one of the greatest mathematicians of the 20th century). Already in 1959, at a conference on the teaching of mathematics in Réalmont, France, Dieudonné rose and hurled the slogans "Down with Euclid!" and "Death to triangles!" —and he maintains his support for these slogans to this day. In numerous speeches on pedagogical topics Dieudonné has repeatedly expressed the wish that secondary-school students (and teachers) forget as soon as possible the very existence of such figures as triangles and circles. Dieudonné's idiosyncratically written book [19] (see especially its Introduction) is entirely devoted to advocacy of the following methodological idea: *Elementary geometry is nothing but linear algebra* —and no other elementary geometry ought to exist (to such a point that the book [19] on elementary geometry is quite without figures, and never mentions the word "triangle"). A different position is taken by A. N. Kolmogorov, whose geometry textbooks are currently in use by all secondary students in the U.S.S.R.; but even these textbooks are arranged in such a way that they almost completely lack substantial geometrical problems. (In this respect they are inferior to American school textbooks, which are generally based on the—also quite ungeometric—axiomatization of G. D. Birkhoff [20].)

[5]It is not at all simple to make a neat division between geometry and algebra; but I think it can be stated without qualification on the basis of contemporary physiological data that geometric representations ("pictures") are among those which enlist the activity of the right half of the human brain, while (sequential) algebraic formulas are controlled by the left hemisphere. From this point of view, maybe people should be divided into natural geometers and natural algebraists according to the predominance in their intellectual life of one or the other hemisphere. Thus I would count Newton (and Hamilton) among geometers, whereas Leibniz (and still more Grassmann) belong rather to the algebraists. (The philological interests of Leibniz and Grassmann are noteworthy here, for it is known that everything related to speech and language relates to the left hemisphere; by contrast, the extramathematical interests of Newton ran to such sharply visual images of world culture as the Apocalypse.) Thus the simultaneous discovery of the calculus by Newton and Leibniz, or of vectors by Hamilton and Grassmann, were made, so to speak, "from different sides."

Thus today "classical" elementary geometry has distinctly lost status. Note too that the subject has virtually disappeared from the problems proposed to competitors in the international mathematical olympiads. But "nature abhors a vacuum"—and the place left vacant by "classical" elementary geometry has been taken with alacrity by the "new" or "contemporary" elementary geometry, the "elementary geometry of the 20th century," to which we now turn.

## 2. Discrete Mathematics and Discrete Geometry

In order to explain the new directions taken in the present period in geometry (and in elementary geometry in particular), it will be necessary to touch on some pervasive tendencies characterizing all of contemporary mathematics.

The first is related to the prominent place occupied, first in applied and then also in pure mathematics, by so-called *optimization problems*, which require the specification of a "best" (or at least "sufficiently good") mode of operation of an individual machine or large system (see the accessible and entirely typical book [21]). The system considered may be, for example, a living organism or a particular part of one (say, the collection of a mammal's visual organs with the parts of the brain which serve them), a factory or any large economic organization (like the European Economic Community), an educational institution, or a line of communication (examples of which would be a single nerve fiber, a television channel, or a whole television network on the scale of a large country). The extreme complexity of such systems, arising from the mutual interaction of an enormous number of separate links, renders their complete mathematical analysis very difficult or even altogether impossible. On the other hand, to set up some satisfactory mode of operation of such a system can be a problem of the utmost importance. It is solved in the case of living organisms, as a rule, by successive self-improvements of biological mechanisms in the course of natural selection; in other systems it may require elaborate and entirely new methods. In the process of seeking solutions to optimization problems in recent decades, there have arisen a whole conglomerate of new directions—unfamiliar mathematical sciences, some of them not yet given definitive formulation, and constantly interacting with each other: *optimal control* and *operations research*, *linear programming* and *dynamic programming*, *theory of games* and *theory of coding of information*. Their importance for practice is well attested by the way most of these disciplines produced their first textbooks and their first courses of lectures before the disciplines took final form—and some of the disciplines, say operations research, have still a rather diffuse character, quite aside from the vastness of the literature (including the school textbook literature) devoted to them.

The second trend of contemporary mathematics which must be mentioned here involves profound shifts in our attitudes on the place in mathematics of the *finite* and the *infinite*, of the *discrete* and the *continuous*. The mathematical revolution of the 17th century, whose central figures were Isaac Newton (1643–1727) and Gottfried Wilhelm Leibniz (1646–1716), consisted largely in replacing the "finite" mathematics of the ancients—arithmetic and geometry—by the *calculus*, dealing with continuous functions which described processes in the course of which the object considered would pass through an infinite set of states (say, the process of motion of a material point or 3-dimensional body). The methods of investigation of these functions are based on the notions of limit and passage to the limit. "If all completed exact computations, the only sort admitted by the ancients, have to be studied by contemporary mathematicians, still their practical significance has much diminished, and sometimes entirely disappeared": so wrote H. L. Lebesgue, one of the greatest mathematicians of the time, in the early thirties of this century [22]—and at that time this point of view was almost universal. But then, beginning at the end of the forties in connection with the advent of *electronic digital computing machines*, and with the establishment of directions of scientific thought largely arising from electronic computers, sometimes called by the catch-all name *cybernetics*, the situation began to change profoundly. The nature of these machines is discrete in principle (we may even say, finite), as is emphasized by the adjective "digital" in their name, and this to some degree opened our eyes to the discrete character of a multitude of phenomena in the world around us. If in the preceding century this world appeared to scientists as an accurately working "machine of continuous action," on the model of a steam or electric engine, today we regard it quite differently.[6] As a typical and important example of discrete phenomena we today cite the *higher neural activity* of humans and animals, which appears in first approximation to be composed of the action of an enormous number of individual neural links—the *neurons* of the brain—each of which will be at any given moment in one or the other of two possible states, "excited" and "passive." Still more important is the delicate mechanism of *heredity*, controlled by the long polymeric molecules called *deoxyribonucleic acid*, now known to the whole world somewhat familiarly as simply DNA (as the great popularity of *Marilyn Monroe* or *Brigitte Bardot* was reflected in their being abbreviated MM and BB); these molecules can be looked at as "words" written in a 4-letter "alphabet" of alternating bases (which are most often designated by only their initials A, G, C, and T, rather than writing *adenine*, *guanine*, *cytosine*, and *thymine*).

All of this has obliged mathematicians to make a thoroughgoing reconsideration of their view of the place within mathematical science of its "discrete" branches; and today Lebesgue's categorical pronouncement of the relatively recent past, quoted above, should be considered utterly outdated. These days we see the flowering of many intrinsically "finite" branches of mathematics—from *mathematical logic*, once regarded as one of the most abstract and far removed from practice of all branches of theoretical mathematics, but now studied by innumerable technicians and engineers, to *combinatorics*, which very recently tended to be put almost with mathematical recreations, but is now among the

[6] From this point of view it is in harmony with the times that the interesting book [23] should maintain that the majority of the functions encountered in real life are nonsmooth, nowhere differentiable, so that the apparatus of classical analysis is not applicable to them.

sciences having the greatest significance for applications.[7] Some of the new mathematical sciences discussed above are of "finite" nature too, for example *game theory*[8] or *coding theory*; these fields sometimes use quite delicate parts of number theory or general algebra.[9]

The changed relationship between "continuous" and "discrete" mathematics could be illustrated by many examples. Take the changed interrelation of *differential equations* and *difference equations*: formerly mathematics students always regarded difference equations as a sort of "toy model" of differential equations, and mathematicians always tried to reduce difference equations to differential equations, which were so much more familiar and better studied; whereas now, with the use of electronic computers, the main method of solution of differential equations everywhere is to reduce them to difference equations (and this is reflected in school textbooks). But these circumstances, well known to all mathematicians, need not detain us.[10] Let us return at once to our present subject, geometry.

What bearing does all of this have on geometry? the impatient reader might ask. The most intimate, I reply, for geometry is part of mathematics and "nothing mathematical is alien to it." The circumstances which have been recalled (and others) have brought about a partial and most likely temporary eclipse of classical *differential geometry*, which is based on the concepts and methods of calculus and which in the first half of this century was firmly entrenched as the "principal" discipline within geometry. But beside that, and more significant for present purposes, the same factors made possible the "entrance center stage" of a number of branches of geometry formerly considered secondary or altogether inessential.

It is appropriate to speak first of all of the rather unexpected development of the subject called *finite geometries*, which deals with "geometries" consisting of only finitely many constituent elements, such as points and lines. Geometries of this sort have been studied for a long time; but they used to be looked at as a sort of "geometrical toy," having some interest for foundations of geometry (and that pedagogical rather than scientific), but not aspiring to any serious significance—still less to applications. Therefore it might seem puzzling, and counter to

[7] One is struck by the explosive growth of the literature on combinatorics, and by the creation of a specialized international journal devoted to it (the *Journal of Combinatorial Theory*, appearing since the early sixties), at a time when a relatively old discipline like projective geometry does not have "its own" journal and can hardly expect it. We mention also the vigorous growth of *graph theory* (closely allied to combinatorics), which also now has an enormous literature.

[8] We will not consider here the so-called *theory of differential games*, which combines in a novel way traditional methods of discrete mathematics with procedures taken from calculus.

[9] This explains the title of a survey [24] of coding theory, attacking the assertion made by the eminent Godfrey Harold Hardy (1877–1948) about the poverty of ideas and primitivity of applied mathematics [24a].

[10] Let me just mention the popular manual [25] of finite mathematics for beginners, reflecting the system of mathematics instruction adopted by Dartmouth College, U.S.A. Under this system, students in the first two years of college are to take two one-year courses, calculus and finite mathematics, but they are free to choose which of them to take first and which second. Note also the collection [26] of papers on combinatorial mathematics, intended for those interested in applied mathematics.

the general trend of contemporary mathematics which has been discussed, that there should be a sudden sharp burst of interest in finite geometries. In the last few years there have appeared in various countries some dozens of books on this subject, of which it will perhaps suffice here to mention the detailed survey [27] by the eminent German geometer Peter Dembowski, which is included in the famous series *Ergebnisse der Mathematik und ihre Grenzgebiete* from J. Springer. This survey came out in 1968 and is already rather out of date; the bibliography accompanying it has about 1500 titles of books and articles, of which 1300 are works published in the fifties and sixties. The key to this anomaly is that finite geometries, along with the rest, have in recent years unexpectedly assumed major significance in applied mathematics (they are used, for instance, in some questions of mathematical statistics and coding theory).

Meanwhile there was a sharp rise in interest in another branch of geometry, which had appeared in the 19th century, and which also is purely discrete in nature—which indeed is so named: *discrete geometry*. The founders of discrete geometry were eminent specialists in number theory, the only serious discipline in 19th-century mathematics of discrete character: the German mathematician Hermann Minkowski (1864–1909), the Russian Georgiĭ Fedos′evič Voronoĭ (1868–1908), the Norwegian Axel Thue (1863–1922). In fact the new field had its first serious applications in number theory. The founding of discrete geometry marked the birth of a new subfield of "higher arithmetic" (number theory), the *geometric theory of numbers*, which consists in using geometric procedures to solve number-theoretic problems; it is no coincidence that H. Minkowski's book, central to this whole development, was called *Geometrie der Zahlen*. Throughout a long period of time mathematicians tended to classify discrete geometry with number theory rather than with geometry, to which the methods of Minkowski, Voronoĭ, and Thue still seemed utterly foreign. So matters stood right up to the second half of this century, when this circle of questions attracted the interest of scientists belonging to several schools which were originally purely geometric: the important English school led by the late Harold Davenport, the younger Claude Ambrose Rogers, and Harold Scott Macdonald Coxeter; the Hungarian school of László Fejes Tóth, which concentrates its attention on somewhat more "recreational" but certainly quite geometrical problems; and the Moscow group of the recently deceased Boris Nikolaevič Delone, which started from problems of mathematical crystallography. Now just as in the case of finite geometries, the renaissance in discrete geometry which occurred in recent decades can be related not only to the general trend which has been noted, but also to the fact that specific achievements of discrete geometry turned out to have uses in several areas of applied mathematics, above all in coding theory (see, for example, the trail-blazing work [28]; we observe that in recent years literally every new result on close packing of multidimensional spheres has been taken over for the arsenal of the communications engineer), and in computational or "computer" mathematics. On the other hand, the investigations of Carl Ludwig Siegel, one of the successors of Minkowski and Voronoĭ, intimately connected some of the problems of discrete geometry with the most current questions of modern number theory and algebra.

Just what is discrete geometry? Essentially it deals with three rather simply stated geometric problems. The first two are the problems of the *densest filling* of the plane or space or some subset thereof by nonintersecting equal figures ("packing" by the figures), and of *sparsest covering* by equal figures, i.e., placement of the figures so that they completely cover the plane (or space, or a subset thereof); in the case of packing the total number of figures is to be as large as possible; in the case of covering it is to be as small as possible. The results obtained concerning these problems by the English school of Davenport, Rogers, and Coxeter are well set forth in the little book [29]; results worked out by the Hungarian school, closely related to this group of problems, are the subject of a detailed survey [30]. Finally, the third class of problems, treated especially by B. N. Delone and his students, concerns *decompositions* of the plane, space, or subsets thereof, i.e., positions of figures which are at once packings and coverings. The best-known problem of this sort originates with the famous Russian crystallographer Evgraf Stepanovič Fedorov (1853–1919): to find all types of *parallelohedron* into copies of which the plane or space may be decomposed, in such a way that the individual figures are oriented parallel to each other, i.e., are obtained from any one by parallel translation[11] (with the supplementary requirement that two of the parallelohedra intersect only in a whole face, not in part of one). It is not hard to see that in the plane there are only two types of parallelohedron (here called *parallelogons*): parallelograms and centrally symmetric hexagons; while in space, as E. S. Fedorov already showed, there are five types of parallelohedron (aside from parallelopipeds and prisms with centrally symmetric hexagonal bases, there are two 12-faced and one 14-faced parallelohedra). In 1929 B. N. Delone showed that in 4-space there are 51 types of parallelohedron; in spaces of 5 or more dimensions they have still not been enumerated.

## 3. Combinatorial Geometry: The Elementary Geometry of the Second Half of the 20th Century

The main problems of discrete geometry concern *infinite* systems of figures, for example, packings or coverings of the whole plane (or space); therefore they surely do not belong to elementary geometry. But in connection with discrete

[11] This restriction, which is sometimes imposed also in packing or covering by equal figures, may be generalized: one may fix a particular group $G$ of motions and require that the figures used be obtained one from another by motions from $G$; the configurations of figures obtained may be called *$G$-configurations*. (Thus, for example, let $S$ be the group of parallel translations and central symmetries. Then to the class of *$S$-parallelogons* will belong, along with the two classes of $T$-parallelogons listed below for the group $T$ of translations, also the triangles and isosceles trapezoids.) Another possibility in discrete geometric problems is to relax the requirement that all figures be equal, substituting some milder restriction. One final "typical restriction" in such problems comes from their connection with number theory, but to the geometer's eye is simply evidence of our helplessness, our inability to cope with problems in their general form: it consists in requiring "lattice" placement of the figures, meaning that some discrete group of motions carries the whole configuration onto itself.

geometry there has arisen in recent decades another direction of geometrical investigation: so-called *combinatorial geometry*, the study of "optimization problems" involving *finite* choices of points or figures (most often convex figures), i.e., problems about finite configurations of points or figures which are in some sense optimal (or anyway sufficiently good).

In contrast to discrete geometry, combinatorial geometry so far has no serious practical applications; in this respect it resembles "classical" elementary geometry, which considered properties of triangles and circles which, beautiful though they were, were scientifically blind alleys—leading nowhere, giving nothing to science at large. Still, "19th-century elementary geometry" was closely bound up with what might be called the "scientific atmosphere" of those years (with projective geometry, and with non-Euclidean geometry, one of whose natural starting points is the so-called "Poincaré models" of hyperbolic and elliptic geometry, via the study of systems of circles); and just so does combinatorial geometry arise from today's serious scientific concerns and reflect the general nature of the "optimization" problems so important for practice. For exactly this reason it seems useful to take up combinatorial geometry in school or university mathematical clubs. This in itself would constitute a significant "application" of combinatorial geometry: a pedagogical application, the only sort of application that was found for "classical" elementary geometry.[12]

The birth of combinatorial geometry may be dated in 1955, with the appearance of the article [35] of the Swiss mathematician Hugo Hadwiger, whose title is apparently the first use of the term; its subsequent growth had the same explosive nature as we noted above for combinatorics and graph theory. By now there are dozens of books devoted to combinatorial geometry (the number is especially large in Russian, it seems to me—both original books and translations; as a characteristic example let me cite here the recent book [36]); the number of articles on it has long since passed 1000. Thus, for instance, the bibliography of the (amplified) Russian translation of the survey [37] lists literature up to 1968 related to just one area within combinatorial geometry, and comprises about 500 books and articles.

To give an idea of the problems typical of combinatorial geometry, of their elementary-geometrical nature, and of their connection with discrete geometry, let us consider some specific examples. The first is intimately related with the problem (basic for discrete geometry) of closest packing of spheres, while the others are more artificial ("recreational"). It is clear that in the plane a given circle $K$ can touch no more than 6 nonintersecting circles the same size as $K$. This leads to the famous *problem of 13 spheres*, which calls for finding the

[12] Beginning in the late forties the present author, with his comrades who like him were connected with the activities of the all-Moscow school mathematics club at Moscow University, worked on collections of "olympiad" problems for students in the higher classes of secondary school (English translations exist of the first of these, in two versions [31], and also of a book [32] continuing this line). The geometry problem book in this series which came out in 1952 was devoted for the most part to traditional construction and proof problems. However, in preparing the new edition of this book I felt it necessary to change its character sharply; this is manifested not only in the new version of that book, but also in the appearance as sequels in the same series of a collection of geometric optimization problems [33] and one on combinatorial geometry [34].

greatest possible number of equal material (nonintersecting) spheres which can touch a given like sphere (this number turns out to equal 12). In contrast to the plane case, this stereometric problem proved quite difficult; the 3-dimensional case has a long history, at the beginning of which stand the great Johann Kepler (1571–1640) and Isaac Newton, while the higher-dimensional cases remain unsolved; see the vivid account by H. S. M. Coxeter [38] or the introductory book [39]. Related to this is the general problem of estimating the Newton numbers of one or another figure, where by the *Newton number* $n(F)$ of a figure $F$ we mean the largest possible number of nonintersecting copies of $F$ which can touch $F$ (see [40]). Thus letting $M_k$ denote a regular $k$-gon, one has $n(M_3) = 12$, $n(M_4) = 8$, and $n(M_k) = 6$ for $k \geqslant 5$ (in proving $n(M_5) = 6$ an electronic computer was resorted to). Just as in the case of problems of discrete geometry discussed above, there is a variant problem, the estimating of *G-Newton numbers* $n_G(F)$ of figures $F$, where $n_G(F)$ is defined similarly to $n(F)$ with the additional restriction that now the (nonintersecting) figures touching $F$ must be obtained from $F$ by motions belonging to a given group $G$. The problem attracting the greatest attention here (H. Hadwiger, B. Grünbaum) has been that of finding $T$-Newton numbers $n_T(F)$, where $T$ is the group of parallel translations; for this the results obtained are almost definitive (see [41]).

A problem having a certain relationship to these is to seek figures of a fixed (narrow) class of convex bodies—say, triangles in the plane or tetrahedra in space—which can be arranged so that no two of them overlap but any two make (essential) contact: Thus for example, the largest number of nonoverlapping tetrahedra which can be so disposed in space that every two have (2-dimensional) contact of their boundaries, is either 8 or 9 (it is almost surely 8; see [42]). And the latter problem is intimately related to the most famous problem of combinatorial geometry, the renowned *four-color problem*, recently solved with the aid of computers in its classical formulation. Even so, there remain an abundance of open problems: take the question of the smallest number of colors needed for a proper coloring of an arbitrary map on two globes, a planet and its satellite, where each country on the satellite (each "colony") is to be colored with the same color as the corresponding country on the primary planet (its "metropolis").

For combinatorial problems involving arrangements of points, we may take as sufficiently typical the problem of Erdös which asks for the largest number $f(n)$ of points which can be arranged in $n$-dimensional (Euclidean) space so that every 3 of them are the vertices of an isosceles triangle. It is clear that $f(1) = 3$; it is relatively simple to establish that $f(2) = 6$; it is harder to prove that $f(3) = 8$. (See [43], [44]. The configuration which serves in the plane consists of the 5 vertices of a regular pentagon and its center; in space, the same 6 points together with two points of the perpendicular to this plane erected at the pentagon's center, at a distance equal to the radius of the circle circumscribed about the pentagon.) For $n \geqslant 4$ the value of $f(n)$ remains unknown.

Different in nature is the difficult problem of *estimating the angles* determined by $k$ points of $n$-dimensional space (this problem goes back to L. Blumenthal

[45]). In full form, this calls for describing the set of those points in $N$-space, where $N = k(k-1)(k-2)/2$, whose coordinates can occur as the tuple of angles $\alpha_1, \alpha_2, \ldots, \alpha_N$ determined by our $k$ points for some way of placing them (it may be assumed that $\alpha_1 \geqslant \alpha_2 \geqslant \cdots \geqslant \alpha_N$). There is, however, one form of the problem which is transparent; indeed, in the case $k = n+1$ obviously $\pi \geqslant \alpha_i \geqslant \pi/3$ for $1 \leqslant i \leqslant (n^3-n)/6$, $\pi/2 > \alpha_j \geqslant 0$ for $(n^3-n)/6 < j \leqslant (n^3-n)/3$, and $\pi/3 \leqslant \alpha_l \leqslant 0$ for $(n^3-n)/3 < l \leqslant (n^3-n)/2$, and all restrictions binding the remaining angles clearly follow from the elementary theorem on the sum of the angles of a triangle. For moderate values of $k$ we are led to very pretty elementary-geometric problems—thus for 4 points in the plane (the case $k=4$, $n=2$) it is not hard to show that $\pi \geqslant \alpha_1 \geqslant \pi/2$, $\pi \geqslant \alpha_2 \geqslant 2\pi/5$, $\pi \geqslant \alpha_3, \alpha_4 \geqslant \pi/3$, $\pi/2 > \alpha_5, \alpha_6, \alpha_7, \alpha_8 \geqslant 0$, $\pi/3 > \alpha_9, \alpha_{10} \geqslant 0$, $\pi/4 \geqslant \alpha_{11}, \alpha_{12} \geqslant 0$. However, for arbitrary (or even just for big enough) values of the number $k$ of points, the picture is hard to describe. For example, for $k$ points in the plane it is fairly easy to show that

$$\alpha_N = \max_{\mathcal{P}_k} \min_{i,j,l} \angle A_i A_j A_l = \frac{\pi}{k},$$

where $i, j, l = 1, 2, \ldots, k$ (and $i \neq j \neq l \neq i$), while $\mathcal{P}_k$ can be an arbitrary configuration of $k$ points $A_1, A_2, \ldots, A_k$ in the plane; on the other hand, Erdös has conjectured that

$$\inf_{\mathcal{P}_k} \max_{i,j,l} \angle A_i A_j A_l = \frac{m-1}{m}\pi$$

for $2^{m-1} < k \leqslant 2^m$ and $k \geqslant 6$, this bound being attained only for $k = 6$, and this conjecture has still not been proved (in this connection see [46], [47]).[13]

One more typical example of a problem on arrangement of (arbitrary) figures, interesting in particular for the relative ease of producing a general solution, is the *problem of patches on jeans* (due, it seems, to E. B. Dynkin). Let the total surface of a pair of blue jeans be 1, and let each of $n$ patches have area $\geqslant \sigma$; then what is the least possible value $f_n^{(2)}(\sigma)$ of the largest area of intersection of any two patches? [Clearly $f_n^{(2)}(\sigma) = \min_{\mathcal{Q}_n} \max_{i,j} M_{ij}$, for $i, j = 1, 2, \ldots, n$ and $i \neq j$, where $M_{ij}$ is the area of intersection of "patches" $M_i$ and $M_j$, i.e., of two of the (measurable) subsets $M_1, M_2, \ldots, M_n$ with area $\geqslant \sigma$ arranged within the figure $M$ (the "jeans") of unit area (say, a square), and $\mathcal{Q}_N$ can be any positioning of the "patches" $M_i$ within $M$.] And what under the same condition is the least possible value $f_n^{(l)}(\sigma)$ of the largest $l$-fold intersection of patches? And if all $k$-fold intersections of patches are $\geqslant \sigma$ in area, what is the least possible value $f_n^{(k,l)}(\sigma)$ of the largest $l$-fold intersection of patches (where, of course, $1 \leqslant k \leqslant l \leqslant n$)? This problem, which can also be given a probability-theoretical motivation, admits an exact solution; the function $f_n^{(k,l)}(\sigma)$ is piecewise linear: we have in

[13] It is instructive to compare this and the following problem with the ubiquitous problem in contemporary applied mathematics of finding "minimaxes" and "maximins."

general

$$f_n^{(k,l)}(\sigma) = \frac{\binom{r-1}{l-1}}{\binom{n}{l}\binom{r-1}{k-1}}\left[\binom{n}{k}\sigma - \frac{l-k}{l}\binom{r}{k}\right]$$

$$\text{for}\quad \binom{r-1}{k}\Big/\binom{n}{k} \leqslant \sigma \leqslant \binom{r}{k}\Big/\binom{n}{k}, \quad r = 1, 2, \ldots, n$$

(see [34], [48], [49], where various solutions of this problem are given).

It seems to me that the whole range of problems of these kinds fully merit the designation of "elementary geometry of our times."

## References


[1] Johnson, R. A., *Advanced Euclidean Geometry*. Dover, New York 1960.

[2] Yaglom, I. M., *Complex Numbers in Geometry*. Academic Press, New York 1968.

[3] Golovina, L. I. and Yaglom, I. M., *Induction in Geometry*, Heath, Boston 1963.

[4] Coxeter, H. S. M. and Greitzer, S. L., *Geometry Revisited*, Random House, New York 1967.

[5] Efremov, D., *Novaya geometriya treugol'nika*. Matezis, Odessa 1903.

[6] Coolidge, J. L., *A Treatise on the Circle and the Sphere*. Clarendon Press, Oxford 1916.

[7] Simon, M., *Über Entwicklung der Elementargeometrie im XIX Jahrhundert*. Berlin 1906.

[8] Bieberbach, L., *Theorie der geometrischer Konstruktionen*. Birkhäuser Verlag 1952.

[9] Yaglom, I. M., *Geometric Transformations III*. Random House, New York 1973.

[10] Klein, F., *Vorlesungen über nicht-euklidische Geometrie*. J. Springer, Berlin 1928.

[11] Coxeter, H. S. M., *Introduction to Geometry*. Wiley, New York 1969.

[12] Pedoe, D., *A Course of Geometry for Colleges and Universities*. University Press, Cambridge 1970.

[13] Baer, R., *Linear Algebra and Projective Geometry*. Academic Press, New York 1952.

[14] Bachmann, F., *Aufbau der Geometrie aus dem Spiegelungsbegriff*. J. Springer, Berlin 1973.

[15] Prenowitz, W. and Jantosciak, J., *Join Geometries*. Springer, New York 1979.

[16] Bachmann, F. and Schmidt, E., *n-Ecke*. Bibliographisches Institut, Mannheim 1970.

[17] Rouché, E. and Comberousse, Ch., *Traité de géométrie*. Gauthier-Villars, Paris 1899.

[18] Hadamard, J., *Leçons de géométrie élémentaire*, I, II. Gauthier-Villars, Paris 1937.

[19] Dieudonné, J., *Algèbre linéaire et géométrie élémentaire*. Hermann, Paris 1968.

[20] Birkhoff, G. D. and Beatley, R., *Basic Geometry*. Chelsea, New York 1959.

[21] Koo, D., *Elements of Optimization*. Springer, New York 1977.

[22] Lebesgue, H., *La mesure des grandeurs*. Université, Genève 1956.

[23] Mandelbrot, B., *Fractals*. Freeman, San Francisco 1977.

[24] Levinson, N., Coding theory: a counterexemple to G. H. Hardy's conception of applied mathematics. *American Math. Monthly* **77** (No. 3, 1970), 249–258.

[24a] Hardy, G. H., *A Mathematician's Apology*. University Press, Cambridge 1941.

[25] Kemeny, J. G., Snell, J. L., and Thompson, G. L., *Introduction to Finite Mathematics*. Prentice-Hall, Englewood Cliffs, N.J. 1957. Kemeny, J. G., Mirkil, H., Snell, J. L., and Thompson, G. L., *Finite Mathematical Structures*. Prentice-Hall 1959.

[26] Beckenbach, E. E., (editor) *Applied Combinatorial Mathematics*. Wiley, New York 1964.

[27] Dembowski, P., *Finite Geometries*. Springer, Berlin 1968.

[28] Shannon, C., Communication in the presence of noise. *Proc. IRE* **37** (No. 1, 1949), 10–21.

[29] Rogers, C. A., *Packing and Covering*. University Press, Cambridge 1964.

[30] Fejes Tóth, L., *Lagerungen in der Ebene, auf ker Kugel und im Raum*. Springer, Berlin 1972.

[31] Shklarsky, D. O., Chentzov, N. N., and Yaglom, I. M., *The USSR Olympiad Problem Book*. Freeman, San Francisco 1962. Shklyarsky, D., Chentsov, N., and Yaglom, I., *Selected Problems and Theorems in Elementary Mathematics*. Mir, Moscow 1979.

[32] Yaglom, A. M. and Yaglom, I. M., *Challenging Mathematical Problems with Elementary Solutions*, I, II. Holden Day, San Francisco 1964, 1967.

[33] Šklyarskiĭ, D. O., Čencov, N. N., and Yaglom, I. M., *Geometričeskie neravenstva i zadači na maksimum i minimum*. Nauka, Moscow 1970.

[34] Šklyarskiĭ, D. O., Čencov, N. N., and Yaglom, I. M., *Geometričeskie ocenki i zadači iz kombinatornoĭ geometrii*. Nauka, Moscow 1974.

[35] Hadwiger, H., Eulers Charakteristik und kombinatorische Geometrie. *J. reine angew. Math.* **194** (1955), 101–110.

[36] Boltyanskiĭ, V. G. and Soltan, P. S., . *Kombinatornaya geometriya različnyh klassov vypuklyh množestv*. Štinca, Kišenev 1978.

[37] Danzer, L., Grünbaum, B., and Klee, V., Helly's theorem and its relatives. In *Convexity*, Proceedings of Symposia in Pure Mathematics, Vol. VII, edited by V. Klee. American Math. Soc., Providence, R.I. 1963; pp. 101–180. Russian translation: Dancer, L., Gryunbaum, B., and Kli, V., *Teorema Helli i ee primeneniya*. Mir, Moscow 1968.

[38] Coxeter, H. S. M., An upper bound for the number of equal nonoverlapping spheres that can touch another of the same size. In *Convexity* (see [37]). Also in Coxeter, H. S. M., *Twelve Geometrical Essays*. London 1968.

[39] Yaglom, I. M., *Problema trinadcati šarov*. Višča Škola, Kiev 1975.

[40] Fejes Tóth, L., On the number of equal discs that can touch another of the same kind. *Studia Scient. Math. Hungar.* **2** (1967), 363–367.

[41] Grünbaum, B., On a conjecture of Hadwiger. *Pacific J. Math.* **11** (1961), 215–219.

[42] Baston, V. J. D., *Some Properties of Polyhedra in Euclidean Space*. Oxford 1965.

[43] Croft, H., 9-point and 7-point configurations in 3-space. *Proc. London Math. Soc.* **12** (No. 3, 1962), 400–424; **13** (1963), 384.

[44] Harazišvili, A. B., *Izbrannye voprosy geometrii Yevklidovyh prostranstv*. Tbilisskiĭ Universitet, Tbilisi 1978.

[45] Blumenthal, L., *Theory and Applications of Distance Geometry*. University Press, Oxford 1953.

[46] Szekeres, G., On an extremal problem in the plane. *Amer. J. Math.* **63** (1941), 208–210.

[47] Erdös, P. and Szekeres, G., On some extremum problems in elementary geometry. *Annales Universitates Scientiarum Budapestinesis de Rolando Eötvös Nominantae* **3–4** (1960/61), 53–62.

[48] Yaglom, I. M., and Faĭnberg, E. I., Ocenki dlya veroyatnosteĭ složnyh sobytiĭ. In *Trudy VI Vsesoyuznogo Soveščaniya po Teorii Veroyatnostei i Matematičeskoi Statistike*. Vilnius 1962; pp. 297–303.

[49] Pirogov, S. A., Veroyatnosti složnyh sobytiĭ i lineĭnoe programmirovanie. *Teoriya Veroyatnosteĭ i ee Primeneniya* **13** (No. 2, 1968), 344–348.

# Some Researches Inspired by H. S. M. Coxeter

L. Fejes Tóth*

Let me start with extremum properties of the regular solids, pointing out how fertile a remark of Professor Coxeter turned out to be in this field. The researches started thirty years ago by Coxeter's remark are still in progress.

Many extremum properties of the regular polygons are known, some of which go back to S. Lhuilier and J. Steiner. One of the best known of these properties is the isoperimetric property of the regular $n$-gon, i.e., the fact that among the $n$-gons of given perimeter the regular $n$-gon has the greatest area. Let me mention four further extremum properties which I sum up in the following theorem: Among the $n$-gons contained in (containing) a given circle the regular $n$-gon inscribed in (circumscribed about) the circle has the greatest (least) area as well as the greatest (least) perimeter.

As to the polyhedra, it was the isoperimetric problem which attracted the greatest interest. The isoperimetric property of the regular tetrahedron, according to which among the tetrahedra of given surface area the regular one has the greatest volume, was known to Lhuilier. Steiner proved, with his famous process of symmetrization, that among the polyhedra topologically isomorphic with the regular octahedron the regular octahedron is the best regarding the isoperimetric problem, and he conjectured that the same is true for the rest of the Platonic solids. Although the isoperimetric problem for polyhedra was investigated by such mathematicians as L. Lindelöf, H. Minkowski, and E. Steinitz, Steiner's conjecture concerning the cube and the dodecahedron was proved comparatively late, and the problem concerning the icosahedron is still open. Let me observe that Steinitz considered Steiner's conjecture, especially concerning the dodecahedron and icosahedron, as unjustified, and he advised great precaution in making similar conjectures. Steinitz's attitude is comprehensible because in the earlier

*Mathematical Institute of the Hungarian Academy of Sciences, Reáltanoda u. 13–15, 1053 Budapest, Hungary.

literature no extremum property of the regular icosahedron or dodecahedron occurs.

Steiner's conjecture for the cube and the dodecahedron has been confirmed under much more general conditions. I proved [7] the inequality

$$\frac{F^3}{V^2} \geqslant 54(n-2)\tan\omega_n(4\sin^2\omega_n - 1), \qquad \omega_n = \frac{n}{n-2}\,\frac{\pi}{6},$$

which holds for the surface area $F$ and the volume $V$ of any convex polyhedron having $n$ faces, with equality only for the regular tetrahedron, hexahedron, and dodecahedron. Thus the cube and the regular dodecahedron are the best not only among the respective isomorphic polyhedra but also among all convex polyhedra with six faces and twelve faces, respectively. An incomplete proof for this inequality was given previously by Goldberg [19].

Originally I was led to extremum properties of the regular solids by packing and covering problems on the sphere. Let us consider e.g. the problem of the thinnest covering of the sphere with a given number $n > 3$ of equal circles. If a sphere of radius $R$ is covered with $n$ circles of angular radius $\rho$, then the basic planes of the circles determine a polyhedron with $n$ faces contained in the sphere of radius $R$ and containing a concentric sphere of radius $r = R\cos\rho$. The problem of finding the least value of $\rho$ such that $n$ circles of radius $\rho$ cover a sphere is equivalent to finding the greatest value of $r$. In other words, the problem is to find the least value of the quotient $R/r$ such that the boundary of a conveniently chosen convex polyhedron with $n$ faces can be imbedded into a spherical shell with radii $R$ and $r$. A polarity with respect to a unit sphere concentric with the spherical shell carries the polyhedron into a polyhedron with $n$ vertices whose boundary is contained in a spherical shell with outer radius $1/r$ and inner radius $1/R$. Thus the problem for polyhedra with $n$ faces is equivalent to the problem for polyhedra with $n$ vertices. A lower bound for $R/r$ is given by the following theorem: If the boundary of a convex polyhedron with $n$ faces or $n$ vertices is contained in a spherical shell with outer radius $R$ and inner radius $r$, then

$$R/r \geqslant \sqrt{3}\,\tan\omega_n.$$

Equality holds only for the regular trihedral polyhedra with $n$ faces and the regular trigonal polyhedra with $n$ vertices.

Let me present the surprisingly simple proof. We prove the inequality for polyhedra with $n$ vertices assuming that $R = 1$. We may suppose that all faces of the polyhedron are triangles, because a face with more than three sides can be decomposed by not intersecting diagonals into triangles. Projecting the faces of the polyhedron radially onto the circumsphere, we obtain a spherical tiling consisting, as a simple consequence of Euler's theorem, of $2n-4$ triangles. Thus there is a triangle of area $\geqslant 4\pi/(2n-4)$. The circumradius of this triangle is not less than the circumradius $\rho_n$ of an equilateral spherical triangle of area $4\pi/(2n-4)$. Thus we have, in accordance with the inequality to be proved,

$$r \leqslant \cos\rho \leqslant \cos\rho_n = \cot(\pi/3)\cot\omega_n.$$

The case of equality is obvious.

I had proved several further inequalities for polyhedra with a given number of faces as well as for polyhedra with a given number of vertices in which equality occurs for the regular trihedral and the regular trigonal polyhedra, respectively, when Professor Coxeter called my attention to his formula

$$\frac{R}{r} = \tan\frac{\pi}{p}\tan\frac{\pi}{q},$$

which holds for the circumradius $R$ and the inradius $r$ of any Platonic solid $\{p,q\}$. This nice formula suggested establishing the following conjecture: In a convex polyhedron with circumradius $R$ and inradius $r$, let $p$ be the average number of the sides of the faces and $q$ the average number of the edges meeting at the vertices. Then

$$\frac{R}{r} \geqslant \tan\frac{\pi}{p}\tan\frac{\pi}{q}, \tag{1}$$

with equality only for the five Platonic solids.

At the same time, Coxeter's formula made it clear that a "real" analogue of an extremum property of the regular polygons arises not by considering polyhedra with a given number of faces or polyhedra with a given number of vertices, but polyhedra with data $p$ and $q$, i.e. polyhedra where both the number of faces $f$ and the number of vertices $v$ is prescribed.

In the theory of convex polyhedra three so-called fundamental gauges play a central part: the volume $V$, the surface area $F$, and the edge curvature $M$ defined by $M = \frac{1}{2}\sum \alpha l$, where $l$ is the length of an edge, $\alpha$ is the angle of the outer normals of the faces meeting at this edge, and the summation extends over all edges of the polyhedron. Thus six problems analogous to those solved by the above theorem for polygons presented themselves: For polyhedra with $f$ faces, $v$ vertices and inradius $r$ (circumradius $R$), find the minimum (maximum) of $V$, $F$, and $M$.

Of course, it seemed to be hopeless to solve any of these problems for all possible values of $f$ and $v$. But it has been conjectured that for values of $f$ and $v$ corresponding to a regular solid, in all six problems the regular solid will yield the solution. Thus it has been expected that for any values of $f$ and $v$ six inequalities can be given, each expressing an extremum property of the regular solids:

$$4r^3e\,\underline{V}(p,q) \leqslant V \leqslant 4R^3e\overline{V}(p,q),$$

$$4r^2e\,\underline{F}(p,q) \leqslant F \leqslant 4R^2e\overline{F}(p,q),$$

$$4re\,\underline{M}(p,q) \leqslant M \leqslant 4Re\overline{M}(p,q).$$

Here $e = f + v - 2$ is the number of edges, $p = 2e/f$, $q = 2e/v$, and the functions of $p$ and $q$ are defined as follows.

Projecting the faces of the polyhedron from an inner point $O$ onto a unit sphere with center $O$, we obtain a spherical tiling. Decompose each face of the tiling into triangles spanned by a point of a face, a vertex of this face, and a point of a side of this face issuing from this vertex. There are altogether $4e$ such triangles. Since the sums of the angles of these triangles at the vertices lying in the faces, in the vertices, and on the edges of the tiling are equal to $2\pi f$, $2\pi v$, and $2\pi e$, respectively, a triangle $ABC$ with angles $A = 2\pi f/4e = \pi/p$, $B = 2\pi v/4e = \pi/q$, and $C = 2\pi e/4e = \pi/2$ can be considered as representing the average of all triangles. This triangle is called the characteristic triangle of the polyhedron. Let the projections of $B$ and $C$ from $O$ onto the plane intersecting the line $OA$ perpendicularly at $A$ be $B'$ and $C'$. Again, let $A'$ and $C''$ be the projections of $A$ and $C$ from $O$ onto the plane passing through $B$ perpendicularly to $OA$. Finally, let $\varphi$ be the inner dihedral angle of the tetrahedron $OAB'C'$ at the edge $B'C'$ (which is, of course, equal to the inner dihedral angle of $OA'BC''$ at $BC''$). Then $\underline{V}$ is the volume of $OAB'C'$, $\overline{V}$ is the volume of $OA'BC''$, $\underline{F}$ is the area of the triangle $AB'C'$, $\overline{F}$ is the area of $A'BC''$, $\underline{M}$ is equal to $\frac{1}{2}(\pi/2 - \varphi)\overline{B'C'}$, and $\overline{M}$ is equal to $\frac{1}{2}(\pi/2 - \varphi)\overline{BC''}$.

We shall refer to these inequalities as $(V, r)$, $(V, R)$, etc.

Soon after Coxeter's remark I succeeded in proving the inequality $(V, r)$ [8]. This inequality can be deduced from a more general theorem. On the unit sphere, let $du$ be the area element at the point $U$, $U_O$ a fixed point, $D$ a domain, and $g(x)$ a function defined for $0 \leqslant x \leqslant \pi$. The integral

$$m(D, U_0) = \int_D g(U_0 U)\, du$$

is called the momentum of $D$ with respect to $U_0$. Let $D_1, \ldots, D_f$ be the faces of a spherical tiling with $f$ faces, $v$ vertices, and $e$ edges, $U_1, \ldots, U_f$ $f$ points on the sphere, and $g(x)$ strictly increasing. The theorem says that

$$\sum_{i=1}^{k} m(D_i, U_i) \geqslant 4em(\Delta, A), \tag{2}$$

where $\Delta = ABC$ is the characteristic triangle of the tiling, defined as for polyhedra. Equality holds only if the tiling is regular and the points $U_1, \ldots, U_f$ are the centers of the faces.

This theorem implies also the inequality $(F, r)$.

Florian [14] proved the inequality $(V, R)$, confirming thereby also the above conjecture concerning $R/r$. By combining the inequalities $(V, r)$ and $(V, R)$ we obtain the inequality (1).

The remaining inequalities $(F, R)$, $(M, r)$, and $(M, R)$ were proved long ago [8, 15] under the condition that the feet of the perpendiculars drawn from the center of the insphere or the center of the circumsphere, respectively, to the face planes and edge lines lie on the corresponding faces and edges. Florian [17] proved the inequality $(M, r)$ for trihedral polyhedra. But after these partial results the researches seemed to come to a dead end. Recently Linhart [22] succeeded in proving the inequality $(M, R)$ by attacking the problem in a new way. Let me outline his ingenious proof.

Let the circumsphere $S$ of the polyhedron $P$ have radius $R = 1$ and center $O$. Using the same symbol for a point $X$ and the vector $OX$, the supporting function $t(U)$ of $P$ at the point $U$ of $S$ is defined by

$$t(U) = \max_{X \in P} (X, U),$$

where $(X, U)$ denotes the inner product of $X$ and $U$. The proof is based on the well-known formula

$$M = \int_S t(U)\,du.$$

Let $V_1, \ldots, V_v$ be the vertices of $P$, $U_1, \ldots, U_v$ the projections of the vertices from $O$ onto $S$, $P^*$ the polyhedron polarly conjugate to $P$ with respect to $S$, and $S_1, \ldots, S_v$ the projections of the faces of $P^*$. Then

$$M = \sum_{i=1}^{v} \int_{S_i} t(U)\,du = \sum_{i=1}^{v} \int_{S_i} |V_i| \cos \widehat{U_i U}\,du \leqslant \sum_{i=1}^{v} \int_{S_i} \cos \widehat{U_i U}\,du.$$

Since $\cos x$ is a decreasing function in the interval $0 \leqslant x \leqslant \pi$, we can apply the inequality (2), obtaining

$$M \leqslant 4e \int_\Delta \cos \widehat{AU}\,du,$$

where $\Delta$ is the characteristic triangle $ABC$ of $P^*$, i.e. a spherical triangle with angles $A = \pi/q$, $B = \pi/p$, and $C = \pi/2$.

This inequality is equivalent to the inequality $(M, R)$. For the integral on the right side is equal to the area of the normal projection $\Delta'$ of $\Delta$ onto a plane perpendicular to $OA$. $\Delta'$ is the sector of an ellipse, and it can easily be checked that the area of $\Delta'$ is equal to $\overline{M}(p,q)$.

The idea of representing $M$ by the above formula and considering the polarly conjugate polyhedron gave also the clue to the proof of the inequality $(M, r)$. The proof, which required further consideration and rather complicated computations, is contained in a recent paper by Florian and Linhart [18].

So far all attempts to prove the inequality $(F, R)$ have failed. Thus, writing $\overline{F}$ in terms of $f$ and $v$, we have the problem: Prove or disprove the conjecture that

$$F \leqslant e \sin \frac{\pi f}{e}\left(1 - \cot^2 \frac{\pi f}{2e} \cot^2 \frac{\pi v}{2e}\right) R^2.$$

The only thing we know in connection with this problem is a nice proof of Heppes [20] for the fact that among the tetrahedra contained in a sphere the regular tetrahedron inscribed in the sphere has the greatest surface area.

Besides the problem concerning $F$ and $R$ there are various analogous unsolved problems involving other data of the polyhedron. Special attention is due to the isoperimetric problem. In spite of the warning of Steinitz, I risk making the conjecture that

$$F^3/V^2 \geqslant 36e\,\underline{F}(p,q).$$

This would imply that the regular icosahedron is the best one not only among the isomorphic polyhedra, but also among all polyhedra with twelve vertices.

Further problems arise in non-Euclidean spaces. The inequality (2) implies the validity of the inequalities $(V, r)$ and $(F, r)$ also in non-Euclidean spaces. What can be said about the inequalities $(V, R)$ and $(F, R)$? The variety and attractiveness of such problems is increased by the fact that in a particular problem difficulties of different kind can occur in the elliptic and in the hyperbolic space.

I proved the isoperimetric property of the regular tetrahedra in the hyperbolic space [12], but was not able to prove the same in the elliptic space.

Inspired by the nice presentation of the theory of the regular star-polyhedra in Coxeter's *Regular Polytopes*, I observed [10] that, conveniently interpreted, the inequalities $(V,r)$ and $(F,r)$ express extremum properties of all nine regular polyhedra. Similar results were obtained also by Florian [16], but systematic researches in this direction remain to be done.

Let me now turn to another field of research.

Confirming a conjecture of mine, Besicovitch and Eggleston [1] showed that among all convex polyhedra of given inradius the cube has the least total edge length. A further conjecture of mine, that among the trigonal polyhedra of given inradius the regular tetrahedron and the regular octahedron have minimal total edge length, was proved by Linhart [21]. Coxeter [4] called attention to the analogous problems in non-Euclidean spaces, pointing out the interesting results which are to be expected there. In a joint paper [6], Coxeter and I considered the problem of finding among the trigonal polyhedra of inradius $r$ that one whose total edge length is minimal. We proved that in spherical space for $r = \arcsin \frac{1}{4}$ the tetrahedron $\{3,3\}$ and the trigonal dihedron $\{3,2\}$ yield the solution. In the hyperbolic space we gave the solutions for two particular values of $r$, namely for $r_1 = 0.364\ldots$ and for $r_2 = 0.828\ldots$. For $r_1$ the solution is given by the regular octahedron and for $r_2$ by the regular icosahedron. Tomor [24] gave intervals $I_1$ and $I_2$ containing $r_1$ and $r_2$, respectively, such that also for any $r \in I_1$ or $r \in I_2$ the solution is $\{3,4\}$ or $\{3,5\}$. But most of the problems suggested by Coxeter's paper are still open.

The general problem may be formulated as follows: In an $n$-dimensional space of constant curvature with $n > 2$ let $P$ be a convex polytope. Let $V_k(P)$ be the sum of the $k$-dimensional volumes of the $k$-dimensional cells of $P$. For a value of $k$ with $1 \leqslant k \leqslant n-2$, find the infimum of $V_k(P)$ extended over all polytopes (or over a special class of polytopes) of inradius $r$.

The method used in our joint paper seems to be suitable for giving lower bounds for $V_{n-2}(P)$, which are especially good in the case of simplicial polytopes, and exact for particular values of $r$ [13]. But no method is known to me to attack the problem for $k < n-2$.

Finally, let us consider the problem of the densest packing of equal spheres in spaces of constant curvature, raised independently by Professor Coxeter [3] and me [9, 11]. In an $n$-dimensional space of constant curvature consider $n+1$ spheres of radius $r$ mutually touching one another. Let $d$ be the density of the spheres in the simplex spanned by the centers of the spheres. It has been conjectured that the density of a packing of spheres of radius $r$ never exceeds $d$. Rogers [23] proved the correctness of this conjecture in the case of Euclidean spaces. In the non-Euclidean spaces especially interesting consequences of the conjecture were pointed out by Coxeter and partly by me. Let me emphasize a paper by Coxeter [5] in which he gave good upper bounds for the Newton number of an $n$-dimensional sphere under the hypothesis of the correctness of the conjecture.

This was a challenge to prove the conjecture. Böröczky [2] observed that in hyperbolic spaces the vaguely phrased conjecture needs a modification and proved that in a packing of spheres of radius $r$ the density of each sphere in its Voronoï cell is less than or equal to the above defined density $d$.

## References

[1] Besicovitch, A. S. and Eggleston, H. G., The total length of the edges of a polyhedron. *Quart. J. Math. Oxford (2)* **8** (1957), 172–190.

[2] Böröczky, K., Packing of spheres in spaces of constant curvature. *Acta Math. Acad. Sci. Hungar.* **32** (1978), 243–261.

[3] Coxeter, H. S. M., Arrangement of equal spheres in non-Euclidean spaces. *Acta Math. Acad. Sci. Hungar.* **4** (1954), 263–274.

[4] Coxeter, H. S. M., The total length of the edges of a non-Euclidean polyhedron. Studies math. anal. related topics, 62-69. Stanford, Calif. 1962.

[5] Coxeter, H. S. M., An upper bound for the number of equal nonoverlapping spheres that can touch another of the same size. *Proc. Sympos. Pure Math. Amer. Math. Soc.*, Providence, R.I., **7** (1963), 53–71.

[6] Coxeter, H. S. M. and Fejes Tóth, L., The total length of the edges of a non-Euclidean polyhedron with triangular faces. *Quart. J. Math. Oxford (2)* **14** (1963), 273–284.

[7] Fejes Tóth, L., The isepiphan problem for *n*-hedra. *Amer. J. Math.* **70** (1948), 174–180.

[8] Fejes Tóth, L., Extremum properties of the regular polyhedra. *Canad. J. Math.* **2** (1950), 22–31.

[9] Fejes Tóth, L., On close-packings of spheres in spaces of constant curvature. *Publ. Math. Debrecen* **3** (1953), 158–167.

[10] Fejes Tóth, L., Characterization of the nine regular polyhedra by extremum properties. *Acta Math. Acad. Sci. Hungar.* **7** (1956), 31–48.

[11] Fejes Tóth, L., Kugelunterdeckungen und Kugelüberdeckungen in Räumen konstanter Krümmung. *Arch. Math.* **10** (1959), 307–313.

[12] Fejes Tóth, L., On the isoperimetric property of the regular hyperbolic tetrahedra. *Publ. Math. Inst. Hungar. Acad. Sci.* **8** (1963), 53–57.

[13] Fejes Tóth, L., On the total area of the faces of a four-dimensional polytope. *Canad. J. Math.* **17** (1965), 93–99.

[14] Florian, A., Eine Ungleichung über konvexe Polyeder. *Monatsh. Math.* **60** (1956), 130–156.

[15] Florian, A., Ungleichungen über konvexe Polyeder. *Monatsh. Math.* **60** (1956), 288–297.

[16] Florian, A., Ungleichungen über Sternpolyeder. *Rend. Sem. Mat. Univ. Padova* **27** (1957), 16–26.

[17] Florian, A., Eine Extremaleigenschaft der regulären Dreikantpolyeder. *Monatsh. Math.* **70** (1966), 309–314.

[18] Florian, A. and Linhart, J., Kantenkrümmung und Inkugelradius konvexer Polyeder. (unpublished manuscript).

[19] Goldberg, M., The isoperimetric problem for polyhedra. *Tohoku Math. J.* **40** (1935), 226–236.

[20] Heppes, A., An extremal property of certain tetrahedra (Hungarian). *Mat. Lapok* **12** (1961), 59–61.

[21] Linhart, J., Über die Kantenlängensumme von Dreieckspolyedern. *Monatsh. Math.* **83** (1977), 25–36.

[22] Linhart, J., Kantenkrümmung und Umkugelradius konvexer Polyeder. *Studia Sci. Math. Hungar.* **11** (1976), 457–458.

[23] Rogers, C. A., The packing of equal spheres. *Proc. London Math. Soc. (3)* **8** (1958), 609–620.

[24] Tomor, B., An extremum property of the regular polyhedra in spaces of constant curvature (Hungarian). *Magyar Tud. Akad. Mat. Fiz. Oszt. Közl.* **15** (1965), 263–271.

# Some Problems in the Geometry of Convex Bodies

C. A. Rogers*

In this note we discuss four problems. The first three remain totally intractable; the fourth has recently yielded some interesting results that are as yet incompletely understood.

The first problem is the equichordal problem stated by Blaschke, Rothe, and Weizenböck in 1917 [2]. A point $E$ in the interior of a convex domain $K$ in the plane is said to be an equichordal point, if each chord of $K$ through $E$ has the same length. The problem is whether or not a convex domain in the plane can have two distinct equichordal points. If $K$ has distinct equichordal points $E$, $E'$ at distance $2c$ apart, then $K$ is fairly easily seen to be symmetrical about the line joining $E$ and $E'$ and about the perpendicular bisector of the line segment $EE'$. Without loss of generality we may take all the chords of $K$ through $E$ and $E'$ to have length 2. One approach to the problem is to introduce bipolar coordinates $(r, r')$, $r$ being the distance of the point from $E$ and $r'$ being the distance of the point from $E'$. One easily finds that the points $(1, (1+4c^2)^{1/2})$, $((1+4c^2)^{1/2}, 1)$ lie on the curve $C$ bounding $K$, and then that the points

$$(r_n, 2 - r_{n-1}), \qquad n \geqslant 1,$$

lie on the curve, with

$$r_0 = 1,$$

$$r_1 = (1+4c^2)^{1/2},$$

and $r_n$, $n \geqslant 2$, defined inductively by the recurrence relation

$$r_{n+1} r_{n+2}^2 + (2 - r_{n+1})(2 - r_n)^2 - 2r_{n+1}(2 - r_{n+1}) = 8c^2. \tag{1}$$

For any fixed $c > 0$ that is not too small, a computer can calculate many points

*Department of Mathematics, University College London, Gower Street, London WC1E 6BT, England.

on the curve $C$. In practice, after a time, the values of $r_n$ for $n$ odd sensibly exceed the values of $r_n$ for $n$ even, and it follows that there is no equichordal curve with the given value of $c$. Unfortunately the size of the oscillations seems to tend to zero exponentially as $c$ tends to zero. In 1958 E. Wirsing [15] managed to prove that such equichordal curves can only exist for at most countably many values of $c$ (with 0 as their only possible limit point), and produced (but did not publish) almost convincing numerical evidence that such curves cannot exist. In 1969 G. J. Butler [5] confirmed the first of these results and at the same time proved the existence of convex domains that nearly have equichordal points, in that there are points $E, E'$ that are (in the obvious sense) equichordal for all chords, making a sufficiently small angle with the line $EE'$. The problem appears to be most intractable. If you are interested in studying the problem, my first advice is "Don't," my second is "If you must, do study the work of Wirsing and Butler," and third is "You may well have to develop a sophisticated technique for obtaining uniform and extremely accurate asymptotic expansions for the solutions of the recurrence relation (1) with its initial conditions; the first terms of such an expansion might be $1 + c \tanh 2nc$."

By way of light relief, let me mention a quite different and quite easy equichordal problem. If $E, E'$ are in the interior of a convex domain $K$ that is symmetrical about the midpoint of the segment $EE'$, then each chord of $K$ through $E$ is of the same length as the parallel chord through $E'$. Recently P. McMullen asked for a proof of the converse. I was able to supply a simple proof [13]; D. G. Larman and N. Tamvakis [11] have obtained an $n$-dimensional generalization.

T. Bang [1] proves a conjecture made by Tarski in 1932 showing, by beautiful and ingenious arguments, that if a convex body is covered by a finite number of "slabs" (i.e., regions that are bounded by a pair of parallel $(n-1)$-dimensional hyperplanes), then the sum of the widths of the slabs is at least as large as the width of the convex body that they cover. Bang asks for a proof of an affine-invariant version of the result asserting that the sum of the widths of the slabs, covering $K$, measured relative to $K$, must be at least 1. I have looked at this problem on and off for many years without ever feeling that I was gaining any insight. There is one very special case which is of interest. Suppose that $K$ and each of the slabs used to cover $K$ are centrally symmetric with the origin as center. Does the result hold in this very special case? This may be a very much easier problem, but I do not see how to make effective use of the extra information. Another interesting special case is when $K$ is a simplex. If the result is true in this case, it is certainly only just true, as there are an immense variety of different ways of covering a simplex with slabs, with the sum of their relative widths equal to 1. If one works with the $n$-simplex in $\mathbb{R}^{n+1}$ defined by

$$x_0 \geqslant 0, \quad x_1 \geqslant 0, \ldots, \quad x_n \geqslant 0, \quad x_0 + x_1 + \cdots + x_n = 1,$$

this is covered by the system of $n+1$ slabs defined by

$$0 \leqslant x_i \leqslant \tau_i, \qquad 0 \leqslant i \leqslant n,$$

for any system of $\tau_0, \tau_1, \ldots, \tau_n$ with

$$\tau_0 > 0, \quad \tau_1 > 0, \ldots, \quad \tau_n > 0, \quad \tau_0 + \tau_1 + \cdots + \tau_n = 1,$$

and the sum of the relative widths of the slabs in $\tau_0 + \tau_1 + \cdots + \tau_n = 1$. But the simplex is also covered by the system of $\frac{1}{2}n(n+1)$ slabs defined by

$$|x_i - x_j| \leqslant \frac{2}{n(n+1)}, \qquad 0 \leqslant i < j \leqslant n,$$

and again the sum of the relative widths of these slabs is precisely 1. There are many ways of combining these types of coverings to give further coverings with the sum of the relative width of the slabs precisely equal to 1. Perhaps some ingenious covering will yield a counterexample. The existence of so many cases of equality may so constrain the possible methods of proof that one is led to construct a proof in this special case (and this might generalize to give a proof of Bang's conjecture). The simple special case when $K$ is a square was reformulated by H. Davenport as the following intriguing problem. Let $n$ straight lines cross a square of side 1 in $\mathbb{R}^2$; is it always possible to find a small square of side $1/(n+1)$, in and homothetic to the square of side 1, that is crossed by none of the lines?

The third problem that I want to discuss is Borsuk's problem. Borsuk conjectured that any set of diameter 1 in $\mathbb{R}^n$ can be partitioned into a system of $n+1$ sets each of diameter less than 1. Hadwiger [8] proved this conjecture in the case when the set is a convex body that has a unique tac-plane at each of its boundary points. It has been proved in $\mathbb{R}^3$ by Eggleston [6], Grünbaum [7], and Heppes [9]. It remains outstanding for $\mathbb{R}^n$ with $n \geqslant 4$. Even the apparently very special case of a finite set in $R^n$, $n \geqslant 4$, has not been solved. I hope that some of those who have made advances in the study of combinatorial geometry will study this problem for finite sets. It is quite possible that some well-known configuration of points provides a counterexample to Borsuk's conjecture. I am only acquainted with a limited circle of finite configurations, and I have discovered no counterexample. This may be because I can only effectively study configurations that have a high degree of symmetry, and Borsuk's conjecture *does* hold for any set in $\mathbb{R}^n$ that is invariant under the symmetry group of the regular $n$-simplex in $\mathbb{R}^n$ (see C. A. Rogers [12]). It also holds for any finite set that is centrally symmetric.

A convex body in $\mathbb{R}^n$ will be called projectively homogeneous if it is invariant under a group of projective transformations that act transitively on its interior. In 1965 H. Busemann [3] asked for geometrical descriptions of all projectively homogeneous convex bodies. In 1967, in his investigation of timelike spaces [4], he was able to discuss some special cases. Earlier E. B. Vinberg (in a series of *Doklady* notes [14], appearing from 1960 to 1962, followed up by detailed articles [15] appearing in 1963 and 1965) gave a theory of projectively homogeneous convex bodies. He gives an algebraic description of all such bodies in terms of his concept of a $T$-algebra, and shows how all $T$-algebras can be generated. He does not throw as much light on the geometry of these bodies as one would wish.

Recently D. G. Larman, P. Mani, and C. A. Rogers [10] have studied this problem from a geometric point of view. For each $m \geqslant 1$, let

$$\eta_{ij}, \qquad 1 \leqslant i \leqslant j \leqslant m,$$

be homogeneous coordinates with respect to some simplex of reference in $\mathbb{R}^k$,

$k = \frac{1}{2}m(m+1) - 1$. Let $K_m$ be the set of points $(\eta_{ij})_{1 \leqslant i \leqslant j \leqslant m}$, for which the symmetric matrix

$$(\eta_{ij})_{1 \leqslant i \leqslant m,\, 1 \leqslant j \leqslant m}$$

is the matrix of a positive semidefinite or positive definite quadratic form. Then $K_m$ is easily seen to be a convex body. Further, $K_m$ is clearly invariant under the group of projective transformations of the form

$$(\eta'_{ij})_{1 \leqslant i \leqslant m,\, 1 \leqslant j \leqslant m} = B^T(\eta_{ij})_{1 \leqslant i \leqslant m,\, 1 \leqslant j \leqslant m} B,$$

where $B$ is any nonsingular $m \times m$ matrix. What is more, this group acts transitively on the interior of $K_m$, which consists of points corresponding to positive definite quadratic forms. So each $K_m$ is a projectively homogeneous convex body. It turns out that each projectively homogeneous convex body in $\mathbb{R}^n$ occurs as a section of some $K_m$ with $m \leqslant 2^n + 1$. Further, the projectively homogeneous convex bodies in $\mathbb{R}^n$ can, at least in principle, be constructed from a knowledge of all the bodies in $\mathbb{R}^n$ with $1 \leqslant r < n$.

In $\mathbb{R}^2$ we have the circle

$$\zeta\xi - \eta^2 = 0,$$

and the triangle

$$\xi_0\xi_1\xi_2 = 0.$$

In $\mathbb{R}^3$ we have the sphere

$$\zeta\xi - \eta_1^2 - \eta_2^2 = 0,$$

the circular cone

$$\xi_0(\zeta\xi_1 - \eta^2) = 0,$$

and the tetrahedron

$$\xi_0\xi_1\xi_2\xi_3 = 0.$$

In $\mathbb{R}^4$ we have the sphere

$$\zeta\xi - \eta_1^2 - \eta_2^2 - \eta_3^2 = 0,$$

two sorts of cone

$$\xi_0(\zeta\xi_1 - \eta_1^2 - \eta_2^2) = 0,$$

$$\xi_0\xi_1(\zeta\xi_2 - \eta^2) = 0$$

(the first based on a 3-sphere and the second on a 2-sphere), and the simplex

$$\xi_0\xi_1\xi_2\xi_3\xi_4 = 0.$$

But we also find a body

$$(\zeta\xi_0 - \eta_0^2)(\zeta\xi_1 - \eta_1^2) = 0$$

that is the intersection of two (unbounded) cones, and a body

$$\zeta\xi_0\xi_1 - \xi_1\eta_1^2 - \xi_0\eta_2^2 = 0$$

that is the convex cover of two ellipses whose planes meet in a single point common to the two ellipses.

Omitting those bodies that are not bounded by a single algebraic surface, the only bodies in dimensions 5 and 6 are

$$\zeta\xi - \eta_1^2 - \eta_2^2 - \eta_3^2 - \eta_4^2 = 0,$$

$$\zeta\xi_0\xi_1 - \xi_1\eta_1^2 - \xi_1\eta_2^2 - \xi_0\eta_3^2 = 0,$$

$$\zeta\xi_0\xi_1 + 2\eta_1\eta_2\xi_2 - \zeta\xi_2^2 - \xi_1\eta_1^2 - \xi_0\eta_2^2 = 0,$$

$$\zeta\xi - \eta_1^2 - \eta_2^2 - \eta_3^2 - \eta_4^2 - \eta_5^2 = 0,$$

$$\zeta\xi_0\xi_1 - \xi_1\eta_1^2 - \xi_1\eta_2^2 - \xi_0\eta_3^2 - \xi_0\eta_4^2 = 0,$$

$$\zeta\xi_0\xi_1 - \xi_1\eta_1^2 - \xi_1\eta_2^2 - \xi_1\eta_3^2 - \xi_0\eta_4^2 = 0,$$

$$\zeta\xi_0\xi_1\xi_2 - \xi_1\xi_2\eta_1^2 - \xi_0\xi_2\eta_2^2 - \xi_1\xi_2\eta_3^2 = 0.$$

In each case the equation arises in determinantal form and the transitive group of projectivities can be easily described. For example, the body in $\mathbb{R}^4$ given by the equation

$$(\zeta\xi_0 - \eta_0^2)(\zeta\xi_1 - \eta_1^2) = 0$$

is the set of points with homogeneous coordinates $(\xi_0, \xi_1, \eta_0, \eta_1, \zeta)$ for which the matrix

$$\Xi = \begin{bmatrix} \zeta & \eta_0 & 0 & 0 \\ \eta_0 & \xi_0 & 0 & 0 \\ 0 & 0 & \zeta & \eta_1 \\ 0 & 0 & \eta_1 & \xi_1 \end{bmatrix}$$

is the matrix of a positive definite or nonnegative semi-definite quadratic form. The body is left invariant by the projective transformations

$$\Xi' = X^T \Xi X,$$

with $X$ any matrix of the form

$$X = \begin{bmatrix} z & 0 & 0 & 0 \\ y_0 & x_0 & 0 & 0 \\ 0 & 0 & z & 0 \\ 0 & 0 & y_1 & x_1 \end{bmatrix}$$

with $zx_0x_1 \neq 0$, and these transformations act transitively on the interior of the body. The body can be described as the least convex cover of the twisted 2-dimensional surface described parametrically by

$$\xi_0 = \rho^2, \quad \xi_1 = \sigma^2, \quad \eta_0 = \rho\tau, \quad \eta_1 = \sigma\tau, \quad \zeta = \tau^2,$$

with $(\rho, \sigma, \tau) \neq (0, 0, 0)$.

The body in $\mathbb{R}^5$ given by the equation

$$\zeta\xi_0\xi_1 + 2\eta_1\eta_2\xi_2 - \zeta\xi_2^2 - \xi_1\eta_1^2 - \xi_0\eta_2^2 = 0$$

is the set of all points $(\xi_0, \xi_1, \xi_2, \eta_1, \eta_2, \zeta)$ for which the matrix

$$\Xi = \begin{bmatrix} \zeta & \eta_1 & \eta_2 \\ \eta_1 & \xi_0 & \xi_2 \\ \eta_2 & \xi_2 & \xi_1 \end{bmatrix}$$

is the matrix of a positive definite or semidefinite quadratic form. This is the body $K_3$ discussed above. It may be described as the convex cover of the twisted 2-dimensional surface given parametrically by

$$\xi_0 = \rho^2, \quad \xi_1 = \sigma^2, \quad \zeta = \tau^2,$$
$$\eta_1 = \rho\tau, \quad \eta_2 = \sigma\tau, \quad \xi_2 = \rho\sigma.$$

## References


[1] Bang, T., A solution of the "plank problem." *Proc. Amer. Math. Soc.* **2** (1951), 990–993.

[2] Blaschke, W. Rothe, and Weizenböck, R., Aufgabe 552. *Arch. Math. Phys.* **27** (1917), 82.

[3] Busemann, H., Problem 3. In *Colloquium on Convexity*, edited by W. Fenchel. Copenhagen 1965.

[4] Busemann, H., Timelike spaces. Dissert. Math., 53. Warsaw 1967.

[5] Butler, G. J., On the "equichordal curve" problem and a problem of packing and covering. Thesis, London 1969.

[6] Eggleston, H. G., Covering a three-dimensional set with sets of smaller diameter. *J. London Math. Soc.* **30** (1955), 11–24.

[7] Grünbaum, B., A simple proof of Borsuk's conjecture in three dimensions. *Proc. Cambridge Phil. Soc.* **53** (1957), 776–778.

[8] Hadwiger, H., Überdeckung einer Menge durch Mengen kleineren Durchmessers. *Comment. Math. Helv.* **18** (1945/46), 73–75. Mitteilung betreffend meiner Note: Überdeckung einer Menge durch Mengen kleineren Durchmessers. *Ibid.* **19** (1946/47), 161–165.

[9] Heppes, A., On the splitting of point sets in three space into the union of sets of smaller diameter (in Hungarian). *Magyar Tud. Akad. Mat. Fiz. Oszt. Kögl.* **7** (1957), 413–416.

[10] Larman, D. G., Mani, P., and Rogers, C. A., Projectively homogeneous convex bodies. (unpublished manuscript)

[11] Larman, D. G. and Tamvakis, N., A characterisation of centrally symmetric convex bodies in $E^n$. *Geometriae Dedicata*, **10** (1981), 161–176.

[12] Rogers, C. A., Symmetrical sets of constant width and their partitions. *Mathematika* **18** (1971), 105–111.

[13] Rogers, C. A., An equi-chordal problem. *Geometriae Dedicata* **10** (1981), 73–78.

[14] Vinberg, È. B., Homogeneous cones. *Soviet Mathematics (Doklady)* **1** (1960), . 787–790. The Morozov-Borel theorem for real Lie groups. *Ibid.* **2** (1961), 1416–1419. Convex homogeneous domains. *Ibid.* **2** (1961), 1470–1473. Automorphisms of homogeneous convex cones. *Ibid.* **3** (1962), 371–374.

[15] Vinberg, È. B., The theory of convex homogeneous cones. *Trans. Moscow Math. Soc.* **12** (1963), 340–403. The structure of the group of automorphisms of a homogeneous convex cone. *Ibid.* **13** (1965), 63–93.

[16] Wirsing, E., Zur Analytizität von Doppelspeichenkurven. *Arch. Math.* **9** (1958), 300–307.

# On an Analog to Minkowski's Lattice Point Theorem

J. M. Wills*

For a convex body $K \subset E^d$ let $V(K)$ be its volume and $G(K) = \text{card}(K \cap \mathbb{Z}^d)$ $[G^0(K) = \text{card}(\text{int}\, K \cap \mathbb{Z}^d)]$ be the lattice point number of $K$ [int $K$]. If $K$ is central symmetric ($K = -K$), then by Minkowski's fundamental theorem

$$G^0(K) = 1 \quad \Rightarrow \quad V(K) \leqslant 2^d. \tag{1}$$

One can easily see that there is no direct analogue for the surface area or any other of Minkowski's quermass integrals $W_i$. On the other hand there is an analogue by Minkowski [3, p. 77]:

$$G^0(K) = 1 \quad \Rightarrow \quad G(K) \leqslant 3^d. \tag{2}$$

The question arises, why $V$ is the only quermass integral with this property similar to $G$. The answer is simple if one restricts to the set $\mathfrak{P}^d$ of lattice polytopes $P \subset E^d$, i.e. the convex hulls of lattice points $\in \mathbb{Z}^d$. Then one gets discrete functions $G_i$ defined on $\mathfrak{P}^d$ by Ehrhart's formulae:

$$\left.\begin{aligned} G(nP) &= \sum_{i=0}^{d} n^i G_i(P), \\ G^0(nP) &= \sum_{i=0}^{d} (-1)^{d-i} n^i G_i(P), \end{aligned}\right\} \quad n = 0, 1, 2, \ldots . \tag{3}$$

In particular $G_0 = 1$, $G_d = V$. The $G_i$ are additive, homogeneous of degree $i$, and dimension-invariant. On the other hand they are not monotone (except $G_0$ and $G_d$) and not positive definite (except $G_0, G_{d-1}, G_d$) [5]. The main property is that they are invariant under unimodular transformations (the $W_i$, $i = 1, \ldots, d-1$ are not).

By a result of Betke [1] for $P \in \mathfrak{P}^d$,

$$a_{id} + b_{id} G_i(P) \leqslant G_d(P), \qquad i = 1, \ldots, d-1, \tag{4}$$

* Dept. of Mathematics, University of Siegen, D-5900 Siegen, Federal Republic of Germany.

where the $a_{id}$, $b_{id}$ depend only on $d$, so that by (1) all the $G_i$ have the above property. Moreover by another theorem of Betke [1], the $G_i$ form a basis in the space of the additive unimodular functionals, so that all these functionals have the mentioned property.

In the following we give some simple partial results.

**Theorem.** *Let $P \subset E^d$ be a central-symmetric lattice polytope with $G^0(P) = 1$. Then*:

(i) *For all $d$,*

$$G_{d-1}(P) \leqslant d2^{d-1} \tag{5}$$

*with equality not in all cases of* (1).

(ii) *For $d = 3$,*

$$G_1(P) \leqslant 6 \tag{6}$$

*with equality not in all cases of* (1).

(iii) *For $d = 4$,*

$$G_1(P) < 21, \qquad G_2(P) < 31,$$

*and $G_2(P) \leqslant G_1(P) + 16$.*

(iv) *For $d = 5$,*

$$G_2(P) < 101.$$

*Remarks.*

(a) For the cube $C^d = \{x \in E^d \,\big|\, |x_i| \leqslant 1\}$ we have

$$G_i(C^d) = \binom{d}{i} 2^i, \qquad i = 0, 1, \ldots, d,$$

so (5) and (6) are best possible.

(b) (iii) and (iv) are far from the expected values $\binom{d}{i}2^i$, but better than the results obtained with (4).

*Proof.* (i) follows directly from (1) and (7). Examples (a) and (b) at the end show that equality in (1) does not imply equality in (5).

(ii): For $d = 3$, $G^0 = G_3 - G_2 + G_1 - 1 = 1 \Rightarrow G_1 = G_2 - G_3 + 2$. From (7), $G_2 \leqslant \frac{3}{2} G_3$, so $G_1 \leqslant \frac{1}{2} G_3 + 2 \leqslant 6$.

(iii): For $d = 4$, $G^0 = G_4 - G_3 + G_2 - G_1 + 1 = 1 \Rightarrow G_2 = G_1 + G_3 - G_4$.

From (7), $G_3 \leqslant 2G_4$, so $G_2 \leqslant G_1 + G_4 \leqslant G_1 + 16$.

From (3), $G - G^0 = \dot{G} = 2(G_3 + G_1)$ and $G + G^0 = \dot{G} + 2 = 2(G_4 + G_2 + 1)$.

From (2), (7), and (9),

$$G_1 \leqslant \frac{1}{2}\dot{G} - G_3 \leqslant \frac{1}{4}\dot{G} + \frac{5}{6} < 21,$$

$$G_2 \leqslant \frac{1}{2}\dot{G} - G_4 \leqslant \frac{1}{2}\dot{G} - \frac{1}{2}G_3 \leqslant \frac{3}{8}\dot{G} + \frac{5}{12} < 31.$$

(iv): For $d=5$, $G - G^0 = \dot{G} = 2(G_4 + G_2 + 1)$. From (2) and (9),

$$G_2 \leqslant \tfrac{1}{2}\dot{G} - G_4 - 1 \leqslant \tfrac{5}{12}\dot{G} - \tfrac{5}{8} < 101. \qquad \square$$

**Lemma 1.** *Let $P$ be a lattice polytope with $G^0(P) > 0$. Then*

(i) $G_{d-1}(P) \leqslant \dfrac{d}{2} G_d(P)$. (7)

(ii) $G_{d-1}(P) = \dfrac{d}{2} G_d(P) \Rightarrow G^0(P) = 1$. (8)

(iii) *For $d \geqslant 3$, $G_{d-1}(P) = (d/2)G_d(P) \not\Leftarrow G^0(P) = 1$.*

(iv) *For $d = 2$, $G_1(P) = G_2(P) \Leftrightarrow G^0(P) = 1$.*

*Proof.* (i): Let $O \in \operatorname{int} P$. For a facet $F_i$ of $P$ let $v_i$ be its $(d-1)$-volume, $h_i$ the distance from aff $F_i$ to $O$, and $\Delta_i$ the lattice determinant of aff $F_i$. Then

$$G_d(P) = \frac{1}{d}\sum_i h_i v_i$$

and

$$G_{d-1}(P) = \frac{1}{2}\sum_i \Delta_i^{-1} v_i,$$

where the sums run over all facets. From $h_i\Delta_i \geqslant 1$ we have

$$G_d(P) = \frac{1}{d}\sum_i (h_i\Delta_i)(\Delta_i^{-1}v_i) \geqslant \frac{2}{d} G_{d-1}(P).$$

(ii): If $G^0(P) > 1$, then at least one $h_i\Delta_i > 1$.
(iii): See examples (a), (b), (c); similarly for $d \geqslant 4$.
(iv): Trivial by $G^0 = G_2 - G_1 + 1$. $\square$

**Lemma 2.** *For a proper lattice polytope $P \subset E^d$,*

$$G_{d-1}(P) \geqslant \frac{1}{2(d-2)!}\dot{G}(P) - \frac{(d+1)(d-2)}{2(d-1)!}.$$

*Proof.* We dissect each facet of $P$ into simplices with the only lattice points at their vertices. By this we obtain a refinement of the boundary complex of $P$ with $f_0 = \dot{G}$ vertices and $f_{d-1}$ $(d-1)$-simplices as facets. For each of these simplices we have $v_i/\Delta_i \geqslant 1/(d-1)!$, so

$$G_{d-1}(P) \geqslant \frac{1}{2}\,\frac{1}{(d-1)!}\,f_{d-1}.$$

By the lower-bound conjecture for simplicial polytopes proved by Barnette [6] we have $f_{d-1} \geqslant (d-1)f_0 - (d-1)(d-2)$. This proves the lemma. $\square$

The following examples $\subset \mathbb{E}^3$ are needed for the theorem and Lemma 1:

(a) $X = \operatorname{conv}\{U, -U\}$, where

$$U = \{(1,0,2),\quad (0,1,2),\quad (1,2,2),\quad (2,1,2)\}.$$

(b) $Y = \operatorname{conv}\{V, -V\}$, where

$$V = \{(1,-1,1),\quad (-1,1,1),\quad (2,0,1),\quad (0,2,1)\}.$$

(c) $Z = \operatorname{conv}\{(1,0,2), (0,1,2), (1,1,2), (-1,-1,-3)\}$.

In all cases $(0,0,0)$ is the only interior lattice point and at least one $h_i\Delta_i > 1$. On the other hand $X$ and $Y$ are central symmetric with $V(X) = V(Y) = 2^3$, so that (1) holds, but not (5) and (6).

Concluding remarks:

(1) Van der Corput's generalization of (1) [2, p. 44]: If $K = -K$ and $k \in \mathbb{N}$, then

$$G^0(K) = 2k - 1 \quad \Rightarrow \quad V(K) \leqslant k\,2^d$$

has the analog for central-symmetric lattice polytopes:

$$G^0(P) = 2k - 1 \quad \Rightarrow \quad G_{d-1}(P) \leqslant kd\,2^{d-1}.$$

For $k > 1$ no equality holds, by (8).

(2) If one omits central symmetry, one gets similar problems for lattice polytopes (but not for general convex bodies), which are solved for $d = 2$ by Scott [4]. Some results for general $d$ will appear in a common paper with M. Perles and J. Zaks.

## References

[1] Betke, U., Gitterpunkte und Gitterpunktfunktionale, to appear.

[2] Lekkerkerker, C. G., *Geometry of Numbers*. Wolters-Noordhoff, Groningen 1969.

[3] Minkowski, H., *Geometrie der Zahlen*. Leipzig 1910.

[4] Scott, P. R., On convex lattice polygons. *Bull. Austral. Math. Soc.* **15** (1976), 395–399.

[5] Wills, J. M., Gitterzahlen und innere Volumina, *Comm. Math. Helv.* **53** (1978), 508–524.

[6] Barnette, D., A proof of the lower bound conjecture for convex polytopes. *Pacific J. Math.* **46** (No. 2, 1973), 349–354.

# Intersections of Convex Bodies with Their Translates

P. R. Goodey*
M. M. Woodcock*

## 1. Introduction

It has been shown by Fujiwara [4] and Bol [2] that if $K$ is a planar convex body which is not a disk, then it is possible to find a congruent copy $K'$ of $K$ such that $K$ and $K'$ have more than two points in common on their boundaries. This result was used by Yanagihara [9] to show that if $K$ is a 3-dimensional convex body with the property that for any congruent copy $K'$ of $K$, the boundaries of $K$ and $K'$ intersect in a planar curve (assuming they do, in fact, meet but do not coincide), then $K$ is a ball.

The above results can be thought of in a slightly different context following an article by Peterson [7]. If $K_1$, $K_2$ are two planar convex bodies which are not coincident, externally tangent, or disjoint, we define $\alpha(K_1, K_2)$ to be the number of connected components of the intersection of the boundaries of $K_1$ and $K_2$. It was conjectured in [7] that if $K$ has the property that $\alpha(K, C)$ is even or infinite for every disk $C$ of diameter $w$, then $S$ has constant width $w$. This conjecture was verified by the present authors in [5]. Since that time we have in fact shown that if $K$ is such that $\alpha(K, K')$ is even or infinite for every congruent copy $K'$ of $K$, then $K$ is a set of constant width. In this context it is easy to show that the Fujiwara–Bol result is equivalent to showing that a ball is the only convex body $K$ with the property that $\alpha(K, K') = 2$ for all congruent copies $K'$ of $K$.

In the present work, instead of intersecting congruent copies we shall consider only intersections of translates. We shall show that if $K_1, K_2$ are such that $\alpha(K_1, K_2') = 2$ for all translates $K_2'$ of $K_2$, then $K_1$ and $K_2$ are translates of one another. In three dimensions we show that if $K_1, K_2$ are convex bodies such that the boundaries of $K_1$ and $K_2'$ meet in a planar curve (assuming they do meet but do not coincide) for all translates $K_2'$ of $K_2$, then $K_1$ and $K_2$ are both translates of

* Mathematics Department, Royal Holloway College, Englefield Green, Surrey, United Kingdom.

the same ellipsoid. It will be seen that these results are both analogues and generalizations of the Fujiwara–Bol and Yanagihara characterizations. Curiously, even though the planar analogue is apparently much weaker than the original Fujiwara–Bol result, our spatial result is precisely what one expects from Yanagihara's work.

## 2. Results

To prove our results we shall use some integral-geometric techniques (see [8] for example). If $K$ is a planar convex body and $P = (x, y)$ is a point fixed in $K$, then the measure of a set of translates of $K$ is defined to be the (Lebesgue) measure of the set of points which $P$ occupies when it is subjected to these translates. The corresponding density $dK$ is given by the exterior product $dK = dx \wedge dy$. We shall denote by $V(K)$ the area of $K$, and by $V(K_1, K_2)$ the mixed area of the bodies $K_1$, $K_2$ (see [3] for example). The following two lemmas will be used in the proofs of our results. The first is an obvious result, and the proof is included for completeness. The second is an analogue of Poincare's formula; see [8, p. 111].

**Lemma 1.** *Let $K_1$, $K_2$ be two planar convex bodies, and let $K_2^*$ denote the reflection of $K_2$ in the origin. The measure of the set translates of $K_2$ which meet $K_1$ is*

$$V(K_1) + V(K_2) + 2V(K_1, K_2^*).$$

*Proof.* Let $P_1, P_2$ be points fixed in $K_1, K_2$, and let $H_1, H_2$ be the support functions of $K_1$, $K_2$ relative to $P_1$, $P_2$ respectively. Then it is clear that if a translate of $K_2$ is to meet $K_1$, then the point $P_2$ must lie within the convex body whose support function relative to $P_1$ is given by

$$H(\theta) = H_1(\theta) + H_2(\theta + \pi) \quad \text{for } 0 \leqslant \theta \leqslant 2\pi.$$

But $H^*(\theta) = H_2(\theta + \pi)$ is the support function of the reflection of $K_2$ in $P_2$. Thus, because of the translation invariance of mixed volumes,

$$\begin{aligned} V(K_1 + K_2^*) &= V(K_1) + V(K_2^*) + 2V(K_1, K_2^*) \\ &= V(K_1) + V(K_2) + 2V(K_1, K_2^*). \end{aligned}$$

□

**Lemma 2.** *Let $K_1, K_2, K_2^*$ be as in Lemma 3, and for each translate $K_2'$ of $K_2$ let $n(K_2')$ denote the number of points which the boundaries of $K_1$ and $K_2'$ have in common. Then*

$$\int n \, dK_2 = 4(V(K_1, K_2) + V(K_1, K_2^*))$$

*where the integration is carried out over all translates of $K_2$.*

*Proof.* Consider any line segment $U = [\mathbf{a}, \mathbf{b}]$. By Lemma 1 the measure of the set of translates of $K_2$ which meet $U$ is

$$V(K_2) + 2V(U, K_2^*) = V(K_2) + l(U)v(K_2^*; U),$$

where $l(U)$ denotes the length of $U$ and $v(K_2^*; U)$ is the length of the projection of $K_2^*$ onto a direction orthogonal to $U$; see [3, p. 45]. Now the measure of the set of translates of $K_2$ which contain $\mathbf{a}$ is $V(K_2)$, and so the measure of the set of translates of $K_2$ which meet $U$ and do not contain $\mathbf{a}$ is

$$l(U)v(K_2^*; U). \tag{1}$$

Let $P_0$ be any polytope with sides $U_1, U_2, \ldots, U_m$ listed in a clockwise sense and assume $U_i = [\mathbf{a}_i, \mathbf{b}_i]$. Denoting the boundary of $P_0$ by $\partial P_0$, we see that the measure of the set of translates $K_2'$ of $K_2$ for which $\partial P_0 \cap \partial K_2'$ has odd or infinite cardinality is zero, since this can only occur if $K_2'$ and $P_0$ share a support line at some point of $\partial P_0 \cap \partial K_2'$. So in the ensuing argument we shall disregard all such translations, as well as those translations for which $\partial K_2'$ contains a vertex of $P_0$ (these are also clearly of measure zero). Then the cardinality of $\partial P_0 \cap \partial K_2'$ is $2k$ if and only if there are precisely $k$ of the edges $U_1, U_2, \ldots, U_m$ for which $\partial K_2' \cap U_i \neq \emptyset$ and $\mathbf{a}_i \notin K_2'$. Consequently, if $n_0(K_2')$ denotes the cardinality of $\partial P_0 \cap \partial K_2'$, we have, from (1),

$$\int n_0 \, dK_2 = 2 \sum_{i=1}^{m} l(U_i) v(K_2^*; U_i). \tag{2}$$

Now let $\mathbf{u}_i$ be the unit outward normal vector to $P_0$ at $U_i$, and let $H_2, H_2^*$ be the support functions of $K_2, K_2^*$ respectively. Clearly

$$v(K_2^*; U_i) = H_2^*(\mathbf{u}_i) + H_2^*(-\mathbf{u}_i) = H_2^*(\mathbf{u}_i) + H_2(\mathbf{u}_i),$$

and so, from (2),

$$\begin{aligned} \int n_0 \, dK_2 &= 2 \sum_{i=1}^{m} \{ H_2^*(\mathbf{u}_i) + H_2(\mathbf{u}_i) \} l(U_i) \\ &= 4\{ V(K_2^*, P_0) + V(K_2, P_0) \}; \end{aligned} \tag{3}$$

see [3, p. 116]. This is the required result when $K_1$ is a polygon.

To extend (3) to arbitrary convex bodies we shall use a technique due to Maak [6]. First we let $U = [\mathbf{a}, \mathbf{b}]$ be a short line segment and find an upper bound for the measure of the set of translates $K_2'$ of $K_2$ for which the cardinality of $U \cap \partial K_2'$ exceeds one. For convenience we shall assume that the origin $\mathbf{0}$ is an interior point of $K_2$ and that the disk $B(\mathbf{0}, r)$ lies in $K_2$. We denote by $S$ the line-segment center $\mathbf{0}$ parallel to $U$ of length $2r$ and assume that $l(U) < 2r$. Now $\mathbf{y} \in \mathbf{x} + K_2$ if and only if $\mathbf{x} \in \mathbf{y} + K_2^*$, and so we require an upper bound for the measure of the set of points in $K_2^* + U$ which are not in $(K_2^* + \mathbf{a}) \cup (K_2^* + \mathbf{b})$. Let $\mathbf{p}, \mathbf{q} \in K_2^*$ lie on support lines of $K_2^*$ parallel to $U$, and let $T$ be the convex hull of $\mathbf{p}$, $\mathbf{q}$, and $S$. Then it suffices to find an upper bound for the measure of the sets of points in $T + U$ which are not in $(T + \mathbf{a}) \cup (T + \mathbf{b})$. If we denote by $D$ the diameter of $K_2^*$, it is easy to see that $\frac{1}{4} r^{-1} (l(U))^2 D$ is such an upper bound.

Now let $K_1$ be an arbitrary convex body, and for $i = 1, 2, \ldots$ let $P_i$ be a polygon inscribed in $K_i$ in such a way that (a) every vertex of $P_i$ is a vertex of $P_{i+1}$; (b) $P_{i+1}$ has one more side than $P_i$; (c) the maximum length of a side of $P_i$ approaches zero as $i \to \infty$; (d) $l(\partial P_i) \to l(\partial K_1)$ as $i \to \infty$. For each $i$ let $U_1^i$, $U_2^i, \ldots, U_{m(i)}^i$ be the sides of $P_i$, and as usual we shall consider only translates

$K_2'$ of $K_2$ for which $\partial K_2'$ does not contain a vertex of any $P_i$. For each admissible translate we let $f_i(K_2')$ denote the number of sides $U_1^i, U_2^i, \ldots, U_{m(i)}^i$ which $\partial K_2'$ intersects precisely once, and let $n_i(K_2')$ denote the cardinality of $\partial P_i \cap \partial K_2'$. Then if $i$ is so large that

$$\max\{l(U_j^i) : 1 \leqslant j \leqslant m(i)\} < 2r,$$

we have

$$\int n_i \, dK_2 \geqslant \int f_i \, dK_2 \geqslant \int n_i \, dK_2 - 2 \sum_{j=1}^{m(i)} \tfrac{1}{4} r^{-1} \big(l(U_j^i)\big)^2 D.$$

Using (3), this gives

$$4(V(K_2^*, P_i) + V(K_2, P_i)) \geqslant \int f_i \, dK_2$$

$$\geqslant 4(V(K_2^*, P_i) + V(K_2, P_i)) - \tfrac{1}{2} r^{-1} D \max\{l(U_j^i) : 1 \leqslant j \leqslant m(i)\} \sum_{j=1}^{m(i)} l(U_j^i).$$

Letting $i \to \infty$, we have

$$\lim_{i \to \infty} \int f_i \, dK_2 = 4(V(K_2^*, K_1) + V(K_2, K_1)).$$

But for almost all translates $K_2'$ of $K_2$ we have

$$f_i(K_2') \leqslant f_{i+1}(K_2') \quad \text{and} \quad \lim_{i \to \infty} f_i(K_2') = n(K_2').$$

So Lebesgue's monotone convergence theorem gives

$$\int n \, dK_2 = 4(V(K_1, K_2) + V(K_1, K_2^*)),$$

as required. □

Our proof of the following theorem is based to some extent on the proof in [8, p. 120] of the Fujiwara–Bol result.

**Theorem 1.** *Let $K_1, K_2$ be planar convex bodies. Then $\alpha(K_1, K_2') = 2$ for all translates $K_2'$ of $K_2$ if and only if $K_2$ is a translate of $K_1$.*

*Proof.* First we assume that $K_2 = \mathbf{a} + K_1$ and that $\operatorname{int} K_1 \cap \operatorname{int} K_2 \neq \emptyset$. Then if $\mathbf{x}, \mathbf{y}, \mathbf{z}$ are three points of $\partial K_1 \cap \partial K_2$, then there must be three parallel chords of $K_1$ through $\mathbf{x}, \mathbf{y}, \mathbf{z}$ all of the same length. Thus $\mathbf{x}, \mathbf{y}, \mathbf{z}$ all lie on the same support line of $K_1$ and thus in the same component of $\partial K_1 \cap \partial K_2$. Hence $\alpha(K_1, K_2) = 2$ as required.

Conversely, we assume that $\alpha(K_1, K_2') = 2$ for all translates $K_2'$ of $K_2$. First we note that if $K_2' \subset K_1$ or $K_1 \subset K_2'$ then $K_1 = K_2'$. For otherwise, if $K_2' \subset K_1$ then $\partial K_2' \cap \partial K_1$ has precisely two components. If there are two points in different components of $\partial K_2' \cap \partial K_1$ at which the support lines to $K_1$ are not parallel, then any small translation of $K_2'$ along one of the support lines towards its intersection with the other will produce more than two boundary components. So the support lines to $K_1$ at the components of $\partial K_2' \cap \partial K_1$ must be parallel. If necessary we translate $K_2'$ the shortest possible distance parallel to these support lines so as to give a $K_2''$ with an extreme point of $K_1$ in at least one of the components of $\partial K_2'' \cap \partial K_1$. If $K_2''$ is not inside $K_1$, then in moving from $K_2'$ to $K_2''$ we must have encountered a translate yielding more than two boundary components. So $K_2'' \subset K_1$, and now any further small translate in the same direction will produce more than two components. The case $K_1 \subset K_2'$ can be dealt with using the same argument.

Now put

$$R = \inf\{\lambda : \lambda K_2' \supset K_1 \quad \text{for some translate} \quad K_2' \text{ of } K_2\}$$

and

$$r = \sup\{\lambda : \lambda K_2' \subset K_1 \quad \text{for some translate} \quad K_2' \text{ of } K_2\}.$$

For $r \leqslant \lambda \leqslant R$ let $m_i(\lambda)$ denote the measure of the set of translates of $\lambda K_2$ which meet $K_1$ and yield precisely $i$ boundary components in common with $K_1$. Then for all $n$ we have as before $m_{2n+1}(\lambda) = m_\infty(\lambda) = 0$. Also since $r \leqslant \lambda \leqslant R$ we have $m_0(\lambda) = 0$. Thus Lemmas 1 and 2 give

$$\sum_{n=1}^{\infty} m_{2n}(\lambda) = V(K_1) + \lambda^2 V(K_2) + 2\lambda V(K_1, K_2^*)$$

and

$$\sum_{n=1}^{\infty} 2n m_{2n}(\lambda) = 2\lambda( V(K_1, K_2) + V(K_1, K_2^*)).$$

This gives us a quadratic satisfying

$$2\lambda V(K_1, K_2) - V(K_1) - \lambda^2 V(K_2) = \sum_{n=1}^{\infty} (n-1) m_{2n}(\lambda) \geqslant 0, \tag{4}$$

for $r \leqslant \lambda \leqslant R$. Now the hypothesis of the theorem gives $m_{2n}(1) = 0$ for $n > 1$, and so $\lambda = 1$ is a root of the quadratic. The inequality (4) shows that $r$ and $R$ must lie between the roots of the quadratic. But $r \leqslant 1 \leqslant R$ since otherwise we would have either $K_2' \subset \operatorname{int} K_1$ or $\operatorname{int} K_2' \supset K_1$ for some translate $K_2'$ of $K_2$. Consequently, either $r = 1$ or $R = 1$, which gives a translate such that either $K_2' \subset K_1$ or $K_2' \supset K_1$ respectively. But we have seen that these both imply $K_1 = K_2'$ as required. □

**Corollary 1.** *Let $K_1, K_2$ be planar convex bodies. Then $\alpha(K_1, K_2') = 2$ for all congruent copies $K_2'$ of $K_2$ if and only if $K_1$ and $K_2$ are congruent discs.*

*Proof*. It follows from Theorem 1 that $K_1$ is a translate of each rotation of $K_2$. Let $\mu(K_1; \cdot)$ and $\mu(K_2; \cdot)$ denote the surface area measures of $K_1$ and $K_2$ (see [1] for example). Let $\omega$ be a Borel subset of $[0, 2\pi]$, and denote by $K_2(\theta)$ a rotation of $K_2$ through angle $\theta$. Then the translation invariance of surface-area measures gives us

$$\mu(K_1; \omega) = \mu(K_2(\theta); \omega) = \mu(K_2; \omega - \{\theta\}) = \mu(K_1; \omega - \{\theta\})$$

for all $\theta \in [0, 2\pi]$. Hence $\mu(K_1; \cdot)$ is a translation invariant Borel measure on $[0, 2\pi]$, and so is a multiple of Lebesgue measure. Consequently $K_1$ is a disk and $K_2$ is a translate of it. $\square$

We now give our analogue of Yanagihara's result.

**Theorem 2.** *Let $K_1$ and $K_2$ be convex bodies in $\mathbb{E}^3$. Then $\partial K_2' \cap \partial K_1$ is a planar curve for all translates $K_2'$ of $K_2$ (except if $K_2' = K_1$ or $K_2' \cap K_1 = \emptyset$) if and only if $K_1$ and $K_2$ are both translates of the same ellipsoid.*

*Proof.* In one direction the result is clear. So we shall assume that for all translates of $K_2$ we know that $\partial K_2' \cap \partial K_1$ is a planar curve. Let $\pi$ denote projection onto an arbitrary plane $P$. Our first aim is to show that we can have neither

$$\pi K_2' \subset \operatorname{relint} \pi K_1 \quad \text{nor} \quad \pi K_1 \subset \operatorname{relint} \pi K_2';$$

here "relint" denotes the interior of a 2-dimensional convex body lying in $P$. For otherwise, we assume without loss of generality that $\pi K_2' \subset \operatorname{relint} \pi K_1$ and consider the infinite cylinder $C$ generated by $K_1$ and a direction $\mathbf{u}$ perpendicular to $P$. Then $\operatorname{int} C$ contains two disjoint regions $R_1, R_2$ of $\partial K_1$, both of them disjoint from the shadow boundary of $K_1$ in direction $\mathbf{u}$. We now consider translates of $K_2'$ in the direction $\pm\mathbf{u}$. From our assumption we see that all such translates lie in $\operatorname{int} C$. Also the hypothesis of the theorem shows that no such translate can lie in $\operatorname{int} K_1$. So there must be a translate $K_2''$ say, such that $\partial K_2'' \cap R_1 \neq \emptyset$ and $\partial K_2'' \cap R_2 \neq \emptyset$. But any planar section of $K_1$ which meets $R_1$ and $R_2$ must also meet the shadow boundary of $K_1$ in direction $\mathbf{u}$. Thus $K_2'' \not\subset \operatorname{int} C$, and so we have the required contradiction.

Our next objective is to show that $\pi K_1$ is a translate of $\pi K_2$. Then since $\pi$ was an arbitrary projection, it follows from [1] that $K_1$ is a translate of $K_2$. If $\pi K_1$ is not a translate of $\pi K_2$, it follows from the above argument and Theorem 1 that we can assume that $\alpha(\pi K_1, \pi K_2) > 2$. So we can choose $\mathbf{x}, \mathbf{y} \in \partial \pi K_2 \cap \partial \pi K_1$ in such a way that

$$[\mathbf{x}, \mathbf{y}] \cap \operatorname{relint} \pi K_2 \cap \operatorname{relint} \pi K_1 \neq \emptyset. \tag{5}$$

Let $Q$ be the plane containing $\mathbf{x}$ and $\mathbf{y}$ which is parallel to $\mathbf{u}$. We note that the sections $Q \cap K_1$ and $Q \cap K_2$ both have width $|\mathbf{x} - \mathbf{y}|$ in the direction perpendicular to $\mathbf{u}$. So if $K_2'$ is a translate of $K_2$ in a direction parallel to $Q$, then we cannot have either $Q \cap K_1 \subset \operatorname{relint} Q \cap K_2'$ or $Q \cap K_2' \subset \operatorname{relint} Q \cap K_1$. Thus Theorem 1 shows that if $Q \cap K_1$ and $Q \cap K_2$ are not translates of one another, we can translate $K_2$ parallel to $Q$ so as to obtain $\alpha(Q \cap K_1, Q \cap K_2') > 2$. But then the common points of the relative boundaries of $Q \cap K_1$ and $Q \cap K_2'$ uniquely define the planar section $Q$ containing $\partial K_2' \cap \partial K_1$, and so $Q \cap K_1 = Q \cap K_2'$. Hence $Q \cap K_1$ is a translate of $Q \cap K_2$, and we can let $K_2'$ be a translate of $K_2$ in a direction parallel to $Q$ chosen so that

$$Q \cap K_1 = Q \cap K_2' = \operatorname{conv}(\partial K_2' \cap \partial K_1),$$

where "conv" denotes the convex hull. We let $Q^+$ and $Q^-$ denote the open half spaces determined by $Q$ in such a way that

$$Q^+ \cap K_1 \subset \operatorname{int}(Q^+ \cap K_2') \quad \text{and} \quad Q^- \cap K_2' \subset \operatorname{int}(Q^- \cap K_1).$$

Thus

$$\pi Q^+ \cap \pi K_1 \subset \operatorname{relint}(\pi Q^+ \cap \pi K_2') \quad \text{and} \quad \pi Q^- \cap \pi K_2' \subset \operatorname{relint}(\pi Q^- \cap \pi K_1),$$

and so

$$\partial \pi K_1 \cap \partial \pi K_2' \subset \partial \pi K_1 \cap \pi Q = \{\mathbf{x}, \mathbf{y}\},$$

which shows that $\mathbf{x}, \mathbf{y} \in \partial \pi K_2'$. Now $\pi K_2'$ is obtained from $\pi K_2$ by means of a translation parallel to $\mathbf{x} - \mathbf{y}$. But since $\mathbf{x}, \mathbf{y} \in \partial \pi K_2'$, we see from (5) that $\pi K_2' = \pi K_2$ and so

$$\partial \pi K_1 \cap \partial \pi K_2 = \{\mathbf{x}, \mathbf{y}\},$$

which contradicts our original assumption that $\alpha(\pi K_1, \pi K_2) > 2$. Hence $K_1$ is a translate of $K_2$.

To complete the proof of the theorem it suffices to show that $K_1$ is an ellipsoid. We shall do this by proving that every shadow boundary of $K_1$ is planar (see [3, p. 142]). Choose an arbitrary direction $\mathbf{v}$, and for $d \in \mathbb{R}$ put $K_1 + d\mathbf{v} = K_1(d)$. Then for all small nonzero values of $d$, $\partial K_1(d) \cap \partial K_1$ is a planar convex curve. We denote by $\boldsymbol{\nu}(d)$ a unit normal to this plane and choose a nonzero sequence $(d_n)_{n=1}^{\infty}$ such that $d_n \to 0$ and $\boldsymbol{\nu}(d_n) \to \mathbf{w}$ as $n \to \infty$. We denote by $H$ the plane which is the limit of the planes $H_n$ containing $\partial K_1(d_n) \cap \partial K_1$.

Next we observe that there are no line segments in $\partial K_1$. For, if $[\mathbf{a}, \mathbf{b}]$ is a maximal line segment in $\partial K_1$, then $K_1' = \frac{1}{2}(\mathbf{b} - \mathbf{a}) + K_1$ is a translate of $K_1$ such that $\partial K_1 \cap \partial K_1'$ is a planar curve. Clearly this plane must contain $[\mathbf{a}, \mathbf{b}]$; but this is impossible, since $\mathbf{a} \in K_1 \setminus K_1'$.

Finally, we show that the shadow boundary $B(\boldsymbol{\nu})$ of $K_1$ in the direction $\boldsymbol{\nu}$ is a planar curve. If $\mathbf{p} \in H_n \cap \partial K_1$, then since $H_n \cap K_1 = H_n \cap K_1(d_n)$, there is a $\mathbf{q} \in \partial K_1$ with $\mathbf{p} = \mathbf{q} + d_n \boldsymbol{\nu}$. Thus $\mathbf{p} \notin B(\boldsymbol{\nu})$, since otherwise $\mathbf{q} \in B(\boldsymbol{\nu})$ and then $[\mathbf{p}, \mathbf{q}]$ is a line segment in $\partial K_1$, contradicting our earlier observation. Now $B(\boldsymbol{\nu})$ separates two regions $T_1, T_2$ of $\partial K_1$, and any line in direction $\boldsymbol{\nu}$ which meets $K_1$ either meets $B(\boldsymbol{\nu})$ or else intersects $\partial K_1$ at two points, one in $T_1$ and the other in $T_2$. So we can assume that $H_n \cap \partial K_1 \subset T_1$ and $J_n \cap \partial K_1 \subset T_2$, where $J_n$ is the plane $H_n - d_n \boldsymbol{\nu}$. Now $B(\boldsymbol{\nu})$ lies between these parallel planes $H_n$ and $J_n$, and so if $\mathbf{s} \in B(\boldsymbol{\nu})$ we have $\mathbf{s} = \mathbf{p}_n + \alpha_n \boldsymbol{\nu}$ for each $n$ where $\mathbf{p}_n \in H_n$ and $|\alpha_n| < |d_n|$. Thus $\mathbf{p}_n \to \mathbf{s}$ as $n \to \infty$, and so $\mathbf{s} \in H$. Consequently $B(\boldsymbol{\nu}) = H \cap \partial K_1$, as required. □

**Corollary 2.** *Let $K_1$ and $K_2$ be convex bodies in $\mathbb{E}^3$. Then $\partial K_2' \cap \partial K_1$ is a planar curve for all congruent copies $K_2'$ of $K_2$ (except if $K_2' = K_1$ or $K_2' \cap K_1 = \emptyset$) if and only if $K_1$ and $K_2$ are congruent balls.*

*Proof.* It follows from Theorem 2 that $K_1$ is an ellipsoid which is a translate of each of its rotations. Hence $K_1$ is a ball, and the same is clearly true of $K_2$. □

As a final remark we observe that it is easy to inductively extend the results of Theorem 2 and its Corollary to $\mathbb{E}^n$.

## References

[1] Aleksandrov, A. D., Zur Theorie der gemischten Volumina von konvexen Körpern. *Mat. Sbornik N.S.* **2** (1937), 947–972, 1205–1238; *Mat. Sbornik N.S.* **3** (1938), 27–46, 227–251.

[2] Bol, G., Zur kinematischen Ordnung ebener Jordan-Kurven. *Abh. Math. Sem. Univ. Hamburg* **11** (1936), 394–408.

[3] Bonnesen, T. and Fenchel, W., *Theorie der konvexen Körper*. Chelsea, New York 1948.

[4] Fujiwara, M., Ein Satz über konvexe geschlossene Kurven. *Sci. Repts. Tôhoku Univ.* **9** (1920), 289–294.

[5] Goodey, P. R. and Woodcock, M. M., A characterization of sets of constant width. *Math. Ann.* **238** (1978), 15–21.

[6] Maak, W., Schnittpunktanzahl rektifizierbarer und nichtrektifizierbarer Kurven. *Math. Ann.* **118** (1942), 299–304.

[7] Peterson, B., Do self-intersections characterize curves of constant width? *Amer. Math. Monthly* **79** (1972), 505–506.

[8] Santalo, L. A., *Integral Geometry and Geometric Probability*. Addison-Wesley Reading, Mass. 1976.

[9] Yanagihara, K., A theorem on surface. *Tôhoku Math. J.* **8** (1915), 42–44.

# An Extremal Property of Plane Convex Curves—P. Ungar's Conjecture[1]

Ignace I. Kolodner*

## 1. Introduction

Let $L$ be a simple closed, rectifiable, plane curve of perimeter $4p$. Using a continuity argument, one can prove that there exist on $L$ four consecutive points, $A$, $A'$, $B$, and $B'$, which divide the perimeter in four equal parts while the segments $AB$ and $A'B'$ are orthogonal. In March 1956, according to my recollection, Peter Ungar of CIMS (then the NYU Institute for Mathematics and Mechanics) conjectured, while studying properties of quasiconformal mappings, that:

*If $L$ is convex, then $\overline{AB} + \overline{A'B'} \geqslant 2p$, with equality iff $L$ is a rectangle.*

It soon became obvious at IMM that the conjecture will be true in general if it can be shown that it is true for quadrilaterals. (A formal proof of this reduction appears in [1].) The latter part is a problem of elementary mathematics, and was thought at first to be trivial. However, early attempts to supply a "quick" proof failed, and to this day, according to my knowledge, and that of Professor Ungar, no proof has been constructed. The objective of this work is to supply a proof.

With arbitrary $L$, the segments $[AB]$ and $[A'B']$ need not intersect. However, if $L$ is convex, then both have positive length and intersect at a point $O$ which is *interior* to both. In such a case, I call, following the terminology in [1], $[AB] \cup [A'B']$ a cross Cr, $A, A', B, B'$ its vertices, and $\overline{AB} + \overline{A'B'}$ its length; and I say that $L$ is *around* Cr. In the sequel I will have to deal with nonconvex quadrilaterals, but will always assume that they are around a cross. Given a cross Cr and a positive number $p$, $Q(\mathrm{Cr}, p)$ will denote a quadrilateral of perimeter $4p$ around

[1] Research partially supported by NSF Grant MCS76-07567.

* Department of Mathematics, Carnegie-Mellon University, Pittsburgh, PA.

Cr. Without loss of generality one may scale the problem to crosses of length 2. With this proviso, used from now on, Ungar's conjecture asserts:

**Theorem.** *If $Q(\mathrm{Cr}, p)$ exists and is convex then $p \leqslant 1$, with equality iff $Q$ is a rectangle.*

The strategy of the proof is outlined in Section 4, following the discussion of some basic results and of notation in Sections 2 and 3. The proofs of the crucial Lemmas 4 and 5 appear in Sections 5 and 6. Section 7 contains further comments about the problem.

In 1963, Chandler Davis [1] considered this problem and an analogous extremal problem involving areas. He supplied a proof of the reduction of both to cases of quadrilaterals, disclaiming originality, and gave a complete proof for the second problem, using analytic geometry. Doing the latter took about two printed pages: one for the setup, and one for the arguments. (These arguments can be really abbreviated; see Section 7.) He concluded tersely with the statement that "for the first problem to be solved in an analogous way would presumably require much messier manipulations." How right he was!

I had been aware of the problem ever since it was conjectured, and on occasion tried, when there was nothing better to do, to construct a proof by some synthetic argument. More serious attempts date from a few years ago, when I settled for less and proved Lemmas 1 and 3. The first seemed to lead nowhere; the other confirmed the conjecture in a special case when two vertices of the cross lie on one edge of the quadrilateral, but the calculation was messy due to an inappropriate notation. After reading [1], I thought of using analytic geometry but felt that this would lead to unending juggling of formulae. However, about three years ago I found an elegant way of computing the necessary and sufficient condition for the existence of $Q(\mathrm{Cr}, p)$; see Lemma 4. This again was in a notation not suitable for discussion and led nowhere. The final impetus was created when I tried to explain to a friend the frustrations occurring in mathematical research.

## 2. Preliminary Results

Refer to Figure 1, showing the cross $[A,B] \cup [A',B']$ together with an auxiliary triangle $[E,F,E']$, with lowercase letters denoting the lengths of the arms of the cross, etc. Assume that

$$0 < b \leqslant a, \qquad 0 < b' \leqslant a', \qquad p \geqslant c := \overline{AA'}\,. \tag{2.1}$$

Exclude from all further considerations the trivial case $a = b$, $a' = b'$, and $p = c$. In that case, $Q(\mathrm{Cr}, p)$ is a rhombus with its diagonals forming the cross, and

$$p = (a^2 + a'^2)^{1/2} < a + a' = 1$$

for a cross of length 2. We note that this exclusion is equivalent to the

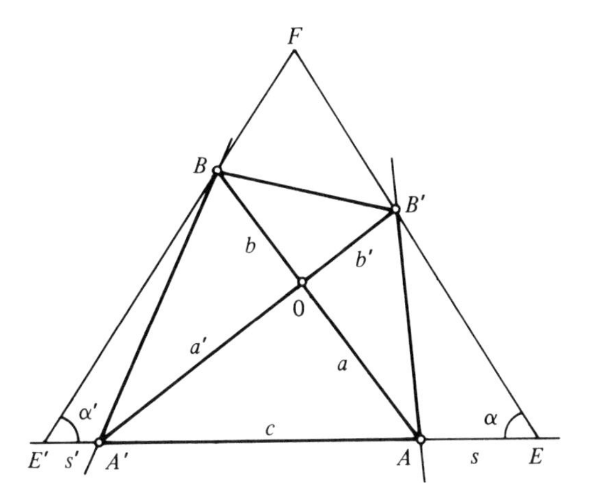


**Figure 1**

assumption

$$p > c' := \overline{BB'} .$$

The triangle $[E, F, E']$ with $A, A' \in [E, E']$, $B \in [E', F]$, $B' \in [E, F]$ is defined by the requirements

$$\overline{AE} + \overline{EB'} = p, \qquad \overline{A'E'} + \overline{E'B} = p.$$

The assumption $p > c'$ assures that the lines through $E', B$ and $E, B'$ are not parallel, so they intersect at a finite point $F$. Define

$$\Delta := p - (\overline{BF} + \overline{B'F}). \tag{2.2}$$

In the sequel we will employ the following results:

**Fact 1.** *If* $Q(\mathrm{Cr}, p)$ *exists and is convex then* $\Delta \leqslant 0$.

**Fact 2.** *If* $Q(\mathrm{Cr}, p)$ *exists and* $p > c$ *then* $\Delta \neq 0$.

These two facts are undoubtedly well known, but in any case their proof is very simple, obtained by tracking the edges of $Q(\mathrm{Cr}, p)$, beginning with a suitable vertex, and checking where the "opposite" vertex must lie. For Fact 1 begin with a vertex in $[E, E']$ if $p = c$, or with the vertex separating $A, A' \in Q(\mathrm{Cr}, p)$ if $p > c$, and conclude that the opposite vertex lies in the *closure* of the triangle $[B, F, B']$. For Fact 2 one assumes that $p > c$, $\Delta = 0$, and shows that the further assumption "$Q(\mathrm{Cr}, p)$ exists" results in a contradiction. If $Q$ is convex, start with the vertex separating $A, A'$ and deduce that the opposite vertex lies in the *interior* of the triangle $[B, F, B']$. For nonconvex $Q$'s we start with the vertex of a reentrant angle and the verification has to be made separately for the four possible locations of this vertex with respect to the vertices of the cross. Here we employ $\Delta_2$, $\Delta_3$, $\Delta_4$ defined analogously to $\Delta_1 = \Delta$ by using for the base the other edges of qu$[A, A', B, B']$, and the following fact:

**Fact 3.** *If* $\Delta_1 \leqslant 0$ *then* $\Delta_i \leqslant 0$ *for all* $i = 1, 2, 3, 4$.

Fact 3 is a consequence of our calculations in Section 3. All other details concerning Facts 1 and 2 are left to the reader.

*Remark*. Proceeding in a similar way we can prove: (i) *if* $p > c$, $\Delta < 0$, *and* $Q(\mathrm{Cr}, p)$ *exists, then* $Q(\mathrm{Cr}, p)$ *is convex*; (ii) *if* $p = c$ *then* $Q(\mathrm{Cr}, p)$ *exists iff* $\Delta \leqslant 0$; *furthermore,* $Q$ *is convex, and it is a triangle—the triangle* $[E, F, E']$—*iff* $\Delta = 0$. However, these observations have no bearing on our proof of Ungar's conjecture.

## 3. Notation

In addition to (2.1) assume that

$$\begin{gathered} d := a + b \geqslant a' + b' =: d', \\ d + d' = 2, \quad \text{the length of the cross.} \end{gathered} \tag{3.1}$$

It will be convenient to employ two additional notations for the elements of the cross:

$$\begin{aligned} a &= a_1 = \tfrac{1}{2}(1 + \rho) + \delta x \sin\theta, \\ a' &= a_2 = \tfrac{1}{2}(1 - \rho) + \delta x \cos\theta, \\ b &= a_3 = \tfrac{1}{2}(1 + \rho) - \delta x \sin\theta, \\ b' &= a_4 = \tfrac{1}{2}(1 - \rho) - \delta x \cos\theta, \end{aligned} \tag{3.2}$$

with

$$\delta := (1 - \rho^2)/4, \qquad \theta \in [0, \pi/2]. \tag{3.3}$$

Together with the second notation, we will occasionally label the vertices of the quadrangle qu$[A, A', B, B']$ with $A_1$, $A_2$, $A_3$ and $A_4$.

If the cross is specified in the first notation, then $\rho$, $\theta$, $x$ are determined by

$$\begin{gathered} \rho := d - 1 = 1 - d', \\ x \sin\theta := \frac{a - b}{2\delta}, \qquad x \cos\theta := \frac{a' - b'}{2\delta}. \end{gathered} \tag{3.4}$$

It follows then that

$$\begin{aligned} &\rho \in [0, 1[, \qquad \delta \in ]0, \tfrac{1}{4}], \\ &\theta \in [0, \pi/2], \\ &x \in [0, \bar{x}[, \end{aligned} \tag{3.5}$$

where

$$\bar{x} := \min\left\{ \frac{2}{(1-\rho)\sin\theta}, \frac{2}{(1+\rho)\cos\theta} \right\} > 1. \tag{3.6}$$

Conversely, each choice of $\rho$, $\theta$, $x$ satisfying (3.5, 6) produces a cross satisfying (2.1), (3.1).

The original notation is convenient for the calculation of $\Delta$ later in this section. The second notation is convenient for the calculations in Section 5. The third notation is used in the main part of the proof, Section 6. The proof has to be carried out for all crosses. We will carry the discussion "at fixed $\rho$, $\theta$." That is, we will pick an arbitrary $(\rho, \theta) \in [0, 1[ \times [0, \pi/2]$ and will scan through all the $x$'s in $[0, \bar{x}[$. The case $x = 0$ corresponds to symmetric crosses ($a = b$, $a' = b'$); in that case the choice of $\theta$ is immaterial.

In the sequel we will also employ the lengths of the edges of qu$[A, A', B, B']$. We have, employing the second notation,

$$c_i^2 := a_i^2 + a_{i+1}^2 \quad \text{(subscripts mod 4)}, \tag{3.7}$$

and note that

$$c := c_1 = \sqrt{a^2 + a'^2} \geqslant c_2, c_4 \geqslant c_3 = \sqrt{b^2 + b'^2} =: c'. \tag{3.8}$$

Setting

$$q := \tfrac{1}{2}\left[(1+\rho)\sin\theta + (1-\rho)\cos\theta\right], \tag{3.9}$$

we find that

$$c^2 = 1 - 2\delta + 2\delta q x + \delta^2 x^2 \tag{3.10}$$

and

$$0 < q \leqslant \bar{q} := (1 - 2\delta)^{1/2} < 1. \tag{3.11}$$

Next we turn to $Q(\mathrm{Cr}, p)$, the quadrilateral around the cross with perimeter $4p$. Defining $y$ by

$$y := (p^2 - 1)/2\delta, \tag{3.12}$$

this quadrilateral is identified (if it exists), for chosen $\rho$, $\theta$, by the pair $(x, y)$. Using $Q(x, y)$ as an alternate notation, Ungar's conjecture can be restated as:

*If $Q(x, y)$ exists and is convex, then $y \leqslant 0$. If furthermore $y = 0$, then $x = 0$ and $Q(0, 0)$ is a rectangle.*

We now proceed with a number of calculations.

### 3.1. Domain of Study

Our problem will be studied in the $(x, y)$ plane. However, certain restrictions on the pairs $(x, y)$ are immediately obvious. Firstly, the cross is defined only if $x \in [0, \bar{x}[$. Secondly, if $Q(x, y)$ exists then $p \geqslant c$. In view of (3.10) and (3.12) this

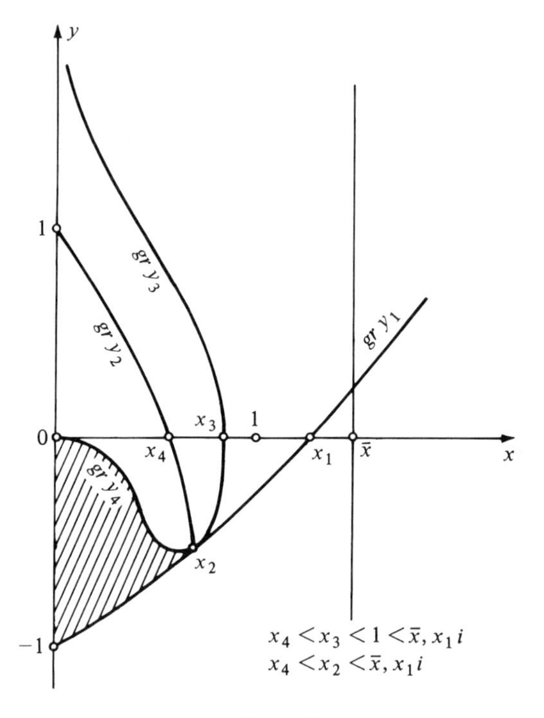


**Figure 2**

condition is equivalent to

$$y \geqslant y_1(x) := -1 + qx + \delta x^2/2. \tag{3.13}$$

Finally, the excluded trivial case corresponds to the point $(0, -1)$. Thus we are led to restrict our considerations to the set $\mathfrak{D}$ defined by

$$\mathfrak{D} := \{(x, y) \mid x \in [0, \bar{x}[, y \geqslant y_1(x)\} \setminus \{(0, -1)\}. \tag{3.14}$$

For illustrations of $\mathfrak{D}$, see Figure 2. In this figure the point where the graph of $y_1$ crosses the $x$-axis has the abscissa

$$x_1 = \frac{2}{q + \sqrt{q^2 + 2\delta}} \geqslant \frac{2}{\bar{q} + \sqrt{\bar{q}^2 + 2\delta}} = \frac{2}{1 + \sqrt{1 - 2\delta}} > 1. \tag{3.15}$$

This point need not belong to $\mathfrak{D}$, as $x_1$ and $\bar{x}$ do not compare the same way for all $\rho, \theta$. Nevertheless it will be convenient to employ this point in Section 3.4 below. Note that at this point, $p = c = 1$.

### 3.2. A Formula for $\Delta$

We define $g_1$ by

$$g_1 := (c + p)(cp - ab - a'b')\Delta \tag{3.16}$$

with $\Delta$ defined by (2.2). Since

$$cp - ab - a'b' \geqslant cp - cc' = c(p - c') > 0 \quad (\text{by C.S.}),$$

we have

$$\text{sg } g_1 = \text{sg}\,\Delta. \tag{3.17}$$

The computation of $\Delta$ and thus of $g_1$ is somewhat tedious, but is routine plane geometry, most easily achieved by employing the first notation. One gets

$$g_1 = (c + p)\left[cp^2 - 2(ab + a'b')p + c'^2c\right] - (ab' + a'b)dd'. \tag{3.18}$$

For the reader who might want to check the derivation we note that

$$\Delta = 3p + c - \overline{EFE'E},$$

$$\overline{EFE'E} = \frac{2(c + s + s')}{1 - \tan\frac{\alpha}{2}\tan\frac{\alpha'}{2}},$$

$$2s = c(p^2 - a^2 - b'^2)/\sigma,$$

$$\tan\frac{\alpha}{2} = \frac{ad'}{\sigma},$$

with

$$\sigma := c(c + p) - a'd'.$$

Formulae for $s'$ and $\tan(\alpha'/2)$ are obtained from those for $s$ and $\tan(\alpha/2)$ by interchanging $a$ with $a'$, $b$ with $b'$; the meaning of $s$, $s'$, $\alpha$, $\alpha'$ is clear from Figure 2.

### 3.3. Sign of $\Delta_i$

In order to prove Fact 2.3 we compute $g_2$, $g_3$, $g_4$, associated with $\Delta_2$, $\Delta_3$, $\Delta_4$ respectively. One obtains $g_2$ from $g_1$ by interchanging $a$ with $b$; $g_3$ by interchanging $a$ with $b$ and $a'$ with $b'$, and $g_4$ by interchanging $a'$ with $b'$. Thus we get

$$g_3 = (c' + p)\left[c'p^2 - 2(ab + a'b')p + c^2c'\right] - (ab' + a'b)dd',$$

$$g_2 = (c_2 + p)\left[c_2^2p^2 - 2(ab + a'b')p + c_4^2c_2\right] - (aa' + bb')dd'.$$

$g_4$ will be also obtained from $g_2$ by interchanging $c_2$ with $c_4$ but need not be written out here. We now compare the $g_i$ with $g_1$ and obtain, after some manipulations,

$$g_1 - g_3 = p(c - c')\left[(p - c)(2c + c' + p) + 2a(a - b) + 2a'(a' - b')\right] \geqslant 0,$$

$$g_1 - g_2 = p(c - c_2)\left[(p - c)(2c + c_2 + p) + 2a(a - b) + (a' - b')^2\right] \geqslant 0.$$

Likewise, it follows that $g_1 - g_4 \geqslant 0$. We conclude:

*If $g_1 \leqslant 0$ then $g_i \leqslant 0$ for all $i$.*

This proves Fact 2.3, since sg $g_i = \text{sg}\,\Delta_i$ for all $i = 1, 2, 3, 4$.

### 3.4. Where Is $\Delta \leqslant 0$?

In order to carry out the proof of the conjecture we will need, among other things, a description of the subset of $\mathfrak{D}$ on which $\Delta \leqslant 0$. For this purpose we express $g_1$ in the third notation and define the function $(x, y) \mapsto g(x, y)$ by

$$g(x, y) := g_1/4\delta^2 = p(p+c)x^2 - 2 + \delta x^2 \sin 2\theta + (p^2 - c^2)\frac{4\delta q x + c(p-c)}{4\delta^2}, \tag{3.19}$$

where $p, c$ are expressed in terms of $x, y$.

In view of (3.17), $\operatorname{sg}\Delta = \operatorname{sg} g$. Of course, we consider only $g$ on $\mathfrak{D}$; however, the formula (3.19) makes sense at all points where $p$ and $c$, given by (3.10) and (3.12), are nonnegative, in particular for all $y \geqslant -1/2\delta$, $x \geqslant 0$.

In order to study $g$, first compute its first partial derivatives,

$$\frac{\partial g}{\partial x}(x, y) = \left[(2p+c)x^2 + \frac{p(4\delta q x + c(p-c))}{2\delta^2} + \frac{(p^2-c^2)c}{4\delta^2}\right]\frac{\delta}{p}, \tag{3.20}$$

$$\frac{\partial g}{\partial y} = \frac{qp(p-c)(p+3c)}{4\delta c} + x\varphi, \tag{3.21}$$

where

$$\varphi := p(p^2 - c^2)/4c + (q + \delta x)\delta x p/c + \delta^2 x^2 + 3p(p+c)/2 + \rho\cos 2\theta.$$

$\partial g/\partial x$ is obviously positive in $\mathfrak{D}$. The same is seen concerning $\partial g/\partial y$ after observing that

$$3p(p+c)/2 \geqslant 3c^2 \geqslant 3(1-2\delta) \geqslant \tfrac{3}{2}, \qquad \rho\cos 2\theta > -1.$$

(Both derivatives vanish at the excluded point $(0, -1)$.)

Next, consider $g$ with $x = 0$,

$$g(0, y) = \frac{c(p^2 - c^2)(p-c)}{4\delta^2} - 2 = \frac{(y+1)^2}{1 + p/c} - 2, \tag{3.22}$$

where

$$\frac{p}{c} = \left[\frac{1 + 2\delta y}{1 - 2\delta}\right]^{1/2}.$$

Since $1 \leqslant p/c \leqslant \sqrt{2+y}$ for all $\delta \in ]0, \frac{1}{4}]$, we get

$$\frac{(y+1)^2}{1 + \sqrt{2+y}} - 2 \leqslant g(0, y) \leqslant \frac{(y+1)^2 - 4}{2} \tag{3.23}$$

for all $\delta$, $y \geqslant -1$. From this it follows that

$$g(0, 1) \leqslant 0, \qquad g(0, 1.383) > 0.000036,$$

so $g(0, \cdot)$ has a zero, $\eta_0$, with $\eta_0 \in [1, 1.383]$ for each $\delta$. Since $g(0, \cdot)$ is strictly increasing, $\eta_0$ is its only zero and we conclude:

$$g(0, y) \gtreqless 0 \quad \text{iff} \quad y \gtreqless \eta_0.$$

Finally consider $g$ with $y = y_1(x)$, i.e. at points of $\mathfrak{D}$ where $p = c$. From (3.19) we get

$$w(x) := g(x, y_1(x)) = 2c^2x^2 - 2 + \delta x^2 \sin 2\theta. \tag{3.24}$$

Now $w(0) = -2$, and

$$w(x_1) = g(x_1, 0) = 2(x_1^2 - 1) + \delta x_1^2 \sin 2\theta > 0,$$

since $c(x_1) = 1$ and $x_1 > 1$; see (3.15). Since $w$ is strictly increasing, there exists then a unique $x_2 \in ]0, x_1[$ such that $\operatorname{sg} w(x) = \operatorname{sg}(x - x_2)$. Furthermore, $y_1(x_2) < y_1(x_1) = 0$, since $y_1$ is strictly increasing.

Combining the above information and employing the implicit-function theorem, we now conclude:

*There exists a smooth, strictly decreasing function $y_2 : [0, x_2] \to \mathbb{R}$ with $y_2(0) = \eta_0 \in [1, 1.4]$, $y_2(x_2) = y_1(x_2) \in ]-1, 0[$, such that $g(x, y) = 0$ iff $y = y_2(x)$, and $g(x, y) < 0$ iff $y < y_2(x)$ with $x < x_2$.*

*Remark.* We observed previously that $x_1$ will not be in the domain of the cross for all choices of $\rho$, $\theta$. This, of course, has no bearing on the proof just given, but so far we do not know whether the cross is defined for $x = x_2$. One can check, however, with some effort, that $w(\bar{x}) > 0$ also, and this implies that $x_2 \in [0, \bar{x}[$.

The graph of $y_2$ is indicated on Figure 2. Note that $y_2$ vanishes at exactly one point $x_4 < x_2$.

## 4. Organization of the Proof

Our proof will be a consequence of the following five lemmas.

**Lemma 1.** *If $Q(x, y)$ exists and is convex, then $x \leqslant x_2$ and $y \leqslant y_2(x)$.*

**Lemma 2.** *If $Q(x, y)$ exists (but is not necessarily convex) and $p > c$, then $y \neq y_2(x)$.*

Lemmas 1 and 2 follow from Facts 2.1, 2.2 and the conclusion in Section 3.4. Lemma 2 is employed only in the proof of Lemma 5 below.

The next lemma asserts the truth of the Ungar's conjecture in a special case.

**Lemma 3.** *If $p = c$ and $Q(x, y)$ exists and is convex, then $y < 0$.*

*Proof.* The premises and Lemma 1 imply that $y = y_1(x)$ and $x \leqslant x_2$. Since $y_1$ is increasing, $y_1(x) \leqslant y_1(x_2) < 0$. □

*Remark.* Since $y_2(x) < 0$ for all $x \in ]x_4, x_2]$, Lemma 1 implies the truth of the conjecture for all such $x$'s.

The last two lemmas employ yet another function $h : \mathfrak{D} \to \mathbb{R}$ defined by

$$h(x, y) := z(ky^3 + ly^2 + my + n/4) - y(y+2), \tag{4.1}$$

where $z$ and the coefficients are

$$\begin{aligned} z &:= x^2, \\ k &:= 2\delta, \\ l &:= 1 + 4\delta + 2\delta^2 z, \\ m &:= 2 - \delta + \delta z(2 + \rho\cos 2\theta) + \delta^3 z^2/2, \\ n &:= \delta^2 z^2 + 2z\left(1 + \rho\cos 2\theta - \tfrac{1}{2}\delta^2\cos^2 2\theta\right) - 2(1 + \rho\cos 2\theta). \end{aligned} \tag{4.2}$$

**Lemma 4.** *Assume $p > c$. Then $Q(x, y)$ exists (but is not necessarily convex) iff $h(x, y) \geqslant 0$. If $h(x, y) = 0$, then $Q(x, y)$ is uniquely determined by the pair $(x, y)$. (If $h(x, y) > 0$, then there exist exactly two such $Q$'s.)*

**Lemma 5.** *If $p > c$, $x > 0$, $y \leqslant y_2(x)$, and $h(x, y) \geqslant 0$, then $y < 0$.*

The proofs of Lemmas 4 and 5 appear in Sections 5 and 6.

*Proof of Ungar's Conjecture.* Lemma 3 asserts its truth in case $p = c$. Assume then that $p > c$ and $Q(x, y)$ exists and is convex. Lemma 1 implies that $x \leqslant x_2$ and $y \leqslant y_2(x)$; Lemma 4 implies that $h(x, y) \geqslant 0$. If $x > 0$ also, then all premises of Lemma 5 are satisfied, so $y < 0$.

In the simple case $x = 0$ (symmetric cross), $h(0, y) = -y(y+2)$ with $p > c$, i.e. $y > -1$, so $Q(x, y)$ exists iff $y \in (-1, 0]$, implying $y \leqslant 0$. In the extreme case $y = 0$, we have $h(0,0) = 0$, so $Q(0,0)$ is uniquely determined—obviously a rectangle with sides parallel to the arms of the cross. □

## 5. Determination of $h$

Refer to Figure 3, which illustrates a method for constructing a quadrilateral around the cross Cr, and gives the notation used here. Starting at some vertex of Cr, say $A_1$, we pick a ray emanating from $A_1$, making an angle $\alpha_1$ with the ray from $A_1$ to $A_2$, and determine on it a point $V_1$ by the requirement $\overline{A_1V_1} + \overline{V_1A_2} = p$; this point exists uniquely under the assumption $p > c$. The construction determines the angle $\beta_1$, and in succession the angle $\alpha_2 = \pi - \beta_1 - 2\theta_1$, denoted by $M_1(\alpha_1)$. (The orientations indicated in Figure 3 are so chosen that $\operatorname{sg}\alpha_1 = \operatorname{sg}\beta_1$.) Proceeding the same way around the cross, we get $\alpha_3 = M_2(\alpha_2)$, $\alpha_4 = M_3(\alpha_3)$. In case $M_4(\alpha_4) = \alpha_1$, our construction produces a quadrilateral around the cross. From this we conclude that *the quadrilaterals $Q(\mathrm{Cr}, p)$ are in one-to-one correspondence with the fixpoints of $M = M_4 \circ M_3 \circ M_2 \circ M_1$.*

Considering the number of parameters and the complexity of $M$, the program just proposed for the construction of $Q(\mathrm{Cr}, p)$ seems completely impractical. There are however, a few fortunate breaks which result in involved, yet manageable calculations.

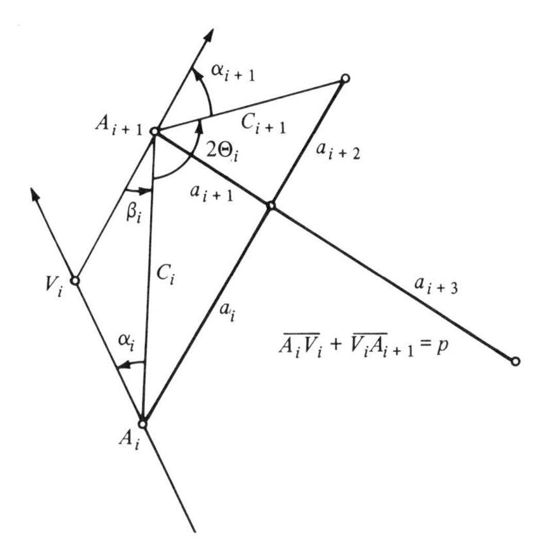


**Figure 3**

### 5.1. Existence of $Q(\mathrm{Cr}, p)$

First note that in handling problems of this type it is better to employ the tangents of half angles,

$$t_i := \tan\frac{\alpha_i}{2}, \qquad s_i := \tan\frac{\beta_i}{2} \tag{5.1}$$

in place of the angles themselves. This imposes a limitation, since it excludes the possibilities $\alpha_i$, $\beta_i = 0$ or $\pi$; we assume this for the moment, and will resolve the difficulty later.

With $\alpha_i \neq 0, \pi$, the requirement $\overline{A_iV_i} + \overline{V_iA_{i+1}} = p$ implies that $\beta_i$ satisfies

$$t_is_i = h_i := \frac{p - c_i}{p + c_i}. \tag{5.2}$$

Since $\alpha_{i+1} = \pi - (2\theta_i + \beta_i)$, we now get

$$t_{i+1} = \cot(\theta_i + \beta_i/2) = \frac{1 - s_i\tan\theta_i}{s_i + \tan\theta_i} = \frac{t_i - h_i\tan\theta_i}{t_i\tan\theta_i + h_i}. \tag{5.3}$$

Recognizing that the mapping $t_i \mapsto t_{i+1}$ is fractional linear (with determinant $h_i/\cos^2\theta_i \neq 0$), we opt to view $t_i$ not as a number, but as a point in the real projective one-space represented by nontrivial pairs $(t_i^1, t_i^2)$ in $\mathbb{R}^2$ with $t_i^1/t_i^2 = t_i$ if $t_i^2 \neq 0$. Using this view, and writing $t_i$ again for the defining pair, we see that the assignment (5.3) is equivalent to.

$$t_{i+1} = T_it_i, \qquad T_i := \begin{pmatrix} (p + c_i)\cos\theta_i & -(p - c_i)\sin\theta_i \\ (p + c_i)\sin\theta_i & (p - c_i)\cos\theta_i \end{pmatrix} \tag{5.4}$$

Note that the previously excluded values $0, \pi$ for $\alpha_i$ are now included by taking for the domain of $t_i$ the whole projective one-space, with $(0, 1)$ corresponding to $\alpha_i = 0$, and $(1,0)$ to $\alpha_i = \pi$. The reader may check that these values of $t_i$ result in correct values of $t_{i+1}$.

Proceeding as with $M$, we are now led to consider the fixpoint equation

$$t = Tt, \qquad T = T_4 T_3 T_2 T_1 \tag{5.5}$$

in the real projective one-space. This equation has solutions iff $T$ has *real* eigenvalues, and the solutions are the corresponding eigenvectors. This now leads to the following conclusion concerning $Q(\mathrm{Cr}, p)$ expressed in terms of

$$\Delta := (\operatorname{tr} T)^2 - 4 \det T: \tag{5.6}$$

*Assume that $p > c$. Then $Q(\mathrm{Cr}, p)$ exists iff $\Delta \geqslant 0$; furthermore, there is exactly one such $Q$ if $\Delta = 0$, and exactly two if $\Delta > 0$.*

It remains to compute $\Delta$. It will be shown that up to a positive factor, $\Delta$ is the same as $h$ specified by (4.1, 2). This will constitute then the proof of Lemma 4.

## 5.2. Computation of det $T$

To compute $\Delta$ we employ the second (subscript) notation, which is already used in the description of the matrices $T_i$, and introduce yet another notation for temporary purposes, setting

$$a_1 =: \lambda + \sigma, \quad a_2 =: \mu + \tau, \quad a_3 =: \lambda - \sigma, \quad a_4 =: \mu - \tau, \tag{5.7}$$

$$\begin{aligned} r &:= \sigma^2 + \tau^2 = \delta^2 x^2, \\ v &:= p^2 - \lambda^2 - \mu^2 - r = 2\delta(y + 1 - \delta x^2/2). \end{aligned} \tag{5.8}$$

(For the reduced cross, $\lambda + \mu = 1$. However, it is preferable not to use this simplification, in order not to destroy the homogeneity of the expressions that occur.) In the new notation we have

$$c_i^2 = \lambda^2 + \mu^2 + r + 2(\pm\lambda\sigma \pm \mu\tau), \tag{5.9}$$

with the choice of signs depending on $i$.

The determinant det $T$ is easily computed. We get

$$\begin{aligned} \det T &= \prod_i \det T_i = \prod_i (p^2 - c_i^2) \\ &= v^4 - 8(\lambda^2\sigma^2 + \mu^2\tau^2)v^2 + 16(\lambda^2\sigma^2 - \mu^2\tau^2)^2, \end{aligned} \tag{5.10}$$

after some simplifications.

## 5.3. Computation of tr $T$

To compute tr $T$, at last we have to multiply out the $T_i$'s. However, even this can be done rather neatly, and several shortcuts will occur also. Although $T$ itself is not invariant under circular permutations of the $a_i$'s, its trace, of course, will be invariant.

First observe that $T_i$ can be factored in the form

$$T_i = R(\theta_i)(pI + c_i R), \tag{5.11}$$

where

$$\begin{aligned} R(\theta) &:= \begin{pmatrix} \cos\theta & -\sin\theta \\ \sin\theta & \cos\theta \end{pmatrix} \quad \text{is a rotation matrix,} \\ R &:= \begin{pmatrix} 1 & 0 \\ 0 & -1 \end{pmatrix} \quad \text{is a reflection matrix.} \end{aligned} \tag{5.12}$$

Noting then the product relations

$$R(\theta)R(\varphi) = R(\theta + \varphi), \qquad R(\theta)R = RR(-\theta), \qquad R^2 = I, \tag{5.13}$$

we now deduce, by expanding the product $T$ as a polynomial in $p$ and moving $R$ always to the right, that

$$T = \sum_{\nu=0}^{4} p^{\nu} S_{\nu} R^{\nu}. \tag{5.14}$$

Since, as follows from (5.11), the $S_\nu$ are linear combinations of rotation matrices, they are all of form

$$S_\nu = u_\nu I + v_\nu R(\pi/2), \tag{5.15}$$

whence

$$\operatorname{tr} S_\nu = 2u_\nu, \qquad \operatorname{tr}(S_\nu R) = 0.$$

It follows then that

$$\operatorname{tr} T = 2(u_0 + u_2 p^2 + u_4 p^4), \tag{5.16}$$

and so we have to evaluate just the three coefficients, $u_0$, $u_2$, and $u_4$.

The matrices $S_\nu$ are best evaluated by "moving fingers" and employing the product relations (5.13), also remembering that $\sum \theta_i = \pi$. Thus we obtain

$$\begin{aligned} S_0 &= \Big(-\prod_i c_i\Big) R(2(\theta_2 + \theta_4)). \\ S_2 &= -\Bigg[\sum_{i=1}^{3} c_i c_{i+1} R(-2\theta_i) + c_4 c_1 R(2\theta_4) \\ &\qquad + c_1 c_3 R(-2(\theta_1 + \theta_2)) + c_2 c_4 R(-2(\theta_2 + \theta_3))\Bigg], \\ S_4 &= -I. \end{aligned} \tag{5.17}$$

Thus, $u_4 = -1$. To compute $u_0$, $u_2$, first obtain the formulae

$$\begin{aligned} \cos 2\theta_i &= \frac{c_i^2 + c_{i+1} - (a_i + a_{i+2})^2}{2 c_i c_{i+1}} = \frac{a_{i+1}^2 - a_i a_{i+2}}{c_i c_{i+1}}, \\ \sin 2\theta_i &= \frac{(a_i + a_{i+2}) a_{i+1}}{c_i c_{i+1}}. \end{aligned} \tag{5.18}$$

Employing these, we finally obtain

$$\begin{aligned} u_0 &= a_1a_3(a_2+a_4)^2 + a_2a_4(a_1+a_3)^2 - (a_1a_3+a_2a_4)^2 \\ &= 8\lambda^2\mu^2 + 4(\lambda^2\sigma^2+\mu^2\tau^2) - (\lambda^2+\mu^2+r)^2, \\ u_2 &= 4(a_1a_3+a_2a_4) - \sum_i a_i^2 = 2(\lambda^2+\mu^2) - 6r. \end{aligned} \tag{5.19}$$

Substitution in (5.16) now yields

$$\operatorname{tr} T = -2\left[v^2 + 8rv - 8\lambda^2\mu^2 - 4(\lambda^2\sigma^2+\mu^2\tau^2) + 8r(\lambda^2+\mu^2+r)\right]. \tag{5.20}$$

### 5.4. Conclusion

The remainder of the calculation is tedious. We substitute (5.10) and (5.20) in (5.6), simplify as much as possible (note that the terms in $v^4$ cancel out), and pass to the third notation. The result is $\Delta = 256\delta^4 h(x, y)$, with $h$ as specified in (4.1) and (4.2).

## 6. Proof of Lemma 5

In order to prove Lemma 5 we need some information about the subset of $\mathfrak{D}$ on which $h \geqslant 0$. Since $h$ is a rather complicated polynomial of degree 4 in $(y, z)$ with coefficients depending on two parameters, we are unable to describe this set completely. However, we will obtain sufficient information for the proof of the Lemma. Our result, proven in Sections 6.1–3 below, is:

*There exists a number* $x_3 \in ]0, 1]$ *and a smooth, strictly decreasing function* $y_3 : ]0, x_3[ \to \mathbb{R}$ *with* $\lim_0 y_3 = \infty$, $\lim_{x_3} y_3 = 0$ *such that if* $x \in ]0, x_3[$ *and* $y \geqslant 0$, *then* $\operatorname{sg} h(x, y) = \operatorname{sg}(y - y_3(x))$.

*Proof of the Lemma*. Refer to $y_2$ defined and described in Section 3.4. We first show that $x_4 < x_3$ and on $]0, x_4]$ we have $y_2(x) < y_3(x)$. Since $\lim_0 y_3 = \infty$, we have $y_3(\xi) > 1.5 > y_2(\xi)$ with some $\xi < x_3, x_4$, as $y_2(x) < 1.5$ always. Now the possibility $y_2(\eta) = y_3(\eta)$ is excluded, since this would imply that $Q(\eta, y_3(\eta))$ exists (as $p > c$ and $h(\eta, y_3(\eta)) = 0$), while Lemma 3 asserts then that $y_3(\eta) \neq y_2(\eta)$. Thus $y_3 - y_2$ does not vanish anywhere (on the intersection of their domains), and since it is continuous, it is strictly positive. Since $y_2(x_4) = 0$, $\lim_{x_3} y_3 = 0$, we conclude that $x_4 < x_3$.

Assume now the premises of the lemma. If $x > x_4$, the conclusion follows already from Lemma 1. Assume then that $x \in ]0, x_4]$. In view of our result about $y_3$, $h(x, y) \geqslant 0$ implies that either $y < 0$ or $y \geqslant y_3(x)$. Since $y \leqslant y_2(x) < y_3(x)$, it follows that $y < 0$. □

For illustration of the above result see Figure 2.

### 6.1. Sign of $h(x, 0)$

We now turn to study of $h$ with $x > 0$, $y \geqslant 0$. Consider $z \mapsto n(z)$, Equation (4.2), with $z \geqslant 0$. Since $\delta^2 > 0$ and

$$1 + \rho \cos 2\theta - \tfrac{1}{2}\delta^2 \cos^2 2\theta \geqslant 1 - \rho - \tfrac{1}{2}\delta^2 \geqslant 7(1-\rho)/8 > 0,$$

$n(\cdot)$ increases with $z$. Since $n(0) < 0$ and $n(1) = \delta^2 \sin^2 2\theta \geqslant 0$, there exists a unique $x_3 \in ]0, 11]$ such that $n(x_3^2) = 0$ while $n(z) < 0$ if $z < x_3^2$. This is the number $x_3$ employed in the assertion at the beginning of this section. We now restrict the study of $h$ to $x$'s in $]0, x_3[$ and note that

$$\begin{aligned} &h(x, 0) < 0 \quad \text{for all } x \in ]0, x_3[, \\ &h(x_3, 0) = 0. \end{aligned} \tag{6.1}$$

Since $\bar{x} > 1$ we have $x_3 < \bar{x}$, so our restriction is to subset of $\mathfrak{D}$.

### 6.2. Determination of $y_3$

Fix $x \in ]0, x_3[$ and consider the function

$$y \mapsto \varphi(y) := h(x, y), \qquad y \geqslant 0. \tag{6.2}$$

This is a cubic polynomial satisfying, in view of (6.1),

$$\varphi(0) < 0. \tag{6.3}$$

Now

$$\begin{aligned} \varphi''(0) - \varphi'(0) &= (2zl - 2) - (zm - 2) = z(2l - m) \\ &= \delta z\big(9 + 2\delta z - z(2 + \rho \cos 2\theta) - \delta^2 z^2/2\big) \\ &\geqslant 5\delta z > 0, \end{aligned} \tag{6.4}$$

employing the fact that $z = x^2 < x_3^2 \leqslant 1$. Thus $\varphi''(0) > \varphi'(0)$, and this implies that the point $y = 0$ lies to the right of the point where $\varphi$ could reach its local maximum if it had one. (To the left of this point we have $\varphi'' < 0 < \varphi'$.) Since the coefficient of $y^3$ in $\varphi$ is positive, either $\varphi$ increases from $\varphi(0) < 0$ to $\infty$, or $\varphi$ first decreases to a local minimum and then increases to $\infty$. We conclude then that *there exists uniquely a number $y_3 > 0$ such that* $\operatorname{sg} \varphi(y) = \operatorname{sg}(y - y_3)$ *if $y \geqslant 0$ and that $\varphi'(y_3) > 0$.*

Proceeding the same way at each $x \in ]0, x_3[$, one creates the function $x \mapsto y_3(x) : ]0, x_3[ \to \mathbb{R}$ with the following property: *if $y > 0$ then*

$$\operatorname{sg} h(x, y) = \operatorname{sg}(y - y_3(x)) \quad \text{and} \quad \frac{\partial h}{\partial y}(x, y_3(x)) > 0.$$

### 6.3. Conclusion of Proof

We just noted that

$$\frac{\partial h}{\partial y}(x, y_3(x)) > 0.$$

On inspecting (4.1, 2) we see that

$$\frac{\partial h}{\partial x}(x, y) > 0 \quad \text{if } x > 0, y \geqslant 0.$$

From this we can easily deduce, by employing the implicit-function theorem and the uniqueness of $y_3$, that $y_3$ *is smooth and strictly decreasing*.

It is obvious now that $\lim_{x_3} y_3 = 0$. To show that $\lim_0 y_3 = \infty$ it will suffice to show that $y_3$ takes on arbitrarily large positive values. But this is obvious from (4.1), since for each $\eta \in \mathbb{R}^+$ we have $-\eta(\eta+2) < 0$, so we can achieve $h(\xi, \eta) < 0$ by picking $\xi$ sufficiently small, and this implies that $y_3(\xi) > \eta$.

## 7. Comments and Addenda

Our proof is now complete, and one wonders whether it can be simplified in an essential way. Of course, we can cut out a line here and there by waving hands more vigorously. I do not believe that the computation of $h$ in Section 5—the crux of this work—can be simplified materially. Also the notation used here seems to be optimal for our strategy of proof.

However, the above refers to details, not to the strategy of proof. In proving an extremal property, one usually starts with the assumed extremal situation and proceeds to look at what happens when one deviates slightly from this situation. In the present problem the extreme situation occurs at the points $(0,0) \in \mathfrak{D}$. When trying to move away from it, I found the problem intractable. Even if this were successful, our pains would not be over, considering the conjectured shape of the set of pairs $(x, y)$ on which $Q(x, y)$ exists and is convex—the shaded region in Figure 2.

We conclude with a number of observations.

### 7.1. Triangles Around a Cross

We can consider an analogous problem for triangles around a cross. This much simpler problem can be considered separately. However, our study includes this problem—$Q(\mathrm{Cr}, p)$ is a triangle iff $p = c$ and $\Delta = 0$—and shows that for each $(\rho, \theta)$ there exists exactly one cross with a triangle around it: the cross is specified by $x_2$, the unique solution of $w(x) = 0$ (see Equation (3.24)), and its perimeter corresponds to $y = y_1(x_2)$. We have shown that $y_1(x_2) < 0$, but it varies with $\rho$, $\theta$, and one can easily check that it can have values arbitrarily close to 0. Thus $y_1(x_2)$ does not attain a maximum. Does it attain a minimum? Even this question cannot be settled trivially, since $w$ does not depend monotonely on $\rho, \theta$, although it is increasing with respect to $x$. However, using continuity arguments one can show quite easily that a minimum is attained. This happens—I am (almost) sure —in a symmetric situation resulting in an isosceles triangle. There are two possibilities: (i) $a' = a$ and $b' = b$, implying $\rho = 0$, $\theta = \pi/4$, and (ii) $b = a$, implying that one vertex of the cross is a vertex of the triangle, and $\theta = 0$ or $\pi$. In case (i) we employ $w$ to find $x_2$, and then $p$ can be computed. Even in this simple case we have to put up with a quartic equation, but with the little hand computers readily available one finds in a few minutes that $x_2 = 0.9838$, leading to $p = 0.9530$. In case (ii) one still has to determine $\rho$, and it is altogether simpler to bypass our setup and deal directly. For the cross we find $a = 0.7395$, $a' = 2a - 1 = 0.4791$, $b' = 3 - 4a = 0.0419$; for the triangle one gets base = 1.6073, leg = 0.9581, and $p = 0.8811$—the minimum.

*Remark*. The minimum problem for *all* quadrilaterals around a cross of length 2 is a trivial problem, since by C. S., $p \geqslant 1/\sqrt{2}$. In the present setup the minimum at a given $\rho$ occurs at $(0, -1)$, and $Q(0, -1)$ is a rhombus, with $p^2 = 1 - 2\delta$. The absolute minimum is attained for a square ($\rho = 0$, $x = 0$) with the cross formed by its diagonals, and we have then $p = 1/\sqrt{2}$.

### 7.2. Domain of Existence of $Q(x, y)$

In this paper we have been concerned only with proving Ungar's conjecture. In the process of proof we had to consider the possibility of construction of an $Q(x, y)$. For those who are interested in such constructions, the problem has not been resolved, since we did not determine the subset $\mathbb{S}$ of $\mathfrak{D}$ on which $Q(x, y)$ exists and is convex. On the basis of the partial evidence that follows, I conjecture that $\mathbb{S}$ looks like the shaded part of Figure 2 when $c > c_i$ for $i > 1$.

(a) It has been established already that if $(x, y) \in \mathbb{S}$, then $y < 0$ if $x > 0$, and even $y < y_2(x)$ for $x \in [x_4, x_2[$; $\overline{\mathfrak{B}}_0 \subset \mathbb{S}$ and $\overline{\mathfrak{B}}_1 \subset \mathbb{S}$, where $\overline{\mathfrak{B}}_0$ is the closure of

$$\mathfrak{B}_0 := \{0\} \times [-1, 0[ \quad \text{(a segment of the } y\text{-axis)},$$

and $\overline{\mathfrak{B}}_1$ is the closure of

$$\mathfrak{B}_1 := \{(x, y_1(x)) \,|\, x \in [0, x_2[\} \quad \text{(an arc of gr } y_1).$$

(b) We show that $\mathbb{S}$ contains a neighborhood of $\mathfrak{B}_0 \cup \mathfrak{B}_1$. Since $h(0, y) = -y(y + 2) > 0$ if $y \in [-1, 0[$, $h > 0$ on some neighborhood of $\mathfrak{B}_0$. The remainder will follow if we show that $h > 0$ on $\mathfrak{B}_1$. (Although $h$ loses its significance on $\mathfrak{B}_1$, it is still defined there.) For this purpose we obtain a suitable representation for $h$.

Since $p^2 - c^2 = 2\delta(y - y_1(x))$, Equation (5.10) implies that

$$\det T = 2\delta(y - y_1(x)) \prod_{i=1}^{3} (p^2 - c_i^2),$$

with the last product *strictly positive* on $\mathfrak{D}$. Next, noting that $v = 2\delta(y - y_1(x) + qx)$ (see Equation (5.8)), we obtain from (5.20), after a relatively simple calculation,

$$\operatorname{tr} T = -8\delta^2\big[(y - y_1(x))(y + 1 + qx + 7\delta x^2/2) + w(x)\big],$$

where $w$ is defined by (3.24). Combining these, we deduce that

$$h(x, y) = \tfrac{1}{4}\big[w(x) + (y - y_1(x))\varphi(x, y)\big]^2 - (y - y_1(x))\psi(x, y), \qquad (7.1)$$

with $\varphi, \psi$ being polynomials that remain positive in $\mathfrak{D}$. Thus

$$h(x, y_1(x)) = (w(x)/2)^2. \qquad (7.2)$$

It was shown in Section 3.4 that $w$ vanishes only at $x_2$, so (7.2) implies that $h > 0$ on $\mathfrak{B}_1$.

(c) Using what has already been proven, one now concludes that for each $x \in ]0, x_2[$, $h(x, \cdot)$ vanishes at a unique point $y_4(x)$ satisfying $y_1(x) < y_4(x) < y_2(x)$. Employing the implicit-function theorem we can now prove that $y_4$ is a smooth function. Finally we obtain the slopes of the graph of $y_4$ by considering $h(x, y_4(x)) = 0$. The slope at 0 turns to be 0, while (7.1) shows that $\lim_{x_2} y_4 = y_1(x_2)$ and at $(x_2, y_1(x_2))$ the graph of $y_2$ is tangent to the graph of $y_1$.

### 7.3. A Limiting Case

It may be instructive to consider the limiting case $\rho = 1$, i.e. $\delta = 0$. Then $q = \sin\theta$, and we get, after some manipulations,

$$g(x, y) = \tfrac{1}{2}(y + 1 + x\sin\theta)^2 - 2(1 - x^2\cos^2\theta),$$

$$h(x, y) = (x^2 - 1)(y^2 + 2y + x^2\cos^2\theta).$$

Assume also that $\theta \neq 0$, or $\pi/2$. Then:

$$y_1(x) = -1 + x\sin\theta, \qquad x_1 = 1/\sin\theta,$$

$$y_2(x) = -1 - x\sin\theta + 2\sqrt{1 - x^2\cos^2\theta}\,, \qquad x \in [0, 1],$$

$$y_4(x) = -1 + \sqrt{1 - x^2\cos^2\theta}\,, \qquad x \in [0, 1],$$

while the counterpart of the graph of $y_3$ is now the line $\{1\} \times [-1 + \sin\theta, \infty[$. All these graphs meet at exactly one point, $(1, -1 + \sin\theta)$; they form a pretty picture: see Figure 4. It should be noted that this picture, obtained by letting $\rho \to 1$ in the formulae for $g$ and $h$, has no bearing on the situation with a degenerate cross, with $a' = b' = 0$; for such a "cross" we need a new notation.

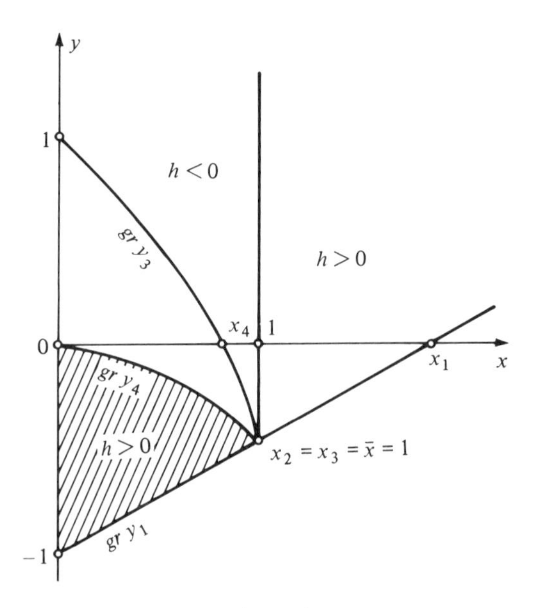


**Figure 4**

### 7.4. Comparison with Work of C. Davis

Up to a point the strategy of our proof is the same as that employed by Davis in [1] in settling fully the analogous problem with areas. We will now review his (rather simple) proof and will explain why that kind of simplicity cannot be expected in our case.

The problem with areas concerns a simple closed plane curve $L$ bounding a region with area $\mathcal{Q}(L)$, which is now around the cross $[AB] \cup [A'B']$ dividing the area in four equal parts. The conjecture is:

*If $L$ is convex, then $\overline{AB} \times \overline{A'B'} \geqslant \mathcal{Q}(L)$, with equality iff $L$ is a rectangle.*

The truth of this conjecture is implied by the corresponding theorem for convex quadrilaterals. Since ratios of areas remain invariant under affine transformations, it suffices to consider crosses with $\overline{AB} = \overline{A'B'} = 2$. In that case we denote by $Q(\mathrm{Cr}, A)$ the quadrilateral of area $4A$ around Cr and have to prove the following.

**Theorem.** *If $Q(\mathrm{Cr}, A)$ exists and is convex, then $A \leqslant 1$, with equality iff $Q$ is a square.*

The notation employed by Davis is the alternate notation given here in (5.7) with $\lambda = \mu = 1$, $\sigma, \tau \in [0, 1[$; there are only two parameters to contend with (instead of three in our case), but this is only a minor simplification. His strategy consists of constructing $Q(\mathrm{Cr}, A)$ in a manner analogous to that explained here at the beginning of Section 5, with reference to Figure 3. In our case it was natural to employ the angles $\alpha_i$ to fix the vertices $V_i$; Davis employs different parameters, $\lambda_i$, with a geometrical significance, and deduces that $Q(\mathrm{Cr}, A)$, *with exactly one vertex in each quadrant of the cross, exists iff the $\lambda$'s are nonnegative and satisfy the equations* (2) *in* [1, p. 184]. (While there exist $Q$'s with no vertex in one of the quadrants and two vertices in another, such $Q$'s are taken care of by the reduction of the problem from convex curves to convex quadrilaterals.)

It is not necessary to copy these equations here. It suffices to note that they are of the form

$$f_i(\lambda_i, \lambda_{i+1}) = 0 \quad \text{(subscripts mod 4)}, \tag{7.3}$$

and constitute a counterpart of our equations (5.3). These equations, though not entirely trivial, are sufficiently simple to entice Davis to proceed immediately with his final arguments, which take about a printed page, to squeeze out the inequality $A \leqslant 1$. There is no such temptation with our equations (5.3), as they contain the $a_i$'s in a much more complicated way.

Thus we are led to inspect (7.3) more closely with a hope of discovering a better approach to (5.3). By a simple rearrangement of terms, one obtains from (7.3)

$$P_i \lambda_i + \frac{P_{i+1}}{\lambda_{i+1}} = 2, \tag{7.3$'$}$$

where

$$P_i = \frac{2A}{(1 \pm \sigma)(1 \pm \tau)} - 1, \tag{7.4}$$

with the choice of $+/-$ depending on $i$. (The $P_i$'s were also used by Davis.) Now the proof follows in a surprisingly simple way. Assume that the equations (7.3′) have a solution with the $\lambda_i$'s positive, so that $Q(\mathrm{Cr}, A)$ exists. We have then

$$\frac{8A}{(1-\sigma^2)(1-\tau^2)} - 4 = \sum_i P_i \leqslant \frac{1}{2} \sum_i P_i\left(\lambda_i + \frac{1}{\lambda_i}\right)$$

$$= \frac{1}{2} \sum_i \left(P_i \lambda_i + \frac{P_{i+1}}{\lambda_{i+1}}\right) = 4, \tag{7.5}$$

whence $A \leqslant (1-\sigma^2)(1-\tau^2) \leqslant 1$ with equality iff $\sigma = \tau = 1$.

Could the same method work in our case? Up to a point. Consider the equations (5.3) and put

$$\begin{aligned} \mu_i &:= \sqrt{h_i}\,/(t_i \sin\theta_i + h_i \cos\theta_i), \\ Q_i &:= \sqrt{h_i}\,/\sin\theta_i, \\ R_i &:= \cot\theta_i + h_{i+1}\cot\theta_{i+1}. \end{aligned} \tag{7.6}$$

One may easily verify that the equations (5.3) are equivalent to

$$Q_i \mu_i + Q_{i+1}/\mu_{i+1} = R_i, \qquad \mu_i > 0, \tag{7.7}$$

the analogues of (7.3′). Proceeding as with (7.3′), we conclude: *if* $Q(\mathrm{Cr}, p)$ *exists, then*

$$\sum_i \frac{\sqrt{h_i}}{\sin\theta_i} \leqslant \frac{1}{2} \sum_i (1 + h_i)\cot\theta_i. \tag{7.8}$$

The inequality (7.8) gives just a necessary condition for the existence of $Q(\mathrm{Cr}, p)$. Besides being very complicated, it is meaningless for our purposes. It just happens—a fortunate accident—that the inequality (7.5) holds whether $Q(\mathrm{Cr}, A)$ is convex or not; that is, the assumption of convexity of $Q(\mathrm{Cr}, A)$ in the statement of the theorem (for quadrilaterals) is not needed. On the other hand, we know now that there exist nonconvex $Q(\mathrm{Cr}, p)$ with $p > 1$, so (7.8), which holds for *all* $Q(\mathrm{Cr}, p)$, *convex or not*, could not possibly imply that $p \leqslant 1$.

Thus we are back to the drawing board, Equation (5.3). With no quick method available, we are resigned to computing of $\hat{\Delta}$, Equation (5.6). It is the computation of $\operatorname{tr} T$ that is forbidding, but, fortunately, the factoring of the $T_i$ and the subsequent observations reduce this computation to manageable proportions. A new question now arises. Can we supply a proof by considering $\hat{\Delta}$ alone, and thus stay within the strategy employed by Davis, with just the "messier computations" that he predicted? This, of course, could not cover the special case $p = c$, which anyhow is settled here separately by Lemma 3. In order to avoid Lemmas 1, 2, and 4, which hinge on the study of $\Delta$ defined by (2.2), we would need more complete information about the subset of $\mathfrak{D}$ on which $\hat{\Delta} \geqslant 0$.

Even after the fact, our information about this set is not yet complete, although the remarks in Sections 7.2, 3 describe its shape quite well. (Further evidence can be gathered from the special case $\rho = 0$, $\theta = \pi/4$ by discovering that then $h(x, -2) = 0$, so that $h(x, y)$ can be factored explicitly.) To complete this information could require further study, and perhaps $\Delta$ would have to be brought into the picture. Since the computation of $\Delta$ is anyhow required for Lemma 3, we found it just pleasant to note the implication:

*If* $\Delta \leqslant 0$ *and* $\hat{\Delta} \geqslant 0$ *then* $p \leqslant 1$.

## References

[1] Davis, Chandler, An extremal problem for plane convex curves. In *Convexity*, Proceedings of the 7th symposium of the American Mathematical Society, Providence, R.I., 1973.

# Part III: Geometric Transformations

# Polygons and Polynomials

J. C. Fisher*
D. Ruoff*
J. Shilleto*

## 1. Three Theorems About $n$-gons

In this paper an algebraic method will be developed to deal with geometry problems of apparently varied nature. We use as examples the following three theorems.

**Theorem 1** (Napoleon, Barlotti). *Let* $\mathbf{A} = A_0A_1 \ldots A_{n-1}$ *be an n-gon in the plane and* $\mathbf{B} = B_0B_1 \ldots B_{n-1}$ *be the n-gon whose vertices are the centers of regular n-gons all erected externally* (*or all internally*) *on the sides of* $\mathbf{A}$. *Then* $\mathbf{B}$ *is regular if and only if* $\mathbf{A}$ *is affinely regular*. (*See* Figure 1).

**Theorem 2.** *Given the square* $\mathbf{A} = A_0A_1A_2A_3$, *define* $B_j$ *to be the summit of the equilateral triangle interior to* $\mathbf{A}$ *with base* $A_jA_{j+1}$ ($j = 0, 1, 2, 3$ *and* $A_4 := A_0$). *Then the point set consisting of the midpoints of* $B_jB_{j+1}$, $A_jB_j$, *and* $A_{j+1}B_j$ ($j = 0, 1, 2, 3$) *comprises the vertices of a regular* 12-*gon* (Figure 2).

**Theorem 3.** *Any pentagon in three dimensions that is both equilateral and equiangular lies in a plane.*

Theorem 1 is a generalization of a fact about triangles attributed to Napoleon [5, p. 23]. The general form was first stated and proved by Barlotti [3], whose arguments involved intricate trigonometric calculations. Theorem 2 was considered sufficiently challenging to appear as a question on the 1977 International Mathematics Olympiad (see [12, pp. 185–188]). Theorem 3 was inspired by a problem in molecular chemistry and was proved by van der Waerden [18]. His

* Department of Mathematics, University of Regina, Regina, Canada S4S 0A2.

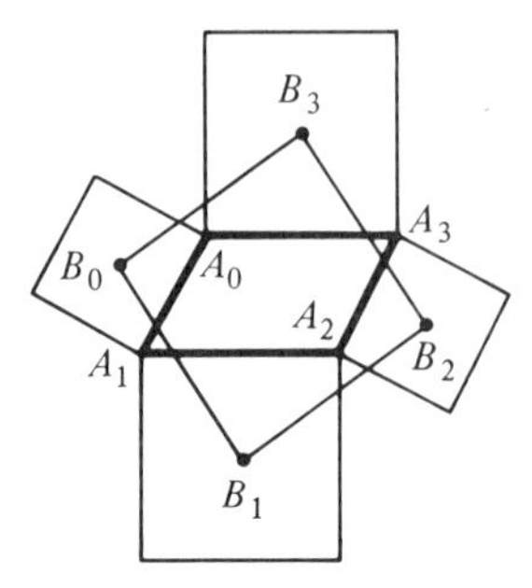


**Figure 1.** The Napoleon–Barlotti Theorem for $n = 4$.

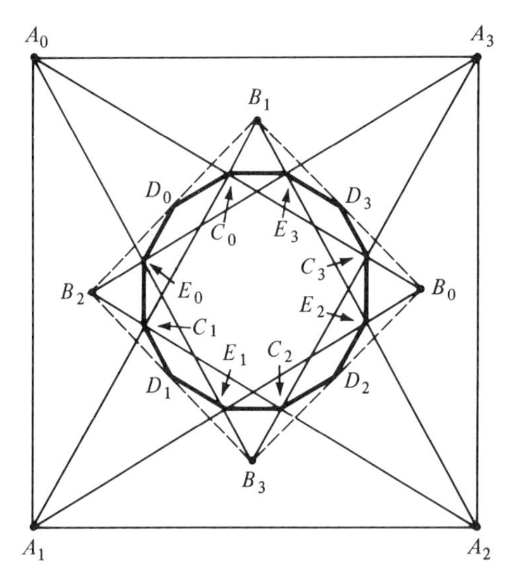


**Figure 2.** Theorem 2.

rather complicated proof has given way to several simpler ones (see *Math. Reviews* **48**, #12235 and *Zbl.* **416**, #51008 for references).

The approach we shall use originated with F. Bachmann and E. Schmidt [2]. It was simplified and expanded in [17]; a related treatment can be found in [8]. The theory enables one to shuttle between geometric properties of $n$-gons and related polynomial equations. The procedure is to translate a given geometric statement into an algebraic form and then perform some routine manipulations to obtain another equation with a resulting new geometric content.

Throughout the paper the exposition is elementary, relying only on basic concepts involving complex numbers, vectors and polynomials. Many readers may wish to skim over our definitions and elementary examples in Section 2, arriving quickly at 2.6; there we summarize the ideas that are required for the proofs of Section 3.

## 2. The Algebraic Treatment of $n$-gons

### 2.1. $n$-gons

An *$n$-gon* **A** in Euclidean space is a sequence $A_0, A_1, \ldots, A_{n-1}$ of $n$ not necessarily distinct points. The points of **A** are called *vertices*, and the segments joining consecutive vertices $A_j$, $A_{j+1}$ are called *sides* or *edges*. Note we do not require an $n$-gon to be either convex or planar. We also allow any number of vertices to coincide. An $n$-gon will be denoted by a boldface letter, its vertices by the corresponding capital Latin letter with subscripts.

### 2.2. Recursive Formulas of $n$-gons

The theory begins with the observation that $n$-gons can be classified by linear interrelations among their vertices. For example, a parallelogram is a quadrangle whose opposite sides are parallel; that is,

$$A_3 - A_2 = A_0 - A_1.$$

The formula remains true if the indices are cyclically permuted. Thus

$$A_{j+3} - A_{j+2} = A_j - A_{j+1}, \qquad j = 0, 1, 2, 3, \quad A_{j+4} := A_j.$$

We can write these four equations compactly as follows:

$$(A_3, A_0, A_1, A_2) - (A_2, A_3, A_0, A_1) = (A_0, A_1, A_2, A_3) - (A_1, A_2, A_3, A_0).$$

Note the elegance of the formula if a cyclic shift is indicated by the multiplier $x$, a double shift by $x^2$, etc. Then the above formula reads simply

$$(x^3 - x^2) \cdot \mathbf{A} = (1 - x) \cdot \mathbf{A},$$

where $\mathbf{A} = (A_0, A_1, A_2, A_3)$.

**Notation.** Multiplication of an $n$-gon by $x^k$ shall denote a cyclic shift of the $n$-gon $k$ places to the left, i.e.,

$$x \cdot (A_0, A_1, \ldots, A_{n-1}) = (A_1, \ldots, A_{n-1}, A_0),$$
$$x^2 \cdot (A_0, A_1, \ldots, A_{n-1}) = (A_2, \ldots, A_0, A_1),$$

etc.

A variety of geometric properties can be expressed by equations of the form

$$p(x) \cdot \mathbf{A} = \mathbf{0}, \qquad \mathbf{0} = (0, 0, \ldots, 0), \tag{$*$}$$

where 0 is the origin. For $p(x) = x^d - c_{d-1}x^{d-1} - \cdots - c_0$, ($*$) is an abbreviation for

$$A_{j+d} = c_{d-1}A_{j+d-1} + c_{d-2}A_{j+d-2} + \cdots + c_0 A_j, \tag{$**$}$$

where $j < n$ and $A_{j+n} := A_j$. Thus the vertices $A_d, A_{d+1}, \ldots, A_{n-1}$ are defined recursively by ($**$) starting from $A_0, A_1, \ldots, A_{d-1}$.

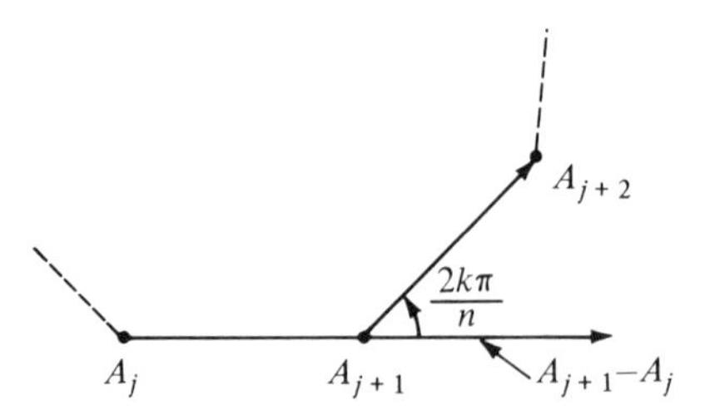


**Figure 3.** Regular $n$-gons (Example (c)).

EXAMPLES.

(a) *Parallelograms* are quadrangles **A** satisfying $(x^3 - x^2 + x - 1) \cdot \mathbf{A} = \mathbf{0}$ (see above).

(b) *$n$-gons* **A** *which repeat after $m$ steps*, where $m \mid n$, satisfy $(x^m - 1) \cdot \mathbf{A} = \mathbf{0}$. When $m = n$ we have $(x^n - 1) \cdot \mathbf{A} = \mathbf{0}$, which is satisfied by every $n$-gon, since a cyclic shift $n$ places is the identity.

When working in the Euclidean plane, we can identify vectors $(a, b)$ with complex numbers $a + bi$. Equation$(*)$ is then a condition on $n$-tuples of complex numbers. In fact, interesting geometric properties can be expressed in this form when $p(x)$ is permitted to have complex coefficients. This is because the rotation of a vector $a + bi$ through an angle of $\theta$ corresponds to the multiplication of it by $e^{i\theta}$.

(c) *Regular plane $n$-gons* **A** satisfy $\omega(x \cdot \mathbf{A} - \mathbf{A}) = x^2 \cdot \mathbf{A} - x \cdot \mathbf{A}$, where $\omega = e^{2k\pi i/n}$ and $k$ is any integer. This says $\omega(\mathbf{A}_{j+1} - A_j) = A_{j+2} - A_{j+1}$; that is, the vector $A_{j+1} - A_j$ when rotated through an angle $2\pi k/n$ becomes $A_{j+2} - A_{j+1}$ (see Figure 3). If $k = \pm 1$ then **A** is convex; otherwise **A** is a regular star $n$-gon (when $k$ is relatively prime to $n$), or it is a regular polygon with repeated vertices.

(d) *Affinely regular $n$-gons* satisfy $x^3 \cdot \mathbf{A} - \mathbf{A} = (1 + c)(x^2 \cdot \mathbf{A} - x \cdot \mathbf{A})$, where $c = 2\cos(2k\pi/n)$; they have $n$ distinct vertices when $k$ is relatively prime to $n$. Recall that an affinity is the composition of a nonsingular linear transformation and a translation [5, §13.3]. An affinely regular $n$-gon is defined to be the image of a regular $n$-gon under an affinity. Since affinities preserve ratios of parallel segments ([5, p. 443, #14] or [6, §2]) and since the regular $n$-gon in (c) satisfies $A_{j+3} - A_j = (1 + c)(A_{j+2} - A_{j+1})$, the above equation holds (see Figure 4).

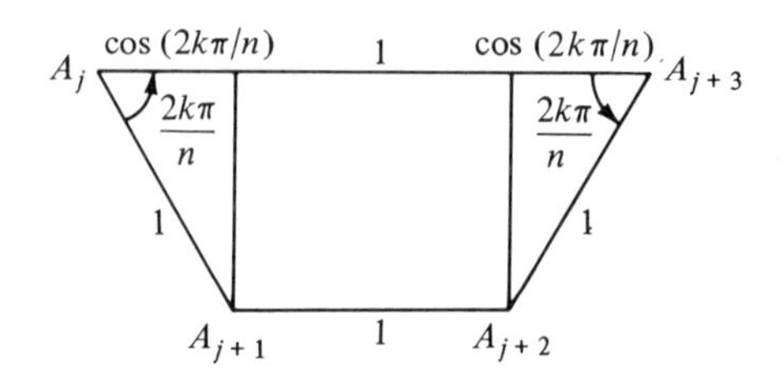


**Figure 4.** $A_jA_{j+3} : A_{j+1}A_{j+2} = 1 + 2\cos(2k\pi/n) : 1$.

An affinely regular $n$-gon $\mathbf{A}$ also may be characterized as the two-dimensional orbit of an affinity $\alpha$ with period $n$; that is, $\mathbf{A} = (A, \alpha A, \alpha^2 A, \ldots, \alpha^{n-1} A)$ and $\alpha^n A = A$ [6, §3].

### 2.3. Some Elementary Applications

(a) For a very simple first example let us prove algebraically the (obvious) fact that *regular n-gons are affinely regular*. By Section 2.2(c), $\mathbf{A}$ is a regular $n$-gon iff $(x-1)(x-\omega) \cdot \mathbf{A} = \mathbf{0}$, where $\omega = e^{2k\pi i/n}$ and $k$ is relatively prime to $n$. Multiply both sides of this equation by $x - \omega^{-1}$ and let $c = \omega + \omega^{-1} = 2\cos(2k\pi/n)$ to obtain $(x-1)(x^2 - cx + 1) \cdot \mathbf{A} = \mathbf{0}$. This is condition (d) of Section 2.2.

(b) *The midpoints of the sides of a quandrangle* $\mathbf{A}$ *form the vertices of a parallelogram* $\mathbf{B}$. The midpoint figure $\mathbf{B}$ is just $\frac{1}{2}(x+1) \cdot \mathbf{A}$. Now $(x^3 - x^2 + x - 1) \cdot \mathbf{B} = (x-1)(x^2+1) \cdot \mathbf{B} = (x-1)(x^2+1)\frac{1}{2}(x+1) \cdot \mathbf{A} = \frac{1}{2}(x^4 - 1) \cdot \mathbf{A} = \mathbf{0}$. By Section 2.2 (a), $\mathbf{B}$ is a parallelogram.

(c) *For any* $n \geqslant 3$ *the set of regular, convex, counterclockwise-oriented n-gons in the Euclidean plane is a two-dimensional vector space over the complex numbers.* The conditions specify all $n$-gons $\mathbf{A}$ which satisfy $(x-1)(x-\omega) \cdot \mathbf{A} = \mathbf{0}$, where $\omega = e^{2k\pi/n}$ (see Section 2.2 (c)). These $n$-gons form a vector space, since $(x-1)(x-\omega) \cdot (c \cdot \mathbf{A} + d \cdot \mathbf{B}) = c(x-1)(x-\omega) \cdot \mathbf{A} + d(x-1)(x-\omega) \cdot \mathbf{B}$. A basis for this vector space consists of any two such $n$-gons with different centers, for example $\mathbf{A} = (1, \omega, \omega^2, \ldots, \omega^{n-1})$ and $\mathbf{B} = (1, \omega, \omega^2, \ldots, \omega^{n-1}) + (1, 1, 1, \ldots, 1)$. This is true because any given $\mathbf{C}$ is of the form $(c, c\omega, c\omega^2, \ldots, c\omega^{n-1}) + (d, d, d, \ldots, d)$; hence $\mathbf{C} = (c-d) \cdot \mathbf{A} + d \cdot \mathbf{B}$.

This theorem can be generalized by replacing the polynomial $(x-1)(x-\omega)$ by an arbitrary one. In its present form it has an immediate consequence in the theorem of Finsler and Hadwiger [10, p. 324],

(d) *Suppose two squares* $ABCD$ *and* $AB'C'D'$ *share a vertex* $A$ *and are similarly oriented. Then the four points formed by the centers of the squares and the midpoints of* $BD'$ *and* $B'D$ *can be ordered to form the vertices of a square* (see Figure 5).

For the proof observe that if $A$ is the first vertex of one square and the third of the other, then the new figure is half the sum of the two squares and is therefore a square itself.

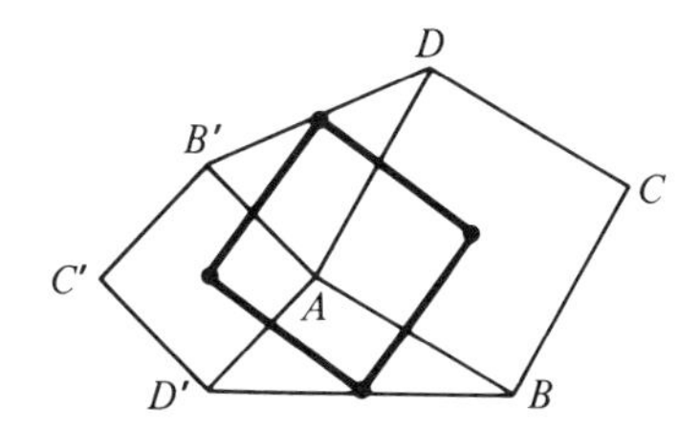


**Figure 5.** The Finsler–Hadwiger theorem (part (d)).

A good source of more examples is [2]. The theory can also be successfully applied to the exercises of §13.2 in [5] and to [16, §2].

### 2.4. Decomposition of $n$-gons

Factoring a polynomial corresponds to a decomposition of an $n$-gon in the following sense.

**Theorem.** *Given the n-gon* $\mathbf{A}$ *and relatively prime polynomials* $p(x)$ *and* $q(x)$ *such that* $p(x)q(x) \cdot \mathbf{A} = \mathbf{0}$; *then there are unique n-gons* $\mathbf{B}$ *and* $\mathbf{C}$ *such that* $\mathbf{A} = \mathbf{B} + \mathbf{C}$ *and* $p(x) \cdot \mathbf{B} = q(x) \cdot \mathbf{C} = \mathbf{0}$.

The proof is easy if one looks at the theorem properly (as involving a module over a PID) [17, Theorem 2], but it will not be given here.

EXAMPLES.

(a) Every $n$-gon $\mathbf{A} = \mathbf{B} + \mathbf{C}$, where $(x^{n-1} + \cdots + x + 1) \cdot \mathbf{B} = (x-1) \cdot \mathbf{C} = \mathbf{0}$, because $(x^n - 1) \cdot \mathbf{A} = \mathbf{0}$. Therefore $\mathbf{C}$ is a single, repeated vertex $C$ and $\mathbf{B}$ has center 0; i.e. $(1/n)(B_{n-1} + \cdots + B_1 + B_0) = 0$. So $C$ must be the center of $\mathbf{A}$, and $\mathbf{B}$ is just the translated figure. More generally if $(x-1)p(x) \cdot \mathbf{A} = \mathbf{0}$ and $x - 1 \nmid p(x)$, then $\mathbf{B}$, the translate of $\mathbf{A}$ with center 0, satisfies $p(x) \cdot \mathbf{B} = \mathbf{0}$.

(b) *Any triangle* $\mathbf{A}$ *is the sum of two regular triangles*. Without loss of generality assume $\mathbf{A}$ has center 0, so that $(x^2 + x + 1) \cdot \mathbf{A} = \mathbf{0}$. For $\omega = e^{2\pi i/3}$, $x^2 + x + 1 = (x - \omega)(x - \omega^{-1})$. So by the decomposition thereom, $\mathbf{A} = \mathbf{B} + \mathbf{C}$, where $\mathbf{B}$ and $\mathbf{C}$ are regular, oppositely oriented triangles, i.e. $(x - \omega) \cdot \mathbf{B} = (x - \omega^{-1}) \cdot \mathbf{C} = \mathbf{0}$.

Similarly, *a parallelogram can be split up into two squares*, since $x^2 + 1 = (x - i) \cdot (x + i)$ (see Section 2.2 (a) and (c)). In general, *an affinely regular polygon is the sum of two regular polygons*, because $x^2 - cx + 1 = (x - \omega)(x - \omega^{-1})$ for $c = 2\cos(2k\pi/n)$ (see Section 2.2 (c) and (d)).

### 2.5. Three-Dimensional Affinely Regular Polygons

Although the idea of a three-dimensional polygon may seem unnatural at first, it has some interesting uses. For example, our third main opening theorem reduces to the question of the existence and exact nature of *three-dimensional regular* and *three-dimensional affinely regular n-gons*. Transplanting the planar situation, we define the former as orbits of periodic isometries that span 3-space [7, §1.5], and the latter as orbits of periodic affinities in 3-space. Again, the affinely regular $n$-gons are the nondegenerate affine images of the regular ones.

**Theorem.** *Every properly three-dimensional affinely regular n-gon has an even number of vertices.*

*Proof.* (We consider here the affine version of 1.7.2 in [7, p. 6].) The generating affinity of the given polygon is finite and so has a fixed point (coinciding with the center of the polygon). Our affinity is not equivalent to a rotation, because then the polygon would lie in a plane. Hence it is equivalent to a rotatory reflection, a motion that is composed of a rotation—possibly the identity—and a reflection in the plane perpendicular to the axis [5, 7.41 and 7.52]. This makes the vertices of the polygon lie alternately in one or the other of two parallel planes; hence the number of vertices is even. □

EXAMPLE. Let **A** be an $n$-gon whose vertices form a hexagonal antiprism; that is, let **A** be affinely equivalent to a zigzag whose vertices lie alternately on two congruent circles in planes perpendicular to the axis joining their centers. The affinity which generates **A** is equivalent to the commutative product of a rotation through $2\pi/12$ about the axis and a reflection in the plane equidistant from the two centers. Such an $n$-gon is called *antiprismatic* in [7, p. 6] and a *12-prismatoid* in [17, §4]. According to the decomposition theorem of Section 2.4, **A** can be considered as the sum of **B** and **C**, where **B** is an affinely regular 12-gon and **C** is a 6-fold repeated segment. One can picture **B** as lying in a horizontal plane, so that the vertices of **A** lie at the ends of vertical segments attached to the 12-gon alternately above and below **B**. From this follows that **A** satisfies the equation $(x-1)(x+1)(x^2-\sqrt{3}\,x+1)\cdot\mathbf{A}=\mathbf{0}$.

*Remark*. More generally, if the three-dimensional polygon **A** is the orbit of an affinity of period $n$, it satisfies $(x-1)(x+1)(x^2-cx+1)\cdot\mathbf{A}=\mathbf{0}$, where $c=2\cos(2k\pi/n)$. When $k$ is relatively prime to $n$, then **A** shares its vertices with an $n/2$-gonal prism or an $n$-gonal antiprism. A similar result is obtained for orbits of affinities in higher dimensions [17, §4].

### 2.6. Summary

We conclude our theoretical discussion with an example that demonstrates explicitly how to obtain geometric information from an equation involving a polygon. This is followed by a dictionary of geometric and algebraic equivalents.

Consider a 12-gon **A** that satisfies $(x^4-x^2+1)\cdot\mathbf{A}=\mathbf{0}$. (Observe that $x^4-x^2+1\,|\,x^{12}-1$.) **A** is affinely regular when four-dimensional. This is so because an affinity $\alpha$ is specified by the conditions $\alpha(A_j)=A_{j+1}$ on the simplex $A_0$, $A_1,\ldots,A_4$; applying the recursive formula $(x^4-x^2+1)\cdot\mathbf{A}=\mathbf{0}$. we obtain for all $j$, $\alpha(A_j)=A_{j+1}$. The hexagon **B** formed by taking the even vertices of **A** is an affinely regular plane figure. This is true because $A_{j+4}-A_{j+2}+A_j=0$ implies (by a replacement of $j$ with $2j$) that $B_{j+2}-B_{j+1}+B_j=0$. So **B** is annihilated by $x^2-x+1$ and is affinely regular (see Section 2.2 (d)). Similarly the odd vertices of **A** form an affinely regular plane hexagon **C**, and one alternates between **B** and **C** in going around **A**. Also $((x^3)^2+1)\cdot\mathbf{A}=(x^2+1)(x^4-x^2+1)\cdot\mathbf{A}=\mathbf{0}$, so each of the three quadrangles formed by taking every third vertex forms a parallelogram (see Section 2.2 (a)). Because $A_{j+6}+A_j=0$, we have $A_{j+7}-A_{j+6}=-(A_{j+1}-A_j)$, and so opposite sides of **A** are parallel and equal, and **A** is symmetric about its center.

We now list in parallel columns some algebraic equivalents of geometric statements about $n$-gons.

| Geometric statement | Algebraic equivalent |
|---|---|
| The vertices of **A** repeat after $m$ steps | $(x^m - 1) \cdot \mathbf{A} = \mathbf{0}$ |
| **A** is a regular $n$-gon (with $n$ distinct vertices) | $(x - 1)(x - \omega) \cdot \mathbf{A} = \mathbf{0}$, where $\omega = e^{2k\pi i/n}$ (with $k$ relatively prime to $n$) (Section 2.2 (c)) |
| **A** is an affinely regular $n$-gon | $(x - 1)(x^2 - cx + 1) \cdot \mathbf{A} = \mathbf{0}$, where $c = \omega + \omega^{-1}$ (Section 2.2 (d)) |
| Special case: **A** is a parallelogram | Special case: $c = 0$ |
| **B** is the $n$-gon formed from the midpoints of the sides of **A** | $\mathbf{B} = \frac{1}{2}(x + 1) \cdot \mathbf{A}$ (Section 2.3 (b)) |
| **A** has center 0 | $p(x) \cdot \mathbf{A} = \mathbf{0}$ implies $x - 1 \nmid p(x)$ for some $p(x)$ |
| **B** is formed by taking every $m$th vertex of the $n$-gon **A**, where $m \mid n$, and its vertices repeat after $n/m$ steps. | If $p(x^m) \cdot \mathbf{A} = \mathbf{0}$, then $p(x) \cdot \mathbf{B} = \mathbf{0}$ (Section 2.6) |
| The number $n$ of vertices of **A** is even, opposite sides of **A** are parallel and equal, and **A** is consequently symmetric about its center. | $(x - 1)(x^{n/2} + 1) \cdot \mathbf{A} = \mathbf{0}$ (Section 2.6) |

## 3. Proofs of the Three Theorems

### 3.1. Proof of Theorem 1 and a Theorem of B. H. Neumann

The assumptions of Theorem 1 imply that $\omega \cdot (A_{j+1} - B_j) = A_j - B_j$, where $\omega = e^{2\pi i/n}$ and $j < n$ (see Figure 6). Therefore, in our notation, $\omega \cdot (x \cdot \mathbf{A} - \mathbf{B}) = \mathbf{A} - \mathbf{B}$, and so $(x - \omega^{-1}) \cdot \mathbf{A} = (1 - \omega^{-1}) \cdot \mathbf{B}$. Multiplying both sides by $x - \omega$ gives $(x - \omega)(x - \omega^{-1}) \cdot \mathbf{A} = \mathbf{0}$ iff $(x - \omega) \cdot \mathbf{B} = \mathbf{0}$. That is, **A** is affinely regular (with center 0) iff **B** is regular (with center 0) according to Section 2.2 (c) and (d). □

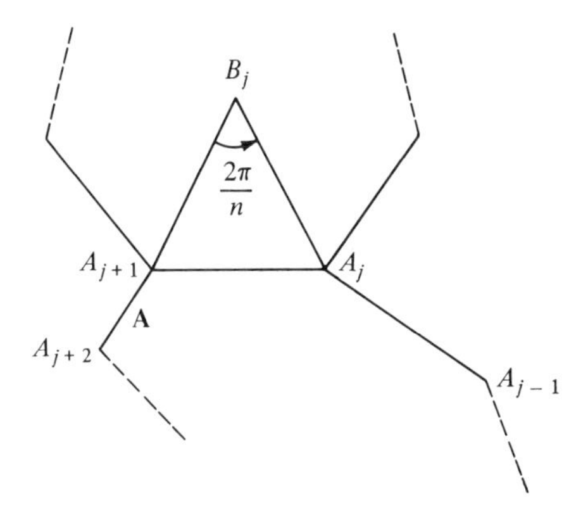


**Figure 6.** $B_j$ is the center of a regular $n$-gon erected on $A_jA_{j+1}$.

The proof of Theorem 1 applies to any $n$th root of unity. For example when $\omega = e^{-2\pi i/n}$, the vertices of $\mathbf{B}$ are the centers of regular $n$-gons erected internally on the sides of $\mathbf{A}$. We are thus led to the notion of a *C-operator*, a slight modification of a concept of B. H. Newmann [15, p. 233 ff. See also note (2) in the Appendix below.]. Such an operator assigns to an $n$-gon $\mathbf{A}^0$ a sequence of $n$-gons $\mathbf{A}^0, \mathbf{A}^1, \mathbf{A}^2, \ldots$ . The inductive definition of $\mathbf{A}^k$ from $\mathbf{A}^{k-1}$ involves a regular $n$-gon that is generated by a rotation through the angle $2k\pi/n$; let us call such an $n$-gon *$\omega_k$-regular*. To carry out the $k$th step of the construction, we erect on each side $A_j^{k-1}A_{j+1}^{k-1}$ of the previous $n$-gon $\mathbf{A}^{k-1}$ an $\omega_k$-regular $n$-gon and take its center to be the vertex $A_j^k$ of $\mathbf{A}^k$ (Figure 7).

**Theorem** (Douglas, Neumann). *The $n$-gon $\mathbf{A}^{n-2}$ in the sequence $\mathbf{A}^0, \mathbf{A}^1, \mathbf{A}^2, \ldots$ obtained from the $n$-gon $\mathbf{A}^0$ by applying the C-operator (described above) is regular.*

*Proof.* Just as in Theorem 1, $(x - \omega_k^{-1}) \cdot \mathbf{A}^{k-1} = (1 - \omega_k^{-1}) \cdot \mathbf{A}^k$. So $A^{n-2} = c(x - \omega_1^{-1})(x - \omega_2^{-1}) \ldots (x - \omega_{n-2}^{-1}) \cdot \mathbf{A}^0$, where $c$ is the constant $[(1 - \omega_1^{-1})$

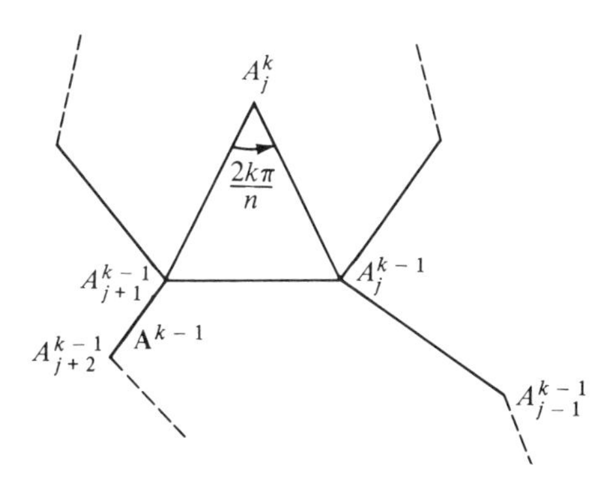


**Figure 7.** $A_j^k$ is the center of an $\omega$-regular $n$-gon erected on $A_j^{k-1}A_{j+1}^{k-1}$.

$(1-\omega_2^{-1})\ldots(1-\omega_{n-2}^{-1})]^{-1}$. Therefore $(x-1)(x-\omega_1)\cdot\mathbf{A}^{n-2}=(x-1)(x-\omega_{n-1}^{-1})\cdot\mathbf{A}^{n-2}=c(x^n-1)\cdot\mathbf{A}^0=\mathbf{0}$. □

Corresponding to a rearrangement of the factors $x-\omega_k^{-1}$ in the above formula for $\mathbf{A}^{n-2}$, there is a different construction of this polygon. Instead of first erecting $\omega_1$-regular polygons on the sides of the given polygon, then $\omega_2$-regular polygons on the sides of the obtained one, etc., we can use the roots of unity in any order. This fact sometimes can be expressed in more familiar geometric terms. Let us illustrate the various aspects of the Theorem in the case $n=4$.

Note first that an $\omega_1$-regular or $\omega_3$-regular 4-gon is a square—with vertices taken counterclockwise and clockwise respectively. An $\omega_2$-regular $n$-gon is a line segment (or more precisely, a 2-fold repeated line segment).

(a) We first erect $\omega_2$-regular, then $\omega_1$- (or $\omega_3$-) regular 4-gons. Since $\mathbf{A}^1=\frac{1}{2}(x+1)\mathbf{A}^0$ is a parallelogram (Section 2.3 (b)), the Napoleon–Barlotti theorem implies the Douglas–Neumann theorem in this case; i.e. $(1-i)^{-1}(x-i)\mathbf{A}^1$ (or $(1+i)^{-1}(x+i)\mathbf{A}^1$) is a square (see Figure 1). In other words, *a square is formed by the centers of squares whose bases join the midpoints of consecutive sides of a quadrangle.*

(b) We begin with $\omega_1$ (or $\omega_3$) and then continue with $\omega_2$. The same square equals the product of the polynomials applied to $\mathbf{A}^0$ in the reversed order. Therefore *the summits of right-angled isosceles triangles all erected externally* (*or all internally*) *on the sides of a quadrangle form a polygon whose midpoint figure is a square.*

(c) Neumann [15, 3.5] points out that the theorem of Finsler and Hadwiger (Section 2.3 (d)) follows immediately from the above one. To see this, apply (b) to the degenerate quadrangle $ACAC'$ in Figure 5.

(d) First $\omega_1$ is used, then $\omega_3$. We interpret geometrically the equation $(x^2-1)(x+i)(x-i)\cdot\mathbf{A}^0=(x^4-1)\cdot\mathbf{A}^0=\mathbf{0}$. The first and third points of the quadrangle $(1+i)^{-1}(x+i)(1-i)^{-1}(x-i)\cdot\mathbf{A}^0$ are equal because it is annihilated by $x^2-1$. Therefore, *when isosceles right triangles with summits $B$, $D$, $B'$, $D'$ are*

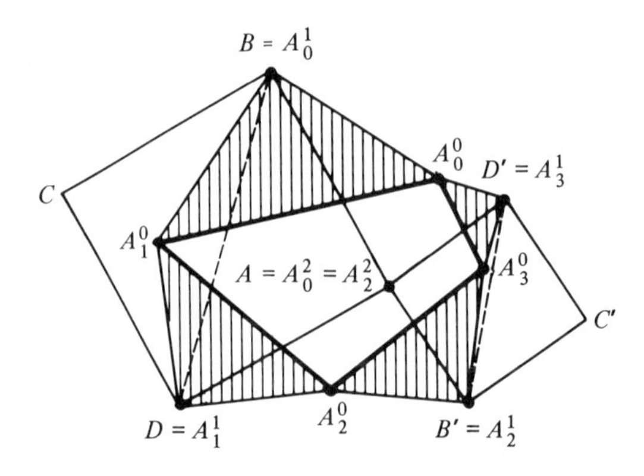


**Figure 8.** The Douglas–Neumann theorem (Section 3.1 (d)).

*erected on the sides of a quadrangle* $A_0^0A_1^0A_2^0A_3^0$ *(all exterior or all interior), the segments BD and B′D′ form diagonals of squares that share a vertex* (Figure 8).

For other interesting generalizations of Napoleon's theorem, see [11, p. 40].

### 3.2. Proof of Theorem 2 and an Affine Version of It

The midpoints mentioned in Theorem 2 will be denoted by $C_j$, $D_j$, and $E_j$. Thus $C_j = \frac{1}{2}(A_j + B_j)$, $D_j = \frac{1}{2}(B_{j+1} + B_{j+2})$, $E_j = \frac{1}{2}(A_j + B_{j+3})$, and form squares (2.3(c)) that satisfy the equations $\mathbf{C} = \frac{1}{2}(\mathbf{A} + \mathbf{B})$, $\mathbf{D} = \frac{1}{2}(x + x^2) \cdot \mathbf{B}$, and $\mathbf{E} = \frac{1}{2}(\mathbf{A} + x^3 \cdot \mathbf{B})$. We must prove that $\mathbf{F} := C_0D_0E_0C_1D_1E_1C_2D_2E_2C_3D_3E_3$ is a regular 12-gon. This can be expressed by the equations $\mathbf{D} = \omega \cdot \mathbf{C}$, $\mathbf{E} = \omega \cdot \mathbf{D}$, and $x \cdot \mathbf{C} = \omega \cdot \mathbf{E}$, where $\omega = e^{2\pi/12} = \frac{1}{2}(\sqrt{3} + i)$. An easy calculation shows that $\mathbf{A} = (1 + \sqrt{3})\mathbf{D}$ (see Figure 2); so $\mathbf{A} = \frac{1}{2}(1 + \sqrt{3})(x + x^2) \cdot \mathbf{B}$. Now $\mathbf{E} - \omega \cdot \mathbf{D} = \frac{1}{2}[(\mathbf{A} + x^3 \cdot \mathbf{B}) - \omega\ (x + x^2) \cdot \mathbf{B}] = \frac{1}{2}[\frac{1}{2}(1 + \sqrt{3})(x + x^2) + x^3 - \omega(x + x^2)] \cdot \mathbf{B} =: p(x) \cdot \mathbf{B}$.

We have to prove that $p(x) \cdot \mathbf{B} = \mathbf{0}$. To do so we need only show that $p(x)$ contains the factor $x - i$ (since $\mathbf{B}$ is a square). This is indeed the case, as $p(i) = \frac{1}{2}[\frac{1}{2}(1 + \sqrt{3})(i - 1) - i - \frac{1}{2}(\sqrt{3} + i)(i - 1)] = 0$. From this and the symmetry in the construction of $\mathbf{C}$ and $\mathbf{E}$, it follows that $\mathbf{D} = \omega \cdot \mathbf{C}$. So a rotation through the angle of 30° moves the square $\mathbf{C}$ into the square $\mathbf{D}$ and the square $\mathbf{D}$ into the square $\mathbf{E}$. (Briefly, since 0 is the common center $(x - 1 \nmid p(x))$, we have shown that $D_j0E_j = 30°$ and, by symmetry, $C_j0D_j = 30°$. Furthermore, since $D_j0D_{j+1} = 90°$, it follows that $E_j0C_{j+1} = 30°$ also; see Figure 2). Therefore $\mathbf{F}$ must be a regular 12-gon. □

**Affine Version.** We take the equation $\mathbf{A} = \frac{1}{2}(1 + \sqrt{3})(x + x^2) \cdot \mathbf{B}$ as the starting point of the affine version of this theorem. *Given the parallelogram* $\mathbf{B}$, *let* $\mathbf{A}$ *be obtained from the midpoint figure of* $\mathbf{B}$ (actually $x \cdot \mathbf{B}$) *by dilating from its center by a factor of* $1 + \sqrt{3}$. *Construct the* 12-*gon* $\mathbf{F}$ *as in Theorem* 2. *Then* $\mathbf{F}$ *is affinely regular and, in fact,* $(x^2 - \sqrt{3}\,x + 1) \cdot \mathbf{F} = \mathbf{0}$. This is true because $\mathbf{A}$, $\mathbf{B}$ and $\mathbf{F}$ are all images, under an affinity $\alpha$, of regular polygons by Theorem 2.

*Remark.* A straightforward calculation along the same lines as in Theorem 2 proves if $\mathbf{A}$ *and* $\mathbf{B}$ *are any two parallelograms with the same center, then* $\mathbf{F}$ *as defined above satisfies* $(x^4 - x^2 + 1) \cdot \mathbf{F} = \mathbf{0}$. Thus $F$ is the 12-gon of Example 2.6.

### 3.3. Proof of Theorem 3 and a Generalization

The hypotheses on the pentagon $\mathbf{A}$ imply that the sides $A_0A_1, A_1A_2, \ldots, A_4A_0$ are congruent, as are the diagonals $A_0A_2, A_1A_3, \ldots, A_4A_1$. Thus $\mathbf{A}$ is congruent to $x \cdot \mathbf{A}$, so that there exists an isometry taking $A_j$ to $A_{j+1}$. Since $\mathbf{A}$ is the orbit of an isometry in three dimensions, it is two-dimensional by the Theorem of Section 2.5. □

**Generalization.** The proof in Section 3.3 actually shows more than we claimed [13, 14]. In the terminology of Branko Grünbaum, we define the $n$-gon **A** in three dimensions to be *3-equilateral* (*with parameters* $c_r$, $r = 1,2,3$) if for each vertex $A_j$, $A_jA_{j+r} = c_r$. It turns out that *if* **A** *is* 3-*equilateral and n is odd, then* **A** *is planar*; *if n is even, then* $A_1A_3 \ldots A_{n-1}$ *and* $A_0A_2 \ldots A_{n-2}$ *are both regular plane* $n/2$-*gons*. Similar results hold in any odd-dimensional Euclidean space ([14]; or see [17, Theorem 7]).

# Appendix

Since May 1979, when this paper was presented, a number of examples and references have been brought to our attention.

(1) The theorem (d) in Section 2.3, attributed to Finsler and Hadwiger, is an immediate consequence of results that have been around for a long time. Murray S. Klamkin has kindly supplied us with a list of references collected by him and Leon Bankoff: *Amer. Math. Monthly*, 1932, pp. 46, 291, 535, 559; 1933, pp. 36, 157; 1934, pp. 330, 370; 1937, p. 525; 1943, p. 64; 1969, p. 698; *Math. Mag.*, 1966, p. 166. See the solution to problem 464 in [*Crux Mathematicorum* **6** (1980), 185–187] for a more complete discussion.

(2) The theorem in Section 3.1 was discovered independently in 1940 by B. H. Neumann [15] and Jesse Douglas [Geometry of polygons in the complex plane. *J. Math. Physics M.I.T.* **19** (1940), 93–130]. See also *J. London Math. Soc.* **17** (1942), 162–166 for alternative treatments of these results.

(3) Here are two other results in the theory of $n$-gons that we learned of recently. Both follow more or less routinely from our theory.

**Theorem A** (Jesse Douglas). *Let* **A** *be any pentagon in three dimensions, and* $B_j = \frac{1}{2}(A_{j+2} + A_{j-2})$ *be the midpoint of the side opposite* $A_j$. *If* $C_j = B_j + (1/\sqrt{5})(B_j - A_j)$ *and* $D_j = B_j - (1/\sqrt{5})(B_j - A_j)$, *then* **C** *and* **D** *are plane, affinely regular pentagons,* **C** *convex, and* **D** *star-shaped.*

**Theorem B.** *Define a sequence* $\mathbf{A}^1, \mathbf{A}^2, \mathbf{A}^3, \ldots$ *of n-gons as follows*: $\mathbf{A}^1$ *is an arbitrary n-gon, and* $\mathbf{A}^{k+1}$ *is the midpoint figure of* $\mathbf{A}^k$ *enlarged by a suitable scaling factor. Then as* $k \to \infty$, $\mathbf{A}^{2k}$ *approaches a plane affinely regular n-gon.*

For a nice proof of Theorem A using finite Fourier series and a discussion of its consequences (with further references), see the paper by I. J. Schoenberg in this volume.

The earliest mention of Theorem B that we have located is [15, p. 233], although Neumann has told us of a German reference from the late 1920s. The theorem and its many variants and elaborations have been rediscovered many times since. For further references see *Math. Reviews* 15-55, **31**, #3925, **40**, #7940, **40**, #7941, **42**, #6716, **43**, #1037, **48**, #12291, **49**, #3683, **53**, #9037;

*Elem. Math.* **16** (1961), 73–78; *Math. Mag.* **52** (1979), 102–105; *Amer. Math. Monthly* **88** (1981), 145–146.

(4) Theorem 1 has been rediscovered [Leon Gerber, *Amer. Math. Monthly* **87** (1980), 644–648]. That article contains a bibliography of related $n$-gon references.

## REFERENCES

[1] Bachmann, Friedrich, $n$-gons. *Ed. Studies in Math.* **3** (1971), 288–309.

[2] Bachmann, Friedrich and Schmidt, E., $n$-gons. Math. Expositions No. 18, University of Toronto Press 1975. (This is a translation by Cyril W. L. Garner of $n$-Ecke. Bibliographisches Institut, Mannheim 1970.)

[3] Barlotti, Adriano, Una proprietà degli $n$-agoni che si ottengono trasformando in una affinità un $n$-agono regolare. *Boll. Un. Mat. Ital. (3)* **10** (1955), 96–98.

[4] Coxeter, H. S. M., *Regular Polytopes* (2nd ed.). Collier-Macmillan, New York 1963.

[5] Coxeter, H. S. M., *Introduction to Geometry* (2nd ed.). Wiley, New York, 1969.

[6] Coxeter, H. S. M., Affinely regular polygons. *Abh. Math. Sem. Univ. Hamburg* **34** (1969), 38–58.

[7] Coxeter, H. S. M., *Regular Complex Polytopes*. Cambridge Univ. Press 1974.

[8] Davis, Philip J., Cyclic transformations of polygons and the generalized inverse. *Canad. J. Math.* **29** (1977), 756–770.

[9] Dunitz, J. D. and Waser, J., The planarity of the equilateral, isogonal pentagon. *Elem. Math.* **27** (1972), 25–32.

[10] Finsler, P. and Hadwiger, H., Einige Relationen im Dreieck. *Comment. Math. Helv.* **10** (1937), 316–326.

[11] Forder, H. G., *The Calculus of Extension*. Chelsea, New York 1960.

[12] Greitzer, Samuel L., *International Mathematical Olympiads 1959–1977*. Math. Assoc. Amer. (New Math. Library #27), Washington, D.C. 1978.

[13] Korchmáros, Gabriele, Poligoni regolari. *Riv. Mat. Univ. Parma (4)* **1** (1975), 45–50.

[14] Lawrence, Jim, $k$-equilateral $(2k + 1)$-gons span only even dimensional spaces. In *The Geometry of Metric and Linear Spaces* (Proc. Conf. Michigan State U., East Lansing, Mich., 1974), Lecture Notes in Math., Vol. 490, Springer, Berlin, 1975.

[15] Neumann, B. H., Some remarks on polygons. *J. London Math. Soc.* **16** (1941), 230–245.

[16] Schoenberg, I. J., The finite Fourier series and elementary geometry. *Amer. Math. Monthly* **57** (1950), 390–404.

[17] Ruoff, D. and Shilleto, J., Recursive polygons. *Boll. Un. Mat. Ital. (5)* **15-B** (1978), 968–981.

[18] van der Waerden, B. L., Ein Satz über räumliche Fünfecke. *Elem. Math.* **25** (1970), 73–78; Nachtrag. *Elem. Math.* **27** (1972), 63.

# Algebraic Surfaces with Hyperelliptic Sections[1]

W. L. Edge*

## 1. The Plane Map

Surfaces whose prime (i.e. hyperplane) sections are hyperelliptic were studied and classified by Castelnuovo (2). If the sections have genus $p$, no surface can have order greater than $4p + 4$, and any of lesser order is a projection of a normal surface $\Phi$ in a projective space $S$ of $3p + 5$ dimensions. There is a pencil of conics, none of them singular, on $\Phi$; through each point of $\Phi$ passes one of the conics and their planes generate a threefold $V$ of order $3p + 3$ (2, §5; as the paper was later republished with different pagination, it may be advisable to refer to it by sections).

If $V$ has no directrix (i.e. curve meeting each generating plane once) of lower order than $p + 1$, Castelnuovo calls $\Phi$ "of the first kind"; it is rational, with prime sections mapped on a plane $\pi$ by those curves of order $p + 3$ with a fixed node $Y$ and a fixed $(p + 1)$-fold point $X$ (2, §7). Suppose henceforward that $\Phi$ is of this first kind: it still awaits attention. It is intended, in the following pages, to describe its nests of tangent spaces, exhibit the polarity that it induces in $S$, and mention some projectivities, notably involutory ones, under which it is invariant.

It suffices to take $p = 2$ for the detailed work. The properties of surfaces having $p > 2$ are sufficiently analogous to those of surfaces with $p = 2$ to permit a summary account of them, although there is a sharp distinction, to be emphasized in its place in §8, between $p$ even and $p$ odd.

$\Phi$, then, has order 12 and is immersed in a projective space $S$ of dimension 11. It is mapped on $\pi$ by quintics with a node $Y$ and a triple point $X$; it contains a pencil of conics $\gamma$, whose planes generate a $V_3^9$, and a pencil of twisted cubics $\delta$; these $\delta$ are among the minimum directrices of $V_3^9$. The conics are mapped by the

[1] An addendum to Castelnuovo.

* Inveresk House, Musselburgh, Scotland.

lines through $X$, the cubics by the lines through $Y$; through any point of $\Phi$ passes a single $\gamma$ and a single $\delta$.

An interesting by-product (§6) of the geometry is the appearance, as a projection of $\Phi$, of a doubly covered del Pezzo quintic surface with two nodes. The nonsingular del Pezzo sextic surface likewise appears (§9) when $p = 3$.

The surface mentioned by Semple and Roth in their comprehensive book is not of the first but of the second kind, where $V_3^9$ is specialized to have a directrix conic (6, p. 155, Example 30; p. 222, Example 5).

If $X$, $Y$ are two of the vertices of the triangle of reference for homogeneous coordinates $(\xi, \eta, \zeta)$ in $\pi$, the equations of the mapping quintics do not involve $\xi$ to higher power than its square or $\eta$ to higher power than its cube; hence, if $x_i, y_i, z_i$ are homogeneous coordinates in $S$, the mapping is

$$\begin{array}{llll} x_0 = \xi^2\eta^3, & x_1 = \xi^2\eta^2\zeta, & x_2 = \xi^2\eta\zeta^2, & x_3 = \xi^2\zeta^3, \\ y_0 = \xi\eta^3\zeta, & y_1 = \xi\eta^2\zeta^2, & y_2 = \xi\eta\zeta^3, & y_3 = \xi\zeta^4, \\ z_0 = \eta^3\zeta^2, & z_1 = \eta^2\zeta^3, & z_2 = \eta\zeta^4, & z_3 = \zeta^5. \end{array} \tag{1.1}$$

Label vertices of the simplex of reference in $S$ by capitals: e.g. $Y_2$ is the vertex opposite $y_2 = 0$. But $X$, $Y$, $Z$ without suffixes always denote points in $\pi$. It is legitimate to speak of "the point $(\xi, \eta, \zeta)$ on $\Phi$" at least when $\zeta \neq 0$, but every point on $XY$ maps the same point on $\Phi$, namely $X_0$. The $\gamma$ through $X_0$ is mapped by the points in the first neighborhood of $Y$. For if $\zeta, \xi$ are infinitesimal, (1.1) indicates that, in the limit, all coordinates other than $x_0$, $y_0$, $z_0$ are zero while $x_0 : y_0 : z_0 = \xi^2 : \xi\zeta : \zeta^2$; this is a conic through $X_0$ and $Z_0$, the tangents there meeting at $Y_0$. The $\delta$ through $X_0$ is mapped by the points in the first neighborhood of $X$; if $\eta, \zeta$ are infinitesimal, all coordinates approach zero save for $x_0 : x_1 : x_2 : x_3 = \eta^3 : \eta^2\zeta : \eta\zeta^2 : \zeta^3$; this is the standard form of a twisted cubic through $X_0$ and $X_3$, with $X_1$ and $X_2$ each on a tangent at one of $X_0$, $X_3$ and the osculating plane at the other.

The point $(0, 0, 1)$ on $\Phi$ is $Z_3$. The $\gamma$ through it is $\eta = 0$ and contains $X_3$; the $\delta$ through it is $\xi = 0$ and contains $Z_0$.

$\Phi$ is homogeneous in the sense of having the same geometrical attributes at every one its points.

## 2. The Nest of Tangent Spaces

One now introduces the tangent spaces $\Omega$ of $\Phi$. A projective space of dimension $n$ will be labeled $[n]$, a standard usage going back as far as Schubert; $S$ is an $[11]$. Spaces $[1]$, $[2]$, $[3]$ are lines, planes, solids; an $[n-1]$ in $[n]$ is a *prime* (1, p. 257).

A surface $F$ in $[3]$ has, at each nonsingular point $P$, a tangent plane $\Omega_2(P)$; this contains the tangents at $P$ to all branches of curves on $F$ that pass through $P$, and the section of $F$ by $\Omega_2(P)$ has a double point at $P$.

But if $F$ is in higher space it has, as explained by del Pezzo (4), a nest of tangent spaces at each nonsingular point $P$. In general their dimensions $0, 2, 5, 9, \ldots$ are one less than the triangular numbers; $\Omega_0$ is $P$ itself, $\Omega_2$ the tangent plane. $\Omega_{(k-1)(k+2)/2}$ contains the osculating $[k-1]$ at $P$ of every branch of curves on $F$ passing through $P$, and every prime through it meets $F$ in a curve having multiplicity at least $k$ at $P$. This is the general situation. But in certain circumstances, as indeed with $\Phi$, these $\Omega$ may have lower dimensions. One effect of this is to afford the nest a longer sequence, since it continues so long as the dimension of $\Omega$ is less than that of the ambient space.

As $\Phi$ is homogeneous, one obtains geometrical information relevant to any of its points by examining the circumstances at $X_0$; here $\zeta = 0$, and $x_0$ is the only one of the twelve coordinates to be nonzero. $\Omega_2(X_0)$ joins $X_0$ to those points whose coordinates involve $\zeta$ only to the first order; it is the plane $X_0X_1Y_0$. Since the degree in $\zeta$ of an entry in (1.1) is constant along any line sloping up from left to right, one incorporates one more such line of entries as one expands from one tangent space to the next. Thus $\Omega_5(X_0)$ is $X_0X_1X_2Y_0Y_1Z_0$. On the next move the dimension increases again by 3, not more, so that the next tangent space is $\Omega_8$—not, as it would be for a nonspecialized surface, $\Omega_9$. The next move produces $\Omega_{10}$, which, being a prime, closes the sequence. The nest of tangent spaces at a point on $\Phi$ is

$$\Omega_0 \subset \Omega_2 \subset \Omega_5 \subset \Omega_8 \subset \Omega_{10}. \tag{2.1}$$

$\Omega_{10}$ may fittingly be called an osculating prime of $\Phi$.

The tangent spaces at $X_0$ are indicated by the partitioning

$$X_0 \,|\, X_1Y_0 \,|\, X_2Y_1Z_0 \,|\, X_3Y_2Z_1 \,|\, Y_3Z_2 \,|\, Z_3 \tag{2.2}$$

of the vertices of the simplex of reference: read from left to right, the passage across each vertical barrier is from one $\Omega(X_0)$ to the next larger.

This same partitioning read from right to left gives the tangent spaces at $Z_3$, as could have been argued directly from (1.1) and the fact that $Z_3$ is determined by $\xi = \eta = 0$. If $\xi, \eta$ are, momentarily, regarded as infinitesimals, each sloping line involves them jointly to the same order, increasing by 1 with each move leftwards.

## 3. Duality

The duality implied by the run of suffixes in (2.1) is no accident. Since, in accordance with del Pezzo's criteria, $\Omega_{10}(\xi,\eta,\zeta)$ meets $\Phi$ in a curve with a quintuple point at $(\xi,\eta,\zeta)$, and since the only quintic curve in $\pi$ with a quintuple point at $A$, a triple point at $X$, and a node at $Y$ consists of $AX$ thrice and $AY$ twice, the section of $\Phi$ by $\Omega_{10}$ consists of $\gamma$ thrice and $\delta$ twice; $(\xi',\eta',\zeta')$ on $\Phi$ is in $\Omega_{10}(\xi,\eta,\zeta)$ when

$$(\eta\zeta' - \eta'\zeta)^3(\xi\zeta' - \xi'\zeta)^2 = 0. \tag{3.1}$$

If the osculating prime at $(\xi,\eta,\zeta)$ contains $(\xi',\eta',\zeta')$, so does the osculating prime at $(\xi',\eta',\zeta')$ contain $(\xi,\eta,\zeta)$. This correspondence on $\Phi$ is subordinate to a null polarity $N$ in $S$. After expansion and multiplication in (3.1), and subsequent replacement of all the quintic monomials in accordance with (1.1), one finds $N$ to be

$$x_0z_3' - x_0'z_3 - 3(x_1z_2' - x_1'z_2) + 3(x_2z_1' - x_2'z_1) - (x_3z_0' - x_3'z_0) \\ - 2(y_0y_3' - y_2'y_3) + 6(y_1y_2' - y_1'y_2) = 0. \tag{3.2}$$

$\Omega_{10}$ is the polar prime in $N$ of its point of osculation.

If $P$ is on $\Phi$, $\Omega_2(P)$ and $\Omega_8(P)$ are polar spaces in $N$, while $\Omega_5(P)$ is self-polar, every two of its points being conjugate in $N$. For if, as is permissible, the point mapping $P$ in $\pi$ is taken for the third vertex $Z$ of the triangle of reference, then $Z_3$ is at $P$ and $\Omega_2(P)$ is $Z_3Z_2Y_3$ with, by (3.2), polar [8] $x_0 = x_1 = y_0 = 0$. But because all primes through $\Omega_8(Z_3)$ meet $\Phi$ in curves with (at least) a quadruple point at $Z_3$, and because quintics with multiplicities 4, 3, 2 at $Z, X, Y$ consist necessarily of $ZX$ twice, $ZY$, and a conic through $X$, $Y$, $Z$, they are linearly dependent on

$$\eta^2\xi \cdot \eta\zeta = 0, \qquad \eta^2\xi \cdot \zeta\xi = 0, \qquad \eta^2\xi \cdot \xi\eta = 0.$$

Thus, by (1.1), the [8] $\Omega_8(Z_3)$ common to the primes whose sections of $\Phi$ are mapped by these composite curves is

$$y_0 = x_1 = x_0 = 0.$$

As for $\Omega_5(Z_3)$, it has already been identified in Section 2 as $Z_3Z_2Z_1Y_3Y_2X_3$, or

$$x_0 = x_1 = x_2 = y_0 = y_1 = z_0 = 0.$$

But if $P, P'$ both satisfy these six equations, they are conjugate by (3.2).

The planes $\Omega_2(P)$ at the different points $P$ of a conic $\gamma$ on $\Phi$ all lie in the same [5]. This becomes clear on mapping $\gamma$ in $\pi$ by $\eta = 0$, for then every $\Omega_2(P)$ lies in that space for which all coordinates are zero whose entries in (1.1) contain $\eta$ to a power higher than the first, namely

$$x_0 = y_0 = z_0 = x_1 = y_1 = z_1 = 0 \tag{3.3}$$

or $X_2X_3Y_2Y_3Z_2Z_3$. So long as $P$ is on $\gamma$, every $\Omega_8(P)$ contains this [5]. Analogous reasoning shows that, if $P$ is on $\gamma$, $\Omega_5(P)$ always lies in the [8]

$$x_0 = y_0 = z_0 = 0. \tag{3.4}$$

If, on the other hand, $P$ moves on a cubic $\delta$, then $\Omega_2(P)$ lies in a [7]; for if $\delta$ is mapped in $\pi$ by $\xi = 0$, the planes all lie in

$$x_0 = x_1 = x_2 = x_3 = 0, \tag{3.5}$$

or $Y_0Y_1Y_2Y_3Z_0Z_1Z_2Z_3$.

$\Omega_8$ generates a primal (for this nomenclature see (1, p. 257)) or hypersurface when its contact traces $\Phi$. Quintics in $\pi$ with multiplicities 4, 3, 2 at $(\xi', \eta', \zeta'), X, Y$ are linearly dependent on (3.1) and

$$(\eta\zeta' - \eta'\zeta)^3(\xi\zeta' - \xi'\zeta)\zeta = 0, \qquad (\eta\zeta' - \eta'\zeta)^2(\xi\zeta' - \xi'\zeta)^2\zeta = 0. \tag{3.6}$$

Replace, in the three equations, the quintic monomials in $\xi, \eta, \zeta$ in accord with (1.1). The equation of the primal is the outcome of eliminating $\xi', \eta', \zeta'$ from the equations so arising.

## 4. Projection from $\Omega_8$

The mapping of $\Phi$ on $\pi$ has been used to establish the existence of the self-dual nest of tangent spaces $\Omega$. It can now be seen that *Castelnuovo's plane mapping is obtainable by direct projection from any* $\Omega_8$. Take the third vertex $Z$ of the triangle of reference in $\pi$ to be at the point mapping the contact of $\Omega_8$; then this contact is $Z_3$. Place $\pi$ on any plane in $S$ skew to $\Omega_8(Z_3)$; since this space is, by (2.2), $Z_3Z_2Y_3Z_1Y_2X_3Z_0Y_1X_2$, $\pi$ could be $X_0X_1Y_0$. Since every $\Omega_2$ whose contact is on the $\delta$ through $Z_3$ lies in the [7] (3.5), it is contained in the [9] joining $\Omega_8(Z_3)$ to $Y_0$: every point of $\delta$ is projected into the same point $Y_0$ of $\pi$. Since every $\Omega_2$ whose contact is on the $\gamma$ through $Z_3$ lies, by (3.3), in $\Omega_8(Z_3)$, the projection of a point $A$ on $\gamma$ is the intersection of $\pi$ with the join of $\Omega_8(Z_3)$ to $\Omega_5(A)$. But, by (3.4), $\Omega_5(A)$ lies in $x_0 = y_0 = z_0$ and so in the [9] $x_0 = y_0 = 0$ joining $\Omega_8(Z_3)$ to $X_1$: every point of $\gamma$ is projected into the same point $X_1$ of $\pi$.

A prime through $\Omega_8(Z_3)$ meets $\Phi$ in a curve including $\delta$ once and $\gamma$ twice, so that, $\Phi$ being of order 12, the residue is a quintic. Hence an arbitrary prime section $C$ of $\Phi$ is projected into a quintic in $\pi$ which, since $C$ meets $\delta$ thrice and $\gamma$ twice, has a triple point at $Y_0$ and a node at $X_1$.

Any $\delta$ and any $\gamma$ may be chosen on $\Phi$ as the ones to be mapped on the first neighborhoods of $Y_0$ and $X_1$.

## 5. Invariance under Harmonic Inversions

The equations (1.1) imply

$$x_0z_3 = x_1z_2 = x_2z_1 = x_3z_0 = y_0y_3 = y_1y_2 = \xi^2\eta^3\zeta^5, \tag{5.1}$$

so that when each of $\xi, \eta, \zeta$ is replaced by its reciprocal, the ratios of the six sums

$$x_0 + z_3, x_1 + z_2, x_2 + z_1, x_3 + z_0, y_0 + y_3, y_1 + y_2$$

are unchanged, as also are those of the six differences

$$x_0 - z_3, x_1 - z_2, x_2 - z_1, x_3 - z_0, y_0 - y_3, y_1 - y_2.$$

The replacement is a standard quadratic transformation in $\pi$ with four fixed points $(\pm 1, \pm 1, 1)$; its three fundamental points $X$, $Y$, $Z$ are the diagonal points of the quadrangle of fixed points.

Perhaps as convenient an approach as any is the following. Take, on $\Phi$, any two $\gamma$ and any two $\delta$; they have intersections

$$A: \gamma_1 \wedge \delta_1, \quad B: \gamma_1 \wedge \delta_2, \quad A': \gamma_2 \wedge \delta_2, \quad B': \gamma_2 \wedge \delta_1. \tag{5.2}$$

These are mapped in $\pi$ by $a, b, a', b'$; $ab$ and $a'b'$ meet at $X$, $ab'$ and $a'b$ at $Y$. Take $Z$ at the intersection of $aa'$ and $bb'$; take the unit point at $a$. Then

$$a(1,1,1), b(-1,1,1), a'(-1,-1,1), b'(1,-1,1) \tag{5.3}$$

are fixed points for the quadratic transformation

$$\xi' : \eta' : \zeta' = \eta\zeta : \zeta\xi : \xi\eta.$$

When each of $\xi, \eta, \zeta$ is replaced in (1.1) by the product of the other two, every resulting monomial is, since $\xi$ never occurs in (1.1) to higher power than its square, divisible by $\xi^3$; it is also, since $\eta$ never occurs in (1.1) to higher power than its cube, divisible by $\eta^2$. When, after the replacement, $\xi^2\eta^2$ is canceled, the remaining quintics are seen to be (1.1) subjected to the sextuple transposition

$$(x_0z_3)(x_1z_2)(x_2z_1)(x_3z_0)(y_0y_3)(y_1y_2).$$

This involutory projectivity in $S$ is the harmonic inversion $\mathcal{H}$ in the pair of skew [5]'s

$$\Sigma: x_0 - z_3 = x_1 - z_2 = x_2 - z_1 = x_3 - z_0 = y_0 - y_3 = y_1 - y_2 = 0,$$
$$\Sigma': x_0 + z_3 = x_1 + z_2 = x_2 + z_1 = x_3 + z_0 = y_0 + y_3 = y_1 + y_2 = 0.$$

$\Phi$ is met by $\Sigma$ at $A$ and $B$, by $\Sigma'$ at $A'$ and $B'$. The transversal from any other point of $\Phi$ to $\Sigma$ and $\Sigma'$ is a chord of $\Phi$, its intersections with $\Phi$ being harmonic to those with $\Sigma$ and $\Sigma'$. The tangent lines to $\Phi$ at $A$ and $B$ all meet $\Sigma'$; those at $A'$ and $B'$ all meet $\Sigma$. The tangent planes $\Omega_2(A')$ and $\Omega_2(B')$ meet $\Sigma$ in lines $d_1$, $d_2$ through the intersection $O$ of the tangents to $\gamma_2$ at $A'$ and $B'$, and similarly with primed and unprimed letters transposed.

## 6. A Doubly Covered Nodal del Pezzo Quintic

The projection $f$ of $\Phi$ from $\Sigma'$ onto $\Sigma$ is the intersection of $\Sigma$ with chords of $\Phi$ and is covered twice. Curves on $\Phi$ are paired by $\mathcal{H}$, members of a pair being projected into the same curve on $f$; but certain curves on $\Phi$ are self-paired and so are projected themselves into curves covered twice on $f$. These self-paired curves include, as their maps in $\pi$ guarantee, $\gamma_1$, $\gamma_2$, $\delta_1$, $\delta_2$; but every point of $\gamma_2$ is projected into $O$, the center of the involution set up on $\gamma_2$ by $\mathcal{H}$. The joins of the involution set up on $\delta_1$ form a regulus, two of them being the tangents at $A$ and $B'$; $\Sigma'$ and $\Sigma$ contain lines $b_1'$, $b_1$ of the complementary regulus, and the projection of $\delta_1$ is $b_1$ covered twice; $b_1$ passes through $A$ and meets $d_2$ at the intersection of $\Sigma$ with the tangent of $\delta_1$ at $B'$. The projection of $\delta_2$ is, similarly, a line $b_2$ covered twice and meeting $d_1$ at the intersection of $\Sigma$ with the tangent of $\delta_2$ at $A'$; $b_2$, of course, contains $B$. Finally $\gamma_1$, with the joins of the involution set up thereon being concurrent on $\Sigma'$, is projected into $AB$ covered twice. So there is a skew pentagon of lines on $f$.

Take an arbitrary [4] $\sigma$ in $\Sigma$; the section $C$ of $\Phi$ by the prime $\sigma\Sigma'$ has order 12 and is self-paired; its chords transversal to $\sigma$ and $\Sigma'$ generate a scroll $R$ and include the tangents of $C$ at $A'$ and $B'$. No other generator of $R$ touches $C$, because its intersections with $C$ are harmonic to its intersections with $\sigma$ and $\Sigma'$ and so cannot coincide save at a point of one of these two spaces. Since $C$ has genus 2, and the (1, 2) correspondence between a prime section of $R$ and $C$ has two coincidences on $C$, $R$ is elliptic by Zeuthen's formula (8, p. 107). The projection of $C$ from $\Sigma'$ onto $\Sigma$ is the section of $R$ by $\sigma$, covered twice; as $C$ meets $\Sigma'$ twice, the order of this projection is $\frac{1}{2}(12 - 2) = 5$; the prime sections of $f$ are elliptic quintics.

Suppose, however, that $\sigma$ contains $A$ and thereby forces $\sigma\Sigma'$ to contain the whole tangent plane $\Omega_2(A)$. Then $C$, the section of $\Phi$ by a prime containing $\Omega_2(A)$, has a node at $A$: both nodal tangents meet $\Sigma'$ and are generators of $R$, which thus has a node at $A$—as, therefore, does the section $f$ of $R$ by $\sigma$. Likewise at $B$. So it appears that the projection of $\Phi$ from $\Sigma'$ onto $\Sigma$ is, covered twice, *a del Pezzo quintic with two nodes*. Nodal del Pezzo surfaces, of orders between 3 and 8 inclusive, were catalogued by Timms. The labeling of the lines on $f$ by the letters $b$ and $d$ has been chosen to accord with his (7, p. 232).

Since the choices of $\gamma_1, \gamma_2, \delta_1, \delta_2$ in the pencils of conics and cubics on $\Phi$ were free, there is a quadruple infinity of such harmonic inversions $\mathcal{H}$ under which $\Phi$ is invariant.

## 7. Invariance Again

Cremona transformations in $\pi$ that turn the linear system of mapping quintics into itself map self-projectivities of $\Phi$; so, therefore, in particular, do projectivities in $\pi$ for which both $X$ and $Y$ are fixed points. Among such are the homologies $h$ with $XY$ for axis, or line of fixed points, and center at any point $Z$ off $XY$. If the third vertex of the triangle of reference is put at $Z$, $h$ is effected by multiplying $\zeta$ by some factor $k \neq 1$. Each entry in (1.1) is then multiplied by that power of $k$ equal to the power of $\zeta$ appearing there, and in entries on a slope upwards from left to right this power is the same. The invariant points of the projectivity $T$ induced in $S$ by $h$ consist therefore of $X_0, Z_3$, the points of the lines $X_1Y_0$ and $Z_2Y_3$, and the points of the planes $X_2Y_1Z_0$ and $Z_1Y_2X_3$. This is the general situation, but there would be pointwise invariant spaces of larger dimensions if there were equalities among $1, k, k^2, k^3, k^4, k^5$. The six spaces that are pointwise invariant for $T$ are seen, on referring to (2.2), to be

$$X_0,\quad \Omega_2(X_0) \wedge \Omega_{10}(Z_3),\quad \Omega_5(X_0) \wedge \Omega_8(Z_3),$$
$$Z_3,\quad \Omega_{10}(X_0) \wedge \Omega_2(Z_3),\quad \Omega_8(X_0) \wedge \Omega_5(Z_3).$$

Any two points of $\Phi$ can be shown to take the places $X_0$ and $Z_3$ for such transformations $T$.

Although it would be too long a digression to describe the geometry in any detail, one may just remark that when $k$ is a complex cube root of 1 the fixed points of $T$ fill the three solids

$$X_0X_3Y_2Z_1,\quad X_1Y_0Y_3Z_2,\quad X_2Y_1Z_0Z_3.$$

## 8. The General Situation

In conclusion, one summarizes the geometry of the surfaces whose sections are hyperelliptic and have genus $p > 2$. The ambient space $S$ has dimension $3p + 5$; $\Phi$, of order $4p + 4$, is mapped in $\pi$ by curves of order $p + 3$ with $X$ of multiplicity $p + 1$ and $Y$ a node; $\Phi$ contains a pencil of conics $\gamma$ and a pencil of curves $\delta$ of order $p+1$. The equations of the mapping, generalizing (1.1), are

$$x_i=\xi^2\eta^{p+1-i}\zeta^i,\quad y_i=\xi\eta^{p+1-i}\zeta^{i+1},\quad z_i=\eta^{p+1-i}\zeta^{i+2}\quad (i=0,1,\ldots,p+1). \tag{8.1}$$

These $3p+6$ relations may helpfully be visualized as strung along 3 rows of $p+2$ entries, the power of $\zeta$ being unvaried along any line sloping up from left to right.

$X_0, Z_0, X_{p+1}, Z_{p+1}$ are on $\Phi$; two "sides" of this quadrangle are $\gamma$, two are $\delta$.

The nest of tangent spaces is

$$\Omega_0\subset\Omega_2\subset\Omega_5\subset\Omega_8\subset\ldots\subset\Omega_{3p-1}\subset\Omega_{3p+2}\subset\Omega_{3p+4},$$

the dimension rising by 3 at each step from 2 to $3p+2$, but only by 2 at the first and last steps. An osculating prime $\Omega_{3p+4}(\xi',\eta',\zeta')$ meets $\Phi$ in a curve having a multiple point of order $p+3$ at $(\xi',\eta',\zeta')$; its map can only be the join of $(\xi',\eta',\zeta')$ to $X$, reckoned $p+1$ times, together with its join to $Y$ reckoned twice. Thus each of $(\xi,\eta,\zeta)$ and $(\xi',\eta',\zeta')$ lies in the osculating prime of $\Phi$ at the other when

$$(\eta\zeta'-\eta'\zeta)^{p+1}(\xi\zeta'-\xi'\zeta)^2=0.$$

This symmetrical pairing of the points of $\Phi$ is subordinate to a polarity in $S$: a null polarity if $p$ is even, reciprocation in a quadric if $p$ is odd. If $p=3$ the quadric is

$$x_0z_4+x_4z_0-4(x_1z_3+x_3z_1)+6x_2z_2-2y_0y_4+8y_1y_3-6y_2^2=0;$$

each $\Omega_2$ has the corresponding $\Omega_{11}$, each $\Omega_5$ the corresponding $\Omega_8$, for its polar. Every $\Omega_5$ lies on the quadric.

One is forcibly reminded here of the geometry of the rational normal curve $\Gamma$, of order $n$ in $[n]$, because of the polarity, signalized by Clifford (3, p. 313), that it induces. If $n$ is odd, $\Gamma$ induces a null polarity in which the osculating spaces of $\Gamma$, at any of its points, that are complementary (i.e. whose dimensions sum to $n-1$) are polars of each other. If $n$ is even, $\Gamma$ induces a polarity with respect to the unique quadric that contains every osculating $[\frac{1}{2}n-1]$ of $\Gamma$, and here too complementary osculating spaces at the same point of $\Gamma$ are polars of each other.

The primal generated by the $\Omega_{3p+2}$ is obtainable from (3.1) and (3.5) when the powers of $\eta'\zeta-\eta\zeta'$ are all raised by $p-2$: replace the monomials, all of degree $p+3$, in $\xi,\eta,\zeta$ by using (8.1), and then eliminate $\xi',\eta',\zeta'$.

The mapping of $\Phi$ on $\pi$ is the outcome of projection from any $\Omega_{3p+2}$.

## 9. Conclusion

One has, analogously to (5.1),

$$x_iz_{p+1-i}=y_iy_{p+1-i}=\xi^2\eta^{p+1}\zeta^{p+3},$$

and the standard quadratic transformation in $\pi$, as is seen by the procedure of Section 5, maps the action on $\Phi$ of the harmonic inversion $\mathcal{H}$ whose fundamental spaces are

$$\left.\begin{array}{ll}\Sigma': & x_i + z_{p+1-i} = y_i + y_{p+1-i} = 0 \\ \Sigma: & x_i - z_{p+1-i} = y_i - y_{p+1-i} = 0\end{array}\right\}. \qquad (i = 0, 1, \ldots, p+1), \qquad (9.1)$$

where it is to be noted that, if $P$ is odd, $\Sigma$ has dimension one higher than does $\Sigma'$ because of the evanescence of the last difference when $i = \frac{1}{2}(p+1)$. If $p = 2m$ both spaces have dimension

$$6m + 5 - (2m + 2) - (m + 1) = 3m + 2;$$

but if $p = 2m + 1$, $\Sigma'$ has dimension

$$6m + 8 - (2m + 3) - (m + 1) - 1 = 3m + 3,$$

and $\Sigma$ dimension $3m + 4$. In this latter contingency all four fixed points $A$, $B$, $A'$, $B'$ on $\Phi$ are in $\Sigma$; in the former, as in Section 5, two are in $\Sigma$, two in $\Sigma'$. This is verified by substituting from (5.3) in (8.1) and then in (9.1).

A word should be said about the geometry when $p = 3$. The [7] $\Sigma$ contains $A$, $B$, $A'$, $B'$; every other point of $\Phi$ is on a chord transversal to $\Sigma$ and $\Sigma'$, but the tangents at $A$, $B$, $A'$, $B'$, all meet $\Sigma'$, so that $\Omega_2(A)$, $\Omega_2(B)$, $\Omega_2(A')$, $\Omega_2(B')$ meet $\Sigma'$ in lines $\alpha$, $\beta$, $\alpha'$, $\beta'$.

The projection $f$ of $\Phi$ from $\Sigma$ onto the [6] $\Sigma'$ is covered twice; since $\Sigma$ is quadrisecant to $\Phi$, whose order is 16, the order of $f$ is $\frac{1}{2}(16 - 4) = 6$.

The conics $\gamma_1$, $\gamma_2$ of (5.2) shrink to the centers $O_1$, $O_2$ of the involutions induced on them by $\mathcal{H}$; $O_1$ is the intersection of $\alpha$ and $\beta$, $O_2$ of $\alpha'$ and $\beta'$.

The scroll of joins of pairs of the involution induced by $\mathcal{H}$ on the quartic $\delta_1$ through $A$ and $B'$ (cf. (5.2)) includes its tangents there; the projection of $\delta_1$ is a line, covered twice, meeting $\alpha$ and $\beta'$. Likewise the projection of $\delta_2$ is a line, covered twice, meeting $\alpha'$ and $\beta$.

So there is a skew hexagon of lines on $f$, which is the sextic surface of del Pezzo (7, p. 225). That its curve sections are elliptic is a consequence of their being in (1,2) correspondence, with 4 branch points, with a curve of genus 3 (a prime section of $\Phi$).

Projectivities that leave $\Phi$ invariant and are induced by a homology in $\pi$ with $Z$ for center and $XY$ for axis have for pointwise invariant spaces the two points $Z_{p+1}$ and $X_0$, the two lines

$$\Omega_2(Z_{p+1}) \wedge \Omega_{3p+4}(X_0) \quad \text{and} \quad \Omega_{3p+4}(Z_{p+1}) \wedge \Omega_2(X_0),$$

and $p$ planes

$$\Omega_{3j+2}(Z_{p+1}) \wedge \Omega_{3(p-j)+5}(X_0) \qquad (j = 1, 2, \ldots, p).$$

## References


(1) Baker, H. F., *Principles of Geometry* 4. Cambridge 1940.

(2) Castelnuovo, G., Sulle superficie algebriche le cui sezione piane sono curve iperellitiche. *Rendiconti del Circolo Matematico di Palermo* **4** (1890). Also in *Memorie Scelte*. Bologna 1937.

(3) Clifford, W. K., *Mathematical Papers*, Macmillan, London 1882.

(4) del Pezzo, P., Sugli spazi tangenti ad una superficie o ad una varietà immersa in uno spazio di più dimensioni. *Rendiconti Acc. Napoli* **25** (1886), 176–180.

(5) Segre, C., Mehrdimensionale Räume. In *Encyklopädie der Math. Wissenschaften*, III, p. C7.

(6) Semple, J. G. and Roth, L., *Algebraic Geometry*. Oxford 1949.

(7) Timms, G., The nodal cubic surfaces and the surfaces from which they are derived by projection. *Proc. Royal Soc. (A)* **119** (1928), 213–248.

(8) Zeuthen, H. G., *Lehrbuch der Abzählenden Methoden der Geometrie*. Teubner, Leipzig, 1914.

# On the Circular Transformations of Möbius, Laguerre, and Lie[†]

I. M. Yaglom*

In line with the elementary-geometric character of this paper, our primary concern is with circular transformations in the (Euclidean) plane (thought of differently for different types of transformations—see, for example, [1]). The (easy) extension of the various constructions to $n$-dimensional space is discussed at the end of the paper. Our aim is to show that it is possible to develop entirely analogous elementary theories of the circular *point transformations* of Steiner and Möbius [8], the circular *axial transformations* of Laguerre [5], and the circular *contact transformations* of Lie [6, 7].

## 1. The Circular Transformations of Möbius

The following well-known development of the theory of circular point transformations, presented in many texts, apparently goes back to Jacob Steiner. Consider the set of points of the inversive plane (the Euclidean plane supplemented by a single point at infinity, $\omega$). Let $A$ be a point different from $\omega$, $S$ a circle (not a line) of finite radius with center $Q$, $a$ a line with $a \ni A$, and $a \cap S = \{M, N\}$. Then the product $\overline{AM} \cdot \overline{AN}$ of the (lengths of) the oriented segments $\overline{AM}$ and $\overline{AN}$ is called the *power* of $A$ with respect to $S$ and is denoted by $po(A, S)$. The number

$$po(A, S) = d^2 - r^2, \qquad d = AQ, \tag{1.1}$$

depends on $A$ and $S$ but not on $a$. Also, $po(A, S) = t\,(A, S)^2$, where $t(A, S)$ is the tangential distance from $A$ to $S$. $t(A, S)$ is real only if $A$ is on or outside $S$. It is

[†]Translated by Abe Shenitzer.
* 1-ĭ Goncharnyĭ per. 7, apt. 17, Moscow 109172, USSR.

obvious that $po(A,S) > 0$, $= 0$, $< 0$, according as $A$ is in the interior of $S$, on $S$, or outside $S$.

It is clear that the set $\{A \mid po(A,S) = k\}$ for fixed $S$ and $k$ is a circle; the set $\{A \mid po(A,S_1) = po(A,S_2)\}$ for fixed $S_1$ and $S_2$ is a line (the *radical axis* of $S_1$ and $S_2$); and the set

$$\{A \mid po(A,S_1) = po(A,S_2) = po(A,S_3)\}$$

for fixed $S_1$, $S_2$, and $S_3$ is a point (the *radical center* of $S_1$, $S_2$, and $S_3$). The set

$$\mathfrak{A} = \mathfrak{A}(Q,k) = \{S \mid po(Q,S) = k\} \tag{1.2}$$

for fixed $Q$ and $k$ is a *bundle* of circles with radical center $Q$ and power $k$; here it is natural to assume that $\mathfrak{A}$ includes all lines $a$ such that $a \ni Q$. It is natural to include in the class of bundles the set of circles perpendicular to some line $l$.

The intersection

$$\mathfrak{a} = \mathfrak{A}_1 \cap \mathfrak{A}_2 = \{S \mid S \in \mathfrak{A}_1 \quad \text{and} \quad S \in \mathfrak{A}_2\} \tag{1.3}$$

with fixed $\mathfrak{A}_1$ and $\mathfrak{A}_2$ is called a *pencil* of circles. In general, a pencil can be defined as a (maximal) set of circles such that any two have the same radical axis (axis of the pencil), and a bundle as a (maximal) set of circles such that any three have the same radical center (center of bundle).

Let $\mathfrak{A}$ be a bundle and $M$ a point. The circles

$$\{S \mid S \in \mathfrak{A} \quad \text{and} \quad S \ni M\} \tag{1.4}$$

define a second point $M'$ (possibly coincident with $M$) common to the circles in (1.4). The map

$$i_{\mathfrak{A}} : M \to M' \qquad (\text{or} \quad i_{\mathfrak{A}}(M) = M') \tag{1.5}$$

is called an *inversion* (or *point inversion*) with respect to the bundle $\mathfrak{A}$. If $\mathfrak{A} = \mathfrak{A}(Q,k)$ $(k \neq 0)$, $i_{\mathfrak{A}}$ is the inversion with center $Q$ and power $k$. (In some treatments it is called an *inversion* when $k > 0$, and an *antiinversion* when $k < 0$.) If $\mathfrak{A}$ is the bundle of circles perpendicular to $l$, $i_{\mathfrak{A}}$ is the *reflection* in $l$.

It is not difficult to show that an inversion $i_{\mathfrak{A}}$ is a point circular transformation, that is, that it maps a circle (of finite or infinite radius—a line) onto a circle:

$$i_{\mathfrak{A}} : S \to S' \qquad (\text{or} \quad i_{\mathfrak{A}}(S) = S'); \tag{1.5a}$$

and that it is conformal, that is, it preserves angles. It can also be shown that every circular transformation can be written as a product of at most four point inversions.

## 2. The Circular Transformations of Laguerre

Next we consider the less well-known but equally simple analogous theory of circular axial transformations (circular transformations of Laguerre). Consider the set of lines (*axes* or spears) in the plane; here circles (of finite radius) will be regarded as oriented and viewed as sets of lines—their (oriented) tangents. (No orientation is assigned to a circle of zero radius—a point—viewed as the totality of lines through that point.) Let $a$ be a directed line, $S$ a circle with center $Q$ and radius $r$ (viewed as a class of lines—see above), $A \in a$ a point, and $A \cap S = \{m, n\}$; then

$$po(a, S) = \tan(\tfrac{1}{2}\angle(a, m))\tan(\tfrac{1}{2}\angle(a, n)) \tag{2.1}$$

depends only on $a$ and $S$ but not on $A$. This is apparent from the alternative expression for $po(a, S)$:

$$po(a, S) = \frac{r-d}{r+d} \quad \left(= \tan^2(\tfrac{1}{2}\angle(a, S)\right), \qquad d = (Q, a). \tag{2.1a}$$

Here $(Q, a)$ is the signed distance from $Q$ to $a$, and $\angle(a, S)$ is the angle between $a$ and $S$. The latter is real only if $a$ intersects $S$ or touches it. The quantity $po(a, S)$ is called the *power of the line a with respect to the circle S*. It is clear that $po(a, S) > 0$ if $a$ intersects $S$; $po(a, S) < 0$ if $a$ does not intersect $S$, in which case $po(a, S) = -\tanh^2 \frac{1}{2}p$, where $p$ is the *inversive distance* between $a$ and $S$ [2, p. 394]; $po(a, S) = 0$ if $a$ is tangent to $S$; $po(a, S) = \infty$ if $a$ is "antitangent" to $S$ (that is, $a$ and $S$ are oppositely oriented); $po(a, S) = -1$ iff $S$ is a point (if $S$ is a point and $a \in S$, then $po(a, S)$ is undefined); $po(a, S) = 1$ iff $a \ni Q$ (equivalently, $a \perp S$).

It is clear that the set $\{a \mid po(a, S) = k\}$ for fixed $S$ and $k$ is a circle; the set $\{a \mid po(a, S_1) = po(a, S_2)\}$ is a point, the *center of similitude* of $S_1$ and $S_2$; the set $\{a \mid po(a, S_1) = po(a, S_2) = po(a, S_3)\}$ is a line, the *axis of similitude* of $S_1$, $S_2$, and $S_3$. The set

$$\mathfrak{B} = \{S \mid po(q, S) = k\} \tag{2.2}$$

for fixed $q$ and $k$ is a *net* of circles with axis $q$ and power $k$. It is natural to regard all circles of fixed radius $a$ (which may be $> 0$, $= 0$, or $< 0$) as a (singular) net. Thus the net of circles with axis $q$ and positive power $k$ is the set of circles cutting $q$ at the constant angle $2\arctan\sqrt{k}$. Similarly, the net with axis $q$ and negative power $k$ is the set of circles at constant inversive distance from $q$, that is, circles which subtend a constant angle $\frac{1}{2}\pi - 2\arctan\sqrt{-k}$ from their nearest point on $q$. (This is a "singular" net when the constant angle is zero and $q$ is at infinity.) The intersection

$$\mathfrak{b} = \mathfrak{B}_1 \cap \mathfrak{B}_2 = \{S \mid S \in \mathfrak{B}_1 \quad \text{and} \quad S \in \mathfrak{B}_2\} \tag{2.3}$$

of two nets is called a *row* of circles. For instance, we speak of a row of congruent circles with collinear centers. In general, a row of circles is a (maximal) set of circles any two of which have the same center of similitude (the center of the net), and a net is a (maximal) set of circles any three of which have the same axis of similitude (the axis of the net).

If $\mathfrak{B}$ is a fixed net and $m$ a fixed (oriented) line, then the set of circles

$$\{S \mid S \in \mathfrak{B} \quad \text{and} \quad S \ni m\} \tag{2.4}$$

defines a second line $m'$ (which may coincide with $m$) tangent to all these circles. The mapping

$$i_{\mathfrak{B}} : m \to m' \qquad (\text{or} \quad i_{\mathfrak{B}}(m) = m') \tag{2.5}$$

is called a *Laguerre inversion* or an *axial inversion* generated by the net $\mathfrak{B}$. If $\mathfrak{B}$ is a net of circles of fixed radius $a$, then $i_{\mathfrak{B}}$ is called a *special inversion* or *axial dilatation* by $a$. If $\mathfrak{B} = \mathfrak{B}(q,k)$ is a net with axis $q$ and power $k$, then $i_{\mathfrak{B}}$ is said to be an *inversion with axis $q$ and power $k$.*

It is possible to show that an inversion $i_{\mathfrak{B}}$ is a circular transformation, that is, it carries every circle $S$ (of finite or zero radius) to a circle $S'$:

$$i_{\mathfrak{B}} : S \to S' \qquad (\text{or} \quad i_{\mathfrak{B}}(S) = S'). \tag{2.5a}$$

An inversion $i_{\mathfrak{B}}$ is an equidistantial transformation, that is, it preserves the tangential distance $t(\Gamma_1, \Gamma_2)$ between two curves $\Gamma_1$ and $\Gamma_2$ (the length of the segment of the common tangent to $\Gamma_1$ and $\Gamma_2$). It is easy to show that every circular (axial) equidistantial transformation is representable as a product of at most five axial inversions, and every circular (axial) transformation differs by at most a similarity from an equidistantial transformation.[1]

## 3. The Circular Transformations of Lie

The (Steiner) concept of power of a point with respect to a circle as defined by the equation

$$po(A,S) = d^2 - r^2 = (t(A,S))^2 \tag{1.1}$$

admits of the following natural generalization:

$$po(S_1,S_2) = d^2 - (r_1 - r_2)^2 = (t(S_1,S_2))^2, \qquad d = Q_1Q_2, \tag{3.1}$$

where $t(S_1,S_2)$ is the tangential distance of the (oriented) circles $S_1$ and $S_2$ with centers $Q_1$ and $Q_2$ and radii $r_1$ and $r_2$ (which is real only if $S_1$ and $S_2$ have a common tangent, that is, if $d \geqslant |r_1 - r_2|$). It is natural to call $po(S_1,S_2)$ the *power of the circle $S_1$ with respect to the circle $S_2$*.[2] It is clear that if $S_1$ is a circle of radius zero, that is, if $S_1$ reduces to a point $A$, and $S_2 = S$, then $po(S_1,S_2) = po(A,S)$, where the right-hand side is defined as in Section 1. On the other hand, the Laguerre definition of the power of a line with respect to a circle is not obtainable in a similar manner. Indeed, if we regard a line as the limit of a circle ("a circle of infinite radius"), and if $S_1$ or $S_2$ is a line, then it might seem natural to put $po(S_1,S_2) = \infty$.[3]

[1] Another parallel treatment of point and axial circular transformations, which is not applicable to circular contact transformations of Lie and does not carry over to circular transformations in space but is helpful in the study of (point and axial) circular transformations of non-Euclidean planes, involves the use of complex coordinates of points ($z = x + iy$, $x, y \in \mathbb{R}$, $i^2 = -1$) and dual coordinates of lines ($w = u + \epsilon v$, $\epsilon^2 = 0$). The circular transformations are described in terms of linear fractional transformations of the appropriate coordinates (see [9], [10]).

[2] Another variant of the concept of the power of a circle with respect to a circle is due to G. Darboux (cf. [4]); Darboux put $\mathrm{pg}(S_1,S_2) = d^2 - r_1^2 - r_2^2$. $\mathrm{pg}(S_1,S_2)$ is called the *Darboux power of $S_1$ with respect to $S_2$*.

[3] Certain considerations dictate the following definitions. If $S_1$ is a line $l$ and $S_2$ is a circle $S$ with center $Q$ and radius $r$, then $po(l,S) = 2(r - d)$, where $d = (Q,l)$. If $S_1$ and $S_2$ are lines $l_1$ and $l_2$, then $po(l_1,l_2) = 2\sin^2(\frac{1}{2}\angle(l_1,l_2))$. Here it must be borne in mind that the quantities $po(S_1,S_2)$ (see (3.1)), $po(l,S)$, and $po(l_1,l_2)$ are not comparable—are "measured in different units"—in the sense that the dimension of $po(S_1,S_2)$ is that of the square of a length, the dimension of $po(l,S)$ is that of length, and $po(l_1,l_2)$ is dimensionless.

The following is another definition of $po(S_1, S_2)$ which is equivalent to (3.1). If $S_1$ is a point $A$ and $S_2 = S$, then we put $po(S_1, S_2) = po(A, S)$, where $po(A, S)$ is given by (1.1). If $S_1$ is a circle of finite radius, then we map it to a point $S_1'$ by means of an (equidistantial) axial circular transformation (Laguerre transformation) $\lambda$. If $\lambda(S_2) = S_2'$, then we put $po(S_1, S_2) = po(S_1', S_2')$, where the right-hand side is defined as in Section 1. Since the map is equidistantial, the value of $po(S_1, S_2)$ is independent of the choice of $\lambda$.[4]

It is clear that $po(S_1, S_2) > 0$, $= 0$, $< 0$ according as $S_1$ and $S_2$ have two, one, or no common tangents. $po(S_1, S_2) = po(S_2, S_1)$. Clearly, $po(S_1, S_2)$ is a "Laguerre invariant" of the circles $S_1$ and $S_2$.

The set $\{S \mid po(S, S_1) = po(S, S_2)\}$ for fixed $S_1$ and $S_2$ is a net of circles (by analogy with the concept of a radical axis, this net could be called the *axial net* of $S_1$ and $S_2$, and its axis could be called simply the *axis* of $S_1$ and $S_2$). The set $\{S \mid po(S, S_1) = po(S, S_2) = po(S, S_3)\}$ for fixed $S_1$, $S_2$, and $S_3$ is a row (the *central row*) of $S_1$, $S_2$, and $S_3$ whose center might be called the *center* of $S_1$, $S_2$, and $S_3$. Finally, the set $\{S \mid po(S, S_1) = po(S, S_2) = po(S, S_3) = po(S, S_4)\}$ for fixed $S_1$, $S_2$, $S_3$, and $S_4$ is a single circle—the *equidistantial circle* of $S_1$, $S_2$, $S_3$, and $S_4$.

Let $\Sigma$ be a circle and $k$ a number. The set of circles

$$\mathfrak{C} = \{S \mid po(\Sigma, S) = k\} \tag{3.2}$$

may be called a *bunch* of circles with central circle $\Sigma$ and power $k$. $\mathfrak{C}$ may be characterized as the set of circles $S$ such that $t(S, \Sigma) = \kappa$ (where $\kappa = \sqrt{k}$ may be imaginary; For $\kappa = 0$, the circles $S$ are tangent to $\Sigma$). The bunch includes all lines tangent to $\Sigma$. It is natural to regard as bunches all nets of circles and the bundle

[4]It is possible to use a "dual" definition of the power of a circle with respect to a circle. If $S_1$ is a line $a$ and $S_2 = S$, then we put $po(S_1, S_2) = po(a, S)$, where the right-hand side is defined as in Section 2. If $S_1$ is a circle of finite radius, then we choose a circular point transformation (Möbius transformation) $\mu$ such that $\mu(S_1) = S_1'$ is a line, and put $po(S_1, S_2) = po(S_1', S_2')$, where $S_2' = \mu(S_2)$ and $po(S_1', S_2')$ is defined as above. The number

$$po(S_1, S_2) = \frac{d^2 - (r_1 - r_2)^2}{(r_1 + r_2)^2 - d^2} = \tan^2(\tfrac{1}{2}\angle(S_1, S_2)),$$

where $\angle(S_1, S_2)$ is real only if the circles $S_1$ and $S_2$ intersect, is independent of $\mu$ and is a "Möbius invariant" of the two circles. For nonintersecting circles $S_1$ and $S_2$,

$$po(S_1, S_2) = -\tanh^2(\tfrac{1}{2}p),$$

where $p$ is the *inversive distance* between the two circles (see [3, pp. 130, 176, Exercise 3]). This number admits of a simple geometric interpretation not connected with $\angle(S_1, S_2)$ and independent of whether or not $S_1$ and $S_2$ intersect (see [9, pp. 237–239]). $po(S_1, S_2)$ is obviously symmetric ($po(S_1, S_2) = po(S_2, S_1)$) and has many remarkable properties (for example, $po(S_1, S_2) = 0$ iff $S_1$ and $S_2$ are tangent; $po(S_1, S_2) = 1$ iff $S_1 \perp S_2$; $po(S_1, S_2) = -1$ iff $S_1$ or $S_2$ is a point; $po(S_1, S_2) = \infty$ iff $S_1$ and $S_2$ are antitangent). Here $\{S \mid po(S, S_1) = po(S, S_2)\}$ is a bundle, and $\{S \mid po(S, S_1) = po(S, S_2) = po(S, S_3)\}$ is, in general, a pencil.

$$\{S \mid po(S, S_1) = po(S, S_2) = po(S, S_3) = po(S, S_4)\}$$

is, in general, a single circle either cutting $S_1$, $S_2$, $S_3$, $S_4$ all at the same angle or situated at equal inversive distances from the four circles.

$po(S_1, S_2)$ can be used to define a variant of the Lie contact inversion, but this approach to the theory of circular contact transformation is somewhat less convenient than our approach.

of all lines in the plane. The intersection

$$\mathfrak{c} = \mathfrak{C}_1 \cap \mathfrak{C}_2 = \{S \mid S \in \mathfrak{C}_1 \quad \text{and} \quad S \in \mathfrak{C}_2\} \tag{3.3}$$

of two bunches of circles may be called a *chain* of circles. Examples of chains are the set of all circles tangent to two fixed circles (hyperbolic chain); the set of all circles which are tangent to one another at a given point (parabolic chain); and chains no three of whose circles have a common tangent circle (elliptic chains). (In general, a bunch can also be described as a (maximal) set of circles every four of which have the same equidistantial circle, and a chain as a (maximal) set of circles every three of which have the same axial pencil.)

Let $\mathfrak{C}$ be a bunch and $s$ a circle. Consider the set of circles

$$\{S \mid S \in \mathfrak{C} \quad \text{and} \quad S \text{ tangent to } s\}. \tag{3.4}$$

It can be shown that the circles in (3.4) other than $s$ are tangent to another circle $s'$ (which may coincide with $s$—if $s \in \mathfrak{C}$, then it is natural to put $s' = s$). Thus a bunch $\mathfrak{C}$ generates a map

$$i_{\mathfrak{C}} : s \rightarrow s' \qquad (\text{or} \quad i_{\mathfrak{C}}(s) = s') \tag{3.5}$$

on the set $\mathbb{S}$ of oriented circles in the plane (including points and lines as special cases of circles). It is natural to call the map $i_{\mathfrak{C}} : \mathbb{S} \rightarrow \mathbb{S}$ a *contact inversion* or *Lie inversion* generated by the bunch $\mathfrak{C}$. If $\mathfrak{C} = \mathfrak{C}(\Sigma, k)$ is a bunch with central circle $\Sigma$ and power $k$, it is natural to call $i_{\mathfrak{C}}$ a (contact) inversion with central circle $\Sigma$ and power $k$.

It is easy to see that a contact inversion preserves tangency of circles: if the circles $s_1$ and $s_2$ touch, then so do the circles $s_1' = i_{\mathfrak{C}}(s_1)$ and $s_2' = i_{\mathfrak{C}}(s_2)$. This fact enables us to regard an inversion $i_{\mathfrak{C}}$ as a map not only on the set of circles in the plane, but also on the set of linear elements in the plane (a linear element is defined as a point and an (oriented) line passing through that point). The latter map is a contact transformation in the sense of Lie [7].

It is clear that the Möbius point inversion and the Laguerre inversion are special cases of the Lie contact inversion—the first corresponds to the case when the central circle of the bunch $\mathfrak{C}$ is a point, and the second to the case when $\mathfrak{C}$ is a net. The general contact inversion $i = i(\Sigma, k)$ with central circle $\Sigma$ and power $k$ can be characterized by the fact that if $i(S) = S'$, then the circles $S$, $S'$, and $\Sigma$ belong to the same row and

$$po(\Sigma, S) \cdot po(\Sigma, S') = k^2. \tag{3.6}$$

Finally, it can be shown that every circular contact transformation (Lie contact transformation mapping circles on circles) is representable as a product of at most eleven contact inversions.

## 4. Extension to Spheres in *n*-dimensional Space

The above constructions are easily extended to spheres in $n$-dimensional space. Here we shall limit ourselves to a brief sketch pertaining to circular contact transformations of Lie. If $S_1$ and $S_2$ are two (oriented) spheres in (real) $n$-space with centers $Q_1(x_1, x_2, \ldots, x_n)$ and $Q_2(y_1, y_2, \ldots, y_n)$ and radii $r_1$ and $r_2$ (which may be positive or not), then it is natural to call the quantity

$$(t(S_1,S_2))^2 = d^2 - (r_1 - r_2)^2$$
$$= (x_1 - y_1)^2 + (x_2 - y_2)^2 + \cdots + (x_n - y_n)^2 - (r_1 - r_2)^2, \quad (4.1)$$

$d = Q_1 Q_2$, the *tangential distance* of $S_1$ and $S_2$; clearly, $t(S_1, S_2)$ is real only if $d \geqslant |r_1 - r_2|$. In particular, two spheres touch if $t(S_1, S_2) = 0$, or

$$(x_1 - y_1)^2 + (x_2 - y_2)^2 + \cdots + (x_n - y_n)^2 - (r_1 - r_2)^2 = 0. \quad (4.1')$$

We use the equation

$$X_0 : X_1 : X_2 : \cdots : X_n : X_{n+1} : X_{n+2}$$
$$= 1 : x_1 : x_2 : \cdots : x_n : \left[ \left( x_1^2 + x_2^2 + \cdots + x_n^2 \right) - r^2 \right] \quad (4.2)$$

to introduce on the sphere $S$ with center $Q(x_1, x_2, \ldots, x_n)$ and radius $r$ (homogeneous, redundant) *polyspheric coordinates* $X_0, X_1, X_2, \ldots, X_{n+2}$ connected by the relation

$$X_1^2 + X_2^2 + \cdots + X_n^2 - X_{n+1}^2 - X_0 X_{n+2} = 0. \quad (4.3)$$

In terms of these coordinates, the criterion (4.1′) of tangency of two spheres $S_1(X_0, X_1, X_2, \ldots, X_{n+2})$ and $S_2(Y_0, Y_1, Y_2, \ldots, Y_{n+2})$ takes the form

$$X_{n+2}Y_0 + X_0 Y_{n+2} - 2X_1Y_1 - 2X_2Y_2 - \cdots - 2X_nY_n + 2X_{n+1}Y_{n+1} = 0 \quad (4.1'a)$$

and Equation (4.1) takes the form

$$X_{n+2}Y_0 + X_0 Y_{n+2} - 2X_1Y_1 - 2X_2Y_2 - \cdots$$
$$- 2X_nY_n + 2X_{n+1}Y_{n+1} - (t(S_1, S_2))^2 X_0 Y_0 = 0. \quad (4.1a)$$

We note that the case $r = 0$ (that is, $X_{n+1} = 0$) is included in the above considerations. If the role of the sphere $S$ is played by the (oriented) (hyper)plane given (in terms of rectangular coordinates) by the equation

$$\cos\alpha_1 x_1 + \cos\alpha_2 x_2 + \cdots + \cos\alpha_n x_n - p = 0,$$

where $\cos^2\alpha_1 + \cos^2\alpha_2 + \cdots + \cos^2\alpha_n - p = 1$, then the polyspheric coordinates $X_0, X_1, X_2, \ldots, X_{n+2}$ of this "sphere of infinite radius" are determined by the equation

$$X_0 : X_1 : X_2 : \cdots : X_n : X_{n+1} : X_{n+2} = 0 : \cos\alpha_1 : \cos\alpha_2 : \cdots : \cos\alpha_n : 1 : 2p \quad (4.2')$$

(and continue to be linked by Equation (4.3)). As before, the condition (4.1′a) characterizes the case of tangency of the spheres $S_1(X_0, X_1, X_2, \ldots, X_{n+2})$ and $S_2(Y_0, Y_1, Y_2, \ldots, Y_{n+2})$ of finite, zero, or infinite radius (two (hyper)planes are said to be tangent if they are parallel and have the same orientation, that is, they have the same outer unit normals $(\cos\alpha_1, \cos\alpha_2, \ldots, \cos\alpha_n)$ and $(\cos\beta_1, \cos\beta_2, \ldots, \cos\beta_n)$).

We call the (positive or nonpositive) quantity

$$po(S_1, S_2) = (t(S_1, S_2))^2 \quad (4.4)$$

the *power of the sphere* $S_1$ *with respect to the sphere* $S_2$. We have

$$po(S_1, S_2) = X_0 Y_0 - X_{n+2} Y_0 - X_0 Y_{n+2} + 2X_1 Y_1 + 2X_2 Y_2 + \cdots + 2X_n Y_n - 2X_{n+1} Y_{n+1}. \tag{4.1b}$$

(cf. (4.1a). Fix a sphere $\Sigma(A_0, A_1, A_2, \ldots, A_{n+2})$ and a number $k$, and call the set of spheres

$$\mathfrak{C} = \{ S \mid po(\Sigma, S) = k \}$$

a *bunch* of spheres. It is clear that the equation of a bunch in polyspheric coordinates is the linear equation

$$F(Y) = (A_{n+2} - kA_0 Y_0) - 2A_1 Y_1 - 2A_2 Y_2 - \cdots - 2A_n Y_n + 2A_{n+1} Y_{n+1} + A_0 Y_{n+2} = 0. \tag{4.5'}$$

Fix a sphere $s(X_0, X_1, X_2, \ldots, X_{n+2})$. The tangency condition of the spheres $S$ and $s$ has the form (4.1′a). In the present case this reduces to

$$G(Y) = X_{n+2} Y_0 + X_0 Y_{n+2} - 2X_1 Y_1 - 2X_2 Y_2 \cdots - 2X_n Y_n + 2X_{n+1} Y_{n+1} = 0. \tag{4.1''a}$$

The spheres $\{ S \mid S \in \mathfrak{C}$ and $S$ tangent to $s\}$ satisfy (4.5′) and (4.1″a) and thus

$$tF(Y) + G(Y) = 0 \tag{4.6}$$

for all real $t$. If we compare (4.6) and (4.1″a), then we see that (4.6) may be viewed as the condition of tangency of the spheres under consideration and the sphere $s'(X_0', X_1', X_2', \ldots, X_{n+2}')$, where

$$X_0' = X_0 + tA_0, \qquad X_1' = X_1 + tA_1, \ldots, \qquad X_{n+2}' = X_{n+2} + tA_{n+2}. \tag{4.7}$$

Here the coordinates (4.7) must satisfy the quadratic equation (4.3). Substitution of the coordinates (4.7) in (4.3) yields a quadratic equation in $t$. One root of this equation is $t = 0$, and the other is

$$t = \frac{-kA_0^2}{kA_0 X_0 + 2A_1 X_1 + 2A_2 X_2 + \cdots + 2A_n X_n - 2A_{n+1} X_{n+1} - A_0 X_{n+2} - A_{n+2} X_0}. \tag{4.7'}$$

The first of these roots corresponds to the sphere $s$, and the second to a sphere $s'$ which is also tangent to all our spheres $S$. Thus each bunch $\mathfrak{C}$ $(\Sigma, k)$ of spheres (with central sphere $\Sigma$ and power $k$) defines on the set of oriented spheres a map

$$i_{\mathfrak{C}} : s(X_0, X_1, X_2, \ldots, X_{n+2}) \to s'(X_0', X_1', X_2', \ldots, X_{n+2}') \tag{4.8}$$

called a *contact inversion* with central sphere $\Sigma$ and power $k$; the coordinates of the sphere $s'$ are given by Equations (4.7)–(4.7′).

It is easy to show that the mapping (4.8) preserves tangency of spheres and that every tangency-preserving mapping on the set of (oriented) spheres in $n$-space (including spheres of zero radius (points) and spheres of infinite radius (hyperplanes))—that is, every spherical contact transformation of Lie—can be represented as a product of at most $n + 9$ contact inversions.

## References

[1] Blaschke, W., *Vorlesungen über Differentialgeometrie III; Differentialgeometrie der Kreise und Kugeln.* J. Springer, Berlin 1929.

[2] Coxeter, H. S. M., Parallel lines, *Can. Math. Bulletin* **21** (1978), 385–397.

[3] Coxeter, H. S. M. and Greitzer, S. L., *Geometry Revisited,* Math. Assoc. of America 1975.

[4] Darboux, G., *Principes de Géométrie Analytique*, Gauthiers-Villars, Paris 1917.

[5] Laguerre, E., Transformations par semidroites reciproques, *Nouv. Annales de Math.*, 1882, 542–556. Also in *Oeuvres*, Vol. II. Gauthiers-Villars, Paris 1905.

[6] Lie, S., Über Complexe, insbesondere Linien- und Kugelcomplexe, mit Anwendungen auf die Theorie partieller Differentialgleichungen. *Math. Annalen*), 145–256. Also in Gesammelte Abhandlungen, Vol. II, Part 1. H. Aschehoug–B. G. Teubner, Oslo–Leipzig 1935.

[7] Lie, S. and Scheffers, G., *Geometrie der Berührungtransformationen*. Teubner, Leipzig 1896.

[8] Möbius, A. F., Theorie der Kreisverwandschaften in rein geometrischer Darstellung. *Abhandlungen Königl. Sachs. Gessellschaft der Wissenschaften, Math.-Phys. Klasse* **2** (1855), 529–595. Also in *Gesammelte Werke*, Vol. II. Hirzel, Leipzig 1886. Über eine neue Verwandschaft zwischen ebenen Figuren. *Journal Reine Angew. Math.* **52** (1856), 218–228. Also in *Gesammelte Werke*, Vol. II.

[9] Yaglom, I. M., *Complex Numbers in Geometry*, Academic Press, New York 1968.

[10] Yaglom, I. M., *A Simple Non-Euclidean Geometry and Its Physical Basis*, Springer, New York Heidelberg Berlin, 1979.

# The Geometry of Cycles, and Generalized Laguerre Inversion†

J. F. Rigby*

## 1. Introduction

This paper on the geometry of cycles (oriented circles and lines) consists of a new look at some old ideas, using mainly synthetic methods. The figures are used for the communication of essentially simple geometrical ideas, and algebraic calculations are kept to a minimum.

In Chapter X of [1] Coolidge gives a method of representing cycles by points of a quadric in complex projective 4-space. Since we can equally well use real projective 4-space, this seems preferable as we are dealing with real cycles (Section 2). On p. 357 of [1] Coolidge remarks: "It is now time to take up the analytic treatment of oriented lines and circles, as thus, naturally, we shall obtain a far greater wealth of results than from purely geometric methods." This may ultimately be true, but it is fascinating to see what elegant synthetic interpretations of cycle theorems can be given in 4-space (Sections 3 and 4); there is not space here to discuss in this way all the cycle theorems mentioned by Coolidge.

In [8] and [9] Pedoe describes and discusses Laguerre inversions [6]. Laguerre's original concept relies heavily on rays (oriented lines), but we shall consider a more general type of mapping of cycles to cycles that can be defined in the inversive plane (where no distinction is made between circles and lines) (Section 5). These transformations are further discussed in Sections 6 and 7, and a study of the representations of Laguerre inversions on the quadric in 4-space (Sections 8 and 9) leads to a further generalization.

A *cycle* is an oriented circle, and a *ray* an oriented line (not to be confused with the term "ray" in ordered geometry, which denotes a half line). As usual in circle geometry, we adjoin a single point at infinity to the Euclidean plane, lying

†Research supported partly by a grant from the National Research Council of Canada.

*Department of Pure Mathematics, University College, Cardiff, Wales.

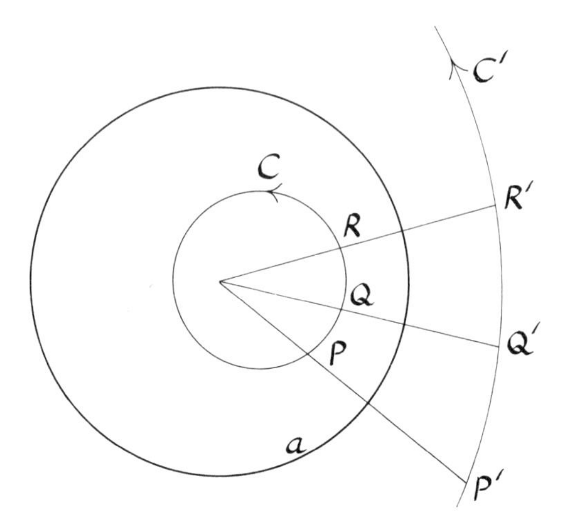


**Figure 1**

on every ray, and we regard rays as cycles through the point at infinity. Thus the term "cycle" includes rays when appropriate. A cycle of zero radius is a point or *point-cycle*. A *proper cycle* is a cycle or ray other than a point-cycle. The cycle *opposite* to the cycle $C$ (i.e., oriented in the opposite direction) is denoted by $\overline{C}$. We do not distinguish between a point-cycle and its opposite, so we may say that a point-cycle has no orientation.

Two cycles *touch*, or are *tangent* ("properly tangent" according to Coolidge), if they are oriented in the same direction at their point of tangency. Cycles $C$ and $S$ *antitouch*, or are *antitangent*, if $C$ and $\overline{S}$ touch. *Parallel* and *antiparallel* rays are defined in the obvious way.

The techniques of ordinary inversion can clearly be applied to cycles. In Figure 1, if a point moves in the positive direction round the cycle $C$, taking up the positions $P, Q, R, \ldots,$ then the inverse point with respect to the circle $a$ traces out a cycle $C'$. It seems natural to say that the inverse of $C$ is $C'$, and such a definition is suitable in many circumstances, but we shall find it more satisfactory in Section 6 to say that the inverse of $C$ is $\overline{C}'$; ordinary inversion then becomes a special case of Laguerre inversion.

## 2. Representation of Cycles by Points on a Quadric in $S^4$

The cycle $S$ with equation

$$a(x^2+y^2)-2bx-2cy+2d=0, \tag{1}$$

where $a \neq 0$, has center $(b/a, c/a)$ and radius $r$ given by

$$r^2=(b^2+c^2-2ad)/a^2$$

We shall write $b^2+c^2-2ad=e^2$ and take $e/a$ to be positive or negative according as the cycle is described in the positive or negative direction. We represent $S$ by the point $\tilde{S}$ in $S^4$ (real projective 4-space) with coordinates $(a,b,c,d,e)$. This point lies on the quadric $\mathbf{Q}$ with equation $y^2 + z^2 - 2xu - v^2 = 0$, using $(x, y, z, u, v)$ as coordinates in $S^4$. The cycle $S$ is a point if and only if $e = 0$. (Coolidge uses the coordinates $(a, -b, -c, d, ie)$, but we prefer to have a real representation.) We shall say that the point $(x, y, z, u, v)$ of $S^4$ lies *inside* or *outside* $\mathbf{Q}$ according as $y^2 + z^2 - 2xu - v^2 < 0$ or $> 0$. The polar of a point $P$ cuts $\mathbf{Q}$ in a nonruled or ruled quadric according as $P$ lies inside or outside $\mathbf{Q}$.

If $a = 0$ but $b$ and $c$ are not both zero, then (1) is the equation of a ray. We represent this ray by the point $(0,b,c,d,e)$, where $e^2 = b^2 + c^2 \neq 0$; we shall determine the sign of $e$ later. The single point at $\infty$ will be represented by $(0,0,0,1,0)$.

Two cycles $\tilde{S}(a,b,c,d,e)$, $\tilde{S}'(a',b',c',d',e')$ touch iff

$$bb' + cc' - ad' - a'd - ee' = 0, \tag{2}$$

i.e. iff $\tilde{S}, \tilde{S}'$ are conjugate points with respect to $\mathbf{Q}$, each lying on the pole of the other. This condition holds also for the ray $(0,b,c,d,e)$ and the point at $\infty$, $(0,0,0,1,0)$, which touch each other; it also holds for the parallel rays $(0,b,c,d,e)$ and $(0,b,c,d',e)$, which touch at $\infty$. We should like it to hold also for rays touching cycles.

For the ray $-2bx - 2cy + 2d = 0$, with coordinates $(0,b,c,d,e)$ where $e^2 = b^2 + c^2$, we can write $b = -e\sin\theta$, $c = e\cos\theta$, where $\theta$ is the *directed* angle that the ray makes with the positive $x$-axis. This determines the sign of $e$. (The opposite ray has the same values of $b$, $c$, $d$, but $\theta$ is increased by $\pi$, so $e$ changes sign.) It is easily checked that, with this convention for the sign of $e$, the condition (2) holds also for rays touching cycles.

Thus *cycles touching a given cycle $S$ are represented by points of* $\mathbf{Q}$ *lying on the polar of* $\tilde{S}$; this polar meets $\mathbf{Q}$ in a quadric cone with vertex $\tilde{S}$. Hence if $S$ and $T$ touch each other, then $\tilde{S}\tilde{T}$ is a generator of this quadric cone, and hence a generator of $\mathbf{Q}$. All points of the generator represent cycles touching $S$ and $T$; thus any set of cycles all touching each other at a point is represented by points on a generator of $\mathbf{Q}$, and conversely.

The points of the inversive plane are represented by points of $\mathbf{Q}$ of the form $(a,b,c,d,0)$, i.e. points on the intersection of $\mathbf{Q}$ with the 3-space $v = 0$. This 3-space is the polar of $V(0,0,0,0,1)$; $V$ lies inside $\mathbf{Q}$, so $v = 0$ meets $\mathbf{Q}$ in the nonruled quadric $\Pi : y^2 + z^2 - 2xu = 0,\ v = 0$. We shall call $v = 0$ the *point-space*.

The points lying on (i.e. touching) a given cycle $S$ are represented by the intersection of $\Pi$ with the polar of $\tilde{S}$; this polar meets the point-space in a plane, so concyclic points are represented by coplanar points of $\Pi$, and conversely. Here we have the usual representation of the inversive plane by points and circles on a sphere, except that the sphere is replaced by a nonruled quadric.

We see from the previous paragraph that if $S, \bar{S}$ are opposite cycles, then the polars of $\tilde{S}, \tilde{\bar{S}}$ meet the point-space $v = 0$ in the same plane, since $S, \bar{S}$ contain the same points. This follows also from the fact that $\tilde{S}$, $\tilde{\bar{S}}$, $V$ are collinear, their coordinates being $(a,b,c,d,e)$, $(a,b,c,d,-e)$, $(0,0,0,0,1)$: *two points of* $\mathbf{Q}$ *representing opposite cycles are collinear with* $V$.

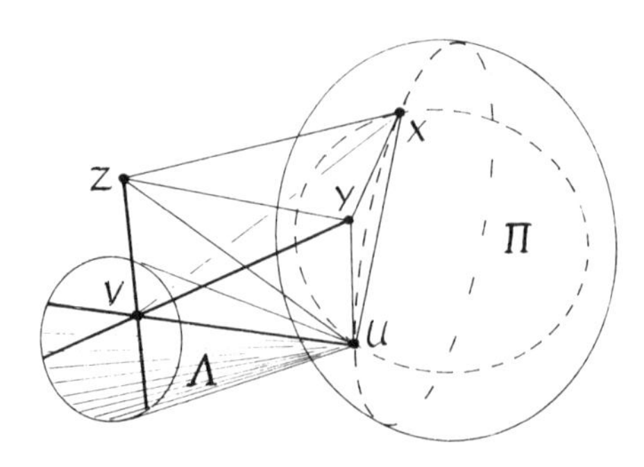


**Figure 2**

Rays are represented by points of **Q** of the form $(0, b, c, d, e)$, i.e. points of **Q** in the 3-space $x = 0$, the polar of $U(0,0,0,1,0)$, which we shall call the *ray-space;* $U$ lies on **Q**, so $x = 0$ meets **Q** in the quadric cone $\Lambda : y^2 + z^2 - v^2 = 0,\ x = 0$, with vertex $U$ (representing the point at $\infty$).[1] See Figure 2.

The rays touching a cycle $S$ are represented by the intersection of $\Lambda$ with the polar of $\tilde{S}$; this polar meets the ray-space in a plane, so rays touching a cycle are represented by coplanar points of $\Lambda$. The plane does not pass through $U$, assuming that $S$ itself is not a ray. Conversely, points of $\Lambda$ on a plane not through $U$ represent rays touching a cycle..

The cycles $\tilde{S}(a, b, c, d, e)$ and $\tilde{S}'(a', b', c', d', e')$ are orthogonal iff $bb' + cc' - ad' - da' = 0$. This means that the planes where the polars of $\tilde{S}$ and $\tilde{S}'$ meet the point-space $v = 0$ are conjugate with respect to $\Pi$.

We shall not be much concerned here with coaxial circles, but it can be shown that the circles of a coaxial system (or rather, the pairs of opposite cycles of which the circles are composed) are represented by points of **Q** on a plane through $V$, and conversely. The polar line $l$ of such a plane lies in the point-space, and the circles of the coaxial system are represented also by sections of $\Pi$ by planes through $l$. For the orthogonal system we take planes in the point-space through the polar line of $l$ with respect to $\Pi$.

As a special case, two cycles are concentric iff the two cycles and the point at $\infty$ belong to a coaxial system. Hence concentric cycles are represented by points of **Q** coplanar with $U$ and $V$, and conversely. This can also be deduced analytically: concentric cycles have coordinates of the form $(a, b, c, d, e)$, $(a, b, c, d', e')$.

## 3. Applications of the Representation on Q

Coolidge does not seem to make full use of the representation on **Q** to obtain elegant synthetic proofs of results about cycles. We shall consider such proofs in this section and the next, after defining a special type of mapping of cycles to cycles in the next paragraph.

[1] $\Lambda$ stands for "line"; the use of a Greek P for "ray" might be confusing.

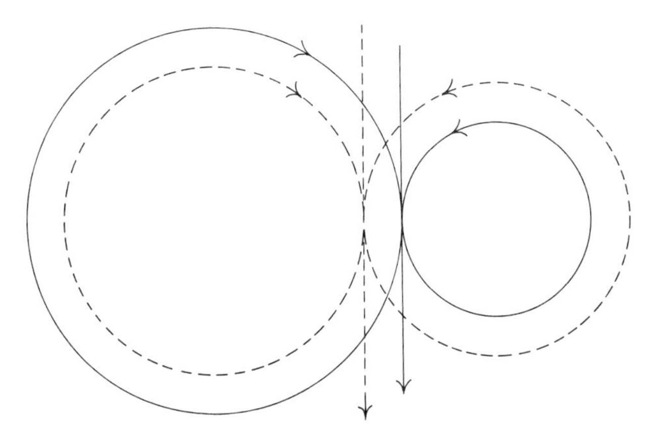

**Figure 3**

In Section 6 we shall study mappings of cycles to cyles that preserve tangency but do not usually map point-cycles to point-cycles or opposite cycles to opposite cycles. It is useful here to consider a very simple mapping of this type.[2] We choose a fixed real number $\delta$, and we map the cycle with center $C$ and radius $\rho$ to the cycle with center $C$ and radius $\rho + \delta$, remembering that $\rho$ is negative for a cycle described in the negative sense. We map the ray $R$ to the ray obtained by translating $R$ perpendicular to itself through a distance $|\delta|$, to the right when looking along $R$ if $\delta$ is positive, to the left otherwise. Denote this mapping by $\tau(\delta)$. Clearly $\tau(\delta)$ preserves tangency, as is shown in Figure 3.

Our first theorem is proved analytically by Coolidge, using the Frobenius identity [1, p. 360].

**3.1.** *Suppose that four cycles, taken in a cyclic order, are such that each touches the next one; then the four common tangents at the points of tangency either are parallel two by two or touch another cycle* (Figure 4).

*First Proof.* If the common tangents $l_1, l_2, l_3, l_4$ are not parallel two by two, then three of them, $l_1, l_2, l_3$ w.l.o.g., touch a cycle $C$ say. We can apply a mapping $\tau(\delta)$, for suitable $\delta$, to map $C$ to a point $C'$, and $l_i$ to $l_i'$. Then $l_1', l_2', l_3'$ are concurrent (Figure 5), and we show by elementary geometry that $l_4'$ is concurrent with them. Now apply $\tau(\delta)$ in reverse. □

*Second Proof.* Represent the cycles by points $P_1, P_2, P_3, P_4$ of $\mathbf{Q}$. Since each cycle touches the next, $P_1P_2$, $P_2P_3$, $P_3P_4$, $P_4P_1$ are generators of $\mathbf{Q}$. The points $L_1, L_2$, $L_3, L_4$ representing the common tangents lie on these generators (Figure 6). Now $L_1, \ldots, L_4$ lie in the 3-space $P_1P_2P_3P_4$. They also lie in the ray-space. Hence they are coplanar. If this plane does not pass through the vertex $U$ of $\Lambda$, they represent rays touching a cycle. If the plane passes through $U$, they lie on two

[2] Note however that the present mapping is not a Laguerre inversion, even in the extended sense of Section 8.

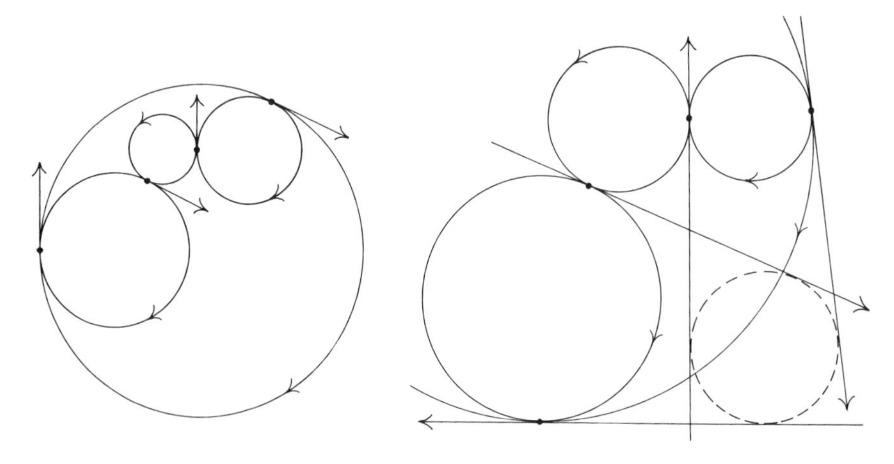

**Figure 4**

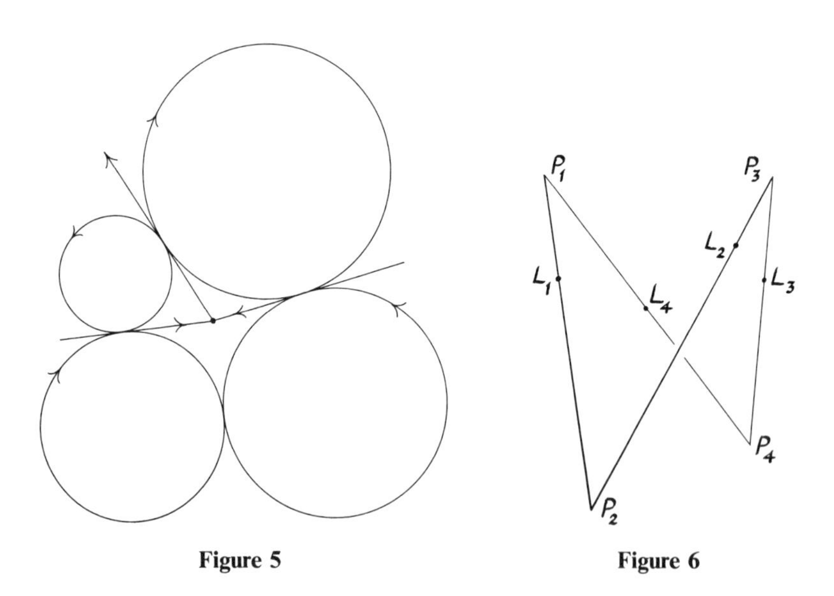


**Figure 5** **Figure 6**

generators of $\Lambda$, two on each. (No three of $L_1, \ldots, L_4$ can be collinear, as we see from Figure 6.) Hence $l_1, \ldots, l_4$ are parallel two by two. □

Figure 4 suggests a further result, not mentioned by Coolidge.

**3.2.** *In the situation of Theorem* 3.1, (*a*) *if the common tangents are parallel two by two, then the four points of contact are collinear, and* (*b*) *in the other case the four points of contact lie on a circle concentric with the cycle touched by the common tangents.*

*Proof.* Once this result has been observed, a proof by elementary geometry is fairly obvious assuming 3.1. With a little ingenuity it is then possible to provide an elementary proof of both 3.1 and 3.2, but it is of more interest here to see how the representation in $S^4$ helps us.

The points $A_1, \ldots, A_4$ of **Q** representing the points of contact of the four cycles lie on the four generators in Figure 6, just as $L_1, \ldots, L_4$ do. Thus they lie in the 3-space $P_1P_2P_3P_4$, which we shall call $\rho$, and in the point-space (on $\Pi$ in fact). Hence they are coplanar on $\Pi$, so they represent concyclic (or collinear) points.

In case (a) we see from the 2nd proof of 3.1 that $\rho$ passes through $U$; $U$ represents the point at $\infty$, so $A_1, \ldots, A_4$ are concyclic with the point at $\infty$, i.e. collinear.

In case (b), let $S, T$ denote the cycle touching $l_1, \ldots, l_4$ and either of the cycles passing through the four points of contact. Let the planes $L_1L_2L_3L_4$ and $A_1A_2A_3A_4$ meet in the line $l$. Then the polars of $\tilde{S}$, $\tilde{T}$, $U$, $V$ all contain $l$, so $\tilde{S}$, $\tilde{T}$, $U$, $V$ lie on the polar plane of $l$. Hence $S, T$ are concentric (see last paragraph of Section 2). $\square$

**3.3.** *Let $H_1, H_2$ be two cycles touching at $P$, and let $K$ be a cycle not touching $H_1, H_2$. Then there exists just one cycle touching $H_1$ and $H_2$ (at $P$) and $K$. (If $K$ passes through $P$, then $P$ is the unique cycle touching $H_1, H_2, K$.)*

This is easily proved by inversion, mapping $P$ to the point at $\infty$. Another proof involves a generator and a quadric cone on **Q**.

The next result was first proved by Tyrrell and Powell ([11]; see also [5]), having been conjectured earlier by accurate drawing.

**3.4.** *Let $R$, $S$, $T$ be three cycles* (Figure 7). *Let $U_1$ be any cycle touching $R, S$, and let cycles $U_2, \ldots, U_6$ be defined* (*uniquely, by* 3.3) *as follows*: *$U_2$ touches $S$, $T$, and $U_1$; $U_3$ touches $T$, $R$, and $U_2$; $U_4$ touches $R$, $S$, and $U_3$; $U_5$ touches $S$, $T$, and $U_4$; $U_6$ touches $T$, $R$, and $U_5$. Then $U_6$ touches $u_1$ also.*

It would appear from Figure 7 that something more is true:

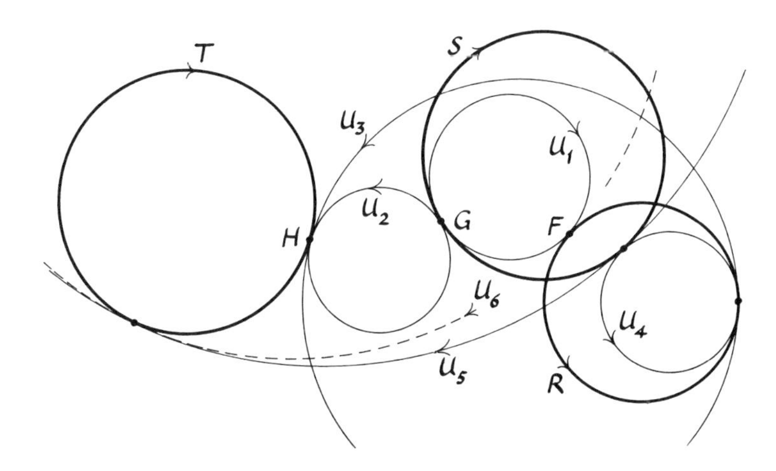


**Figure 7**

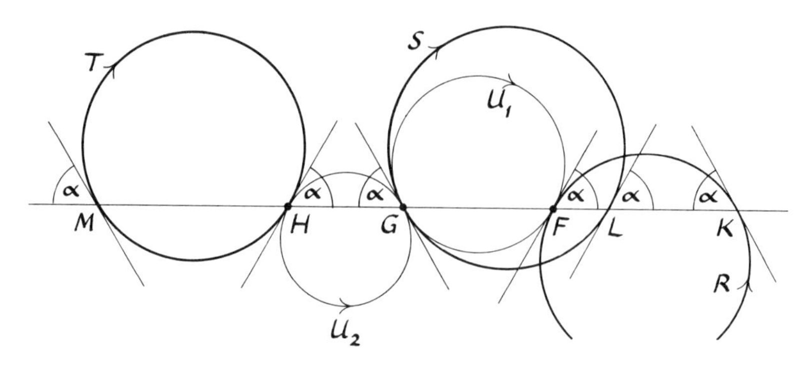


**Figure 8**

**3.5.** *The six points of tangency of the cycles in* 3.4 *are concyclic.*

This observation suggests a method of proof of 3.4 and 3.5 by elementary geometry. We shall also give a proof using the representation on **Q**.

*First Proof of* 3.4 *and* 3.5. Invert Figure 7 so that $F, G, H$ become collinear (Figure 8). Let $R, S, T$ meet the line $FGH$ again at $K, L, M$. Clearly all the angles marked $\alpha$ are equal. Hence $U_3$ touches the tangents at $H$ and $K$, $U_4$ at $K$ and $L$, $U_5$ at $L$ and $M$, and $U_6$ at $M$ and $F$. □

*Second Proof of* 3.4 *and* 3.5. The points $\tilde{S}$, $\tilde{U}_1$, $\tilde{U}_2$ lie on a generator of **Q** (see Figure 9, in which $\tilde{U}_1, \ldots, \tilde{U}_6$ are denoted by $1, \ldots, 6$), and $\tilde{T}$, $\tilde{U}_2$, $\tilde{U}_3$ lie on a generator. These generators are coplanar, and together with $\tilde{R}$ they determine a 3-space meeting **Q** in a ruled quadric, on which the two generators are of opposite systems, $\tilde{S}\tilde{U}_1\tilde{U}_2$ of the first system, $\tilde{T}\tilde{U}_2\tilde{U}_3$ of the second. Then $\tilde{R}\tilde{U}_3\tilde{U}_4$ is a generator of the first system, $\tilde{S}\tilde{U}_4\tilde{U}_5$ of the second, $\tilde{T}\tilde{U}_5\tilde{U}_6$ of the first. Now $\tilde{R}\tilde{U}_1$ is a generator of the second system, and so is $\tilde{R}\tilde{U}_6$; these must coincide, and hence $R$, $U_1$, $U_6$ all touch each other.

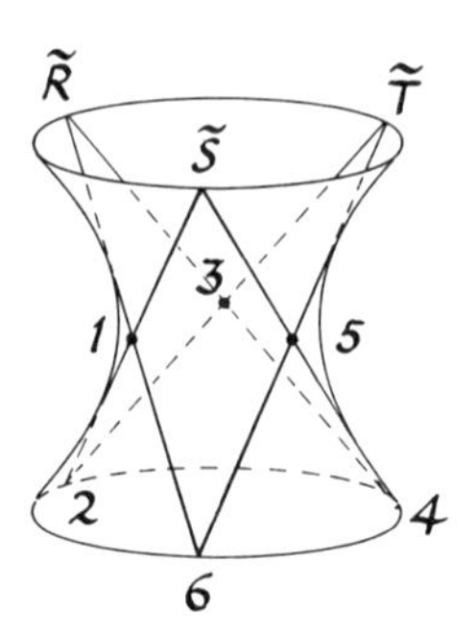


**Figure 9**

The six points of contact of the cycles are represented on **Q** by six points (one on each generator) lying in the above 3-space, and also lying in the point-space. Hence the points are coplanar, so they represent concyclic points. □

The nine cycles can be arranged in a square array

$$\begin{matrix} R & U_1 & U_6 \\ U_4 & S & U_5 \\ U_3 & U_2 & T \end{matrix};$$

the three cycles in any row or column touch each other. We see that $R$, $S$, $T$, and $U_1$ do not play a special role: we can build up the complete configuration starting from any three mutually nontangent cycles and a cycle touching two of them.

The existence of a similar configuration of nine cycles, in which each cycle *antitouches* four others according to the above square array (each cycle antitouching the other cycles in the same row and column), is much more difficult to prove. The original proof by Tyrrell and Powell [11] involves elliptic functions; the author has given a more geometrical proof, together with other properties of the configuration [10].

## 4. The Cone of Rays, Λ

We saw in Section 2 that the rays touching a cycle $S$ (not itself a ray) are represented on **Q** by points of $\Lambda$ lying in a plane $s$ not passing through the vertex $U$ of $\Lambda$. We can say that *$s$ represents $S$ in the ray-space.*

The simplex of reference for the ray-space is $YZUV$ (Figure 2), and we can conveniently represent the ray-space in Euclidean space by taking $YZU$ as the plane at infinity and $V$ as origin. More precisely, the ray $-e\sin\theta \cdot x + e\cos\theta \cdot y - d = 0$ $(e \neq 0)$, i.e. $-\sin\theta \cdot x + \cos\theta \cdot y - d/e = 0$, is represented on **Q** by the point $(0, -\sin\theta, \cos\theta, d/e, 1)$ of $\Lambda$, which we now represent in Euclidean space $E^3$ by the point $(-\sin\theta, \cos\theta, d/e)$ on the cylinder $y^2 + z^2 = 1$, using $(y, z, u)$ as coordinates in $E^3$. This cylinder is a cone with vertex $U$ at $\infty$. The ray $-\sin\theta \cdot x + \cos\theta \cdot y - d/e = 0$ makes an angle $\theta$ with the positive $x$-axis and touches the cycle of radius $-d/e$ with center at the origin; the point of tangency is $(-d/e\sin\theta,\ d/e\cos\theta)$.

Points on a generator of the cylinder represent parallel rays; points on the circular section by the plane $t = d/e$ represent rays touching the cycle of radius $-d/e$, center the origin; points that are reflections of each other in the origin represent opposite rays.

This representation of rays can be made without any excursion into four dimensions.

In Chapter X of [1], Coolidge frequently refers to a *duality* between points and circles on the one hand and rays and cycles on the other. This is also mentioned by Pedoe [8]. In particular, there is a duality between theorems about incidence of points and circles and theorems about tangency of rays and cycles, but this

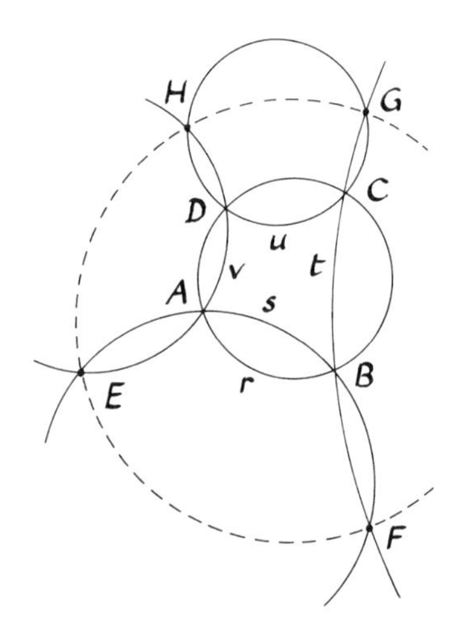


Figure 10

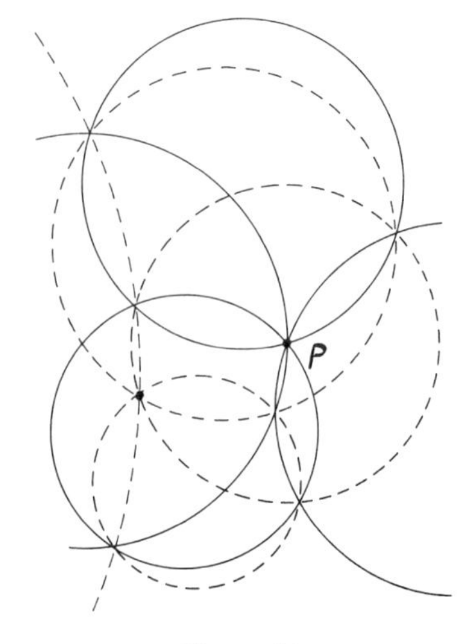


Figure 11

duality is observed after the results have been proved. Is there a *principle of duality* whereby theorems about tangency of rays and cycles may be immediately deduced from those about incidence of points and circles?

The duality is most clearly seen when theorems are represented on $\Pi$ and $\Lambda$, and in our attempt to answer this question we consider two well-known theorems about circles, and their duals.

**4.1** (Miquel [1, p. 86]). *Let $A, B, C, D$ be points on a circle $r$. Let $s, t, u, v$ be circles through $AB$, $BC$, $CD$, $DA$ respectively. Define $E, F, G, H$ as in* Figure 10. *Then $E, F, G, H$ are concyclic.*

**4.2** (Clifford [2, p. 262]). *Let $p_1, \ldots, p_4$ be circles through a point $P$. Let $p_i, p_j$ meet again at $P_{ij} = P_{ji}$. Let $p_{ijk}$ denote the circle $P_{ij}P_{jk}P_{ki}$. Then the circles $p_{234}, p_{341}, p_{412.}, p_{123}$ are concurrent, at $P_{1234}$ say* (Figure 11).

The duals of these theorems are:

**4.1*** [1, p. 363]. *Let $a, b, c, d$ be rays touching a cycle $R$. Let $S, T, U, V$ be cycles touching $a, b$; $b, c$; $c, d$; $d, a$ respectively. Let $e$ be the second common tangent of $S, T$, and define rays $f$, $g$, $h$ similarly as in* Figure 12. *Then $e$, $f$, $g, h$ touch a cycle or are parallel in pairs.*

**4.2*** [1, p. 365]. *Let $P_1, \ldots, P_4$ be four cycles touching a ray $p$. Let $p_{ij} = p_{ji}$ be the second common tangent of $P_i, P_j$. Let $P_{ijk}$ denote the cycle touching $p_{ij}, p_{jk}, p_{ki}$. Then the cycles $P_{234}, P_{341}, P_{412}, P_{123}$ (if they exist) have a common tangent ray.*

Now 4.1, interpreted on $\Pi$, is a theorem about eight points on a nonruled quadric, certain sets of four points being coplanar; 4.1* is a precisely similar

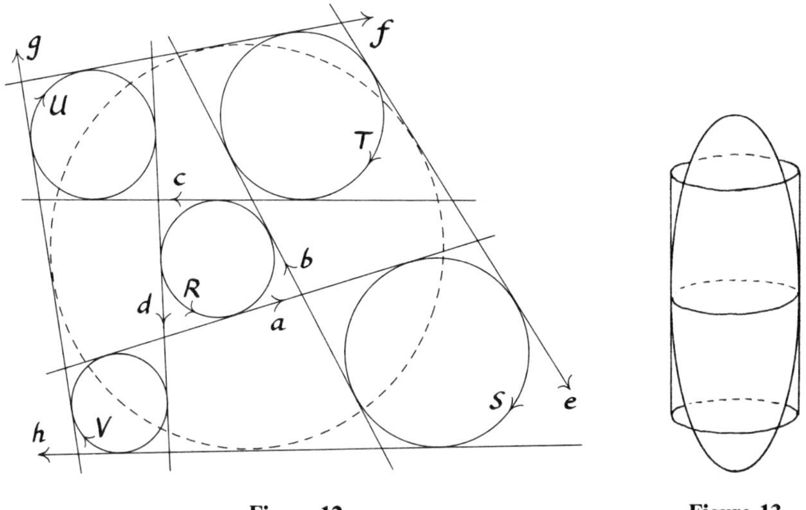


Figure 12

Figure 13

theorem about points on the quadric cone $\Lambda$, which has now become a circular cylinder. A similar remark applies to 4.2 and 4.2*. Given any similar theorem about concyclic sets of points, we can interpret it as a theorem about coplanar sets of points on *any nonruled quadric*; can we then say that the same theorem is true on the *quadric cone* $\Lambda$? We can take the nonruled quadric to be a prolate spheroid fitting neatly inside the cylinder (Figure 13). Then, intuitively, we stretch the spheroid vertically, at the same time moving the relevant points on the spheroid so that they remain close to the equator. In the limit we obtain the required theorem on the cylinder. If a plane section of the spheroid becomes vertical in the limit, it cuts the cylinder in two generators, the points of which correspond to two sets of parallel rays.

The chains of theorems due to Clifford and Grace [1, pp. 90, 92] have duals about rays and cycles that can be deduced by this method [1, pp. 365, 366], but we have not gained much here: 4.2 and the chains of theorems of Clifford and Grace can all be deduced from 4.1, and in the same way all the dual theorems can be deduced from 4.1*.

This attempt to provide a principle of duality is so far inconclusive. It may be that, when we dualize certain theorems, too many sets of four "concyclic" rays degenerate unavoidably into two pairs of parallel rays. We need to study more examples, which will not be done here.

For direct proofs of 4.1, 4.2, and their duals, we can use the well-known result about eight associated points:

**4.3** [2, p. 259; 7, p. 420]. *If three quadrics meet in eight points, then any other quadric through seven of these points passes through the eighth also.*

To prove 4.1 (or 4.1*) we have the quadric $\Pi$ (or $\Lambda$) and the plane-pairs $\tilde{A}\tilde{B}\tilde{E}\tilde{F}$, $\tilde{C}\tilde{D}\tilde{G}\tilde{H}$ and $\tilde{A}\tilde{D}\tilde{E}\tilde{H}$, $\tilde{B}\tilde{C}\tilde{F}\tilde{G}$ meeting in the eight points $\tilde{A}, \tilde{B}, \ldots, \tilde{H}$; the plane-pair $\tilde{A}\tilde{B}\tilde{C}\tilde{D}$, $\tilde{E}\tilde{F}\tilde{G}$ passes through seven of these points, so it must pass through $\tilde{H}$ also.

A similar proof holds for 4.2 and 4.2*, but we must now use 4.3 twice. We end up with eight points that can be regarded in four ways as two Möbius tetrahedra [2, p. 258; 7, pp. 268, 423] each inscribed in the other: e.g. $\tilde{P}\tilde{P}_{23}\tilde{P}_{24}\tilde{P}_{34}$ and $\tilde{P}_{12}\tilde{P}_{13}\tilde{P}_{14}\tilde{P}_{1234}$.

## 5. The Angle Between Two Cycles

The theorems in this section are required for the subsequent definition and study of Laguerre inversion.

If two proper cycles meet each other, their radii being $\rho$ and $\sigma$ and the distance between their centers $\varepsilon$, then their angle of intersection $\theta$ is given by

$$\cos\theta = \frac{\rho^2 + \sigma^2 - \varepsilon^2}{2\rho\sigma}. \tag{3}$$

If a proper cycle of radius $\rho$ meets a ray whose distance from its center $C$ is $\delta$, then their angle of intersection is given by $\cos\theta = \delta/\rho$; we must regard $\delta/\rho$ as positive if the cycle and the ray "rotate in the same direction about $C$" and negative otherwise (Figure 14). Without needing to define the notion of imaginary-valued angles between nonintersecting cycles and rays, we can use the above formulae to define the *cosine* $\cos(r,s)$ between two proper cycles $r$ and $s$. Two cycles are tangent, orthogonal, or antitangent according as their cosine is 1, 0, or $-1$. We can say that *the cycles $r$ and $r'$ make the same angle with the cycle $s$* when $\cos(r,s) = \cos(r',s)$, even when $r$ and $r'$ do not meet $s$.

It is well known that the angle between two intersecting cycles is unaltered by inversion. A simple algebraic calculation shows that the same is true of the cosine between any two cycles.[3] It is clear from the definition of cosines that two cycles

[3] When the cycles $r$ and $s$ do not meet, then

$$|\cos(r,s)| = \cosh\phi,$$

where $\phi$ is the *inversive distance* between $r$ and $s$ [3, 4].

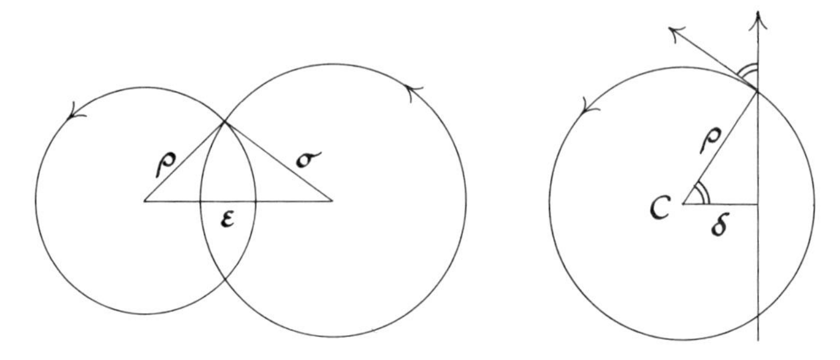


**Figure 14**

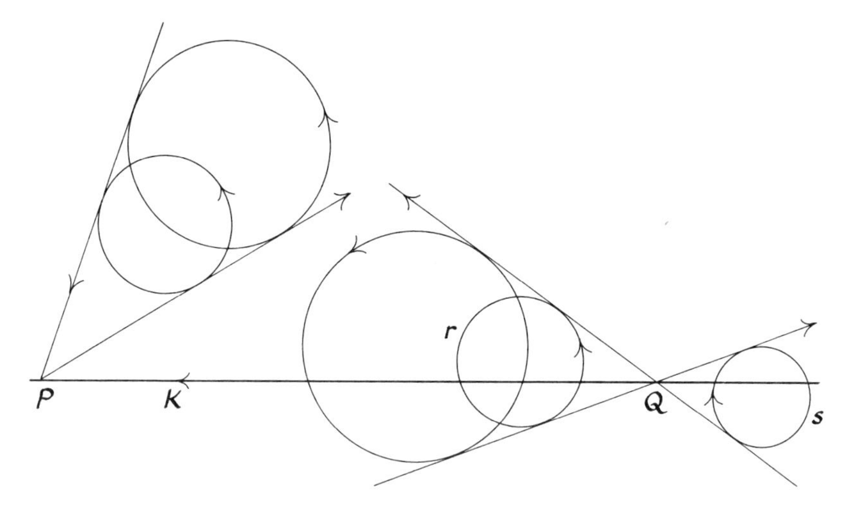


**Figure 15**

make the same angle with a ray $K$ if and only if their center of similitude lies on $K$ (Figure 15). There is then a dilatation that fixes $K$ and maps one cycle to the other (but if $P$ in Figure 15 is at infinity, then the dilatation becomes a translation). In Figure 15 one's first reaction is to say that there is a negative dilatation with center $Q$ mapping $r$ to $\bar{s}$, but we must agree that a negative dilatation reverses the expected sense of the image cycle (otherwise the image of $K$ would be $\bar{K}$ and not $K$) so that $r$ is mapped to $s$.

**5.1.** *Let $S, S', K$ be coaxial cycles, $K$ being a proper cycle. Then the cycles touching $S$ and $S'$ all make the same angle with $K$.*

*Proof*. If $S, S'$ are proper cycles, we invert them to intersecting or parallel lines, or to concentric circles. If one of them, $S$ say, is a point, invert it to infinity. The result is then obvious in all cases, and is illustrated in Figure 16(a)–(f). □

**5.2.** *If the proper cycles $S$ and $T$ make the same angle with the proper cycle $K$, then $S$ and $T$ cannot touch each other except at a point of $K$.*

*Proof*. If $S$ and $T$ touch at $P$, invert $P$ to infinity; then $S$ and $T$ become parallel rays, and only a ray can make the same angle with both of them. Hence $K$ passes through $P$ also. □

Now let $K$ be a proper cycle, and let $\lambda$ be a real number distinct from 1 and $-1$. Let $\mathcal{C}$ be the set of all cycles $C$ such that $\cos(K, C) = \lambda$. Thus all the cycles in $\mathcal{C}$ make the same angle with $K$, and they do not touch or antitouch $K$, since $\lambda \neq \pm 1$. We shall refer to the cycles in $\mathcal{C}$ as *$\mathcal{C}$-cycles*. Since the radius of a $\mathcal{C}$-cycie tends to zero as the center of the cycle approaches $K$, we shall include all the points of $K$ in $\mathcal{C}$ also.

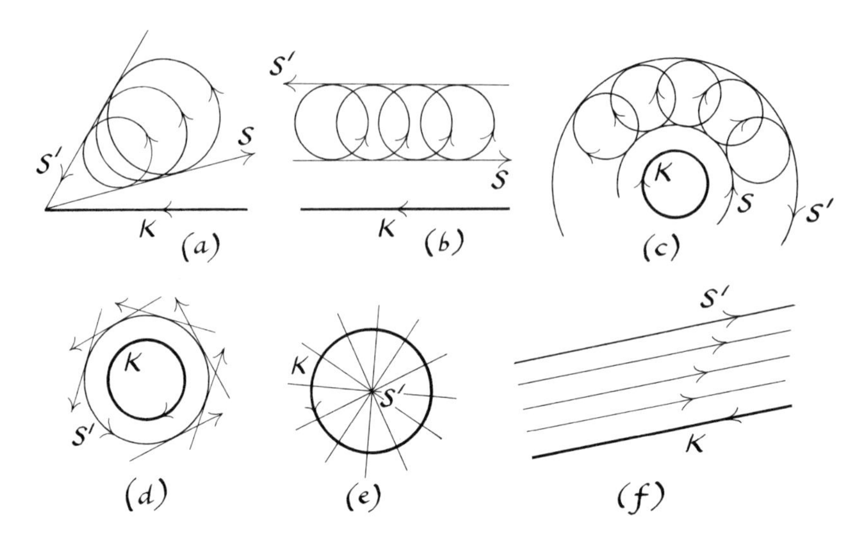


Figure 16

**5.3.** *Let $S$ be any cycle not in $\mathcal{C}$. Then there is an infinite number of $\mathcal{C}$-cycles touching $S$, and they all touch a second cycle $S'$ distinct from $S$. Also $K, S, S'$ are coaxial.*

*Proof*. Suppose first that $S \neq K$. If we invert $K$ to a ray, we can show that there is at least one $\mathcal{C}$-cycle touching $S$, using the method illustrated in Figure 17. (From a point of $K$ we draw rays $r$ and $s$ touching a $\mathcal{C}$-cycle $D$. Then at each point of $r$ there is a $\mathcal{C}$-cycle touching $r$. If we perform a translation parallel to $K$ so that $r$ becomes $r'$ touching $S$, then one of these $\mathcal{C}$-cycles will be translated to a $\mathcal{C}$-cycle touching $S$. If $S$ is a ray, a simpler translation argument can be used.)

Now invert $K$ and $S$ to intersecting lines, or concentric circles, etc., as in the proof of 5.1, and we can then easily reconstruct Figure 16 (16(f) is inapplicable here). It only remains to show that *all* $\mathcal{C}$-cycles touching $S$ touch $S'$ also. Since we already have, in Figure 16(a)–(c), a $\mathcal{C}$-cycle touching $S$ at each point of $S$,

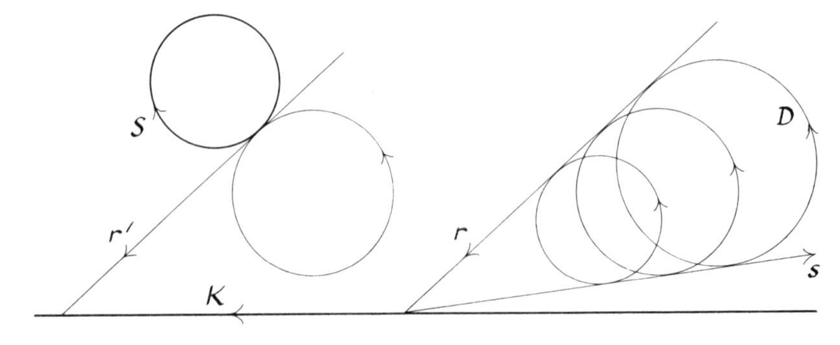


Figure 17

any other $\mathcal{C}$-cycle touching $S$ would contradict 5.2. The result is obvious in Figure 16(d), (e), when $S$ is a point.

If $S = K$, the $\mathcal{C}$-cycles touching $K$ are just the points of $K$, and these touch a second cycle, namely $\bar{K}$, so $S' = \bar{K}$. □

## 6. Generalized Laguerre Inversions

For given $K$ and $\lambda$ ($\neq \pm 1$) we now define a mapping of cycles to cycles, called the (*generalized*) *Laguerre inversion with base cycle $K$ and constant* $\lambda$, in the following way: (i) a cycle $S$ that is not a $\mathcal{C}$-cycle is mapped to $S'$ as given by 5.3, and (ii) all $\mathcal{C}$-cycles are mapped to themselves.

The following points should be noted:

(a) in Laguerre's original definition $K$ is a ray (and also the $\mathcal{C}$-cycles meet $K$ twice, but many of his theorems are true without this restriction); when $K$ is a ray we shall use the term "*axial* Laguerre inversion";

(b) point-cycles are not in general mapped to point-cycles;

(c) opposite cycles are not in general mapped to opposite cycles;

(d) although the points of $K$, being $\mathcal{C}$-cycles, are invariant, $K$ itself is not invariant but is mapped to $\bar{K}$. (See the last paragraph of the proof of 5.3.)

Since the relation between $S$ and $S'$ in 5.3 is symmetrical, we immediately have:

**6.1.** *Laguerre inversions are of order* 2.

**6.2.** *Laguerre inversions map tangent cycles to tangent cycles.*

*Proof.* Let $S$ and $T$ be distinct tangent cycles.

(a) If $S$ and $T$ are both $\mathcal{C}$-cycles, then they are invariant.

(b) If one of them, $T$ say, is a $\mathcal{C}$-cycle, then by 5.3 $T$ touches $S'$, i.e. $T'$ touches $S'$.

(c) If neither $S$ nor $T$ is a $\mathcal{C}$-cycle or a point-cycle, invert their point of contact to infinity, so that they become parallel rays. If $K$ is then a cycle (not a ray), we have Figure 18, in which the dotted line is a line of (anti)-symmetry of $S$, $T$ and $K$. There is just one $\mathcal{C}$-cycle that is a ray $C$ parallel to $S$, i.e. touching $S$. Some of the other $\mathcal{C}$-cycles touching $S$ are also illustrated to emphasize the symmetry. Because of this symmetry, $S'$ (which touches $C$) must touch $C$ at $P$. Now $C$ touches $T$ also, so by the same argument $T'$ touches $C$ at $P$. Hence $S'$ and $T'$ touch at $P$.

If $S$ and $T$ are parallel rays and $K$ is a ray parallel or antiparallel to them, then $S'$ and $T'$ are antiparallel to $S$ and $T$, and hence $S'$ and $T'$ touch at infinity (Figure 19).

If $S$ and $T$ are parallel rays and $K$ is a ray meeting them, we have Figure 20, which speaks for itself: again $S'$ and $T'$ are parallel.

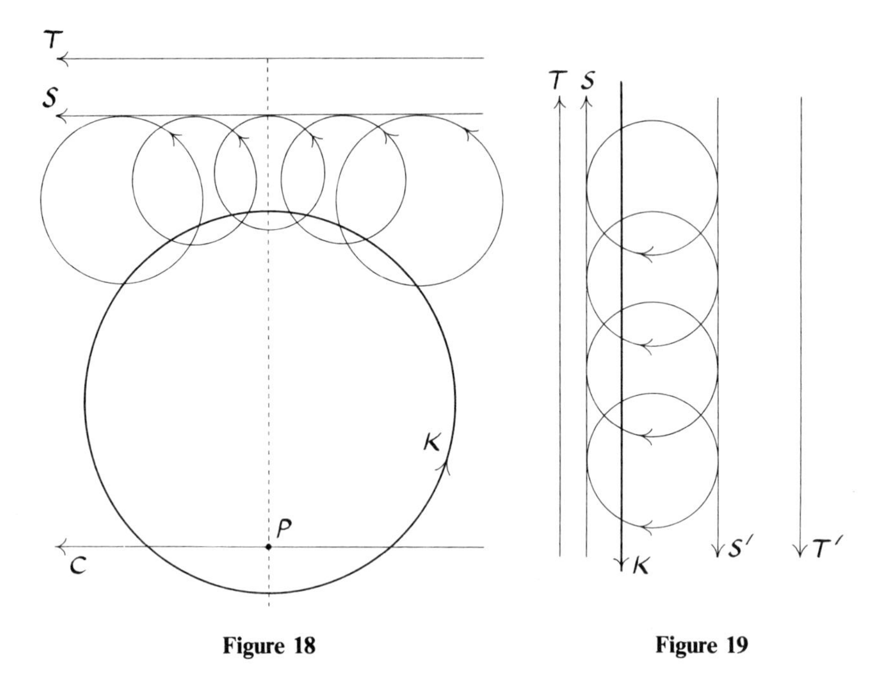


Figure 18 Figure 19

(d) If one of $S$ and $T$ is a point-cycle, invert this point to infinity. The reader will easily provide a proof of this final case. □

If $\lambda = 0$, the $\mathcal{C}$-cycles are orthogonal to $K$, and the corresponding Laguerre inversion is ordinary inversion in $K$, defined as in the last paragraph of Section 1. We can interpret the case $\lambda = \infty$ as the limiting case in which $\mathcal{C}$ consists of all point-cycles. Then the $\mathcal{C}$-cycles touching a cycle $S$ are just the points of $S$, and the other cycle touching these points is $\overline{S}$; hence the corresponding Laguerre inversion is just the *reversal mapping*, which maps each cycle to the opposite cycle.

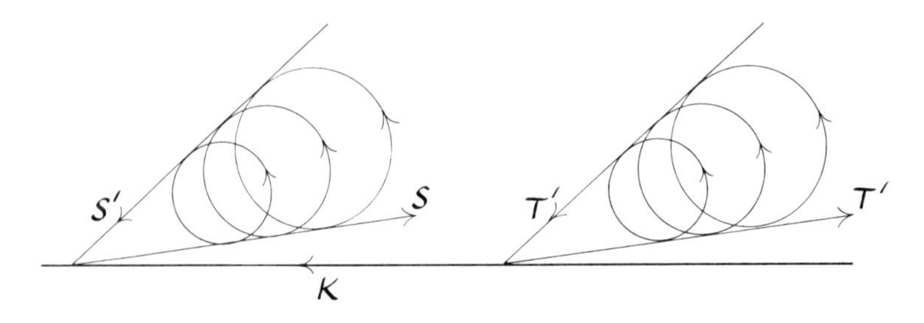


Figure 20

## 7. Mapping Cycles to Points

Except in the case of ordinary inversion, Laguerre inversions map points to proper cycles (apart from the points of $K$, which are always invariant).

**7.1.** *If $S$ is the image of a point $P$ not on $K$, then $S$ does not meet $K$.*

*Proof.* If $S$ meets $K$, invert the point of meeting to infinity. Then $S$ and $K$ are lines, and we have seen previously that $S'$ is then a line also (Figure 16(a) and (b)) and not a point.

Alternatively, $S$, $K$, and $P$ are coaxial (5.3), and the circles of a coaxial system containing points do not meet each other. □

**7.2.** *The images of the point-cycles in a Laguerre inversion all make the same angle with $K$. Conversely, if $\mathfrak{D}$ is a set of cycles all making the same angle with a cycle $K$, and if the $\mathfrak{D}$-cycles do not meet $K$, then there exist two Laguerre inversions with base cycle $K$ mapping the $\mathfrak{D}$-cycles to points.*

*Proof.* Invert the figures so that $K$ becomes a line. Let the images of the points $P$ and $Q$ be cycles $S$ and $T$. There is a dilatation (or a translation) fixing $K$ and mapping $P$ to $Q$. This dilatation maps $\mathcal{C}$ to itself and $S$ to $S^*$ say. There is an infinity of $\mathcal{C}$-cycles touching $P$ and $S$; hence there is an infinity of $\mathcal{C}$-cycles touching $Q$ and $S^*$. Hence $S^*$ is the image of $Q$ in the Laguerre inversion. Hence $S^* = T$. Hence the result, since a dilatation preserves angles.

For the converse, let $S$ be a $\mathfrak{D}$-cycle. Since $S$ does not meet $K$, the coaxial system determined by $S$ and $K$ contains two points. Let $P$ be one of these points; the same argument will hold for the other point, giving a second Laguerre inversion. The cycles touching $S$ and $P$ all make the same angle with $K$ (5.1), and hence they determine a complete set of $\mathcal{C}$-cycles. We now use the same type of proof as in the previous paragraph, using a dilatation or a translation to map $S$ to any other $\mathfrak{D}$-cycle. □

The next result does not form an essential part of the present discussion, so we omit the proof.

**7.3.** *Let $C$ be a fixed cycle of a Laguerre inversion with base cycle $K$, and let $D$ be a cycle that is mapped to a point. Write* $\cos(C, K) = \lambda$, $\cos(D, K) = \mu$. *Then*

$$\frac{\mu - 1}{\mu + 1} = \left(\frac{\lambda - 1}{\lambda + 1}\right)^2.$$

Laguerre's final theorem in [6] states that

(A) *if $S_1, S_2, S_3$ are cycles, then there exists an axial Laguerre inversion $\alpha$ mapping the cycles to points.*

(See also [8, §4] and [9].) He uses this theorem to show that

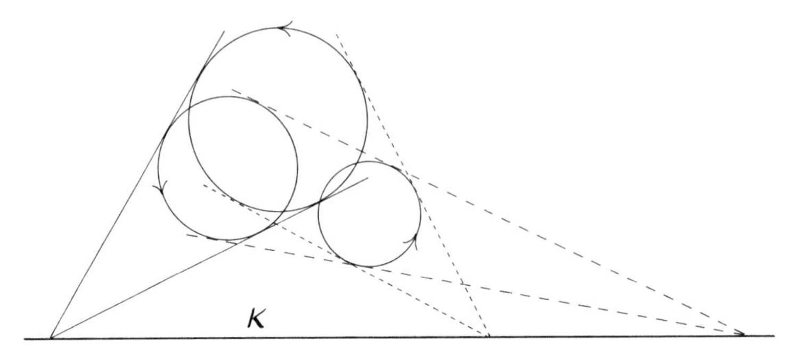


**Figure 21**

(B) *there exist two cycles touching* $S_1, S_2, S_3$,

in the following way. Suppose $\alpha$ maps $S_1$, $S_2$, $S_3$ to the points $P_1, P_2, P_3$. There exist just two (opposite) cycles, $S$ and $\bar{S}$ say, touching (i.e. passing through) $P_1$, $P_2$, $P_3$. Then by 6.2, $S\alpha$ and $\bar{S}\alpha$ are the only two cycles touching $S_1$, $S_2$, $S_3$.

Let us analyze Theorem (A) and its consequence Theorem (B). From 7.2 and a remark in Section 5 we see that, if an axial Laguerre inversion maps $S_1, S_2$ to points, then the axis $K$ must pass through the center of similitude of $S_1, S_2$. The centers of similitude of $S_1$, $S_2$, $S_3$ taken in pairs are well known to be collinear (Figure 21). Let $K$ be the line on which they lie; then $K$ is the only possible axis for an axial Laguerre inversion mapping $S_1$, $S_2$, $S_3$ to points. But such an axis must not meet $S_1$, $S_2$, $S_3$ (7.1). Hence *Theorem* (A) *is true only if $K$ does not meet* $S_1$, $S_2$, $S_3$. However, Figure 22 shows three cycles whose "line of similitude" meets all three cycles but which are touched by two cycles. On the other hand, there exist sets of three cycles that have no common tangent cycle (Figure 23). Hence

(i) *Theorem* (A) *is not always true*;
(ii) *Theorem* (B) *can be true when Theorem* (A) *is not*;
(iii) *Theorem* (B) *is not always true*.

If we now ask under what circumstances three cycles can be mapped to points, we obtain a complete reversal of the above situation by proving the following result.

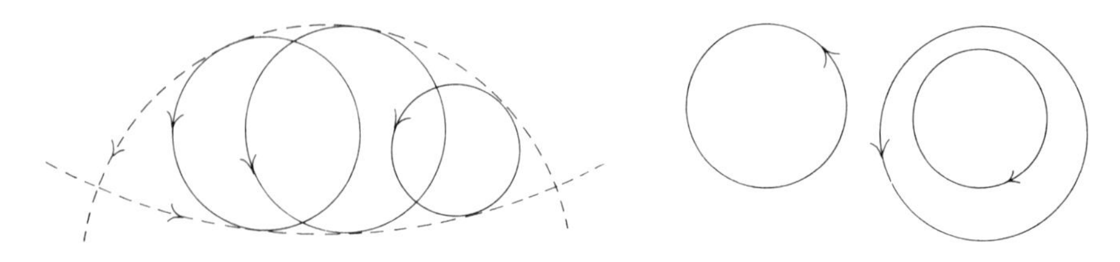

**Figure 22** **Figure 23**

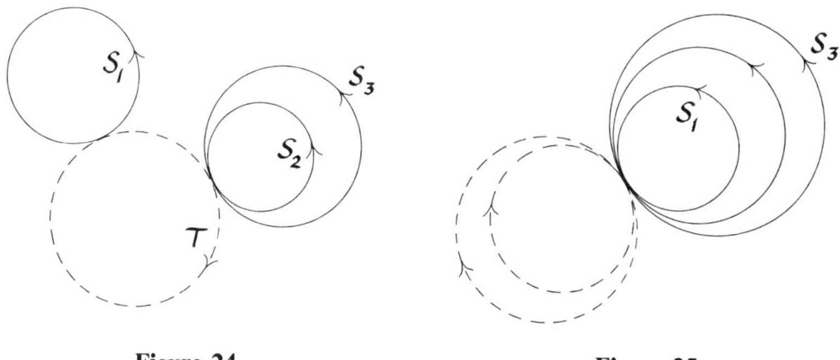


Figure 24

Figure 25

**7.4.** *If three cycles have a common tangent cycle, and if no two of them touch each other, then there exist Laguerre inversions that map the three cycles to points.*

First we prove a lemma.

**7.5.** *Let $S_1$, $S_2$, $S_3$ be distinct cycles with a common tangent cycle $T$. Then* (a) *if no two of $S_1$, $S_2$, $S_3$ touch each other, they have a second common tangent cycle, distinct from $T$*; (b) *if two of them touch each other, they have no other common tangent cycle* (Figure 24); (c) *if all three touch each other, they have an infinite number of common tangent cycles* (Figure 25).

*Proof.* Apply an axial Laguerre inversion, followed by an ordinary inversion, to map $T$ to the point at infinity. (This can also be done by a single nonaxial Laguerre inversion). Then $S_1$, $S_2$, $S_3$ become, in case (a), three nonparallel rays, which have a unique common tangent cycle (other than the point at infinity) by elementary geometry. (Two of the lines may be antiparallel; this does not matter.) By reversing the mapping we obtain the result. Cases (b) and (c) are dealt with similarly. □

*Proof of* 7.4. Denote the three cycles by $S_1$, $S_2$, $S_3$. By 7.5 they have two common tangent cycles, $T$ and $U$ say. It is clear from the proof of 5.1 that many of the cycles coaxial with $T$ and $U$ do not meet $S_1$, $S_2$, $S_3$, and by 5.1 all such cycles make the same angle with $S_1$, $S_2$, $S_3$. Hence by 7.2 any such cycle can be used as the base cycle of a Laguerre inversion mapping $S_1$, $S_2$, $S_3$ to points. □

Pedoe [8, p. 266] gives a neat proof of 3.4, using a Laguerre inversion to map $R$, $S$, $T$ to points. Unfortunately, in Figure 7 we cannot do this, since $R$, $S$, $T$ have no common cycle.

## 8. Representation of Laguerre Inversions on the Quadric Q

Let $A$ and $P$ be proper cycles, other than rays, represented on **Q** by $\tilde{A}(a,b,c,d,e)$ and $\tilde{P}(p,q,r,s,t)$, where $a$, $e$, $p$, $t$ are nonzero. Then $A$ has center $(b/a, c/a)$ and radius $e/a$, whilst $P$ has center $(q/p, r/p)$ and radius $t/p$. We find that

$$\cos(A, P) = \frac{bq + cr - dp - as}{et} \tag{4}$$

(making use of the equations $e^2 = b^2 + c^2 - 2ad$ and $t^2 = q^2 + r^2 - 2ps$). It is easily checked that (4) remains true if we allow $A$ and $P$ to be rays, so that (4) is true for all proper cycles.

Hence $\cos(A, P) = \lambda$ if and only lf

$$bq + cr - dp - as - \lambda et = 0, \tag{5}$$

and the condition (5) remains true if $P$ is a point-cycle (with $t = 0$), since (5) is then the condition for $P$ to lie on (i.e. to touch) $A$. Hence we have the following result.

**8.1.** *The cycle $P$ makes an angle $\lambda$ with the proper cycle $A$, represented on* $\mathbf{Q}$ *by $\tilde{A}(a, b, c, d, e)$, if and only if $\tilde{P}$ lies in the 3-space whose equation is*

$$by + cz - dx - au - \lambda ev = 0.$$

Let us denote the point $(a, b, c, d, \lambda e)$ by $\lambda\tilde{A}$.

**8.2.** *Using the notation of* 8.1, *the Laguerre inversion with base cycle $A$ and constant $\lambda$ is represented on* $\mathbf{Q}$ *by the harmonic homology whose center is $\lambda\tilde{A}$ and whose axis is the polar space of $\lambda\tilde{A}$ with respect to* $\mathbf{Q}$.

(The three-dimensional analog is represented in Figure 26: the line joining $P$ to $\lambda\tilde{A}$ meets the polar at $Q$, and $(\lambda\tilde{A}Q, PP')$ is a harmonic set; the harmonic homology maps $P$ on the quadric to $P'$ on the quadric.)

*Proof.* Denote the harmonic homology by $\tilde{\alpha}$; it is a collineation of the 4-space mapping $\mathbf{Q}$ onto itself, and hence it maps generators of $\mathbf{Q}$ to generators of $\mathbf{Q}$. Hence points representing tangent cycles are mapped to points representing tangent cycles. Now suppose $S$ is not a $\mathcal{C}$-cycle, and let $C$ be any $\mathcal{C}$-cycle touching $S$. The 3-space in 8.1 is the polar of $\lambda\tilde{A}$; hence $\tilde{C}$ lies in the polar but $\tilde{S}$ does not. Hence $\tilde{C}\tilde{\alpha} = \tilde{C}$ but $\tilde{S}\tilde{\alpha} \neq \tilde{S}$; suppose $\tilde{S}\tilde{\alpha}$ represents the cycle $T \neq S$, so that $\tilde{S}\tilde{\alpha} = \tilde{T}$. Then $T$ touches $C$ by the remark made above. This is true for every $\mathcal{C}$-cycle $C$ touching $S$. Hence $T = S'$, the image of $S$ in the Laguerre inversion. □

The cycle $\bar{A}$ opposite to $A$ is represented by $\tilde{\bar{A}}(a, b, c, d, -e)$, so $\lambda\tilde{A} = (-\lambda)\tilde{\bar{A}}$, and we see either geometrically or by 8.2 that the Laguerre inversion with base cycle $\bar{A}$ and constant $-\lambda$ is the same as that with base cycle $A$ and constant $\lambda$. Also $\tilde{A}$, $\tilde{\bar{A}}$, $\lambda\tilde{A}$, and $V(0,0,0,0,1)$ are collinear; $\tilde{A}$ and $\tilde{\bar{A}}$ are distinct, since $A$ is not a point-cycle. The situation is illustrated in Figure 27: the polar space of $V$ is the point-space, which meets $\mathbf{Q}$ in the quadric $\Pi$ (illustrated here by an ellipse).

Let $B$ be any point not on $\mathbf{Q}$, so that $B$ does not lie in its polar space $b$, and let $\tilde{\beta}$ be the harmonic homology with center $B$ and axis $b$. When does $\tilde{\beta}$ represent a

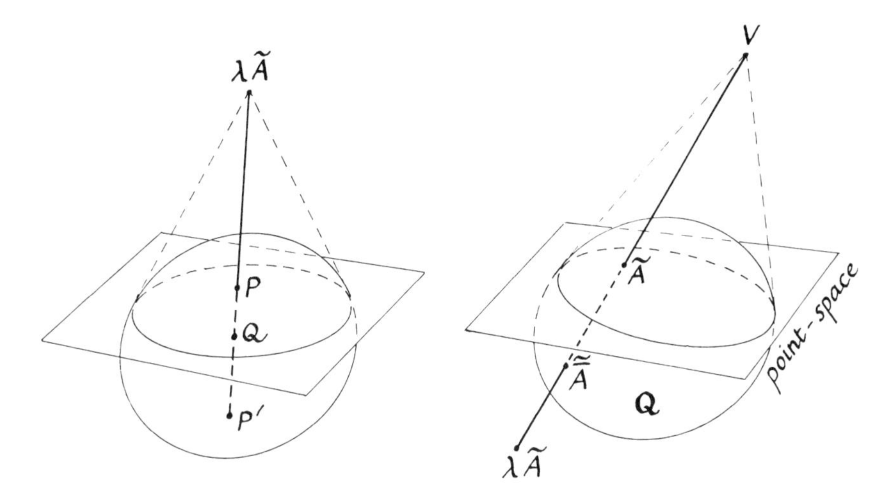


**Figure 26** **Figure 27**

Laguerre inversion? The answer is provided by Figure 27: if $VB$ meets **Q** in the *distinct* points $\tilde{A}$, $\bar{A}$, then $\tilde{\beta}$ represents a Laguerre inversion with axis $A$ or $\bar{A}$. In this situation $VB$ lies inside the quadric cone $y^2 + z^2 - 2xu = 0$ joining $V$ to the quadric $\Pi$. In the special case when $B = V$, $\beta$ represents the special Laguerre inversion that maps each cycle to its opposite cycle.

What does $\tilde{\beta}$ represent when $VB$ does not meet **Q** in two distinct points? The answer is suggested by the notion of *antiinversion* [4, p. 230] or inversion in an imaginary circle with real center $O$ and imaginary radius $i\rho$, which maps a point $P$ to $P'$ where $O$, $P$, $P'$ are collinear and $OP, OP' = -\rho^2$. This antiinversion has no fixed points; its fixed circles are the real circles orthogonal to the imaginary circle of inversion, orthogonality being defined by taking $\cos\theta = 0$ in formula (3) of Section 5, with $\rho$ replaced by $i\rho$. Thus we are led to consider a Laguerre inversion whose $\mathcal{C}$-cycles make a fixed angle with an imaginary cycle.

The imaginary cycle $A$ with real center $(b/a, c/a)$ and radius $ie/a$ has equation $a(x^2 + y^2) - 2bx - 2cy + 2d = 0$, where $b^2 + c^2 - 2ad = (ie)^2 = -e^2$. Let $P$ be a real cycle represented by $(p, q, r, s, t)$. Then

$$\cos(A, P) = \frac{bq + cr - dp - as}{iet};$$

hence $\cos(A, P) = -i\mu$ if and only if

$$bq + cr - dp - as - \mu et = 0.$$

Hence we can say that the Laguerre inversion with imaginary base cycle $A$ and constant $-i\mu$ is represented by the harmonic homology whose center is $B(a, b, c, d, \mu e)$ and whose axis is the polar of $B$. We now have $b^2 + c^2 - 2ad = -e^2 < 0$, so $VB$ does not meet **Q**: $VB$ lies outside the quadric cone of Figure 27.

The harmonic homology $\tilde{\beta}$ represents an inversion or antiinversion if and only if $B$ lies in the point-space, the polar of $V$.

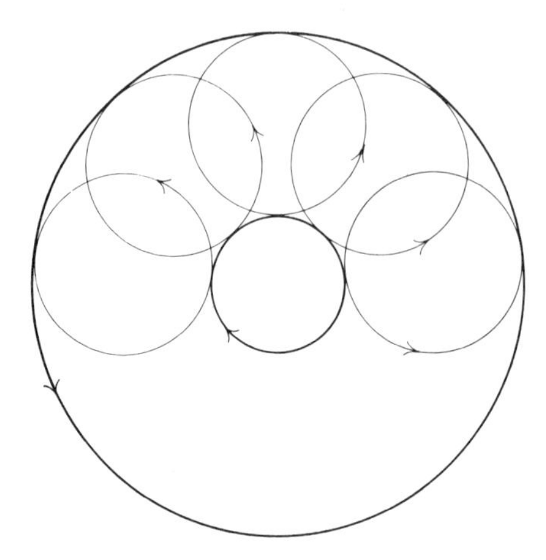

**Figure 28**

Finally, what if $B$ lies *on* the quadric cone $y^2 + z^2 - 2xu = 0$? Consider the particular case when $B$ is $(0, 0, 0, f, 1)$, a point on $UV$; $P$ is a fixed cycle of the mapping represented by $\tilde{\beta}$ if and only if

$$-fp - t = 0,$$

i.e. if and only if $t/p$, the radius of $P$, is equal to $-f$. Hence the fixed cycles of this mapping (which preserves tangency) all have the same radius. The mapping (which we shall denote by $\beta$) is illustrated in Figure 28; it is seen to be a mapping of the type described in Section 3 followed by the reversal mapping.

Now let $VN$ be any other generator of the quadric cone, $N$ lying on $\mathbf{Q}$, and let $C$ be any point of $VN$ distinct from $V$ and $N$ (Figure 29). Then $BC$ meets the point-space at a point $L$ on $UN$, and by choosing $B$ suitably we can ensure that $L$ lies inside the quadric cone. Denote the harmonic homologies determined by $C$ .

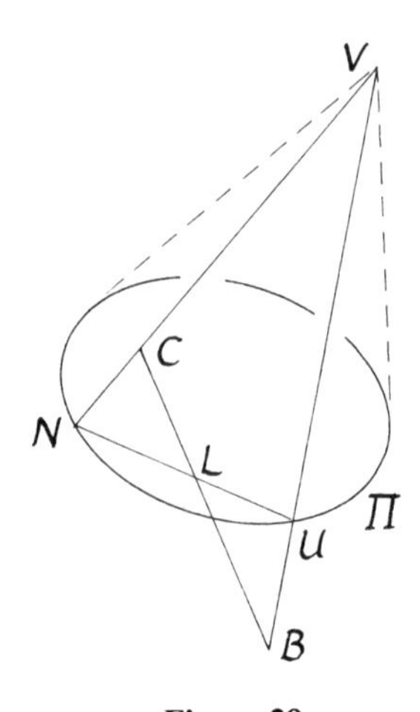


**Figure 29**

and $L$ (and their polars) by $\tilde{\gamma}$ and $\tilde{\lambda}$. Then $\tilde{\lambda}$ maps $U$ to $N$ and fixes $V$ (since $V, L$ are conjugate with respect to $\mathbf{Q}$). Hence $B\tilde{\lambda} = C$; hence $\tilde{\gamma} = \tilde{\lambda}^{-1}\tilde{\beta}\tilde{\lambda} = \tilde{\lambda}\tilde{\beta}\tilde{\lambda}$. Denote the mappings represented by $\tilde{\gamma}$ and $\tilde{\lambda}$ by $\gamma$ and $\lambda$. Then $\gamma = \lambda^{-1}\beta\lambda$, and the fixed cycles of $\gamma$ are the images under the inversion $\lambda$ of the fixed cycles of $\beta$. The point at infinity is the only fixed point of $\beta$, so $\gamma$ has just one fixed point, namely the center of $\lambda$.

The sets of rays and points are not invariant under Laguerre inversions (just as the set of lines is not invariant under ordinary inversions). Hence we can regard all points, rays, and cycles as having equal status in the *Laguerre plane*, where there is a relation of *tangency* between cycles but no relation of being opposite. In the Laguerre plane all Laguerre inversions are of the same type; the distinctions among the three types in Section 8 arise from our using a particular representation of the Laguerre plane in the Euclidean plane.

## 9. Representing Axial Laguerre Inversions in the Ray-Space

Let $A$ be the base ray of an *axial* Laguerre inversion $\alpha$ with constant $\lambda$. Then $\tilde{A}$ lies in the ray-space; so does $V$, and hence $\lambda\tilde{A}$ lies in the ray space also. Let $C$ be a fixed cycle of $\alpha$, not itself a ray; then $\tilde{C}$ lies in the polar space of $\lambda\tilde{A}$. Hence the polar of $\tilde{C}$ meets the ray-space in a plane through $\lambda\tilde{A}$ but not through $U$ (since $C$ is not a ray). This plane represents $C$ in the ray-space (Section 4). The converse also holds; thus we have proved:

**9.1.** *If $\lambda\tilde{A}$, in the ray-space, is the center of the harmonic homology representing an axial Laguerre inversion $\alpha$, then the fixed cycles (not rays) of $\alpha$ are represented by the planes through $\lambda\tilde{A}$ but not through $U$.*

If the center of $\tilde{\alpha}$ lies on $UV$ (in the ray-space), then $\alpha$ is a Laguerre inversion of the type depicted in Figure 28; hence we must regard such a Laguerre inversion as a limiting case of axial Laguerre inversion.

The proof of the next result is left to the reader.

**9.2.** *The fixed rays of $\alpha$ are represented by the points lying on the two generators along which the tangent planes through the line $\lambda\tilde{A}U$ meet $\Lambda$.*

We end with a synthetic proof of another theorem given by Coolidge [1, p. 362], slightly rephrased.

**9.3.** *Let $l_{12}, l_{13}, l_{14}, l_{23}, l_{24}, l_{34}$ be six rays, no two parallel; let $C_1$ denote the cycle touching $l_{12}, l_{13}, l_{14}$, and define $C_2, C_3, C_4$ similarly. Let $D_1$ denote the cycle touching $l_{23}, l_{24}, l_{34}$, and define $D_2, D_3, D_4$ similarly.*

*Suppose that $C_1, C_2, C_3, C_4$ are fixed cycles of an axial Laguerre inversion. Then $D_1, D_2, D_3, D_4$ are fixed cycles of another axial Laguerre inversion.*

*Proof*. Representing the rays and cycles by points of $\Lambda$ and planes in the ray-space, and using 9.1, we see that 9.3 is equivalent to the following result in projective 3-space: *if four alternate faces of an octahedron are concurrent, then so are the remaining four*. This is equivalent to the well-known theorem on Möbius tetrahedra (see also Section 4): *if* $ABCD$, $A'B'C'D'$ *are tetrahedra such that* $A$, $B$, $C$, $D$, $A'$, $B'$, $C'$ *lie in the planes* $B'C'D'$, $C'D'A'$, $D'A'B'$, $A'B'C'$, $BCD$, $CDA$, $DAB$ *respectively, then* $D'$ *lies in* $ABC$ [2, p. 258; 7, pp. 268, 423]. To obtain the tetrahedra from the octahedron, use any face of the octahedron and the three adjacent faces as the faces of one tetrahedron. □

## References


[1] Coolidge, J. L., *A Treatise on the Circle and the Sphere*. Oxford 1916.

[2] Coxeter, H. S. M., *An Introduction to Geometry*. Wiley, New York 1961.

[3] Coxeter, H. S. M., Inversive distance. *Annali di Mat. (4)* **71** (1966), 73–83.

[4] Coxeter, H. S. M., The inversive plane and hyperbolic space. *Abh. Math. Sem. Univ. Hamburg* **29** (1966), 217–241.

[5] Evelyn, C. J. A., Money-Coutts, G. B., and Tyrrell, J. A., *The Seven Circles Theorem and Other New Theorems*. Stacey International, London 1974.

[6] Laguerre, E., pp. 608–619 in *Oeuvres de Laguerre*, Tome II. Paris 1905.

[7] Pedoe, D., *A Course of Geometry for Colleges and Universities*, Cambridge University Press, London 1970.

[8] Pedoe, D., A forgotten geometrical transformation. *L'Enseignement Math.* **28** (1972), 255–267.

[9] Pedoe, D., Laguerre's axial transformation. *Math. Mag.* **48** (1975), 23–30.

[10] Rigby, J. F., On the Money-Coutts configuration of nine anti-tangent cycles *Proc. London Math. Soc. (3)* **43** (1981), 110–132.

[11] Tyrrell, J. A. and Powell, M. T., A theorem in circle geometry. *Bull. London Math. Soc.* **3** (1971), 70–74.

# Inversive Geometry[1]

J. B. Wilker*

## 0. Preface

Let me begin by describing one of the gems of classical mathematics which first stirred my own enthusiasm for inversive geometry. It illustrates the elegance of the subject and provides a point of interest which we shall glimpse again in the closing chapters of this account.

Imagine a unit circle $A$ with a line $D$ through its centre and, inside $A$, two smaller circles $B$ and $C_0$ with their centres on $D$. The circles $B$ and $C_0$, which may be of different sizes, should be externally tangent to each other and internally tangent to $A$ so that together the three circles bound two congruent curvilinear triangles. Choose one of these triangles and consider the sequence of successively tangent circles $C_0, C_1, C_2, \ldots$ which lie inside this triangle tangent to $A$ and $B$. Let $r_n$ denote the radius of $C_n$ and let $y_n$ denote the distance of its centre from $D$. Then $y_n$ and $r_n$ are related by the remarkably simple formula $y_n = 2nr_n$.

The figure is called Pappus' Arbelos and the formula relating $y_n$ to $r_n$ seems to be part of Greek mathematics. However, the following proof of this formula is much more recent because it is based on a nineteenth-century invention, the transformation called inversion in a circle.

Consider inversion in the circle orthogonal to $C_n$ with its centre at the point where $A$ touches $B$. The elementary properties of inversion in a circle show that

[§]I would like to thank H. S. M. Coxeter, J. C. Fisher, E. Honig, W. Israel, B. Salzberg, L. Southwell, J. F. Rigby, and B. Wilker for their encouraging enthusiasm. My interest in inversive geometry goes back more than ten years, and for much of this time I have received financial support from Canadian NRC Grant A8100. The opportunity to write the final version of this paper came while I was enjoying a sabbatical year as Visiting Fellow at the Institute of Advanced Studies, The Australian National University.

*Department of Mathematics, University of Toronto, Toronto, Ontario, M5S 1A1, Canada.

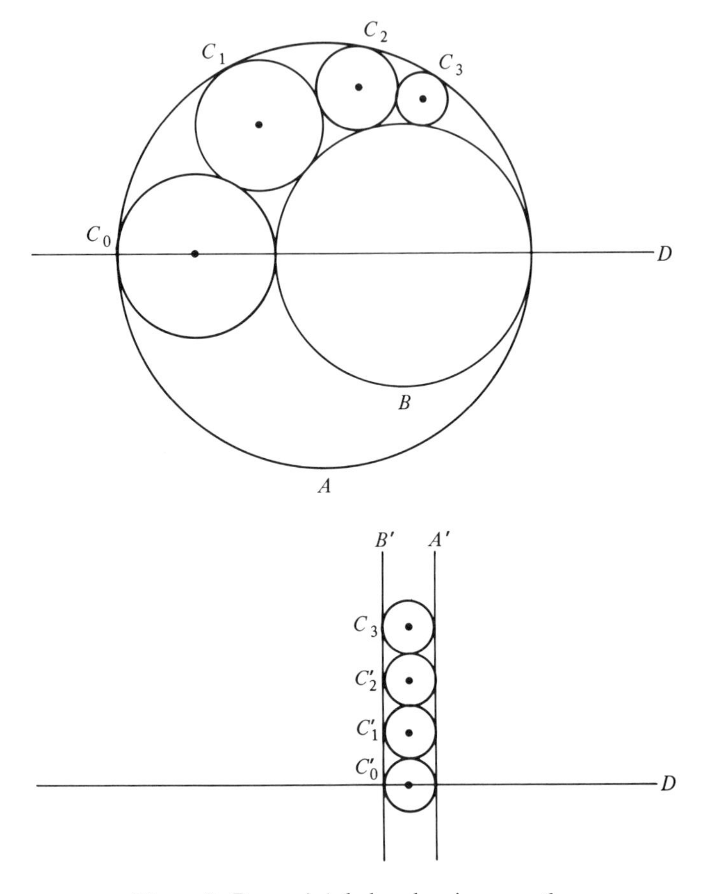


**Figure 0.** Pappus' Arbelos showing $y_3 = 6r_3$.

this transformation fixes $C_n$ and $D$ hence $r_n$ and $y_n$ and maps the original sequence into a column of congruent circles $C_n, C'_{n-1}, C'_{n-2}, \ldots, C'_0$ sandwiched between parallel lines $A'$ and $B'$ which are perpendicular to $D$. Since $C'_0$ is orthogonal to $D$, its centre lines on $D$ and we can read off from the transformed figure that $y_n = 2nr_n$.

I think that is a lovely proof and quite a few people must agree because a good many geometry texts devote a chapter to inversion in a circle and Pappus' Arbelos is one of the most popular applications. Some of these texts go on to consider the group generated by inversions in circles of the extended plane $\mathbb{R}^2 \cup \{\infty\}$ and to establish that this group is equal to the group of fractional linear transformations

$$z \to \frac{az + b}{cz + d} \quad \text{and} \quad z \to \frac{a\bar{z} + b}{cz + d} \qquad (ad - bc = 1).$$

These transformations include the Euclidean similarities, the isometries of the Riemann sphere and the isometries of the Poincaré model of the hyperbolic plane. This suggests that inversive geometry can serve as the common foundation of all three classical geometries. In the spirit of Klein, these geometries belong to subgroups of the group generated by all inversions.

My present thesis is threefold. First of all, I believe that inversive geometry is not much harder in $n$-dimensions with $n$ arbitrary than it is in 2-dimensions. Since there are many interesting problems in higher dimensions, I have followed my inclination to give a presentation which is $n$-dimensional from the outset. Incidentally, if any dimension is troublesome it is $n = 1$ where the $(n - 1)$-sphere is not connected and the angles are not defined. To help compensate for this difficulty I have drawn most of the figures for the case $n = 1$.

Second, there is an intimate connection between inversive geometry and linear algebra, and I have emphasized this connection throughout. Curriculum designers who are looking for a lively course to follow introductory linear algebra might consider inversive geometry as well as convexity and linear programming. Better still, why not follow through with both courses?

Finally, the connection between inversive geometry and the classical geometries is in no way dependent on a complex variable. I demonstrate this for the sake of better understanding in the 2-dimensional situation as well as for the sake of including $n$-dimensional results. Both these aspects are important in making this kind of geometry a more useful tool for research in such modern areas as automorphic forms group representations and the classification of 3-manifolds.

So much for the motivation and intended spirit of this paper. The introduction gives fairly precise details about the material which is actually covered. In order to make the presentation as self contained as possible I have avoided using any references in the first twelve chapters. Then in the last three chapters I have tried to give credit where it is due and to indicate how the story of inversive geometry continues beyond the scope of this paper.

One interesting area deals with the properties of $(n - 1)$-spheres in Euclidean and non-Euclidean $n$-space. I have marked references in this area with an asterisk whenever I felt the author's treatment could be improved by using the present techniques. One of my own papers is [95]* and chapter 13 explains how I would have written that paper if I had known this material in 1968. Readers who want an entertaining project might try a similar reformulation and generalization of other papers marked with an asterisk.

# 1. Introduction

Let $\Sigma$ be the unit $n$-sphere lying in Euclidean $(n + 1)$-space, $\Sigma = \{x \in \mathbb{R}^{n+1} : \|x\| = 1\}$. The $(n - 1)$-spheres which lie on $\Sigma$ are the sections of $\Sigma$ by $n$-flats which contain more than one point of $\Sigma$. To each such $(n - 1)$-sphere $\gamma$ there corresponds a bijection $\bar{\gamma} : \Sigma \to \Sigma$ called inversion in $\gamma$. The inversion $\bar{\gamma}$ is a transformation of period 2 which fixes the points of $\gamma$ and interchanges other points in pairs whose members are separated by $\gamma$. If $\gamma$ is a great or equatorial $(n - 1)$-sphere, $\bar{\gamma}$ is induced by reflection in the $n$-flat which cuts $\gamma$ from $\Sigma$ (Figure 1). If

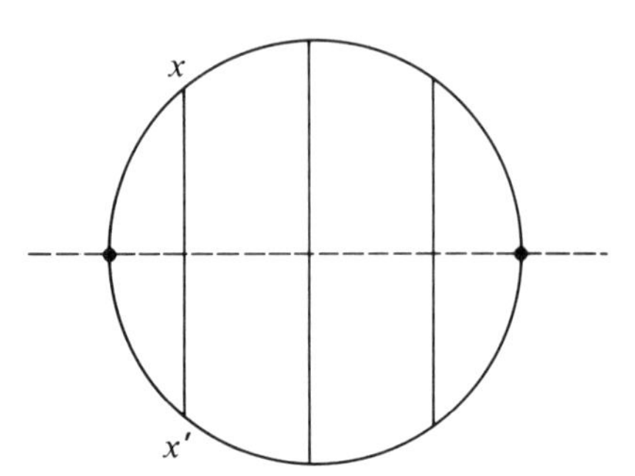


**Figure 1.** Inversion in a great $(n-1)$-sphere is induced by reflection in an $n$-flat $[n = 1]$.

$\gamma$ is not an equatorial $(n-1)$-sphere, there is a unique point $x_\gamma \in \mathbb{R}^{n+1}$ whose tangents to $\Sigma$ touch it along $\gamma$ (Figure 2). In this case $\bar{\gamma}$ interchanges two points $x$ and $x'$ provided the secant $xx'$ passes through $x_\gamma$.

The bijections of $\Sigma$ form a group, and those that can be expressed as a product of inversions form a subgroup. This subgroup is called the $n$-dimensional Möbius group and denoted $\mathfrak{M}_n$. The elements of this group are called Möbius transformations.

Let $x_0$ be a point of $\Sigma$, and let $\Pi_0'$ be the $n$-flat tangent to $\Sigma$ at $x_0$ (Figure 3). Let $\Pi'$ be an arbitrary $n$-flat parallel to $\Pi_0'$ and lying in the same half space as $\Sigma$. The lines through $x_0$ which do not lie in $\Pi_0'$ establish a 1-1 correspondence between the points of $\Sigma - \{x_0\}$ and the points of $\Pi'$. It can be extended to a 1-1 correspondence between $\Sigma$ and $\Pi = \Pi' \cup \{\infty\}$ by pairing $x_0$ with the new symbol, $\infty$. The extended correspondence is called stereographic projection, and through it the action of $\mathfrak{M}_n$ can be transferred to $\Pi$.

A model for $n$-dimensional inversive geometry is obtained by taking either $\Sigma$ or $\Pi$ under the action of $\mathfrak{M}_n$. Our object is to study inversive geometry, and it is useful to keep both models in mind. We investigate geometric invariants of $\mathfrak{M}_n$, show that $\mathfrak{M}_n$ is a linear group, relate the geometric invariants to an invariant

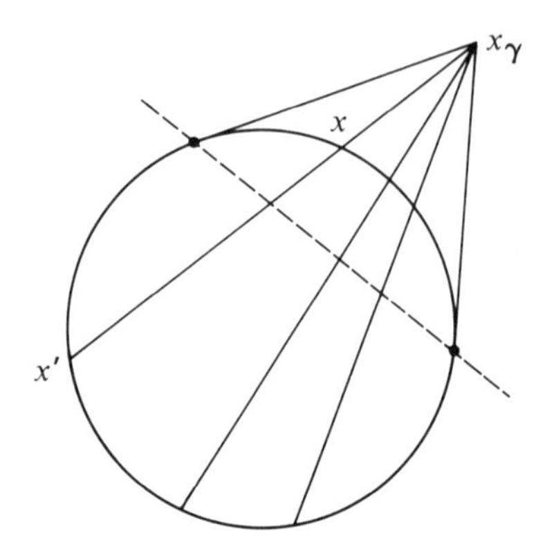


**Figure 2.** Inversion in a small $(n-1)$-sphere is induced by projection from a point $[n = 1]$.

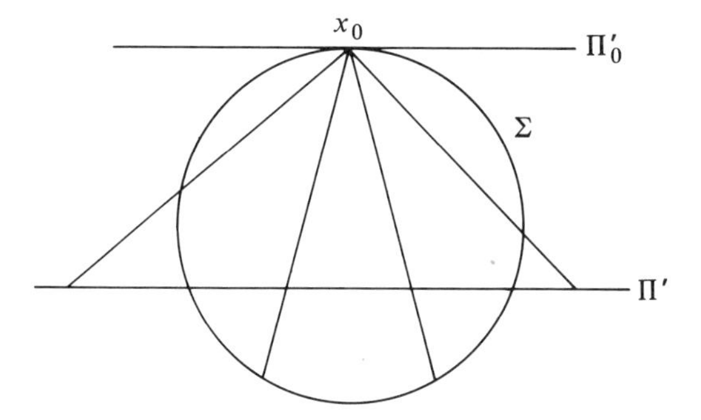


**Figure 3.** Stereographic projection to any $n$-flat $\Pi'$ parallel to $\Pi'_0$ [$n = 1$].

bilinear form, and carry out a synthetic description of the conjugacy classes of $\mathfrak{M}_n$ for $n = 1, 2, 3$. We apply our results in the case $n = 2$ to classify the transformations

$$z \to \frac{az + b}{cz + d} \quad \text{and} \quad z \to \frac{a\bar{z} + b}{c\bar{z} + d}, \qquad ad - bc \neq 0,$$

and to deduce the existence and full particulars of the famous 2-1 homomorphism of $SL(2, \mathbb{C})$ onto the proper orthochroneous Lorentz group. We indicate how to apply our results for general $n$ to study arrays of spheres in Euclidean and non-Euclidean geometry. In the last section, we give a survey of related literature.

## 2. Stereographic Projection and the $\Pi$-Model

We want to prove that when stereographic projection maps the $n$-sphere $\Sigma$ to an $n$-flat $\Pi = \mathbb{R}^n \cup \{\infty\}$ it carries the $(n-1)$-spheres of $\Sigma$ to the $(n-1)$-spheres of $\mathbb{R}^n$ together with the $(n-1)$-flats of $\mathbb{R}^n$, each including $\infty$. Moreover we want to show that stereographic projection conjugates inversions on $\Sigma$ into inversions or reflections on $\Pi$. These are defined as follows. Reflection in the $(n-1)$-flat $(x - a) \cdot n = 0$ acts on $\Pi$ by fixing $\infty$ and interchanging $x$ and $x'$ where

$$x' = x - 2\frac{n \cdot (x - a)}{n \cdot n} n.$$

Inversion in the $(n-1)$-sphere $\|x - a\| = k$ acts on $\Pi$ by interchanging $a$ and $\infty$ and interchanging $x$ and $x'$ where

$$x' = a + \frac{k^2}{\|x - a\|^2}(x - a).$$

To show that stereographic projection has the desired properties we step up one dimension and describe how inversion in an $n$-sphere, $n \geqslant 1$, acts on $\mathbb{R}^{n+1} \cup \{\infty\}$. Notice that "inversion" has two distinct meanings depending on whether the context is $\Sigma$ or $\Pi$. These meanings are *a priori* unrelated.

**Theorem 1.** *Let $\Gamma$ be the n-sphere, $\|x - a\| = k$, in $\mathbb{R}^{n+1} \cup \{\infty\}$. Inversion in $\Gamma$ has the following properties*:

(i) *it fixes n-flats through a and acts on them by inversion in an $(n-1)$-sphere with center a and radius k*;

(ii) *it interchanges n-spheres through a with n-flats not through a and does this by stereographic projection from a* (Figure 4);

(iii) (a) *it interchanges n-spheres not through a with others of the same description, and*

(b) *in particular, it fixes n-spheres $\Sigma$ which are orthogonal to $\Gamma$ and acts on them by inversion in the $(n-1)$-sphere $\Gamma \cap \Sigma$* (Figure 5);

(iv) *it preserves the set of m-spheres and m-flats, $0 \leqslant m \leqslant n$*;

(v) *it preserves angles*;

(vi) *it distorts distance in $\mathbb{R}^{n+1} - \{a\}$ but preserves each of the six cross ratios determined by four distinct points of this set.*

*Proof*. Without loss of generality assume that $a = 0$, so the inversion is given by $x' = (k/\|x\|)^2 x$. Part (i) is clear, and the fact that $x'$ is on the ray from the center of $\Gamma$ through $x$ helps with parts (ii) and (iii)(b).

The general equation of an $n$-sphere or $n$-flat is

$$c\|x\|^2 + y \cdot x + d = 0.$$

This passes through the center of $\Gamma$ precisely when $d = 0$ and represents an $n$-flat precisely when $c = 0$. Its image under the inversion is given by

$$c\left(\frac{k}{\|x\|}\right)^4 \|x\|^2 + \left(\frac{k}{\|x\|}\right)^2 y \cdot x + d = 0,$$

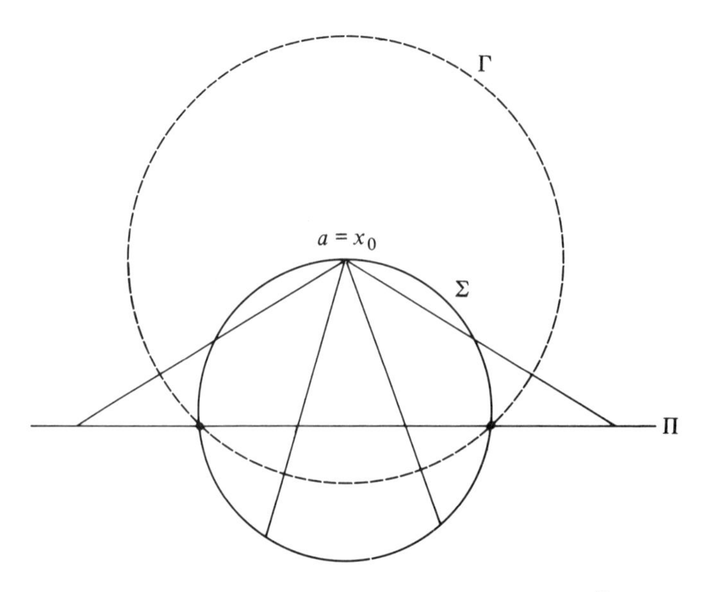


**Figure 4.** Stereographic projection is induced by $\overline{\Gamma}$.

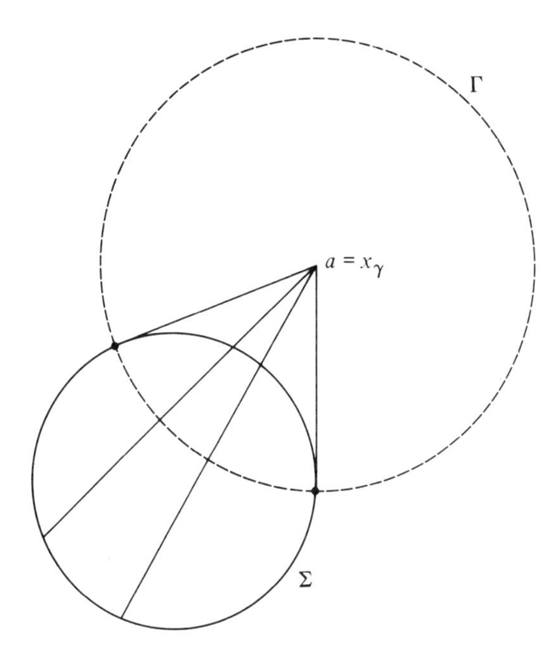


**Figure 5.** Inversion in $\gamma$ is induced by $\bar{\Gamma}$.

or

$$ck^4 + k^2y \cdot x + d\|x\|^2 = 0.$$

The exchange of roles between $c$ and $d$ completes the proof of parts (ii) and (iii)(a). Since the $n$-spheres $\|x\| = k$ and $c\|x\|^2 + y \cdot x + d = 0$ are orthogonal when $c = 1$ and $d = k^2$, the above calculation completes the proof of (iii)(b) as well.

Taken together, parts (i)–(iii) prove that inversion in an $n$-sphere preserves the set of $n$-spheres and $n$-flats. Appropriate intersections of these represent the $m$-spheres and $m$-flats $0 \leqslant m \leqslant n - 1$, and (iv) follows because the image of an intersection is exactly the intersection of the images.

To see that angles are preserved, notice that the $n$-sphere $\|x\| = r$ is mapped to the $n$-sphere $\|x\| = k^2/r$ just as it would be by the dilation $x \to (k/r)^2x$. Since the derivative of $t \to k^2/t$ is $t \to -(k/t)^2$, the rescaling factor along orthogonal rays matches its value on these $n$-spheres and the mapping is therefore conformal.

Two points $x_1$ and $x_2$ a distance $\|x_1 - x_2\|$ apart get mapped to points $x_1' = (k/\|x_1\|)^2x_1$ and $x_2' = (k/\|x_2\|)^2x_2$ whose distance apart works out to be

$$\|x_1' - x_2'\| = \frac{k^2}{\|x_1\|\,\|x_2\|}\,\|x_1 - x_2\|$$

because triangles $0x_1x_2$ and $0x_2'x_1'$ are similar. This means that given four points $x_1, x_2, x_3, x_4$, any ratio of the form

$$\|x_1 - x_2\|\,\|x_3 - x_4\|\big[\,\|x_1 - x_3\|\,\|x_2 - x_4\|\,\big]^{-1}$$

will be invariant. To see that there are six ratios of this type, think of the points as vertices of a 3-simplex. A numerator or denominator is the product of the lengths of a pair of opposite edges, and the three possible pairs give six nontrivial ratios. □

**Corollary 1.** *Inversion in an n-sphere* $\Gamma$ *and reflection in an n-flat* $\Gamma$ *can be given a unified definition. In either case the transformation* $\overline{\Gamma}$ *fixes the points of* $\Gamma$ *and maps a point* $x \notin \Gamma$ *to the second point of intersection of any two circles or lines through* $x$ *perpendicular to* $\Gamma$.

*Proof.* Both types of transformation fix the points of $\Gamma$ and interchange other points in pairs whose members are separated by $\Gamma$ (Figure 6). Both types of transformation fix circles and lines orthogonal to $\Gamma$. The result follows because the image of an intersection is the intersection of the images. □

**Corollary 2.** *Inversion in an n-sphere conjugates inversion [reflection] in an n-sphere [n-flat]* $\Gamma$ *into inversion or reflection in its image* $\Gamma'$.

*Proof.* The image set $\Gamma'$ is an $n$-sphere or an $n$-flat. The result follows because the circles and lines orthogonal to $\Gamma$ are mapped to the circles and lines orthogonal to $\Gamma'$, and therefore mates under $\overline{\Gamma}$ are mapped to mates under $\overline{\Gamma'}$. □

**Corollary 3.** *Stereographic projection has the desired property of conjugating inversions on* $\Sigma$ *[described in Section 1] into inversions and reflections on* $\Pi$ *[described earlier in this section].*

*Proof.* Stereographic projection from the $n$-sphere $\Sigma$ to an $n$-flat $\Pi$ is induced by inversion in an $n$-sphere whose center lies on $\Sigma$ (Figure 7). An arbitrary $(n-1)$-sphere $\gamma$ on $\Sigma$ can be written $\gamma = \Sigma \cap \Gamma$ for a suitable $n$-sphere or $n$-flat $\Gamma$ orthogonal to $\Sigma$. Then $\bar{\gamma}$ is induced by $\overline{\Gamma}$.

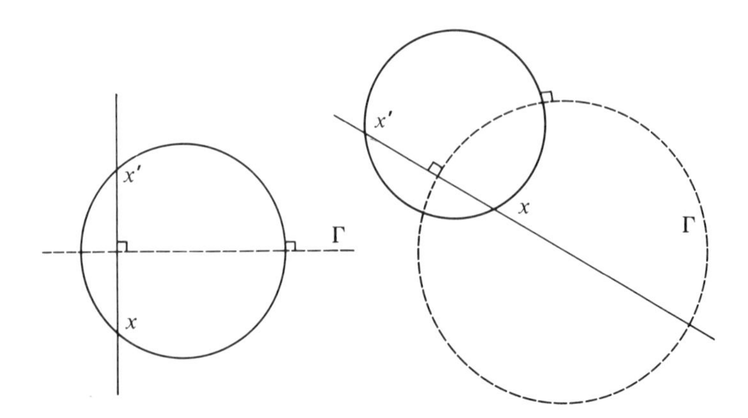


**Figure 6.** The unified definition of reflection in an $n$-flat and inversion in an $n$-sphere $n \geqslant 1$ $[n = 1]$.

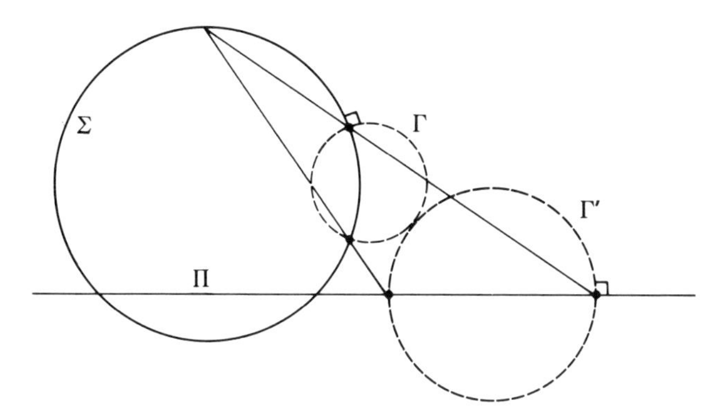


**Figure 7.** Stereographic projection conjugates inversions on $\Sigma$ into inversions on $\Pi$.

Stereographic projection carries $\gamma$ to an $(n-1)$-sphere or $(n-1)$-flat $\gamma'$ in $\Pi$ given by $\gamma' = \Pi \cap \Gamma'$. Since $\Gamma'$ is orthogonal to $\Pi$, $\overline{\gamma'}$ is induced by $\overline{\Gamma'}$. Finally, since $\overline{\Gamma'}$ is conjugate to $\overline{\Gamma}$, it follows that $\overline{\gamma'}$ is conjugate to $\bar{\gamma}$. □

*Remark* 1. The proof of Corollary 3 shows that $\mathfrak{M}_{n+1}$ contains one copy of $\mathfrak{M}_n$ for each $n$-sphere or $n$-flat and that these subgroups are mutually conjugate.

*Remark* 2. The fact that stereographic projection is induced by an inversion shows that every conceivable $\mathfrak{M}_n$-invariant can be measured equally well in $\Sigma$ and in $\Pi$.

*Remark* 3. The $\Sigma$-model shows that any four distinct points can be used to form a cross ratio and the resulting invariant depends continuously on its arguments. This justifies the $\Pi$-model calculation

$$\frac{\|x_1 - x_2\| \, \|x_3 - \infty\|}{\|x_1 - x_3\| \, \|x_2 - \infty\|} = \lim_{y \to \infty} \frac{\|x_1 - x_2\| \, \|x_3 - y\|}{\|x_1 - x_3\| \, \|x_2 - y\|} = \frac{\|x_1 - x_2\|}{\|x_1 - x_3\|} .$$

*Remark* 4. When $n \geqslant 2$ the unified definition of inversion and reflection in the $\Pi$-model can be lifted to give an intrinsic definition of inversion in the $\Sigma$-model. Inversion in an $(n-1)$-sphere $\gamma$ of $\Sigma$ fixes the points of $\gamma$ and maps other points to the second meeting of circles through them perpendicular to $\gamma$.

## 3. Linearization of $\mathfrak{M}_n$

Recall that $\Sigma$ is the unit $n$-sphere lying in Euclidean $(n+1)$-space,

$$\Sigma = \{x \in \mathbb{R}^{n+1} : \|x\| = 1\}.$$

Let $C$ be a closed $n$-cap on $\Sigma$ (Figure 8) with center $c \in \Sigma$ and angular radius $\theta$, $0 < \theta < \pi$,

$$C = \{x \in \Sigma : x \cdot c \geqslant \cos\theta\}.$$

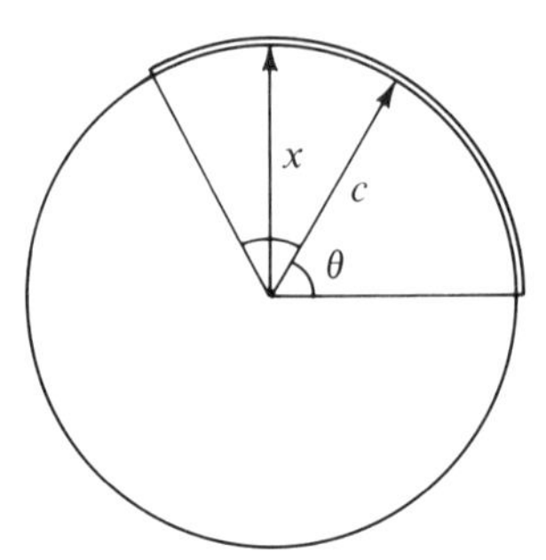


**Figure 8.** A cap on $\Sigma$.

The condition for a point $x$ of $\Sigma$ to belong to $C$ can be rewritten in the form

$$x \cdot c - 1 \cdot \cos\theta \geqslant 0$$

or better still, since $\sin\theta > 0$,

$$x \cdot \frac{c}{\sin\theta} - 1 \cdot \frac{\cos\theta}{\sin\theta} \geqslant 0.$$

This suggests that we should represent the point by the $(n + 2)$-vector

$$X = (x, 1)$$

and the cap by the $(n + 2)$-vector

$$C = (\csc\theta\, c, \cot\theta).$$

Then if we introduce the indefinite bilinear form

$$U * V = U^{(1)}V^{(1)} + U^{(2)}V^{(2)} + \cdots + U^{(n+1)}V^{(n+1)} - U^{(n+2)}V^{(n+2)},$$

the condition for the point $X$ to belong to the cap $C$ is that

$$X * C \geqslant 0.$$

With these coordinates a vector represents a cap if and only if it satisfies $C * C = 1$, and a vector represents a point if and only if it satisfies $X * X = 0$ and $X^{(n+2)} = 1$. There is some utility in allowing the coordinates of a point to be positive homogeneous so that if $\lambda > 0$, $X$ and $\lambda X$ are allowable names for the same point.

If a cap $C = (\csc\theta\, c,\ \cot\theta)$ has center $c$ and angular radius $\theta$, then the "complementary" cap $C'$ has center $-c$ and angular radius $\pi - \theta$. This means that the coordinate for $C'$ is just $-C$. A point $X$ belongs to the $(n - 1)$-sphere which forms the common boundary of $C$ and $C'$ if and only if $X * C = 0$.

**Theorem 2.** *With the coordinates described above, inversion in the common boundary of $C$ and $-C$ is given by the linear transformation*

$$U \to U - 2(C * U)C.$$

*This formula describes the action of the inversion on points and caps alike.*

*Proof*. The linear transformation described in the theorem preserves the bilinear form, since

$$\begin{aligned} U' * V' &= \left[ U - 2(C * U)C \right] * \left[ V - 2(C * V)C \right] \\ &= U * V - 4(C * U)(C * V) + 4(C * U)(C * V)(C * C) \\ &= U * V. \end{aligned}$$

This proves that it maps caps to caps. Once it has been shown that it maps points properly, the invariance of $U * V$ will imply that incidence is preserved and therefore that the transformation maps caps properly as well.

If $C^{(n+2)} = 0$, then $C = (c, 0)$ is a hemisphere and $c$ is the unit normal to the $n$-flat whose reflection induces our inversion. If $X = (x, 1)$ is a point of $\Sigma$, this reflection is described in terms of $(n + 1)$-vectors by

$$x \to x - 2(c \cdot x)c.$$

The desired description of the inversion in terms of $(n + 2)$-vectors is immediate.

If $C^{(n+2)} = \cot\theta \neq 0$ and $C = (\cos\theta\, c, \cot\theta)$, then $\sec\theta\, c \in \mathbb{R}^{n+1}$ (Figure 9) is the point of concurrence of the joins of mates in the inversion. If $X = (x, 1)$ represents a point of $\Sigma$, its image under the transformation of the theorem can be written in the form

$$X' = \lambda(x', 1) = \lambda\left(\lambda^{-1}x + (1 - \lambda^{-1})\sec\theta\, c, 1\right)$$

where

$$\lambda = \left[ 1 + \cos^2\theta - 2(c \cdot x)\cos\theta \right] \csc^2\theta.$$

Since $\lambda > 0$ and $X' * X' = X * X = 0$, $X'$ represents a point $x'$ on $\Sigma$. Since $x'$ is on the straight line joining $x$ and $\sec\theta\, c$, it really is the image of $x$ under inversion in the boundary of $C$. □

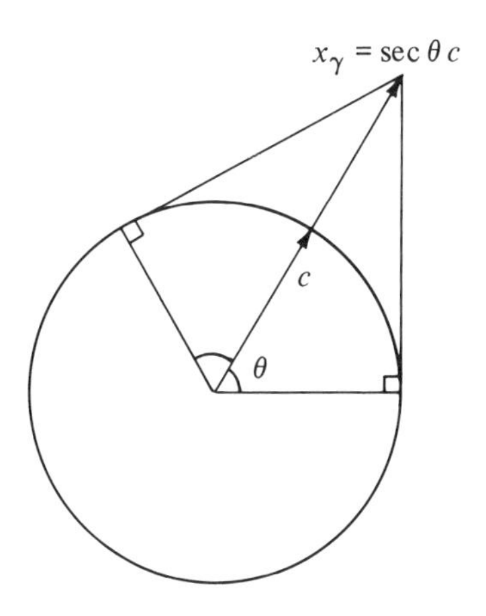


**Figure 9.** The point of projection corresponding to inversion in the boundary of a cap.

**Corollary.** *The n-dimensional Möbius group* $\mathfrak{M}_n$ *is isomorphic to a subgroup of* $\mathfrak{L}_{n+2}^{\uparrow}$, *the* $(n+2)$*-dimensional linear group which preserves the bilinear form* $U * V$ *and the sign of* $U^{(n+2)}$ *on the cone* $U * U = 0$.

*Proof.* $\mathfrak{M}_n$ is defined to be the group generated by inversions in the $(n-1)$-spheres of $\Sigma$, and we have just found that these inversions are given by linear transformations which preserve $U * V$ and the sign of $U^{(n+2)}$ on the cone $U * U = 0$. □

In the next two sections we shall prove that $\mathfrak{M}_n$ is isomorphic to the full group $\mathfrak{L}_{n+2}^{\uparrow}$. We conclude this section with a few preliminary remarks about tangent caps and their coordinates.

*Remark* 1. It is possible to have a set of $n+2$ $n$-caps on the $n$-sphere with the property that any two of them are externally tangent. When a set like this is ordered, we call it a cluster. To obtain an example of a cluster, let $c_1$, $c_2, \ldots, c_{n+2}$ be any ordering of the vertices of a regular $(n+1)$-simplex inscribed in $\Sigma$. These points are equally spaced, and the angular separation between any pair of them is $2\psi$, where $\sec 2\psi = -(n+1)$. It follows that the congruent $n$-caps $C_j$ $(j = 1, 2, \ldots, n+2)$ with centers $c_j$ and angular radius $\psi$ form a cluster.

*Remark* 2. If $C = (\csc\theta\, c, \cot\theta)$ and $D = (\csc\psi\, d, \cot\psi)$ are externally tangent, then $c \cdot d = \cos(\theta + \psi)$ and $C * D = -1$. The caps of a cluster $\mathcal{C} = (C_1, C_2, \ldots, C_{n+2})$ therefore satisfy

$$c_{ij} \equiv C_i * C_j = \begin{cases} 1 & \text{if } i = j, \\ -1 & \text{if } i \neq j. \end{cases}$$

The cluster matrix $M = (c_{ij})$ is given by $M = 2I - J$, where $I$ is the $(n+2) \times (n+2)$ identity matrix and $J$ is the $(n+2) \times (n+2)$ matrix of 1's. Since $J^2 = (n+2)J$, it is easy to verify that $\frac{1}{2}[I - (1/n)J] = M^{-1}$. Since $M$ is nonsingular, the vectors in $\mathcal{C}$ are linearly independent and form a basis for $(n+2)$-space.

*Remark* 3. Any element in $\mathfrak{L}_{n+2}^{\uparrow}$ induces a mapping of the $n$-sphere $\Sigma$ which takes points to points and caps to caps in such a way as to preserve incidence. In particular it maps clusters to clusters. Since the vectors of a cluster form a basis, there is at most one transformation in $\mathfrak{L}_{n+2}^{\uparrow}$ which maps a given cluster to another. In order to complete the proof that $\mathfrak{M}_n \cong \mathfrak{L}_{n+2}^{\uparrow}$ it remains to show that $\mathfrak{M}_n$ is transitive on clusters, and this is done in Section 5 using information about the isometries and similarities of $\mathbb{R}^n$ which is assembled in Section 4.

## 4. Isometries and Similarities of $\mathbb{R}^n$

An isometry is a mapping $f: \mathbb{R}^n \to \mathbb{R}^n$ such that $\|x^f - y^f\| = \|x - y\|$. We shall see that any isometry can be written as the product of at most $n + 1$ reflections, and so the isometries form a subgroup of $\mathfrak{M}_n$. A similarity with scale factor $k > 0$ is a mapping $g: \mathbb{R}^n \to \mathbb{R}^n$ such that $\|x^g - y^g\| = k\|x - y\|$. The similarity $g$ can be written $g = fd$ where $f$ is an isometry and $d$ is the dilatation $x \to kx$. Since $d$ can be expressed as the product of any two inversions $x \to (k_1/\|x\|)^2 x$ and $x \to (k_2/\|x\|)^2 x$ satisfying $k_2 k_1^{-1} = k^{1/2}$, the similarity $g = fd$ lies in $\mathfrak{M}_n$. We shall see that the similarities constitute the subgroup of $\mathfrak{M}_n$ which stabilizes $\infty$.

The mapping $m \to m + 1$ shows that the isometries of the positive integers do not form a group. Warned by this simple example, we take some care with our assertions about the isometries and similarities of $\mathbb{R}^n$.

Two subsets $S = \{x_i\}$ $(i \in I)$ and $S' = \{x_i'\}$ $(i \in I)$ of $\mathbb{R}^n$ are similar [congruent] if there is a constant $k > 0$ $[k = 1]$ such that $\|x_i' - x_j'\| = k\|x_i - x_j\|$ $(i, j \in I)$.

**Lemma 1.** *If $S$ and $S'$ are congruent $m$-point subsets of $\mathbb{R}^n$, there is an isometry $f: \mathbb{R}^n \to \mathbb{R}^n$ which satisfies $S^f = S'$ and can be written as the product of at most $m$ reflections.*

*Proof.* Advance the points of $S$ in turn from their current positions to their desired final positions by reflections (Figure 10). Once a point reaches its desired final position, it automatically lies on the mirror of each subsequent reflection and is not disturbed. □

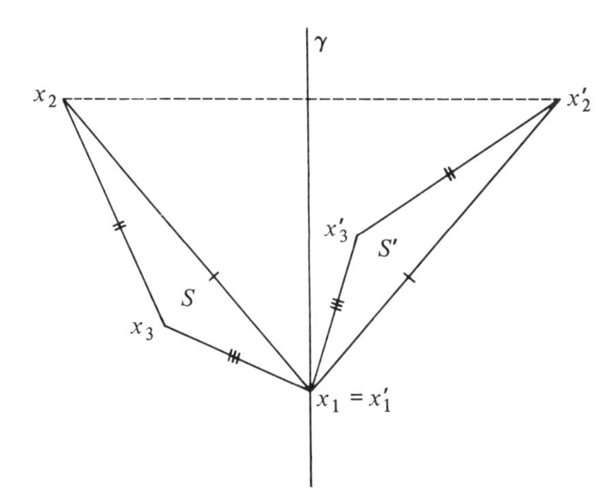


**Figure 10.** The $(n - 1)$-flat $\gamma$ such that $x_2^{\gamma} = x_2'$ automatically passes through $x_1 = x_1'$ $[n = 2]$.

An $n$-simplex $S = \{x_i\}$ $(i = 1, 2, \ldots, n + 1)$ is a set of $n + 1$ points of $\mathbb{R}^n$ which do not lie on an $(n-1)$-flat. Since reflections preserve $n$-simplices, any set congruent to an $n$-simplex is an $n$-simplex. In particular, isometries map $n$-simplices to congruent $n$-simplices.

**Lemma 2.** *An isometry $f: \mathbb{R}^n \to \mathbb{R}^n$ is determined by its effect on an arbitrary $n$-simplex. It can be written as the product of at most $n + 1$ reflections.*

*Proof.* Let $S$ be an $n$-simplex and $f_1$ an isometry such that $S^{f_1} = S^f = S'$. If $f_1 \neq f$, there is a point $x$ such that $x^{f_1} \neq x^f$. But then $S'$ lies in the $(n-1)$-flat equidistant from $x^{f_1}$ and $x^f$, and this contradicts the fact that $S'$ is an $n$-simplex. It follows that $f$ is determined by its effect on an $n$-simplex, and the rest is clear from Lemma 1.

**Lemma 3.** *The isometries of $\mathbb{R}^n$ form a group, and this group is sharply transitive on any class of mutually congruent $n$-simplices.*

*Proof.* An arbitrary isometry can be written as a product of reflections and is therefore a bijection. It is clear that distance-preserving bijections form a group. If $S$ and $S'$ are congruent $n$-simplices, then Lemma 1 shows that there is an isometry $f$ with $S^f = S'$, and Lemma 2 shows that there is only one such $f$.

**Lemma 4.** *The similarities of $\mathbb{R}^n$ form a group with the mapping to scale factors as a homomorphism and its kernel, the isometries, as a normal subgroup. The similarities are sharply transitive on any class of mutually similar $n$-simplices.*

*Proof.* If $g_1$ has scale factor $k_1$, and $g_2$ has scale factor $k_2$, then $g_1 g_2$ has scale factor $k_1 k_2$. In particular if $g$ has scale factor $k$, and $d$ is the dilatation $x \to kx$, then $gd^{-1} = f$ is an isometry and $g = fd$. This shows that similarities are bijections, and it is clear that distance rescaling bijections form a group.

If $S$ and $S'$ are similar $n$-simplices, there is a dilatation $d$ such that $S^d$ and $S'$ are congruent, and then an isometry $f$ such that $S^{df} = S'$. The similarity $g = df$ such that $S^g = S$ is unique, because if $S^{g_1} = S'$, then $S^{gg_1^{-1}} = S$ and $gg_1^{-1}$ is a similarity fixing an $n$-simplex, hence an isometry fixing an $n$-simplex, and hence the identity. □

**Lemma 5.** *A proper similarity can be written as the commuting product of an isometry with a fixed point and a dilatation with the same fixed point.*

*Proof.* If $\|x^g - y^g\| = k\|x - y\|$ with $k \neq 1$, then $g$ or $g^{-1}$ is a contraction mapping and Banach's theorem implies that it has a unique fixed point $x_0$. If $d$ is the dilatation $x \to k(x - x_0) + x_0$ with fixed point $x_0$ and scale factor $k$, then $gd^{-1} = f$ is an isometry with fixed point $x_0$. It follows that $g = fd = df$, where the last equality is seen to hold by factoring $f$ into reflections in $(n-1)$-flats through $x_0$ and $d$ into inversions in $(n-1)$-spheres centered at $x_0$, and observing that the reflections commute with the inversions. □

Some of the information in Lemmas 1–5 is for immediate use, and some is related information for later use. We now return to our consideration of $\mathfrak{M}_n$.

## 5. The Complete Description of $\mathfrak{M}_n$

We begin by proving that $\mathfrak{M}_n$ is transitive on clusters.

**Lemma 6.** *Let $\mathcal{C} = (C_1, C_2, \ldots, C_{n+2})$ and $\mathcal{C}' = (C_1', C_2', \ldots, C_{n+2}')$ be any two clusters on the n-sphere $\Sigma$. Then there is a sequence of at most $n + 2$ inversions whose product maps $\mathcal{C}$ to $\mathcal{C}'$.*

*Proof.* We pass to a $\Pi$-model with $\infty$ at the point of contact of $C_{n+1}'$ and $C_{n+2}'$ (Figure 11). This means that $C_{n+1}'$ and $C_{n+2}'$ are parallel half spaces and $C_1', C_2', \ldots, C_n'$ are congruent $n$-balls sandwiched between them. The centers of $C_1', C_2', \ldots, C_n'$ are the vertices of a regular $(n-1)$-simplex $(x_1', x_2', \ldots, x_n')$ lying in the $(n-1)$-flat midway between the boundaries of $C_{n+1}'$ and $C_{n+2}'$. We complete this $(n-1)$-simplex to an $n$-simplex $S' = (x_1', x_2', \ldots, x_n', x_{n+1}')$ by adding the point of contact of $C_1'$ and $C_{n+1}'$.

Let $x_0$ be the point where $C_{n+1}$ touches $C_{n+2}$. There are two cases to consider, depending on whether $x_0 = \infty$ or $x_0 \neq \infty$.

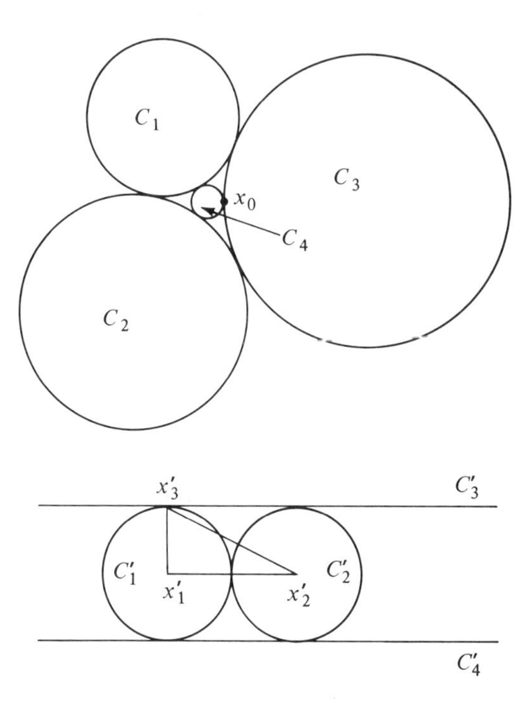


**Figure 11.** Lemma 6 with $n = 2$ and $x_0 \neq \infty$.

If $x_0 \neq \infty$, we begin by inverting in an $(n-1)$-sphere $\gamma$ with center $x_0$ and radius $k$. This maps $C_{n+1}$ and $C_{n+2}$ to parallel half spaces separated by a gap whose width is proportional to $k^2$. We choose $\gamma$ so that the width of the gap between $C_{n+1}^{\bar{\gamma}}$ and $C_{n+2}^{\bar{\gamma}}$ is equal to that between $C'_{n+1}$ and $C'_{n+2}$. Then we introduce an $n$-simplex $S = (x_1, x_2, \ldots, x_n, x_{n+1})$ congruent to $S'$ by using $C_1^{\bar{\gamma}}, C_2^{\bar{\gamma}}, \ldots, C_{n+2}^{\bar{\gamma}}$ in place of $C'_1, C'_2, \ldots, C'_{n+2}$. The isometry which carries $S$ to $S'$ maps $\mathcal{C}^{\bar{\gamma}}$ to $\mathcal{C}'$. This isometry requires at most $n+1$ reflections, so counting these together with the initial inversion, we have proved the lemma in the case $x_0 \neq \infty$.

If $x_0 = \infty$, we can immediately define a simplex $S$ which is similar to $S'$. If $S$ is actually congruent to $S'$, we proceed as above. If $S$ is not congruent to $S'$, we apply the dilatation $d$ which maps the $n$-ball $C_1$ to the $n$-ball $C'_1$. This dilatation costs two inversions, but it makes $S^d$ and $S'$ congruent and gives them at least one vertex in common, namely $x_1^d = x'_1$. Then we can map $S^d$ to $S'$ by a product of at most $n$ reflections. Since this completes the mapping of $\mathcal{C}$ to $\mathcal{C}'$, we have proved the lemma in this case as well. □

**Theorem 3.** *The $n$-dimensional Möbius group $\mathfrak{M}_n$ is isomorphic to the $(n+2)$-dimensional linear group $\mathfrak{L}_{n+2}^{\uparrow}$. The group $\mathfrak{M}_n$ is sharply transitive on clusters, and its most general element can be written as the product of at most $n+2$ inversions.*

*Proof.* The Corollary and Remarks following Theorem 2 show that $\mathfrak{M}_n$ is a subgroup of $\mathfrak{L}_{n+2}^{\uparrow}$, and $\mathfrak{L}_{n+2}^{\uparrow}$ contains at most one transformation mapping one cluster to another. On the other hand, Lemma 6 shows that given any two clusters, there is a transformation in $\mathfrak{M}_n$ mapping one to the other, and this transformation can be written as the product of at most $n+2$ inversions. □

**Corollary 1.** *When $\mathfrak{M}_n$ acts on the $\Pi$-model, its most general element is either a similarity of $\mathbb{R}^n$ or the product of an inversion and an isometry of $\mathbb{R}^n$.*

*Proof.* Take a cluster $\mathcal{C}'$ with $\infty$ as the point of contact of $C'_{n+1}$ and $C'_{n+2}$. Let $\mathcal{C}$ be the image of this cluster under the inverse of a given transformation $h \in \mathfrak{M}_n$. Then $\mathcal{C}^h = \mathcal{C}'$, and the analysis of Lemma 6 applies to $h$. If $x_0 = \infty$, $h$ is a similarity, and if $x_0 \neq \infty$, $h$ is the product of an inversion and an isometry. □

**Corollary 2.** *An element of $\mathfrak{M}_n$ which fixes a point of $\Sigma$ can be considered as a Euclidean similarity.*

*Proof.* If $h$ fixes $x_0 \in \Sigma$, choose a $\Pi$-model with $x_0 = \infty$, and $h$ will act as a similarity on $\Pi$.

**Corollary 3.** *An element of $\mathfrak{M}_n$ which acts without fixed points on $\Sigma$ is conjugate to an isometry of $\Sigma$.*

*Proof.* Let $\sum = \{x \in \mathbb{R}^{n+1} : \|x\| = 1\}$. A mapping $h \in \mathfrak{M}_n$ which acts on $\Sigma$ can be extended to a mapping $\tilde{h} \in \mathfrak{M}_{n+1}$ which acts on $\mathbb{R}^{n+1} \cup \{\infty\}$ and maps the ball $\|x\| \leqslant 1$ continuously onto itself. The extension is obtained by factoring $h$ into inversions in $(n-1)$-spheres $\gamma$ of $\Sigma$ and then inducing these by inversions in $n$-spheres [or reflections in $n$-flats] $\Gamma$ orthogonal to $\Sigma$ along the $\gamma$'s. If $h$ has no fixed points, then $\tilde{h}$ has no fixed points on $\Sigma$, and by the Brouwer fixed-point theorem, $\tilde{h}$ must have a fixed point $x_0$ with $\|x_0\| < 1$.

Let $\Gamma_0$ be the $n$-sphere orthogonal to $\Sigma$ with the property that $x_0^{\overline{\Gamma}_0} = 0$. The conjugate transformation $\tilde{h}' = \overline{\Gamma}_0 \tilde{h} \overline{\Gamma}_0$ fixes 0 and $\Sigma$, and hence $0^{\overline{\Sigma}} = \infty$. It follows that $\tilde{h}'$ is a Euclidean isometry and that it factors into a product of reflections in $n$-flats through 0.

Finally, if $\gamma_0 = \Gamma_0 \cap \Sigma$, then the conjugate transformation $h' = \bar{\gamma}_0 h \bar{\gamma}_0$ is the restriction of $\tilde{h}'$ to $\Sigma$. This means that $h'$ factors into a product of inversions in equatorial $(n-1)$-spheres and is an isometry of $\Sigma$. $\square$

The $n$-sphere $\Sigma$ is orientable, and inversion in an $(n-1)$-sphere $\gamma$ reverses this orientation. It follows that the product of an even number of inversions preserves orientation, and the product of an odd number of inversions reverses orientation. We refer to a Möbius transformation as direct or opposite depending on whether it preserves or reverses orientation. This geometrical distinction is reflected in the linear algebra of $\mathfrak{L}_{n+2}^{\uparrow}$.

**Corollary 4.** *A Möbius transformation is direct or opposite depending on whether its matrix in $\mathfrak{L}_{n+2}^{\uparrow}$ has determinant $+1$ or $-1$.*

*Proof.* It is enough to show that if $\gamma$ is an $(n-1)$-sphere, then the linear transformation $\bar{\gamma}$ has determinant $-1$. Let $C_1$ be an $n$-cap bounded by $\gamma$, and let $C_2, C_3, \ldots, C_{n+2}$ be $n$-caps completing a cluster. Since inversion in $\gamma$ is given by

$$U \to U - 2(C_1 * U)C_1,$$

we have $C_1 \to -C_1$ and $C_j \to C_j + 2C_1$ $(j = 2, 3, \ldots, n+2)$. It is immediate to write down the matrix of $\bar{\gamma}$ relative to the cluster basis and to see that its determinant is $-1$.

**Corollary 5.** *The group $\mathfrak{L}_{n+2}^{\uparrow}$ depends on $\binom{n+2}{2}$ continuous real parameters.*

*Proof.* The elements of this group are labeled by the cluster $\mathcal{C}'$ to which they move a fixed cluster $\mathcal{C}$. We consider the parameters involved when a possible $\mathcal{C}'$ is constructed in the $\Pi$-model. The center and radius of the first $n$-ball account for $n+1$ continuous parameters. Thereafter each successively chosen $n$-ball has one less degree of freedom than its predecessor until the $(n+1)$st $n$-ball is chosen with just a single degree of freedom. This accounts for a total of

$$(n+1) + n + (n-1) + \cdots + 1 = \binom{n+2}{2}$$

continuous real parameters. When $n \geqslant 2$, the $(n+2)$nd $n$-ball can then be added in one of two positions, and these account for determinant $+1$ and determinant $-1$. The case $n = 1$ is somewhat different in this last respect. Here the position of the third 1-ball is determined by those of the first two. The orientation of the cluster is fixed by the position of the second 1-ball relative to the first, and this determines whether the transformation has determinant $+1$ or $-1$. $\square$

## 6. Cross Ratio

Cross ratio is the fundamental inversive invariant. If $\Pi = \mathbb{R}^n \cup \{\infty\}$, then a mapping $h : \Pi \to \Pi$ belongs to $\mathfrak{M}_n$ if and only if it preserves cross ratio. The phrase "preserves cross ratio" implies that $h$ maps sets of four distinct points to sets of four distinct points, and in particular it implies that $h$ is injective. However, it does not imply *a priori* that $h$ is surjective. This makes the result more striking and shows the strength of the material developed in Section 4.

**Theorem 4.** *A mapping $h : \Pi \to \Pi$ belongs to $\mathfrak{M}_n$ if and only if it preserves cross ratio.*

*Proof.* If $h$ belongs to $\mathfrak{M}_n$, it preserves cross ratio. It remains to show that if $h$ preserves cross ratio, then it belongs to $\mathfrak{M}_n$. If $h$ does not fix $\infty$, then it can be composed with an inversion which restores $\infty$. It is therefore sufficient to prove that if $h$ preserves cross ratio and fixes $\infty$, then it is a similarity.

Let $x_1, x_2, x, y$ be four distinct points of $\mathbb{R}^n$. Using Remark 3 of Section 2, we obtain

$$\frac{\|x - x_2\|}{\|x_2 - x_1\|} = \frac{\|x - x_2\|\,\|x_1 - \infty\|}{\|x_2 - x_1\|\,\|x - \infty\|} = \frac{\|x^h - x_2^h\|\,\|x_1^h - \infty\|}{\|x_2^h - x_1^h\|\,\|x^h - \infty\|} = \frac{\|x^h - x_2^h\|}{\|x_2^h - x_1^h\|},$$

and similarly

$$\frac{\|x - y\|}{\|x - x_2\|} = \frac{\|x^h - y^h\|}{\|x^h - x_2^h\|}.$$

By multiplying these equal ratios we obtain

$$\frac{\|x - y\|}{\|x_2 - x_1\|} = \frac{\|x^h - y^h\|}{\|x_2^h - x_1^h\|}, \quad \text{or} \quad \|x^h - y^h\| = \frac{\|x_2^h - x_1^h\|}{\|x_2 - x_1\|}\|x - y\|,$$

and if this last equality is read with $x_1$ and $x_2$ fixed and $x$ and $y$ variable, it implies that $h$ is a similarity with scale factor $k = \|x_2^h - x_1^h\|[\|x_2 - x_1\|]^{-1}$.

**Corollary.** *Given two subsets $S = \{x_i\}$ $(i \in I)$ and $S' = \{x_i'\}$ $(i \in I)$ lying in $\Pi$, there is a Möbius transformation $h$ such that $S^h = S'$ if and only if every cross ratio taken in $S$ is matched by an equal cross ratio involving the corresponding points of $S'$. Given that $h$ exists, it is unique if and only if $S$ contains $n + 2$ points not all on the same $(n-1)$-sphere or $(n-1)$-flat.*

*Proof.* The necessity of the condition follows from the invariance of cross ratio. To prove its sufficiency we invert $x_1$ and $x_1'$ to $\infty$, thereby transforming $S$ to

$T = \{y_i\}$ and $S'$ to $T' = \{y_i'\}$. It follows that there exists a Möbius transformation satisfying $S^h = S'$ if and only if there exists a similarity satisfying $T^g = T'$ if and only if there exists a constant $k > 0$ such that $\|y_i' - y_j'\| = k\|y_i - y_j\|$ $(i, j \in I)$. Since a cross ratio in $T$ matches the corresponding one in $S$, hence in $S'$, and hence in $T'$, the argument of the theorem shows that the last condition holds.

To complete the proof of the corollary we note that the following statements are equivalent: $h$ is unique; $g$ is unique; $T$ contains an $n$-simplex; $S$ contains $n + 2$ points not all on the same $(n - 1)$-sphere or $(n - 1)$-flat. □

*Remark* 1. The conditions of the corollary are satisfied vacuously if $S$ and $S'$ are triples, and this implies that $\mathfrak{M}_n$ is at least 3-transitive in every dimension. In point of fact $\mathfrak{M}_1$ and the direct subgroup of index 2 in $\mathfrak{M}_2$ are each sharply 3-transitive.

*Remark* 2. Let $x_1, x_2, x_3, x_4$ be four distinct points in $\Pi = \mathbb{R}^n \cup \{\infty\}$. We think of them as the vertices of a 3-simplex even when $n = 1$. A double interchange such as $x_1 \leftrightarrow x_2$, $x_3 \leftrightarrow x_4$ fixes one pair of opposite edges and interchanges the members of the other two pairs. It follows that $S = (x_1, x_2, x_3, x_4)$ and $S' = (x_2, x_1, x_4, x_3)$ have the same cross ratios. The corollary therefore implies that the double interchange is induced by a Möbius transformation. Moreover, when $n = 1$ or when $n = 2$ and the points do not lie on a circle or a line, this Möbius transformation is unique and an involution.

*Remark* 3. Let $x_1, x_2, x_3, x_4$ be the vertices of a "3-simplex" in $\Pi$, and let $\alpha, \beta, \gamma$ with $\gamma = \max\{\alpha, \beta, \gamma\}$ be the products of the lengths of its opposite edges. The six cross ratios of the vertices are $\lambda = \alpha\gamma^{-1}$ and $\mu = \beta\gamma^{-1}$, each $\leqslant 1$; $\lambda^{-1}$ and $\mu^{-1}$, each $\geqslant 1$; and the reciprocal pair $\lambda\mu^{-1}$, $\mu\lambda^{-1}$. Ptolemy's theorem [usually stated for the vertices of a quadrangle] asserts that $\alpha, \beta, \gamma$ satisfy the triangle inequality $\alpha + \beta \geqslant \gamma$ with equality if and only if $x_1, x_2, x_3, x_4$ lie on a circle or a line. An equivalent statement is that the cross ratios $0 < \lambda \leqslant 1$ and $0 < \mu \leqslant 1$ satisfy the additional inequality $\lambda + \mu \geqslant 1$ with equality if and only if $x_1, x_2, x_3, x_4$ lie on a circle or a line.

A direct proof of the last assertion is to invert $x_4$ to $\infty$ and thereby transform $x_1, x_2, x_3$ to triangle $x_1'x_2'x_3'$ with sides $a, b, c$ satisfying $c = \max\{a, b, c\}$. Then $a + b \geqslant c$, with equality if and only if $x_1', x_2', x_3'$ lie on a line, and hence if and only if $x_1, x_2, x_3, x_4$ lie on a circle or a line. Moreover $\lambda = ac^{-1}$ and $\mu = bc^{-1}$, and therefore $\lambda + \mu \geqslant 1$ with the condition of equality as stated before.

*Remark* 4. If $x_1, x_2, x_3, x_4$ are four distinct points in $\Pi = \mathbb{R}^n \cup \{\infty\}$ and if $p$ and $q$ are transformations in $\mathfrak{M}_n$ such that $x_3^p = x_4^q = \infty$, then the triangles $x_1^p x_2^p x_4^p$ and $x_2^q x_1^q x_3^q$ must be similar because they summarize the same cross-ratio information. If $g$ is a similarity mapping $x_1^p x_2^p x_4^p$ to $x_2^q x_1^q x_3^q$, then the Möbius transformation $h = pgq^{-1}$ induces the double interchange $x_1 \leftrightarrow x_2$, $x_3 \leftrightarrow x_4$ described in Remark 2.

## 7. The Product of Two Inversions

Let $(\alpha, \beta)$ and $(\alpha', \beta')$ be two pairs of $(n-1)$-spheres lying on $\Sigma$. We say that $(\alpha, \beta)$ and $(\alpha', \beta')$ are equivalent under $\mathfrak{M}_n$ if there is an element $h \in \mathfrak{M}_n$ such that $\alpha^h = \alpha'$ and $\beta^h = \beta'$. We shall see that $(\alpha, \beta)$ and $(\alpha', \beta')$ are equivalent if and only if the Möbius transformations $\bar{\alpha}\bar{\beta}$ and $\bar{\alpha}'\bar{\beta}'$ are conjugate. In particular we shall see that $(\alpha, \beta)$ and $(\beta, \alpha)$ are always equivalent and therefore the inverse transformations $\bar{\alpha}\bar{\beta}$ and $\bar{\beta}\bar{\alpha}$ are always conjugate.

Two $(n-1)$-spheres $\alpha$ and $\beta$ are tangent, intersecting, or disjoint depending on whether they have a single point in common, an $(n-2)$-sphere in common, or no points in common. In the ambiguous case when $n = 1$ and $\alpha \cap \beta = \emptyset$, we define $\alpha$ and $\beta$ to be intersecting if their points are interlaced and disjoint if they are not. This agrees with the behavior of the circles [or lines] of $\mathbb{R}^2 \cup \{\infty\}$ whose inversions [or reflections] induce the one-dimensional transformations $\bar{\alpha}$ and $\bar{\beta}$.

If $\alpha$ and $\beta$ are small $(n-1)$-spheres, then $\bar{\alpha}$ and $\bar{\beta}$ are given by projection from points $x_\alpha$ and $x_\beta$ of $\mathbb{R}^{n+1}$. We observe (Figure 12) that $\alpha$ and $\beta$ are tangent, intersecting or disjoint depending on whether the line $x_\alpha x_\beta$ is a tangent, nonsecant, or secant of $\Sigma$. This observation is helpful because every equivalence class under $\mathfrak{M}_n$ can be represented by a pair of small $(n-1)$-spheres.

When $n \geqslant 2$ it is meaningful to speak of the circles perpendicular to $\alpha$ and $\beta$. We refer to this set of circles as the pencil perpendicular to $\alpha$ and $\beta$, and denote it by $[\alpha, \beta]^\perp$. Most points of $\Sigma$ lie on exactly one circle belonging to $[\alpha, \beta]^\perp$. It is also meaningful to speak of the $(n-1)$-spheres perpendicular to every circle in $[\alpha, \beta]^\perp$. We refer to this set of $(n-1)$-spheres as the pencil including $\alpha$ and $\beta$, and denote it by $[\alpha, \beta]$. Most points of $\Sigma$ lie on exactly one $(n-1)$-sphere belonging to $[\alpha, \beta]$. When $n = 2$, $\alpha$ and $\beta$ are themselves circles and there is a possibility of confusion between $[\alpha, \beta]$, $[\alpha, \beta]^\perp$, and the family of $(n-1)$-spheres perpendicular to $\alpha$ and $\beta$ [especially since the latter two are identical and dual to the first in the familiar theory of coaxial circles]. For this reason the 3-dimensional situation is a more helpful guide to the general situation.

We shall discuss the transformation $h = \bar{\alpha}\bar{\beta}$ in the three cases (i) $\alpha$ and $\beta$ tangent, (ii) $\alpha$ and $\beta$ intersecting, and (iii) $\alpha$ and $\beta$ disjoint. In the appropriate $\Pi$-model $h$ will appear as a translation, rotation, or dilatation. At the same time

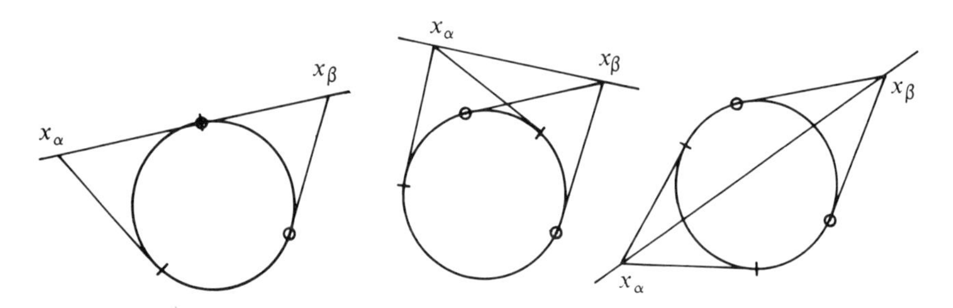


**Figure 12.** Two $(n-1)$-spheres $\alpha$ and $\beta$ are tangent, intersecting, or disjoint according to whether the line $x_\alpha x_\beta$ is a tangent, nonsecant, or secant of the $n$-sphere $\Sigma$ $[n = 1]$.

$[\alpha, \beta]$ and $[\alpha, \beta]^{\perp}$ will assume very simple forms, and the extent to which they afford foliations of $\Sigma$ will become apparent. We shall obtain an inversively significant description of $h$ as a flow along the circles of $[\alpha, \beta]^{\perp}$ with wave fronts belonging to $[\alpha, \beta]$. Our procedure is only appropriate to $n \geqslant 2$, but the missing case $n = 1$ is actually included on every circle of $[\alpha, \beta]^{\perp}$.

(i) *$\alpha$ and $\beta$ tangent.* If $\alpha$ and $\beta$ are tangent (Figure 13), we pass to a $\Pi$-model with $\alpha \cap \beta = \infty$. Then $\alpha$ and $\beta$ appear as parallel $(n-1)$-flats, $[\alpha, \beta]$ is the set of all $(n-1)$-flats parallel to $\alpha$, and $[\alpha, \beta]^{\perp}$ is the set of all lines perpendicular to $\alpha$. Each point of $\Pi$ except $\infty$ is on exactly one member of $[\alpha, \beta]$ and one member of $[\alpha, \beta]^{\perp}$, while $\infty$ is on every member of $[\alpha, \beta]$ and every member of $[\alpha, \beta]^{\perp}$.

The transformation $h = \bar{\alpha}\bar{\beta}$ acts on $\Pi$ by fixing $\infty$ and translating the points of $\mathbb{R}^n$ along the lines of $[\alpha, \beta]^{\perp}$. The length of the translation is twice the width of the gap between $\alpha$ and $\beta$, and its sense is from $\alpha$ towards $\beta$. This description shows that $h$ can be refactored as $h = \bar{\gamma}\bar{\delta}$ for any $\gamma, \delta \in [\alpha, \beta]$ which have the appropriate gap and sense.

The pair $(\alpha, \beta)$ is equivalent to the pair $(\beta, \alpha)$ by reflection in the $(n-1)$-flat midway between them. More generally any two tangent pairs are equivalent. To see this we recall that $\mathfrak{M}_n$ is 3-transitive and note that $(\alpha, \beta)$ is determined by the three distinct points $\alpha \cap \beta$, $a = \alpha \cap l$, and $b = \beta \cap l$, where $l$ is any member of $[\alpha, \beta]^{\perp}$.

(ii) *$\alpha$ and $\beta$ intersecting.* If $\alpha$ and $\beta$ are intersecting (Figure 14), $\sigma = \alpha \cap \beta$ is an $(n-2)$-sphere and has at least 2 points. We pass to a $\Pi$-model with one point of $\sigma$ at $\infty$. Then $\alpha$ and $\beta$ appear as $(n-1)$-flats intersecting in the $(n-2)$-flat $\sigma$,

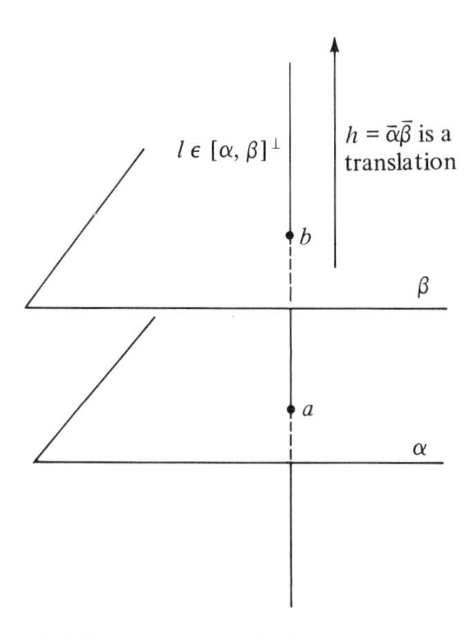


**Figure 13.** The canonical form when $\alpha$ and $\beta$ are tangent $(n-1)$-spheres $[n = 3]$.

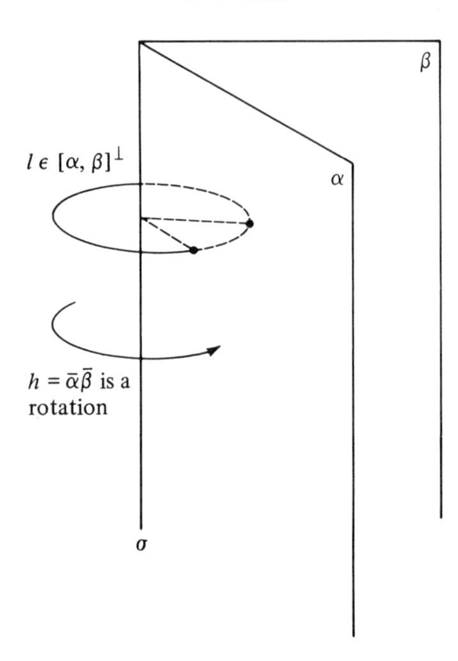


**Figure 14.** The canonical form when $\alpha$ and $\beta$ are intersecting $(n-1)$-spheres $[n = 3]$.

and $[\alpha, \beta]$ is the set of all $(n-1)$-flats through $\sigma$. Each point of $\Pi - \sigma$ lies in a unique 2-flat which is perpendicular to $\sigma$ and meets it in a single point. The circles of $[\alpha, \beta]^{\perp}$ lie in these 2-flats, and those in a given 2-flat are concentric about its single point of $\sigma$. Each point of $\Pi - \sigma$ is on exactly one member of $[\alpha, \beta]$ and one member of $[\alpha, \beta]^{\perp}$, while each point of $\sigma$ is on every member of $[\alpha, \beta]$ and on no member of $[\alpha, \beta]^{\perp}$.

The transformation $h = \bar{\alpha}\bar{\beta}$ acts on $\Pi$ by fixing the points of $\sigma$ including $\infty$ and rotating the other points of $\Pi$ about $\sigma$ along the circles of $[\alpha, \beta]^{\perp}$. The angle from $\alpha$ to $\beta$ is either $\theta$ or $\pi - \theta$, depending on the sense in which it is measured. If the rotation is applied in the same sense, its angle is either $2\theta$ or $2(\pi - \theta)$. These are equivalent descriptions, and they show that $h$ can be refactored as $h = \bar{\gamma}\bar{\delta}$ for any $\gamma, \delta \in [\alpha, \beta]$ which have the appropriate dihedral angle and sense.

The pair $(\alpha, \beta)$ is equivalent to the pair $(\beta, \alpha)$ by reflection in either of the two $(n-1)$-flats through $\sigma$ which bisect a dihedral angle between $\alpha$ and $\beta$. More generally, any two intersecting pairs which meet at the same dihedral angles are equivalent. This is obvious, since they can differ by at most a rotation or reflection once their common $(n-2)$-spheres are brought into coincidence.

(iii) *$\alpha$ and $\beta$ disjoint.* If $\alpha$ and $\beta$ are disjoint $(n-1)$-spheres (Figure 15), the secant $x_{\alpha}x_{\beta}$ cuts $\Sigma$ at two points which are exchanged by each of $\bar{\alpha}$ and $\bar{\beta}$ and therefore fixed by $h = \bar{\alpha}\bar{\beta}$. If we pass to a $\Pi$-model in which these fixed points are 0 and $\infty$, then $\alpha$ and $\beta$ are $(n-1)$-spheres with center 0 and radii $k_1$ and $k_2$, $[\alpha, \beta]$ is the set of all $(n-1)$-spheres with center 0, and $[\alpha, \beta]^{\perp}$ is the set of all lines through 0. Each point of $\Pi$ except 0 and $\infty$ is on exactly one member of

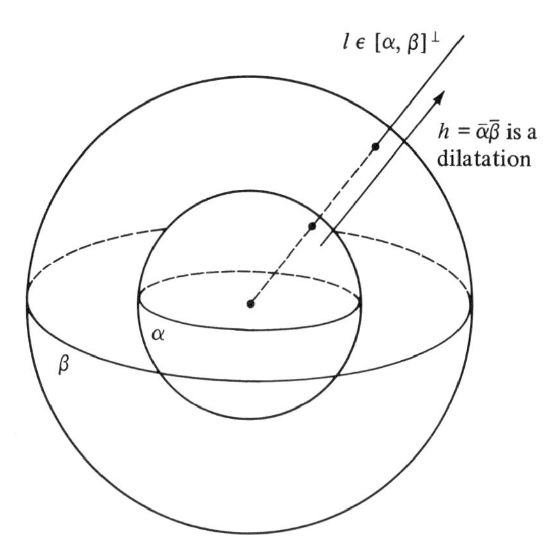


**Figure 15.** The canonical form when $\alpha$ and $\beta$ are disjoint $(n-1)$-spheres $[n = 3]$.

$[\alpha, \beta]$ and one member of $[\alpha, \beta]^\perp$, while 0 and $\infty$ are on no members of $[\alpha, \beta]$ and every member of $[\alpha, \beta]^\perp$. The points 0 and $\infty$ are called the limiting points of $[\alpha, \beta]$, and $[\alpha, \beta]^\perp$ can be described as the set of all circles and lines through these limiting points.

The transformation $h = \bar{\alpha}\bar{\beta}$ acts on $\Pi$ by fixing 0 and $\infty$ and pushing the other points of $\mathbb{R}^n$ along the lines of $[\alpha, \beta]^\perp$ by the dilatation $x \to (k_2/k_1)^2 x$. This description shows that $h$ can be factored as $h = \bar{\gamma}\bar{\delta}$ for any $\gamma, \delta \in [\alpha, \beta]$ which have the same ratio of radii, the ratio being formed with attention to the order of the $(n-1)$-spheres.

The pair $(\alpha, \beta)$ is equivalent to the pair $(\beta, \alpha)$ by inversion in the $(n-1)$-sphere with center 0 and radius $(k_1 k_2)^{1/2}$. More generally, two nonintersecting pairs $(\alpha, \beta)$ and $(\alpha', \beta')$ can both be transformed by $\mathfrak{M}_n$ so that their limiting points are 0 and $\infty$, $\alpha = \alpha'$, and $\beta$ and $\beta'$ are each larger. The inversive distance between $\alpha$ and $\beta$ is defined to be $\delta = \log(k_2/k_1)$ $(k_2 > k_1)$. We shall soon see that the inversive distance between disjoint $(n-1)$-spheres is an inversive invariant like the angle between intersecting $(n-1)$-spheres. Granting this, two disjoint pairs are equivalent precisely when they have the same inversive distance.

**Lemma 7.** *Let $\alpha$ and $\beta$ be two $(n-1)$-spheres on $\Sigma$ which intersect at an angle $\theta$ or are disjoint and separated by an inversive distance $\delta$. Suppose a circle of $[\alpha, \beta]^\perp$ meets $\alpha$ at $a$ and $a'$ and $\beta$ at $b$ and $b'$ where $a$ and $b$ are adjacent. Then the cross ratio*

$$\frac{\|a - b'\| \, \|b - a'\|}{\|a - a'\| \, \|b - b'\|} = \begin{cases} \cos^2 \dfrac{\theta}{2} & \text{if } \alpha \text{ and } \beta \text{ are intersecting,} \\[2ex] \cosh^2 \dfrac{\delta}{2} & \text{if } \alpha \text{ and } \beta \text{ are disjoint.} \end{cases}$$

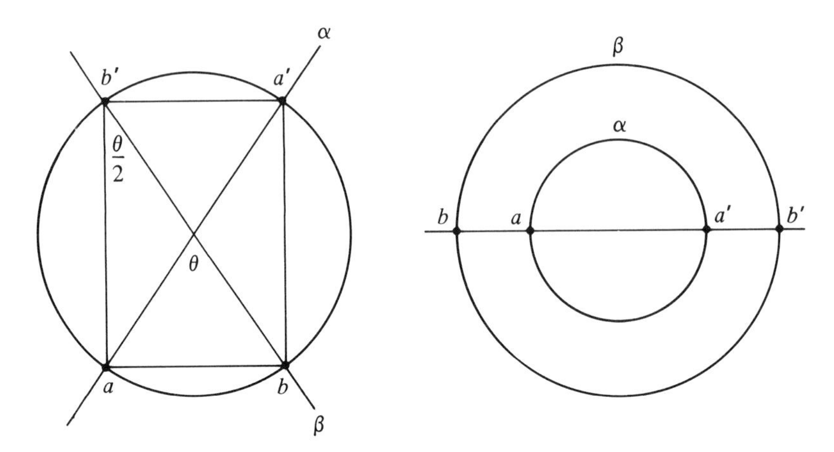


**Figure 16.** Cross ratio measures angle and inversive distance.

*Proof.* We work in the 2-space through the circle $abb'$ and perpendicular to $\alpha$ and $\beta$ (Figure 16). We pass to the $\Pi$-model, where $\alpha$ and $\beta$ can be regarded as intersecting lines or concentric circles.

In the intersecting case $\alpha$ and $\beta$ are diameters of the circle $aba'b'$, and $ab$ subtends an angle of $\theta$ at the center of this circle and therefore an angle of $\theta/2$ at $b'$ on its circumference. It follows that

$$\frac{\|a-b'\|\,\|b-a'\|}{\|a-a'\|\,\|b-b'\|}=\left(\frac{\|a-b'\|}{\|b-b'\|}\right)^2=\cos^2\frac{\theta}{2}\,.$$

In the disjoint case $\alpha$ and $\beta$ are concentric circles with radii $k_1$ and $k_2$ such that $\delta=\log(k_2/k_1)$. The orthogonal circle $baa'b'$ appears as a common diameter, and

$$\frac{\|a-b'\|\,\|b-a'\|}{\|a-a'\|\,\|b-b'\|}=\frac{(k_1+k_2)^2}{4k_1k_2}=\left|\frac{(k_1/k_2)^{1/2}+(k_2/k_1)^{1/2}}{2}\right|^2=\cosh^2\frac{\delta}{2}\,.$$

□

*Remark* 1. In the intersecting case the labeling of $b$ given $a$ is ambiguous. This corresponds to the fact that $\alpha$ and $\beta$ determine two dihedral angles $\theta$ and $\pi-\theta$.

*Remark* 2. In Section 6 we saw that four points lie on a circle if and only if their independent cross ratios satisfy $\lambda+\mu=1$. Lemma 7 shows us that when four points lie on a circle, their six cross ratios are the squares of the trigonometric functions of a certain angle and we can write $\lambda=\sin^2(\theta/2)$, $\mu=\cos^2(\theta/2)$.

*Remark* 3. Lemma 7 expresses angle and inversive distance in terms of the fundamental inversive invariant, cross ratio. This proves that inversive distance is an inversive invariant. It also provides an intrinsic definition of the angle between intersecting 0-spheres in dimension $n=1$.

We summarize the results of Section 7 in

**Theorem 5.** *Let $\alpha$ and $\beta$ be $(n-1)$-spheres on $\Sigma$. Then the transformation $h = \bar{\alpha}\bar{\beta}$ lies in one of the following conjugacy classes: translation [$\alpha$ and $\beta$ tangent], rotation $\theta$ [$\alpha$ and $\beta$ intersect at an angle of $\theta$], or dilatation $\delta$ [$\alpha$ and $\beta$ are disjoint and separated by an inversive distance $\delta$]. In all cases the orbits of $h$ lie in $[\alpha, \beta]^{\perp}$ and the $(n-1)$-spheres of $[\alpha, \beta]$ are advanced coherently. The transformation $h = \bar{\alpha}\bar{\beta}$ can be refactored with either the first or second mirror arbitrary in $[\alpha, \beta]$.*

**Corollary** (Three-inversion theorem). *Let $\alpha$, $\beta$, and $\gamma$ be three $(n-1)$-spheres in a pencil on $\Sigma$. Then there is a fourth $(n-1)$-sphere $\delta$ in the pencil such that $\bar{\alpha}\bar{\beta}\bar{\gamma} = \bar{\delta}$.*

*Proof.* The transformation $\bar{\beta}\bar{\gamma}$ can be refactored as $\bar{\alpha}\bar{\delta}$ for a suitable $\delta$ in the pencil. Then

$$\bar{\alpha}\bar{\beta}\bar{\gamma} = \bar{\alpha}(\bar{\beta}\bar{\gamma}) = \bar{\alpha}(\bar{\alpha}\bar{\delta}) = \bar{\alpha}^2\bar{\delta} = \bar{\delta}. \qquad \square$$

## 8. The Conjugacy Classes of $\mathcal{M}_n$ $(n = 1, 2, 3)$

An arbitrary element of $\mathcal{M}_n$ can be written as the product of at most $n+2$ inversions. If it has a fixed point, it is conjugate to a Euclidean similarity, and if it does not, it is conjugate to a spherical isometry. We use this information to build our understanding of the conjugacy classes of $\mathcal{M}_n$ for $n = 1, 2, 3$. We refer to the conjugacy classes by the name of their simplest representative, and we designate them by symbols which remind us of their canonical arrangement of mirrors. For example, the classes encountered in Section 7 are reflection [—], translation [||], rotation $\theta$ $[\times, \theta]$, and dilatation $\delta$ $[\circledcirc, \delta]$. As in Section 7, it is best to regard the case $n = 1$ as imbedded in the case $n = 2$.

(i) *Euclidean isometries*, $n = 2$. Here we have a product of at most three reflections in lines. A single reflection is [—], and a product of two must be [||] or $[\times, \theta]$ because two lines are either parallel or intersecting.

The Corollary to Theorem 5 tells us that if a transformation $\bar{\gamma}_1\bar{\gamma}_2\bar{\gamma}_3$ requires three reflections, its mirrors cannot be parallel or concurrent. It follows that $\gamma_2$ meets $\gamma_1$ or $\gamma_3$ at a point not on the other mirror (Figure 17). If $\gamma_1$ meets $\gamma_2$, we rewrite the rotation $\bar{\gamma}_1\bar{\gamma}_2$ as $\bar{\gamma}'_1\bar{\gamma}'_2$ with $\gamma'_2$ perpendicular to $\gamma_3$. Then we rewrite the half turn $\bar{\gamma}'_2\bar{\gamma}_3$ as $\bar{\gamma}''_2\bar{\gamma}'_3$ with $\gamma''_2$ parallel to $\gamma'_1$ and therefore $\gamma'_3$ perpendicular to both $\gamma'_1$ and $\gamma''_2$. A similar consideration when $\gamma_2$ meets $\gamma_3$ shows that the product of three reflections is always a glide reflection [⫲], the commuting product of a translation and a reflection.

(ii) *Similarities*, $n = 2$. A proper similarity is the commuting product of a dilatation and an isometry with the same fixed point. Since the isometries with a fixed point are reflection and the rotations, we obtain the following classes: dilatation $\delta$ $[\circledcirc, \delta]$, dilative reflection $\delta$ $[\ominus, \delta]$, and dilative rotation $\theta, \delta$ $[\otimes, \theta, \delta]$.

(iii) *Spherical isometries*, $n = 2$. Here we have the product of at most three inversions in great circles. We have already listed [—] and $[\times, \theta]$, and since two great circles always intersect, the list is complete up to two inversions. We have

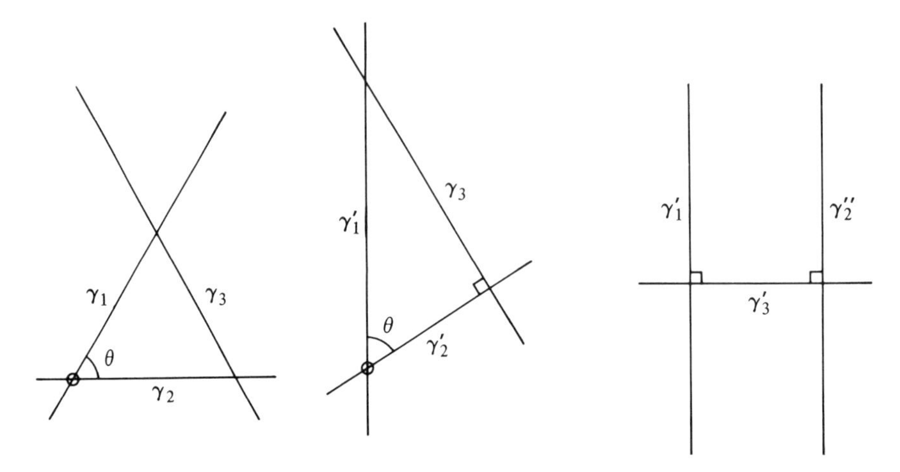


**Figure 17.** The product of reflections in three lines which do not belong to a pencil is a glide reflection.

therefore shown that every direct isometry of the sphere is a rotation, and this gives Euler's theorem: the product of two rotations [of the sphere] is a rotation.

A product of three inversions can be simplified in exactly the same way as the product of three reflections treated in (i). It turns out to be a rotatory reflection $\theta$ $[\boxtimes\ ,\theta]$, the commuting product of a rotation and a reflection. These transformations have no fixed points. The special case $[\boxtimes\ ,\pi/2]$ is the involution which exchanges every point with its antipode.

(iv) *Euclidean isometries*, $n = 3$. These isometries are the product of at most four reflections in planes. A single reflection is [—], and the product of two must be [||] or $[\times,\theta]$ because two planes are either parallel or meet in a line.

If a transformation $\bar{\gamma}_1\bar{\gamma}_2\bar{\gamma}_3$ really requires three reflections, its mirrors cannot be parallel or coaxial. It follows that $\gamma_2$ meets $\gamma_1$ or $\gamma_3$ in a line, and this line either is parallel to the remaining plane or intersects it in a single point. In the first case, the mirrors form a "vertical" prism and the transformation acts on every "horizontal" plane as [$\#$]. In the second case, the three mirrors have a single point in common and the transformation acts on every sphere about this point as $[\boxtimes\ ,\theta]$.

If $h$ requires four reflections, it cannot have a fixed point. Let $t$ be the translation which restores some point 0. Then $ht = r$ is direct and therefore a rotation about 0. Write $h = rt^{-1} = rt_\perp t_\parallel$, where $t_\perp$ is a translation perpendicular to the axis of $r$ and $t_\parallel$ is a translation parallel to this axis. An easy manipulation of mirrors shows that $r_1 = rt_\perp$ is a rotation through the same angle as $r$ about a parallel axis. It follows that $h = r_1 t_\parallel$ is a twist $\theta$ $[\sqsubset,\theta]$, the commuting product of a rotation $\theta$ and a translation along its axis.

(v) *Similarities*, $n = 3$. The isometries with a fixed point are reflection, rotation $\theta$, and rotatory reflection $\theta$. These lead to the classes dilatation $\delta$ $[\odot,\delta]$, dilative reflection $\delta$ $[\ominus,\delta]$, dilative rotation $\theta,\delta$ $[\otimes,\theta,\delta]$, and dilative rotatory reflection $\theta,\delta$ $[—,\theta,\delta]$. The abbreviated symbol for this last class must serve to

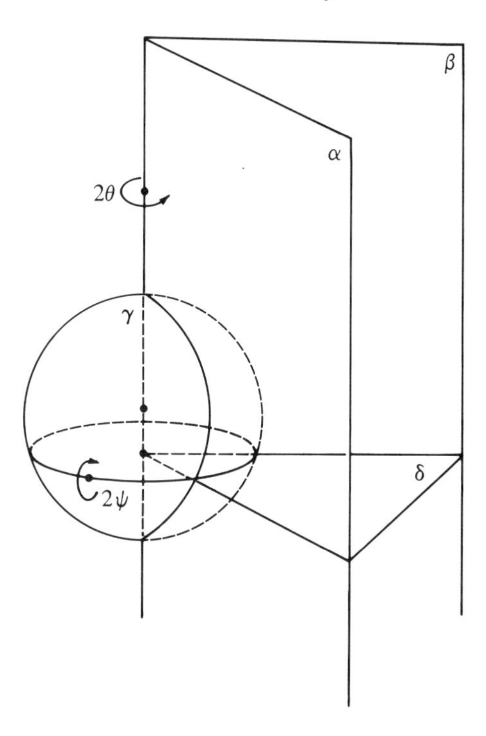


**Figure 18.** The double rotation $[\theta, \psi]$ as the product of a [—], $\bar{\gamma}$, and a [⊠, $\theta$], $\bar{\alpha}\bar{\beta}\bar{\delta}$.

remind us of five mirrors: two concentric spheres for [◎, $\delta$], two meridian planes for $[\times, \theta]$ and an equatorial plane for [—].

(vi) *Fixed point free in* $\mathfrak{M}_3$. Such a transformation is conjugate to an isometry of the 3-sphere and can therefore be written as the product of at most four inversions. The possibilities $m = 1$ and $m = 2$ have obvious fixed points. When $m = 3$ we consider the $\Pi$-representation as an inversion followed by an isometry which must be [|||] or $[\times, \theta]$. In either case there is a plane perpendicular to all three mirrors. Since it and the half spaces into which it divides $\Pi$ are mapped onto themselves, we can find a fixed point for the transformation by conjugating either half space to a ball and applying Brouwer's theorem. When $m = 4$ we again consider the $\Pi$-representation as an inversion followed by an isometry, which must now be [╫] or [⊠ , $\theta$]. The Brouwer argument eliminates the first case. An instance of the second case (Figure 18) is the double rotation $[\theta, \psi]$ which arises if the sphere of inversion is centered on the axis of rotation and meets the plane of reflection at an angle $\psi$. This transformation is fixed-point-free and can be described as the commuting product of a rotation through $2\theta$ about a line and a rotation through $2\psi$ about a circle centered on the line and lying in a plane perpendicular to the line.

If we grant for the moment that the double infinity of conjugacy classes $[\theta, \psi]$ are all that can arise as the fixed-point-free product of a [—] and a [⊠ , $\theta$], then we can summarize our results.

**Theorem 6.** *The conjugacy classes of the groups* $\mathfrak{M}_n$ $(n = 1,2,3)$ *are as shown in the following table. The rows of the table are labeled to indicate the number of mirrors required. The conjugacy symbols are suffixed to indicate the first dimension in which they appear.*

| | | | |
|---|---|---|---|
| $m = 1$ | $[-]_1$ | | |
| $m = 2$ | $[\circledcirc, \delta]_1$ | $[\times, \theta]_1$ | $[\parallel]_1$ |
| $m = 3$ | $[\ominus, \delta]_1$ | $[\boxtimes, \theta]_2$ | $[\nparallel]_2$ |
| $m = 4$ | $[\otimes, \theta, \delta]_2$ | $[\theta, \psi]_3$ | $[\sqsubset, \theta]_3$ |
| $m = 5$ | $[-, \theta, \delta]_3$ | | |

*Remark.* The square of an opposite transformation is direct. Since the component transformations in our canonical factorizations commute, we can identify the squares of the opposite transformations as follows: $[-]^2 = e$, $[\ominus, \delta]^2 = [\circledcirc, 2\delta]$, $[\otimes, \theta]^2 = [\times, 2\theta]$, $[\nparallel]^2 = [\parallel]$ and $[-, \theta, \delta]^2 = [\otimes, 2\theta, 2\delta]$. Two special cases should be noted: $[\boxtimes, \pi/2]^2 = e$ and $[-, \pi/2, \delta]^2 = [\circledcirc, 2\delta]$.

Now we return to the product of an inversion $\bar{\gamma}_1$ and a $[\boxtimes, \theta]$ given by the rotation $\bar{\gamma}_2\bar{\gamma}_3$ and the reflection $\bar{\gamma}_4$. We can assume that $\gamma_2$ passes through the center of $\gamma_1$, and this allows us to regroup our transformation as the product of a half turn about a circle, $\bar{\gamma}_1\bar{\gamma}_2$, and a half turn about a line, $\bar{\gamma}_3\bar{\gamma}_4$. We obtain our desired result through a theorem of independent interest.

**Theorem 7.** *Every direct transformation in* $\mathfrak{M}_3$ *can be written as the product of half turns about two circles* [*two lines or a circle and a line*]. *The incidence properties of the circles determine the conjugacy class of the transformation in accordance with the following table.*

| Pair of circles | Transformation |
|---|---|
| *Touch once* | $[\sqsubset, \theta]$ |
| *Touch once and lie on a sphere* [*plane*] | $[\parallel]$ |
| *Touch twice* [*and lie on a sphere* [*plane*]] | $[\times, \theta]$ |
| *Disjoint* | $[\otimes, \theta, \delta]$ |
| *Disjoint and lie on a sphere* [*plane*] | $[\circledcirc, \delta]$ |
| *Interlocked* | $[\theta, \psi]$ |

*Proof.* The paragraph preceding this theorem shows that the "unknown" direct transformations can be written as a product of half turns. Our case-by-case considerations will reveal the nature of these unknown transformations. It will also turn up all the "known" direct transformations and thereby prove that every direct transformation is the product of two half turns.

If the circles touch once, we invert their common point to $\infty$ and obtain two skew lines $l_1$ and $l_2$ with a unique perpendicular transversal $m$. For $i = 1,2$, let $\gamma_i$ be the plane through $l_i$ with normal direction $m$, and let $\gamma_{i+2}$ be the plane spanned by $l_i$ and $m$. Then the product of two half turns is $(\bar{\gamma}_1\bar{\gamma}_3)(\bar{\gamma}_2\bar{\gamma}_4) = \bar{\gamma}_1\bar{\gamma}_2\bar{\gamma}_3\bar{\gamma}_4$, and this is a $[\sqsubseteq, \theta]$.

If the circles touch once and lie on a sphere, then $\gamma_3 = \gamma_4$ and the transformation reduces to a $[||]$. If the circles touch twice, they must lie on a sphere, $\gamma_1 = \gamma_2$, and the transformation reduces to a $[\times, \theta]$.

If the circles are disjoint, we consider them as a line $l$ and a circle $c$. Swell the line to a cylinder which touches $c$; let $\gamma$ be the plane tangent to the cylinder at its point of contact with $c$; let $\gamma_1$ be the plane through $l$ parallel to $\gamma$, and let $\gamma_2$ be the sphere through $c$ tangent to $\gamma$. The half turn about $l$ interchanges the sides of $\gamma_1$, and the half turn about $c$ interchanges the "sides" of $\gamma_2$. In all, the transformation maps the ball bounded by $\gamma_2$ into itself, and the Brouwer theorem gives a fixed point. The image of this fixed point under either half turn must be a second fixed point.

We consider the $\Pi$-representation with these fixed points at 0 and $\infty$. Since a half turn about a circle interchanges its center and $\infty$, the axes of our half turns are now circles $c_1$ and $c_2$ with common center 0. For $i = 1, 2$, let $\gamma_i$ be the sphere with center 0 containing $c_i$, and let $\gamma_{i+2}$ be the plane of $c_i$. The product of half turns is $(\bar{\gamma}_1\bar{\gamma}_3)(\bar{\gamma}_2\bar{\gamma}_4) = \bar{\gamma}_1\bar{\gamma}_2\bar{\gamma}_3\bar{\gamma}_4$, and this is a $[\otimes, \theta, \delta]$.

If the circles are disjoint and lie on a sphere, $\gamma_3 = \gamma_4$ and the transformation reduces to a $[\circledcirc, \delta]$.

If the circles are interlocked, we consider them first as a line $l$ and a circle $c$. If $l$ cuts the plane of $c$ at right angles to the radius vector from the center of $c$ (Figure 19), we factor the half turn about $l$ as $\bar{\gamma}_1\bar{\gamma}_3$, where $\gamma_1$ passes through the center of $c$, and we factor the half turn about $c$ as $\bar{\gamma}_2\bar{\gamma}_4$, where $\gamma_4$ is the sphere whose equator is $c$. It follows that $\gamma_2$ and $\gamma_3$ are perpendicular planes, and the transformation can be written as $(\bar{\gamma}_1\bar{\gamma}_3)(\bar{\gamma}_2\bar{\gamma}_4) = (\bar{\gamma}_1\bar{\gamma}_2)(\bar{\gamma}_3\bar{\gamma}_4)$. This is the commuting product of a rotation about the line $\gamma_1 \cap \gamma_2$ and a rotation about the circle $\gamma_3 \cap \gamma_4$.

If $l$ is not perpendicular to the radius vector from the center of $c$, we seek a circle $d$ which is centered on $l$ and meets $c$ and $l$ twice at right angles. For then we obtain the desired configuration by inverting either of the points $l \cap d$ to $\infty$.

The existence of $d$ follows from a continuity argument. Any point $x$ on $l$ is the center of a unique circle $d(x)$ which meets $c$ and $l$ twice. Let $l(x)$ be the line of intersection of the plane of $d(x)$ and the plane of $c$. The angle between $d(x)$ and $c$ varies continuously, and is $< \pi/2$ when $l(x)$ is a diameter of $c$ and $> \pi/2$ when $l(x)$ is the projection of $l$ on the plane of $c$ (Figure 20). $\square$

*Remark* 1. Points of $\mathbb{R}^4$ can be represented by pairs of complex numbers, $(z_1, z_2)$. Those points on the unit sphere satisfy $|z_1|^2 + |z_2|^2 = 1$. The conjugacy class $[\theta, \psi]$ can be represented on $\Sigma$ by the double rotation $(z_1, z_2) \to (e^{2i\theta}z_1, e^{2i\psi}z_2)$. The special case $[\pi/2, \pi/2]$ can be represented by the antipodal map $(z_1, z_2) \to (-z_1, -z_2)$.

*Remark* 2. Our double rotations are closely related to the Hopf fibration of the 3-sphere into great circles continuously indexed by the points of the 2-sphere.

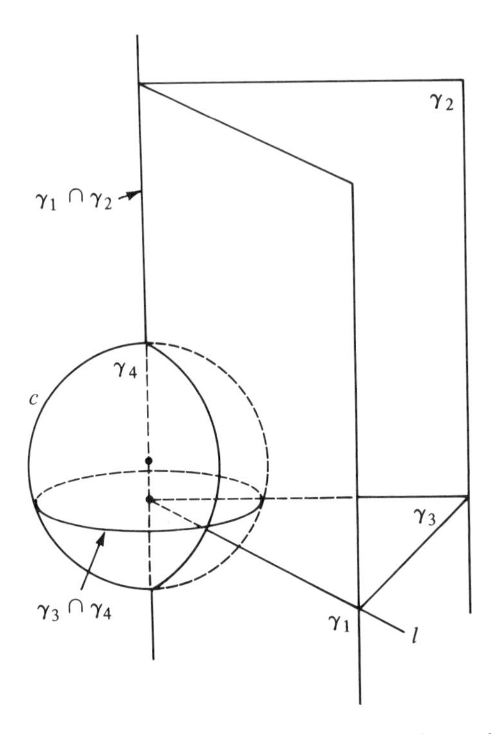


**Figure 19.** The line $l$ cuts the plane of the circle $c$ at right angles to the radius vector from the center of $c$.

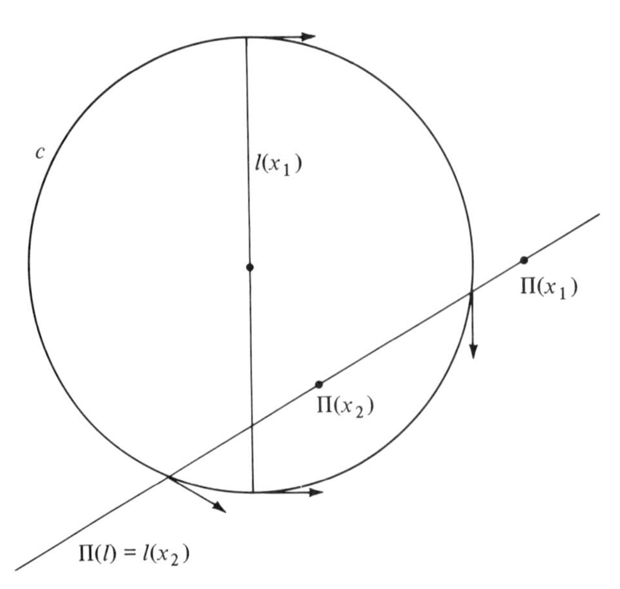


**Figure 20.** If $\Pi$ is projection to the plane of $c$, $d = d(x)$ for $x$ between $x_1$ and $x_2$.

The great circles are of the form $(e^{it}z_1, e^{it}z_2)$, $0 \leqslant t \leqslant 2\pi$, and they are indexed by $z_1 z_2^{-1}$.

## 9. Coordinates and Invariants

We begin by extending our use of $(n+2)$-tuple coordinates from $\Sigma$ to $\Pi$. The points of $\Pi$ are named by $n$-tuple coordinates $x$ and the single symbol $\infty$. The points of $\Pi - \{\infty\}$ can be imbedded as $(x, 0)$ in the $\mathbb{R}^{n+1}$ containing $\Sigma$. Stereographic projection (Figure 21) from $(0, 1)$ carries $\infty$ to $(0, 1)$ and $(x, 0)$ to the point where the line $(0,1) + \lambda(x, -1)$ cuts $\Sigma$. By imposing the condition $\|(\lambda x, 1 - \lambda)\| = 1$ we find that this point is $(2x/(\|x\|^2 + 1), (\|x\|^2 - 1)/(\|x\|^2 + 1))$. Passing to $(n+2)$-tuple coordinates and making use of the fact that they are positive homogeneous, we find that we can name

$\infty$ by $(0, 1, 1)$ and
$x$ by $(2x, \|x\|^2 - 1, \|x\|^2 + 1)$.

The important subsets of $\Sigma$ are $n$-caps bounded by $(n-1)$-spheres. The corresponding subsets of $\Pi$ are

(i) half spaces, $(x - a) \cdot n \geqslant 0$, bounded by $(n-1)$-flats,
(ii) $n$-balls, $\|x - a\| \leqslant r$, bounded by $(n-1)$-spheres, and
(iii) improper $n$-balls, $\|x - a\| \geqslant r$, bounded by $(n-1)$-spheres.

In order to determine coordinates for these sets, we take a cap

$$C = (c, C^{(n+1)}, C^{(n+2)}) \quad \text{satisfying } C * C = 1$$

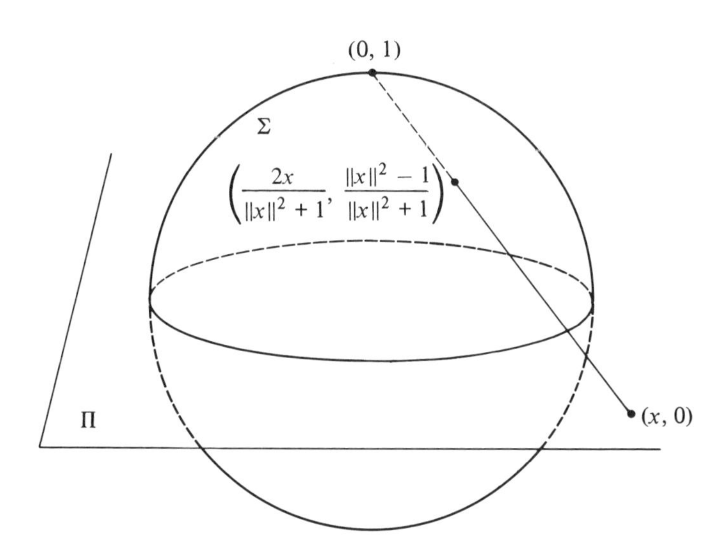


**Figure 21.** Stereographic projection from the $n$-sphere $\Sigma$ to the $n$-flat $\Pi$ $[n = 2]$.

and match

$$(2x, \|x\|^2 - 1, \|x\|^2 + 1) * (c, C^{(n+1)}, C^{(n+2)}) \geqslant 0,$$

or

$$2x \cdot c - (C^{(n+2)} - C^{(n+1)})\|x\|^2 - (C^{(n+2)} + C^{(n+1)}) \geqslant 0,$$

to the above equations.

In case (i) we divide by 2 and read off $c = n$ and $C^{(n+1)} = C^{(n+2)} = a \cdot n$. This is already properly normalized if $n$ is the unit normal pointing into the half space.

For case (ii) it is helpful to rewrite $\|x - a\| \leqslant r$ as

$$2x \cdot a - \|x\|^2 - (\|a\|^2 - r^2) \geqslant 0.$$

Then $c = a$, $C^{(n+1)} = \frac{1}{2}(\|a\|^2 - r^2 - 1)$, and $C^{(n+2)} = \frac{1}{2}(\|a\|^2 - r^2 + 1)$. Since the resulting $(n+2)$-vector satisfies $C * C = r^2$, the properly normalized coordinate of the $n$-ball can be written

$$\frac{1}{2r}(2a, \|a\|^2 - 1 - r^2, \|a\|^2 + 1 - r^2).$$

An effective memory aid is that this is the standard coordinate of the center modified by subtracting $r^2$ from the last two components and dividing the whole vector by $2r$ with $r > 0$. Case (iii) is the complement of case (ii), and its coordinate differs by a minus sign. We incorporate this into $r$ by adopting the convention that improper $n$-balls have negative radius.

The Π-model affords a convenient setting for interpreting the algebraic invariant $U * V$ of Section 3 in terms of the geometric invariants: cross ratio, angle, and inversive distance.

Suppose that $X_1 = (2x_1, \|x_1\|^2 - 1, \|x_1\|^2 + 1)$ and $X_2 = (2x_2, \|x_2\|^2 - 1, \|x_2\|^2 + 1)$ are standard coordinates for two points. Then a simple calculation shows that $X_1 * X_2 = -2\|x_1 - x_2\|^2$. The fact that distance is not an invariant corresponds to the fact that point coordinates are positive homogeneous. Thus, with arbitrary coordinates $X_1$ and $X_2$, the best we can say is that $X_1 * X_2 = -2\lambda_1\lambda_2\|x_1 - x_2.\|^2$ for $\lambda_1, \lambda_2 > 0$. Nevertheless this shows that $U * V$ can be used to express the cross ratio of four distinct points:

$$\frac{\|x_1 - x_2\|\,\|x_3 - x_4\|}{\|x_1 - x_3\|\,\|x_2 - x_4\|} = \left(\frac{(X_1 * X_2)(X_3 * X_4)}{(X_1 * X_3)(X_2 * X_4)}\right)^{1/2}.$$

If $X$ and $C$ are coordinates for a point and a set [half space, $n$-ball, or improper $n$-ball], then $X * C$ is positive, negative, or zero as $X \in \text{interior}\, C$, $X \in \text{complement}\, C$, or $X \in \text{boundary}\, C$. The positive homogeneity of $X$ implies that only the sign of $X * C$ is significant.

Finally we turn to $A * B$ where $A$ and $B$ name sets. There are three cases, depending on whether the boundaries of these sets are tangent, intersecting, or disjoint. We examine the sets with their boundaries $\alpha$ and $\beta$ in the standard positions considered in Section 7.

Tangent boundaries correspond to parallel $(n-1)$-flats, and we have

$$A = (n_1, d_1, d_1),$$
$$B = (n_2, d_2, d_2).$$

In this case $A * B = n_1 \cdot n_2 = \pm 1$, where the sign is "+" if the half spaces are nested and "−" if they are not.

Intersecting boundaries correspond to intersecting $(n-1)$-flats, and we have

$$A = (n_1, 0, 0),$$
$$B = (n_2, 0, 0).$$

In this case $A * B = n_1 \cdot n_2 = \cos\theta$ where $\theta$ is the dihedral angle between $\alpha$ and $\beta$ measured in $A \backslash B$ or $B \backslash A$.

Disjoint boundaries correspond to concentric $(n-1)$-spheres, and we have

$$A = \left(0, \frac{-1 - r_1^2}{2r_1}, \frac{1 - r_1^2}{2r_1}\right),$$

$$B = \left(0, \frac{-1 - r_2^2}{2r_2}, \frac{1 - r_2^2}{2r_2}\right).$$

In this case $A * B = (r_1^2 + r_2^2)/2r_1r_2 = \pm\cosh\delta$, where the sign is "+" if the $n$-balls are nested [both proper or both improper, so that $r_1r_2 > 0$] and "−" if they are not nested [one proper and the other improper, so that $r_1r_2 < 0$].

An $(n-1)$-sphere on $\Sigma$ bounds the complementary $n$-caps $C$ and $-C$. The relative position of two $(n-1)$-spheres $\alpha$ and $\beta$ with coordinates $\pm A$ and $\pm B$ is determined by $|A * B|$. We have seen that such $(n-1)$-spheres are tangent, intersecting, or disjoint as $|A * B|$ is equal to 1, less than 1, or greater than 1. On the other hand, the relative position of two $n$-caps $A$ and $B$ is not fully determined even by the signed invariant $A * B$. For example, if $A * B = -1$ it is not clear whether $A \cup B$ is all of $\Sigma$ or a proper subset of $\Sigma$.

In order to complete our classification of pairs of $n$-caps we introduce the size functional $s$. If a cap $C$ has $C = (\csc\theta\, c,\ \cot\theta)$ as its $(n+2)$-tuple coordinate, then $s(C) = \cot\theta$. Since $\cot\theta$ decreases from $+\infty$ to $-\infty$ as $\theta$ increases from 0 to $\Pi$, the actual size of a cap bears an inverse relation to $s(C)$. It is appropriate to think of $s(C)$ as a curvature. While $s(C)$ is certainly not an invariant, the following theorem shows that for certain pairs of caps the sign of $s(A) \pm s(B) = s(A \pm B)$ is an invariant.

**Theorem 8.** *Table 1 gives a complete classification of ordered pairs of distinct caps on $\Sigma$. The group $\mathfrak{M}_n$ is transitive on pairs in the same class.*

*Proof.* The group $\mathfrak{M}_n$ is transitive on pairs of $(n-1)$-spheres which are tangent, intersect at a given angle $\theta$, or are disjoint and separated by a given inversive distance $\delta$.

The verbal descriptions (see also Figure 22) show that the tangent and disjoint cases each split into four subcases which cannot be related to one another by

**Table 1.** Classification in Theorem 8

| Verbal description | | | | $A * B$ | $s$ |
|---|---|---|---|---|---|
| | Boundaries | Interiors | $A, B$ | | |
| (i) | tangent | nested | $A \supset B$ | $1$ | $s(A-B)<0$ |
| (ii) | tangent | nested | $B \supset A$ | $1$ | $s(A-B)>0$ |
| (iii) | tangent | disjoint | | $-1$ | $s(A+B)>0$ |
| (iv) | tangent | | $A \cup B = \Sigma$ | $-1$ | $s(A+B)<0$ |
| (v) $\theta$ | intersecting | | | $\cos\theta$ | |
| (vi) $\delta$ | disjoint | nested | $A \supset B$ | $\cosh\delta$ | $s(A-B)<0$ |
| (vii) $\delta$ | disjoint | nested | $B \supset A$ | $\cosh\delta$ | $s(A-B)>0$ |
| (viii) $\delta$ | disjoint | disjoint | | $-\cosh\delta$ | $s(A+B)>0$ |
| (ix) $\delta$ | disjoint | | $A \cup B = \Sigma$ | $-\cosh\delta$ | $s(A+B)<0$ |

elements of $\mathfrak{M}_n$. We remark that the pairs of cases (i) and (ii), (iii) and (iv), (vi) and (vii), and also (viii) and (ix) can be interchanged by the set maps $(A,B) \to (-A,-B)$. However, this kind of interchange can be induced by an element of $\mathfrak{M}_n$ only in case (v).

It remains to show that the verbal descriptions are equivalent to the corresponding conditions on $s$. Because of the last paragraph it is enough to consider cases (i), (iii), (vi), and (viii). In cases (i) and (vi) $A \supset B$ if and only if $\theta_A > \theta_B$ if and only if $s(A) < s(B)$ or $s(A-B) < 0$. In cases (iii) and (viii) the fact that $A \cup B$ does not cover $\Sigma$ is equivalent to $\theta_A + \theta_B < \pi$, $\theta_A < \pi - \theta_B$, $s(A) > s(-B)$, and finally to $s(A+B) > 0$. □

**Corollary.** *Two n-caps A and B have disjoint interiors if and only if* $A * B \leqslant -1$ *and* $s(A+B) > 0$.

*Proof.* These conditions characterize the relevant cases (iii) and (viii). □

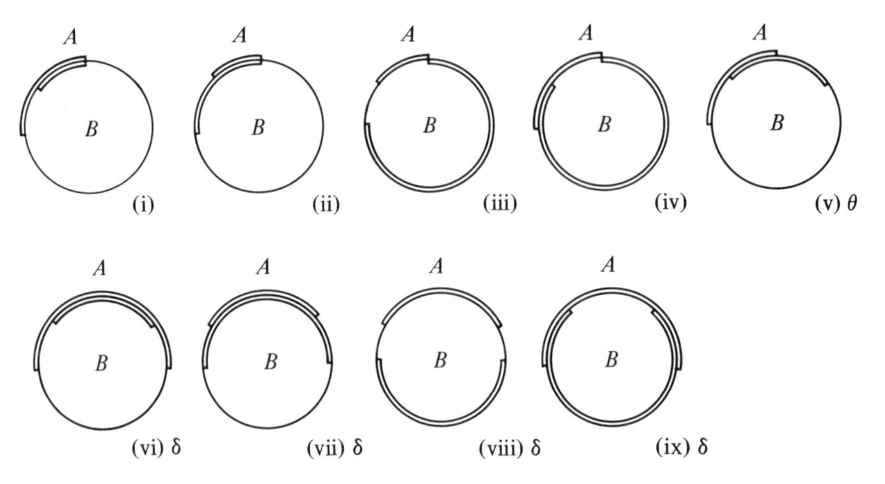


**Figure 22.** The pairs of $n$-caps on the $n$-sphere $[n=1]$.

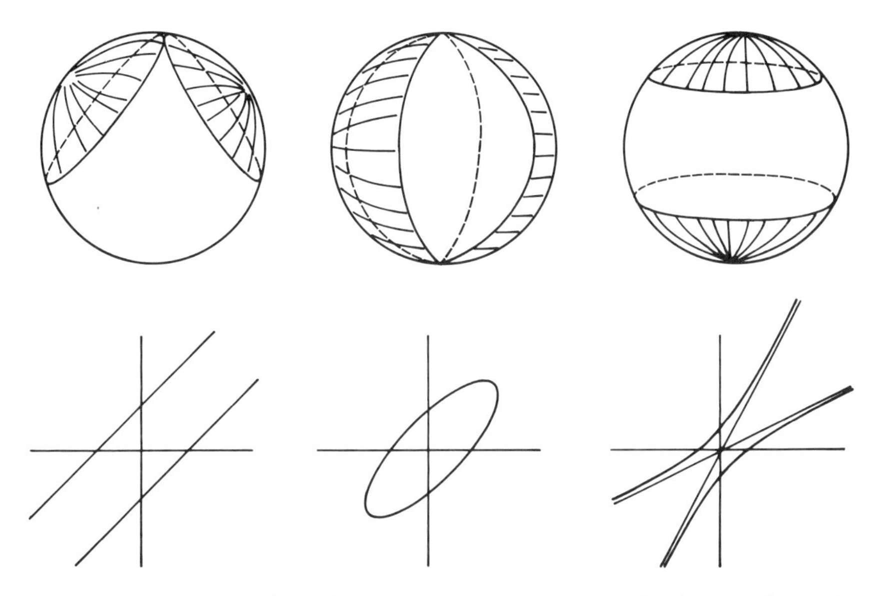

**Figure 23.** Two $n$-caps determine a tangent, intersecting, or disjoint pencil $[n = 2]$.

We close this section by reconsidering the pencil $[\alpha, \beta]$ from an analytic point of view (Figure 23). If $\alpha$ and $\beta$ bound caps $\pm A$ and $\pm B$, then $\gamma \in [\alpha, \beta]$ if and only if it bounds caps $\pm C$, where $C = aA + bB$. This much is clear by considering the coordinates of $A$, $B$, and $C$ when the pencil is in its canonical form. Of course $C = aA + bB$ represents a cap only if it admits the normalization

$$C * C = a^2 + b^2 + 2ab\, A * B = 1,$$

and this represents a conic in the $(a, b)$-parameter plane.

If $\alpha$ and $\beta$ are tangent, then $|A * B| = 1$ and we can choose $A$ and $B$ with $A * B = -1$ and $s(A + B) > 0$ so that the $n$-caps $A$ and $B$ are tangent with disjoint interiors. Then $C * C = (a - b)^2$. If $a = b$, normalization is impossible and it follows that $A + B$ names the common point of the tangent pencil. If $a \neq b$, normalization is possible and gives the pair of lines $a - b = \pm 1$. If $a - b = 1$, then $C$ is nested with $A$, and its boundary $\gamma$ is contained in $A$, contained in $B$, or contained in neither as $a \geqslant 1$, $a \leqslant 0$, or $0 < a < 1$. In particular the boundary of $\frac{1}{2}(A - B)$ lies between $A$ and $B$ and inverts one into the other.

If $\alpha$ and $\beta$ are intersecting, then $A * B = \cos\theta$, $0 < \theta < \pi$. The normalization constraint gives the ellipse $a^2 + b^2 + 2ab\cos\theta = 1$, and there are no vectors in the pencil which cannot be normalized to caps. Inversion in the common boundary of $\pm[2(1 - \cos\theta)]^{-1/2}(A - B)$ interchanges $A$ and $B$, while inversion in the common boundary of $\pm[2(1 + \cos\theta)]^{-1/2}(A + B)$ interchanges $A$ and $-B$.

If $\alpha$ and $\beta$ are disjoint, then $|A * B| = \cosh\delta > 1$, and we can choose $A$ and $B$ with $A * B = -\cosh\delta$ and $s(A + B) > 0$ so that the $n$-caps $A$ and $B$ are completely disjoint. The normalization constraint gives the hyperbola $a^2 + b^2 -$

$2ab\cosh\delta = 1$. On the branch with $b < a$, $C = aA + bB$ is nested with $A$, and its boundary $\gamma$ is contained in $A$, contained in $B$, or contained in neither as $a \geqslant 1$, $a \leqslant 0$, or $0 < a < 1$. In particular the boundary of $[2(1+\cosh\delta)]^{-1/2}(A - B)$ lies between $A$ and $B$ and inverts one into the other. The limiting points of the pencil are obtained by setting $(A + bB) * (A + bB) = 0$. These turn out to be $A + e^{-\delta}B \in A$ and $A + e^{\delta}B \in B$ and correspond to asymptotes of the normalization hyperbola.

## 10. Two Dimensions and a Complex Coordinate

In the two-dimensional $\Pi$-model it is natural to coordinatize each point of $\mathbb{R}^2 = \Pi - \{\infty\}$ with a complex number $z$. When this is done, reflection in the real axis is given by $z \to \bar{z}$, and inversion in the unit circle centered at the origin is given by $z \to 1/\bar{z}$. More generally, reflection in the line $\operatorname{Im} z = v$ is given by $z \to \bar{z} + 2iv$, reflection in the line which crosses the real axis at a point $u$ with an angle $\theta$ is given by $z \to e^{2i\theta}\bar{z} + (1 - e^{2i\theta})u$, and inversion in the circle $|z - z_0| = k$ is given by

$$z \to \frac{z_0\bar{z} + \left(k^2 - |z_0|^2\right)}{\bar{z} - \bar{z}_0}.$$

These can be put in the form

$$z \to \frac{a\bar{z} + b}{c\bar{z} + d} \quad \text{with } ad - bc = 1,$$

and when this is done the matrices $\begin{pmatrix} a & c \\ b & d \end{pmatrix}$ are, respectively,

$$\pm\begin{pmatrix} 1 & 0 \\ 2iv & 1 \end{pmatrix}, \quad \pm\begin{pmatrix} e^{i\theta} & 0 \\ -2i\sin\theta\, u & e^{-i\theta} \end{pmatrix}, \quad \text{and} \quad \pm\begin{bmatrix} iz_0k^{-1} & ik^{-1} \\ i\left(k^2 - |z_0|^2\right)k^{-1} & -i\bar{z}_0k^{-1} \end{bmatrix}.$$

**Observation 1.** These matrices all satisfy $\overline{G}G = I$.

**Observation 2.** Each of these matrices is related to a real 4-vector by the formula

$$\begin{pmatrix} a & c \\ b & d \end{pmatrix} \to \tfrac{1}{2}\big(i(d-a), -(d+a), i(b+c), i(b-c)\big).$$

The unimodular condition, $ad - bc = 1$, guarantees that these 4-vectors satisfy $C * C = 1$. Closer inspection reveals that the matrices $\pm G$ which are related to a given inversion $\bar{\gamma}$ give rise to 4-vectors $\pm C$ which name the complementary caps bounded by $\gamma$. For example, the third pair of matrices gives rise to

$$\pm\frac{1}{2}\left((z_0 + \bar{z}_0)k^{-1}, (z_0 - \bar{z}_0)(ik)^{-1}, \left(|z_0|^2 - 1 - k^2\right)k^{-1}, \left(|z_0|^2 + 1 - k^2\right)k^{-1}\right)$$

$$= \pm\frac{1}{2k}\left(2\operatorname{Re} z_0, 2\operatorname{Im} z_0, |z_0|^2 - 1 - k^2, |z_0|^2 + 1 - k^2\right).$$

We temporarily leave aside this connection with 4-vector coordinates and take up the point that inversions can be represented in the form $z \to (a\bar{z} + b)/(c\bar{z} + d)$. This means that the product of an even number of inversions is represented by a homography $z \to (az + b)/(cz + d)$ $[ad - bc \neq 0]$, and the product of an odd number of inversions is represented by an antihomography $z \to (a\bar{z} + b)/(c\bar{z} + d)$ $[ad - bc \neq 0]$.

**Lemma 8.** *The Möbius group* $\mathfrak{M}_2$ *is equal to the full group of homographies and antihomographies.*

*Proof.* Since the antihomography $z \to (a\bar{z} + b)/(c\bar{z} + d)$ is equal to the reflection $z \to \bar{z}$ followed by the homography $z \to (az + b)/(cz + d)$, it remains to show that every homography represents an element of $\mathfrak{M}$.

If $c = 0$, the homography is of the form $z \to az + b$, and this represents a dilative rotation about the origin with scale factor $|a|$ and rotation angle $\arg a$ followed by a translation with displacement $b$. If $c \neq 0$ and $ad - bc = 1$, the given homography is the product of four simpler homographies which clearly belong to $\mathfrak{M}_2$: $z_1 = cz + d$, $z_2 = -cz_1$, $z_3 = 1/z_2$, and $z_4 = z_3 + ac^{-1}$. □

If the homography $z \to (a_1 z + b_1)/(c_1 z + d_1)$ with matrix

$$H_1 = \begin{pmatrix} a_1 & c_1 \\ b_1 & d_1 \end{pmatrix}$$

is followed by the homography $z \to (a_2 z + b_2)/(c_2 z + d_2)$ with matrix $H_2$, then the product homography has matrix $H_1 H_2$. Notice that when these matrices are multiplied they appear from left to right in the order in which the transformations are applied. It is useful to think of them as acting on homogeneous row vectors $z \sim (z, 1)$ and $\infty \sim (1, 0)$.

We have already used the fact that the matrix of a homography can be normalized by setting $ad - bc = 1$. The remaining twofold ambiguity cannot be systematically eliminated. For example, the halfturn, $z \to -z$, is represented by $\pm\begin{pmatrix} i & 0 \\ 0 & i \end{pmatrix}$, and both of these matrices square to $-I$ rather than $I$. The situation is best described by saying that there is a 2-1 homomorphism of the special linear group $SL(2, \mathbb{C})$ onto the homographies given by

$$\pm\begin{pmatrix} a & c \\ b & d \end{pmatrix} \to z \to \frac{az + b}{cz + d}.$$

The matrix calculations which this homomorphism allows can be extended to include the antihomographies if we let

$$^{\#} \pm I \to z \to \bar{z}.$$

Here $I$ is the $2 \times 2$ identity matrix and $^{\#}$ is an involutary operator which shifts to the left in any calculation and effects complex conjugation $H \to \bar{H}$ on those matrices which it passes over. For example,

$$(^{\#}H_1)H_2(^{\#}H_3) = \bar{H}_1 \bar{H}_2 H_3 \quad \text{and}$$

$$H_1(^{\#}H_2)(^{\#}H_3)(^{\#}H_4) = {}^{\#}\bar{H}_1 H_2 \bar{H}_3 H_4.$$

**Lemma 9.** *Let $h$ be a homography with canonical matrix $H$. Then* trace $H$ *is determined up to sign and complex conjugation by the conjugacy class of $h$ in $\mathfrak{M}_2$.*

*Proof.* The ambiguity in the sign of $H$ gives an ambiguity to the sign of its trace.

If $h$ is conjugate to $h' = \bar{\gamma}h\bar{\gamma}$ by inversion in a circle, then $h'$ can be represented by the matrix $H' = ({}^{\#}G)H({}^{\#}G) = \overline{G}\overline{H}G$. It follows from Observation 1 that

$$\operatorname{trace} H' = \operatorname{trace} G\overline{G}\overline{H} = \operatorname{trace} \overline{H} = \overline{\operatorname{trace} H}\ .$$

The lemma holds because any conjugating element can be written as a product of inversions. □

We can now use the algebraic properties of homographies and antihomographies to distinguish between the conjugacy classes of $\mathfrak{M}_2$ described in Theorem 6.

**Theorem 9.** *The direct transformations in $\mathfrak{M}_2$ are represented by the homographies $z \to (az+b)/(cz+d)$ $[ad - bc = 1]$. A homography is a translation, rotation $\theta$, dilatation $\delta$, or dilative rotation $\theta, \delta$ depending on whether $\frac{1}{2}(a+d)$ belongs to $\{\pm 1\}$, $\{\pm\cos\theta\}$, $\{\pm\cosh\delta\}$, or $\{\pm\cosh(\delta+i\theta), \pm\overline{\cosh(\delta+i\theta)}\}$. The opposite transformations in $\mathfrak{M}_2$ are represented by antihomographies $z \to (a\bar{z}+b)/(c\bar{z}+d)$ $[ad-bc=1]$. An antihomography is an inversion or glide reflection, rotatory reflection $\theta$, or dilative reflection $\delta$ depending on whether $\frac{1}{2}(|a|^2+|d|^2+2\operatorname{Re} b\bar{c})$ is equal to* 1, $\cos 2\theta$, *or* $\cosh 2\delta$. *The fact that inversions are involutions serves to distinguish them from the glide reflections.*

*Proof.* Table 2 shows the even conjugacy classes in $\mathfrak{M}_2$, the simplest representative transformation of each class, and the corresponding canonical matrices and their traces. It follows from Lemma 9 that these traces are inversive invariants of the classes, and it is clear by inspection that they split the classes. The first three classes have real invariants, and as to the fourth, we note that $\cosh(\delta - i\theta) = \overline{\cosh(\delta + i\theta)}$.

**Table 2.** Even Conjugacy Classes

| Class | Transformation | $H$ | $\frac{1}{2}$ trace $H$ |
|---|---|---|---|
| [‖\|] | $z \to z+1$ | $\pm\begin{pmatrix}1 & 0\\ 1 & 1\end{pmatrix}$ | $\pm 1$ |
| $[\times, \theta]$ | $z \to e^{2i\theta}z$ | $\pm\begin{pmatrix}e^{i\theta} & 0\\ 0 & e^{-i\theta}\end{pmatrix}$ | $\pm\cos\theta$ |
| $[\odot, \delta]$ | $z \to e^{2\delta}z$ | $\pm\begin{pmatrix}e^{\delta} & 0\\ 0 & e^{-\delta}\end{pmatrix}$ | $\pm\cosh\delta$ |
| $[\otimes, \theta, \delta]$ | $z \to e^{2(\delta+i\theta)}z$ | $\pm\begin{pmatrix}e^{\delta+i\theta} & 0\\ 0 & e^{-(\delta+i\theta)}\end{pmatrix}$ | $\pm\cosh(\delta+i\theta)$ |

**Table 3.** Odd Conjugacy Classes

| Class | Transformation | Matrix | $\frac{1}{2}(\lvert a\rvert^2+\lvert d\rvert^2+2\operatorname{Re} b\bar{c})$ |
|---|---|---|---|
| [—] | $z\to\bar{z}$ | $^{\#}\pm\begin{pmatrix}1&0\\0&1\end{pmatrix}$ | 1 |
| [⧺] | $z\to\bar{z}+1$ | $^{\#}\pm\begin{pmatrix}1&0\\1&1\end{pmatrix}$ | 1 |
| [⊠, $\theta$] | $z\to\dfrac{e^{2i\theta}}{\bar{z}}$ | $^{\#}\pm\begin{pmatrix}0&ie^{-i\theta}\\ie^{i\theta}&0\end{pmatrix}$ | $\cos 2\theta$ |
| [⊖, $\delta$] | $z\to e^{2\delta}\bar{z}$ | $^{\#}\pm\begin{pmatrix}e^{\delta}&0\\0&e^{-\delta}\end{pmatrix}$ | $\cosh 2\delta$ |

If an odd transformation in $\mathfrak{M}_2$ is represented by $^{\#}\pm\begin{pmatrix}a&c\\b&d\end{pmatrix}$, then its square is represented by

$$H=\begin{pmatrix}\bar{a}&\bar{c}\\\bar{b}&\bar{d}\end{pmatrix}\begin{pmatrix}a&c\\b&d\end{pmatrix},$$

and $\frac{1}{2}\operatorname{trace} H=\frac{1}{2}(|a|^2+|d|^2+2\operatorname{Re} b\bar{c})$ is an invariant of the conjugacy class. Table 3 shows the odd conjugacy classes in $\mathfrak{M}_2$, the simplest representative transformation of each class, the corresponding canonical matrix, and the invariant $\frac{1}{2}(|a|^2+|d|^2+2\operatorname{Re} b\bar{c})$. It is noteworthy that this invariant distinguishes between the involutions [—] and [⊠, $\pi/2$]. It is only a minor shortcoming that it fails to distinguish between [—] and [⧺].

As a further exercise in the use of our matrix notation for the elements of $\mathfrak{M}_2$ we prove the following lemma.

**Lemma 10.** *An element of $\mathfrak{M}_2$ is an isometry of $\Sigma$ if and only if it commutes with the antipodal map. The rotations of $\Sigma$ are given by the homographies*

$$z\to\frac{az+b}{-\bar{b}z+\bar{a}}.$$

*Proof.* An isometry of $\Sigma$ can be written as a product of inversions in great circles and therefore commutes with the antipodal map.

On the other hand, Lemma 7 shows that the usual distance on $\Sigma$ can be expressed in terms of the antipodal map $a$ by the formula

$$\cos^2\frac{d(x,y)}{2}=\frac{\|x-y^a\|\,\|y-x^a\|}{\|x-x^a\|\,\|y-y^a\|}.$$

This shows that an element of $\mathfrak{M}_2$ that commutes with the antipodal map is an isometry of $\Sigma$.

The antipodal map $z\to-1/\bar{z}$ has matrix $^{\#}\left(\begin{smallmatrix}0&1\\-1&0\end{smallmatrix}\right)$, and a homography $z\to(az+b)/(cz+d)$ commutes with it if and only if

$$^{\#}\begin{pmatrix}0&1\\-1&0\end{pmatrix}\begin{pmatrix}a&c\\b&d\end{pmatrix}{}^{\#}\begin{pmatrix}0&1\\-1&0\end{pmatrix}=\begin{pmatrix}-\bar{d}&\bar{b}\\\bar{c}&-\bar{a}\end{pmatrix}=\pm\begin{pmatrix}a&c\\b&d\end{pmatrix}.$$

The condition "+" leads to $d = -\bar{a}$, $c = \bar{b}$, and the contradiction $-|a|^2 - |b|^2 = 1$. The condition "−" leads to $d = \bar{a}$, $c = -\bar{b}$, as stated in the lemma. □

*Remark.* Similar calculations can be applied to determine the elements of $\mathfrak{M}_2$ which fix the upper half plane $\operatorname{Im} z > 0$ or the unit disk $|z| < 1$.

## 11. The Homomorphism of $SL(2,\mathbb{C})$ onto $\mathfrak{L}_4^{\uparrow +}$

If the 2-sphere $\Sigma$ is coordinatized by $\mathbb{C} \cup \{\infty\}$, then $\mathfrak{M}_2$ is given by the group of homographies and antihomographies. On the other hand, if $\Sigma$ is coordinatized by real 4-vectors, then $\mathfrak{M}_2$ is given by $\mathfrak{L}_4^{\uparrow}$. These two groups are therefore isomorphic, and the isomorphism between them can be computed by a change of coordinates. This isomorphism must carry the homographies onto the elements of $\mathfrak{L}_4^{\uparrow}$ which have determinant $+1$, because it is these elements which represent the direct transformations of $\mathfrak{M}_2$. The obvious 2-1 homomorphism of $SL(2,\mathbb{C})$ onto the homographies therefore gives rise to a 2-1 homomorphism $SL(2,\mathbb{C}) \to \mathfrak{L}_4^{\uparrow +}$. We describe the details of this homomorphism in the last theorem.

**Theorem 10.** *The mapping given below is a 2-1 homomorphism of the special linear group $SL(2,\mathbb{C})$ onto the proper orthochroneous Lorentz group $\mathfrak{L}_4^{\uparrow +}$:*

$$\pm\begin{pmatrix} a & c \\ b & d \end{pmatrix} \to \begin{bmatrix} \operatorname{Re}(a\bar{d} + b\bar{c}) & \operatorname{Im}(a\bar{d} + b\bar{c}) & \operatorname{Re}(a\bar{b} - c\bar{d}) & \operatorname{Re}(a\bar{b} + c\bar{d}) \\ \operatorname{Im}(-a\bar{d} + b\bar{c}) & \operatorname{Re}(a\bar{d} - b\bar{c}) & \operatorname{Im}(-a\bar{b} + c\bar{d}) & \operatorname{Im}(-a\bar{b} - c\bar{d}) \\ \operatorname{Re}(a\bar{c} - b\bar{d}) & \operatorname{Im}(a\bar{c} - b\bar{d}) & \frac{1}{2}(|a|^2 - |b|^2 - |c|^2 + |d|^2) & \frac{1}{2}(|a|^2 - |b|^2 + |c|^2 - |d|^2) \\ \operatorname{Re}(a\bar{c} + b\bar{d}) & \operatorname{Im}(a\bar{c} + b\bar{d}) & \frac{1}{2}(|a|^2 + |b|^2 - |c|^2 - |d|^2) & \frac{1}{2}(|a|^2 + |b|^2 + |c|^2 + |d|^2) \end{bmatrix}.$$

*The trace of the real $4 \times 4$ matrix is equal to $|a + d|^2$, and this is 4, $4\cos^2\theta$, $4\cosh^2\delta$, or $4(\sinh^2\delta + \cos^2\theta)$ depending on whether $z \to (az + b)/(cz + d)$ represents a translation, rotation $\theta$, dilatation $\delta$, or dilative rotation $\theta, \delta$.*

*Proof.* The point $z$ has 4-vector coordinate

$$(2\operatorname{Re} z, 2\operatorname{Im} z, |z|^2 - 1, |z|^2 + 1) = \big(z + \bar{z}, -i(z - \bar{z}), z\bar{z} - 1, z\bar{z} + 1\big).$$

Under the given transformation $z$ is mapped to $w = (az + b)/(cz + d)$, and this point can be represented by the 4-vector coordinate

$$|cz + d|^2\big(w + \bar{w}, -i(w - \bar{w}), w\bar{w} - 1, w\bar{w} + 1\big).$$

The first component in this vector is

$$\begin{aligned}(az + b)(\bar{c}\bar{z} + \bar{d}) + (\bar{a}\bar{z} + \bar{b})(cz + d) \\ = (a\bar{d} + c\bar{b})z + (b\bar{c} + d\bar{a})\bar{z} + (a\bar{c} + c\bar{a})z\bar{z} + b\bar{d} + d\bar{b}.\end{aligned}$$

The first two terms of this sum can be written

$$\tfrac{1}{2}(a\bar{d}+c\bar{b}+b\bar{c}+d\bar{a})(z+\bar{z})+\tfrac{1}{2}(a\bar{d}+c\bar{b}-b\bar{c}-d\bar{a})(z-\bar{z})$$
$$=\operatorname{Re}(a\bar{d}+b\bar{c})(z+\bar{z})+\operatorname{Im}(-a\bar{d}+b\bar{c})\left[-i(z-\bar{z})\right].$$

The last two terms in the above sum can be written

$$\tfrac{1}{2}(a\bar{c}+c\bar{a}-b\bar{d}-d\bar{b})(z\bar{z}-1)+\tfrac{1}{2}(a\bar{c}+c\bar{a}+b\bar{d}+d\bar{b})(z\bar{z}+1)$$
$$=\operatorname{Re}(a\bar{c}-b\bar{d})(z\bar{z}-1)+\operatorname{Re}(a\bar{c}+b\bar{d})(z\bar{z}+1).$$

Together, these four coefficients account for the first column in our $4\times 4$ matrix.

The analogous calculation of the other three components of the image vector give the last three columns in our $4\times 4$ matrix. To check that this matrix is not in error by a positive factor we must check that the norm of the first row vector is one. Rather than do this directly, we illustrate another technique for computing the homomorphism.

The first row of our $4\times 4$ matrix is the image of $(1,0,0,0)$. This vector names the right half plane, and inversion in its boundary is given by $z\to -\bar{z}$, which has the matrix ${}^{\#}\left(\begin{smallmatrix} i & 0\\ 0 & -i\end{smallmatrix}\right)$. The given transformation conjugates this inversion into

$$\begin{pmatrix} a & c\\ b & d\end{pmatrix}^{-1}\left[{}^{\#}\begin{pmatrix} i & 0\\ 0 & -i\end{pmatrix}\right]\begin{pmatrix} a & c\\ b & d\end{pmatrix}={}^{\#}\begin{pmatrix} \bar{d} & -\bar{c}\\ -\bar{b} & \bar{a}\end{pmatrix}\begin{pmatrix} ia & ic\\ -ib & -id\end{pmatrix}$$
$$={}^{\#}\begin{pmatrix} ia\bar{d}+ib\bar{c} & ic\bar{d}+id\bar{c}\\ -ia\bar{b}-ib\bar{a} & -ic\bar{b}-id\bar{a}\end{pmatrix}.$$

According to Observation 2 made at the beginning of Section 10, the circle of this conjugated inversion bounds the caps $\pm C$, where

$$C=\tfrac{1}{2}\big(i(-ic\bar{b}-id\bar{a}-ia\bar{d}-ib\bar{c}),(ic\bar{b}+id\bar{a}-ia\bar{d}-ib\bar{c}),$$
$$i(-ia\bar{b}-ib\bar{a}+ic\bar{d}+id\bar{c}),i(-ia\bar{b}-ib\bar{a}-ic\bar{d}-id\bar{c})\big)$$
$$=\big(\operatorname{Re}(a\bar{d}+b\bar{c}),\operatorname{Im}(a\bar{d}+b\bar{c}),\operatorname{Re}(a\bar{b}-c\bar{d}),\operatorname{Re}(a\bar{b}+c\bar{d})\big).$$

Thus $\pm C$ is the image of $(1,0,0,0)$ under $z\to(az+b)/(cz+d)$, and it must therefore appear in the first row of the $4\times 4$ matrix related to $\left(\begin{smallmatrix} a & c\\ b & d\end{smallmatrix}\right)$. Since $+C$ matches the first row of the matrix we have calculated via $z$, this matrix is correct as it stands and does not need positive rescaling.

The trace of the $4\times 4$ matrix is obviously $|a+d|^2$, and Theorem 9 gives the information required to compute its possible values. □

*Remark* 1. The trace of the $4\times 4$ matrix does not split the transformations into conjugacy classes. This prompts the question of how to recover the representation $z\to(az+b)/(cz+d)$ from a given matrix $L\in\mathfrak{L}_4^{\uparrow +}$.

The points, 0, 1, $\infty$ have 4-vector coordinates $(0,0,-1,1),(1,0,0,1),(0,0,1,1)$, and for each of these values of $X$ we can compute $Y=XL$. A 4-vector $Y=(Y^{(1)},Y^{(2)},Y^{(3)},Y^{(4)})$ which satisfies $Y*Y=0$ and $Y^{(4)}>0$ names the point

$$z = \frac{Y^{(1)} + iY^{(2)}}{Y^{(4)} - Y^{(3)}}.$$

We can therefore determine the images of 0, 1, $\infty$ and use the resulting three equations to determine the homography which corresponds to $L$.

*Remark* 2. According to Lemma 10 a homography represents a rotation of $\Sigma$ if and only if it satisfies $d = \bar{a}$ and $c = -\bar{b}$. It is reassuring to observe that under these conditions the matrix $L$ of Theorem 10 reduces to a $3 \times 3$ block [which must be an orthogonal matrix] and a diagonal entry $L_{44} = 1$.

*Remark* 3. In special relativity, with the speed of light set equal to 1, the points of $\Sigma$ correspond to photon trajectories and therefore $\Sigma$ itself can be regarded as a celestial sphere. The fact that the elements of $\mathfrak{M}_2$ are angle-preserving transformations of $\Sigma$ can be given the physical interpretation that Lorentz-related observers have conformally related celestial spheres.

## 12. Inversive Models of the Classical Geometries

Up to this point Euclidean geometry of $n$ and $n+1$ dimensions has provided the framework for our discussion of $n$-dimensional inversive geometry and its group $\mathfrak{M}_n$. In this section we regard the inversive geometry as fundamental and show how it can be used to model the classical geometries, including Euclidean geometry. Our object is therefore to show how cross ratio, an $\mathfrak{M}_n$-invariant, can be used to construct metrics on $\Sigma$ or appropriate subsets of $\Sigma$. In each case we describe the subgroup of $\mathfrak{M}_n$ which belongs to the given geometry. We also provide an expression for the metric in terms of the invariant bilinear form $U * V$.

### 12.1 Euclidean $n$-space

We fix a point $x_0 \in \Sigma$. If $a$, $b$, $c$, $d$ are four points of $\Sigma - \{x_0\}$ and $a'$, $b'$, $c'$, $d'$ are their images in a $\Pi$-model obtained by stereographic projection from $x_0$, then

$$\begin{aligned}\frac{\|a'-b'\|}{\|c'-d'\|} &= \frac{\|a'-b'\|\,\|c'-\infty\|}{\|a'-\infty\|\,\|b'-c'\|} \cdot \frac{\|b'-c'\|\,\|d'-\infty\|}{\|b'-\infty\|\,\|c'-d'\|} \\ &= \frac{\|a-b\|\,\|c-x_0\|}{\|a-x_0\|\,\|b-c\|} \cdot \frac{\|b-c\|\,\|d-x_0\|}{\|b-x_0\|\,\|c-d\|}.\end{aligned}$$

Thus the ratio of Euclidean distances is given by a product of cross ratios. Since the stabilizer of $x_0$ in $\mathfrak{M}_n$ is the full group of Euclidean similarities, it is quite natural that there is no absolute unit of distance.

Since $\mathfrak{M}_n$ is transitive on points, there is no loss of generality in taking $x_0$ to be the point of $\Sigma$ whose $(n+2)$-vector coordinates are of the form $(0, 0, \ldots, 0,$

$\lambda, \lambda)$ $[\lambda > 0]$. Then, if we set

$$E = (0, 0, \ldots, 0, -1, -1),$$

a proper or improper $n$-ball, $C = (1/2r)(2a, \|a\|^2 - 1 - r^2, \|a\|^2 + 1 - r^2)$, satisfies

$$C * E = \frac{1}{r},$$

and a half space, $D = (n, d, d)$, satisfies

$$D * E = 0.$$

Moreover if $X = \lambda(2x, \|x\|^2 - 1, \|x\|^2 + 1)$ and $Y = \mu(2y, \|y\|^2 - 1, \|y\|^2 + 1)$ are the positive homogeneous names of two Euclidean points $x$ and $y$, then

$$\frac{X * Y}{(X * E)(Y * E)} = -\frac{1}{2}\|x - y\|^2.$$

Thus in terms of the linear group $\mathcal{L}_{n+2}^{\uparrow}$, the Euclidean isometries fix $E$, and the proper similarities map $E$ to a positive multiple of itself.

The vector $E$ satisfies $E * E = 0$ and $s(E) < 0$, and any other such vector will define an equivalent inversive model of Euclidean $n$-space.

## 12.2 Spherical $n$-space

The usual definition of distance between two points $x$ and $y$ on the $n$-sphere $\Sigma$ is that $d(x, y) = \theta$, where $\theta$ is the angle which they subtend at the center of $\Sigma$. This obviously gives a metric. If $a$ denotes the antipodal map $x \to x^a = -x$, then this metric can be expressed through cross ratio by the formula

$$\cos^2 \frac{d(x, y)}{2} = \frac{\|x - y^a\| \, \|y - x^a\|}{\|x - x^a\| \, \|y - y^a\|}.$$

Let us consider the $(n + 2)$-vector

$$S = (0, -1).$$

Then if $U$ is an arbitrary $(n + 2)$-vector, the value of the size functional at $U$ is given by $s(U) = U * S$. In particular, if $C = (\csc\theta \, c, \cot\theta)$ names a cap on $\Sigma$,

$$C * S = \cot\theta.$$

Moreover, if $X = \lambda(x, 1)$ and $Y = \mu(y, 1)$ are the positive homogeneous names of two spherical points $x$ and $y$, then

$$\frac{X * Y}{(X * S)(Y * S)} = x \cdot y - 1 = -2\sin^2 \frac{d(x, y)}{2}.$$

In Lemma 10 we saw that the isometries of $\Sigma$ are precisely the elements of $\mathfrak{M}_n$ which commute with the antipodal map. Since the antipodal map can be expressed in terms of $S$ by the formula

$$U \to -\left[ U + 2(S * U)S \right],$$

the spherical isometries can be described in terms of the linear group $\mathfrak{L}_{n+2}^{\uparrow}$ as the stabilizer of $S$. The formula for the antipodal map also gives us an independent derivation of the $U * V$-formula for spherical distance. We have

$$\begin{aligned}\cos^2 \frac{d(x, y)}{2} &= \frac{\|x - y^a\| \, \|y - x^a\|}{\|x - x^a\| \, \|y - y^a\|} \\ &= \left\{ \frac{(X * Y^a)(Y * X^a)}{(X * X^a)(Y * Y^a)} \right\}^{1/2} \\ &= \left\{ \frac{(X * [Y + 2(S * Y)S])(Y * [X + 2(S * X)S])}{(X * [X + 2(S * X)S])(Y * [Y + 2(S * Y)S])} \right\}^{1/2} \\ &= \frac{X * Y + 2(X * S)(Y * S)}{2(X * S)(Y * S)} \\ &= 1 + \frac{X * Y}{2(X * S)(Y * S)}\end{aligned}$$

and therefore

$$\frac{X * Y}{(X * S)(Y * S)} = -2 \sin^2 \frac{d(x, y)}{2}.$$

The preceding discussion of an inversive model for spherical $n$-space is not yet inversively invariant. However, this defect is easily remedied. In Theorem 3, Corollary 3, we saw that an element of $\mathfrak{M}_n$ which acts without fixed points is conjugate to an ordinary isometry of $\Sigma$. It is an easy extension to see that every fixed-point-free involution in $\mathfrak{M}_n$ is conjugate to the antipodal map. Thus, just as there is an inversive model of Euclidean $n$-space for every point $x_0 \in \Sigma$, so there is an inversive model of spherical $n$-space for every fixed-point-free involution $a : \Sigma \to \Sigma$. These involutions are given by vectors $S$ which satisfy $S * S = -1$ and $s(S) < 0$, and each such vector labels a different model of spherical $n$-space. The formulae which we derived for the standard model and its $S$ conjugate to identical formulae for the other models and their $S$'s.

### 12.3 Hyperbolic $n$-space

Here we generalize the Poincaré model, which is usually defined, for $n = 2$, in the unit disk $|z| < 1$ or the upper half plane $\operatorname{Im} z > 0$.

Let $H$ be an arbitrary $n$-cap on the $n$-sphere $\Sigma$, and let $\gamma$ be the $(n-1)$-sphere which bounds $H$. The points of the model are the points in the interior of $H$. If $x$ and $y$ are points of the model, then we define the distance between them to be $h(x, y)$, where

$$\cosh^2 \frac{h(x, y)}{2} = \frac{\|x - y^{\bar{\gamma}}\| \, \|y - x^{\bar{\gamma}}\|}{\|x - x^{\bar{\gamma}}\| \, \|y - y^{\bar{\gamma}}\|}.$$

This makes a striking parallel with the definition of spherical distance, but it is not as clear that it gives a metric. We therefore put down a second expression for

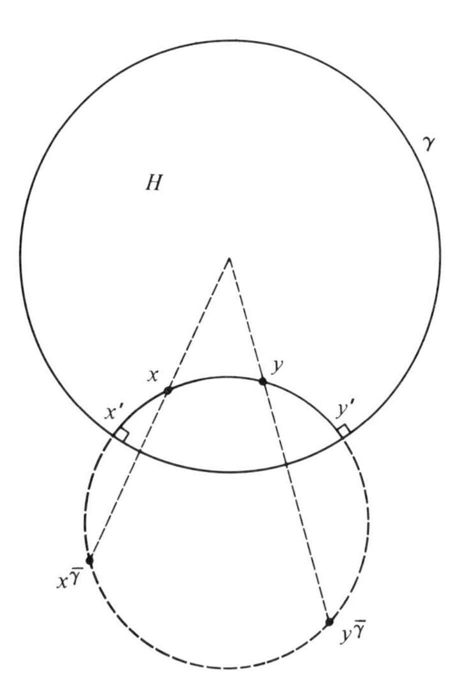


**Figure 24.** The points involved in our two equivalent definitions of the metric on hyperbolic $n$-space $[n = 2]$.

$h$ which we will prove to be equivalent (see Figure 24). There is a unique circle through $x$ and $y$ which is perpendicular to $\gamma$. If this circle meets $\gamma$ at $x'$ beyond $x$ and $y'$ beyond $y$, then

$$h(x, y) = \log \frac{\|x - y'\| \, \|y - x'\|}{\|x - x'\| \, \|y - y'\|}.$$

Since our expressions for $h$ are invariant under the action of $\mathfrak{M}_n$, we can assume that $H$ is the unit ball $\|z\| \leqslant 1$ [in the $\Pi$-model of inversive $n$-space], $x = 0$, and $y$ is situated a distance $0 < r < 1$ away from $x$ on a diameter of $\gamma$. Then

$$\cosh^2 \frac{h(x, y)}{2} = \frac{\|x - y^{\bar{\gamma}}\| \, \|y - x^{\bar{\gamma}}\|}{\|x - x^{\bar{\gamma}}\| \, \|y - y^{\bar{\gamma}}\|}$$

$$= \frac{r^{-1} \cdot \infty}{\infty \cdot (r^{-1} - r)} = \frac{1}{1 - r^2},$$

where the "canceling of infinities" is justified, as it has been before, by Remark 3 following Theorem 1. Using the other expression for $h$, we compute

$$h(x, y) = \log \frac{\|x - y'\| \, \|y - x'\|}{\|x - x'\| \, \|y - y'\|} = \log \frac{1 + r}{1 - r}$$

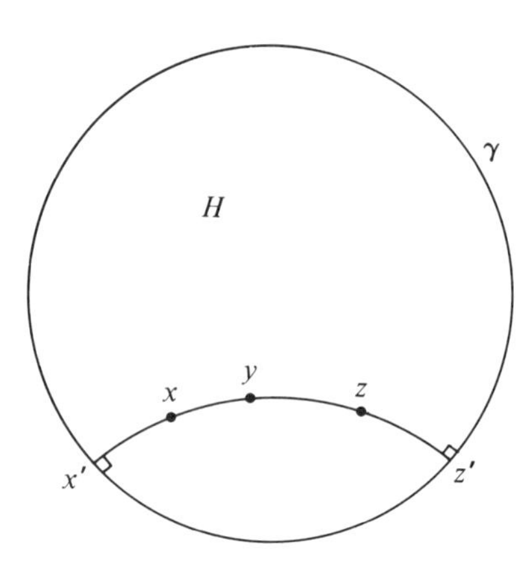


**Figure 25.** There is equality in the triangle inequality if $x$, $y$, and $z$ lie in that order on a circle perpendicular to $\gamma$.

and verify that

$$\cosh^2\frac{h(x,y)}{2}=\left[\frac{\left(\frac{1+r}{1-r}\right)^{1/2}+\left(\frac{1-r}{1+r}\right)^{1/2}}{2}\right]^2=\frac{1}{1-r^2}.$$

This shows that the two definitions of $h$ are equivalent.

The function $h$ is positive and symmetric. In order to complete the proof that $h$ is a metric, we consider the triangle inequality in two cases. If $x$, $y$, and $z$ lie in that order on a circle perpendicular to $\gamma$ (Figure 25), then

$$\begin{aligned}h(x,y)+h(y,z)&=\log\frac{\|x-z'\|\,\|y-x'\|}{\|x-x'\|\,\|y-z'\|}+\log\frac{\|y-z'\|\,\|z-x'\|}{\|y-x'\|\,\|z-z'\|}\\&=\log\frac{\|x-z'\|\,\|z-x'\|}{\|x-x'\|\,\|z-z'\|}\\&=h(x,z).\end{aligned}$$

This also shows that if $x$, $y$, and $z$ lie in a different order on such a circle, $h(x,y)+h(y,z)>h(x,z)$. If $x$, $y$, and $z$ do not lie on a circle perpendicular to $\gamma$, let $c_{xy}$, $c_{yz}$, and $c_{zx}$ be the three circles perpendicular to $\gamma$ which are determined by pairs of these points (Figure 26). Without loss of generality we can assume $y=0$, so that $c_{xy}$ and $c_{yz}$ are diameters of $\gamma$. Then

$$\begin{aligned}h(x,y)+h(y,z)&=\log\frac{1+\|x\|}{1-\|x\|}+\log\frac{1+\|z\|}{1-\|z\|}\\&=\log\frac{(1+\|x\|)(1+\|z\|)}{(1-\|x\|)(1-\|z\|)}\\&>\log\frac{\|x-z'\|\,\|z-x'\|}{\|x-x'\|\,\|z-z'\|}\\&=h(x,z).\end{aligned}$$

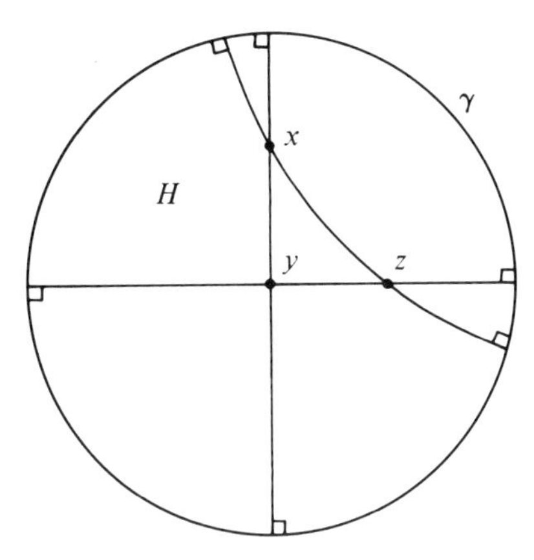


**Figure 26.** There is strict inequality in the triangle inequality if $x$, $y$, and $z$ do not lie in that order on a circle perpendicular to $\gamma$.

This completes the proof that $h$ is a metric. Morever it shows that equality holds in the triangle inequality precisely when $x$, $y$, and $z$ lie in that order on a circle perpendicular to $\gamma$. Naturally we refer to these circles as lines of the hyperbolic geometry.

There is an inversive model of hyperbolic $n$-space for every $n$-cap $H$ on $\Sigma$. The $(n+2)$-vector $H$ satisfies $H * H = 1$, and any such vector names a cap and indexes a model. The vector $H$ is to hyperbolic geometry as the vectors $E$ and $S$ are to Euclidean and spherical geometry. In particular, the hyperbolic distance between points $x$ and $y$ of the model is given by

$$\frac{X * Y}{(X * H)(Y * H)} = -2\sinh^2\frac{h(x, y)}{2}.$$

This formula can be derived from

$$\cosh^2\frac{h(x, y)}{2} = \frac{\|x - y^{\bar{\gamma}}\| \, \|y - x^{\bar{\gamma}}\|}{\|x - x^{\bar{\gamma}}\| \, \|y - y^{\bar{\gamma}}\|}$$

by using the familiar fact that $\bar{\gamma}$ is given by $U \to U - 2(H * U)H$. Alternatively, there is no loss of generality in verifying the formula from the special case with $H = (0, 0, \ldots, 0, -1, 0)$ $[\|z\| \leqslant 1]$, $X = \lambda(0, 0, \ldots, 0, -1, 1)$ $[x = 0]$, and $Y = \mu(0, 0, \ldots, 2r, r^2 - 1, r^2 + 1)$ $[\|y\| = r]$.

A hyperbolic $(n-1)$-sphere with center $x$ and radius $\rho$ is the set of points of the geometry given by $\{z : h(x, z) = \rho\}$. If we take $x = 0$ in the $\Pi$-model with $H = \{z : \|z\| < 1\}$, then this $(n-1)$-sphere is given by by the points $z$ which satisfy

$$\log\frac{1 + \|z\|}{1 - \|z\|} = \rho, \quad \text{or } \|z\| = r = \frac{e^\rho - 1}{e^\rho + 1}.$$

This shows that hyperbolic $(n-1)$-spheres are represented by inversive $(n-1)$-spheres and their radii $\rho$ can be arbitrarily large. A hyperbolic $n$-ball of radius $\rho$

corresponds to an inversive $n$-ball $C$ which is nested within $H$ and satisfies

$$C * H = \cosh\left(\log \frac{1}{r}\right) = \frac{1 + r^2}{2r} = \coth \rho.$$

A hyperbolic $(n-1)$-flat is the set of points equidistant from two given points $\{z : h(z, z_1) = h(z, z_2)\}$. If we take $z_2 = -z_1$ in the model with $H = \{z : \|z\| < 1\}$, then the hyperbolic $(n-1)$-flat is a section of a Euclidean $(n-1)$-flat. Since any pair of points in the model can be moved to this position by suitable inversions, the most general hyperbolic $(n-1)$-flat is a section of an inversive $(n-1)$-sphere perpendicular to $\gamma$, and a hyperbolic half space corresponds to an inversive $n$-ball $C$ which satisfies

$$C * H = 0.$$

Inversions in the $(n-1)$-spheres perpendicular to $\gamma$ fix $H$ and commute with $\bar{\gamma}$, so the formula

$$\cosh^2 \frac{h(x, y)}{2} = \frac{\|x - y^{\bar{\gamma}}\| \, \|y - x^{\bar{\gamma}}\|}{\|x - x^{\bar{\gamma}}\| \, \|y - y^{\bar{\gamma}}\|}$$

shows that they are hyperbolic isometries. On the other hand, these inversions can be described within the model as reflections in hyperbolic $(n-1)$-flats. A discussion of hyperbolic isometries parallel to that of Euclidean isometries shows that any hyperbolic isometry is the product of at most $n+1$ reflections. It follows that these inversions generate the full group of hyperbolic isometries and this group can be described as the subgroup of $\mathfrak{L}_{n+2}^{\uparrow}$ which fixes $H$. By considering the action of this group on $\gamma$, we see that it is an $\mathfrak{M}_{n-1}$ lying in $\mathfrak{M}_n$ in the way described in Remark 1 following Theorem 1.

From an inversive point of view a pencil of $(n-1)$-spheres can be disjoint or tangent or intersecting (Figures 27–29). If such a pencil includes the boundary $\gamma$ of a hyperbolic model $H$, then the circles of the orthogonal pencil will represent lines of the hyperbolic model. The lines will be intersecting, parallel, or ultraparallel depending on whether the pencil is disjoint, tangent, or intersecting.

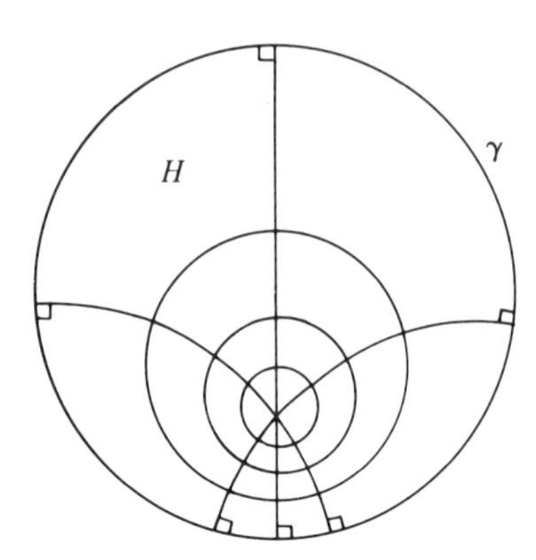


**Figure 27.** A nonintersecting pencil including $\gamma$ represents a nest of concentric hyperbolic $(n-1)$-spheres $[n = 2]$.

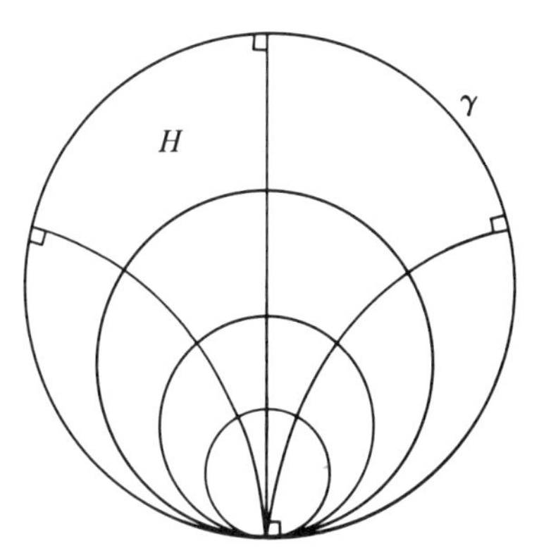


**Figure 28.** A tangent pencil including $\gamma$ represents a nest of concentric horocycles.

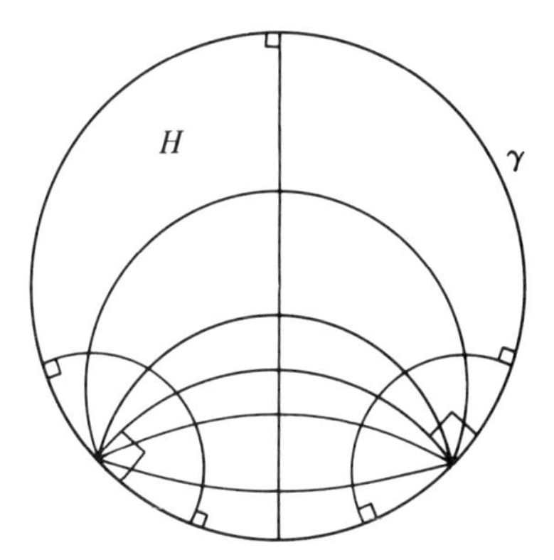


**Figure 29.** An intersecting pencil including $\gamma$ represents an $(n-1)$-flat and its related equidistant surfaces $[n=2]$.

In the first case the $(n-1)$-spheres interior to $H$ which lie in the pencil with $\gamma$ represent a nest of concentric hyperbolic $(n-1)$-spheres. In the second case the $(n-1)$-spheres interior to $H$ which lie in the pencil with $\gamma$ represent a nest of concentric horospheres. If one of these has the inversive coordinate $\pm C$, it will satisfy

$$C * H = \pm 1.$$

In the third case, one of the $(n-1)$-spheres in the pencil with $\gamma$ will be perpendicular to $\gamma$ and represent a hyperbolic $(n-1)$-flat. The rest will meet $\gamma$ at various angles $\theta$ and will represent equidistant surfaces a constant hyperbolic distance $\beta = \beta(\theta)$ away from this $(n-1)$-flat (Figure 30). If one of these has inversive coordinate $\pm C$, it will satisfy

$$C * H = \pm \cos\theta = \pm \tanh \beta.$$

In order to verify that equidistant surfaces deserve their name and furthermore that the above formula holds, we let $y$ be a point on one of the equidistant

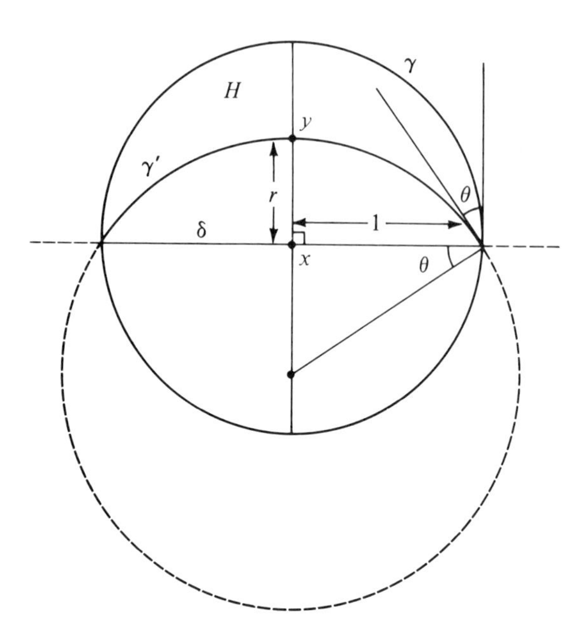


**Figure 30.** The $(n-1)$-sphere $\gamma'$ meets $\gamma$ at an angle $\theta$ and represents an equidistant surface lying a distance $\beta$ above its hyperbolic $(n-1)$-flat $\delta$ $[n = 2]$.

surfaces and let $x$ be the foot of the perpendicular from $y$ to the hyperbolic $(n-1)$-flat in question. We work in the Π-model with $H$ given by $\|z\| < 1$ and apply a hyperbolic isometry which takes $x$ to the center of the model. This means that the hyperbolic $(n-1)$-flat appears as an equatorial $(n-1)$-flat, and the point $y$ lies a Euclidean distance $r$ above it on a diameter of $\gamma$. The equidistant surface is represented by an $(n-1)$-sphere of radius $\sec\theta$ whose center lies a distance $\tan\theta$ below the equatorial $(n-1)$-flat. This gives

$$\beta = \log\frac{1+r}{1-r}$$

and

$$r = \sec\theta - \tan\theta;$$

hence

$$\tanh\beta = \cos\theta.$$

## 13. Formulae for Families of Spheres

The references include approximately 30 papers marked with an asterisk, e.g. [95]*. These provide formulae related to families of spheres in Euclidean and non-Euclidean spaces. The material treated in the present paper furnishes a common inversive foundation for these results. This means that many of the formulae can be given an expanded meaning and, at the same time, a simpler

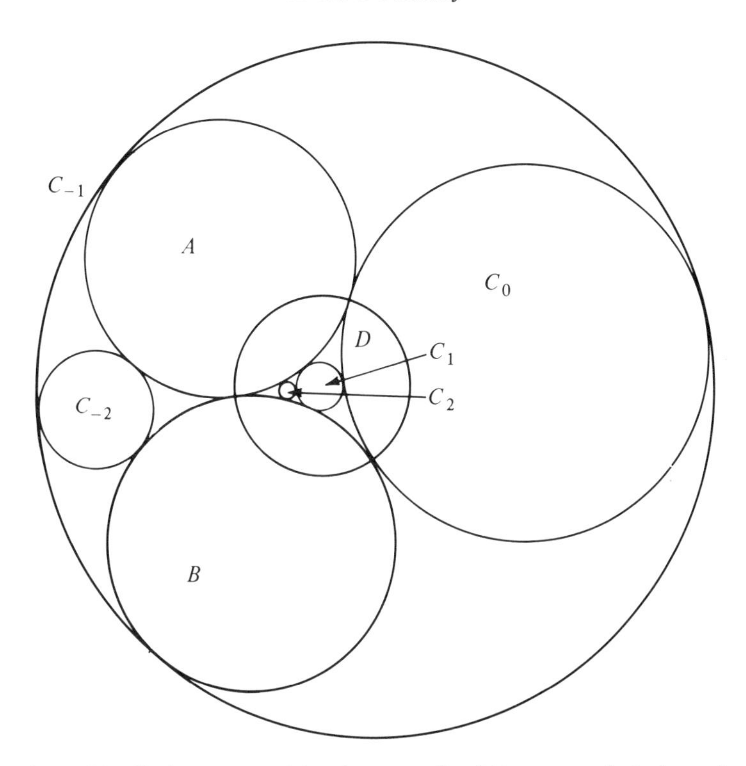


**Figure 31.** Circles governed by the generalized Descartes circle formula.

proof. The details of this reworking will be given in another paper. Here we indicate the flavor of this future paper by considering one example.

In [95] I proved the "generalized Descartes circle formula",

$$C_n = C_0 + 2nD + n^2(A + B).$$

In the original context (Figure 31), $A$ and $B$ were tangent circles lying in the Euclidean plane; $C_0, C_1, C_2, \ldots$ was a sequence of successively tangent circles, each tangent to $A$ and $B$, and $D$ was the circle through the three points of contact of $A$, $B$, and $C_0$. The formula was interpreted as a relation between the curvatures of the circles named, and "curvature" was counted 0 for a line and negative for internal contacts.

In the present context $A$ and $B$ are tangent caps lying in the inversive plane; $C_0, C_1, C_2, \ldots$ is a sequence of successively tangent caps, each tangent to $A$ and $B$, and $D$ is the cap which contains $C_1$ and is bounded by the circle through the three points of contact of $A$, $B$, and $C_0$. The formula is interpreted as a vector relation involving the 4-vectors which name the caps.

The vector relation includes the Euclidean relation in the form

$$\begin{aligned} C_n * E &= \left[ C_0 + 2nD + n^2(A + B) \right] * E \\ &= C_0 * E + 2nD * E + n^2(A * E + B * E). \end{aligned}$$

Moreover there is an immediate generalization to spherical or hyperbolic geometry obtained by replacing $E$ with $S$ or $H$ respectively. We recall from Section 12 that $C * S = \cot\theta$, where $\theta$ is the angular radius of the spherical cap $C$ and $C * H = \coth\rho$, 1, $\tanh\beta$, or 0 depending on whether $C$ represents a hyperbolic disk of radius $\rho$, a horocyclic disk, a region bounded by an equidistant curve running a distance $\beta$ from its line, or a hyperbolic half space. A negative value in the hyperbolic case indicates the complement of one of the above regions, just as a negative curvature in the Euclidean case indicates an improper disk.

Inversive geometry has allowed us to unify the Euclidean and non-Euclidean aspects of the generalized Descartes circle formula. Inversive geometry also helps us to prove the formula. In fact it helps so much that the proof becomes utterly trivial.

Any configuration of caps to which the formula applies can be transformed by an element of $\mathfrak{M}_2$ to the canonical form (Figure 32) with $A = (0,1,1,1)$ $[y \geqslant 1]$, $B = (0,-1,1,1)$ $[y \leqslant -1]$, $D = (1,0,0,0)$ $[x \geqslant 0]$, and $C_n = (2n, 0, 2n^2 - 1, 2n^2)$ $[(x-2n)^2 + y^2 \leqslant 1]$. Since the formula holds in this case and is linear, it holds in general.

For an application of the Generalized Descartes' Circle Formula we return to Pappus' Arbelos, which was described in the first paragraphs of the Preface. The Descartes Formula applies with

$$A = -\frac{1}{2}(0,0,-2,0) \qquad \left[x^2 + y^2 \geqslant 1\right]$$

$$B = \frac{1}{2r}(2b, 0, b^2 - 1 - r^2, b^2 + 1 - r^2) \qquad \left[(x-b)^2 + y^2 \leqslant r^2\right]$$

$$C_n = \frac{1}{2r_n}\left(2x_n, 2y_n, x_n^2 + y_n^2 - 1 - r_n^2, x_n^2 + y_n^2 + 1 - r_n^2\right)$$

$$\left[(x - x_n)^2 + (y - y_n)^2 \leqslant r_n^2\right]$$

$$D = (0,1,0,0) \qquad \left[y \geqslant 0\right].$$

If we take the second component in the Descartes Formula we obtain our old Arbelos formula

$$\frac{y_n}{r_n} = \frac{0}{r_0} + 2n + n^2(0+0) \quad \text{or} \quad y_n = 2nr_n.$$

Similarly, the first component yields

$$\frac{x_n}{r_n} = \frac{x_0}{r_0} + n^2\frac{b}{r},$$

the curvature equation yields

$$\frac{1}{r_n} = \frac{1}{r_0} + n^2\left(\frac{1}{r} - 1\right) = \frac{1}{r_0} + n^2\frac{b}{r},$$

and these combine to yield

$$1 - x_n = \frac{r_n}{r_0}(1 - x_0).$$

This is quite reasonable, since obviously $r_n \to 0$ and $(x_n, y_n) \to (1,0)$ as $n \to \infty$.

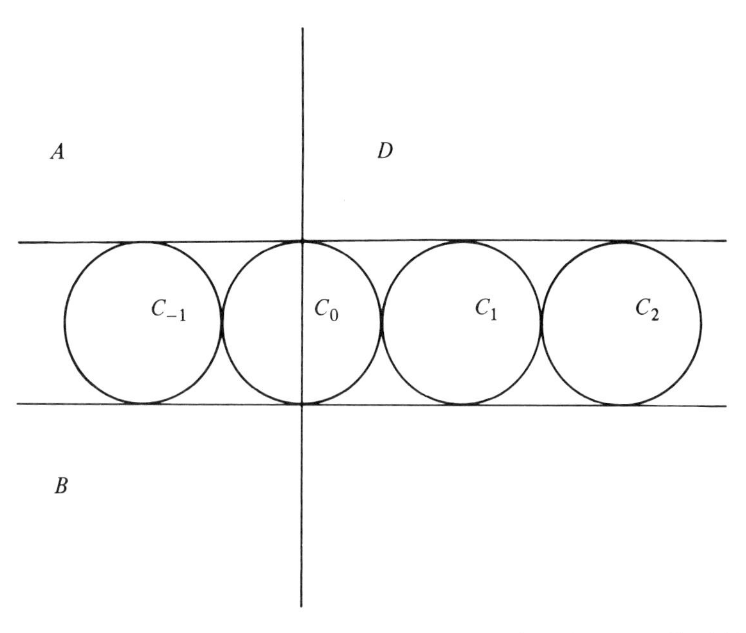


**Figure 32.** The canonical form of the configuration related to the generalized Descartes circle formula.

If we write the last equation in the form

$$\frac{1 - x_n}{r_n} = \frac{1 - x_0}{r_0},$$

it suggests an attractive geometric interpretation. The "lollipops" obtained by dropping a perpendicular from the centre of $C_n$ to the common tangent of $A$ and $B$ are all similar.

## 14. Packings and Inversive Crystals

Let $U$ denote the unit disk lying in the Euclidean plane, and let $D_1$, $D_2$, $D_3$ be three mutually tangent disks which lie inside $U$ and are tangent to its boundary. These three disks divide their complement in $U$ into four curvilinear triangles, and each of these contains a unique disk of maximal radius inscribed within it. The maximal disk in each triangle divides its complement in that triangle into three new curvilinear triangles, and in this way we establish an induction which defines an infinite sequence of disks $P = \{D_n\}_{n=1}^{\infty}$, $D_n$ of radius $r_n$. These disks give a solid packing of $U$ in the sense that the residual set, $R(P) = U \setminus \cup_{n=1}^{\infty} D_n$, has Lebesgue measure 0. In order to measure the asymptotic behavior of the sequence of packing radii, Melzak [73] introduced the moment sum

$$M(P, \alpha) = \sum_{n=1}^{\infty} r_n^{\alpha}$$

and defined the exponent

$$e(P) = \inf\{\alpha : M(P,\alpha) < \infty\}.$$

He showed that $1 < e(P) < 2$, and his paper prompted further efforts to bound $e(P)$ and determine its value exactly. My own papers [97] and [98] describe the significance of this exponent, the contributions towards determining its value, and the reasons for believing it to be minimal among exponents for arbitrary solid disk packings of $U$. The latest result in the quest for $e(P)$ is Boyd's algorithm [3, 4], which can determine it to any desired degree of accuracy and has been used to show that

$$1.300 < e(P) < 1.315.$$

There is a close connection between this packing problem and our present subject. We begin to describe this connection by giving a scheme for naming the disks of the packing.

Let us denote the improper disk complementary to $U$ by $D_4$. Then $D_1$, $D_2$, $D_3$, $D_4$ form a cluster whose disks can conveniently be named by the shorthand [1], [2], [3], [4]. If $\{1,2,3,4\} = \{i,j,k,l\}$, then the curvilinear triangle bounded by $[i]$, $[j]$, $[k]$ can be called $l$, and the maximal disk which sits in this triangle can also be called $l$. (See Figure 33.) This is the beginning of an inductive procedure

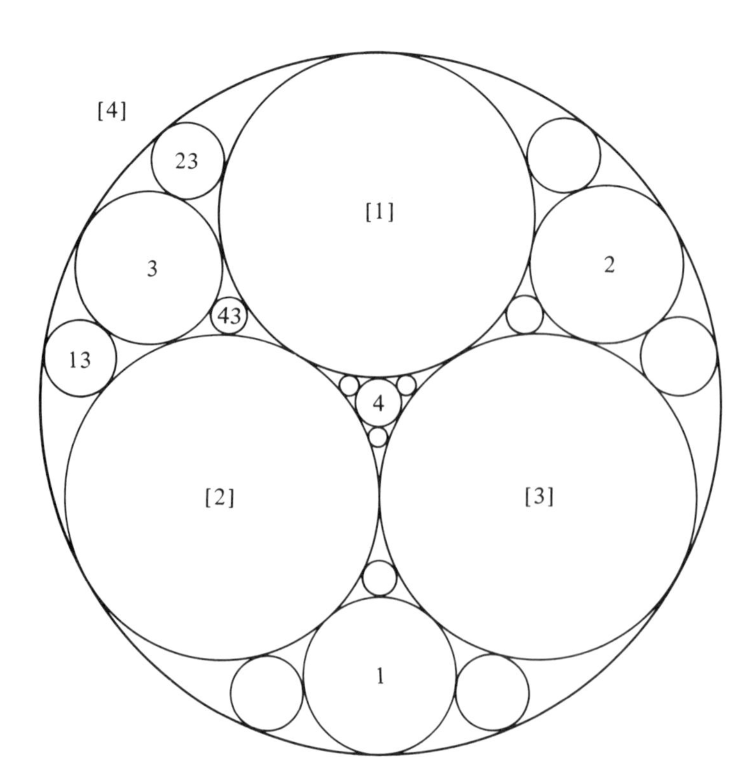


**Figure 33.** The disk labeled 43 belongs to family 4 and sits inside a triangle bounded by [1] from family 1, [2] from family 2, and 3 from family 3.

for naming the packing disks. It automatically divides them into four classes such that a member of the $l$th class is named either by $[l]$ or by a finite string of symbols from $\{1,2,3,4\}$ which begins with $l$ and contains no immediate repetitions.

It is a property of this naming scheme that a yet unnamed triangle is bounded by disks of families $i, j$, and $k$ with the property that the longest of their names includes the other two as terminal segments unless they are of the form $[m]$. This triangle and the new disk which it contains are then named by prefixing this longest name with $l$. Since the name of the new disk begins with $l$ and includes the names of its three ancestral neighbors from families $i, j$, and $k$, the three new triangles which it helps to bound must satisfy the inductive property mentioned at the beginning of this paragraph.

Now we reconsider the cluster [1], [2], [3], [4] and name the circle perpendicular to the boundaries of $[i]$, $[j]$, $[k]$ by the symbol $l$. No confusion will result from our double use of the symbol $l$, and in fact we go on to give it a third meaning. We denote inversion in the circle $l$ by writing $l$ as a superscript, and we note that this inversion has the following effect on the disks of the cluster:

$$[i]^l=[i],\quad [j]^l=[j],\quad [k]^l=[k],\quad [l]^l=l.$$

A product of these inversions is represented by a string of digits from $\{1,2,3,4\}$ which contains no immediate repetitions. This is because inversions are involutions and such a repetition would lead to cancellation. If a sequence of inversions is applied in the natural left-to-right order of the string which names it and if this string begins with $l$, then the effect of the product on $[l]$ will be to move it to the disk which is named by the string. (See Figure 34.) Since every possible name is used once and only once in naming the disks, this shows that the group generated by our four inversions is the "free group" on four involutions. Since the packing $P$ has all the symmetry of this large group, we refer to it as an inversive crystal.

The vectors which name the disks of a cluster form a basis for the coordinatizing 4-space, and satisfy $[i]*[i]=1$ and $[i]*[j]=-1$. Since $l$ is tangent to $[i]$, $[j]$, $[k]$ and satisfies $l*l=1$, it is easy to deduce that

$$l=[l]^l=-[l]+2([i]+[j]+[k]).$$

Thus the matrix which represents the inversion $l$ relative to the basis $[i]$, $[j]$, $[k]$, $[l]$ is obtained by altering the $l$th row of the $4\times 4$ identity matrix by changing the diagonal 1 to $-1$ and the off-diagonal 0's to 2's. Let us denote this matrix by $M_l$.

The four matrices $M_1$, $M_2$, $M_3$, $M_4$ generate a faithful representation of the free group on four involutions. The row vectors that occur in the matrices of this group are the coordinates of the packing disks relative to the cluster basis. The curvature of the disk named by the vector $V$ is the value of a linear form on $V$, and this form is effectively the sum of the components of $V$.

To verify this last remark, lift the packing from the plane $\Pi$ to the sphere $\Sigma$ by stereographic projection. There are constants $0<k_1<k_2$ such that if $D\neq D_4$ is a disk of the packing with radius $r$ which lifts to a cap $D'$ with angular radius $\theta$,

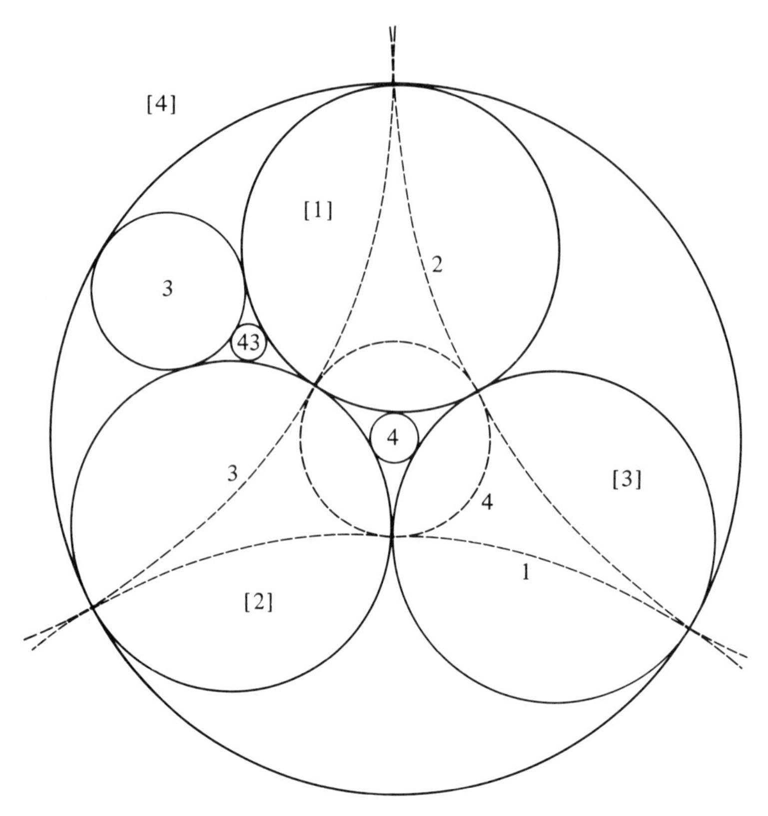


**Figure 34.** The disk labeled 43 is obtained from the improper disk labeled [4] by inverting in circle 4 and then in circle 3.

then

$$k_1 \cot\theta < r^{-1} < k_2 \cot\theta.$$

A further adjustment on $\Sigma$ can make $D_1'$, $D_2'$, $D_3'$, $D_4'$ congruent while maintaining the inequality.

Now let $\mathcal{V}$ denote the set of vectors $V$ which occur as row vectors in the matrix group generated by $M_1$, $M_2$, $M_3$, $M_4$. For each positive integer $n$, let $f(n)$ denote the number of vectors in $\mathcal{V}$ whose components sum to $n$. Then the convergence properties of $M(P,\alpha) = \sum_{n=1}^{\infty} r_n^{\alpha}$ are the same as those of the series $\sum_{n=1}^{\infty} f(n)/n^{\alpha}$, and therefore

$$e(P) = \inf\left\{ \alpha : \sum_{n=1}^{\infty} \frac{f(n)}{n^{\alpha}} < \infty \right\}.$$

So far this approach has not shed light on the value of $e(P)$. On the other hand, machinery like this has allowed Boyd [5–7] to discover other inversive crystals in dimensions $n = 2, 3, 4, 5, 9$.

## 15. Literature and Selected Problems

The first order of business is to mention several places where the work of other authors is used in the preceding sections. The unified definition of inversion and reflection is taken from Coxeter [16, p. 80]. This book also inspired our discussion of Euclidean isometries and similarities [16, Chapters 3, 5, 7]. The notion of inversive distance was first introduced by Coxeter [17] and is also described in Coxeter and Greitzer [23, pp. 123–131]. Finally we point out that our discussion of the twist $\theta$ in Euclidean 3-space is taken directly from L. Fejes Tóth [31, p. 57].

The next item of business is to discuss areas where our work runs parallel to that of other authors.

Our $(n+2)$-vector coordinates and especially those referred to a cluster basis are related to the polyspherical coordinates used by Clifford [11], Darboux [24; 25, livre II, chapitre VI, pp. 265–284], Lachlan [63], Coolidge [12], and Woods [100]. However, all of these authors obtain their coordinates by algebraic considerations involving the properties of determinants. A concise summary of their approach is given by Boyd [5, §3]. Two other books about inversive geometry are those of Morley and Morley [75] and Lagrange [64]. All of these books are recommended, but all of them are more algebraic than the present account.

Most texts on complex analysis mention the group of homographies, and many consider the full group of homographies and antihomographies. Probably the most complete treatments of this material are those of Carathéodory [10, Vol. 1, Chapters 1–3], and Schwerdtfeger [84]. The novelty in the present treatment is that we view $\mathfrak{M}_2$ as a special case of $\mathfrak{M}_n$. This leads us to give a synthetic treatment of the products of up to 5 inversions which occur in dimension $\leqslant 3$. Algebraic manipulations are not used to obtain our list of the conjugacy classes of $\mathfrak{M}_2$. This has the advantage that when the algebra is brought in to label these conjugacy classes it can be employed more decisively.

The discussion of $\mathfrak{M}_2$ via a complex coordinate as well as via real 4-vectors virtually forces us to notice the homomorphism $SL(2,\mathbb{C}) \to \mathfrak{L}_4^{\uparrow +}$. Some of the history of this homomorphism is given by Coxeter [19] along with an alternative geometric explanation for its existence. Ebner [28] reverses our argument and deduces the existence of the homomorphism from a physical argument that Lorentz-related observers must have conformally related celestial spheres. A modern algebraic treatment of the homomorphism is given in Greub [46, pp. 345–350].

Traditionally, the common foundation for Euclidean and non-Euclidean geometry has been projective geometry. See, for example, Coxeter [15]. In Section 12 we show that a common foundation can also be given in terms of inversive geometry. To compare these approaches we consider two models of the hyperbolic plane—the projective or Beltrami–Klein model, and the inversive or Poincaré model.

Both of these models can be set in the disk $|z| < 1$ in the complex plane. In the projective model lines are represented by line segments within the circle $|z| = 1$, and in the inversive model they are represented by circular arcs perpendicular to the circle $|z| = 1$. The intimate connection between the models can be exhibited by parallel-projecting the Beltrami–Klein model from $\Pi$ onto the southern hemisphere of $\Sigma$ and then stereographically projecting it from $\Sigma$ onto the Poincaré model in $\Pi$ (cf. Coxeter [16, p. 290]). In both the projective and inversive models there is a reasonable description of the hyperbolic metric in terms of invariants of the larger geometry and of the hyperbolic isometries in terms of the group of the larger geometry. Nevertheless there is a decisive factor from outside geometry which favors the inversive model over the projective one. It is the close connection between the Poincaré model and complex function theory.

One instance of this close connection is Pick's generalization of the Schwarz lemma which states that an analytic function $f: H \to H$ defined on a disk or half space $H$ must satisfy the inequality $h(f(z_1), f(z_2)) \leq h(z_1, z_2)$ in terms of the Poincaré metric $h$. Moreover any instance of equality forces $f$ to be a fractional linear transformation that acts as an isometry on $H$. The details appear in Carathéodory [10, Vol. 2, pp. 14–16].

The inversive models of Euclidean spherical and hyperbolic $n$-space make it clear that in each case the isometries fixing a point constitute a group isomorphic to the isometries of spherical $(n-1)$-space. This observation is put in different terms in Sommerville [90, pp. 69, 120]. Another useful isomorphism is the one mentioned in Section 12 between the full Möbius group $\mathfrak{M}_n$ and the isometries of hyperbolic $(n+1)$-space. See Coxeter [18] for further details of the case $n = 2$.

Carathéodory [10, Vol. 1, Chapter 3] and Schwerdtfeger [84, Chapter III] give inversive treatments of non-Euclidean planes using a complex coordinate. Alexander [1] and Ewald [30, Chapter 7] each give inversive treatments which do not involve a complex coordinate and can therefore be generalized to higher dimensions. The present treatment is similar to those of Alexander and Ewald but does not coincide with either of them.

In order to give complete inversive foundations for the classical geometries it is necessary to produce suitable foundations for inversive geometry. One possibility is to establish inversive geometry within projective geometry, instead of within Euclidean geometry as we have done. This might make the subject appear less familiar to some readers, but it would also have a number of advantages. Inversions in $(n-1)$-spheres on $\Sigma$ could be given a uniform definition [inversion in a great $(n-1)$-sphere $\gamma$ on $\Sigma$ would be induced by projection from a point $x_\gamma$ on the $n$-flat at infinity], the $(n+2)$-vector coordinates could be regarded as projective coordinates, albeit with emphasis on an unusual normalization, and finally the vectors $E$, $S$, and $H$ defining models of Euclidean and non-Euclidean geometry could be interpreted as projective points on $\Sigma$, within $\Sigma$, or outside of $\Sigma$ respectively.

Since inversive geometry is important in its own right, we should also consider independent axiomatic foundations. Various possibilities occur in Ewald [29], Mäurer [72], and Volenec [92, 93]. Coxeter and Greitzer [23, pp. 103–107] indicate that the cyclic order of four points on a circle might give a point of entry for inversive axioms. Our own Section 6 indicates that foundations might also be developed in terms of cross ratio.

Once we discuss axioms, we invite generalizations of real inversive geometry. Scherk [83] gives a discussion of inversive geometry over the complex members. Dembowski and Hughes [27] introduce general incidence axioms for inversive planes and then concentrate on those with a finite number of points. Further references are available in Dembowski [26, Chapter 6]. Another subject makes its contact with inversive geometry on the $\mathfrak{L}_{n+2}^{\uparrow}$ side of the picture. It is the general study of linear groups which leave invariant a bilinear form. This is a vast field, and it is beyond the scope of this paper to begin a discussion of its literature.

My own inclination is to concentrate on real inversive geometry, including of course the very important case of dimension $n = 2$, where it is convenient to do real inversive geometry with a complex coordinate. I will close by mentioning some areas of current interest related to this subject but different from the sphere formulae and packing problems touched on in Sections 13 and 14.

There is a close connection between angle-preserving transformations and Möbius transformations. For contrast and comparison we recall the result of Section 6 that in every dimension $n$ a cross-ratio-preserving mapping $h:\Pi\to\Pi$ must be a Möbius transformation.

If $n = 1$, "angle-preserving" does not make sense, so we pass on to $n = 2$. When $n = 2$, Ford [33, p. 3] shows that an analytic bijection $h:\Pi\to\Pi$ must be a Möbius transformation. In the light of results which we will quote in a moment, it is significant to ask for the most general definition of "angle-preserving" which allows us to prove such a transformation is analytic or conjugate analytic. We note that if $U$ is an open subset of $\Pi$, there are analytic mappings $h: U\to\Pi$ which are conformal but not the restrictions of Möbius transformations.

If $n \geqslant 3$ and $U$ and $V$ are open subsets of $\Pi$, then it is a famous result of Liouville [67] that an angle preserving homeomorphism $h: U\to V$ must be the restriction of a Möbius transformation. But what do we mean by "angle-preserving"? Textbooks on differential geometry such as Guggenheimer [47, pp. 224–226] or Spivak [91, pp. 302–313] assume directly or indirectly that the mapping is about four times differentiable. A number of writers have tried to reduce the differentiability hypothesis. There are early papers by Hartman [48, 49], an approach by Nevanlinna [78] which is reexamined by Flanders [32], and a treatment by Phillips [80] which is nonstandard in the sense of Abraham Robinson. All of these writers achieve some success, but none can match the result of Gehring [35–41]. Using the techniques of quasiconformal mapping, he shows that Liouville's theorem holds if "angle-preserving" is interpreted to mean that for each $x_0 \in U$

$$\lim_{r\to 0^+} \frac{\displaystyle\max_{\|x-x_0\|=r} \|h(x)-h(x_0)\|}{\displaystyle\min_{\|x-x_0\|=r} \|h(x)-h(x_0)\|} = 1.$$

Unfortunately, Gehring's result appears near the end of a pair of fairly technical papers [37, 38]. It would be very nice to have a simple expository paper aimed at proving Liouville's theorem on minimal hypotheses. A generous source of related references is Caraman [8]. In particular, potential authors should consult the work of Rešetnjak and especially [81, 82].

Another very interesting approach to the characterization of Möbius transformations was initiated by Carathéodory [9]. He showed that when $n = 2$ a bijection $h: U \to V$ is the restriction of a Möbius transformation if it maps points which lie on a circle or a line to points which lie on a circle or a line. The hypothesis that $h$ is injective cannot be abandoned, because any constant map satisfies the other conditions. However, Gibbons and Webb [43] have recently weakened this hypothesis to the nearly minimal nontriviality condition that every circle in $\Pi$ omits at least two points of $h(U)$, and $h(U)$ contains at least six points. Their methods are applicable when $n \geqslant 3$, and they have announced some fine results in this direction [94].

Older work by de Kerékjártó [59–61] considers the group of homographies acting on the 2-sphere and characterizes it up to conjugation by a homeomorphism as the only sharply 3-transitive homeomorphism group acting on a surface.

When $n = 2$, conformal mapping by homographies and other analytic functions is a well-studied method of producing harmonic functions with prescribed boundary conditions. When $n \geqslant 3$, conformal mapping is still available, but now the only mapping functions are Möbius transformations. The key result on these transformations is the Kelvin inversion formula, which states that if $\varphi$ is harmonic in $U \subset \mathbb{R}^n$ $[n \geqslant 3]$ and if $h$ is inversion in an $(n-1)$-sphere with center $x_0 \notin U$, then

$$\|x - x_0\|^{2-n}\varphi(h(x))$$

is a harmonic function of $x$ in $h(U)$. It is easy to remember the formula because it interchanges constant functions with the point potential $\|x - x_0\|^{2-n}$. Details are discussed in Kellogg [58, pp. 231–236], Courant and Hilbert [13, pp. 242–243], and Helms [50, pp. 36–42].

A second area of applied mathematics where inversive geometry should play a role is in relativity theory. Above and beyond the homomorphism of $SL(2, \mathbb{C})$ onto $\mathfrak{L}_4^{\uparrow +}$, we have the isomorphism of $\mathfrak{M}_2$ and $\mathfrak{L}_4^{\uparrow}$ and of $\mathfrak{M}_1$ and $\mathfrak{L}_3^{\uparrow}$. The calculus of mirrors which we employed to study the conjugacy classes of $\mathfrak{M}_n$ $[n = 1, 2, 3]$ can also be used to compute in these groups, and there must be physical applications of this kind of mental arithmetic in the Lorentz group.

Most of our discussion of finite-dimensional inversive geometry has an infinite-dimensional analogue based on infinite dimensional real Hilbert space. The coordinates on $\Pi$ $[\lambda(2x, \|x\|^2 - 1, \|x\|^2 + 1)$ and $(n, d, d)$ or $(1/2r)(2a, \|a\|^2 - 1 - r^2, \|a\|^2 + 1 - r^2)]$ and on $\Sigma$ $[\lambda(x, 1)$ and $(\csc\theta\, c, \cot\theta)]$ can be used just as in the finite-dimensional case. However, a distinction arises between the group generated by inversions and the strictly larger group of transformations of the coordinate space, which could be denoted $\mathfrak{L}_{\infty+2}^{\uparrow}$. Another difference is indicated by the presence of the unilateral shift, which is an isometry of $\Pi$ and not a bijection.

An inversion can be defined by the formula $x \to \|x\|^{-2}x$, and this remains meaningful in general Banach spaces. However, it is not clear in this context that inversion preserves flats and spheres or cross ratio. Perhaps the amount by which it fails to do this can be used to measure some of the geometrical properties of Banach spaces which are important in functional analysis.

Finally, I hope that the present treatment of inversive geometry and the non-Euclidean geometries will be of interest to workers in automorphic function theory. The books by Fricke and Klein [34], Ford [33], Siegel [85], Lehner [65, 66], Kra [62], Bers and Kra [2], and Magnus [68] all have a geometric side which is concerned in large measure with the discrete subgroups of the hyperbolic isometries in dimensions $n = 2$ and $n = 3$. Here, as in special relativity, the full groups in question are $\mathfrak{M}_1$ and $\mathfrak{M}_2$.

Incidentally, the subgroup of $\mathfrak{M}_2$ which acts as the symmetry group of the inversive crystal described in Section 14 is conjugate to a subgroup of the Picard group. It would be exciting if this fact should suggest an attack on the function $\sum_{n=1}^{\infty} f(n)/n^{\alpha}$ and the related exponent $e(P)$.

## References


[1] Alexander, H. W., Vectorial inversive and non-Euclidean geometry. *Amer. Math. Monthly* **74** (1967), 128–140. MR **35**, #864.

[2] Bers, L. and Kra, I. (eds.), *A Crash Course on Kleinian Groups*. Springer-Verlag, Berlin–Heidelberg–New York 1974. MR **49**, #10878.

[3] Boyd, D. W., The disc-packing constant. *Aequationes Math.* **7** (1972), 182–193. MR **46**, #2557.

[4] Boyd, D. W., Improved bounds for the disc-packing constant. *Aequationes Math.* **9** (1973), 99–106. MR **49**, #5728.

[5]* Boyd, D. W., The osculatory packing of a three-dimensional sphere. *Canad. J. Math.* **25** (1973), 303–322. MR **47**, #9430.

[6]* Boyd, D. W., An algorithm for generating the sphere coordinates in a three-dimensional osculatory packing. *Math. Comp.* **27** (1973), 369–377. MR **49**, #3700.

[7]* Boyd, D. W., A new class of infinite sphere packings. *Pacific J. Math.* **50** (1974), 383–398. MR **50**, #3118.

[8] Caraman, P., *n-dimensional Quasiconformal (QCf) Mappings*. Abacus Press, Tunbridge Wells 1974. MR **38**, #3428 and **50**, #10249.

[9] Carathéodory, C., The most general transformations of plane regions which transform circles into circles. *Bull. Amer. Math. Soc.* **43** (1937), 573–579.

[10] Carathéodory, C., *Theory of Functions of a Complex Variable* (Vols. 1, 2). Chelsea, New York 1954. MR **15**, 612.

[11] Clifford, W. H., On the powers of spheres (1868). In *Mathematical Papers*. Macmillan, London 1882.

[12] Coolidge, J. L., *A Treatise on the Circle and the Sphere*. Clarendon Press, Oxford 1916.

[13] Courant, R. and Hilbert, D., *Methods of Mathematical Physics* (Vol. 2). Wiley, New York 1962. MR **25**, #4216.

[14]* Coxeter, H. S. M., Interlocked rings of spheres. *Scripta Math.* **18** (1952), 113–121. MR **14**, 492.

[15] Coxeter, H. S. M., *Non-Euclidean Geometry* (5th edition). University of Toronto Press, Toronto 1965. MR **19**, 445.

[16] Coxeter, H. S. M., *Introduction to Geometry* (2nd edition), Wiley, New York–London–Sydney–Toronto 1969. MR **23**, #A1251.

[17] Coxeter, H. S. M., Inversive distance. *Ann. Mat. Pura Appl. (4)* **71** (1966), 73–83. MR **34**, #3418.

[18] Coxeter, H. S. M., The inversive plane and hyperbolic space. *Abh. Math. Sem. Univ. Hamburg* **29** (1966), 217–242. MR **33**, #7920.

[19] Coxeter, H. S. M., The Lorentz group and the group of homographies. In *Proc. Internat. Conf. Theory of Groups (Canberra 1965)*. Gordon and Breach, New York 1967. MR **37**, #5768.

[20]* Coxeter, H. S. M., Mid-circles and loxodromes. *Ontario Mathematics Gazette* **5** (1967), 4–15.

[21]* Coxeter, H. S. M., The problem of Apollonius. *Canad. Math. Bull.* **11** (1968), 1–17. Reprinted in *Amer. Math. Monthly* **75** (1968), 5–15. MR **37**, #5767.

[22]* Coxeter, H. S. M., Loxodromic sequences of tangent spheres. *Aequationes Math.* **1** (1968), 104–121. MR **38**, #3765.

[23] Coxeter, H. S. M. and Greitzer, S. L., *Geometry Revisited*. Mathematical Association of America, Washington 1977.

[24] Darboux, G., Sur les relations entre les groupes de points de cercles et de spheres dans le plan et dans l'espace. *Ann. Ecole Norm. Sup.* **1** (1872), 323–392.

[25] Darboux, G., *Théorie Générale des Surfaces* (3rd edition). Chelsea, New York 1972. MR **53**, #79–82.

[26] Dembowski, P., *Finite Geometries*. Springer-Verlag, Berlin–New York 1968. MR **38**, #1597.

[27] Dembowski, P. and Hughes, D. R., On finite inversive planes. *J. London Math. Soc.* **40** (1965), 171–182. MR **30**, #2382.

[28] Ebner, D. W., A purely geometrical introduction of spinors in special relativity by means of conformal mappings on the celestial sphere. *Annalen der Physik (7)* **30** (1973), 206–210.

[29] Ewald, G., Begründung der Geometrie der ebenen Schnitte einer Semiquadrik. *Arch. Math.* **8** (1957), 203–208. MR **19**, 1190.

[30] Ewald, G., *Geometry: An Introduction*. Wadsworth, Belmont, California 1971. MR **51**, #1575.

[31] Fejes Tóth, L., *Regular Figures*. Pergamon Press, London 1964. MR **29**, #2705.

[32] Flanders, H., Liouville's theorem on conformal mapping. *J. Math. Mech.* **15** (1966), 157–161. MR **32**, #1626.

[33] Ford, L. R., *Automorphic Functions* (2nd edition). Chelsea, New York 1951.

[34] Fricke, R. and Klein, F., *Vorlesungen über die Theorie der automorphen Funktionen* (erster Band). Teubner, Stuttgart 1897. Johnson Reprint Corporation, New York 1965.

[35] Gehring, F. W., The Liouville theorem in space. *Notices Amer. Math. Soc.* **7** (1960), 523–524.

[36] Gehring, F. W., Rings and quasiconformal mappings in space. *Proc. Nat. Acad. Sci. U.S.A.* **47** (1961), 98–105. MR **23**, #A3261.

[37] Gehring, F. W., Symmetrization of rings in space. *Trans. Amer. Math. Soc.* **101** (1961), 499–519. MR **24**, #A2677.

[38] Gehring, F. W., Rings and quasiconformal mappings in space. *Trans. Amer. Math. Soc.* **103** (1962), 353–393. MR **25**, #3166.

[39] Gehring, F. W., Quasiconformal mappings in space. *Bull. Amer. Math. Soc.* **69** (1963), 146–164. MR **26**, #2606.

[40] Gehring, F. W., Quasiconformal mappings in $\mathbb{R}^n$. In *Lectures on Quasiconformal Mapping*. Department of Mathematics, University of Maryland, College Park 1975. MR **52**, #14281.

[41] Gehring, F. W., Quasiconformal mappings. In *Complex Analysis and Its Applications* (Vol. II). International Atomic Energy Agency, Vienna 1976.

[42]* Gerber, L., Sequences of isoclinical spheres. *Aequationes Math.* **17** (1978), 53–72.

[43] Gibbons, J. and Webb, C., Circle-preserving functions of spheres. *Trans. Amer. Math. Soc.* **248** (1979), 67–83.

[44]* Gosset, T., The kiss precise. *Nature* **139** (1937), 62.

[45]* Gosset, T., The hexlet. *Nature* **139** (1937), 251.

[46] Greub, W., *Linear Algebra* (4th edition). Springer-Verlag, New York 1967. MR **37**, #221.

[47] Guggenheimer, H. W., *Differential Geometry*. McGraw-Hill, New York–San Francisco–Toronto–London 1963. MR **27**, #6194.

[48] Hartman, P., Systems of total differential equations and Liouville's theorem on conformal mappings. *Amer. J. Math.* **69** (1947), 327–332. MR **9**, 59.

[49] Hartman, P., On isometries and on a theorem of Liouville. *Math. Z.* **69** (1958), 202–210. MR **21**, #7521.

[50] Helms, L. L., *Introduction to Potential Theory*. Wiley, New York–London–Sydney 1969. MR **41**, #5638.

[51]* Iwata, S., Generalizations of Ohara-Iwata's theorem in Wasan to the $n$-dimensional space. *Sci. Rep. Fac. Ed. Gifu Univ. Natur. Sci.* **4** (1970), 243–250. MR **48**, #12234.

[52]* Iwata, S., Generalization of Steiner's contact circle theorem to the $n$-dimensional space. *Sci. Rep. Fac. Ed. Gifu Univ. Natur. Sci.* **4** (1970/71), 349–354. MR **46**, #4343.

[53]* Iwata, S., On a theorem connected with the contact hyperspheres. *Sci. Rep. Fac. Ed. Gifu Univ. Natur. Sci.* **5** (1972), 5–7. MR **48**, #1017.

[54]* Iwata, S., Generalizations of Pappus's tangent circle theorem to the $n$-dimensional space, *Bull. Gifu Coll. Ed.* **1** (1974), 55–58. MR **51**, #6551.

[55]* Iwata, S. and Naito, J., The problem of Apollonius in the $n$-dimensional space. *Sci. Rep. Fac. Ed. Gifu univ. Natur. Sci.* **4** (1969), 138–148. MR **41**, #6038.

[56]* Iwata, S. and Naito, J., A generalization of Wilker's calculation to the $n$-dimensional space. *Sci. Rep. Fac. Ed. Gifu Univ. Natur. Sci.* **4** (1970), 251–255. MR **49**, #3657.

[57]* Iwata, S. and Naito, J., Relations between the radii of successively tangent hyperspheres touching a hyperellipsoid. *Sci. Rep. Fac. Ed. Gifu Univ. Natur. Sci.* **5** (1973), 121–130. MR **48**, #12236.

[58] Kellogg, O. D., *Foundations of Potential Theory*. Ungar, New York 1929.

[59] de Kerékjártó, B., Sur le groupe des homographies et des antihomographies d'une variable complex. *Comment. Math. Helv.* **13** (1940), 68–82. MR **2**, 322–323.

[60] de Kerékjártó, B., Sur le caractère topologique du groupe homographique de la sphere. *Acta Math.* **74** (1941), 311–341. MR **7**, 137.

[61] de Kerékjártó, B., Sur les groupes transitifs de la droite. *Acta. Univ. Szeged. Sect. Sci. Math.* **10** (1941), 21–35. MR **2**, 322.

[62] Kra, I., Automorphic forms and Kleinian groups. Benjamin, Reading, Massachusetts 1972. MR **50**, #10242.

[63] Lachlan, R., On systems of circles and spheres. *Philos. Trans. Roy. Soc. London A* **177** (1886), 481–625.

[64] Lagrange, R., *Produits d'Inversions et Métrique Conforme*. Gauthier-Villars, Paris 1957. MR **19**, 162.

[65] Lehner, J., Discontinuous groups and automorphic functions. American Mathematical Society, Providence 1964. MR **29**, #1332.

[66] Lehner, J., A short course in automorphic functions. *Holt Rinehart and Winston*, New York–Toronto–London 1966. MR **34**, #1519.

[67] Liouville, J., Théorème sur l'equation $dx^2 + dy^2 + dz^2 = \lambda(d\alpha^2 + d\beta^2 + d\gamma^2)$. *J. Math. Pures Appl. (1)* **15** (1850), 103.

[68] Magnus, W., *Noneuclidean Tesselations and Their Groups*. Academic Press, New York–London 1974. MR **50**, #4774.

[69]* Mauldon, J. G., Sets of equally inclined spheres. *Canad. J. Math.* **14** (1962), 509–516. MR **25**, #5425.

[70]* Mauldon, J. G., Equally inclined spheres. *Proc. Camb. Phil. Soc.* **58** (1962), 420–421.

[71]* Mauldon, J. G., Bunches of cones. *Amer. Math. Monthly* **69** (1962), 206–207.

[72] Mäurer, H., Ein axiomatischer Aufbau der mindestens 3-dimensionalen Möbius-Geometrie. *Math. Z.* **103** (1968), 282–305. MR **36**, #7016.

[73] Melzak, Z. A., Infinite packings of discs. *Canad. J. Math.* **18** (1966), 838–852. MR **34**, #3443.

[74]* Morley, F., The hexlet. *Nature* **139** (1937), 72–73.

[75] Morley, F. and Morley, F. V., *Inversive Geometry*. G. Bell and Sons, London 1933.

[76]* Naito, J., Some properties of the contact hyperspheres in the $n$-dimensional space. *Sci. Rep. Fac. Ed. Gifu Univ. Natur. Sci.* **4** (1970), 256–267. MR **49**, #1313.

[77]* Naito, J., A generalization of Malfatti's problem. *Sci. Rep. Fac. Ed. Gifu Univ. Natur. Sci.* **5** (1975), 277–286. MR **52**, #15218.

[78] Nevanlinna, R., On differentiable mappings. In *Analytic Functions*. Princeton University Press, Princeton 1960. MR **22**, #7075.

[79]* Pedoe, D., On a theorem in geometry. *Amer. Math. Monthly* **74** (1967), 627–640. MR **35**, #6012.

[80] Phillips, R., Liouville's theorem. *Pacific J. Math.* **28** (1969), 397–405. MR **42**, #4715.

[81] Rešetnjak, J. G., On conformal mappings of a space. *Dokl. Akad. Nauk. SSSR* **130** (1960), 1196–1198. *Soviet Math. Dokl.* **1** (1960), 153–155. MR **22**, #9935.

[82] Rešetnjak, J. G., Liouville's conformal mapping theorem under minimal regularity hypothesis (Russian). *Sibirsk. Mat. Z.* **8** (1967), 835–840. MR **36**, #1630.

[83] Scherk, P., Some concepts of conformal geometry. *Amer. Math. Monthly* **67** (1960), 1–29.

[84] Schwerdtfeger, H., *Geometry of Complex Numbers*. University of Toronto Press, Toronto 1962. MR **24**, #A2880.

[85] Siegel, C. L., *Topics in Complex Function Theory* (3 vols.). Wiley, New York 1969–73. MR **41**, #1977.

[86]* Soddy, F., The kiss precise. *Nature* **137** (1936), 1021.

[87]* Soddy, F., The hexlet. *Nature* **138** (1936), 958.

[88]* Soddy, F., The bowl of integers and the hexlet. *Nature* **139** (1937), 77–79.

[89]* Soddy, F., The hexlet. *Nature* **139** (1937), 154.

[90] Sommerville, D. M. Y., *The Elements of Non-Euclidean Geometry*, Bell and Sons, London 1914.

[91] Spivak, M., *A Comprehensive Introduction to Differential Geometry*, Vol. 3. Publish or Perish, Boston 1975. MR **51**, #8962.

[92] Volenec, V., An axiomatic foundation of the circle plane with orthogonality. *Glasnik Mat.* **8** (No. 28, 1973), 85–92. MR **47**, #7574.

[93] Volenec, V., Axiomatic foundations of the $n$-dimensional Möbius geometry. *Publ. Math. Debrecen* **23** (1976) 89–102. MR **54**, #11171.

[94] Webb, C. and Gibbons, J., The inversive group of $S^n$. *Notices Amer. Math. Soc.* **24** (1977), A335.

[95]* Wilker, J. B., Four proofs of a generalization of the Descartes circle theorem. *Amer. Math. Monthly* **76** (1969), 278–282. MR **39**, #7511.

[96]* Wilker, J. B., Circular sequences of disks and balls. *Notices Amer. Math. Soc.* **19** (1972), A193.

[97] Wilker, J. B., The interval of disk packing exponents. *Proc. Amer. Math. Soc.* **41** (1973), 255–260. MR **50**, #3120.

[98] Wilker, J. B., Sizing up a solid packing. *Period. Math. Hungar.* **8** (1977), 117–134. MR **58**, #30759.

[99] Wise, M. E., On the radii of five packed spheres in mutual contact. *Philips Res. Reports* **15** (1960), 101. MR **22**, #11293.

[100] Woods, F. S., *Higher Geometry*. Ginn, Boston 1922. Dover, New York 1961. MR **23**, #A2094.

# Absolute Polarities and Central Inversions

Norman W. Johnson*

## 1. Introduction

There are two different geometric transformations that go by the name of "inversion." One is inversion in a point, also called "central" inversion, and the other is inversion in a circle or a sphere, which is the basis of *inversive* geometry. Other than the fact that both of these transformations are of period two, they seem to have little in common except for the name. However, when we extend the concept of inversion in a circle to include inversion in imaginary circles, we find that inversion in an ordinary or ideal point of hyperbolic space can be identified with inversion in an imaginary or real circle at infinity, thus uniting the two meanings of the term. Such a correspondence is possible because the group of isometries of a hyperbolic space of two or more dimensions is isomorphic to the group of circle-preserving transformations of an inversive space of one dimension less. This isomorphism, noted in 1905 by Liebmann [13, p. 54] and subsequently by many others, has been dealt with extensively in two papers by Coxeter [6; 10].

There also turns out to be a very close analogy between the well-known derivation of the classical metric geometries from projective geometry by fixing an *absolute polarity* and a construction I shall give for deriving various spherical geometries from inversive geometry by fixing what is appropriately called a *central inversion* (in both senses). Each of these geometries leads in turn to one or another of the classical geometries through mappings called *central projection* and *polar projection*. As we examine these relationships, we shall also discover convenient ways of classifying circles, both on a sphere and in a plane.

*Wheaton College, Norton, Massachusetts 02766, U.S.A.

## 2. Projective Geometry

In what follows, $F$ may be any field. For the purposes of this paper, however, it can be supposed that $F$ is either the real field **R** or the complex field **C**. Where no field is specified, the real field is to be assumed. Likewise, $n$ may be any positive integer, though detailed consideration will be given only to cases where $n \leqslant 4$.

The group of invertible linear transformations of an $n$-dimensional (left) vector space $\boldsymbol{N}$ over a field $F$ is the *general linear group* $\mathrm{GL}_n(F)$. The center of this group, consisting of the nonzero scalar transformations, is the *general scalar group* $\mathrm{GZ}_n(F)$, which is isomorphic to the multiplicative group of $F$. Each coset of $\mathrm{GZ}_n(F)$ is the *projection* of any of its elements, and the quotient group $\mathrm{GL}_n(F)/\mathrm{GZ}_n(F)$ is the *projective general linear group* $\mathrm{PGL}_n(F)$. Factoring out the center corresponds to mapping each nonzero vector $\boldsymbol{x}$ onto the one-dimensional subspace $\langle \boldsymbol{x} \rangle$ of its scalar multiples or, geometrically, to projecting the nonzero vectors $\lambda \boldsymbol{x}$ of $\boldsymbol{N}$ onto a *point* $\langle \boldsymbol{x} \rangle$ of an $(n-1)$-dimensional *projective space* $\mathrm{P}\boldsymbol{N}$ (at infinity). The same projection takes each vector (linear form) $\hat{\boldsymbol{u}}$ of the dual (right) vector space $\hat{\boldsymbol{N}}$ into a hyperplane $\langle \hat{\boldsymbol{u}} \rangle$ of $\mathrm{P}\boldsymbol{N}$, with $\langle \boldsymbol{x} \rangle$ and $\langle \hat{\boldsymbol{u}} \rangle$ being incident—written $\langle \boldsymbol{x} \rangle \lozenge \langle \hat{\boldsymbol{u}} \rangle$—if and only if $\boldsymbol{x}\hat{\boldsymbol{u}} = 0$.

Because multiplication in a field is commutative, the two vector spaces $\boldsymbol{N}$ and $\hat{\boldsymbol{N}}$ are isomorphic [14, p. 224], and it is possible to define a *double linear transformation* of the pair of spaces $\{\boldsymbol{N}, \hat{\boldsymbol{N}}\}$ into itself. The group of invertible double linear transformations is the *double general linear group* $\mathrm{DGL}_n(F)$, which contains $\mathrm{GL}_n(F)$ as a subgroup of index 2 when $n \geqslant 2$. The group $\mathrm{DGL}_n(F)/\mathrm{GZ}_n(F)$ is the *double projective general linear group* $\mathrm{DPGL}_n(F)$, containing $\mathrm{PGL}_n(F)$ as a subgroup of index 2 when $n \geqslant 2$.

A *collineation* of an $(n-1)$-dimensional projective space is an incidence-preserving transformation that takes points into points and hyperplanes into hyperplanes. A *correlation* is an incidence-preserving transformation that takes points into hyperplanes and hyperplanes into points. A collineation or correlation is *projective* if it is the projection of a linear or double linear transformation. Thus $\mathrm{PGL}_n(F)$ is the group of projective collineations of an $(n-1)$-dimensional projective space over $F$, and $\mathrm{DPGL}_n(F)$ is the group of projective collineations and correlations if $n \geqslant 3$. On a projective line ($n = 2$) points are self-dual, and there is no difference between a collineation and a correlation; if projective, the transformation is called a *projectivity*.

## 3. Harmonic Homologies and Polarities

On a projective line a projectivity of period two is called an *involution*. An involution of the real projective line has either no invariant points or exactly two, being called *elliptic* in the first case and *hyperbolic* in the second [9, p. 47]. More generally, a transformation of period two is said to be *involutory*.

In a projective plane a projective collineation that leaves invariant every line through some point, the *center*, and every point on some line, the *axis*, is a *perspective collineation* and is called an *elation* or a *homology* according as the center and the axis are or are not incident. An involutory perspective collineation is a *harmonic homology* [4, p. 64; 9, pp. 53–55]. In three or more dimensions a perspective collineation leaves invariant every (hyper)plane through its *center* and every point in its *median (hyper)plane*. Again, an involutory perspective collineation is a *harmonic homology*.

In a projective space of two or more dimensions an involutory projective correlation is a *polarity*, with each point being the *pole* of a unique hyperplane and each hyperplane the *polar* of a unique point. Two points (or two hyperplanes) are *conjugate* in a polarity if one is incident with the polar (or pole) of the other [4, p. 67; 9, p. 60].

A polarity of the real projective plane is said to be *hyperbolic* or *elliptic* according as there are or are not any self-conjugate points. A *conic* is the locus of self-conjugate points or the envelope of self-conjugate lines in a hyperbolic polarity. A hyperbolic polarity induces an involution of conjugate points on any line that is not self-conjugate; the induced involution is hyperbolic if the line intersects the conic, elliptic if it does not [4, pp. 68, 80–84; 9, pp. 62, 71–73].[1]

In real projective space of three or more dimensions polarities come in greater variety. In odd dimensions there is a *null* polarity, in which every point lies on its polar hyperplane. Other than this, the locus of self-conjugate points, if any, is a *quadric*. A line that has more than two points in common with a quadric lies in the quadric, and the quadric is then said to be *ruled*. If no line meets a quadric in more than two points, the quadric is *oval*. As before, an *elliptic* polarity is one that has no self-conjugate points. A *hyperbolic* polarity is one whose locus of self-conjugate points is an oval quadric. A hyperbolic polarity induces a polarity in each hyperplane that is not self-conjugate, the induced polarity being hyperbolic if the hyperplane intersects the quadric, elliptic if not [5, pp. 65–70].

For a three-dimensional projective space Veblen [17, p. 259] proved the following:

**3.1. Theorem.** *A harmonic homology whose center is the pole of its median plane with regard to a quadric transforms the quadric into itself.*

**3.2. Theorem.** *A projective collineation of an oval quadric that leaves three points of the quadric invariant is either the identity or a harmonic homology whose center and median plane are polar with respect to the quadric.*

Both of these theorems can be generalized to spaces of higher dimension.

## 4. Hyperbolic and Elliptic Geometry

As was shown by Cayley [2] for the hyperbolic case and Klein [12] for the elliptic, a non-Euclidean metric can be introduced into the real projective plane by fixing an *absolute polarity* and defining an *isometry* to be a projective

[1] In fact, *any* polarity induces an involution on a non-self-conjugate line, but only the hyperbolic case is relevant here.

collineation that commutes with it. Two lines are *perpendicular* if they are conjugate in the absolute polarity, and an *absolute* point or line is one that is self-conjugate. If the absolute polarity is hyperbolic, the conic of absolute points and lines is the *absolute circle*. (The justification for the use of the term "circle" in this context will appear later.) The resulting geometry is that of a *hyperbolic plane*, whose *ordinary* and *ideal* regions are respectively the interior and the exterior of the absolute circle. If the absolute polarity is elliptic, there are no (real) absolute points or lines, and the resulting geometry is that of an *elliptic plane* [17, pp. 350, 371].

In three dimensions the same procedure of fixing an absolute polarity leads to *hyperbolic space* or *elliptic space* [17, pp. 369, 373]. In hyperbolic space the oval quadric defined by the absolute polarity is the *absolute sphere*. Points lying in the interior of the absolute sphere and lines and planes that intersect it are the *ordinary* points, lines, and planes of hyperbolic space. Points on the absolute sphere and lines and planes tangent to it are *absolute* (at infinity), while points exterior to the absolute sphere and lines and planes that do not intersect it are *ideal* (ultrainfinite). Two ordinary planes are said to be *intersecting*, *parallel*, or *ultraparallel*, according as they meet in an ordinary line, an absolute line, or an ideal line [5, pp. 195–196].

It follows from Theorem 3.1 that a harmonic homology whose center is the absolute pole of its median plane leaves the absolute sphere invariant and hence is an isometry of hyperbolic space. Such an isometry is an *ordinary reflection* if the median plane is ordinary, an *ideal reflection* if the median plane is ideal. Since the absolute pole of an ideal plane is an ordinary point, it may be seen that an ideal reflection is the same as an inversion in a point. The ordinary reflections generate the complete group of isometries of hyperbolic space, and every hyperbolic isometry is the product of at most four ordinary reflections [7, pp. 62–63].

Analogous results hold for spaces of higher dimension.

Fixing an absolute elliptic polarity in an $(n-1)$-dimensional real projective space $\mathrm{P}\boldsymbol{N}$ corresponds to defining an inner product on the vector space $\boldsymbol{N}$. If, relative to some basis, vectors $\boldsymbol{x}$ and $\boldsymbol{y}$ have coordinates

$$(x_0, x_1, \ldots, x_{n-1}) = (x) \quad \text{and} \quad (y_0, y_1, \ldots, y_{n-1}) = (y),$$

such an inner product $(\boldsymbol{x}, \boldsymbol{y})$ is given by a symmetric bilinear form $(x)E(y)\check{}$ whose associated quadratic form $(x)E(x)\check{}$ is positive definite.[2] A linear transformation that preserves inner products, i.e., one that leaves the quadratic form invariant, is said to be *orthogonal*. The group of all orthogonal transformations is the *orthogonal group* $\mathrm{O}_n$ [14, pp. 390–401]. The center of this group is the group $\mathrm{S}_2\mathrm{Z}_n$ of order two generated by the *central inversion* that interchanges $\boldsymbol{x}$ and $-\boldsymbol{x}$. The quotient group $\mathrm{O}_n/\mathrm{S}_2\mathrm{Z}_n$ is the *projective orthogonal group* $\mathrm{PO}_n$, which is the group of isometries of elliptic $(n-1)$-space.

Let $(x)H(y)\check{}$ be a symmetric bilinear form whose associated quadratic form $(x)H(x)\check{}$, is indefinite on $\mathbf{R}^n$ but positive definite on some $(n-1)$-dimensional subspace. A linear transformation that leaves this quadratic form invariant may

[2] I use a circumflex to indicate the transpose of a matrix. Thus $(x)$ is a row, and $(y)\check{}$ is a column.

be called a *hyperbolic orthogonal* transformation, and the group of all such transformations is the *hyperbolic orthogonal group* $O_{n-1,1}$. The central quotient group $O_{n-1,1}/S_2Z_n$ is the *projective hyperbolic orthogonal group* $PO_{n-1,1}$, the group of isometries of hyperbolic $(n-1)$-space.

## 5. The Inversive Sphere

The so-called "inversive plane" is commonly represented by the Euclidean plane extended by a single point at infinity that is assumed to lie on every line. An *inversive circle* is either an ordinary circle of finite radius or a line, regarded as a circle of infinite radius. Inversion in a circle (including reflection in a line) is a bijective transformation of the extended plane that interchanges points, taking circles and lines into circles and lines, i.e., taking inversive circles into inversive circles. An equivalent model is the Argand plane of complex numbers, again extended by a point at infinity, which is actually a two-dimensional real representation of the complex projective line [17, pp. 222–226].

A more symmetric model for what is more appropriately called the *inversive sphere* is a sphere in Euclidean space, with no distinction made between great circles and small circles. Since the center of such a sphere has no special significance, we could just as well use an ellipsoid in affine space or an oval quadric in projective space. We shall find the projective model particularly instructive.

There is a unique circle through any three points of the inversive sphere. Points that lie on the same circle will be called *concircular* (or "concylic"). In agreement with the theory of separation of pairs of points on an ordered projective line [17, pp. 44–45], we shall say that two concircular pairs of points are *separating* if they separate each other, are *tangent* if they have one point in common, or are *separated* if they are disjoint and do not separate each other.

For example, given five concircular points named in cyclic order $P_1$, $Q_1$, $P_2 = R_1$, $R_2$, $Q_2$, as in Figure 1, let $\mathrm{P} = \{P_1, P_2\}$, $\mathrm{Q} = \{Q_1, Q_2\}$, and $\mathrm{R} = \{R_1, R_2\}$. Then pairs P and Q are separating, pairs P and R are tangent, and pairs Q and R are separated.

Analogously, two distinct circles will be called *separating*, *tangent*, or *separated*, according as they have two, one, or no points in common. It is possible to

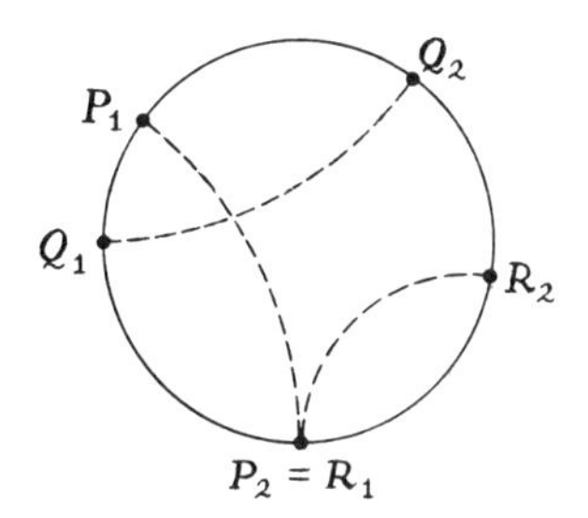


**Figure 1**

regard two separated circles as meeting in an imaginary pair of points; we may then say that any two distinct circles meet in a pair of points, which may be *real* (and distinct), *degenerate* (coincident), or *imaginary*. In like manner, it is sometimes convenient, particularly in analytic treatments, to consider a circle as *real* if it has (at least) two real points, *degenerate* if it has just one real point, or *imaginary* if it has no real points. Two circles are *orthogonal* if one is taken into itself by inversion in the other.

A bijective transformation of the inversive sphere that takes circles into circles will be called a *concirculation*. Because the inversive sphere is orientable, a concirculation may be either direct (sense-preserving) or opposite (sense-reversing). A direct concirculation is called a *homography*; an opposite one, an *antihomography*. All the concirculations form a group, in which the homographies constitute a subgroup of index 2. This group is generated by inversions in (real) circles, and every concirculation can be expressed as the product of at most four inversions [6, p. 230].

Concirculations are more commonly known as *Möbius transformations*, and the group of all concirculations is called the *Möbius group*. The nomenclature used here has the advantage of bringing out more clearly some of the connections that exist between projective and inversive geometry. Just as three or more points lying on a line are *collinear*, so four or more points lying on a circle are *concircular*. A transformation preserving collinearity is a *collineation*; a transformation preserving concircularity is a *concirculation*.

There are three kinds of involutory concirculation: (1) inversion in a real circle, (2) the product of inversions in two orthogonal real circles, and (3) the product of inversions in three mutually orthogonal real circles. The second of these transformations is known as a *Möbius involution* [6, pp. 232–234]. The third may also be regarded as inversion in an imaginary circle and will be called an *elliptic inversion*, with an inversion in a real circle being distinguished as a *hyperbolic* inversion. (This terminology may be compared with that of Du Val [11, p. 7] and Schwerdtfeger [15, pp. 65, 79–80], who call the first and third transformations hyperbolic and elliptic "anti-involutions.")

## 6. The Absolute Sphere

The following theorem was proved by Veblen [17, p. 243] (cf. [6, pp. 234–235]):

**6.1. Theorem.** *If a concirculation leaves invariant each of three distinct points of the inversive sphere, it is either the identity or the inversion in the circle through the three points.*

Comparing this with Theorem 3.2, we see that the group generated by inversions in the real circles of an inversive sphere is isomorphic to the group generated by reflections in the ordinary planes of hyperbolic space. That is, the group of concirculations of the inversive sphere (the Möbius group) is isomorphic to the group of isometries of hyperbolic space, which we have identified as the group $PO_{3,1}$. The subgroup of homographies is isomorphic to the group $PO_{3,1}^{+}$ of direct isometries of hyperbolic space.

Since the complex projective line provides another model for the inversive sphere, we also have $PO_{3,1}^{+} \cong PGL_2(\mathbf{C})$ [17, p. 226]. This isomorphism, however, has no analog in higher space, whereas the geometry of the absolute hypersphere of $(n-1)$-dimensional hyperbolic space is inversive for all $n \geqslant 2$.

The relationship between hyperbolic and inversive geometry may be stated in another way. Just as the geometry at infinity of affine space is projective and the geometry at infinity of Euclidean space is elliptic, so the geometry at infinity of hyperbolic space is inversive [6, p. 221]. Thus the absolute circles in which ordinary hyperbolic planes cut the absolute sphere are the real circles of an inversive sphere; the absolute points at which absolute planes touch the absolute sphere are degenerate circles; and ideal planes, which do not meet the absolute sphere at all, correspond to imaginary circles. An ordinary reflection in hyperbolic space induces a hyperbolic inversion on the absolute sphere, and an ideal reflection induces an elliptic inversion.

Any two ordinary planes of hyperbolic space belong to a unique *pencil of planes* through their common line, the *axis* of the pencil. The pencil is said to be *elliptic*, *parabolic*, or *hyperbolic* according as the axis is ordinary, absolute, or ideal, i.e., according as the given planes are intersecting, parallel, or ultraparallel. Associating each pencil of planes with its trace on the absolute sphere, we see that any two real circles of the inversive sphere belong to a *pencil of circles*, which is *elliptic*, *parabolic*, or *hyperbolic* according as the given circles are separating, tangent, or separated [1, p. 130; 6, p. 223].[3]

If two circles of one pencil are each orthogonal to two circles of another pencil, then every circle of the first pencil is orthogonal to every circle of the second pencil, and the two pencils are *mutually orthogonal*. Each pencil of circles has a unique mutually orthogonal pencil, which is hyperbolic, parabolic, or elliptic according as the given pencil is elliptic, parabolic, or hyperbolic.

Any three ordinary planes of hyperbolic space that do not belong to a pencil have a unique common point. This point is the *center* of a *bundle of planes*, which is said to be *elliptic*, *parabolic*, or *hyperbolic* according as the center is ordinary, absolute, or ideal. The absolute polar of the center is the *median plane* of the bundle. Correspondingly, any three real circles of the inversive sphere belong to a *bundle of circles*, each circle of the bundle being orthogonal to a unique circle (corresponding to the median plane of a bundle of planes). A bundle is *elliptic*, *parabolic*, or *hyperbolic*, according as its orthogonal circle is imaginary, degenerate, or real [17, pp. 256–257].

## 7. Homogeneous Coordinates

Many of the relationships between inversive and hyperbolic geometry can best be seen in analytic form. To this end, let points $\langle \boldsymbol{x} \rangle$ and planes $\langle \hat{\boldsymbol{u}} \rangle$ of real projective space, defined as in Section 2, be given homogeneous coordinates

$$(x) = (x_0, x_1, x_2, x_3) \quad \text{and} \quad [u] = [u_0, u_1, u_2, u_3],$$

[3]Veblen [17, p. 242] interchanged the meanings of "hyperbolic" and "elliptic" as applied to pencils.

with $\langle \boldsymbol{x} \rangle \Diamond \langle \hat{\boldsymbol{u}} \rangle$ if and only if

$$(x)[u] = x_0u_0 + x_1u_1 + x_2u_2 + x_3u_3 = 0. \tag{7.1}$$

Note that if $(x)$ is regarded as a row and $[u]$ as a column, $(x)[u]$ can be calculated as a matrix product.

Let $H$ be a symmetric matrix that is congruent to the diagonal matrix $\backslash -1, 1, 1, 1 \backslash$. Then the double linear transformation

$$(x) \mapsto H(x)\,\check{}, \qquad [u] \mapsto [u]\,\check{}\,H^{-1} \tag{7.2}$$

induces a hyperbolic polarity on the projective space. Using the above condition for a point and a plane to be incident, we see that two points $\langle \boldsymbol{x} \rangle$ and $\langle \boldsymbol{y} \rangle$ or two planes $\langle \hat{\boldsymbol{u}} \rangle$ and $\langle \hat{\boldsymbol{v}} \rangle$ are conjugate in the polarity if

$$(x)H(y)\,\check{} = 0 \quad \text{or} \quad [u]\,\check{}\,H^{-1}[v] = 0. \tag{7.3}$$

The locus of self-conjugate points or the envelope of self-conjugate planes is an oval quadric whose equation is

$$(x)H(x)\,\check{} = 0 \quad \text{or} \quad [u]\,\check{}\,H^{-1}[u] = 0. \tag{7.4}$$

Following Coolidge [3, p. 130] and Coxeter [5, p. 224], let us write $(x\,y)$ for $(x)H(y)\check{}$ and $[u\,v]$ for $[u]\check{}\,H^{-1}[v]$, and take this to be the absolute polarity for hyperbolic space. It is then a simple matter to classify points and planes:

$$\text{A point } \langle \boldsymbol{x} \rangle \text{ is } \begin{cases} \text{ordinary} & \text{if } (x\,x) < 0, \\ \text{absolute} & \text{if } (x\,x) = 0, \\ \text{ideal} & \text{if } (x\,x) > 0. \end{cases} \tag{7.5}$$

$$\text{A plane } \langle \hat{\boldsymbol{u}} \rangle \text{ is } \begin{cases} \text{ordinary} & \text{if } [u\,u] > 0, \\ \text{absolute} & \text{if } [u\,u] = 0, \\ \text{ideal} & \text{if } [u\,u] < 0. \end{cases} \tag{7.6}$$

Two planes $\langle \hat{\boldsymbol{u}} \rangle$ and $\langle \hat{\boldsymbol{v}} \rangle$ are orthogonal if $[u\,v] = 0$.

An ordinary plane $\langle \hat{\boldsymbol{u}} \rangle$ meets the absolute sphere in an absolute circle of points $\langle \boldsymbol{x} \rangle$ satisfying the equations

$$(x)[u] = 0, \qquad (x\,x) = 0. \tag{7.7}$$

If $\langle \hat{\boldsymbol{u}} \rangle$ is an absolute plane, these equations are satisfied by only one real point, the absolute pole of the plane. If $\langle \hat{\boldsymbol{u}} \rangle$ is an ideal plane, the equations have no real solution but may be regarded as representing an imaginary circle. If we now let $\langle \hat{\boldsymbol{u}} \rangle$ denote the circle—real, degenerate, or imaginary—in which the plane $\langle \hat{\boldsymbol{u}} \rangle$ meets the absolute sphere $(x\,x) = 0$, thus passing from hyperbolic to inversive geometry, we have the following classification of inversive circles (cf. [15, p. 113]):

$$\text{A circle } \langle \hat{\boldsymbol{u}} \rangle \text{ is } \begin{cases} \text{real} & \text{if } [u\ \ u] > 0, \\ \text{degenerate} & \text{if } [u\ \ u] = 0, \\ \text{imaginary} & \text{if } [u\ \ u] < 0. \end{cases} \tag{7.8}$$

Two circles $\langle \hat{\boldsymbol{u}} \rangle$ and $\langle \hat{\boldsymbol{v}} \rangle$ are orthogonal if $[u\,v] = 0$.

## 8. Some Useful Formulas

A similar approach to inversive geometry was taken by Coolidge [3, pp. 129–135] and by Veblen [17, pp. 253–256], except that each of them associated an inversive circle with a *point* of hyperbolic or projective space. Alexander [1, pp. 128–132] identified circles and lines of the Euclidean plane with four-dimensional vectors, employing the bilinear form

$$[u\,v] = u_3v_0 + u_1v_1 + u_2v_2 + u_0v_3$$

to obtain a number of interesting results (none dependent on the choice of this particular form), some of which were also given by Coolidge. For example, if the *determinant* $\Delta$ of two real or degenerate circles $\langle\hat{\boldsymbol{u}}\rangle$ and $\langle\hat{\boldsymbol{v}}\rangle$ is defined by

$$\Delta(\hat{\boldsymbol{u}},\hat{\boldsymbol{v}}) = [u\,u][v\,v] - [u\,v]^2, \tag{8.1}$$

we have the following criterion:

$$\text{Circles } \langle\hat{\boldsymbol{u}}\rangle \text{ and } \langle\hat{\boldsymbol{v}}\rangle \text{ are } \begin{cases} \text{separating} & \text{if } \Delta(\hat{\boldsymbol{u}},\hat{\boldsymbol{v}}) > 0, \\ \text{tangent} & \text{if } \Delta(\hat{\boldsymbol{u}},\hat{\boldsymbol{v}}) = 0, \\ \text{separated} & \text{if } \Delta(\hat{\boldsymbol{u}},\hat{\boldsymbol{v}}) < 0. \end{cases} \tag{8.2}$$

The *angle* between separating circles is the same as the angle between the corresponding intersecting planes of hyperbolic space and is given by

$$\cos^{-1}\frac{|[u\,v]|}{\sqrt{[u\,u][v\,v]}} \tag{8.3}$$

[1, p. 130; 3, p. 132] (cf. [5, p. 225]). The *inversive distance* between separated circles is the same as the distance between the corresponding ultraparallel planes of hyperbolic space and is given by

$$\cosh^{-1}\frac{|[u\,v]|}{\sqrt{[u\,u][v\,v]}} \tag{8.4}$$

[5, p. 225; 6, pp. 225–228; 10, pp. 391–392].

Inversion in a real circle $\langle\hat{\boldsymbol{a}}\rangle$ takes each circle $\langle\hat{\boldsymbol{u}}\rangle$ into a circle $\langle\hat{\boldsymbol{u}}\rangle_a$, homogeneous coordinates for which are given by

$$[u] - 2[a]\frac{[u\,a]}{[a\,a]} \tag{8.5}$$

[1, p. 131]. Remarkably, the same formula applies when $\langle\hat{\boldsymbol{a}}\rangle$ is an imaginary circle, so that analytically no distinction need be made between hyperbolic and elliptic inversions.

The treatment of pencils and bundles can also be greatly simplified through the use of coordinates [1, pp. 130–131]. The type of pencil to which two real or degenerate circles $\langle\hat{\boldsymbol{p}}\rangle$ and $\langle\hat{\boldsymbol{q}}\rangle$ belong depends on whether the circles are separating, tangent, or separated:

$$\text{A pencil } \langle \hat{\boldsymbol{p}}, \hat{\boldsymbol{q}} \rangle \text{ is } \begin{cases} \text{elliptic} & \text{if } \Delta(\hat{\boldsymbol{p}}, \hat{\boldsymbol{q}}) > 0, \\ \text{parabolic} & \text{if } \Delta(\hat{\boldsymbol{p}}, \hat{\boldsymbol{q}}) = 0, \\ \text{hyperbolic} & \text{if } \Delta(\hat{\boldsymbol{p}}, \hat{\boldsymbol{q}}) < 0. \end{cases} \tag{8.6}$$

Two pencils $\langle \hat{\boldsymbol{p}}, \hat{\boldsymbol{q}} \rangle$ and $\langle \hat{\boldsymbol{r}}, \hat{\boldsymbol{s}} \rangle$ are mutually orthogonal if

$$[p\,r] = [p\,s] = [q\,r] = [q\,s] = 0. \tag{8.7}$$

Three circles $\langle \hat{\boldsymbol{p}} \rangle$, $\langle \hat{\boldsymbol{q}} \rangle$, $\langle \hat{\boldsymbol{r}} \rangle$ belong to the same pencil if and only if their coordinate vectors $[p]$, $[q]$, $[r]$ are linearly dependent. If the coordinate vectors are linearly independent, then the circles belong to a bundle $\langle \hat{\boldsymbol{p}}, \hat{\boldsymbol{q}}, \hat{\boldsymbol{r}} \rangle$, and there is a unique circle $\langle \hat{\boldsymbol{s}} \rangle$ (real, degenerate, or imaginary) to which every circle of the bundle is orthogonal. The bundle may be classified by reference to this circle:

$$\text{A bundle } \langle \hat{\boldsymbol{p}}, \hat{\boldsymbol{q}}, \hat{\boldsymbol{r}} \rangle \text{ is } \begin{cases} \text{elliptic} & \text{if } [s\,s] < 0, \\ \text{parabolic} & \text{if } [s\,s] = 0, \\ \text{hyperbolic} & \text{if } [s\,s] > 0, \end{cases} \tag{8.8}$$

where $\langle \hat{\boldsymbol{s}} \rangle$ is orthogonal to $\langle \hat{\boldsymbol{p}}, \hat{\boldsymbol{q}}, \hat{\boldsymbol{r}} \rangle$.

## 9. Central Spheres and Median Planes

When the inversive sphere is taken to be the absolute sphere of hyperbolic space, defined by an absolute polarity of projective space, a projective collineation that commutes with the absolute polarity induces a concirculation of the inversive sphere. If the collineation is a harmonic homology, the concirculation is an inversion, hyperbolic or elliptic according as the median plane of the homology is ordinary or ideal.

Let us fix an inversion of the inversive sphere and call it the *central inversion*. Any circle that is orthogonal to the (real or imaginary) *median circle* of this inversion will be called a *great* circle, all others being called *small* circles. A real circle other than the median circle will be said to be *hyperbolic*, *parabolic*, or *elliptic* according as it meets the median circle in a pair of points, a single point, or no points. A point on the median circle is a *mediary* point; other points are *ordinary*. Pairs of points that are interchanged by the central inversion will be said to be *antipodal*.

Any inversive concirculation that commutes with the central inversion, i.e., any transformation that takes great circles into great circles, is an *isometry* of what may be called a *central sphere*. An inversion in a great circle is a *reflection*, and the product of two reflections is a *rotation*. Depending on whether the central inversion is elliptic or hyperbolic, we shall call the central sphere an *elliptic sphere* or a *hyperbolic sphere* and refer to the corresponding geometry as *elliptic spherical* or *hyperbolic spherical*. When a central sphere is regarded as being embedded in projective space, its *center* and *median plane* are the center and median plane of the harmonic homology that induces the central inversion (cf. [11, pp. 8–10]).

The median circle of the elliptic sphere is imaginary, so that all circles great and small are elliptic circles; all degenerate circles are ordinary points. There is more variety on the hyperbolic sphere: all real great circles are hyperbolic, but a real small circle may be hyperbolic, parabolic, or elliptic. A degenerate small circle is an ordinary point; a degenerate great circle is a mediary point.

The central inversion naturally commutes with itself and is therefore an isometry of the central sphere. By definition, the central inversion also commutes with every other isometry. It is apparent that, except for the identity, no other isometry does so, and consequently the subgroup of order two generated by the central inversion is the center of the group of isometries of the central sphere. By factoring out the center, we obtain the corresponding projective (central quotient) group, which is the group of isometries of a projective plane with a fixed polarity. Geometrically, this amounts to identifying antipodal points of the central sphere or, equivalently, projecting the central sphere from its center into its median plane, so that a real pair of antipodal points is taken into an ordinary point and a real great circle is taken into an ordinary line. The latter operation will be called *central projection*.

In this manner the elliptic sphere is projected onto the *elliptic plane*, with great circles being projected into elliptic lines (cf. [8, p. 93]). Likewise, the hyperbolic sphere is projected into the *hyperbolic plane*, with real great circles being projected into ordinary hyperbolic lines and the median circle becoming the absolute circle. This is essentially *Beltrami's model*, with the absolute circle and the ordinary region of the hyperbolic plane being represented by the median circle and its interior (cf. [5, pp. 252, 260–262]).

It may be noted that central projection does not yield any ideal points or lines of the hyperbolic plane. To remedy this defect, let us map the central sphere into the median plane in a different way, using tangent lines and tangent planes. Each projective line that is tangent to the central sphere belongs to a cone of tangent lines whose points of contact lie on a great circle and which all meet the median plane in the same point, the vertex of the cone. (See Figure 2.) Each projective plane that is tangent to the central sphere touches it at a unique point and may be paired with the plane that touches it at the antipodal point; both planes meet

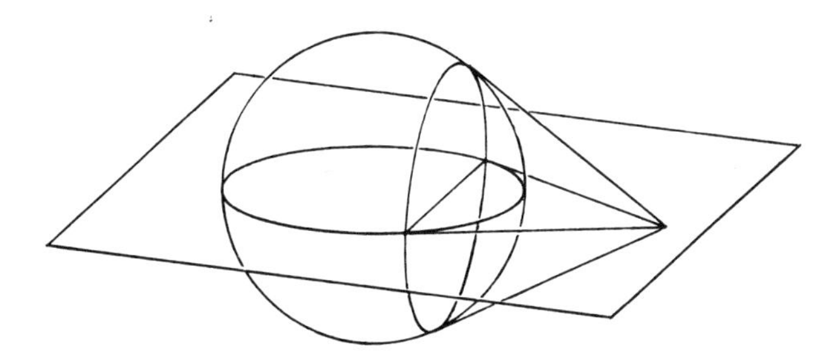

**Figure 2**

the median plane in the same line. Thus real great circles and real pairs of antipodal points of the central sphere are respectively taken into points and lines of its median plane by what may be called *polar projection*.

The image under polar projection of a circle on the central sphere is the polar, with respect to the median circle, of its image under central projection. In other words, the correspondence between the respective images of the same circle is a *polarity*, which is the same as the absolute polarity induced in the median plane by the absolute polarity that defines the central sphere as a quadric in projective space. We see, therefore, that polar projection, like central projection, maps the elliptic sphere onto the elliptic plane, while the hyperbolic sphere is mapped onto the absolute circle and the ideal region of the hyperbolic plane, represented by the median circle and its exterior [7, p. 68]. Combining central projection with polar projection, we obtain the complete hyperbolic plane.

## 10. The Parabolic Sphere

While there are no parabolic inversions, it is possible to define a third kind of central sphere. Again regarding the inversive sphere as the absolute sphere of a hyperbolic polarity of projective space, fix a projective plane tangent to it. This has the effect of fixing a degenerate circle on the inversive sphere, viz., the point of contact. This point is the unique *mediary point*, or degenerate *median circle*, of a *parabolic sphere*, and the tangent plane is its *median plane* (cf. [11, p. 12]). As before, any circle that is orthogonal to the median circle is a *great* circle, and others are *small* circles. (A circle is orthogonal to a degenerate circle when it is incident with it.) Every point will be regarded as being *antipodal* to the mediary point.

Any inversive concirculation that leaves the mediary point fixed, i.e., any transformation that takes great circles into great circles, is a *similarity* of the parabolic sphere. All nondegenerate great circles are parabolic circles, passing through the mediary point, and all real small circles are elliptic circles, which do not pass through the mediary point; a degenerate small circle is an ordinary point. An inversion in a great circle is a *reflection*. A similarity that can be expressed as the product of reflections is an *isometry*. The product of two reflections is a *rotation*.

Removal of the mediary point converts the parabolic sphere into the *Euclidean plane*, with great circles becoming Euclidean lines. This operation may be regarded as the central projection of the parabolic sphere (minus the mediary point) into itself. Alternatively, we may use polar projection and take the section by the median plane of the lines and planes tangent to the parabolic sphere. Each tangent line not in the median plane belongs to a cone of tangent lines whose points of contact lie on a great circle and which all meet the median plane in the same point, the vertex of the cone. Each tangent plane other than the median plane touches the parabolic sphere at a unique ordinary point and meets the median plane in a unique line. If we add the degenerate cone consisting of the central (flat) pencil of lines in the median plane through the mediary point, we obtain the *paratactic plane*. The mediary point of the parabolic sphere is the *absolute* point of the paratactic plane, and lines through it, being tangent to the (degenerate) median circle, are *absolute* lines. All other points and lines are *ideal*.[4]

## 11. Mapping Spheres into Planes

The relationships among the three types of central sphere and the planes obtained from them by central or polar projection may perhaps be better understood through the use of conveniently chosen coordinate systems. Let us define the inversive sphere by means of the matrix

$$H = \backslash -1,1,1,1\backslash = H^{-1},$$

so that when it is represented as an oval quadric in projective space, its locus and envelope equations (7.4) are

$$-x_0^2 + x_1^2 + x_2^2 + x_3^2 = 0 \quad \text{and} \quad -u_0^2 + u_1^2 + u_2^2 + u_3^2 = 0. \tag{11.1}$$

The circle $\langle \hat{\boldsymbol{u}} \rangle$ with plane coordinates $[u] = [u_0, u_1, u_2, u_3]$ is degenerate if $[u\,u] = 0$. A degenerate circle $\langle \hat{\boldsymbol{u}} \rangle$ can be identified with its single real point $\langle \dot{\boldsymbol{u}} \rangle$, which has point coordinates $(\dot{u}) = (-u_0, u_1, u_2, u_3)$. The real points of a circle $\langle \hat{\boldsymbol{a}} \rangle$ are those points $\langle \dot{\boldsymbol{u}} \rangle$ such that $[u\,a] = 0$.

The circles $\langle \hat{\boldsymbol{e}} \rangle$, $\langle \hat{\boldsymbol{p}} \rangle$, and $\langle \hat{\boldsymbol{h}} \rangle$, whose plane coordinates are

$$[e] = [1,0,0,0], \qquad [p] = [1,0,0,-1], \qquad [h] = [0,0,0,1],$$

are respectively imaginary, degenerate, and real. The central spheres whose median planes have the same coordinates are thus respectively elliptic, parabolic, and hyperbolic.

On the elliptic sphere with median circle $\langle \hat{\boldsymbol{e}} \rangle$ the central inversion takes a circle $\langle \hat{\boldsymbol{u}} \rangle$ with coordinates $[u_0, u_1, u_2, u_3]$ into a circle $\langle \hat{\boldsymbol{u}} \rangle_e$ with coordinates $[-u_0, u_1, u_2, u_3]$; $\langle \hat{\boldsymbol{u}} \rangle$ is a great circle if $u_0 = 0$. If $x_0^2 = x_1^2 + x_2^2 + x_3^2$, then $(x_0, x_1, x_2, x_3)$ and $(-x_0, x_1, x_2, x_3)$ are antipodal points. Central projection takes a pair of antipodal points $(\pm x_0, x_1, x_2, x_3)$ on the elliptic sphere into a point $(x_1, x_2, x_3)$ in the elliptic plane $x_0 = 0$, and it takes a great circle $[0, u_1, u_2, u_3]$ into a line $[u_1, u_2, u_3]$, with $(x_1, x_2, x_3)$ and $[u_1, u_2, u_3]$ being incident if and only if

$$x_1 u_1 + x_2 u_2 + x_3 u_3 = 0. \tag{11.2}$$

Polar projection takes a great circle $[0, x_1, x_2, x_3]$ on the elliptic sphere into a point $(x_1, x_2, x_3)$ in the elliptic plane, and it takes a pair of antipodal points $(\pm u_0, u_1, u_2, u_3)$ into a line $[u_1, u_2, u_3]$.

The parabolic sphere with median circle $\langle \hat{\boldsymbol{p}} \rangle$ has the mediary point $(1,0,0,1)$. A circle $\langle \hat{\boldsymbol{u}} \rangle$ with coordinates $[u_0, u_1, u_2, u_3]$ is a great circle if $u_0 + u_3 = 0$. Central projection takes a point

$$\left(\tfrac{1}{2}(x_1^2 + x_2^2 + 1), x_1, x_2, \tfrac{1}{2}(x_1^2 + x_2^2 - 1)\right)$$

[4]It will be seen that a paratactic plane is the two-dimensional dual of a Euclidean plane extended by the customary linear range of points at infinity.

on the parabolic sphere into a point $(x_1, x_2)$ in the Euclidean plane, and it takes a great circle $[u_0, u_1, u_2, -u_0]$ into a line $[u_0, u_1, u_2]$, where $(x_1, x_2)$ is incident with $[u_0, u_1, u_2]$ if and only if

$$u_0 + x_1 u_1 + x_2 u_2 = 0 \tag{11.3}$$

(cf. [1, p. 129; 17, pp. 253–256]). Polar projection takes a great circle $[x_0, x_1, x_2, -x_0]$ on the parabolic sphere into a point $(x_0, x_1, x_2)$ in the paratactic plane $x_0 = x_3$, and it takes an ordinary point $(-u_0, u_1, u_2, u_3)$, where $u_0^2 - u_3^2 = u_1^2 + u_2^2$, into an ideal line $[u_0 + u_3, u_1, u_2]$. The point $(x_0, x_1, x_2)$ and the line $[u_0 + u_3, u_1, u_2]$ are incident if and only if

$$x_0(u_0 + u_3) + x_1 u_1 + x_2 u_2 = 0. \tag{11.4}$$

On the hyperbolic sphere with median circle $\langle \hat{\boldsymbol{h}} \rangle$ the central inversion takes a circle $\langle \hat{\boldsymbol{u}} \rangle$ with coordinates $[u_0, u_1, u_2, u_3]$ into the circle $\langle \hat{\boldsymbol{u}} \rangle_h$ with coordinates $[u_0, u_1, u_2, -u_3]$; $\langle \hat{\boldsymbol{u}} \rangle$ is a great circle if $u_3 = 0$. If $x_3^2 = x_0^2 - x_1^2 - x_2^2$, then the points $(x_0, x_1, x_2, x_3)$ and $(x_0, x_1, x_2, -x_3)$ are antipodal. Central projection takes a pair of antipodal points $(x_0, x_1, x_2, \pm x_3)$ on the hyperbolic sphere into the ordinary point $(x_0, x_1, x_2)$ in the hyperbolic plane $x_3 = 0$, and it takes a great circle $[u_0, u_1, u_2, 0]$ into the ordinary line $[u_0, u_1, u_2]$, with $(x_0, x_1, x_2)$ and $[u_0, u_1, u_2]$ being incident if and only if

$$x_0 u_0 + x_1 u_1 + x_2 u_2 = 0. \tag{11.5}$$

Polar projection takes a great circle $[-x_0, x_1, x_2, 0]$ on the hyperbolic sphere into the ideal point $(x_0, x_1, x_2)$ in the hyperbolic plane, and it takes a pair of antipodal points $(-u_0, u_1, u_2, \pm u_3)$ into the line $[u_0, u_1, u_2]$.

## 12. Circles in a Plane

The great circles of a central sphere, being orthogonal to the median circle, belong to a bundle of circles. This is an elliptic bundle on the elliptic sphere, a parabolic bundle on the parabolic sphere, a hyperbolic bundle on the hyperbolic sphere. Any two great circles belong to a pencil of circles, and all the pairs of antipodal points on a real great circle belong to a *pencil of pairs*, which is *elliptic*, *parabolic*, or *hyperbolic* according as the pairs are separating, tangent, or separated, i.e., according as the central sphere is elliptic, parabolic, or hyperbolic.

On each kind of central sphere an inversion in a great circle has been defined to be a reflection, and every isometry of the central sphere can be expressed as the product of reflections. Central projection transforms a reflection in a great circle of the sphere into a reflection in an ordinary line of an elliptic, Euclidean, or hyperbolic plane; the line is called the *axis* of the reflection. Polar projection transforms a reflection in a great circle into an inversion in an ordinary point of an elliptic plane or an ideal point of a paratactic or hyperbolic plane; the point is called the *center* of the inversion.

The product of reflections in two tangent great circles of the parabolic sphere becomes the product of reflections in two parallel lines of the Euclidean plane, which is a *translation*, or the product of inversions in two parallel points of a paratactic plane, which is a *transvection*. In all other cases the product of two induced reflections or inversions is a *rotation* of the plane, the point of intersection of two axes of reflection being the *center* of the rotation, the line joining two centers of inversion being the *axis* of the rotation. While the axis of a reflection is required to be an ordinary line, the axis of a rotation may be ordinary, absolute, or ideal [17, pp. 352–353].

In the elliptic and hyperbolic planes a reflection in an ordinary line is the same as an inversion in the absolute pole of the line, so that every rotation has both a center and an axis. The center of a rotation being the pole of the axis, any line through the center is perpendicular to the axis. Thus if the axis is an ordinary line, the product of a reflection in the axis and a reflection in an ordinary line through the center is an involutory rotation, or *half-turn*, that leaves invariant both the center and the axis of the given rotation. Any such half-turn also leaves invariant the set of absolute points on the axis and the set of absolute lines through the center.

In Figure 3 the ordinary line $o$ is the axis of a rotation in the hyperbolic plane. The ideal point $O$, the pole of $o$, is the center of the rotation. An ordinary line $p$ passing through $O$ is perpendicular to $o$; the pole of $p$ is an ideal point $P$. Lines $o$ and $p$ meet in an ordinary point $Q$, whose polar is an ideal line $q = OP$. The product of reflections in the lines $o$ and $p$ is a half-turn about $Q$. This half-turn, a harmonic homology, leaves invariant every line through $Q$, including the axis $o$ of the rotation, and every point on $q$, including the center $O$ of the rotation. It interchanges the absolute points $M$ and $N$ and likewise the absolute lines $m$ and $n$.

For any given point in an elliptic, Euclidean, or hyperbolic plane or any given line in an elliptic, paratactic, or hyperbolic plane, all the rotations with that point as center or that line as axis form a group. This group may be the same as, or a

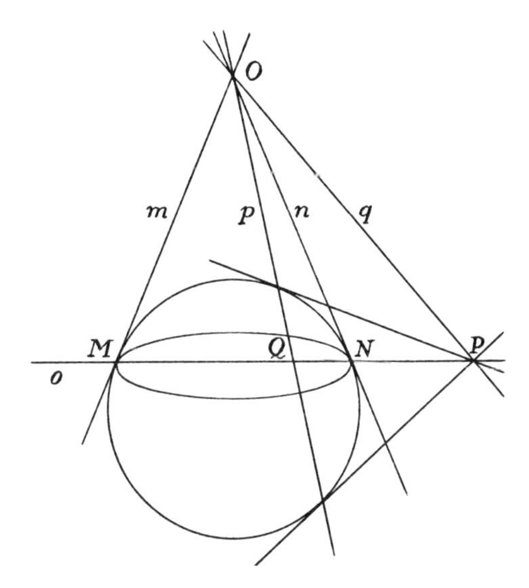


**Figure 3**

proper subgroup of, the group of rotations that leave the center and/or axis invariant. The orbit in the latter group of any point not the center and not lying on the axis or on an absolute line through the center, together with any absolute points on the axis, is a *circle* in the plane. In the former group the orbit of such a point is either a circle or a *branch* of a circle. The center and/or axis are the *center* and/or *axis* of the circle [5, pp. 213–214; 17, pp. 354–356]. For the sake of completeness, the absolute point may be regarded as the center of every circle in the paratactic plane, and the line at infinity may be regarded as the axis of every circle in the extended Euclidean plane. A circle is a conic [5, pp. 117, 215–217] and so can also be identified with its envelope of tangent lines.

In the elliptic and hyperbolic planes a circle can also be defined as a conic that has two (real or imaginary, coincident or distinct) points of contact with or common tangent lines with the absolute circle [16, p. 136]. In any case a circle in a plane is *elliptic* if it has no real absolute points, *parabolic* if it has just one, and *hyperbolic* if it has two; if it has more than two, which can happen only in the hyperbolic plane, it coincides with the absolute circle.

It is a simple matter to classify circles in the elliptic, Euclidean, and paratactic planes: in each case every circle is an elliptic circle. Moreover, in each case every circle in the plane can be obtained from a real (small) circle on the central sphere by a suitable projection. The situation is quite different in the hyperbolic plane, as we shall see in Section 15.

## 13. Circles in the Elliptic Plane

Taking the elliptic case first, and using the same coordinates as before, we see that the (imaginary) absolute circle of the elliptic plane has the equation

$$x_1^2 + x_2^2 + x_3^2 = 0. \tag{13.1}$$

The general conic is $(x)A(x)^{\check{}} = 0$, or

$$a_{11}x_1^2 + a_{22}x_2^2 + a_{33}x_3^2 + 2a_{12}x_1x_2 + 2a_{13}x_1x_3 + 2a_{23}x_2x_3 = 0.$$

If $\det A \neq 0$, the conic is real provided that the conditions

$$a_{11}a_{22} - a_{12}^2 > 0 \quad \text{and} \quad a_{11}\det A > 0 \tag{13.2}$$

are not both true [17, p. 205]. Eliminating $x_3$ between the above equations, we find that the absolute points $(x_1, x_2, x_3)$ of the conic must satisfy the following fourth-degree equation:

$$\begin{aligned}&\left[(a_{11} - a_{33})^2 + 4a_{13}^2\right]x_1^4 + 4\left[(a_{11} - a_{33})a_{12} + 2a_{13}a_{23}\right]x_1^3x_2\\ &\quad + 2\left[2a_{12}^2 + (a_{11} - a_{33})(a_{22} - a_{33}) + 2\left(a_{13}^2 + a_{23}^2\right)\right]x_1^2x_2^2\\ &\quad + 4\left[(a_{22} - a_{33})a_{12} + 2a_{13}a_{23}\right]x_1x_2^3 + \left[(a_{22} - a_{33})^2 + 4a_{23}^2\right]x_2^4 = 0.\end{aligned} \tag{13.3}$$

In order for the conic to be a circle, this equation must have two double roots or one quadruple root. Label the bracketed quantities as follows[5]:

$$A = (a_{11} - a_{33})^2 + 4a_{13}^2,$$
$$B = 2a_{12}^2 + (a_{11} - a_{33})(a_{22} - a_{33}) + 2(a_{13}^2 + a_{23}^2),$$
$$C = (a_{22} - a_{33})^2 + 4a_{23}^2,$$
$$M = (a_{11} - a_{33})a_{12} + 2a_{13}a_{23},$$
$$N = (a_{22} - a_{33})a_{12} + 2a_{13}a_{23}.$$

Then $(x)A(x)^{\wedge} = 0$ is the equation of a circle only if

$$(AC + 2MN)^2 - ACB^2 = 0, \tag{13.4}$$

this being the condition for a quartic

$$Ax_1^4 + 4Mx_1^3x_2 + 2Bx_1^2x_2^2 + 4Nx_1x_2^3 + Cx_2^4$$

to be a square $(ax_1^2 + bx_1x_2 + cx_2^2)^2$. If $a_{13}^2 + a_{23}^2 \neq 0$, this condition is sufficient for the four (imaginary) absolute points of the conic to coincide in pairs. However, if $a_{13} = a_{23} = 0$, it is also necessary to have

$$B^2 = AC \quad \text{or} \quad B^2 = 9AC. \tag{13.5}$$

Since the absolute circle has no real points, it is apparent that all circles in the elliptic plane must be elliptic.

Central projection takes a small circle $\langle \hat{\boldsymbol{a}} \rangle$ on the elliptic sphere into an ordinary circle

$$(x)A(x)^{\wedge} = 0 \quad \text{or} \quad [u]^{\wedge}\tilde{A}[u] = 0 \tag{13.6}$$

in the elliptic plane, while polar projection takes $\langle \hat{\boldsymbol{a}} \rangle$ into an ordinary circle

$$[u]^{\wedge}A[u] = 0 \quad \text{or} \quad (x)\tilde{A}(x)^{\wedge} = 0. \tag{13.7}$$

If $\langle \hat{\boldsymbol{a}} \rangle$ has coordinates $[a_0, a_1, a_2, a_3]$, the matrices $A$ and $\tilde{A}$ are defined by

$$A = \begin{bmatrix} a_1^2 - a_0^2 & a_1a_2 & a_1a_3 \\ a_1a_2 & a_2^2 - a_0^2 & a_2a_3 \\ a_1a_3 & a_2a_3 & a_3^2 - a_0^2 \end{bmatrix}, \quad \tilde{A} = \begin{bmatrix} a_0^2 - a_2^2 - a_3^2 & a_1a_2 & a_1a_3 \\ a_1a_2 & a_0^2 - a_1^2 - a_3^2 & a_2a_3 \\ a_1a_3 & a_2a_3 & a_0^2 - a_1^2 - a_2^2 \end{bmatrix}.$$

It can be shown that a matrix of either of these forms represents a circle and that every circle in the elliptic plane corresponds to such a pair of matrices.

## 14. Euclidean and Paratactic Circles

In the Euclidean plane the general conic has an equation of the form

$$Ax_1^2 + 2Bx_1x_2 + Cx_2^2 + 2Dx_1 + 2Ex_2 + F = 0.$$

[5]The $A$ here is, of course, not the matrix $A$.

This is the equation of a real circle provided that

$$A = C > 0, \qquad B = 0, \quad \text{and} \quad D^2 + E^2 - AF > 0. \tag{14.1}$$

Since the Euclidean plane contains no real absolute points, all circles in the plane must be elliptic. Every circle in the extended Euclidean plane passes through the two (imaginary) circular points at infinity [17, p. 120].

A small circle $\langle \hat{\boldsymbol{a}} \rangle$ on the parabolic sphere with coordinates $[a_0, a_1, a_2, a_3]$ is taken by central projection into an ordinary circle in the Euclidean plane with equation

$$\left(x_1 + \frac{a_1}{a_0 + a_3}\right)^2 + \left(x_2 + \frac{a_2}{a_0 + a_3}\right)^2 = \frac{-a_0^2 + a_1^2 + a_2^2 + a_3^2}{(a_0 + a_3)^2}. \tag{14.2}$$

The same small circle is taken by polar projection into an ideal circle

$$[u]\check{\ } \dot{A}[u] = 0 \quad \text{or} \quad (x)\ddot{A}(x)\check{\ } = 0 \tag{14.3}$$

in the paratactic plane. The matrices $\dot{A}$ and $\ddot{A}$ are defined by

$$\dot{A} = \begin{bmatrix} a_0 - a_3 & -a_1 & -a_2 \\ -a_1 & a_0 + a_3 & 0 \\ -a_2 & 0 & a_0 + a_3 \end{bmatrix}, \quad \ddot{A} = \begin{bmatrix} (a_0 + a_3)^2 & a_0a_1 + a_1a_3 & a_0a_2 + a_2a_3 \\ a_0a_1 + a_1a_3 & a_0^2 - a_2^2 - a_3^2 & a_1a_2 \\ a_0a_2 + a_2a_3 & a_1a_2 & a_0^2 - a_1^2 - a_3^2 \end{bmatrix}.$$

By definition, all circles in the paratactic plane have the same center $(1, 0, 0)$, the absolute point. Since no circle passes through the absolute point, every circle in the paratactic plane is elliptic.

## 15. Circles in the Hyperbolic Plane

The hyperbolic plane presents considerably more variety. Circles may be elliptic, parabolic, or hyperbolic, and their nonabsolute points and tangent lines may be either ordinary or ideal. A circle will be termed *ordinary* if its points and tangent lines (with one or two exceptions in the case of a parabolic or hyperbolic circle) are ordinary, *ideal* if they are ideal.

The absolute circle of the hyperbolic plane has the equation

$$-x_0^2 + x_1^2 + x_2^2 = 0. \tag{15.1}$$

Combining this with the equation of the general conic $(x)A(x)\check{\ } = 0$ and eliminating $x_0$, we obtain the following equation that must be satisfied by the absolute points $(x_0, x_1, x_2)$ of a real conic:

$$\begin{aligned} &\left[(a_{00} + a_{11})^2 - 4a_{01}^2\right]x_1^4 + 4\left[(a_{00} + a_{11})a_{12} - 2a_{01}a_{02}\right]x_1^3x_2 \\ &\quad + 2\left[2a_{12}^2 + (a_{00} + a_{11})(a_{00} + a_{22}) - 2(a_{01}^2 + a_{02}^2)\right]x_1^2x_2^2 \\ &\quad + 4\left[(a_{00} + a_{22})a_{12} - 2a_{01}a_{02}\right]x_1x_2^3 + \left[(a_{00} + a_{22})^2 - 4a_{02}^2\right]x_2^4 = 0. \end{aligned} \tag{15.2}$$

In order for the conic to be a circle, this equation must have either two double roots or one quadruple root. Set

$$\begin{aligned}
A &= (a_{00} + a_{11})^2 - 4a_{01}^2,\\
B &= 2a_{12}^2 + (a_{00} + a_{11})(a_{00} + a_{22}) - 2\left(a_{01}^2 + a_{02}^2\right),\\
C &= (a_{00} + a_{22})^2 - 4a_{02}^2,\\
M &= (a_{00} + a_{11})a_{12} - 2a_{01}a_{02},\\
N &= (a_{00} + a_{22})a_{12} - 2a_{01}a_{02}.
\end{aligned}$$

If $a_{01}^2 + a_{02}^2 \neq 0$, then $(x)A(x)^{\hat{}} = 0$ is the equation of a circle if and only if

$$(AC + 2MN)^2 - ACB^2 = 0. \tag{15.3}$$

When this condition is met, the above quartic can be written as a square $(ax_1^2 + bx_1x_2 + cx_2^2)^2$. The ratios of $x_1$ to $x_2$ that make this expression equal to zero can be found by the quadratic formula. The results can then be expressed in terms of the original coefficients by means of the relations

$$a^2 = A, \quad ab = 2M, \quad b^2 + 2ac = 2B, \quad bc = 2N, \quad c^2 = C.$$

If we define the parameter $s$ by the rule

$$s = \begin{cases} 1 & \text{if } MN > 0 \text{ or if } MN = 0 \text{ and } B > 0,\\ 0 & \text{if } MN = 0 \text{ and } B = 0,\\ -1 & \text{if } MN < 0 \text{ or if } MN = 0 \text{ and } B < 0, \end{cases} \tag{15.4}$$

we can write the general solution in one or both of the following forms:

$$\begin{aligned}
\frac{x_1}{x_2} &= \frac{-(\operatorname{sgn} M)\sqrt{B - s\sqrt{AC}} \pm \sqrt{B - 3s\sqrt{AC}}}{\sqrt{2A}},\\
\frac{x_2}{x_1} &= \frac{-(\operatorname{sgn} N)\sqrt{B - s\sqrt{AC}} \mp \sqrt{B - 3s\sqrt{AC}}}{\sqrt{2C}}.
\end{aligned} \tag{15.5}$$

It can then be shown that:

$$\text{The circle is } \begin{cases} \text{elliptic} & \text{if } B - 3s\sqrt{AC} < 0,\\ \text{parabolic} & \text{if } B - 3s\sqrt{AC} = 0,\\ \text{hyperbolic} & \text{if } B - 3s\sqrt{AC} > 0. \end{cases} \tag{15.6}$$

If $a_{01} = a_{02} = 0$, the conic will be a circle only if (15.3) holds and also either $B^2 = AC$ or $B^2 = 9AC$. In this case we have the criterion:

$$\text{The circle is } \begin{cases} \text{elliptic} & \text{if } B^2 = AC \neq 0,\\ \text{absolute} & \text{if } A = B = C = 0,\\ \text{hyperbolic} & \text{if } B^2 = 9AC,\ A^2 + C^2 \neq 0. \end{cases} \tag{15.7}$$

One way of obtaining circles in the hyperbolic plane is by central or polar projection of small circles on the hyperbolic sphere. Central projection takes a small circle $\langle \hat{\boldsymbol{a}} \rangle$ on the sphere into an ordinary circle

$$(x)A(x)^{\wedge} = 0 \quad \text{or} \quad [u]^{\wedge}\tilde{A}[u] = 0 \tag{15.8}$$

in the plane, while polar projection takes $\langle \hat{\boldsymbol{a}} \rangle$ into an ideal circle

$$[u]^{\wedge}\ddot{A}[u] = 0 \quad \text{or} \quad (x)\ddot{\tilde{A}}(x)^{\wedge} = 0. \tag{15.9}$$

If $\langle \hat{\boldsymbol{a}} \rangle$ has coordinates $[a_0, a_1, a_2, a_3]$, the matrices $A$, $\tilde{A}$, $\ddot{A}$, and $\ddot{\tilde{A}}$ are defined by

$$A = \begin{bmatrix} a_0^2 - a_3^2 & a_0a_1 & a_0a_2 \\ a_0a_1 & a_1^2 + a_3^2 & a_1a_2 \\ a_0a_2 & a_1a_2 & a_2^2 + a_3^2 \end{bmatrix}, \quad \tilde{A} = \begin{bmatrix} a_1^2 + a_2^2 + a_3^2 & -a_0a_1 & -a_0a_2 \\ -a_0a_1 & a_0^2 - a_2^2 - a_3^2 & -a_1a_2 \\ -a_0a_2 & -a_1a_2 & a_0^2 - a_1^2 - a_3^2 \end{bmatrix},$$

$$\ddot{A} = \begin{bmatrix} a_0^2 - a_3^2 & -a_0a_1 & -a_0a_2 \\ -a_0a_1 & a_1^2 + a_3^2 & a_1a_2 \\ -a_0a_2 & a_1a_2 & a_2^2 + a_3^2 \end{bmatrix}, \quad \ddot{\tilde{A}} = \begin{bmatrix} a_1^2 + a_2^2 + a_3^2 & a_0a_1 & a_0a_2 \\ a_0a_1 & a_0^2 - a_2^2 - a_3^2 & a_1a_2 \\ a_0a_2 & a_1a_2 & a_0^2 - a_1^2 - a_3^2 \end{bmatrix}.$$

In each case the type of circle in the hyperbolic plane is the same as that of the corresponding circle on the hyperbolic sphere. Thus central and polar projection respectively take elliptic small circles into ordinary and ideal elliptic circles, or *proper circles*, parabolic small circles into ordinary and ideal parabolic circles, or *horocycles*, and hyperbolic small circles into ordinary and ideal hyperbolic circles, or *equidistant curves*. An equidistant curve has two branches.

Every matrix of one of the above forms represents a circle in the hyperbolic plane that can be obtained by central or polar projection. But not every circle can be so obtained. Whether a circle is obtainable by projection or not, its nonabsolute points are either all ordinary or all ideal, and the same is true of the nonabsolute lines tangent to the circle. If the points are ordinary, so are the lines, and if the lines are ideal, so are the points. It is possible, however, for all but two points to be ideal and all but two lines to be ordinary, yet neither central nor polar projection can yield such a circle, which will therefore be termed *exceptional*.

Figure 4 shows the absolute circle $a$ and three hyperbolic circles with the same axis $MN$ and the same center (the pole of $MN$, not shown): an ordinary circle $o$, an ideal circle $i$, and an exceptional circle $e$.

An exceptional hyperbolic circle can be described as the envelope of lines that all cut a given ordinary line, its axis, at the same acute angle. For this reason it may be called an *equiangular curve*. If the angle is $\pi/4$, the curve is self-polar. For example, consider the circle defined by the equation

$$-x_0^2 + 2x_1x_2 = 0.$$

Its axis is the ordinary line $x_1 = x_2$, or $[0, 1, -1]$, and its center is the ideal point $(0, 1, -1)$. The envelope equation is

$$-u_0^2 + 2u_1u_2 = 0;$$

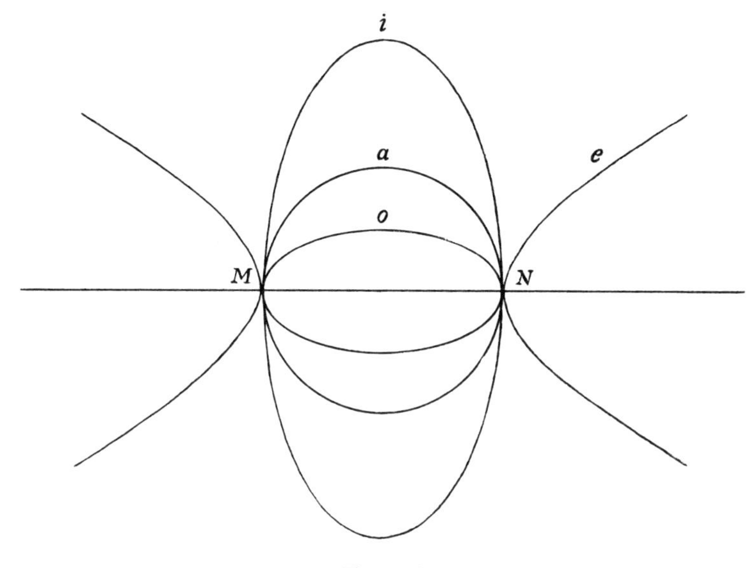


**Figure 4**

the lines $[\pm\sqrt{2u_1u_2}, u_1, u_2]$ are in the envelope. Applying the two-dimensional version of (8.3), we find that the angle between one of these lines and the axis has cosine

$$\frac{|u_1 - u_2|}{\sqrt{2(u_1^2 - 2u_1u_2 + u_2^2)}} = \frac{1}{\sqrt{2}},$$

so that this is an equiangular curve of angle $\pi/4$.

There are thus eight types of circle in the hyperbolic plane: the absolute circle; ordinary elliptic, parabolic, and hyperbolic circles; exceptional hyperbolic circles; and ideal hyperbolic, parabolic, and elliptic circles.

## References

[1] Alexander, H. W., Vectorial inversive and non-Euclidean geometry. *Amer. Math. Monthly* **74** (1967), 128–140.

[2] Cayley, Arthur, A sixth memoir upon quantics. *Philos. Trans. Roy. Soc. London* **149** (1859), 61–90.

[3] Coolidge, J. L., *A Treatise on the Circle and the Sphere*. Clarendon Press, Oxford 1916.

[4] Coxeter, H. S. M., *The Real Projective Plane* (2nd edition). University Press, Cambridge 1955.

[5] Coxeter, H. S. M., *Non-Euclidean Geometry* (5th edition). University of Toronto Press, Toronto 1965.

[6] Coxeter, H. S. M., The inversive plane and hyperbolic space. *Abh. Math. Sem. Univ. Hamburg* **29** (1966), 217–241.

[7] Coxeter, H. S. M., Transformation groups from the geometric viewpoint. In *CUPM Geometry Conference Proceedings*, CUPM Report No. 18, 1967.

[8] Coxeter, H. S. M., *Introduction to Geometry* (2nd edition). Wiley, New York 1969.

[9] Coxeter, H. S. M., *Projective Geometry* (2nd edition). University of Toronto Press, Toronto 1974.

[10] Coxeter, H. S. M., Parallel lines. *Canad. Math. Bull.* **21** (1978), 385–397.

[11] Du Val, Patrick, *Homographies, Quaternions and Rotations*. Clarendon Press, Oxford 1964.

[12] Klein, Felix, Ueber die sogenannte Nicht-Euklidsche Geometrie. *Math. Annalen* **4** (1871), 573–625.

[13] Liebmann, Heinrich, *Nichteuklidische Geometrie*. G. J. Göschen, Leipzig 1905.

[14] MacLane, Saunders and Birkhoff, Garrett, *Algebra*. Macmillan, New York 1967.

[15] Schwerdtfeger, Hans, *Geometry of Complex Numbers*. University of Toronto Press, Toronto 1962.

[16] Sommerville, D. M. Y., *The Elements of Non-Euclidean Geometry*. Bell, London 1914. Dover, New York 1958.

[17] Veblen, Oswald and Young, J. W., *Projective Geometry*, Vol. 2. Ginn, Boston 1918.

# Products of Axial Affinities and Products of Central Collineations[1]

Erich W. Ellers*

## 1. Introduction

In order to obtain information about a mapping, it is often advantageous to factorize it into mappings of a special nature. We shall announce a number of results dealing with the factorization of affinities and projectivities into axial affinities and central collineations, respectively. These mappings are as simple as possible, since they leave all points of a hyperplane fixed. We shall distinguish different types of axial affinities such as shears, affine reflections, and affine hyperreflections, and of central collineations such as elations, projective reflections, and projective hyperreflections. We shall be interested in factorizations with a minimal number of factors for each mapping and also in finding upper bounds for the number of factors needed to express all mappings in a certain group.

It may be of interest that our factorization theorems also lead to characterizations of Pappian geometries among all Desarguesian geometries; see Corollary 8 for projectivities and Theorems 12 and 13 for affinities.

For the proofs, we heavily rely on the knowledge of similar results for the general linear group. Therefore, our account of the geometric consequences will be preceded by a summary of the background in the general linear group. Of special interest here should be Theorems 2 and 3. They have not been stated in this particular form before, but they are clearly contained in Lemmas 1 to 4 of [10]. Special cases of Theorems 2 and 3 have recently been published [2, 7].

[1] Research supported in part by NSERC grant no. A7251.

* Department of Mathematics, University of Toronto, Toronto, Ontario, Canada M5S 1A1.

## 2. The General Linear Group

Let $V$ be a vector space over a field $K$. The dimension of $V$ may be infinite, and the multiplication of $K$ may not be commutative.

We intend to write every element $\pi$ in the general linear group $GL(V)$ of $V$ as a product of simple mappings. For each $\pi \in GL(V)$ we shall also determine the minimal number of factors needed in such a product. An element in $GL(V)$ is simple if it is distinct from the identity, but fixes every vector of some hyperplane in $V$.

With each $\pi \in GL(V)$ we associate two subspaces of $V$, namely $F(\pi) = \{x \in V;\ x^\pi = x\}$, the fix of $\pi$, and $B(\pi) = \{x^\pi - x;\ x \in V\}$, the path of $\pi$. Thus an element $\sigma$ is simple if $F(\sigma)$ is a hyperplane or, equivalently, if $\dim B(\sigma) = 1$. If $\pi_1, \pi_2 \in GL(V)$, then $F(\pi_1\pi_2) \supset F(\pi_1) \cap F(\pi_2)$ and $B(\pi_1\pi_2) \subset B(\pi_1) + B(\pi_2)$.

Let $r \in V\backslash\{0\}$ and $\psi \in V^*\backslash\{0\}$, where $V^*$ denotes the dual space of $V$, such that $r^\psi \neq -1$; then $\sigma: x \to x + x^\psi r$ is a simple mapping in $GL(V)$. Also, $B(\sigma) = Kr$ and $F(\sigma) = \psi^0 = \{x \in V;\ x^\psi = 0\}$. Conversely, if $\sigma \in GL(V)$ is simple, then there are $r$, $\psi$ with these properties. Clearly, for each $\lambda \in K\backslash\{0\}$, $x \to x + x^{\psi\lambda}\lambda^{-1}r$ also describes $\sigma$.

The type of $\sigma$ is the conjugacy class of $1 + r^\psi$: $\operatorname{type}\sigma = \overline{1 + r^\psi}$. Clearly, if $K$ is commutative, then $\operatorname{type}\sigma = \det\sigma$.

Simple transformations $\sigma$ of $\operatorname{type}\sigma = 1$ and $\operatorname{type}\sigma = -1$ are called transvections and reflections, respectively.

**Theorem 1.** *If $\pi \in GL(V)$ and* $\operatorname{codim} F(\pi) = t$, *then there are simple mappings $\sigma_i \in GL(V)$, $i = 1, \ldots, t$, such that $\pi = \sigma_1 \ldots \sigma_t$, and $t$ is the smallest number for which such a factorization of $\pi$ exists.*

For each $\pi \in GL(V)$ we define $\bar{\pi} \in GL(V)/F(\pi)$ by $\bar{\pi}: x + F(\pi) \to x^\pi + F(\pi)$. For $\epsilon$ in the center of $K\backslash\{0\}$, the mapping $\eta_\epsilon: x \to \epsilon x$ is called a homothety.

If we factorize as in Theorem 1, we may even insist that each factor but one have a certain type. If we do so, then occasionally we need one extra factor.

**Theorem 2.** *If $\pi \in GL(V)$,* $\operatorname{codim} F(\pi) = d$, *and $\epsilon_i \in K\backslash\{0\}$ for $i = 1, \ldots, d$, then there are simple mappings $\sigma_i, \delta \in GL(V)$ with* $\operatorname{type}\sigma_i = \bar{\epsilon}_i$, *$i = 1, \ldots, t$, such that $\pi = \sigma_1 \ldots \sigma_t\delta$. If $t$ is the smallest number for which such a factorization exists, then $t = d - 1$ if $\bar{\pi} = \eta_\epsilon$ and $\epsilon_1 = \cdots = \epsilon_{d-1} = \epsilon$, or if $\bar{\pi}$ is not a homothety; $t = d$ otherwise.*

For a special case see [7], where all $\epsilon_i = -1$.

The subgroup of $GL(V)$ that consists of all $\pi \in GL(V)$ with finite codimension of $F(\pi)$, will be denoted by $FL(V)$. For $\pi \in FL(V)$, the dimension of $B(\pi)$ is finite. Therefore, $\det_{B(\pi)}|\pi$ is defined. We put $\det\pi = \det_{B(\pi)}|\pi$. Then $\pi \to \det\pi$ is a homomorphism of $FL(V)$ into $K^*/C(K^*)$, where $K^*$ is the multiplicative group of $K$ and $C(K^*)$ the commutator subgroup of $K^*$.

If $\sigma \in GL(V)$ is simple and type $\sigma = \bar{\epsilon}$ for some $\epsilon \in K^*$, then $\det \sigma = \epsilon C(K^*)$.

In a similar way as in the preceding theorem we obtain a factorization where we prescribe the determinant of each factor but one.

**Theorem 3.** *If $\pi \in GL(V)$, $\operatorname{codim} F(\pi) = d$, and $\gamma_i \in K^*/C(K^*)$, $i = 1, \ldots, d$, then there are simple mappings $\sigma_i, \delta \in GL(V)$ with $\det \sigma_i = \gamma_i$, $i = 1, \ldots, t$, such that $\pi = \sigma_1 \ldots \sigma_t \delta$. If $t$ is the smallest number for which such a factorization exists, then $t = d - 1$ if $\bar{\pi} = \eta_\epsilon$ and $\epsilon \in \gamma_i$ for $i = 1, \ldots, d-1$, or if $\bar{\pi}$ is not a homothety; $t = d$ otherwise. Clearly,* $\det \delta = \det \pi \cdot \prod_{i=1}^{t} \gamma_i^{-1}$.

In case $K$ is commutative, this theorem yields a result of F. S. Cater [2].

Let $\Gamma$ be a cyclic subgroup of $K^*/C(K^*)$, $\gamma$ a generator of $\Gamma$, and $m$ the order of $\Gamma$. We define $G_m = \{\pi \in FL(V);\ \det \pi \in \Gamma\}$, and we say $G_m$ is a hyperreflection group. The group $G_1$ is the special linear group $SL(V)$. A simple transformation $\rho \in G_m$ with $\det \rho = \gamma$ will be called a hyperreflection. If $K$ is commutative, then $\det \rho = \operatorname{type} \rho$. Therefore, a hyperreflection is a reflection if $m = 2$; it is a transvection if $m = 1$. This is in general not true if $K$ is not commutative.

**Theorem 4.** *Let $\pi \in G_m$, $\dim B(\pi) = d$, and $\det \pi = \gamma^k$, $0 < k \leqslant m$. Then $\pi$ is a product of hyperreflections $\rho_i$: $\pi = \rho_1 \ldots \rho_t$. If $t$ is the smallest number for which such a factorization exists, then $t \equiv k \bmod m$ and $d \leqslant t < d + m$ if $\bar{\pi}$ is not a homothety or $\bar{\pi} = \eta_\epsilon$ with $\epsilon \in \gamma$; $d \leqslant t - 1 < d + m$ otherwise. Furthermore, if $B$ is a hyperplane and if $B(\pi) \subset B$, then $B(\rho_i) \subset B$.*

This theorem includes a number of interesting special cases. Namely, for commutative $K$ it determines the minimal number of transvections needed to express any element in $SL(V)$, and similarly it determines the minimal number of reflections needed to express any element in $G_2$. Clearly, $SL(V)$ is contained in $G_2$.

## 3. The Affine Subgroup

We shall turn our attention to a subgroup of $GL(V)$ that yields a geometric interpretation. Let $B$ be a fixed hyperplane of $V$; then $N = \{\pi \in GL(V);\ B(\pi) \subset B\}$ is called the affine subgroup of $GL(V)$.

A simple transformation $\sigma$ in $N$ is called an axial affinity if $F(\sigma) \neq B$, and a translation if $F(\sigma) = B$. An axial affinity $\sigma$ is a shear if $B(\sigma) \subset F(\sigma)$, it is a strain if $B(\sigma) \not\subset F(\sigma)$.

**Theorem 5.** *Assume $\pi \in N$ and $\pi$ is not a translation. If $\dim B(\pi) = d$, then $\pi$ is a product of $d$ (but not fewer) axial affinities.*

A transformation $\pi \in GL(V)$ is called a big $\epsilon$-dilatation if $\bar{\pi} = \eta_\epsilon$.

A product of a translation $\tau$ and a shear $\sigma$ is called a parabolic rotation if $B(\tau) \not\subset F(\sigma)$.

**Theorem 6.** *Let $\pi \in N$ with* $\operatorname{codim} F(\pi) = d$ *and $\epsilon_i \in K\backslash\{0\}$ for $i = 2, \ldots, d+1$. Then there are axial affinities $\rho_1, \ldots, \rho_t$ such that* $\operatorname{type} \rho_i = \bar{\epsilon}_i$ *for $i = 1, \ldots, t$ and $\pi = \rho_1\rho_2 \ldots \rho_t$. Let $t$ be the smallest positive integer for which such a factorization of $\pi$ exists; then $t = d + 1$ if*

(i) *$\pi$ is a translation,*
(ii) *$\pi$ is a parabolic rotation and $\epsilon_2 = 1$, or*
(iii) *$\pi$ is a big $\epsilon$-dilatation and $\epsilon_i \neq \epsilon$ for some $i$ with $1 < i \leqslant r$.*

*In all other cases, $t = r$.*

The group $G'_m = G_m \cap N$ is called an affine hyperreflection group.

If $\pi \in G'_m$ and if $m \neq 1$, then all factors of the factorization of $\pi$ into hyperreflections given in Theorem 4 may be chosen in $G'_m$. Namely, if $B(\rho_i) \subset B$, $F(\rho_i) = B$ implies that $\rho_i$ is a transvection and $\det \rho_i = C(K^*)$.

For $m = 1$, we shall assume that $K$ is commutative. Then $\operatorname{type} \rho = \det \rho$, and an axial affinity $\rho$ with $\det \rho = 1$ is a shear. If $\pi \in G'_1$, i.e. $\pi$ is an equiaffinity, then $\det \pi = 1$. Now we obtain a factorization of $\pi$ from Theorem 6. We assume there that $\det \rho_i = 1$ for $i = 2, \ldots, t$. Then under our assumptions, also $\det \rho_1 = 1$, i.e., we have factorized $\pi$ into shears.

## 4. Projectivities

The projective geometry connected with $V$ will be denoted by $P(V)$.

Let $p \in PGL(V)$; then $F(p)$ denotes the set of fixed points in $P(V)$. In general, $F(p)$ is not a subspace of $P(V)$.

Let $p \in PGL(V)$. Then $\dim F(p)$ is the maximal number that occurs as dimension for any subspace contained in $F(p)$. Also, $\operatorname{codim} F(p)$ is the minimal number that occurs as codimension for any subspace contained in $F(p)$.

If $\operatorname{codim} F(p) = 0$, then $p$ is called a central collineation. If $\operatorname{codim} F(p) = -1$, then $p$ is the identity.

The element $p \in PGL(V)$ will be called exceptional if $\dim F(p) = 0$ and if for every point $P(Kx)$ in $F(p)$ and $p = P(\pi)$, we have $x^\pi = \lambda x$, where $\lambda \notin Z(K)$, the center of $K$.

Let $p \in PGL(V)$ and $p = c_1, \ldots, c_r$, where $c_i$ are central collineations for $i = 1, \ldots, r$. The smallest number $r$ of any decomposition of $p$ of that kind will be called the length of $p$. This length will be denoted by $l(p)$.

**Theorem 7.** *Let $p \in PGL(V)$ and* $\operatorname{codim} F(p) = r - 1 > 0$. *Then $l(p) = r$ if $p$ is not exceptional. If $p$ is exceptional, then* $\dim P(V) = r$ *and $l(p) = r + 1$.*

**Corollary 8.** *An $n$-dimensional Desarguesian geometry is Pappian if and only if every projectivity $p$ with* $\dim F(p) = 0$ *is a product of exactly $n$ central collineations.*

If $\rho$ is a simple transformation and $\operatorname{type} \rho = \bar{\epsilon}$, then we define $\operatorname{type} P(\rho) = \bar{\epsilon}$. If $\rho$ is a transvection, then $P(\rho)$ is an elation. If $\rho$ is a reflection, then $P(\rho)$ is a projective reflection.

Let $p \in PGL(V)$ and $\operatorname{codim} F(p) = r - 1$. Then $p$ is a big $\epsilon$-homology if there is a big $\epsilon$-dilatation $\pi$ such that $P(\pi) = p$ and $\operatorname{codim} F(\pi) = r$.

**Theorem 9.** *Let $p \in PGL(V)$ with $\operatorname{codim} F(p) = r - 1 > 0$, and $\epsilon_i \in K\setminus\{0\}$ for $i = 1, \ldots, r+1$. Then $p = c_1 \ldots c_t$, where $c_1$ is a central collineation and* type $c_i = \bar{\epsilon}_i$ *for $i = 2, \ldots, t$. Let $t$ be the smallest positive integer for which such a factorization exists. Then $t = r$ if*

(i) *$p$ is neither exceptional nor a big $\epsilon$-homology,*
(ii) *$p$ is a big $\epsilon$-homology and* type $c_i = \bar{\epsilon}$ *for $1 \leqslant i \leqslant r$,*
(iii) *$p$ is a big $\epsilon$-homology, $2r = \dim P(V) + 1$, and* type $c_i = \overline{\epsilon^{-1}}$ *for $1 \leqslant i \leqslant r$.*

*In all other cases, $t = r + 1$.*

**Corollary 10.** *Let $p \in PGL(V)$ and $\operatorname{codim} F(p) = r - 1 > 0$. Let $t$ be the number of factors $c_i$ in Theorem 9. Then $t \leqslant \dim P(V) + 1$.*

The upper bound given in Corollary 10 cannot be decreased. We need $\dim P(V) + 1$ factors if $\dim F(p) = -1$ or if $p$ is exceptional. It is easy to verify that $\dim P(V)$ factors are sufficient if $p$ is a big $\epsilon$-homology.

Projective hyperreflection groups will be denoted by $PG_m(V)$. If $\rho$ is a hyperreflection, then $P(\rho)$ is a projective hyperreflection.

**Theorem 11.** *Every projectivity $p \in PG_m(V)$ is a product of at most* $\dim P(V) + m$ *projective hyperreflections.*

## 5. Skew Affinities

Let $B$ be a hyperplane of $V$. Then $P(B)$ is a projective hyperplane. The group $M = \{\pi \in GL(V);\ B^\pi \subset B\}$ is called the preaffine group, and $P(M) = A$ the affine group. Obviously, the group $H(V)$ of homotheties is contained in $M$. Every projectivity $p \in A$ is called an affinity. Clearly, $A = \{p \in PGL(V);\ P(B)^p \subset P(B)\}$. Also, $A$ is isomorphic to $M/H(V)$; namely, $\pi \to P(\pi)$ is a homomorphism of $GL(V)$ onto $PGL(V)$ whose kernel is $H(V)$.

The group $P(N)$ is isomorphic to $N$. This isomorphism relates central collineations in $P(N)$ to simple transformations in $N$. The elements in $P(N)$ corresponding to translations, shears, and strains in $N$ will also be called translations, shears, and strains, respectively.

Every element in $P(N)$ is called a Pappian affinity; every element in $A\setminus P(N)$ is called a skew affinity.

Axial affinities are Pappian affinities.

**Theorem 12.** *The affine group $A$ is isomorphic to the affine subgroup $N$ if and only if $K$ is commutative.*

**Theorem 13.** *Let $p \in A$ be a skew affinity with* $\operatorname{codim} F(p) = r - 1$ *and $\epsilon_i \in K\setminus\{0\}$ for $i = 2, \ldots, r$. Then $p = c_1 c_2 \ldots c_t$, where $c_t$ is a skew affine dilata-*

*tion,* $c_1, \ldots, c_{t-1}$ *are axial affinities, and* type $c_i = \bar{\epsilon}_i$ *for* $1 < i < t$. *Let* $t$ *be the smallest number for which such a factorization exists. Then* $t = r + 1$ *if*

(i) $p$ *is exceptional,*
(ii) ${}_{P(B)}|p$ *is a big* $\epsilon$*-homology and* $\epsilon_i \neq \epsilon$ *for some* $i$ *with* $1 < i \leqslant r - 1$.

*In all other cases,* $t = r$.

## References


[1] Bachmann, F., *Aufbau der Geometrie aus dem Spiegelungsbegriff* (2nd edition). Springer-Verlag, New York–Heidelberg–Berlin 1973.

[2] Cater, F. S., Products of central collineations. *Linear Algebra Appl.* **19** (1978), 251–274.

[3] Coxeter, H. S. M., *Introduction to Geometry* (2nd edition). John Wiley & Sons, New York–Sydney–London–Toronto 1969.

[4] Coxeter, H. S. M., Affinely regular polygons. *Abh. Math. Sem. Univ. Hamburg* **34** (1969), 38–58.

[5] Coxeter, H. S. M., Products of shears in an affine Pappian plane. *Rend. Matematica (1) Ser. VI* **3** (1970), 161–166.

[6] Dieudonné, J., Sur les générateurs des groupes classiques. *Summa Bras. Mathem.* **3** (1955), 149–178.

[7] Djoković, D. Ž. and Malzan, J., Products of reflections in the general linear group over a division ring. *Linear Algebra Appl.* **28** (1979), 53–62.

[8] Ellers, E. W., The length problem for the equiaffine group of a Pappian geometry. *Rend. Matematica (2) Ser. VI* **9** (1976), 327–336.

[9] Ellers, E. W., Decomposition of orthogonal, symplectic, and unitary isometries into simple isometries. *Abh. Math. Sem. Univ. Hamburg* **46** (1977), 97–127.

[10] Ellers, E. W., Decomposition of equiaffinities into reflections. *Geometriae Dedicata* **6** (1977), 297–304.

[11] Ellers, E. W., Factorization of affinities. *Canadian J. Math.* **31** (1979), 354–362.

[12] Ellers, E. W., Projectivities as products of homologies, elations, and projective reflections.

[13] Ellers, E. W., Skew affinities. *Geometriae Dedicata.*

[14] Gruenberg, K. W. and Weir, A. J., *Linear Geometry* (2nd edition). Graduate Texts in Math., Vol. 49. Springer-Verlag, New York–Heidelberg–Berlin 1977.

[15] O'Meara, O. T., *Lectures on Linear Groups*. CBMS Regional Conference Series in Math., No. 22. (1974).

[16] Scherk, P., On the decomposition of orthogonalities into symmetries. *Proc. Amer. Math. Soc.* **1** (1950), 481–491.

# Normal Forms of Isometries

G. Ewald*

## 1. Introduction

Let $\sigma$ be an isometry in a Euclidean vector space $V = V_n(\mathbb{R}, f)$ of dimension $n$, the bilinear form $f$ representing the ordinary inner product. One of the basic theorems proved in linear algebra states that a basis of $V$ can be chosen in such a way that $\sigma$ is represented by a matrix

$$\begin{bmatrix} A_1 & & & & & & & & 0 \\ & A_2 & & & & & & & \\ & & \ddots & & & & & & \\ & & & A_k & & & & & \\ & & & & \pm 1 & & & & \\ & & & & & 1 & & & \\ & 0 & & & & & 1 & & \\ & & & & & & & & 1 \end{bmatrix} \tag{1}$$

where

$$A_j = \begin{pmatrix} \cos\theta_j & \sin\theta_j \\ -\sin\theta_j & \cos\theta_j \end{pmatrix}, \qquad j = 1, \ldots, k.$$

The proof is achieved by using eigenvalue theory.

Let us look at this "normal form" of an isometry in a somewhat more geometrical manner. It is an elementary theorem of planar Euclidean geometry that a rotation $A_j$ can be expressed as a product of two reflections $a_j, b_j$ in lines $\overline{a}_j, \overline{b}_j$ passing through the center $O$ of rotation (Figure 1). If $\overline{a}_j, \overline{b}_j$ are considered

* Department of Mathematics, Ruhr-Universität, Bochum, Germany.

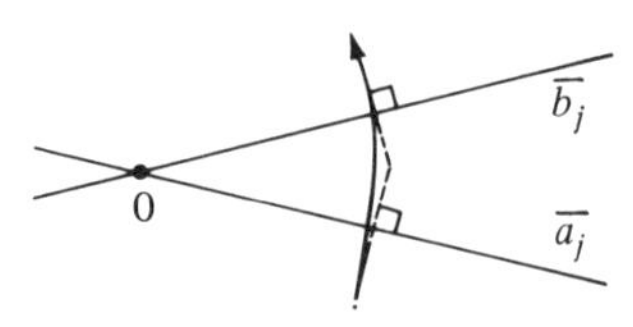


**Figure 1**

lines in $V$, the product $a_jb_j$ of the reflections $a_j, b_j$ in $\overline{a}_j, \overline{b}_j$, respectively, represents the rotation given by a matrix which is obtained from (1) by replacing all $A_l$, $l \neq j$, by unit matrices and $-1$ by 1. Its axis has dimension $n-2$. If we denote by $P_O$ the point symmetry in $O$ and let $P_O^\epsilon = P_O$ if $\epsilon = 1$, $=$ identity if $\epsilon = 0$, we can express $\sigma$ as follows:

$$\sigma = (a_1b_1)(a_2b_2) \dots (a_k, b_k)P_O^\epsilon, \tag{2}$$

where the following condition is satisfied:

Any two planes spanned by pairs $\overline{a}_j$, $\overline{b}_j$ and $\overline{a}_l$, $\overline{b}_l$ of different lines, $j \neq l$, are perpendicular to each other. (3)

Statements (2), (3) can be used to define normal forms of a class of isometries also in case $V$ is a vector space $V(\mathbb{R}, f)$ of arbitrary dimension provided by an inner product $f$ (Hilbert space or pre-Hilbert space). In fact, we need not even refer to a coordinate system; the selection of the lines $\overline{a}_j, \overline{b}_j$ replaces the choice of an appropriate basis of $V$. To be more precise, (2) represents an element of the "small" orthogonal group $O(\mathbb{R}, f)$ of all isometries leaving a subspace of finite dimension or finite codimension (0 included) pointwise fixed (axis). Only in case $V$ is finite-dimensional do $O_n(\mathbb{R}, f)$ and $O(\mathbb{R}, f)$ coincide.

Proceeding from vector-space to ordinary Euclidean, elliptic, or hyperbolic geometry of arbitrary dimension, we consider the "small" group $O(\mathbb{R}, \perp)$ of isometries that corresponds to $O(\mathbb{R}, f)$. If we denote the point symmetry at $\overline{P}$ by $P$, and let $P^\epsilon = P$ or $=$ identity depending on whether $\epsilon = 1$ or $\epsilon = 0$, we can show (using (2), (3)) that any element of $O(\mathbb{R}, \perp)$ is expressible as

$$(a_1b_1)(a_2b_2) \dots (a_kb_k)P^\epsilon Q \tag{4}$$

where now $a_j, b_j$ are reflections in lines $\overline{a}_j, \overline{b}_j$ passing through $P$ and satisfy (3).

We may ask whether (4) together with (3) can be obtained without using eigenvalue theory. Moreover, we wish to know whether an analogous result for more general orthogonal groups can be proved. In presenting an answer we shall not aim at "wild" generalizations, but rather at generalizations which turn out to be appropriate to the problem of normal forms.

## 2. Elliptic Groups

Let us restrict our attention mainly to elliptic spaces. We define an elliptic space (like other spaces) by group-theoretic means, following up the tradition of F. Bachmann's school. The starting point is not to define points and lines, but to

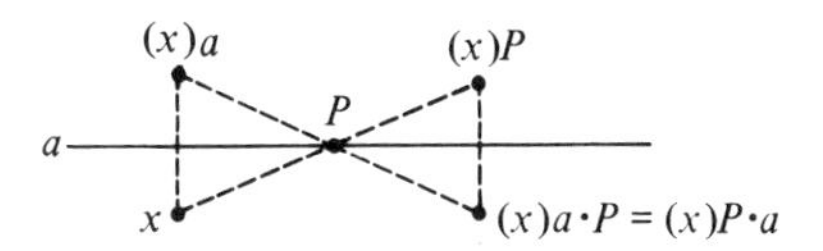


**Figure 2**

consider point reflections and line reflections as (involutoric) generating elements of an abstract group. Any point symmetry $P$ can also be called "point" and any line symmetry $a$ can be called "line," the incidence of $P$ and $a$ being given by requiring $P \cdot a$ to be an involutoric element of the group (in particular $P \cdot a = a \cdot P$; see Figure 2). This illustrates why in (4) we use reflections in lines and points rather than rotations. Every isometry is defined as a product of finitely many such reflections; by a theorem of Cartan and Dieudonné, in the finite-dimensional case $O_n(\mathbb{R}, \perp)$, the full group of isometries is obtained.

If the space is elliptic, the set of generators can even be restricted to the set of reflections in points. So let a group $\mathfrak{G}$ be given generated by a set $\mathfrak{R}$ of involutoric elements, called *points* $P, Q, \ldots$. We assume $\mathfrak{R}$ to be invariant under all inner automorphisms of $\mathfrak{G}$ and to have a trivial center. If a product $\alpha\beta$ of two elements of $\mathfrak{G}$ is involutoric, we write $\alpha \mid \beta$. If $\alpha\beta$ is not involutoric, we write $\alpha \dagger \beta$. In particular, we call $P$ and $P^*$ *conjugate* to each other if $P|P^*$ and call any product $PP^*$ of two conjugate points a *line*. If $Q \mid PP^*$, we say $Q$ *lies on* $PP^*$ (or is incident to $PP^*$, etc.). If $PP^* \mid QQ^*$, we call the lines $PP^*$, $QQ^*$ *perpendicular*. Axioms (Friedlein [3]):

**(E1)** Given two different points $P$, $Q$, there exists a uniquely determined $P^*$ such that $Q \mid PP^*$ and $Q \dagger P, P^*$.

**(E2)** Let $R \mid (P, Q)$ but $R \dagger P, Q$, and let $T \dagger (P, Q)$. Then

(1) $PQR$ is a point $S$.
(2) $(T, P)(T, Q)(T, R)$ is a line $(T, S')$ (Figure 3).

**(E3)** There exist points $P, Q, R$ no two of which are conjugate to each other and such that $R \dagger (P, Q)$. On every line $(P, Q)$ there exists an $R$ not conjugate to $P$ or $Q$.

Under the assumptions (E1)–(E3) we call $\mathfrak{G}$ an *elliptic group*.

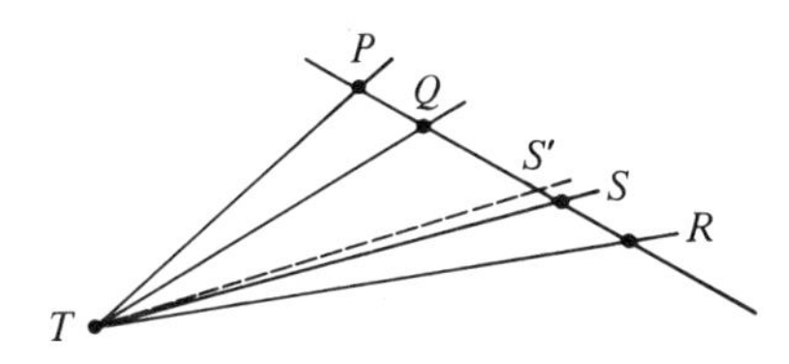


**Figure 3**

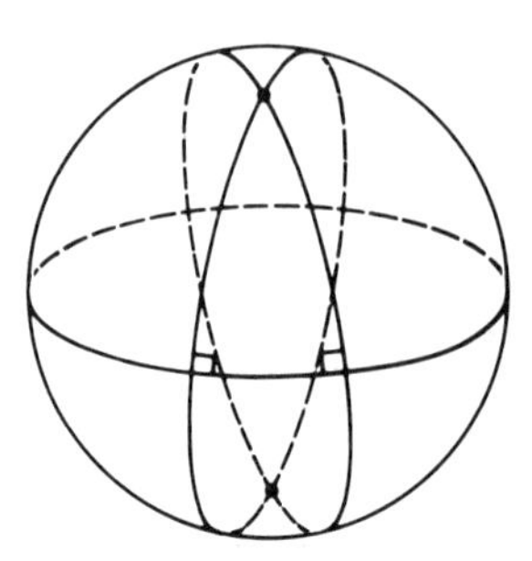

**Figure 4**

As an example we consider the sphere model of an elliptic plane: Points are represented by pairs of antipodal points on the surface of a sphere $S$, lines by great circles of $S$ (Figure 4).

More generally, if, in addition to (E1)–(E3), we require any two lines to intersect, $\mathfrak{G}$ is the group defining an elliptic plane in the sense of F. Bachmann [3]. (Concerning the finite-dimensional case see also J. Ahrens [1] and H. Kinder [11].)

**Theorem 1** (Friedlein [8]). *Any elliptic group $\mathfrak{G}$ is isomorphic to a projective orthogonal group $PO(K, f)$ of all projectivities leaving a nonsingular form $f$ of index 0 invariant and leaving an axis (set of fixed points) of finite dimension or finite codimension.*

An analog of Theorem 1 for "absolute" geometry is found in Ewald [5]. Both papers make use of the work of Lenz [12] in order to show that $\mathfrak{G}$ is isomorphic to a subgroup of $PO(K, f)$. The theorem of Cartan and Dieudonné mentioned above shows that this subgroup is the group $PO(K, f)$ itself.

If an elliptic group $\mathfrak{G}$ is given, the normal form (4), (3) can be written by using point symmetries only:

$$P^{\epsilon}(P_1 Q_1) \dots (P_k Q_k), \quad \text{where } P_i Q_i \mid P_j, Q_j \quad \text{if } i \neq j. \tag{5}$$

It can be shown, however, that not in all elliptic groups can each element be written in a normal form. Let, for example, $K$ be the field $\mathbb{Q}$ of rational numbers and let $V_4(\mathbb{Q}, f)$ be the 4-dimensional metric vector space whose bilinear form is represented by the matrix

$$\begin{bmatrix} 2 & & & 0 \\ & 1 & & \\ & & 1 & \\ 0 & & & 1 \end{bmatrix}$$

The 3-dimensional elliptic space defined by $V_4(\mathbb{Q}, f)$ does not permit all isometries to be represented in a normal form.

**Theorem 2** (Friedlein [10]). *Let $\mathfrak{G}$ be an elliptic group. The following conditions are equivalent*:

(a) *Each element of $\mathfrak{G}$ can be represented in a normal form.*
(b) *Any two nonintersecting lines are intersected by a line perpendicular to both of them.*
(c) *In the representation by $PO(K, f)$ according to Theorem 1, $K$ is Pythagorean (that is, all sums of squares are squares, and $-1$ is not a square), and $f$ restricted to any finite-dimensional subspace can be represented by a matrix*

$$\begin{bmatrix} \alpha & & 0 \\ & \ddots & \\ 0 & & \alpha \end{bmatrix}, \qquad 0 \neq \alpha \in K.$$

Theorem 2 thus shows which generalization of ordinary elliptic space is appropriate to the problem of normal forms: (b) or (c) is not only sufficient but also necessary for the existence of normal forms for all isometries. Furthermore, Friedlein's proof of (a) from (b) contains an affirmative answer to another question posed above: Can (4), (3) be obtained without eigenvalue theory? The basic idea of this proof may be summarized as follows:

According to (5), the pairs $P_i$, $Q_i$ must lie on nonintersecting 1-dimensional sides of what is called a polar simplex: a finite collection of mutually conjugate points and the subspaces spanned by any subset of this collection (Figure 5 for dimension 3 and $k = 2$). If $\sigma = R_1 R_2 \ldots R_m$ is an arbitrary element of $\mathfrak{G}$, one first replaces $\alpha$ by a product $R_1' R_2' \ldots R_k'$ in which the $R_i'$ are in "general position," that is, none of them is contained in the (projective) subspace spanned by the others. This can be done by a theorem in [5] (see also [6]).

Now $R_1' R_2'$ is, according to (E2), replaced by a product $R_1'' R_2''$, where $R_1''$, $R_2''$ lie on $(R_1', R_2')$, in such a way that $R_1'' \mid R_3'$. In this way more and more pairs of adjacent points in the product that represents $\sigma$ are placed in "conjugate position." Proceeding carefully, it turns out that under assumption (b) a normal form can be obtained.

Concerning the equivalence of (b) and (c), a rather intricate calculation must be performed (Friedlein [9]).

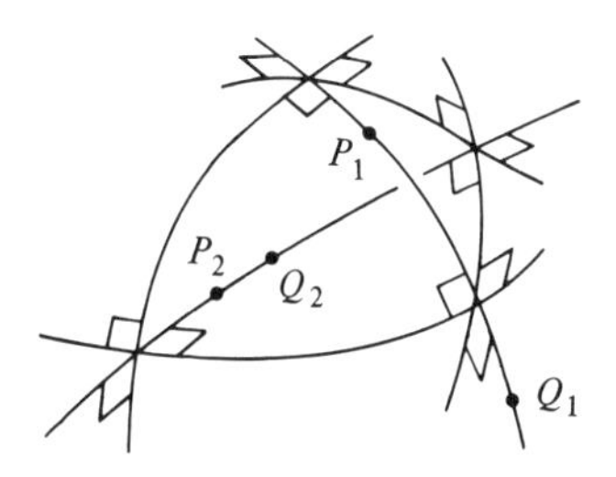


**Figure 5**

## 3. Final Remarks

It is an open question if (a) can be deduced from (c) by using eigenvalue theory. Though the field $K$ and the form $f$ are rather special, there are difficulties involved.

Concerning a generalization of Theorem 2 the following can be said. In [5] general spaces are defined which are the analogs of Bachmann's metric planes for arbitrary dimension, that is, they represent the common basis for Euclidean and non-Euclidean geometry ("absolute" geometry). So far the analog of Theorem 2 is only known for the Euclidean case. However, it seems probable that in all cases the problem of normal forms can be solved in an analogy to Theorem 2.

### References

[1] Ahrens, J., Begründung der absoluten Geometrie des Raumes aus dem Spiegelungsbegriff. *Math. Z.* **71** (1959), 154–185.

[2] Artin, E., *Geometric Algebra*. Interscience Publ., New York 1957.

[3] Bachmann, F., *Aufbau der Geometrie aus dem Spiegelungsbegriff*, Springer-Verlag, Berlin–Göttingen–Heidelberg 1973.

[4] Coxeter, H. S. M., *Introduction to Geometry*, Wiley, New York–London 1961.

[5] Ewald, G., Spiegelungsgeometrische Kennzeichnung euklidischer und nichteuklidischer Räume beliebiger Dimension. *Abh. Math. Sem. Univ. Hamburg* **41** (1974), 224–251.

[6] Ewald, G., Über Bewegungen in der absoluten Geometrie. *J. of Geometry* **9** (1977), 1/2 Birkhäuser Verlag, Basel.

[7] Ewald, G., *Geometry. An Introduction*. Wadsworth Publishing Company, Belmont, California 1971.

[8] Friedlein, H. R., Elliptische Bewegungsgruppen und orthogonale Gruppen vom Index 0. *Abh. Math. Sem. Univ. Hamburg* (to appear).

[9] Friedlein, H. R., Über gemeinsame Lote windschiefer Geraden in elliptischen Räumen. Manuscript 1978.

[10] Friedlein, H. R., Normalformen für Bewegungen elliptischer Räume. Manuscript 1979.

[11] Kinder, H., Elliptische Geometrie endlicher Dimension. *Arch. Math.* **21** (1970), 515–527.

[12] Lenz, H., Über die Einführung einer absoluten Polarität in die projektive und affine Geometrie des Raumes. *Math. Ann.* **128** (1954), 363–372.

# Finite Geometries with Simple, Semisimple, and Quasisimple Fundamental Groups[1]

B. A. Rosenfeld*
N. I. Haritonova†
I. N. Kashirina†

Deep connections between the classical theory of simple Lie groups and the theory of finite simple groups, discovered by C. Chevalley [1], lead to the construction of finite geometries analogous to geometries of classical Lie groups.

## 1. Geometries of Classical Lie Groups

The complex simple Lie groups of the four infinite series $A_n$, $B_n$, $C_n$, and $D_n$ have well-known geometrical meaning. The complex group $A_n$ is the group $SL_{n+1}(i)$ of complex unimodular $(n+1)$-matrices or, locally isomorphic to it, the group of collineations of complex projective $n$-space $P_n(i)$; the complex groups $B_n$ and $D_n$ are respectively the groups $O_{2n+1}(i)$ and $O_{2n}(i)$ of complex orthogonal $(2n+1)$- and $2n$-matrices or, locally isomorphic to them, the groups of motions of complex quadratic elliptic $2n$- and $(2n-1)$-spaces $S_{2n}(i)$ and $S_{2n-1}(i)$; the complex group $C_n$ is the group $Sp_{2n}(i)$ of complex symplectic $2n$-matrices ("*Komplex-Gruppe*," group of a linear complex) or, locally isomorphic to it, the group of symplectic transformations of complex symplectic $(2n-1)$-space $Sp_{2n-1}(i)$. The geometrical meaning of the compact real simple Lie groups is analogous: the compact real group $A_n$ is the group $SU_{n+1}(i)$ of complex unimodular unitary $(n+1)$-matrices or, locally isomorphic to it, the group of motions of complex Hermitian elliptic space $\bar{S}_n(i)$ (not to be confused with quadratic space $S_n(i)$); the compact real groups $B_n$ and $D_n$ are respectively the groups $O_{2n+1}$ and

[1] Section 1 of this paper was written by Boris A. Rosenfeld (Moscow). Section 2 by Nadezhda I. Haritonova (Cheboksary), and Section 3 by Irina N. Kashirina (Tula).

* Institute of History of Natural Science and Technology, Academy of Sciences, Staropanskiĭ per. 1/5, Moscow 103012, USSR.

† c/o B. A. Rosenfeld, same address as in footnote above (*).

$O_{2n}$ of real orthogonal $(2n+1)$- and $2n$-matrices or, locally isomorphic to them, the groups of motions of real quadratic elliptic $2n$- and $(2n-1)$-spaces $S_{2n}$ and $S_{2n-1}$; the compact real group $C_n$ is the group $U_n(i,j)$ of quaternionic unitary $n$-matrices or, locally isomorphic to it, the group of motions of quaternionic Hermitian elliptic $(n-1)$-space $\overline{S}_{n-1}(i,j)$. The noncompact real simple Lie groups $A_n$ are the groups of collineations of the real projective $n$-space $P_n$ and of the quaternionic projective $(n-1)/2$-space $P_{(n-1)/2}(i,j)$ and the groups of motions of the complex Hermitian hyperbolic $n$-spaces $S_n(i)$; the noncompact real simple Lie groups $B_n$ are the real quadratic hyperbolic $2n$-spaces ${}^1S_{2n}$; the noncompact real simple Lie groups $C_n$ are the groups of symplectic transformations of the real symplectic $(2n-1)$-space $Sp_{2n-1}$ and the groups of motions of the quaternionic Hermitian hyperbolic $(n-1)$-space ${}^1\overline{S}_{n-1}(i,j)$; the noncompact real simple Lie groups $D_n$ are the groups of motions of the real quadratic hyperbolic $(2n-1)$-spaces ${}^1S_{2n-1}$ and the group of symplectic transformations of the quaternionic Hermitian symplectic $(n-1)$-space $\overline{Sp}_{n-1}(i,j)$. The space ${}^1S_n$ is the well-known Lobachevskian space. The spaces $\overline{S}_n(i)$ and ${}^1\overline{S}_n(i)$ were discovered by G. Fubini and E. Study at the beginning of the 20th century. All the above-mentioned spaces are described by B. A. Rosenfeld [2], and the last one by L. V. Rumyanceva [3].

The spaces $S_n$, ${}^1S_n$, and $S_n(i)$ are respectively spaces $P_n$ and $P_n(i)$ with absolute quadrics

$$(x,x)=a_{\mu\nu}x^{\mu}x^{\nu}=0, \qquad a_{\mu\nu}=a_{\nu\mu}, \tag{1}$$

reduced in the cases of $S_n$ and $S_n(i)$ to the form $\sum(x^{\mu})^2=0$ and in the case of ${}^1S_n$ to the form $-\sum(x^{a})^2+\sum(x^{u})^2=0$. The spaces $\overline{S}_n(i)$, ${}^1\overline{S}_n(i)$, $\overline{S}_n(i,j)$, and ${}^1\overline{S}_n(i,j)$ are respectively spaces $P_n(i)$ and $P_n(i,j)$ with absolute Hermitian quadrics

$$(x,x)=\bar{x}^{\nu}a_{\mu\nu}x^{\mu}=0, \qquad a_{\mu\nu}=\bar{a}_{\nu\mu}. \tag{2}$$

The spaces $Sp_{2n-1}$ and $Sp_{2n-1}(i)$ are respectively the spaces $P_{2n-1}$ and $P_{2n-1}(i)$ with absolute correlation

$$u_{\mu}=a_{\mu\nu}x^{\nu}, \qquad a_{\mu\nu}=-a_{\nu\mu} \tag{3}$$

reduced to the form $u_{2\mu}=x^{2\mu+1}$, $u_{2\mu+1}=-x^{2\mu}$; the space $\overline{Sp}_n(i,j)$ is the space $P_n(i,j)$ with absolute correlation

$$u_{\mu}=\bar{x}^{\nu}a^{\mu\nu}, \qquad a_{\mu\nu}=-\bar{a}_{\nu\mu} \tag{4}$$

reduced to the form $u_{\mu}=\bar{x}^{\mu}i$.

Each two points $x$ and $y$ of the spaces $S_n$, ${}^1S_n$, $S_n(i)$, $\overline{S}_n(i)$, ${}^1\overline{S}_n(i)$, $\overline{S}_n(i,j)$, and ${}^1\overline{S}_n(i,j)$ have metric invariant

$$\mu=\frac{(x,y)(y,x)}{(x,x)(y,y)} \tag{5}$$

and distance $\delta: \cos^2\delta=\mu$, where $(x,y)$ is the polarized form of $(x,x)$. Each two $m$-planes $X$ and $Y$ of these spaces also have metric invariants—the eigenvalues $\mu_a=\cos^2\delta_a$ of the matrix

$$M=(UX)^{-1}(UY)(VY)^{-1}(VX), \tag{6}$$

where $X = (x_a^\mu)$ is the rectangular matrix of coordinates of the basic points $x_a$ of the plane $X$, $Y = (y_a^\mu)$ is the analogous matrix for the plane $Y$, and $U = (u_\mu^a)$ is the rectangular matrix of tangential coordinates of hyperplanes $u^a$ corresponding to the points $x_a$ in the polarities $u_\mu = a_{\mu\nu}x^\nu$ and $u_\mu = \bar{x}^\nu a_{\mu\nu}$ respective to the absolute quadrics. Each two lines $X$ and $Y$ of the spaces $Sp_{2n-1}$ and $Sp_{2n-1}(i)$ have symplectic invariant defined by the formula (6) where $m = 1$; each two points of the space $\overline{Sp}_n(i,j)$ have symplectic invariant (5), where $(x,x) = \bar{x}^\nu a_{\mu\nu}x^\mu$.

The isomorphisms $A_1 = B_1 = C_1$, $B_2 = C_2$, $D_2 = B_1 \times B_1$, and $A_3 = D_3$ determine the interpretations $\bar{S}_1(i) = S_2$, ${}^1\bar{S}_1(i) = {}^1S_2$ (Poincaré's interpretation of the Lobachevskian plane), $P_1 = {}^1S_2$ (Hesse's "*Übertragungsprinzip*"), $\bar{S}_1(i,j) = S_4$, ${}^1\bar{S}_1(i,j) = {}^1S_4$, $SP_3 = {}^2S_4$, $S_3 = S_2 \times S_2$ (Fubini and Study's interpretation), ${}^1S_3 = S_2(i)$ (Kotelnikov and Study's interpretation), $\overline{Sp}_1(i,j) = S_2 \times {}^1S_2$, $\bar{S}_3(i) = S_5$, ${}^1\bar{S}_3(i) = \overline{Sp}_2(i,j)$, ${}^2\bar{S}_3(i) = {}^2S_5$, $P_3 = {}^3S_5$ (Plücker's interpretation), $P_1(i,j) = {}^1S_5$ (Study's interpretation) [2].

If we define, analogously to the spaces $P_n(i)$, $S_n(i)$, $\bar{S}_n(i)$, $P_n(i,j)$, $\bar{S}_n(i,j)$, and $\overline{Sp}_n(i,j)$, the spaces $P_n(e)$, $S_n(e)$, $\bar{S}_n(e)$, $P_n(i,e)$, $\bar{S}_n(i,e)$, and $\overline{Sp}_n(i,e)$ over the algebra $\mathbf{R}(e)$ of "double numbers" (split complex numbers) $a + be, e^2 = 1$, and $\mathbf{R}(i,e)$ of "antiquaternions" (split quaternions) $a + bi + ce + df$, $i^2 = -1$, $e^2 = 1$, $ie = -ei = f$ ($\mathbf{R}(e)$ is isomorphic to the direct sum $\mathbf{R} \oplus \mathbf{R}$ of two fields $\mathbf{R}$ of real numbers; $\mathbf{R}(i,e)$ is isomorphic to the algebra $\mathbf{R}_2$ of real 2-matrices), we obtain new interpretations $P_n(e) = P_n \times P_n$, $S_n(e) = S_n \times S_n$, $\bar{S}_n(e) = P_n$ (points of $\bar{S}_n(e)$ are images of $O$-couples (points + hyperplanes) of $P_n$), $P_n(i,e) = P_{2n+1}$, $\bar{S}_n(i,e) = Sp_{2n+1}$, $\overline{Sp}_n(i,e) = S_{2n+1}$ (points of the first spaces of the last three equalities are images of lines of the second spaces). The last three interpretations are special cases of the interpretations of the projective, elliptic, and symplectic $n$-spaces over the algebra $\mathbf{R}_m$ of real $m$-matrices as multiplicities of $(m-1)$-planes of, respectively, $P_{nm+m-1}$, $S_{nm+m-1}$, and $Sp_{nm+m-1}$ [2].

The compact group $G_2$ is the group of automorphisms of the alternative skew-field $\mathbf{R}(i,j,l)$ of octaves and the group of motions of $G$-elliptic 6-space $Sg_6$ ($S_6$ with 14-dimensional fundamental group) [4]. The compact group $F_4$ is the group of motions of the octave Hermitian elliptic plane $\bar{S}_2(i,j,l)$ [5]; the compact group $E_6$ is the group of motions of the bioctave Hermitian elliptic space $\bar{S}_2(i,j,l,I)$ [2] (bioctaves are complex octaves). The noncompact group $G_2$ is the group of automorphisms of the alternative algebra $\mathbf{R}(i,j,e)$ of antioctaves (split octaves) and the group of motions of $G$-hyperbolic 6-space ${}^3Sg_6$ (${}^3S_6$ with 14-dimensional fundamental group); the noncompact groups $F_4$ are the groups of motions of the octave Hermitian hyperbolic plane ${}^1\bar{S}_2(i,j,l)$ and of the antioctave Hermitian elliptic plane $\bar{S}_2(i,j,e)$; the noncompact groups $E_6$ are the groups of collineations of octave and antioctave projective planes $P_2(i,j,l)$ [5] and $P_2(i,j,e)$ as well as the groups of motions of the bioctave Hermitian hyperbolic plane ${}^1\bar{S}_2(i,j,l,I)$ and of the biantioctave Hermitian elliptic plane $\tilde{\bar{S}}_2(i,j,e,I)$ [2].

The noncompact simple Lie groups are obtained from the compact ones by means of Cartan's algorithm: if the compact group $G$ has an involutory automorphism, its Lie algebra $\mathbf{G}$ admits the Cartan decomposition $\mathbf{G} = \mathbf{H} \oplus \mathbf{E}$, in which this automorphism has the form $\mathbf{G} \rightarrow \mathbf{H} \oplus (-\mathbf{E})$, then the Lie algebra $\mathbf{G}' = \mathbf{H} \oplus i\mathbf{E}$ is the Lie algebra of the corresponding noncompact Lie group $G'$ having the same complex form $G(i)$ as $G$.

The "pseudo-Cartanian algorithm" $\mathbf{G}' \rightarrow \mathbf{G}^\circ = \mathbf{H} \oplus \epsilon\mathbf{E}$, where $\epsilon^2 = 0$, transforms each simple Lie group $G$ into a semisimple Lie group $G^\circ$. The quasisimple groups are the groups of motions of Euclidean space $R_n$; pseudo-Euclidean space ${}^lR_n$; quasielliptic space $S_n^m$ ($S_n^0 = R_n$, $S_n^{n-1} = R_n^*$ corresponding to $R_n$ in the principle of duality of $P_n$); quasihyperbolic spaces ${}^{kl}S_n^m$ (${}^{0l}S_n^0 = {}^lR_n$, ${}^{l0}S_n^{n-1} = {}^lR_n^*$); Hermitian Euclidean and pseudo-Euclidean spaces $\bar{R}_n(i)$, $\bar{R}_n\,(i, j)$, ${}^l\bar{R}_n(i)$, and ${}^l\bar{R}_n(i, j)$; Hermitian quasielliptic and quasihyperbolic spaces $\bar{S}_n^m(i)$, $\bar{S}_n^m(i, j)$, ${}^{kl}\bar{S}_n^m(i)$, and ${}^{kl}\bar{S}_n^m(i, j)$; spaces over dual numbers $a + b\epsilon$, $\epsilon^2 = 0$, $P_n(\epsilon)$, $S_n(\epsilon)$, ${}^lS_n(\epsilon)$, $\bar{S}_n(\epsilon)$, ${}^l\bar{S}_n(\epsilon_2)$, $Sp_{2n-1}(\epsilon)$, $\bar{S}p_{2n-1}(\epsilon)$; spaces over semiquaternions $a + bi + c\epsilon + d\eta$ ($i^2 = -1$, $\epsilon^2 = 0$, $i\epsilon = -\epsilon i = \eta$) and semiantiquaternions $a + be + c\epsilon + d\eta$ ($e^2 = 1$, $\epsilon^2 = 0$, $e\epsilon = -\epsilon e = \eta$), namely, $P_n(i,\epsilon)$, $\bar{S}_n(i,\epsilon)$, ${}^l\bar{S}_n(i,\epsilon)$, $\bar{S}p_n(i,\epsilon)$, $P_n(e,\epsilon)$, $\bar{S}_n(e,\epsilon)$, $\bar{S}p_n(e,\epsilon)$, and others; and planes over semioctaves and semiantioctaves $P_2(i, j,\epsilon)$, $\bar{S}_2(i, j,\epsilon)$, ${}^1\bar{S}_2(i, j,\epsilon)$, $P_2(i, e,\epsilon)$, $\bar{S}_2(i,e,\epsilon)$, and others [2, 6–7].

## 2. Geometries of Finite Simple and Semisimple Groups

Finite geometries over Galois fields have been studied by many mathematicians. The projective space $P_n(q)$ over the Galois field $GF(q)$ was defined by O. Veblen and W. Bussey in 1906 [8] (see also [9–10]). On the basis of the theories of quadrics and null polarities in $P_n(q)$ and of Hermitian quadrics in $P_n(q^2)$, Yu. G. Sokolova has defined non-Euclidean spaces $S_{2n}(q)$, $S_{2n-1}(q)$, and ${}'S_{2n-1}(q)$ [11], symplectic space $Sp_{2n-1}(q)$ [12], and Hermitian non-Euclidean space $\bar{S}_n(q^2)$ [13]. The spaces $S_{2n}(q)$, $S_{2n-1}(q)$, and ${}'S_{2n-1}(q)$ are respectively the spaces $P_{2n}(q)$ and $P_{2n-1}(q)$ in which the nondegenerate quadric (1) is given. In the case of $S_{2n}(q)$ the equation (1) is reduced to the form

$$\sum x^\nu x^{2n-\nu-1} + (x^{2n})^2 = 0; \tag{7}$$

in the cases of $S_{2n-1}(q)$ and ${}'S_{2n-1}(q)$ the equation (1) is reduced to the forms, respectively,

$$\sum x^\nu x^{2n-\nu-1} = 0 \tag{8}$$

and

$$\sum x^\nu x^{2n-\nu-3} + (x^{2n-2})^2 + \alpha(x^{2n-1})^2 = 0, \tag{9}$$

where $\alpha$ is a nonsquare of the field $GF(q)$. The space $Sp_{2n-1}(q)$ is the space $P_{2n-1}(q)$ in which the correlation (3) is given. The space $\bar{S}_n(q^2)$ is the space $P_n(q^2)$ in which the nondegenerate quadric (2) is given, where $\bar{\alpha} = \alpha^q$.

Each two points $x$ and $y$ of the spaces $S_{2n}(q)$, $S_{2n-1}(q)$, ${}'S_{2n-1}(q)$, and $\bar{S}_n(q^2)$ have metric invariant (5); each two $m$-planes $X$ and $Y$ of these spaces with basic points $x_a$ and $y_a$ have metric invariants eigenvalues of the matrix (6); each two lines $X$ and $Y$ of the space $Sp_{2n-1}(q)$ have symplectic invariant defined by the same formula (6) where $m = 1$.

The groups of collineations of $P_n(q)$, the groups of motions of $S_{2n}(q)$, $S_{2n-1}(q)$, $'S_{2n-1}(q)$, and $\bar{S}_n(q^2)$ (i.e., the groups of collineations respectively of $P_{2n}(q)$, $P_{2n-1}(q)$, and $P_n(q^2)$ conserving the metric invariants $\mu$ of their points), and the group of symplectic transformations of $Sp_{2n-1}(q)$ (i.e., the group of collineations of $P_{2n-1}(q)$ conserving the symplectic invariant $M$ of their lines) are finite simple groups and, together with the cyclic groups $Z_p$, exhaust all infinite series of finite simple groups. These groups are factor groups respectively of the groups $SL_{n+1}(q)$, $O_{2n+1}(q)$, $O_{2n}(q)$, $'O_{2n}(q)$, $SU_{n+1}(q^2)$, and $Sp_{2n}(q)$ by their normal divisors consisting of scalar matrices. These groups are indicated also respectively $A_n(q)$, $B_n(q)$, $D_n(q)$, ${}^2D_n(q)$, ${}^2A_n(q)$, and $C_n(q)$ [14].

The isomorphisms $A_1(q) = B_1(q) = C_1(q)$, $D_2(q) = B_1(q) \times B_1(q)$, $B_2(q) = C_2(q)$, $A_3(q) = D_3(q)$, ${}^2D_2(q) = B_1(q^2)$, and ${}^2A_3(q) = {}^2D_3(q)$ determine the interpretations $\bar{S}_1(q^2) = S_2(q)$, $S_3(q) = S_2(q) \times S_2(q)$, $S_4(q) = Sp_3(q)$, $P_3(q) = S_5(q)$, ${}^1S_3(q) = S_2(q^2)$, and $\bar{S}_3(q^2) = {}'S_5(q)$ analogous to the interpretations mentioned in Section 1.

Finite spaces may be built also over finite rings with divisors of zero. N. I. Haritonova [15] has built projective and non-Euclidean spaces over the "Galois square" $[GF(q)]^2$ isomorphic to the direct sum $GF(q) + GF(q)$. This ring may be called a "double extension" of $GF(q)$; its elements may be written in the form $a + be$, $e^2 = 1, a, b \in GF(q)$. Let us note that the extension of $GF(q)$ with a polynomial modulus $p_2(x)$ is isomorphic to $GF(q^2)$ when $p_2(x)$ has no solution in $GF(q)$, is isomorphic to $[GF(q)]^2$ when $p_2(x)$ has two solutions in $GF(q)$, and is a "dual extension" of $GF(q)$ (i.e., its elements may be written in the form $a + b\epsilon$, $\epsilon^2 = 0, a, b \in GF(q)$) when $p_2(x)$ has a single solution in $GF(q)$. Let us denote the double and dual extensions of $GF(q)$ respectively by $GF(q,e)$ and $GF(q,\epsilon)$.

J. Thas [16] has built projective spaces over the total ring $GF_m(q)$ of $m$-matrices over $GF(q)$ and has proved (independently of [2]) that projective $n$-space over $GF_m(q)$ admits an interpretation as a variety of $(m-1)$-planes of $P_{nm+m-1}(q)$; the groups of collineations of $P_n[GF_m(q)]$ and $P_{nm+m-1}$ are isomorphic.

N. I. Haritonova [15] has proved that the space $P_n(q,e)$ over $GF(q,e)$ admits the interpretation as $P_n(q) \times P_n(q)$, and the spaces $S_{2n}(q,e)$, $S_{2n-1}(q,e)$, $'S_{2n-1}(q,e)$, $Sp_{2n-1}(q,e)$, and $\bar{S}_n(q^2,e)$ admit analogous interpretations as $S_{2n}(q_1) \times S_{2n}(q_1), \ldots, \bar{S}_n(q^2) \times \bar{S}_n(q^2)$; the group of collineations of $P_n(q,e)$ and the groups of motions and symplectic transformations of $S_{2n}(q, e), \ldots, \bar{S}_n(q^2,e)$ are isomorphic to direct products of two groups of collineations of $P_n(q)$ and, respectively, to the direct product of two groups of motions or symplectic transformations of $S_{2n}(q), \ldots, \bar{S}_n(q^2)$; and all cardinalities of $P_n(q, e)$, $S_{2n}(q,e), \ldots, \bar{S}_n(q^2,e)$ are equal to the squares of the corresponding cardinalities of $P_n(q)$, $S_{2n}(q), \ldots, \bar{S}_n(q^2)$. The space $\tilde{S}_n(q,e)$ defined analogously to $\bar{S}_n(q^2)$, but with $\alpha = a + be \to \tilde{\alpha} = a + be$, admits the interpretation as a multiplicity of $O$-couples of $P_n(q)$; the group of motions of $\tilde{S}_n(q,e)$ is isomorphic to the group of collineations of $P_n(q)$.

Analogous spaces may be defined over the ring $GF_m(q)$. These spaces admit interpretations as varieties of $(m-1)$-planes of $S_{nm+m-1}(q)$, $Sp_{nm+m-1}(q)$, etc.; the groups of motions or symplectic transformations of these spaces are isomorphic to the corresponding fundamental groups of $S_{nm+m-1}(q)$, $Sp_{nm+m-1}(q)$, etc.

The spaces over rings with divisors of zero have "adjacent points" incident with more than one line, "adjacent lines" incident with more than one point, and analogous "adjacent $m$-planes."

Quaternionic and antiquaternionic extensions of $GF(q)$ are isomorphic to $GF_2(q)$, and the spaces over these rings are special cases of spaces over $GF_m(q)$.

Very interesting are the octave extensions of $GF(q)$. This alternative ring is isomorphic to the antioctave extension of $GF(q)$ and therefore has divisors of zero. Let us denote this ring by $GF(q,i,j,l)$. The group of automorphisms of this ring is the finite simple group $G_2(q)$, which is also the group of motions of the $G$-elliptic 6-space $Sg_6(q)$. Over this ring are defined the Hermitian elliptic plane $\bar{S}_2(q,i,j,l)$ and projective plane $P_2(q,i,j,l)$ investigated together with octave planes over an arbitrary field by J. R. Faulkner [17]. The fundamental groups of these planes are $F_4(q)$ and $E_6(q)$. There is also the finite simple group ${}^2E_6(q)$, which is the group of motions of the "bi-Hermitian" elliptic plane $\bar{S}_2(q^2,i,j,l)$.

The most general semisimple finite associative ring is the direct sum $GF_{m_1}(q_1) \oplus \ldots \oplus GF_{m_r}(q_r)$; the elements of this ring may be defined as $(\alpha_1, \alpha_2, \ldots, \alpha_r)$, where $\alpha_i \in GF_{m_i}(q_i)$. The points of projective $n$-spaces over this ring are defined as $(x^0, x^1, \ldots, x^n)$, where $x^\nu$ are elements of this ring determined up to multiplication $x^\nu \rightarrow x^\nu k$, $k$ an element of this ring not zero or a divisor of zero. Collineations of this space are transformations $'x^\mu = a^\mu_\nu f(x^\nu)$ where $x \rightarrow f(x)$ is an automorphism of this ring. This space admits the interpretation as $P_n[GF_{m_1}(q_1)] \times \cdots \times P_n[GF_{m_r}(q_r)]$ or as a product of varieties of $(m_i-1)$-planes of $P_{nm_i+m_i-1}(q_i)$. The group of collineations of this space is a direct sum of groups of collineations of $P_{nm_i+m_i-1}(q_i)$ and therefore is a finite semisimple group. Analogously we may define non-Euclidean and symplectic spaces over this ring and give their interpretations; the fundamental groups of these spaces also are finite semisimple groups.

## 3. Geometries of Finite Quasisimple Groups

The finite simple groups $A_n(q)$, $B_n(q)$, $C_n(q)$, $D_n(q)$, ${}^2A_n(q)$, ${}^2D_n(q)$, $G_2(q)$, $F_4(q)$, $E_6(q)$, and ${}^2E_6(q)$ mentioned above may be defined also as groups of automorphisms of corresponding finite Lie algebras. If we apply to these Lie algebras the pseudo-Cartanian algorithm defined in Section 1, we obtain finite quasisimple Lie algebras. The groups of automorphisms of these Lie algebras are finite quasisimple groups. Special cases of these groups are the groups of motions of quasi-non-Euclidean spaces $S^m_n(q)$, $'S^m_n(q)$, and $\bar{S}^m_n(q^2)$ defined by Yu. G. Sokolova [11, 13] and the groups of quasisymplectic transformations of quasisymplectic spaces $Sp^{2m-1}_{2n-1}(q)$ defined by the same author [12].

Another special case of these groups is the group of collineations of projective $n$-space $P_n(q,\epsilon)$ over the dual extension $GF(q,\epsilon)$ of $GF(q)$ studied by S. B.

Kapralova [18]. The ring $GF(q,\epsilon)$ is a Hjelmslev ring, and therefore the space $P_n(q,\epsilon)$ is a Hjelmslev finite space [9, pp. 291–300].

If the number of $m$-planes of $P_n(q)$ is

$$N_n^m(q) = \frac{(q^{n+1}-1)(q^n-1)\dots(q^{n-m+1}-1)}{(q^{m+1}-1)(q^m-1)\dots(q-1)},$$

the number of $m$-planes of $P_n(q^2)$, which is the "complex extension" of $GF(q)$, is

$$N_n^m(q^2) = \frac{(q^{2(n+1)}-1)(q^{2n}-1)\dots(q^{2n-2m+2}-1)}{(q^{2(m+1)}-1)(q^{2m}-1)\dots(q^2-1)}$$

$$= \frac{(q^{n+1}-1)(q^{n+1}+1)(q^n-1)(q^n+1)\dots(q^{n-m+1}-1)(q^{n-m+1}+1)}{(q^{m+1}-1)(q^{m+1}+1)(q^m-1)(q^m+1)\dots(q-1)(q+1)};$$

the number of $m$-planes of $P_n(q,e)$, or the double extension of $P_n(q)$, is

$$N_n^m(q,e) = \frac{(q^{n+1}-1)^2(q^n-1)^2\dots(q^{n-m+1}-1)^2}{(q^{m+1}-1)^2(q^m-1)^2\dots(q-1)^2};$$

so the number of $m$-planes of $P_n(q,\epsilon)$ is

$$N_n^m(q,\epsilon) = \frac{(q^{m+1}-1)q^{n+1}(q^n-1)q^n\dots(q^{n+1}-1)q^{n+1}}{(q^{m+1}-1)q^{m+1}(q^m-1)q^m\dots(q-1)q}.$$

Analogously we define spaces $S_{2n}(q,\epsilon)$, $S_{2n-1}(q,\epsilon)$, $'S_{2n-1}(q,\epsilon)$, $Sp_{2n-1}(q,\epsilon)$, and $\bar{S}_n(q^2,\epsilon)$. Just as, in the cases of the complex and double extensions $GF(q^2)$ and $GF(q,e)$ of $GF(q)$, all cardinalities of spaces over these rings are obtained from corresponding cardinalities for spaces over $GF(q)$ by replacing all factors $q^k-1$ respectively with $(q^k-1)(q^k+1)$ and $(q^k-1)^2$, so, in the case of the dual extension $GF(q,\epsilon)$ of $GF(q)$, all cardinalities of all spaces over this ring are obtained from cardinalities for spaces over $GF(q)$ by replacing all factors $q^k-1$ with products $(q^k-1)q^k$.

## References


[1] Chevalley, C., Sur certains groupes simples. *Tôhoku Math. Journal* **7** (1955), 14–66.

[2] Rosenfeld, B. A., *Neevklidovy geometrii*. Moscow Gos. Izd. Tehn. Teop. Lit. 1955.

[3] Rumyanceva, L. V., Kvaternionnaya simplekticheskaya geometriya. In *Trudy Seminara po Vektornomu i Tenzornomu Analizu pri Moskovskom Gos. Universitete*, **12** (1963), 287–314.

[4] Adamushko, N. N., Geometriya prostyh i kvaziprostyh grupp Li klassa *G. Uchenye Zapiski Moskovskogo Oblastnogo Pedagogicheskogo Instituta* **253** (No. 3, 1969), 23–42.

[5] Freudenthal, H., *Oktaven, Ausnahmegruppen und Oktavengeometrie*. Utrecht, Math. inst. Rijksuniversiteit 1951.

[6] Rosenfeld, B. A., *Neevklidovy prostranstva*. Moscow, Nauka 1969.

[7] Klimanova, T. M., Unitarnye poluellipticheskie prostranstva. *Izvestiya Akademii Nauk Azerbaydzhanskoy SSR Ser. Fiz. Mat. i Tehn. Nauk* (No. 3, 1963), 21–29.

[8] Veblen, O. and Bussey, W., Finite projective geometries. *Transactions Amer. Math. Society* **7** (1906), 241–259.

[9] Dembowski, P., *Finite Geometries*. Berlin–Heidelberg–New York, Springer 1968.

[10] Sokolova, Yu. G., Proektivnye prostranstva Galua. *Uchenye Zapiski Moskovskogo Oblastnogo Pedagogicheskogo Instituta* **173** (No. 2, 1967), 57–66.

[11] Sokolova, Yu. G., Neevklidovy i kvazineevklidovy prostranstva Galua. *Ibid.*, 67–82.

[12] Sokolova, Yu. G., Simplekticheskie i kvazisimplekticheskie prostranstva Galua. *Matematicheskie Zametki* **6** (No. 1, 1969), 119–123.

[13] Sokolova, Yu. G., Ermitovy neevklidovy i kvazineevklidovy prostranstva Galua. In *Uchenye Zapiski Matematicheskih Kafedr Tul'skogo Pedagogicheskogo Instituta*. Tula 1968; pp. 109–114.

[14] Tits, J., Groupes simples et géométries associées. In *Proceedings of the International Congress of Mathematicians*. Stockholm 1964; pp. 197–221.

[15] Haritonova, N. I., Proektivnye i èllipticheskie prostranstva nad kvadratom Galua. In *Aktual'nye problemy geometrii i ee prilozheniy*, Vol. 2. Cheboksary, Churash. Gos. Univ. 1976; pp. 44–49.

[16] Thas, J., The $m$-dimensional projective space $P_m(M_n(GF(q)))$ over the total matrix algebra $M_n(GF(q))$ of the $n \times n$-matrices with elements in the Galois field $GF(q)$. *Rend. Matem.* **4** (No. 3, 1971), 449–452.

[17] Faulkner, J. R., *Octonion Planes Defined by Quadratic Jordan Algebras*. Memoirs Amer. Math. Society, No. 104, Providence 1970.

[18] Kapralova, S. B., Dual'nye prostranstva Galua. In *Trudy Seminara Kafedry Geometrii Kazanskogo Gos. Universiteta*, Vol. 12. Kazan 1979; pp. 38–44.

# Motions in a Finite Hyperbolic Plane

Cyril W. L. Garner*

## 1. Introduction

Let $\mathcal{P}$ be a finite projective plane of arbitrary odd order $n$, and let $\pi$ be a regular polarity of $\mathcal{P}$: that is, a polarity for which there exists an integer $s = s(\pi)$ such that every line containing two or more absolute points of $\pi$ contains $s + 1$ absolute points [11, p. 247]. Baer [1] has shown that the absolute points form an oval when $n$ is odd and nonsquare, and Segre [13] has shown that every oval in a Desarguesian projective plane is a conic. This implies $s = 1$, and just as in the real projective plane, there are two disjoint classes of nonabsolute points:

$$\mathcal{O} = \{\text{outer points, or points having 2 absolute lines}\},$$

$$\mathcal{I} = \{\text{inner points, or points having 0 absolute lines}\};$$

and two disjoint classes of nonabsolute lines:

$$o = \{\text{outer lines, or lines having 0 absolute points}\},$$

$$\iota = \{\text{inner lines, or lines having 2 absolute points}\}.$$

Clearly $\mathcal{I}\pi = o$ and $\iota\pi = \mathcal{O}$.

In analogy with the real projective plane, we might expect the incidence structure $HA(n)$ whose points are $\mathcal{I}$ and lines are $\iota$, with incidence as given in $\mathcal{P}$, to be a finite hyperbolic plane, since intersecting, parallel, and ultraparallel lines can be defined in the obvious way. Baer [1] has shown that this is not true—a simpler, but less general argument [9, p. 316] shows that the incidence structure consisting of $\mathcal{I}$ and $\iota$ involves "parallel points," that is, points which do not determine a common line.

In a previous paper [9], we investigated the results of the simple hypothesis:

* Department of Mathematics, Carleton University, Ottawa, Ontario, Canada K1S 5B6.

(1) No outer point is elliptic
in an arbitrary finite projective plane of order $n$ which is endowed with a regular polarity $\pi$; here we use the term "elliptic" in Baer's original sense, as the intersection of two perpendicular inner lines, where two lines $a$, $b$ are said to be perpendicular ($a \perp b$) with respect to $\pi$ if each passes through the other's pole. Then, as Baer has shown ([1], but using the outline given in [8, p. 155]):
(2) Every inner point is elliptic.
(3) If $P \in l$ with $P$ elliptic and $l \in i$, then the line through $P$ perpendicular to $l$ also $\in i$.
(4) $s(\pi) = 1$.
(5) $n \equiv 3 \pmod 4$.

If we start with a finite projective plane of order $n \equiv 3 \pmod 4$ whose absolute points form a conic (suitably defined in the finite case—see [10]), then no outer point is elliptic [9, p. 318].

However, in this paper we wish to study the geometry of this incidence structure $HA(n)$ where $n \equiv 3 \pmod 4$ over a *Desarguesian* projective plane $\mathcal{P}$. This restriction to Desarguesian (which implies Pappian) finite projective planes is necessary so that we can avail ourselves of the many rich results found in Professor Coxeter's beautiful book [7]; in particular, the consequences of the fundamental theorem will be exploited. Any of the results obtained below can be exhibited in $\mathcal{P}(n)$, the classical projective plane defined over $GF(n)$, for any $n \equiv 3 \pmod 4$. In the figures, a solid dot represents an inner point, an open dot an outer point, and a $\times$ an absolute point; inner lines are represented by solid lines, outer lines by dashed lines, and absolute lines by dotted lines.

The study of motions in such an $HA(n)$ was initiated in [8], but instead of completing the classification of motions, it was merely stated that "we could obtain a classification of motions in $HA(n)$ analogous to that of the classical hyperbolic plane" [5, p. 201], and the three-reflection theorem was investigated. However, further investigation has shown that this classification differs markedly in some respects from that of the classical plane, and some interesting points arise; moreover, this classification is helpful in correcting the proof of one case of the celebrated three-reflection theorem.

## 2. Reflections in $HA(n)$ over a Desarguesian Projective Plane

Let $\mathcal{P}$ be a Desarguesian projective plane of order $n \equiv 3 \pmod 4$, and $\pi$ a regular polarity whose absolute points form a conic; we have seen that the associated $HA(n)$ involves parallel points. Since $\mathcal{P}$ is finite, it is also Pappian [8, p. 160]; since $n$ is odd, $\mathcal{P}$ is also Fanonian, i.e., the diagonal points of a quadrangle are never collinear [12, pp. 190–191]. Thus $\mathcal{P}$ satisfies the axioms of projective geometry as enunciated by Coxeter ([6, p. 230] or [7, p. 25]), so that we may use the results proved there.

Let $\sigma_{A,a}$ denote the involutory homology with center $A$ and axis $a$, the polar of $A$. Since every involutory homology is an harmonic homology [7, p. 55], this means that if $\sigma_{A,a}(X) = Y$, $X$ and $Y$ are harmonic conjugates with respect to $A$ and $a \cdot XY = O_a$. We denote this by $H(XY, AO_a)$.

**Definition 1.** If $a$ is an inner line, $\sigma_{A,a}$ is called the *reflection in the line* $a$, and denoted $\sigma_a$.

If $A$ is an inner point, $\sigma_{A,a}$ is called the *reflection in the point* $A$, and denoted $\sigma_A$.

Clearly, no involutory homology can be both a line reflection and a point reflection, since a pole–polar pair always consists of one inner element and one outer element (cf. Bachmann's axiom $\sim P$ in [2, p. 47] or the English translation [3, pp. 12–14]). Since reflections are defined only for inner lines and points, we shall omit the adjective "inner" when referring to reflections.

The following results proved in [9] will be needed:

**Result 1.** *Given two inner lines which intersect in an inner point, there is no inner line which is perpendicular to both.*

**Result 2.** *The polar of an exterior point $X$ is the line joining the two points of intersection with the conic of the two tangents through $X$.*

**Result 3.** *Reflections in points and lines fix the conic $\mathcal{C}$ of absolute points, but not pointwise or linewise.*

**Result 4.** *Reflections map inner points into inner points, outer points into outer points, and dually; i.e., reflections preserve $HA(n)$.*

**Result 5.** *Let $a, b$ be two perpendicular lines of $HA(n)$ intersecting in the point $C$. Then $\sigma_b\sigma_a = \sigma_C = \sigma_a\sigma_b$.*

**Corollary 1.** *Reflections in perpendicular lines commute.*

**Corollary 2.** *Reflections in an incident point–line pair commute.*

**Corollary 3.** *A point reflection can be represented as the product of line reflections in any two perpendicular lines passing through the point.*

**Result 6.**

(i) *The only fixed inner points of a line reflection $\sigma_a$ are the inner points of $a$.*
(ii) *The only fixed inner lines of a line reflection $\sigma_a$ are the line $a$ and the inner lines perpendicular to $a$.*

**Result 7.**

(i) *The only fixed inner lines of a point reflection $\sigma_A$ are the inner lines through $A$.*
(ii) *The only fixed inner points of a point reflection $\sigma_A$ are the point $A$ and the inner points on $a$.*

To study the group of motions of $HA(n)$ we must consider all products of line reflections. Fortunately, we are able to restrict attention to products of only two reflections, either point or line reflections, because of a result found in a recent book by Brauner [4, Satz 2.4, p. 73]:

*In a classical projective plane, every automorphic projective collineation of a conic k is the product of two automorphic harmonic homologies.*

For "automorphic projective collineation of a conic $k$" we use the classical term "motion."

One result of a combinatorial nature should be pointed out:

**Result 8.** *The inner lines through an inner point are paired under perpendicularity, as are the outer lines. But through an outer point, the unique perpendicular to an inner line is an outer line.*

*Proof.* The first part is simply a rewording of condition (3). The second part follows from the hypothesis (1) that no outer point is elliptic; since half the nonabsolute lines through an outer point (i.e. $(n-1)/2$) are outer and half are inner, and since no two of these inner lines can be perpendicular (or the point would be inner), the perpendicularity relation must pair inner and outer lines. □

This result gives a useful criterion for deciding whether a point is inner or outer.

## 3. Products of Two Reflections in $HA(n)$ over a Desarguesian Projective Plane

Two (inner) lines $a$ and $b$ can intersect in an inner point, an outer point, or an absolute point; they are called intersecting, ultraparallel, or parallel lines respectively. There are thus three motions which arise as products of two line reflections:

**Definition 2.** The product of reflections in two distinct intersecting lines is called a *rotation*.

**Definition 3.** The product of reflections in two ultraparallel lines is called a *translation*.

**Definition 4.** The product of reflections in two parallel lines is called a *parallel displacement*.

These terms are of course adopted from classical hyperbolic geometry [6, p. 300]. Table 1 at the end of the paper shows the (inner) points and lines which are fixed under these motions and others to be mentioned later. In [9, Result 9] we have already shown that a nonidentity translation fixes the common perpendicular of the two generating line reflections, and no points. The analogous results for rotations and parallel displacements are as follows:

**Theorem 1.** *A nonidentity rotation $\sigma_b\sigma_a$ where $a \cdot b = P$, fixes the point $P$ and no other (inner) point, and fixes no (inner) line, unless $a \perp b$, in which case $\sigma_b\sigma_a$ is a point reflection $\sigma_P$.*

*Proof.* Consider $\sigma_b\sigma_a$ where $a \cdot b = P$, an inner point. Now if $X$ is a fixed point, $\sigma_b\sigma_a(X) = X$ implies that $\sigma_b(X) = \sigma_a(X) = X'$ say. If $X = X'$, then $X$ is fixed under $\sigma_a$, so that either $X = A$ or $X \in a$. But $X$ is also fixed under $\sigma_b$, so that either $X = B$ or $X \in b$. Together we have four possibilities:

$X = A = B$, which means $\sigma_a = \sigma_b$, a contradiction;
$X = a \cdot b = P$;
$X = A \in b$ or $X = B \in a$, both of which imply $a \perp b$, so that $\sigma_b\sigma_a$ is a point reflection.

On the other hand, if $X \neq X'$, then $\sigma_a(X) = X'$ implies that $A, X, X'$ are collinear and $H(XX', AO_a)$ where $O_a = a \cdot XX'$. Similarly, $\sigma_b(X) = X'$ implies that $B, X, X'$ are collinear and $H(XX', BO_b)$ where $O_b = b \cdot XX'$. Thus $X, X' \in AB = p$, and we have the two harmonic relations $H(AO_a, XX')$ and $H(BO_b, XX')$. Thus $(A, O_a)$ and $(B, O_b)$ are two pairs in the involution on $AB$ with invariant points $X, X'$ [7, p. 47]. In symbols, $(AO_a), (BO_b) \in (XX)(X'X')$. But $(AO_a)$, $(BO_b)$ are also two pairs of the involution induced on $p$ by the polarity $\pi$ [7, p. 62], and since an involution is uniquely determined by two pairs of mates [7, p. 45], these two involutions are identical. Thus the involution induced on $p$ by $\pi$ is hyperbolic, having two invariant points $X$ and $X'$, and since these invariant points must be absolute points under $\pi$ [7, p. 62], $p$ is an inner line. But $P$ is an inner point, and we have the desired contradiction.

We have exhibited this argument at some length, as it will reappear in abbreviated form in most of the subsequent theorems.

Similarly, if $\sigma_b\sigma_a(x) = x$, then $\sigma_b(x) = \sigma_a(x) = x'$ say. If $x = x'$, then $\sigma_a(x) = x$ implies that $x = a$ or $A \in x$, and $\sigma_b(x) = x$ implies that $x = b$ or $B \in x$. Thus we have four possibilities: $x = a = b$; $x = a \ni B$; $x = b \ni A$; or $x = AB$. The first is an immediate contradiction, the second and third imply that $a \perp b$, and the fourth implies that $AB$ is inner, while its pole $a \cdot b$ is also inner, again a contradiction. On the other hand, if $x \neq x'$, then the argument is the exact dual of the above for fixed points, and so there are no fixed lines. $\square$

**Theorem 2.** *A parallel displacement fixes no (inner) point or line.*

*Proof.* Suppose $\sigma_b\sigma_a(X) = X$, which implies $\sigma_a(X) = \sigma_b(X) = X'$, say. If $X = X'$, then again we have four possibilities: $X = A = B$, a contradiction; $X = a \cdot b$, which is not an inner point but an absolute one; $X = A \in b$; or $X = B \in a$. If $X = A \in b$, then $b = AC$ where $C = a \cdot b$ is an absolute point (see Figure 1). But $C \in a$ implies $A \in c$, and since $C \in c$, $c$ must be the line $AC$. Then $b = c$, an

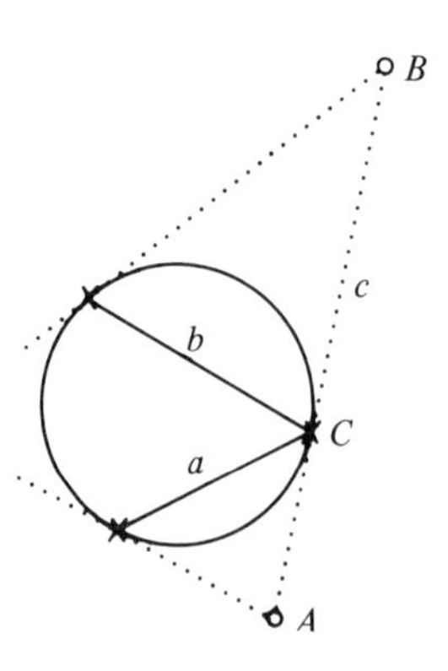


**Figure 1**

absolute line, giving a contradiction. Similarly $X = B \in a$ would imply $c = a$. But if $X \neq X'$, then as before we shall have $A, B, X, X'$ collinear and $H(XX', AO_a)$ and $H(XX', BO_b)$. Once more $X$ and $X'$ must be absolute points on the line $AB$, which is, however, an absolute line, being the polar of $a \cdot b$, and so has only one absolute point.

The dual argument to the above shows that there are no fixed lines. □

Thus far, the motions are strictly analogous to those of the classical hyperbolic plane, and precisely what one would expect. A quick glance at the fixed points and lines in Table 1 (p. 493) clearly indicates that these motions are distinct. However, when we try to combine point reflections and line reflections, some interesting cases arise:

**Definition 5.** The product $\sigma_b \sigma_A$ of a point reflection and a line reflection is called a *glide reflection*. It is a *glide reflection of the first type*, or *glide I*, if $a \cdot b$ is an inner point, and a *glide reflection of the second type* or *glide II*, if $a \cdot b$ is an outer point.

Note that if $A \in b$, then $a \perp b$ and so $a \cdot b$ must be an outer point by Result 8. In this case we have a glide II, and if $c$ is the unique perpendicular to $b$ through $A$ (as guaranteed by (3)), then $\sigma_b \sigma_A = \sigma_b \sigma_b \sigma_c = \sigma_c$, a single line reflection. It is also readily verified that a single line reflection is a glide II.

**Theorem 3.** *A glide reflection $\sigma_b \sigma_A$ which is not a single line reflection admits one fixed point, namely $a \cdot b$, only if it is a glide I; dually, it admits one fixed line or "axis," namely $AB$, only if it is a glide II.*

*Proof.* Suppose $\sigma_b \sigma_A$ is not a line reflection, so that $A \notin b$. Then if $X$ is fixed, $\sigma_b \sigma_A(X) = X$ implies that $\sigma_A(X) = \sigma_b(X) = X'$, say. If $X = X'$, then we have $X = A$ or $X \in a$ and $X = B$ or $X \in b$. The four possibilities are then: $X = A = B$, which is an immediate contradiction; $X = a \cdot b$, which is an inner point only for a glide I; $X = B \in a$, implying $X$ is outer; $X = A \in b$, which implies that $\sigma_b\sigma_A$ is a single line reflection. If $X \neq X'$, then $\sigma_A(X) = X'$ implies $A, X, X'$ collinear and $H(XX', AO_a)$, while $\sigma_B(X) = X'$ implies $B, X, X'$ collinear and $H(XX', BO_b)$. Thus as in Theorems 1 and 2, $X$ and $X'$ are absolute points on the line $AB$. But $AB$ is an inner line only for a glide II, since it is the pole of the outer point $a \cdot b$, and in this case $X, X'$ are absolute points, not inner points.

The property of being a fixed line for a glide reflection is dual to that of being a fixed point, and the proof proceeds dually. □

Finally we must consider products of two point reflections, say $\sigma_B\sigma_A$. If $A, B$ have a common (inner) line $c$, let $a, b$ be the perpendiculars to $c$ through $A, B$ respectively—they are inner lines by (3). Then $\sigma_B\sigma_A = \sigma_b\sigma_c\sigma_c\sigma_a$ (Corollary 3) $= \sigma_b\sigma_a$, where $a$ and $b$, having a common perpendicular $c$, are ultraparallel. Thus in this case $\sigma_B\sigma_A$ is a translation.

If, however, $AB$ is not an inner line, we distinguish two cases, according as $a \perp b$ or $a \not\perp b$.

**Theorem 4.** *If $a \perp b$, then $\sigma_B\sigma_A$ is a point reflection.*

*Proof.* Since $a \perp b$, $C = a \cdot b$ is an inner point (by Result 8) and so $c = AB$ is an outer line. Moreover, $A \in b$ and $B \in a$, since $a \perp b$. Thus $ABC$ is a self-polar triangle whose vertices are inner points and whose sides are outer lines (see Figure 2). But the product of three such involutory homologies $\sigma_{A,a}$, $\sigma_{B,b}$, and $\sigma_{C,c}$ is the identity [11, p. 101], and so $\sigma_B\sigma_A = \sigma_C$. □

Finally we must consider $\sigma_B\sigma_A$ where $AB$ is not an inner line, i.e. $a \cdot b$ is an inner point, and $a \not\perp b$. Calling such a motion a *semirotation*, we have:

**Theorem 5.** *A semirotation $\sigma_B\sigma_A$ has precisely one invariant point $a \cdot b$ and no invariant line.*

The proof is so similar to the above proofs that it can be omitted.

From Table 1, we note that a rotation, glide I, and semirotation have the same fixed point and no fixed line; the obvious question is whether these motions are simply the same motion expressed in three different ways. A partial answer is given by the following:

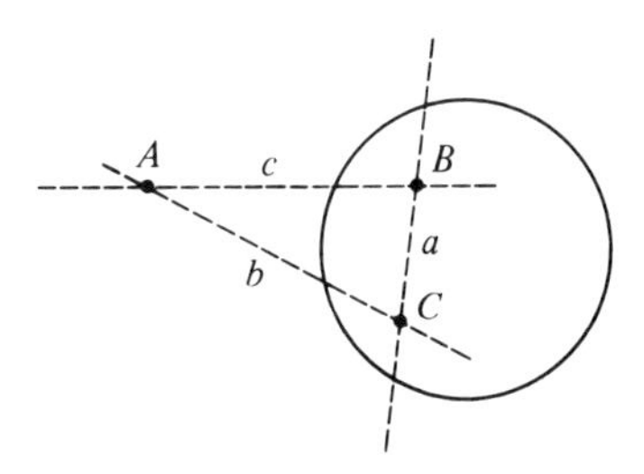


**Figure 2**

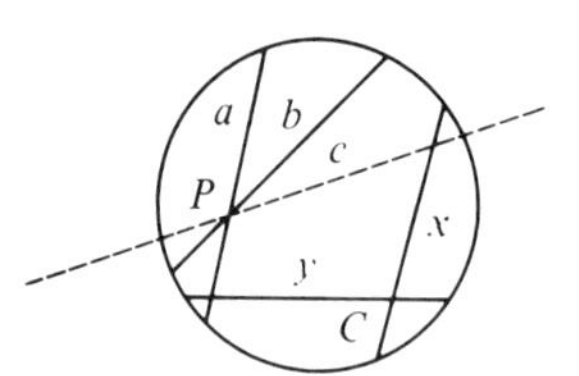


**Figure 3**

**Theorem 6.** *A rotation is not representable as a glide reflection of the first type.*

*Proof.* Suppose there exist (inner) point and lines $a, b, d, C$ such that $\sigma_b\sigma_a$ is a rotation (with $a \cdot b = P$ the fixed inner point), $\sigma_d\sigma_C$ is a glide I (with $c \cdot d$ the fixed inner point), and $\sigma_b\sigma_a = \sigma_d\sigma_C$. Then these fixed points must be identical, and so $c \cdot d = P$. Moreover, $\sigma_b\sigma_a = \sigma_d\sigma_C$ implies $\sigma_d\sigma_b\sigma_a = \sigma_C$.

Since $C$ is an inner point, let $x$ and $y$ be a pair of perpendicular (inner) lines through it; by Corollary 3, $\sigma_C = \sigma_y\sigma_x$. (See Figure 3.)

Now $\sigma_x\sigma_y(P) = \sigma_C(P) = P$, since $P$ is fixed under $\sigma_d$, $\sigma_b$, and $\sigma_a$ and so under their product, which is $\sigma_C$, and so $\sigma_x(P) = \sigma_y(P) = P'$, say. By a now familiar argument this implies (in the case that $P = P'$) that $P = X = Y$ (a contradiction), or $P = X \in y$ or $P = Y \in x$ (a contradiction, since $X$, $Y$ are outer and $P$ inner), or $P = C$, which, since $P \in c$, implies that $C$ is an absolute point, giving a contradiction. If $P \neq P'$, then $\sigma_x(P) = \sigma_y(P) = P'$ implies that $P, P' \in XY = c$ and $H(PP', XO_x)$, $H(PP', YO_y)$. Then as before, $P$ and $P'$ are absolute points on the line $XY = c$, which is an outer line. This contradiction completes the theorem. □

In a similar way, the distinction between a glide I and a semirotation, or between a rotation and a semirotation, could be exhibited.

In summary, we see that an $HA(n)$ defined over a classical projective plane admits eight distinct kinds of motion, in contrast to the classical hyperbolic plane, which admits only six. The classical glide reflection appears as two distinct types of glide, and a new motion with a single invariant point, similar to a rotation but representable as the product of two reflections in points which do not have a common line, arises.

## 4. Addendum

Theorem 6 can be used to complete the proof of Theorem 5 of [9], which states:

*Let a, b, c be three inner lines concurrent in an inner point G. Then $\sigma_c\sigma_b\sigma_a = \sigma_d$ for some inner line d incident with G.*

The proof is correct down to the conclusion $\sigma_c\sigma_b\sigma_a = \sigma_{D,d}$. But to prove that $\sigma_{D,d}$ is a line reflection and not a point reflection (i.e. that $d$ is inner) the incorrect Lemma 3 was used (cf. Theorem 4 above). Instead, if $\sigma_{D,d} = \sigma_D$, then we have $\sigma_c\sigma_b\sigma_a = \sigma_D$ or $\sigma_b\sigma_a = \sigma_c\sigma_D$, which Theorem 6 of this paper has shown cannot be the case. Hence $\sigma_c\sigma_b\sigma_a = \sigma_d$, and the three-reflection theorem is valid in all cases.

**Table 1**

| Motion $\neq$ identity | Symbol | Fixed point(s) | Fixed line(s) |
|---|---|---|---|
| Line reflection | $\sigma_a$ | Inner points of $a$ | $a$, inner lines perpendicular to $a$ |
| Point reflection | $\sigma_A$ | $A$, inner points of $a$ | Inner lines through $A$ |
| Rotation | $\sigma_b\sigma_a$ where $a \cdot b$ is an inner point, $a \not\perp b$ | $a \cdot b$ | None |
| Parallel displacement | $\sigma_b\sigma_a$ where $a \cdot b$ is an absolute point | None | None |
| Translation | $\sigma_b\sigma_a$ where $a \cdot b$ is an outer point | None | Common perpendicular to $a$ and $b$ (polar of $a \cdot b$) |
| Glide I | $\sigma_b\sigma_A$ where $a \cdot b$ is an inner point, $A \notin b$ | $a \cdot b$ | None |
| Glide II | $\sigma_b\sigma_A$ where $a \cdot b$ is an outer point | None | $AB$ |
| Semirotation | $\sigma_B\sigma_A$ where $a \cdot b$ is an inner point, $a \not\perp b$ | $a \cdot b$ | None |

## References


[1] Baer, R., Polarities in finite projective planes. *Bull. Am. Math. Soc.* **51** (1946), 77–93.

[2] Bachmann, F., *Aufbau der Geometrie aus dem Spiegelungsbegriff* (2nd edition). Springer-Verlag 1973.

[3] Bachmann, F., *Hjelmslev-Gruppen* (2nd edition). Kiel University 1974. (English translation of Part I by C. W. L. Garner for the Conference on the Foundations of Geometry, held at the University of Toronto, July–August 1974.)

[4] Brauner, H., *Geometrische Projektiver Raüme* I, II. Bibliographisches Institut, Mannheim 1977.

[5] Coxeter, H. S. M., *Non-Euclidean Geometry* (3rd edition). University of Toronto Press 1957.

[6] Coxeter, H. S. M., *Introduction to Geometry* (2nd edition). Wiley 1969.

[7] Coxeter, H. S. M., *Projective Geometry* (2nd edition). University of Toronto Press 1974.

[8] Dembowski, P., *Finite Geometries*. Springer-Verlag 1968.

[9] Garner, C. W. L., A finite analogue of the classical hyperbolic plane and Hjelmslev groups. *Geom. Ded.* **7** (1978), 315–331.

[10] Garner, C. W. L., Conics in finite projective planes. *J. Geometry* **12** (1979), 132–138.

[11] Hughes, D. R. and Piper, F. C., *Projective Planes*. Springer-Verlag 1973.

[12] Pickert, G., *Projektive Ebenen*, Springer-Verlag 1955.

[13] Segre, B., Ovals in finite projective planes. *Can. J. Math.* **7** (1955), 414–416.

# Part IV: Groups and Presentations of Groups

# Generation of Linear Groups[1]

William M. Kantor*

## 1. Introduction

Let $G$ be a *finite*, primitive subgroup of $GL(V) = GL(n, D)$, where $V$ is an $n$-dimensional vector space over the division ring $D$. Assume that $G$ is generated by "nice" transformations. The problem is then to try to determine (up to $GL(V)$-conjugacy) all possibilities for $G$. Of course, this problem is very vague. But it is a classical one, going back 150 years, and yet very much alive today. The purpose of this paper is to discuss both old and new results in this area, and in particular to indicate some of its history. Our emphasis will be on especially geometric situations, rather than on representation-theoretic ones.

For small $n$, all transformations may be considered "nice" (Sections 2 and 4). For general $n$, the nicest transformations are reflections and transvections (or, projectively, homologies and elations); these occupy Sections 3 and 5. Finally, Section 6 touches on several other types of "nice" transformations.

We will generally regard as equivalent the study of subgroups of $GL(n, D)$ and of the projective group $PGL(n, D)$. It should, however, be realized that this point of view was occasionally not taken by some of the authors cited here.

In general, we will not list the groups in the classifications discussed; nor will we discuss further properties of the groups obtained.

Further historical information may be found in Wiman (1899b) and van der Waerden (1935).

## 2. Characteristic 0: Small Dimensions

While the subject of this paper began in the case of finite $D$, we will start with the possibly more familiar characteristic 0 case. In this section, $D$ will be

[1]Supported in part by NSF Grant MCS-7903130.

*Department of Mathematics, University of Oregon, Eugene, OR 97403, USA.

commutative of characteristic 0—in which case we may take $D = \mathbb{C}$—and $n$ will be small. By a fundamental result of Jordan (1878, 1879), for each $n$ the number of types of primitive subgroups of $SL(n, \mathbb{C})$ is finite.

All finite subgroups of $SL(2, \mathbb{C})$ were first determined by Klein in 1874 (Klein (1876, 1884)). His method was very geometric, based upon regarding the extended complex plane as a sphere in $\mathbb{R}^3$. Of course, the groups he found all arise from regular polygons and regular polyhedra.

Jordan, who had been working on $SL(2, \mathbb{C})$, turned to $SL(3, \mathbb{C})$ (Jordan (1878)). However, he missed two examples (later found by Klein (1879) and Valentiner (1889)). His approach was not at all geometric. He derived information about $G$ by a case-by-case analysis of a diophantine equation he had used successfully in the proof of his general finiteness theorem. (This equation arises by expressing $|G|$ as a sum in terms of the orders of suitable—and especially, maximal—abelian subgroups of $G$ and of the indices of their normalizers, great care being taken with intersections of pairs of such subgroups.) He used the same methods soon afterwards (Jordan (1879)) in order to (attempt to) correct his previous work on $SL(3, \mathbb{C})$, and in order to obtain very preliminary results concerning $SL(4, \mathbb{C})$. His diophantine approach was later used a number of times, especially in the case of finite fields (Moore (1904), Wiman (1899a), Dickson (1900), Mitchell (1911a, 1913), Huppert (1967)).

Valentiner (1889) devised a similar diophantine method in his attempt at $SL(3, \mathbb{C})$. In addition, he proceeded somewhat geometrically, but erred in his treatment of homologies of order 3 (Mitchell (1911b)), thereby missing one example. (He was apparently unaware of Jordan's work on the same problem, where this example is listed.) Valentiner's treatment seems to have otherwise been correct: Wiman (1896) stated that Valentiner's error was easily corrected, and that all examples were known. For further historical discussion up to this point, as well as for properties of these groups, see Wiman (1899b).

Blichfeldt (1904, 1907) was the first to publish a complete proof for $SL(3, \mathbb{C})$. His methods were nongeometric: they involved a careful analysis of eigenvalues in order to obtain precise information concerning $|G|$. A purely geometric proof was later obtained by Mitchell (1911a). In fact, since it is easy to show that a primitive subgroup of $PSL(3, \mathbb{C})$ contains homologies (compare Mitchell (1911a), p. 215), a geometric proof is implicitly contained in Bagnera (1905); for the same reason, Mitchell's proof depends upon homologies (cf. Section 3).

Eigenvalue and order considerations also dominate the determination by Blichfeldt (1905) (also 1917) of all finite primitive subgroups of $SL(4, \mathbb{C})$. At about the same time, Bagnera (1905) gave a geometric solution to this problem when $G$ contains homologies; the case when $G$ does not contain homologies was handled later by Mitchell (1913), thereby providing an alternative, geometric proof of Blichfeldt's result.

At this point, the subject seems to have died, probably because much more sophisticated methods were needed. It was finally revived again by Brauer (1967), who handled $SL(5, \mathbb{C})$. The cases $n = 6$, 7, 8, and 9 have now been completed, by Lindsay (1971), Wales (1969, 1970), Doro (1975), Huffman and Wales (1976, 1978), and Feit (1976). In these results, geometry essentially disappears. It is replaced by representation theory (ordinary and modular) and by simple group classification theorems.

## 3. Characteristic 0: Reflections

Recall that a *reflection* is a diagonalizable transformation having eigenvalue 1 with multiplicity $n - 1$. The corresponding eigenspace is its *axis*; the remaining 1-dimensional eigenspace is its *center*. A *homology* is just a reflection viewed projectively (i.e., as acting on $PG(n-1, D)$). Classification problems concerning reflections or homologies are thus essentially the same, and will generally be identified.

Finite subgroups of $GL(n, \mathbb{R})$ generated by reflections are a very familiar topic. For a discussion of them and their history, we defer to Coxeter (1948) and Bourbaki (1968). However, it is worth mentioning that the classification and study of these groups occupy a far more central role in mathematics than the other groups discussed in this survey. They are the crystals (or rather, "apartments") from which Tits' theory of buildings grows (Tits (1974), Carter (1972)), and hence are central in the theories of algebraic groups (Tits (1966)) and of finite groups (Chevalley (1955), Carter (1972)). Further incredibly varied but fundamental occurrences of them are discussed at length in Hazewinkel et al. (1977).

The determination of all finite primitive subgroups of $GL(n, \mathbb{C})$ generated by reflections is due primarily to Mitchell (1914a). Namely, he dealt with the cases $n \geqslant 5$, the smaller values of $n$ having been handled earlier (as described in Section 2). His method was short, elegant, and very geometric. It involved building up groups, homology by homology and dimension by dimension. Namely, suppose that $W$ is a subspace of $V$, spanned by some of the homology centers for $G$, and for which the induced group generated by these homologies is known—and, hopefully, primitive. Mitchell picked a homology $h$ moving $W$, with center $c$, and studied the group induced on $\langle W, c \rangle$. (Since a homology fixes every subspace containing its center, both the known group and $h$ send $\langle W, c \rangle$ to itself.)

However, Mitchell's result apparently went largely unnoticed. He was clearly far ahead of his time: he handled the complex case several years before all real reflection groups were independently determined by Cartan and Coxeter (cf. Coxeter (1948, p. 209), and Bourbaki (1968, p. 237)). Only very recently has another complete proof of his result appeared (Cohen (1976)). Important special cases have, however, been re-proved (Shepard (1952, 1953); Coxeter (1957), (1974)); namely, those leading to regular complex polytopes.

Shephard and Todd (1954) took the (projective) groups generated by homologies obtained by Klein (1876), Blichfeldt (1904, 1907), Bagnera (1905), and Mitchell (1914a, b), and listed all complex reflection groups giving rise to them. The case $n \geqslant 3$ is implicit in the above papers (and is freely used in Mitchell's proof); the case $n = 2$ is more involved. This list will not be reproduced here. Instead, we will simply make a few comments about the largest example which is not already a real reflection group.

A group $G = 6 \cdot P\Omega^-(6,3) \cdot 2$, having $|Z(G)| = 6$, $|G : G'| = 2$, and $G'/Z(G) \cong P\Omega^-(6,3)$, arises as a subgroup of $GL(6,\mathbb{C})$ generated by involutory reflections. It was discovered by Mitchell (1914a), who wrote down coordinates for its reflecting hyperplanes. Geometric properties of the action on the corresponding projective space $PG(5,\mathbb{C})$ were studied by Hamill (1951) and Hartley (1950). Its reflection centers (dual to the reflecting hyperplanes) determine the $\mathbb{Z}[\omega]$-lattice $\Lambda$ of Coxeter and Todd (1953) (where $\omega$ is a primitive cube root of unity). This lattice consists of all $(x_i) \in \mathbb{Z}[\omega]^6$ such that $\sum_1^6 x_i \equiv 0 \pmod 3$ and $x_i \equiv x_j \pmod \theta$ for all $i, j$ (where $\theta = \omega - \omega^2$ satisfies $\theta^2 = -3$); $\Lambda$ is equipped with the usual hermitian inner product inherited from $\mathbb{C}^6$. Its automorphism group is $G$, generated by the reflections in $GL(6,\mathbb{C})$ preserving $\Lambda$; these are the reflections with centers $\langle\lambda\rangle$ for $\lambda \in \Lambda$ of norm 6. This group induces $\Omega^-(6,3) \cdot 2$ on $\Lambda/\theta\Lambda$, where $\Lambda/\theta\Lambda$ is the natural $GF(3)$-module for $O^-(6,3)$. The 126 reflections in $G$ induce 126 reflections of the orthogonal space $\Lambda/\theta\Lambda$. The remaining 126 reflections of that space are induced by using semilinear automorphisms of $\Lambda$; for example, $-cr$ induces one of them, where $c$ denotes complex conjugation on $\Lambda$, while $r$ is the reflection with center $\langle(1,1,1,1,1,1)\rangle$. On the other hand, the hermitian product on $\Lambda$ induces one on the $GF(4)$-space $\Lambda/2\Lambda$, and reflections in $G$ induce 126 transvections (defined in Section 5) belonging to $SU(6,2)$. This produces an embedding $P\Omega^-(6,3) \cdot 2 < PSU(6,2)$, which is crucial to the existence of the sporadic finite simple groups found by Fischer (1969). Also, the lattice $\Lambda \oplus \Lambda$ is a sublattice of the Leech $\mathbb{Z}[\omega]$-lattice, described in Conway (1971). Similarly, the direct sum of three copies of the 8-dimensional real lattice of type $E_8$ is a sublattice of the Leech lattice itself (Conway (1971)); while the corresponding real reflection group, when embedded in $O^+(8,3)$, also plays a significant role in Fischer's constructions.

The study of small-dimensional complex groups, and of large-dimensional groups generated by reflections, seems to have (temporarily) ended with Blichfeldt (1917) and Mitchell (1914a, b). Mitchell's attitude towards this is indicated on pp. 596–7 of Mitchell (1935). First he states that "comparatively few groups of interest appear to be known in more than four variables." This leads to a discussion of work of Burnside (1912) concerning real reflection groups. Mitchell then turns to his own work on complex reflection groups: "In spite of the more general character of this problem as compared with that solved by Burnside, no restrictions being placed on the character of the coefficients, the results were chiefly negative." Only one new example arose (the 6-dimensional one just discussed). Thus, Mitchell was looking for new groups, or at least new linear groups, and was not entirely happy with the outcome of this work.

It is unfortunate, both for geometry and group theory, that Mitchell (or someone else of his generation) did not pursue reflections further. Certainly, if $D$ is commutative of characteristic 0, then $D$ may be assumed to be a subfield of $\mathbb{C}$. However, reflection groups over the quaternions $\mathbb{H}$ do indeed yield new examples. One 3-dimensional example is (projectively) a simple group discovered in 1967. Its discovery 50 years earlier might have revived the then nearly dead theory of finite groups.

The determination of all finite primitive subgroups of $GL(n,\mathbb{H})$ generated by reflections was made by Cohen (1980), although some of this had been done earlier by Wales and Conway. The groups $G$ obtained which are not complex reflection groups can be described as follows, if $n \geqslant 3$:

(i) $n = 3$, $G = Z_2 \times PSU(3,3)$;
(ii) $n = 3$, $G = 2 \cdot HJ$ (where $HJ$ denotes the Hall–Janko simple group, predicted by Janko in 1967 and constructed by Hall as a permutation group of degree 100 on the cosets of a subgroup $PSU(3,3)$; cf. Hall and Wales (1968);
(iii) $n = 4$, $G/Z(G)$ has an elementary abelian normal subgroup of order $2^6$, modulo which it is one of 3 subgroups of $\Omega^-(6,2)$ (note the similarity to some 4-dimensional complex groups);
(iv) $n = 4$, $G/Z(G) \cong (A_5 \times A_5 \times A_5) \rtimes S_3$ (a wreathed product; compare the situation $(A_5 \times A_5) \rtimes S_2$ for the real reflection group [3, 3, 5]); and
(v) $G = Z_2 \times PSU(5,2)$.

In each case, all reflections turn out to be involutory. Tits has shown that example (ii) is related to a quaternionic version of the real Leech lattice.

Cohen's proof is definitely nongeometric. Quaternionic $n$-space can be regarded as complex $2n$-space (in many ways). When this is done, quaternionic reflections become complex transformations having a $(2n - 2)$-dimensional eigenspace. Results of Huffman and Wales (Huffman (1975); Huffman and Wales (1975); Wales (1978)), to be discussed soon, then provide a list of complex groups; these must be checked to see which arise from quaternionic groups.

It would be desirable to have a new geometric proof of Cohen's result. The present proof is not elegant, using machinery of an overly sophisticated sort. A new proof would presumably proceed along the lines of Mitchell's approach. The case $n = 2$ merely requires knowledge of the finite subgroups of $SL(4,\mathbb{C})$. The case $n = 3$ is probably the hardest and most interesting one, in view of the examples. Starting from these cases, Mitchell's approach should have a good chance of success.

In the papers just cited, Huffman and Wales extended Mitchell's work in quite a different direction. They determined all finite primitive subgroups of $GL(n,\mathbb{C})$ which are generated by transformations having $(n - 2)$-dimensional eigenspaces. The resulting list is too long to reproduce here, but is probably worthy of geometric investigation. It may not be possible to give a direct proof of their result. Their proof relies very heavily on representation theory (ordinary and modular), and on very deep simple group classification theorems. Little geometry is involved. It is precisely for this reason that an alternative approach is needed to Cohen's quaternionic results.

However, there is an obvious advantage to applying group-theoretic classification theorems in geometry: results can be obtained which may otherwise be difficult to prove, or which may later be proved more elegantly. For example, consider the problem of determining all finite primitive reflection groups $G$ in $GL(n, D)$, for $D$ an arbitrary noncommutative division ring of characteristic 0. If $n = 1$, this is just the famous problem solved by Amitsur (1955) (and independently and almost simultaneously by J. A. Green). If $n = 2$ and $G$ is solvable, the problem seems to involve even more difficult number theory than Amitsur used.

But if $n \geqslant 3$, and if simple group classification theorems are thrown at the problem, no new nonsolvable examples arise. Similarly, the Cayley–Moufang projective plane appears not to admit any new examples of finite groups, generated by involutory reflections, which fix no point, line, triangle, or proper subplane, other than ${}^3D_4(2)$.

We have only been discussing the classification of reflection groups. There is, of course, a large body of literature concerning their properties. Their invariants have been of interest for a century (see, e.g., Klein (1876, 1884), and Shephard and Todd (1954), and the papers by Hiller and Solomon in these Proceedings). So have their associated polytopes in the real and complex cases (Coxeter (1948, 1957, 1974); Shephard (1952, 1953)). The case of quaternionic polytopes has recently been begun by Hoggar (1978) (see also his paper in these Proceedings). For remarkable extremal properties of real, complex, and quaternionic examples, see Delsarte, Goethals and Seidel (1975, 1977), Hoggar (1978), and Odlyzko and Sloane (1979).

## 4. Finite $D$: Small Dimensions

The detailed study of the subgroups of $PSL(2, D)$ was begun by Galois in 1832 with the case of a prime field $D$ (cf. Galois (1846), pp. 411–412, 443–444). For prime $q$, all subgroups of $PSL(2,q)$ were first determined by Gierster (1881). Burnside (1894) worked on the case of arbitrary $q$. Finally, all subgroups of $PSL(2,q)$ were determined for all $q$ independently by Moore (1904) and Wiman (1899a). We refer to Kantor (1979b) and references given there for further historical remarks concerning 2-dimensional groups.

The group $PSL(3,q)$ brings us back to Mitchell. The first attempt at determining its subgroups was made by Burnside (1895) in case $q$ and $(q^2+q+1)/(3,q+1)$ are both prime; but he missed the groups $PSO(3,q)$. Dickson (1905) later enumerated all subgroups of order divisible by $q$, when $q$ is prime, using an explicit knowledge of all conjugacy classes of $q$-groups. Both authors relied on group theory and matrices, not on geometry. Veblen suggested to his student Mitchell that he provide a geometric solution to the problem for $PSL(3,5)$ (where, incidentally, $q^2+q+1$ is prime). Mitchell solved the problem for $PSL(3,q)$, first for odd prime $q$, then for arbitrary odd $q$ in his thesis "The subgroups of the linear group $LF(3, p^n)$," written in 1910; the solution appears in Mitchell (1911a). (Another student of Veblen's, U. G. Mitchell, determined the subgroups of $PSL(3,4)$ in his thesis entitled "Geometry and collineation groups of the finite projective plane $PG(2,2^2)$," also written in 1910.) H. H. Mitchell went even further in his paper: he dealt with $PSL(3,\mathbb{C})$ at the same time as $PSL(3,q)$. His approach was very geometric, and highly original. (A very different approach, based on modular characters and simple group classification theorems, was given by Bloom (1967).) It should, in fact, be noted that Mitchell solved problems which Jordan (1878, 1879), Valentiner (1889), Burnside (1895), and Dickson (1905) could not. The maximal subgroups of $PSL(3,q)$, $q$ even, were later determined by Hartley (1926) in his thesis written under Mitchell. By

Mitchell (1911a), $|G|$ must be even here, so Hartley naturally concentrated on the elations $G$ must contain (cf. Section 5).

Mitchell's only other major papers on linear groups were Mitchell (1913), where all subgroups of $PGL(4,\mathbb{C})$ and $PGL(4,q)$ are determined which do not contain nontrivial homologies and have order not divisible by the characteristic of the field; Mitchell (1914a), which was discussed in Section 2; and Mitchell (1914b), in which all maximal subgroups of the symplectic groups $Sp(4,q)$ were found for odd $q$. All four papers rely heavily on geometry. The most important ones are certainly the ones on reflection groups and $PSL(3,q)$. The work of Mitchell and Hartley on $PSL(3,q)$ has been quoted often in recent papers on finite groups, besides providing some motivation for Piper's work on elations of finite projective planes (Piper (1965, 1966b)).

The groups $PSL(n,q)$, $n = 4$ or 5, have been the object of several recent papers. Mwene (1976) and Wagner (1979) enumerated all maximal subgroups when $q$ is even and $n$ is 4 and 5, respectively. The same was done, independently, by Zalesskii (1977). Zalesskii and Suprenenko (1978) handled the case $PSL(4,q)$ when the prime $p$ dividing $q$ is greater than 5, and Mwene (1980) discussed the general case for odd characteristic. $PSL(5,q)$ was handled by Zalesskii (1976) for $p > 5$, and completed for $p \geqslant 3$ by DiMartino and Wagner (1981). All these papers rely heavily on modular representation theory and simple group classification theorems. See Kantor and Liebler (1982) for further discussion and applications of these results.

## 5. Finite *D*: Homologies and Elations

Mitchell (1914a) observed that his work on complex groups generated by homologies applied equally well when the field was $GF(q)$, so long as $q$ is relatively prime to the order of the group. When this condition fails, so does complete reducibility, and the problem becomes considerably harder. As a further indication of its difficulty, note that Mitchell's problem turned out to be a finite one: only finitely many primitive examples exist. However, when $D = GF(q)$ and $q > 2$, infinitely many examples arise, such as orthogonal groups, unitary groups, and $PGL(n,q)$ itself. In addition, complex examples produce examples for suitable odd $q$, simply by passing modulo a suitable prime ideal. Of course, all of the above remarks apply to Section 4 as well.

Primitive subgroups of $PGL(n,q)$ containing a homology of order greater than 2 were determined independently by Wagner (1978) and by Zalesskii and Serezkin (1977). Homologies of order 2 were handled by Serezkin (1976) when $q$ is not a power of 3 or 5. The general case of groups containing involutory homologies was settled by Wagner (1980–1981). All of these papers are highly geometric. The general case was also dealt with independently and nongeometrically by Zalesskii and Serezkin (1980).

Wagner's approach is based on that of Mitchell (1914a). It is direct and reasonably elementary (but long). More than half of the work is devoted to fields of characteristic 3 or 5. The results may be summarized as follows.

Suppose that $G$ contains involutory homologies, but no homologies of higher order and no nontrivial elations (defined below). Then either

(i) $G \trianglerighteq P\Omega^{\pm}(n,q')$ with $GF(q') \subseteq GF(q)$;
(ii) $G = S_{n+2}$ and $(q, n+2) \neq 1$;
(iii) $G$ arises from a complex reflection group; or
(iv) $G = PSL(3,4) \cdot 2$, $n = 4$, and $GF(9) \subseteq GF(q)$.

Example (iv) arises from the embedding $PSL(3,4) \cdot 2 < PSU(4,3) \cdot 2$, which in turn arises from the complex 6-dimensional reflection group discussed in Section 3. The embedding $PSL(3,4) < PSU(4,3)$ was discovered by Hartley (1950) by considering the action of that reflection group on $PG(5,\mathbb{C})$. An alternative proof can be given, by observing that $SL(3,4)$ is induced on any totally isotropic 3-space of the unitary space $\Lambda/2\Lambda$ which is fixed by none of the transvections in the group. This embedding is the basis for the construction by McLaughlin (1969) of his sporadic simple group.

Homologies are not the only collineations inducing the identity on a hyperplane of a projective space. The other type of collineations behaving in this manner are the *elations*. They have order 1 or $p$ if $D$ has characteristic $p \neq 0$. The corresponding linear transformations are *transvections*; such a transformation $t$ satisfies $(t-1)^2 = 0$ and $\dim V(t-1) \leqslant 1$. Then, with respect to some basis, $t$ has the form

$$t = \begin{bmatrix} 1 & 0 & & \alpha \\ & 1 & & \\ & & \ddots & \\ 0 & & & 1 \end{bmatrix} \quad \text{for some } \alpha \in D;$$

if $\alpha$ is allowed to be arbitrary, then the resulting transvections form a group $\cong D^+$, called a *root group*. (This is a special case of root groups of Chevalley groups; cf. Carter (1972).)

McLaughlin (1967, 1969a) determined all irreducible subgroups of $GL(n,D)$ generated by root groups, for any field $D$. His approach is elegant and geometric.

The primitive subgroups of $PSL(n,q)$ generated by elations have also been determined, primarily by Piper (1966b, 1968, 1973) and Wagner (1974) (and, independently, by Zalesskii and Serezkin (1976) for odd $q$). Their arguments are beautifully geometric. Unfortunately, in one characteristic 2 situation simple group classifications were also used (Kantor (1979a)). For $n \geqslant 4$, the possibilities are as follows:

(i) $PSL(n,q')$, $PSp(n,q')$, and $PSU(n,q')$, where $GF(q') \subseteq GF(q)$;
(ii) $PO^{\pm}(n,q')$, where $q'$ is even and $GF(q') \subseteq GF(q)$;
(iii) $S_{n+2}$, where $n$ and $q$ are even; and
(iv) $P\Omega^{-}(6,3) \cdot 2$, where $n = 6$ and $GF(4) \subseteq GF(q)$.

Of course, example (iv) arises from Mitchell's 6-dimensional complex reflection group. An entirely geometric proof of the above result would again be desirable.

Elations appear in several situations. Ever since Galois, they have been involved in the proof of the simplicity of linear groups—not just of $PSL(n,q)$, but also of $PSp(2n, q)$ and $PSU(n, q)$ (Jordan (1870), Dickson (1900), Huppert (1967), as well as implicitly in Carter (1972)). Elations and homologies were used throughout the study of subgroups of $PSL(3, q)$ by Mitchell (1911a) and Hartley (1926). Elations were equally important for $PSL(4, q)$ and $PSL(5, q)$; for example, if $q$ is even, then the Sylow 2-subgroups of a subgroup of $PSL(5, q)$ containing no nontrivial elations have nilpotence class at most 2, a fact which was crucial for Mwene (1976), Wagner (1979), and Zalesskii (1977). Elations also arose in the determination of the 2-transitive permutation representations of $PSL(n, q)$, $PSp(2n, q)$, and $PSU(n, q)$ (Curtis, Kantor, and Seitz (1976)); in particular, McLaughlin's result was essential for $PSp(2n, 2)$. Elations (and involutory reflections) arise throughout the classification of Fischer (1969); and Fischer's work was, in fact, used at one point in the determination of the primitive groups generated by elations. The latter determination was fundamental in bounding from below the degree (among other things) of a primitive permutation representation of $PSL(n, q)$, $PSp(2n, q)$, or $PSU(n, q)$ (Patton (1972), Cooperstein (1978), Kantor (1979b)).

## 6. Other Transformations

We conclude with a brief discussion of subgroups $G$ of $GL(V) = GL(n, q)$ generated by other "nice" types of transformations.

(i) Call $t \in GL(V)$ *quadratic* if $(t-1)^2 = 0$. Clearly, $|t|$ is 1 or the prime $p$ dividing $q$. Transvections are quadratic, and if $p = 2$ then so are all involutions. If $t$ is quadratic and $t \neq 1$, then the subspace $C_V(t)$ of fixed vectors contains the intersection $[V, t] = \{vt - v \mid v \in V\}$ of all fixed hyperplanes. Thus, quadratic transformations can be regarded as generalizations of transvections.

Thompson (1970) classified all irreducible groups generated by quadratic transformations if $p > 3$, at the same time determining all possible modules for each group obtained. The groups are $SL(n, q')$, $Sp(n, q')$, $SU(n, q')$, $\Omega^{\pm}(n, q')$, $G_2(q')$, ${}^3D_4(q')$, $F_4(q')$, $E_6(q')$, ${}^2E_6(q')$, and $E_7(q')$, where $q' \mid q$. (The last six classes of groups are defined in Carter (1972): they are Chevalley and twisted groups.) Some sporadic simple groups arise when $p = 3$; this case has been the subject of a great deal of work by Ho (cf. Ho (1976) and the references given there). Thompson's theorem provided part of the impetus for the remarkable result of Aschbacher (1977) (where no module is present). The latter result to a certain extent supersedes Thompson's, and was a main tool in Ho (1976).

(ii) Dempwolff (1978, 1979) has classified all irreducible subgroups of $SL(n, 2)$ generated by involutions $t$ for which $\dim C_V(t) = n - 2$. His proof uses simple group classification theorems.

(iii) Kantor (1979a) determined all irreducible subgroups of orthogonal groups $\Omega^{\pm}(n, q)$ which are generated by "long root elements." These are analogues of transvections, provided by the theory of Chevalley groups. While they are quadratic transformations, it is the characteristic 2 case that provides the most interesting examples.

The corresponding type of problem for all other Chevalley groups has been settled by Cooperstein (1979, 1981).

Of greater importance is the work recently begun by Seitz concerning the structure of subgroups of Chevalley groups. When specialized to the case of $SL(n,q)$, one of the preliminary applications of his methods (Seitz (1979)) is the determination of all subgroups of $SL(n,q)$ containing all diagonal matrices when $q > 11$ and $q$ is odd. His methods depend upon algebraic groups, not geometry. Further results on generation of yet another type are found in Seitz (1982).

(iv) *Singer cycles* are elements of $GL(n,q)$ of order $q^n - 1$. Their geometric significance was first noticed by Singer (1938). They arise in the special case $k = 1$ of the following construction.

Let $k \mid n$, and write $s = n/k$. Then a $k$-dimensional vector space over $GF(q^s)$ is also an $n$-dimensional vector space over $GF(q)$. Thus, $GL(k,q^s) \leqslant GL(n,q)$. In particular, $GF(q^n)^* \cong GL(1,q^n) \leqslant GL(n,q)$.

Kantor (1980) showed that any subgroup of $GL(n,q)$ generated by Singer cycles is a group $GL(k,q^s)$ (for some $k$ and $s = n/k$) obtained in the above manner. This time, simple group classification theorems are in no way involved in the proof. The proof is geometric, and is based upon the determination (geometrically) of all collineation groups acting 2-transitively on the points of a finite projective space (Cameron and Kantor (1979)).

## References


Amitsur, S. A. (1955), Finite subgroups of division rings. *TAMS* **80**, 361–386.

Aschbacher, M. (1977), A characterization of Chevalley groups over fields of odd order. *Ann. of Math. (2)* **106**, 353–468.

Bagnera, G. (1905), I gruppi finiti di transforminazioni lineari dello spazio che contengono omologie. *Rend. Circ. Mat. Palermo* **19**, 1–56.

Blichfeldt, H. F. (1904), On the order of linear homogeneous groups (second paper). *TAMS* **5**, 310–325.

Blichfeldt, H. F. (1905), The finite, discontinuous primitive groups of collineations in four variables. *Math. Ann.* **60**, 204–231.

Blichfeldt, H. F. (1907), The finite discontinuous primitive groups of collineations in three variables. *Math. Ann.* **63**, 552–572.

Blichfeldt, H. F. (1917), *Finite Collineation Groups*. University of Chicago Press, Chicago.

Bloom, D. M. (1967), The subgroups of $PSL(3,q)$ for odd $q$. *TAMS* **127**, 150–178.

Bourbaki, N. (1968), *Groupes et Algèbres de Lie, Chapters* IV, V, VI. Hermann, Paris.

Brauer, R. (1967), Über endliche lineare Gruppen von Primzahlgrad. *Math. Ann.* **169**, 73–96.

Burnside, W. (1894), On a class of groups defined by congruences. *Proc. LMS* **25**, 113–139.

Burnside, W. (1895), On a class of groups defined by congruences. *Proc. LMS* **26**, 58–106.

Burnside, W. (1912), The determination of all groups of rational linear substitutions of finite order which contain the symmetric group in the variables. *Proc. LMS (2)* **10**, 284–308.

Cameron, P. J. and Kantor, W. M. (1979), 2-transitive and antiflag transitive collineation groups of finite projective spaces. *J. Algebra*, **60**, 384–422.

Carter, R. W. (1972), *Simple Groups of Lie Type*. Wiley, London–New York–Sydney–Toronto.

Chevalley, C. (1955), Sur certains groupes simples. *Tohoku Math. J.* **7**, 14–66.

Cohen, A. M. (1976), Finite complex reflection groups. *Ann. Scient. Ec. Norm. Sup. (4)* **9**, 379–436.

Cohen, A. M. (1980), Finite quaternionic reflection groups. *J. Algebra* **64**, 293–324.

Conway, J. H. (1971), Three lectures on exceptional groups. In *Finite Simple Groups*, edited by M. B. Powell and G. Higman. Academic Press, New York, pp. 215–247.

Cooperstein, B. N. (1978), Minimal degree for a permutation representation of a classical group. *Israel J. Math.* **30**, 213–235.

Cooperstein, B. N. (1979), The geometry of root subgroups in exceptional groups I, *Geom. Dedicata* **8**, 317–381.

Cooperstein, B. N. (1981), Subgroups of exceptional groups of Lie type generated by long root elements, I, II. *J. Algebra* **70**, 270–282, 283–298.

Coxeter, H. S. M. (1948), *Regular Polytopes*. Methuen, London.

Coxeter, H. S. M. (1957), Groups generated by unitary reflections of period two. *Can. J. Math.* **9**, 243–272.

Coxeter, H. S. M. (1967), Finite groups generated by unitary reflections. *Abh. Hamburg* **31**, 125–135.

Coxeter, H. S. M. (1974), *Regular Complex Polytopes*. Cambridge University Press, Cambridge.

Coxeter, H. S. M. and Todd, J. A. (1953), An extreme duodenary form. *Can. J. Math.* **5**, 384–392.

Curtis, C. W., Kantor, W. M., and Seitz, G. M. (1976), The 2-transitive permutation representations of the finite Chevalley groups. *TAMS* **218**, 1–59.

Delsarte, P., Goethals, J. M., and Seidel, J. J. (1975), Bounds for systems of lines, and Jacobi polynomials. Philips Res. Rep. **30**, 91–105.

Delsarte, P., Goethals, J. M., and Seidel, J. J. (1977), Spherical codes and designs. *Geom. Ded.* **6**, 363–388.

Dempwolff, U. (1978), Some subgroups of $GL(n, 2)$ generated by involutions, I. *J. Algebra* **54**, 332–352.

Dempwolff, U. (1979), Some subgroups of $GL(n, 2)$ generated by involutions, II. *J. Algebra* **56**, 255–261.

Dickson, L. E. (1900), *Linear Groups, with an Exposition of the Galois Field Theory*. Reprinted, Dover, New York, 1958.

Dickson, L. E. (1905), Determination of the ternary modular linear groups. *Amer. J. Math.* **27**, 189–202.

DiMartino, L. and Wagner, A. (1981), The irreducible subgroups of $PSL(V_5, q)$, where $q$ is odd. *Resultate der Math.*

Doro, S. (1975), On finite linear groups in nine and ten variables. Thesis, Yale University.

Feit, W. (1976), On finite linear groups in dimension at most 10. In *Proc. Conf. Finite Groups*, edited by W. R. Scott and F. Gross. Academic Press, New York, pp. 397–407.

Fischer, B. (1969), Finite groups generated by 3-transpositions. Preprint. University of Warwick.

Galois, É. (1846), Oeuvres mathématiques, *J. de Math.* **11**, 381–444.

Gierster, J. (1881), Die Untergruppen der Galois'schen Gruppe der Modulargleichungen für den Fall eine primzahligen Transformationsgrades. *Math. Ann.* **18**, 319–365.

Hall, Jr., M. and Wales, D. B. (1968), The simple group of order 604,800. *J. Algebra* **9**, 417–450.

Hamill, C. M. (1951), On a finite group of order 6,531,840. *Proc. LMS (2)* **52**, 401–454.

Hartley, E. M. (1950), Two maximal subgroups of a collineation group in five dimensions. *Proc. Camb. Phil. Soc.* **46**, 555–569.

Hartley, R. W. (1926), Determination of the ternary linear collineation groups whose coefficients lie in the $GF(2^n)$. *Ann. of Math. (2)* **27**, 140–158.

Hazewinkel, M., Hesselink, W., Siersma, D., and Veldkamp, F. D. (1977), The ubiquity of Coxeter–Dynkin diagrams (an introduction to the A–D–E problem). *Nieuw. Archief (3)* **25**, 257–307.

Ho, C. Y. (1976), Chevalley groups of odd characteristic as quadratic pairs. *J. Algebra* **41**, 202–211.

Hoggar, S. G. (1978), Bounds for quaternionic line systems and reflection groups. *Math. Scand.* **43**, 241–249.

Huffman, W. C. (1975), Linear groups containing an element with an eigenspace of codimension two. *J. Algebra* **34**, 260–287.

Huffman, W. C. and Wales, D. B. (1975), Linear groups of degree $n$ containing an element with exactly $n - 2$ equal eigenvalues. *Linear and Multilinear Algebra* **3**, 53–59.

Huffman, W. C. and Wales, D. B. (1976), Linear groups of degree eight with no elements of order 7. *Ill. J. Math.* **20**, 519–527.

Huffman, W. C. and Wales, D. B. (1977), Linear groups containing an involution with two eigenvalues $-1$. *J. Algebra* **45**, 465–515.

Huffman, W. C. and Wales, D. B. (1978), Linear groups of degree nine with no elements of order seven. *J. Algebra* **51**, 149–163.

Huppert, B. (1967), *Endliche Gruppen I*. Springer, Berlin–Heidelberg–New York.

Jordan, C. (1870), *Traité des Substitutions et des Équations Algébriques*. Gauthier-Villars, Paris.

Jordan, C. (1878), Mémoire sur les équations différentielles linéaires à intégrale algébrique. *J. Reine Angew. Math.* **84**, 89–215.

Jordan, C. (1879), Sur la détermination des groupes d'ordre fini contenus dans le group linéaire. *Atti Accad. Napoli* **8**, 1–41.

Kantor, W. M. (1979a), Subgroups of classical groups generated by long root elements. *TAMS* **248**, 347–379.

Kantor, W. M. (1979b), Permutation representations of the finite classical groups of small degree or rank. *J. Algebra* **60**, 158–168.

Kantor, W. M. (1980), Linear groups containing a Singer cycle. *J. Algebra* **62**, 232–234.

Kantor, W. M., and Liebler, R. A. (1982), The rank 3 permutation representations of the finite classical groups. *TAMS* (to appear).

Klein, F. (1876), Ueber binäre Formen mit linearen Transformationen in sich selbst. *Math. Ann.* **9**, 183–208.

Klein, F. (1879), Ueber die Transformationen siebenter Ordnung der elliptischen Functionen. *Math. Ann.* **14**, 428–471.

Klein, F. (1884), *Vorlesungen über das Ikosaeder und die Auflösung der Gleichungen vom fünften Grade*. Teubner, Leipzig.

Lindsay, II, J. H., (1971), Finite linear groups of degree six. *Can. J. Math.* **23**, 771–790.

McLaughlin, J. (1967), Some groups generated by transvections. *Arch. Math.* **18**, 364–368.

McLaughlin, J. (1969), Some subgroups of $SL_n(F_2)$. *Ill. J. Math.* **13**, 108–115.

McLaughlin, J. (1969), A simple group of order 898,128,000. In *Theory of Finite Groups*, edited by R. Brauer and C.-H. Sah. Benjamin, New York, pp. 109–111.

Mitchell, H. H. (1911a), Determination of the ordinary and modular ternary linear groups. *TAMS* **12**, 207–242.

Mitchell, H. H. (1911b), Note on collineation groups. *Bull. AMS* **18**, 146–147.

Mitchell, H. H. (1913), Determination of the finite quaternary linear groups. *TAMS* **14**, 123–142.

Mitchell, H. H. (1914a), Determination of all primitive collineation groups in more than four variables which contain homologies. *Amer. J. Math.* **36**, 1–12.

Mitchell, H. H. (1914b), The subgroups of the quaternary abelian linear groups. *TAMS* **15**, 379–396.

Mitchell, H. H. (1935), Linear groups and finite geometries. *Amer. Math. Monthly* **42**, 592–603.

Moore, E. H. (1904), The subgroups of the generalized finite modular group. *Decennial Publications of the University of Chicago* **9**, 141–190.

Mwene, B. (1976), On the subgroups of the group $PSL_4(2^m)$. *J. Algebra* **41**, 79–107.

Mwene, B. (1982), On some subgroups of $PSL(4, q)$ $q$ odd.

Odlyzko, A. M. and Sloane, N. J. A. (1979), New bounds on the number of unit spheres that can touch a unit sphere in $n$ dimensions, *J. Comb. Theory (A)* **26**, 210–214.

Patton, W. H. (1972), The minimum index for subgroups in some classical groups: a generalization of a theorem of Galois. Thesis, University of Illinois at Chicago Circle.

Piper, F. C. (1965), Collineation groups containing elations, I. *Math. Z.* **89**, 181–191.

Piper, F. C. (1966a), Collineation groups containing elations, II. *Math. Z.* **92**, 281–287.

Piper, F. C. (1966b), On elations of finite projective spaces of odd order. *JLMS* **41**, 641–648.

Piper, F. C. (1968), On elations of finite projective spaces of even order. *JLMS* **43**, 459–464.

Piper, F. C. (1973), On elations of finite projective spaces. *Geom. Ded.* **2**, 13–27.

Seitz, G. M. (1979), Subgroups of finite groups of Lie type. *J. Algebra* **61**, 16–27.

Seitz, G. M. (1982), Generation of finite groups of Lie type. *TAMS* (to appear).

Serezkin, V. N. (1976), Reflection groups over finite fields of characteristic $p > 5$. *Soviet Math. Dokl.* **17**, 478–480. (Correction. *Dokl. Akad. Nauk SSSR* **237** (1977), 504 (Russian).)

Shephard, G. C. (1952), Regular complex polytopes. *Proc. LMS (3)* **2**, 82–97.

Shephard, G. C. (1953), Unitary groups generated by reflections. *Can. J. Math.* **5**, 364–383.

Shephard, G. C. and Todd, J. A. (1954), Finite unitary reflection groups. *Can. J. Math.* **6**, 274–304.

Singer, J. (1938), A theorem in finite projective geometry and some applications to number theory. *TAMS* **43**, 377–385.

Thompson, J. G. (1970), Quadratic pairs. *Actes Cong. Intern. Math.* **1**, 375–376.

Tits, J. (1966), Classification of algebraic semi-simple groups. *AMS Proc. Symp. Pure Math.* **9**, 33–62.

Tits, J. (1974), *Buildings of Spherical Type and Finite BN-pairs*. Springer Lecture Notes 386.

Valentiner, H. (1889), De endelige Transformations-Gruppers Theori. *Kjöb. Skrift. (6)* **5**, 64–235.

van der Waerden, B. L. (1935), *Gruppen von Linearen Transformationen*. Springer, Berlin. Reprinted: Chelsea, New York, 1948.

Wagner, A. (1974), Groups generated by elations. *Abh. Hamburg* **41**, 190–205.

Wagner, A. (1978), Collineation groups generated by homologies of order greater than 2. *Geom. Ded.* **7**, 387–398.

Wagner, A. (1979), The subgroups of $PSL(5, 2^a)$. *Resultate der Math.* **1**, 207–226.

Wagner, A. (1980–1981), Determination of finite primitive reflections groups over an arbitrary field of characteristic not two, I, II, III *Geom. Dedicata* **9**, 239–253; **10**, 191–203, 475–523.

Wales, D. B. (1969), Finite linear groups of degree seven I. *Can. J. Math.* **21**, 1042–1056.

Wales, D. B. (1970), Finite linear groups of degree seven II, *Pacif. J. Math.* **34**, 207–235.

Wales, D. B. (1978), Linear groups of degree $n$ containing an involution with two eigenvalues $-1$, II. *J. Algebra* **53**, 58–67.

Wiman, A. (1896), Ueber eine einfache Gruppe von 360 ebenen Collineationen. *Math. Ann.* **47**, 531–556.

Wiman, A. (1899a), Bestimmung alle Untergruppen einer doppelt unendlichen Reihe von einfachen Gruppen. *Bihang till K. Svenska Vet.-Akad. Handl* **25**, 1–47.

Wiman, A. (1899b), Endliche Gruppen linearer Substitutionen. IB3f in *Encyk. Math. Wiss.*

Zalesskii, A. E. (1976), A classification of the finite irreducible linear groups of degree 5 over a field of characteristic other than $0, 2, 3, 5$ (Russian). *Dokl. Akad. Nauk BSSR* **20**, 773–775, 858.

Zalesskii, A. E. (1977), Classification of finite linear groups of degree 4 and 5 over fields of characteristic 2. *Dokl. Akad. Nauk BSSR* **21**, 389–392, 475 (Russian).

Zalesskii, A. E. and Serezkin, V. N. (1976), Linear groups generated by transvections. *Math. USSR Izvestija* **10**, 25–46.

Zalesskii, A. E. and Serezkin, V. N. (1977), Linear groups which are generated by pseudoreflections. *Izv. Akad. Nauk BSSR*, 9–16 (Russian).

Zalesskii, A. E. and Serezkin, V. N. (1980), Finite linear groups generated by reflections. *Izv. Akad. Nauk SSSR* **44**, 1279–1307 (Russian).

Zalesskii, A. E. and Suprenenko, I. D. (1978), Classification of finite irreducible linear groups of degree 4 over a field of characteristic $p > 5$. *Izv. Akad. Nauk BSSR*, 1978, 9–15 (Russian).

# On Covering Klein's Curve and Generating Projective Groups

Jeffrey Cohen*

An important event in the history of Riemann surface theory was F. Klein's investigation [6] of the principal congruence subgroup at level seven of the modular group and his subsequent discovery of the famous quartic curve $x^3y + y^3z + z^3x = 0$. This genus 3 curve was soon put to good use. A. Hurwitz [5] showed that a compact Riemann surface of genus $g$ has no more than $84(g-1)$ conformal homeomorphisms and cited Klein's curve to show that this bound is attainable. It is an easy consequence of Hurwitz's work that there is a one-to-one correspondence between conformal equivalence classes of compact Riemann surfaces that attain the upper bound and normal subgroups of finite index of

$$(2,3,7) := \langle x, y : x^2 = y^3 = (xy)^7 = 1 \rangle.$$

Factors of $(2,3,7)$ are therefore called *Hurwitz groups*. In [2], all such normal subgroups were obtained whose factor is an extension of an Abelian group by $PSL_2(7)$. In other words all Abelian covers of Klein's surface by compact Riemann surfaces exhibiting Hurwitz's upper bound were obtained. Here a new infinite family of Hurwitz groups is given whose members act on covers of Klein's curve. Further, matrix representations of certain groups from [2] are obtained and used to decide precisely when the extension splits.

Related to $(2,3,7)$ is the classical modular group, i.e. $PSL_2(Z)$. This group is known to be the free product of cyclic groups of orders two and three. In [12], Sinkov computed the precise number of normal subgroups of the modular group with factor isomorphic to $PSL_2(p)$. Here, a more efficient reckoning is achieved by employing the results of [9]. In fact, the same computation is performed for $PGL_2(p)$. The method used extends in a fairly obvious manner to free products of cyclic groups.

*Mathematics and Statistics Department, University of Pittsburgh, Pittsburgh, Pennsylvania 15160, U.S.A.

## 1. Modular Group Factors

**Theorem 1.** *Let $p \geqslant 7$ be a prime number, and let $\mathscr{Q}(G)$ denote the number of abstract definitions of $G$ as a factor of the modular group. Let $K$ be the number of elements of $\{5, 8, 10, 12\}$ which divide either $p+1$ or $p-1$, and let $L$ equal the number of elements of $\{8, 12\}$ which either divide $2(p+1)$ but not $p+1$ or divide $2(p-1)$ but not $p-1$. Then*

(i) (*Sinkov*) $\mathscr{Q}[PSL_2(p)] = \frac{1}{2}(p-3) - K$;
(ii) $\mathscr{Q}[PGL_2(p)] = \frac{1}{2}(p-1) - L$.

*Proof.* By [9, Theorem 1], given $\alpha \in GF(q^2)$, there exist $A, B, C \in SL_2(q^2)$ with

$$\operatorname{tr} A = 0, \qquad \operatorname{tr} B = 1, \qquad \operatorname{tr} AB = \alpha.$$

Further, by Corollary 2 and Theorem 3 of [9], if $n \geqslant 7$, then all such matrix triples differ by an automorphism of $SL_2(q^2)$. Thus once $\alpha$ is specified, the normal subgroup of $(2, 3, \infty)$ is forced. Now since $(2, 3, n)$ has order $\leqslant 60$ for $n \leqslant 5$ and $(2, 3, 6)$ is solvable, it follows that if $\langle \bar{A}, \bar{B} \rangle$ is either of the two groups in question, then $|\overline{AB}| > 6$. (We employ "bar" to denote the natural homomorphism from $GL_2(p)$ to $PGL_2(p)$.) By [9, Theorem 5], if $|\overline{AB}| > 6$, then $PSL_2(p)$ or $PGL_2(p)$ is generated according to whether $|\overline{AB}|$ is the order of an element of $PSL_2(p)$ or not. Suppose $|\overline{AB}| = d$. If $d \neq p$ and is odd, then $|AB| \in \{d, 2d\}$, and since $\alpha = \omega + \omega^{-1}$ where $|\omega| = |AB|$, there are $\frac{1}{2}\phi(d) + \frac{1}{2}\phi(2d)$ possibilities for $\alpha$, where $\phi$ is Euler's function. Passing to the factor group, one obtains $\frac{1}{4}\phi(d) + \frac{1}{4}\phi(2d)$ cosets. Similarly, if $d$ is even, one obtains $\frac{1}{4}\phi(2d)$ cosets. If $d = p$, then $\alpha = \pm 2$, which produces one coset. Therefore

$$\mathscr{Q}\big[PSL_2(p)\big] = 1 + \frac{1}{4} \sum_{S \not\ni d \mid (p+1)} \phi(d) + \frac{1}{4} \sum_{S \not\ni e \mid (p-1)} \phi(e),$$

where $S = \{1, 2, 3, 4, 5, 6, 8, 10, 12\}$. In the sum 1 and 2 appear twice, while 3, 4, and 6 appear once, for a total contribution of 10. Thus

$$\begin{aligned} \mathscr{Q}\big[PSL_2(p)\big] &= 1 + \tfrac{1}{4}\big[(p+1) + (p-1) - 10 - 4K\big] \\ &= \tfrac{1}{2}(p-3) - K. \end{aligned}$$

Let $T$ denote the union of the following three sets:

(1) set of all divisors of $p+1$;
(2) set of all divisors of $p-1$;
(3) $\{4, 8, 12\}$.

Then

$$\mathscr{Q}\big[PGL_2(p)\big] = \frac{1}{4} \sum_{T \not\ni d \mid 2(p+1)} \phi(d) + \frac{1}{4} \sum_{T \not\ni e \mid 2(p-1)} \phi(e).$$

The result follows from the fact that 4 appears once for a contribution of $\phi(4) = 2$. $\square$

For the sake of completeness we include the following easily checked results:

$$\mathscr{C}[PSL_2(p)] = 1 \quad \text{for } p \leqslant 5,$$

$$\mathscr{C}[PGL_2(p)] = \begin{cases} 1 & \text{if } p = 2, \\ 1 & \text{if } p = 3, \\ 0 & \text{if } p = 5. \end{cases}$$

The industrious reader may now reformulate Theorem 1 in terms of congruency classes modulo 120 and 24.

There is a useful result credited to Fricke [7] in [13], where it is restated in terms of Wohlfahrt's new concept of level. In [10] this result is called Wohlfahrt's theorem, and not wishing to decide whose theorem it is, the author [4] has referred to this result as the Fricke–Wohlfahrt theorem. We shall only state a special case of the theorem.

**Theorem 2** [Fricke–Wohlfahrt]. *Suppose the modular group $\Gamma$ is represented as the free product of $\langle x \rangle$ and $\langle y \rangle$ where $|x| = 2$ and $|y| = 3$. Let $G$ be a finite group, $\phi: \Gamma \to G$ a homomorphism, and suppose that the kernel of $\phi$ is a congruence subgroup. Then* $\operatorname{Ker}\phi$ *contains the principal congruence subgroup at level $|\phi(xy)|$.*

Let $\Gamma_n$ denote the principal congruence subgroup at level $n$. Then it is easy to check (see [4]) that the composition series of $\Gamma/\Gamma_r$ consists solely of elementary Abelian $p_1$-groups and $PSL_2(p_1)$ for $p_1 \mid n$. Now suppose $x$, $y$, and $\phi$ are as in the Fricke–Wohlfahrt theorem, $G \in \{PSL_2(p), PGL_2(p)\}$, and $\operatorname{Ker}\phi$ is a congruence subgroup. Then $PSL_2(p)$ must appear among the composition factors of $\Gamma/\Gamma_{|\phi(xy)|}$. Since $|\phi(xy)|$ divides one of $p, p+1, p-1$, it follows that $|\phi(xy)| = p$. In the proof of Theorem 1, we saw that only one definition of $PSL_2(p)$ involved an element of order $p$. This yields a result,

**Corollary 1.** *Let $p$ be any prime number. Then there is precisely one* (*respectively, is no*) *normal congruence subgroup of the modular group whose factor is isomorphic to $PSL_2(p)$* (*respectively, $PGL_2(p)$*).

This corollary and theorem lead to a slight strengthening of the final result in [10].

**Corollary 2.** *Let $p \geqslant 13$ be a prime number. Then there exists a normal noncongruence subgroup of the modular group whose factor is isomorphic to $PSL_2(p)$.*

In [4] this result has been generalized further. It is shown that if $n$ has a prime factor $\geqslant 13$, then there exists a normal noncongruence subgroup of the modular group whose factor is isomorphic to $\Gamma/\Gamma_n$.

There is an open question related to the contents of this section. This also involves the ongoing program to classify all finite simple groups. All known finite simple groups are two-generator groups, and it might be of interest not only to know which of these are factors of the modular group, but to try to classify all simple modular group factors. Garbe has recently shown that $PSL_3(q)$ is a factor of the modular group for $q \neq 4$, and it is known [3] that $PSU_3(q^2)$ is such a factor for $q \notin \{2,5\}$, but clearly, much remains to be done. In fact, one can ask the same question about $(2,3,7)$, and even here little is known about its finite simple factors.

## 2. Hurwitz Groups

In a recent paper [2], all Hurwitz groups which are an extension of an Abelian group by $PSL_2(7)$ were determined and presented by generators and relations. The smallest of these is Sinkov's group of order 1344 which appears in [11]. Several investigators have erroneously believed that Sinkov's group is the holomorph of the elementary Abelian group of order 8 (it is perhaps best not to name anyone), and to combat such tendencies concerning the newly found groups, it seems desirable to decide which are split extensions. The notation of this section is to be found in [2] and [8], and the reader will need to be familiar with both.

Most of the factors of the modular group so far considered have many different definitions as modular group factors. We now present a family of groups with relatively few such definitions. Let $G$ be one of the groups in [2]. Since the matrix $B$ has no eigenvector[1] associated with one, the order of $B$ as an element of $G$ is seven. Now if $\langle x, y\rangle = G$ with $x^2 = y^3 = 1$, then modulo a normal subgroup, $x$ and $y$ generate $PSL_2(7)$, so that $|xy| = 7$. Hence the number of definitions of $G$ as a factor of the modular group equals the number of definitions of $G$ as a factor of $(2,3,7)$. We restate this as

**Proposition 1.** *The number of normal subgroups of the modular group with factor $G$ is $\sigma$.*

N.B. The value of $\sigma$ is 1 for infinitely many $G$.

In order to represent $G$ by matrices we use the $6 \times 6$ matrices in [2] and the fact that as elements of $(2,3,7)$, $R = B^2A^{-1}B^2$ and $S = B^2$. This suggests that one investigate the following $7 \times 7$ matrices with entries in the ring of integers modulo $n$:

$$\bar{R} = \left[\begin{array}{cccccc|c} 0 & 1 & -1 & 0 & 0 & 0 & 0 \\ 0 & 1 & 0 & 0 & 0 & 0 & 0 \\ 0 & 0 & 0 & 0 & 0 & 1 & 0 \\ 0 & 0 & 0 & -1 & 1 & 0 & 0 \\ 0 & 1 & 0 & -1 & 0 & 0 & 0 \\ -1 & 1 & 0 & 0 & 0 & 0 & 0 \\ \hline 0 & 0 & 0 & 0 & 0 & 0 & 1 \end{array}\right],$$

$$\bar{S} = \left[\begin{array}{cccccc|c} 0 & 0 & 0 & 0 & -1 & 1 & 0 \\ 0 & 0 & 0 & 0 & -1 & 0 & 0 \\ 1 & 0 & 0 & 0 & -1 & 0 & -1 \\ 0 & 1 & 0 & 0 & -1 & 0 & 0 \\ 0 & 0 & 1 & 0 & -1 & 0 & 0 \\ 0 & 0 & 0 & 1 & -1 & 0 & 0 \\ \hline 0 & 0 & 0 & 0 & 0 & 0 & 1 \end{array}\right].$$

[1] If $\rho \neq 7$. The remainder of the proof of Proposition 1 uses Theorem 3.

Then $\bar{R}^3 = \bar{S}^7 = (\bar{R}\bar{S})^2 = I$, and $[\bar{R}, \bar{S}]^4$ has the identity matrix for its $6 \times 6$ block, and its $6 \times 1$ block is

$$e_0 = [-6 \quad -2 \quad -2 \quad 0 \quad -2 \quad 0]'.$$

Applying the $6 \times 6$ block of $\bar{S}, \bar{S}^2, \bar{S}^3$ to $e_0$ yields four columns which are linearly independent over $Z_n$ if and only if $n$ is odd. Thus if $n$ is odd, then $\bar{R}$ and $\bar{S}$ generate the group of all matrices of the form

$$\left[\begin{array}{c|c} W & V \\ \hline 0 & 1 \end{array}\right]$$

where $W$ is a word in $\bar{R}$ and $\bar{S}$, and $V$ is any column. This is clearly a split extension of $(Z_n)^6$ by $PSL_2(7)$, and the stated objective has been accomplished.

**Theorem 3.** *$PSL_2(7) \precsim G$ if and only if $p_i$ and $q_j$ are odd for $i = 1, \ldots, s$ and $j = 1, \ldots, t$. Hence the extension splits iff $n$ is odd.*

It is likely that $\bar{R}$ and $\bar{S}$ (and certain similar generating sets) will prove useful in further, more geometric investigations related to the work in [2]. For example, if $n = 4$, $\langle \bar{R}, \bar{S} \rangle$ is Sinkov's group of order 1344. A representation (due to G. A. Miller) is given in [11], but it seems easier to derive group-theoretic (and hence geometric) information from our representation. The term "Sinkov's group" is somewhat a misnomer. Miller actually discovered it in his investigation of transitive subgroups of $S_{14}$. Sinkov's achievement, however, has geometric significance, since he showed that this group is a factor of $(2,3,7)$.

If $\bar{R}$ and $\bar{S}$ are taken over the rational integers, then an infinite factor of $(2,3,7)$ is obtained. This is an extension (which from Theorem 3 does not split) of the free Abelian group of rank 6 by $PSL_2(7)$. Alternatively, if $(2,3,7)/K \simeq PSL_2(7)$, then this group is isomorphic to $(2,3,7)/[K,K]$. One can construct representations of any of the groups $G$ in [2] by working modulo $n$ or $4n$ according as $n$ is even or odd and factoring out an easily obtained $Z_n$-submodule.

*Proof of Theorem* 3. From the remarks preceding the theorem's statement, it need only be established that Sinkov's group of order 1344 is a nonsplit extension. This group has an element of order 8, which has order 4 in the factor. In a split extension, $PSL_2(7)$ acts as a group of $3 \times 3$ matrices on the elementary Abelian group of order 8. Now any element of order 4 in $GL_3(2)$ satisfies the polynomial $x^3 + x^2 + x + 1$. Thus, in a split extension, if $M$ is a matrix of order 4 and $V$ a member of the normal Abelian group, then

$$(V, A)^4 = \left( \sum_{i=0}^{3} A^i V, A^4 \right) = (0, I). \qquad \square$$

## 3. A New Family of Hurwitz Groups

In [4] extensions of a $p$-group by $PSL_2(p^n)$ for all Hurwitz $PSL_2(p^n)$ were obtained with the exception of $p^n = 7$. The Hurwitz $PSL_2(q)$ were described by Macbeath [9]. The case of 7 will now be reexamined; even though such extensions were constructed in [2], we want something that resembles the construction of [4] in a natural way.

Denote by $Z_{7^n}[\alpha]$ the ring $\{\sum_{i=0}^{2} a_i\alpha^i : a_i \in Z_{7^n}\}$ subject to the relation $\alpha^3 + \alpha^2 - 2\alpha - 1 \equiv 0 \pmod{7^n}$. Then

$$Z_{7^n}[\alpha] \simeq Z_{7^n}[x]/(x^3 + x^2 - 2x - 1).$$

For $n = 1$, the relation becomes $(\alpha - 2)^3$ so that there is a natural homomorphism onto $Z_7[x]/((x-2)^2)$. As usual, define

$$PSL_2(Z_{7^n}[\alpha]) = SL_2(Z_{7^n}[\alpha])/\{\pm I\}.$$

For $n \geqslant m$, let $\Pi_m^n$ denote the natural homomorphism

$$\Pi_m^n : PSL_2(Z_{7^n}[\alpha]) \to PSL_2(Z_{7^m}[\alpha]).$$

Clearly

$$\operatorname{Ker} \Pi_{n-1}^n = \{I + 7A : \operatorname{trace} A = 0\},$$

a set of order $7^9$.

**Proposition 2.** *$PSL_2(Z_{7^n}[\alpha])$ is an extension of the elementary Abelian group of order $7^9$ by $PSL_2(Z_{7^{n-1}}[\alpha])$. $PSL_2(Z_7[\alpha])$ is an extension of $(Z_7)^3$ by $PSL_2(Z_7[x]/((x-2)^2))$. $PSL_2(Z_7[x]/((x-2)^2))$ is a split extension of $(Z_7)^3$ by $PSL_2(7)$ and hence is one of the groups described in* [2].

**Corollary 3**

$$|PSL_2(Z_{7^n}[\alpha])| = 2^3 \times 3 \times 7^{9n-2},$$

$$\left|PSL_2\left(\frac{Z_7[x]}{((x-2)^2)}\right)\right| = 57624.$$

For $G$ a group let $\Phi(G)$ denote the Frattini subgroup of $G$.

**Proposition 3.** $\operatorname{Ker} \Pi_{n-1}^n \leqslant \Phi(\operatorname{Ker} \Pi_1^n)$ *for* $n \geqslant 3$.

*Proof.* We shall show that if $\operatorname{tr} A = 0$, then $I + 7^{n-1}A$ is the seventh power of a member of $\operatorname{Ker} \Pi_1^n$. Let $k = (1 - 4 : 7^{2n-4}) \det A$, and note that since $\operatorname{tr} A = 0$,

$$\det(I + 7^{n-2}A) = 1 + 7^{2n-4} \det A. \tag{$*$}$$

Let $B = k(I + 7^{n-2}A)$, so that by $(*)$ $\det B = 1$ and an easy computation gives $B^7 = I + 7^{n-1}A$. $\square$

**Proposition 4.** *In*

$$PSL_2(Z_7[\alpha]) \xrightarrow{\Pi_1} PSL_2\left(\frac{Z_7[x]}{((x-2)^2)}\right) \xrightarrow{\Pi_0} PSL_2(7),$$

*let $\Pi_1$ and $\Pi_0$ be the natural maps. Then $\Phi(\operatorname{Ker} \Pi_0 \circ \Pi_1) = \operatorname{Ker} \Pi_1$.*

*Proof*. Since $PSL_2(7)$ cannot act on a two-dimensional vector space over $GF(7)$, $|\Phi(G)| \in \{7^3, 7^6\}$. Since $\operatorname{Ker} \Pi_1 \circ \Pi_0$ is non-Abelian, $7^6$ is excluded. □

**Proposition 5.** *Suppose that $R$ is a commutative ring with 1 of characteristic $p^2$, that $n \geqslant 3$ and that*

$$A = \begin{pmatrix} 1+pa & 1+pc \\ pb & 1+pd \end{pmatrix}.$$

*Then*

$$A^n = \begin{pmatrix} 1+p\left[na + \binom{n}{2}d\right] & n + p\left[\binom{n}{2}a + \binom{n}{3}b + nc + \binom{n}{2}d\right] \\ npb & 1 + p\left[\binom{n}{2}a + \binom{n}{3}b\right] \end{pmatrix}.$$

*In particular if $(p,b) = 1$, then*

$$A^p = \begin{pmatrix} 1 & p \\ 0 & 1 \end{pmatrix}.$$

*Proof*. Induction on $n$. □

**Theorem 4.** *Suppose $G \leqslant PSL_2(Z_{49}[\alpha])$ and that $\Pi_1^2(G) = PSL_2(Z_7[\alpha])$. Then $G = PSL_2(Z_{49}[\alpha])$.*

*Proof*. Since the restriction of $\Pi_1^2$ to $G$ is epic, $G$ contains an element having the form of $A$ in Proposition 1. Thus

$$I + 7\begin{pmatrix} 0 & 1 \\ 0 & 0 \end{pmatrix} \in G.$$

Let $c, d \in Z_{49}[\alpha]$ be invertible, so that $I + 7B(c,d) \in G$, where

$$B(c,d) = \begin{pmatrix} d & 0 \\ -c & d^{-1} \end{pmatrix}\begin{pmatrix} 0 & 1 \\ 0 & 0 \end{pmatrix}\begin{pmatrix} d^{-1} & 0 \\ c & d \end{pmatrix} = \begin{pmatrix} cd & d^2 \\ -c^2 & -cd \end{pmatrix}.$$

Note that $\operatorname{Ker} \Pi_1^2$ is a nine-dimensional vector space over $GF(7)$. It is now a tedious but straightforward computation to check that the nine $I + 7B(c,d)$ obtained by letting $c$ and $d$ take values in $\{1, \alpha, \alpha^2\}$ are linearly independent. □

Employing Propositions 3 and 4 and Lemma 1, we obtain a result.

**Theorem 5.** *Suppose $G \leqslant PSL_2(Z_{7^n}[\alpha])$ and that $\Pi_1 \circ \Pi_1^n \cap \operatorname{Ker} \Pi_0 \neq \{1\}$. Then $G = PSL_2(Z_{7^n}[\alpha])$.*

Let $t \in Z_{7^n}$ satisfy $t^2 - t + 1 = 0$, and set

$$A = \begin{pmatrix} 0 & t^{-1} \\ -t & 0 \end{pmatrix} \quad \text{and} \quad B = \begin{pmatrix} -t & 0 \\ \alpha t & -t^{-1} \end{pmatrix}.$$

Then $-A^2 = B^3 = (AB)^7 = I$. One easily verifies that $(A^{-1}B^{-1}AB)^4$ represents the identity coset only in $PSL_2(Z)$ so that by Theorem 4, $\langle \bar{A}, \bar{B} \rangle = PSL_2(Z_{7^n}[\alpha])$.

**Corollary 4.** *$PSL_2(Z_{7^n}[\alpha])$ and $PSL_2(Z_7[x]/((x - \alpha)^2))$ are Hurwitz groups.*

Let $\Gamma$ be a cocompact Fuchsian group and $N \nabla \Gamma$ of finite index. In [1] normal subgroups of $\Gamma$ contained in $N$ were constructed. The reader can now use [1] to find further factors of $(2, 3, 7)$.

## References


[1] Cohen, J., Some compact Riemann surfaces via Fuchsian groups (submitted).

[2] Cohen, J., On Hurwitz extensions by $PSL_2(7)$ (to appear).

[3] Cohen, J., On non-Hurwitz groups and non-congruence subgroups of the modular group (to appear).

[4] Cohen, J., Homomorphisms of cocompact Fuchsian groups on $PSL(Z_n[x]/(f(x)))$ (submitted).

[5] Hurwitz, A., Ueber algebraische Gebilde mit eindeutigen Transformationen in sich. *Math. Annalen* **41** (1893), 403–442.

[6] Klein, F., Ueber die Transformationen siebenter Ordnung der elliptischer Funktionen. *Math. Annalen* **14** (1879), 428–471.

[7] Klein, F. and Fricke, R., *Vorlesungen ueber die Theorie der elliptischen Modulfunktionen*, Band I, Leipzig 1890.

[8] Leech, J., Generators for certain normal subgroups of (2, 3, 7), *Proc. Cambridge Phil. Soc.* **61** (1965), 321–332.

[9] Macbeath, A. M., Generators of the linear fractional groups. In *Proc. Sym. of Pure Math. in Number Theory*, Vol. XII. Houston 1967.

[10] Newman, M., Maximal normal subgroups of the modular group. *Proc. Amer. Math. Soc.* **19** (1968), 1138–1144.

[11] Sinkov, A., On the group-defining relations $(2, 3, 7; p)$. *Ann. of Math.* **38** (1937), 577–584.

[12] Sinkov, A., The number of abstract definitions of $LF(2, p)$ as a quotient group of $(2, 3, n)$. *J. Algebra* **12** (1969), 525–532.

[13] Wohlfahrt, K., An extension of F. Klein's level concept, *Illinois J. Math.* **8** (1964), 529–535.

# A Local Approach to Buildings

J. Tits*

The object of this paper is the comparison of two notions of (combinatorial) buildings, that of [14] (or [1]), and an earlier version (cf. e.g. [10]), which has lately regained interest through the work of F. Buekenhout on sporadic groups (cf. [2], [3], [16]).

The origin of the notion lies in [6], [7], [8], and other, related papers, where a method was devised for the geometric study of the semisimple Lie groups (especially the exceptional ones) and, later on, the semisimple algebraic groups. To every such group was associated a certain geometry (the geometry of its "parabolic subgroups," in present-day terminology), and the key observation was that essential properties of these geometries can be easily read from the Coxeter–Witt–Dynkin diagrams of the corresponding groups. In particular, an inclusion relation between two diagrams is reflected by the simple geometric notion of residue (cf. Section 1.2 below). In [10], an axiomatization of those results was proposed: to every Coxeter diagram $M$ (as defined below, in Section 1.1), I associated a certain class of incidence geometries, the *geometries of type* $M$, taking the correspondence between subdiagrams and residues as the crucial axiom. However, examples showed that that axiom—here called (Res) (cf. Section 1.4)—was not sufficient to define the "good" geometries of type $M$, that is, to prove some of the properties which always hold in geometries coming from semisimple algebraic groups (for instance the "linearity" expressed by the property (Int′) of the last remark in this paper). In a footnote of [10], it was suggested that "good" geometries of type $M$ could be characterized by adding to the axiom (Res) a certain condition of "simple connectedness." Later on, I found another, completely different and technically more efficient approach to the "good" geometries, inspired by the fundamental papers [4] and [5] of C. Chevalley (in particular, by his version of Bruhat's lemma). This led, in [11] and [12], to the

*Collège de France, (75231) Paris (Cedex 05) France.

introduction of the objects now called *buildings* (after N. Bourbaki [1]). The main result of the present paper, expressed for instance in Theorem 1(ii) (cf. Section 1.6) and in Corollary 3 (cf. Section 5.3), is a characterization of the buildings in the spirit of [10], that is, by means of the "local" property (Res) completed by a "global" condition, reminiscent of the topological simple connectedness.

The main axiom (Res) being inductive, our "local approach" needs a starting point. In [10], the role of "building blocks" was played by the generalized polygons—i.e. the buildings of rank 2. Here our characterization of buildings requires also some "blocks" of rank 3, namely the buildings of type $\mathbf{C}_3$ and $\mathbf{H}_3$, and that turns out to be inescapable (cf. the comments following the statement of Theorem 1).

The connection established here between buildings and geometries of type $M$ can be used in two ways:

to deduce information on buildings from known results on geometries, for instance to prove the existence of a "great many" buildings of "most" types;

to investigate geometries of certain given types for which buildings are relatively well understood, as in the case of spherical or affine types (cf. Section 6.2 for example).

So far, buildings have always been described as incidence geometries or simplicial complexes. The results of the present paper find their simplest expression in a somewhat more abstract framework, that of *chamber systems*. It should be mentioned here that the idea of founding the theory of buildings on the notion of chamber was first expressed by L. Puig.

The paper is organized as follows. In Section 1, we establish the basic terminology concerning geometries and buildings (viewed as simplicial complexes), and we state Theorem 1, a first version of our main result. Chamber systems are introduced in Section 2, and various characterizations of buildings in terms of those systems are given in Sections 3 and 5, whereas Section 4 essentially has an auxiliary character. In Section 6, we return to geometries with the proof of Theorem 1 and some applications.

# 1. Geometries and Buildings; First Statement of the Main Result

Most definitions in this section are inspired by [10] and [14], but we do not strictly adhere to the terminology of those two papers, which does not quite suit our purpose.

### 1.1. Coxeter Diagrams

*In the whole paper, $I$ denotes a set*, on which no hypothesis is made for the time being. A *Coxeter diagram* over $I$ is a function $M: I \times I \to \mathbb{N} \cup \{\infty\}$ such that, for $i, j \in I$, $M(i,i) = 1$ and $M(i,j) = M(j,i) \geqslant 2$ if $i \neq j$. The elements of $I$ are called the *vertices* of the diagram $M$. We shall also use the familiar pictorial representation of $M$ as a graph whose vertices are indexed by $I$ and whose vertices $i$ and $j$ are joined by an edge of multiplicity $M(i,j)-2$ (no edge if $M(i,j)=2$) or by a simple edge labeled with $M(i,j)$. For instance, if $I=\{1,2,3\}$, the function $M$ whose values is given by the matrix

$$((M(i,j)) = \begin{bmatrix} 1 & 2 & 3 \\ 2 & 1 & 5 \\ 3 & 5 & 1 \end{bmatrix}$$

corresponds to the diagram

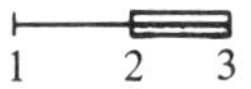

$$\underset{1}{\circ}\!\!-\!\!\!-\!\!\underset{2}{\circ}\!\equiv\!\underset{3}{\circ}$$

or equivalently

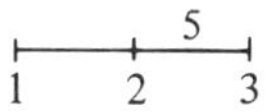

$$\underset{1}{\circ}\!\!-\!\!\!-\!\!\underset{2}{\circ}\!\!\overset{5}{-\!\!\!-}\!\!\underset{3}{\circ}\,.$$

If $J$ is a subset of $I$, the restriction of $M$ to $J\times J$, that is, the subdiagram of $M$ whose set of vertices is $J$, will be denoted by $M_J$. We define the *rank* of a Coxeter diagram as the cardinality of its set of vertices.

### 1.2. Geometries

We give the name of *geometry over $I$* to a system $\Gamma=(V,\tau,*)$ consisting of a set $V$, a map $\tau: V\to I$, and a binary symmetric relation $*$ on $V$ such that for any two elements $x$, $y$ of $V$ whose images by $\tau$ are the same, the relation $x*y$ holds if and only if $x=y$. The relation $*$ is the *incidence relation*, the image by $\tau$ of an element or a subset of $V$ is called its *type*, and the *rank* of $\Gamma$ is defined as the cardinality of $I$.

For example, the set of all proper nonempty linear subspaces of a projective space of dimension $n$ endowed with the dimension function and the symmetrized inclusion as incidence relation is a geometry over $\{0,1,\ldots,n-1\}$, called an *$n$-dimensional projective geometry*. (N.B. By "projective space," we mean a space satisfying the "formal" axioms of projective geometry, concerning intersections and spans; the lines are required to have at least *two* points.)

The above example suggests the following definition: a *flag* of $\Gamma$ is a set of pairwise incident elements of $V$. Two flags are said to be *incident* if their union is a flag. The *rank* (the *corank*) of a flag $X$ is the cardinality of $X$ (of $I-\tau(X)$).

Let $X$ be a flag, and let $Y$ be the set of all elements of $V-X$ incident to $X$. Then the system $(Y,\tau|_Y,*\cap(Y\times Y))$, considered as a geometry over $I-\tau(X)$, is called the *residue of $X$* in $\Gamma$ and denoted by $\Gamma_X$.

With every geometry $\Gamma=(V,\tau,*)$ is associated *its graph*, whose set of vertices is $V$ and whose edges join the incident pairs. We say that $\Gamma$ is *connected* if the associated graph is connected (which, for us, implies nonempty), and that $\Gamma$ is *residually connected* if the residue of every flag of corank $\geqslant 2$ (of corank 1) is connected (nonempty). Note that a residually connected geometry of rank $\geqslant 2$ is connected, since it is the residue of its empty flag.

### 1.3. Morphisms and Quotients

If $\Gamma = (V, \tau, *)$ and $\Gamma' = (V', \tau', *')$ are two geometries over $I$, we define a *morphism* (of geometries over $I$) of $\Gamma'$ into $\Gamma$ as a type-preserving function $\varphi: V' \to V$ mapping incident pairs onto incident pairs. Isomorphisms and automorphisms are defined as usual. The morphism $\varphi$ is called a *covering* if it is surjective and if, for every $x \in V'$, it maps the residue of $x$ in $\Gamma'$ isomorphically onto the residue of $\varphi(x)$ in $\Gamma$. We say that $\Gamma$ is *simply connected* (*residually simply connected*) if it is connected (residually connected) and if every covering by a connected geometry is an isomorphism (if all residues of flags of corank $\geqslant 3$ in $\Gamma$ are simply connected).

Let $A$ be an automorphism group of $\Gamma = (V, \tau, *)$ and let us denote by $V/A$ the set of orbits of $A$ in $V$, by $\tau/A$ the "type mapping" on $V/A$ (the type of an orbit is an element of $I$), and by $*/A$ the incidence relation on $V/A$ defined as follows: two orbits are incident for $*/A$ if they are of the form $Ax, Ay$ where $x$, $y$ are incident for $*$. Then $(V/A, \tau/A, */A)$ is a geometry over $I$ denoted by $\Gamma/A$ and called the *quotient of* $\Gamma$ *by* $A$. The mapping $x \mapsto Ax$ (for $x \in V$) is a morphism of $\Gamma$ onto $\Gamma/A$ called the *canonical projection*.

Later on, we shall be interested in the following condition:

**(Q1)** for every flag $X$ of $\Gamma$, the canonical projection $\pi: \Gamma \to \Gamma/A$ induces an isomorphism of the quotient of the residue $\Gamma_X$ by the stabilizer of $X$ in $A$ onto the residue $(\Gamma/A)_{\pi(X)}$.

This can be diversely translated in terms of properties of the group $A$. For instance, if $I$ is finite, (Q1) is easily seen to be equivalent to the combination of the following two conditions:

**(Q2′)** the elements of an orbit of $A$ in $V$ which are incident to a given flag $X$ form at most one orbit of the stabilizer of $X$ in $A$;

**(Q2″)** if $x$ and $y$ are two incident elements of $V$ and if $X$ is a flag incident to some element of $Ax$ and some element of $Ay$, then there exists $a \in A$ such that $X$ is incident to $ax$ and $ay$.

Note that the following simple condition implies both (Q2′) and (Q2″), and hence also (Q1):

**(Q3)** in the graph of $\Gamma$, the distance between two distinct vertices belonging to the same orbit of $A$ is at least 4.

### 1.4. Geometries of Type $M$

*In this subsection, the set $I$ is supposed to be finite.*

For $m \in \mathbb{N} \cup \{\infty\}$ and $m \geqslant 2$, a geometry of rank 2 is called a *generalized m-gon* if its graph has diameter $m$ and girth (minimum length of a cycle) $2m$ and if every vertex of the graph belongs to at least two edges.

Let $M$ be a Coxeter diagram over $I$. We define a *geometry of type M* as a residually connected geometry over $I$ such that, for $i, j \in I$ and $i \neq j$, the residue of any flag of type $I - \{i, j\}$ is a generalized $M(i, j)$-gon. Clearly,

**(Res)** the residue of a flag of type $J$ $(\subset I)$ in a geometry of type $M$ is a geometry of type $M_{I-J}$,

and that property, together with the residual connectedness and the fact that the geometries of type $\overset{m}{\longmapsto\!\!\dashv}$ are the generalized $m$-gons, characterizes the geometries of type $M$.

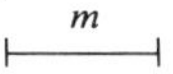

The following trivial observation will be useful:

(1) *Let* $\Gamma$ *be a geometry of type* $M$, *and let* $A$ *be a group of automorphisms of* $\Gamma$ *satisfying the condition* (Q1) (*or, equivalently, the conditions* (Q2′) *and* (Q2″)) *of Section* 1.3 *and operating freely on the set of all flags of corank* 2 *of* $\Gamma$; *then,* $\Gamma/A$ *is a geometry of type* $M$.

EXAMPLES

(a) The geometries of type

$$\mathbf{A}_n \quad \underbrace{\vdash\!\!\!-\!\!\!+\!\!\!-\!\!\!\dashv \;\cdots\; \vdash\!\!\!-\!\!\!\dashv}_{n \text{ vertices}}$$

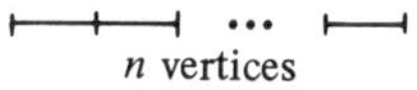

are just the $n$-dimensional projective geometries (cf. Sections 1.2 and 6.1.5 below).

(b) A "polar geometry of rank $n$," that is, a geometry whose elements are the linear subspaces of a polar space of rank $n$ (cf. [14, §7]) is of type

$$\mathbf{C}_n \quad \underbrace{\vdash\!\!\!-\!\!\!+\!\!\!-\!\!\!\dashv \cdots \vdash\!\!\!-\!\!\!+\!\!\!=\!\!\!=}_{n \text{ vertices}}$$

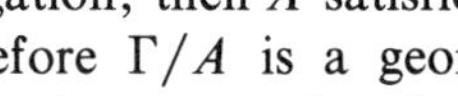

but there are other geometries of that type. Examples are easily obtained by application of (1). For instance, let $\Gamma$ be the geometry of all complex linear subspaces of a hyperquadric defined by a real quadric form of Witt index $\leqslant 1$ in $2n$ or $2n+1$ variables, and let $A$ be the automorphism group of order two of $\Gamma$ generated by the complex conjugation; then $A$ satisfies the conditions (Q2′) and (Q2″) of Section 1.3, and therefore $\Gamma/A$ is a geometry of type $\mathbf{C}_n$. In the language of the introduction, the polar geometries of rank $n$ would be the "good" geometries of type $\mathbf{C}_n$; we now proceed to define the "good" geometries of any type $M$, that is, the buildings.

### 1.5. Buildings

The objects which will be called *buildings* in this paper essentially correspond to the "weak buildings" of [14] (the "buildings" of [14] being of little use here). We shall merely recall the basic definitions, referring the reader to [12], [14], and [16] for further motivation.

A *complex* over $I$ is a system $\Delta = (V, \tau, \mathbb{S})$ consisting of a set $V$, a map $\tau: V \to I$ (called the *type*, as in Section 1.2), and a set $\mathbb{S}$ of subsets of $V$ (the *simplices*), with the following properties: all subsets of $V$ consisting of a single element are simplices, every subset of a simplex $S$ is a simplex (a *face* of $S$), and the restriction of $\tau$ to any simplex is injective. With every geometry $\Gamma = (V, \tau, *)$ over $I$ is associated its *flag complex* $\Delta(\Gamma) = (V, \tau, \mathbb{S})$, where $\mathbb{S}$ is the set of all flags of $\Gamma$. Clearly, the complex $\Delta(\Gamma)$ determines the geometry $\Gamma$, and we shall often make no distinction between them. A *morphism* (of complexes over $I$) of a complex $\Delta' = (V', \tau', \mathbb{S}')$ in a complex $\Delta = (V, \tau, \mathbb{S})$ is defined as a mapping $\varphi: V' \to V$ which is type-preserving (i.e. $\tau' = \tau \circ \varphi$) and maps simplices onto simplices; if $V' \subset V$ and if the inclusion mapping is a morphism, $V'$ is called a *subcomplex* of $V$. We define the *rank* and the *corank* of a simplex $S$ as the cardinalities of the sets $S$ and $I - \tau(S)$.

Let $M$ be a Coxeter diagram, and let $(W; (r_i)_{i \in I})$ be a Coxeter system of type $M$: this means that $W$ is a group—called *Coxeter group*—generated by the set $\{r_i \mid i \in I\}$, and that the relations

$$(r_i r_j)^{M(i,j)} = 1 \quad \text{whenever } M(i,j) \neq \infty$$

form a presentation of $W$ (cf. [1, p. 11]). For $J \subset I$, let $W_J$ denote the subgroup of $W$ generated by $\{r_j \mid j \in J\}$. A *Coxeter complex of type $M$* is defined as a complex over $I$ isomorphic with the complex $\Delta(W) = (V, \tau, \mathbb{S})$, where $V$ is the direct sum of the sets $W / W_{I - \{j\}}$, with $j \in J$, $\tau$ is given by $\tau(W / W_{I - \{j\}}) = \{j\}$, and $\mathbb{S}$ consists of all sets of the form

$$\{wW_{I - \{j\}} \mid j \in J\} \qquad (w \in W, \quad J \subset I).$$

A set $\mathcal{Q}$ of subcomplexes of a complex is called a *system of apartments of type $M$* if the elements of $\mathcal{Q}$, called the *apartments*, are Coxeter complexes of type $M$, and if the following two axioms hold:

**(Ap 1)** every two simplices belong to an apartment;

**(Ap 2)** if two apartments $\Sigma$ and $\Sigma'$ have two simplices $S$ and $S'$ in common, there is an isomorphism (of complexes over $I$) $\Sigma \to \Sigma'$ fixing $S \cup S'$.

We define a *building* of type $M$ as a complex possessing a system of apartments of type $M$; it can be shown that the set of *all* Coxeter subcomplexes of a building is a system of apartments.

### 1.6. Buildings as Geometries of Type $M$

Suppose $I$ finite. By [14, Sections 3.16 and 3.12], the *buildings of type $M$ are special cases of geometries of type $M$ ("identified" with their flag complexes*: cf. Section 1.5) *in which all residues of flags are also buildings*. The following theorem, a first version of our main result, is a partial converse of that statement.

**Theorem 1.** *Let $I$ be a finite set of cardinality $\geqslant 2$, $M$ a Coxeter diagram over $I$ and $\Gamma$ a geometry of type $M$.*

(i) *Suppose that in $\Gamma$, all residues (of flags) of type*

$$\mathbf{C}_3 \quad \textit{or} \quad \mathbf{H}_3$$

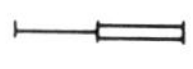

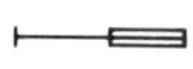

*are covered by buildings* (cf. Section 1.3). *Then there is a building $\Delta$ of type $M$ and a group $A$ of automorphisms of $\Delta$ satisfying the condition* (Q1)*—or, equivalently, the conditions* (Q2′) *and* (Q2″)*—of Section* 1.3 *and operating freely on the set of all flags of corank* 2, *such that $\Gamma$ is isomorphic with $\Delta/A$; the pair $(\Delta, A)$ is unique up to isomorphism. In particular, if $M$ has no subdiagram of type $\mathbf{C}_3$ or $\mathbf{H}_3$, every geometry of type $M$ is isomorphic to a quotient $\Delta/A$ of that kind.*

(ii) *The geometry $\Gamma$ is a building if and only if it is residually simply connected and if all residues of type $\mathbf{C}_3$ or $\mathbf{H}_3$ in $\Gamma$ are buildings.*

Note that (ii) can be viewed as a corollary of (i); but here, the two statements will be proved simultaneously.

In the above theorem, the condition on the residues of type $\mathbf{C}_3$ and $\mathbf{H}_3$ is essential. Indeed, it is possible to construct "free geometries" of type $\mathbf{C}_3$ and $\mathbf{H}_3$, which can have no relation with buildings in view of [14, Sections 8 and 9], and [15]. More generally, we shall show in a forthcoming paper that "free constructions" exist for geometries of type $M$ whenever $M$ has no subdiagram of type $\mathbf{A}_3$ (thus, projective 3-spaces are the only obstructions to the existence of such constructions). If, moreover, $M$ has no subdiagram of type $\mathbf{C}_3$ or $\mathbf{H}_3$, the above theorem then shows that there exist, in a somewhat derived sense, "free buildings" of type $M$. Thus, for diagrams $M$ with no subdiagram of type $\mathbf{A}_3$, $\mathbf{C}_3$, or $\mathbf{H}_3$, the notion of building of type $M$ appears to be quite "soft."

Sections 3 to 5 aim at the proof of Theorem 1, which will be completed in Section 6.1.

## 2. Chamber Systems

### 2.1. Definitions

A *chamber system* over the set $I$ consists of a set $\mathcal{C}$—whose elements are called *chambers*—and a system $(\mathcal{P}_i)_{i \in I}$ of partitions of $\mathcal{C}$ indexed by $I$. *Par abus de langage* the system $(\mathcal{C}, (\mathcal{P}_i)_{i \in I})$ will also be called "the chamber system $\mathcal{C}$." For $J \subset I$, we denote by $\mathcal{P}_J$ the join of the partitions $\mathcal{P}_j$ with $j \in J$, i.e. the finest partition of which all $\mathcal{P}_j$ (with $j \in J$) are refinements. The elements of $\mathcal{P}_I$ are the *connected components* of $\mathcal{C}$, and the system $\mathcal{C}$ is called *connected* if $\mathcal{P}_I = \{\mathcal{C}\}$. Two chambers belonging to the same member of some $\mathcal{P}_j$ (respectively, of a given $\mathcal{P}_i$) are called *adjacent* (respectively, *i-adjacent*). The *rank* of the chamber system $\mathcal{C}$ is defined as the cardinality of $I$.

## 2.2. Relations with Complexes and Geometries

To every complex $\Delta = (V, \tau, \mathbb{S})$ over $I$, and in particular to every geometry over $I$ (cf. Section 1.5), is associated as follows a chamber system $\mathcal{C}(\Delta)$, called *the chamber system of* $\Delta$: the chambers are the simplices of type $I$, and two chambers belong to the same member of $\mathcal{P}_i$ if they have the same face of type $I - \{i\}$.

Conversely, to every chamber system $\mathcal{C}$, one can associate a complex $\Delta(\mathcal{C}) = (V, \tau, \mathbb{S})$ and a geometry $\Gamma(\mathcal{C}) = (V, \tau, *)$: the set of vertices $V$ is the direct sum of the sets $\mathcal{P}_{I-\{i\}}$ (thus, an element of $V$ is a member of some $\mathcal{P}_{I-\{i\}}$), the type function $\tau$ maps $\mathcal{P}_{I-\{i\}}$ onto $\{i\}$, a subset of $V$ is a simplex of type $J$ if $\tau$ bijects it on $J$ and if the intersection of its elements (as subsets of $\mathcal{C}$) is not empty, and similarly, two elements of $V$ are incident if (again as subsets of $\mathcal{C}$) they have a nonempty intersection.

In general, those two constructions are not inverse to the function $\Delta \to \mathcal{C}(\Delta)$. Indeed, it is easy to see that a complex $\Delta$ over $I$ satisfies the relation $\Delta = \Delta(\mathcal{C}(\Delta))$ if and only if every simplex of $\Delta$ is contained in a simplex of type $I$ (chamber) and if, given any two chambers $c, c'$ and a common vertex $v$, there exists a finite sequence $c = c_0, c_1, \ldots, c_m = c'$ of chambers containing $v$ and such that, for $0 < k \leqslant m$, $c_k$ is adjacent to $c_{k-1}$ (i.e. $c_k \cap c_{k-1}$ has corank one: cf. Section 1.5). By [14, Section 3.12], the buildings have that property. It is also readily verified that if $I$ is finite and if $\Gamma$ is a residually connected geometry over $I$ (cf. Section 1.2), the flag complex of $\Gamma$ has the above properties, therefore, $\Gamma = \Gamma(\mathcal{C}(\Gamma)) = \Delta(\mathcal{C}(\Gamma))$ (with the convention of Section 1.5 identifying a geometry with its flag complex). Thus, in particular, buildings and geometries (of finite rank) of type $M$ can legitimately be "identified" with their chamber systems. *Par abus de langage*, the chamber system of a building of type $M$ will also be called a building of type $M$. A more direct definition of buildings as chamber systems will be given in Section 2.5.

## 2.3. Examples

2.3.1. Given a group $G$, a subgroup $B$, and a system $(P_i)_{i \in I}$ of subgroups containing $B$, one can construct a chamber system $\mathcal{C}(G; B, (P_i)_{i \in I})$ as follows: the chambers are the elements of $G/B$, and two chambers are $j$-adjacent if their images by the canonical projection $G/B \to G/P_j$ coincide. If $(W; (r_i)_{i \in I})$ is a Coxeter system (cf. Section 1.5), and if $W_i$ denotes the group $\langle r_i \rangle = \{1, r_i\}$, then $\mathcal{C}(W; \{1\}, (W_i))$ is the chamber system of the corresponding Coxeter complex (cf. [14, Section 2.16]), whose set of chambers is $W = W/\{1\}$. Thus, a Coxeter group can be viewed as a chamber system, which we shall call a *Coxeter chamber system*. If $W$ is a finite or affine reflection group, the corresponding chamber system can be realized geometrically in the usual way, as the system of connected components of the complement of the union of all reflection hyperplanes (cf. [1, Chapter 5, Section 3]). If $G$ is a group, $(B, N)$ a $BN$-pair in $G$, and $(P_i)_{i \in I}$ the system of all subgroups of $G$ containing $B$ properly and minimal with that property, the chamber system $\mathcal{C}(G; B, (P_i))$ is the chamber system of the building associated with the $BN$-pair in question (cf. [14, Section 3.2.6]).

2.3.2. *Direct products.* Let $A$ be an index set, and for $\alpha \in A$ let $I_\alpha$ be a set and let $(\mathcal{C}_\alpha, (\mathcal{P}_{\alpha,i})_{i \in I_\alpha})$ be a chamber system over $I_\alpha$. Let now $\mathcal{C}$ be the direct product of the sets $\mathcal{C}_\alpha$, let $I$ be the direct sum of the sets $I_\alpha$, and for $i \in I_\alpha$ let $\mathcal{P}_i$ be the partition of $\mathcal{C}$ defined as follows: two elements $x, y$ of $\mathcal{C}$ belong to the same element of that partition if and only if their projections in $\mathcal{C}_\beta$ coincide whenever $\beta \neq \alpha$ and belong to the same element of $\mathcal{P}_{\alpha,i}$ if $\beta = \alpha$. Then $(\mathcal{C}, (\mathcal{P}_i)_{i \in I})$ is a chamber system, called the *direct product* of the systems $\mathcal{C}_\alpha$.

## 2.4. Morphisms

We define an *I-morphism*—or simply a *morphism*—of a chamber system $(\mathcal{C}, (\mathcal{P}_i)_{i \in I})$ into another $(\mathcal{C}', (\mathcal{P}'_i)_{i \in I})$ as a mapping of $\mathcal{C}$ into $\mathcal{C}'$ such that the image of every element of $\mathcal{P}_i$ is contained in an element of $\mathcal{P}'_i$ for all $i$. If $\Delta$ and $\Delta'$ are two complexes over $I$ such that $\Delta(\mathcal{C}(\Delta)) = \Delta$ and $\Delta(\mathcal{C}(\Delta')) = \Delta'$ (with the notation of Section 2.2), then the morphisms $\mathcal{C}(\Delta) \to \mathcal{C}(\Delta')$ are "the same" as the morphisms $\Delta \to \Delta'$ defined in Section 1.5. By Section 2.2, this remark applies in particular to buildings and, for $I$ finite, to residually connected geometries.

## 2.5. Buildings as Chamber Systems: A Direct Definition

The following statement, which we shall now prove, is essentially a reformulation of the definition of buildings in terms of chamber systems. As before, $M$ denotes a Coxeter diagram.

*A chamber system $\mathcal{C}$ is a building of type $M$ if and only if there exists a set $\mathcal{G}$ of morphisms of the Coxeter group $W = W(M)$ (viewed as a chamber system*: cf. *Section* 2.3.1) *into $\mathcal{C}$ and, for every $a \in \mathcal{C}$, a morphism $\rho_a : \mathcal{C} \to W$, with the following properties*:

(i) *every two chambers in $\mathcal{C}$ belong to the image of some element of $\mathcal{G}$*;
(ii) *for $\alpha \in \mathcal{G}$ and $a \in \alpha(W)$, the mapping $\rho_a \circ \alpha$ is an automorphism of the chamber system $W$.*

If $\mathcal{C} = \mathcal{C}(\Delta)$ (cf. Section 2.2) is the chamber system of a building $\Delta$ of type $M$, denote by $\mathcal{G}$ the set of all isomorphisms of the Coxeter complex $\Delta(W)$ of Section 1.5 onto the apartments of a system of apartments in $\Delta$, and by $\rho_a$ the retraction $\mathrm{retr}_{\Sigma,a}$ of $\Delta$ onto an apartment $\Sigma$ containing $a$ with center $a$ (cf. [14, Section 3.3]) followed by any isomorphism of $\Sigma$ onto $\Delta(W)$. The properties (i) and (ii) are then obvious.

Conversely, let $\mathcal{G}$ and $\{\rho_a \mid a \in C\}$ be two sets of morphisms satisfying (i) and (ii), and set $\Delta = \Delta(\mathcal{C})$ (with the notation of Section 2.2). It is immediate that each $\alpha \in \mathcal{G}$ (each $\rho_a$) "extends" uniquely to a morphism of complexes over $I$, $\Delta(W) \to \Delta$ $(\Delta \to \Delta(W))$, which will also be called $\alpha$ $(\rho_a)$. From (ii), it readily follows that, for $\alpha \in \mathcal{G}$, the morphism $\alpha : \Delta(W) \to \Delta$ is an isomorphism of $\Delta(W)$ onto a subcomplex of $\Delta$. We claim that the set of all those subcomplexes is a system of apartments (cf. Section 1.5). The axiom (Ap 1) is an immediate consequence of (i). To prove (Ap 2), let us consider two simplices $S_1, S_2$ of $\Delta$ belonging to the images of two morphisms $\alpha_1, \alpha_2 \in \mathcal{G}$. We must show that there is an automorphism $\omega$ of the Coxeter complex $\Delta(W)$ such that, for $k = 1, 2, \alpha_2^{-1}(S_k) = \omega(\alpha_1^{-1}(S_k))$. Let $c_k$ be a chamber of $\alpha_k(\Delta(W))$ containing $S_k$. By (i), $c_1$ and $c_2$ belong to the image of some $\alpha \in \mathcal{G}$. Now, for $k, m \in \{1, 2\}$, the automorphism $(\rho_{c_m} \circ \alpha)^{-1} \circ (\rho_{c_m} \circ \alpha_m)$ of $\Delta(W)$ (cf. (ii)) maps $\alpha_m^{-1}(S_k)$ onto $\alpha^{-1}(S_k)$. Therefore,

$$\omega = (\rho_{c_2} \circ \alpha_2)^{-1} \circ (\rho_{c_2} \circ \alpha) \circ (\rho_{c_1} \circ \alpha)^{-1} \circ (\rho_{c_1} \circ \alpha_1)$$

has the desired property. Thus, $\Delta$ is a building, and to finish our proof we only have to show that $\mathcal{C} = \mathcal{C}(\Delta)$ (with the notation of Section 2.2). By definition of $\Delta$, there is a natural map $\psi$ from $\mathcal{C}$ onto the set of chambers of the complex $\Delta$, and we just have to verify that two chambers $c, c' \in \mathcal{C}$ are distinct (respectively, $j$-adjacent for a given $j \in I$) if and only if $\psi(c)$ and $\psi(c')$ are distinct (respectively $j$-adjacent). Let $\alpha$ be an element of $\mathcal{G}$ whose image contains $c$ and $c'$. If $c$ and $c'$ are distinct (respectively, are not $j$-adjacent), the same is true of $\alpha^{-1}(c)$ and $\alpha^{-1}(c')$, hence of $\rho_c(c)$ and $\rho_c(c')$ (by (ii)), hence of $\rho_c(\psi(c))$ and $\rho_c(\psi(c'))$, and hence finally of $\psi(c)$ and $\psi(c')$. The converse being obvious, our assertion is proved.

### 2.6. Galleries

We denote by $F(I) = F$ the free monoid over the set $I$; its elements are called *words*. For $f = i_1 \ldots i_l \in F$, the number $l$, called the *length* of the word $f$, is denoted by $l(f)$. We define a *gallery of type f* and of length $l$ in a chamber system over $I$ as a sequence of chambers $(c_0, \ldots, c_l)$ such that, for $j \in \{1, \ldots, l\}$, the chambers $c_{j-1}$ and $c_j$ are $i_j$-adjacent; if $c_{j-1} \neq c_j$ for all $j$, we say that the gallery is *simple* (="non-stammering" in the terminology of [14]). The *distance* of two chambers $c, c'$, denoted by $d(c, c')$, is the minimum length of a gallery connecting them. Clearly, a morphism of chamber systems maps a gallery of type $f$ onto a gallery of type $f$ and diminishes the distances.

## 3. Chamber Systems of Type $M$ and Buildings

*From now on, we always assume* Card $I \geqslant 2$.

### 3.1

We choose a Coxeter diagram $M$ over $I$, we denote by $(W; R)$ with $R = (r_i)_{i \in I}$ a Coxeter system of type $M$ (cf. Section 1.5) and by $f \mapsto r_f$ the homomorphism of $F$ onto $W$ which maps $i$ onto $r_i$ for all $i \in I$. By Section 2.3, $W$ is, in a natural way, a chamber system, and, for $w \in W$, we set $l(w) = d(1, w)$; thus, $l(w)$ is the length of $w$ with respect to the generating set $R$ (cf. [1, p. 9]). We recall that a word $f$ is called *reduced* (with respect to $M$) if $l(f) = l(r_f)$.

For $i, j \in I$ and $n \in \mathbb{N}$, we denote by $p_n(i, j)$ the word $\ldots ijij$ of length $n$, and we set $p(i, j) = p_{M(i, j)}(i, j)$ if $M(i, j) \neq \infty$.

### 3.2. Chamber Systems of Type $M$

A chamber system $(\mathcal{C}; (\mathcal{P}_i)_{i \in I})$ is called *of type M* if it is connected and satisfies the following three conditions:

**(CS$_M$1)** every element of each partition $\mathcal{P}_i$ contains at least two chambers;

**(CS$_M$2)** if $i, j$ are two distinct elements of $I$ and if $n$ is a strictly positive integer strictly smaller than $2M(i, j)$, then $\mathcal{C}$ contains no simple closed gallery of type $p_n(i, j)$;

**(CS$_M$3)** if $i, j \in I$ and $M(i, j) \neq \infty$, for every simple gallery of type $p(i, j)$ there is a simple gallery of type $p(j, i)$ having the same origin and the same extremity.

By (CS$_M$2), the gallery of type $p(j, i)$ in (CS$_M$3) is unique. Setting $n = 2$ in (CS$_M$2), one sees that a simple gallery $G$ has only one type, which we shall denote by $\gamma(G)$.

It is readily verified that the chamber system of a geometry of type $M$ (cf. Section 1.3) is a chamber system of type $M$, and that that property characterizes the geometries of type $M$ among the residually connected geometries $\Gamma$ (of finite rank) satisfying the relation $\Gamma(\mathcal{C}(\Gamma)) = \Gamma$ (cf. Section 2.2). In view of [14, Section 3.9], buildings of type $M$ are also chamber systems of type $M$.

*From now on,* $(\mathcal{C}, (\mathcal{P}_i))$ *will always denote a chamber system of type M*, except when the contrary is specified.

### 3.3. A First Characterization of Buildings: Statement of the Theorem

The following theorem, proved in Section 3.7, will be the main object of this section. In order to formulate it conveniently, we first state two properties depending on a chamber $x \in \mathcal{C}$:

**(P$_x$)** if two reduced words $f$ and $f'$ are the types of two simple galleries with common origin $x$ and common extremity, then $r_f = r_{f'}$;

**(Q$_x$)** if two simple galleries have the same origin $x$, the same extremity, and the same reduced type, they coincide.

**Theorem 2.** *The chamber system $\mathcal{C}$ of type $M$ is a building if and only if it has property* (P$_c$) *for some $c \in \mathcal{C}$, in which case* (P$_x$) *and* (Q$_x$) *hold for all $x \in \mathcal{C}$.*

### 3.4. Homotopy

3.4.1. A sequence of two words (respectively two galleries) will be called an *elementary homotopy*—relative to $M$—if it can be written $(f_1 f f_2, f_1 f' f_2)$, where $f = p(i, j)$ and $f' = p(j, i)$ (respectively where $f$ and $f'$ are galleries of types $p(i, j)$ and $p(j, i)$ with common origin and common extremity), for some distinct $i, j \in I$ such that $M(i, j) \neq \infty$. A *homotopy* is a sequence of words or of galleries such that two consecutive terms of the sequence form an elementary homotopy; we also say that the first and last terms of the sequence are *homotopic*.

3.4.2.
**Proposition 1.** *If $G$ is a gallery of type $f$ and if $(f = f_0, f_1, \ldots, f_n)$ is a homotopy of words, there exists a homotopy of galleries $(G = G_0, G_1, \ldots, G_n)$ such that $\gamma(G_k) = f_k$ for all $k$. If $f$ is reduced and $G$ is simple, then all $G_k$ are simple and the homotopy $(G, G_1, \ldots, G_n)$ is unique.*

It is clearly sufficient to consider the case where $n = 1$, $f = p(i, j)$ and $f_1 = p(j, i)$ for some $i, j \in I$. Then the assertion is practically contained in the definition of Section 3.2 if $G$ is simple; otherwise, the proof is straightforward.

3.4.3.
**Proposition 2**

(i) *Every nonreduced word is homotopic with a word containing a repetition.*
(ii) *Two reduced words $f, f'$ such that $r_f = r_{f'}$ are homotopic. If $f$ and $f'$ have a common tail end $f''$ (i.e. if $f = f_1 f''$ and $f' = f_1' f''$), they can be joined by a homotopy all of whose terms end with $f''$.*

These are known properties of Coxeter groups (cf. [13, Théorème 3]). Note that the second assertion of (ii) says nothing more than the first one, applied to $f_1$ and $f_1'$.

3.4.4.
**Corollary 1.** *Given $c, c' \in \mathcal{C}$, the type of any gallery of minimum length joining $c$ and $c'$ is reduced.*

Indeed, by Propositions 1 and 2, if $G$ is a gallery of nonreduced type joining $c$ and $c'$, then there exists a gallery homotopic to $G$ and whose type contains a repetition. But then, there also exists a shorter gallery joining $c$ and $c'$.

3.4.5.
**Corollary 2.** *If two reduced words $f, f'$ are such that $r_f = r_{f'}$, any two chambers which can be joined by a gallery of type $f$ can also be joined by a gallery of type $f'$.*

### 3.5. Two Properties of Coxeter Chamber Systems

We recall that $(W; R)$ denotes a Coxeter system of type $M$ and that, by Section 2.3, $W$ can also be viewed as a chamber system. For $r \in R$, we set $\Phi_r^+ = \{w \in W \mid l(rw) > l(w)\}$ and $\Phi_r^- = W - \Phi_r^+ = \{w \in W \mid l(rw) < l(w)\}$. In the language of [14, Section 1.12], $\Phi_r^+$ and $\Phi_r^-$ are the sets of chambers of two opposite roots.

**Lemma 1.** *Let $r \in R$ and $w, w' \in W$. Suppose that $w$ and $w'$ belong both to $\Phi_r^+$ or both to $\Phi_r^-$. Then $d(w, rw') > d(w, w')$.*

Since the automorphism $x \mapsto rx$ of the chamber system $W$ permutes $\Phi_r^+$ and $\Phi_r^-$, it suffices to consider the case where $w, w' \in \Phi_r^+$. Let $G = (w, w_1, \ldots, w_{d-1}, rw')$ be a gallery of length $d = d(w, rw')$, and let $w_s$ be the last element of $G$ belonging to $\Phi_r^+$. Then, by [14, Section 1], one has $rw_{s+1} = w_s$, and $(w, w_1, \ldots, w_s, rw_{s+2}, \ldots, rw_{d-1}, w')$ is a gallery of length $d - 1$ joining $w$ and $w'$; hence the claim.

**Lemma 2.** *Let $r \in R$, let $X$ be a subset of $W$, and let $\beta: X \to W$ be a mapping decreasing the distances and such that $\beta(x) \in \{x, rx\}$ for all $x \in X$. Suppose that $\beta$ fixes an element of $\Phi_r^+$ and an element of $\Phi_r^-$. Then it is the inclusion mapping.*

That is an immediate consequence of Lemma 1.

### 3.6. A Property of Buildings

**Proposition 3.** *Suppose $\mathcal{C}$ is a building, and let $G = (c_0, \ldots, c_l)$ be a simple gallery of reduced type and length $l$. Then, $d(c_0, c_l) = l$ and $G$ is contained in any apartment containing $c_0$ and $c_l$.*

By [14, Section 3.4], it suffices to prove the first assertion, which we shall do by induction on $l$. Let $\Sigma$ be an apartment of the building containing $c_1$, and let $\rho$ be the retraction of the building onto $\Sigma$ with center $c_1$ (cf. [14, Section 3.3]). The induction hypothesis implies that $d(c_1, c_l) = l - 1$. In view of the theorem of [14, Section 3.3], it follows that $\rho$ maps the gallery $(c_1, \ldots, c_l)$—and hence also the gallery $G$—onto a simple gallery of $\Sigma$. Since the type of $\rho(G)$, equal to that of $G$, is reduced, we have $d(\rho(c_0), \rho(c_l)) = l$. But $\rho$ diminishes the distances. Therefore, $d(c_0, c_l) \geqslant l$, and the opposite inequality is obvious.

### 3.7. Proof of Theorem 2

3.7.1. We first show that

(1) *for $x \in \mathcal{C}$, $(\mathrm{P}_x)$ implies $(\mathrm{Q}_x)$.*

Suppose $(\mathrm{Q}_x)$ does not hold, and let $G, G'$ be two different simple galleries with same origin $x$, same extremity, and same reduced type. Upon shortening them if necessary, we may choose them so that their elements before the last are different, that is, so that $G = (G_1, y, z)$ and $G' = (G_1', y', z)$, with $y \neq y'$. But then, the galleries $(G_1, y)$ and $(G_1', y', y)$ contradict $(\mathrm{P}_x)$.

3.7.2. The fact that $(\mathrm{P}_x)$ and $(\mathrm{Q}_x)$ hold for all chambers $x$ in any building is an immediate consequence of Proposition 3. Therefore, we only have to prove that if $(\mathrm{P}_c)$ holds for some $c \in \mathcal{C}$, which we shall assume from now on, then $\mathcal{C}$ is a building. Our next aim is showing that

(2) *the condition $(\mathrm{P}_c)$ holds for all $c \in \mathcal{C}$.*

By an obvious induction, it suffices to prove that

(3) *if* $j \in I$ *and if* $c, c'$ *are two distinct* $j$*-adjacent chambers, then* $(\mathrm{P}_c)$ *implies* $(\mathrm{P}_{c'})$.

We suppose that $(\mathrm{P}_c)$ holds, and we must show that if $G, G'$ are two simple galleries with common origin $c'$ and common extremity, and such that the words $f = \gamma(G)$ and $f' = \gamma(G')$ are reduced, then $r_f = r_{f'}$. We shall distinguish three cases.

*Case* 1: *Both* $jf$ *and* $jf'$ *are reduced.* Then, the assertion follows from $(\mathrm{P}_c)$ applied to the galleries $(c, G)$ and $(c, G')$.

*Case* 2: *Neither* $jf$ *nor* $jf'$ *is reduced.* Let $f_1, f_1'$ be reduced words such that $r_{jf} = r_{f_1}$ and $r_{jf'} = r_{f_1'}$. Then the words $jf_1$ and $jf_1'$ are reduced (cf. [1, p. 15, Proposition 4]), and, upon replacing $G$ and $G'$ by homotopic galleries and using Proposition 1, we may assume that $f = jf_1$ and $f' = jf_1'$. Set $G = (c', c_1, G_1)$ and $G' = (c', c_1', G_1')$. If $c_1 = c_1' = c$ (respectively if $c_1 \neq c$ and $c_1' \neq c$), the equality $r_f = r_{f'}$ follows from $(\mathrm{P}_c)$ applied to the galleries $(c, G_1)$ and $(c, G_1')$ (respectively, $(c, c_1, G_1)$ and $(c, c_1', G_1')$). If $c_1 = c \neq c_1'$, the property $(\mathrm{P}_c)$ applied to $(c, G_1)$ and $(c, c_1', G_1')$ implies that $r_{f_1} = r_{f'}$, in contradiction with the fact that $jf_1$ is reduced whereas $jf'$ is not. A contradiction is derived in a similar way from the assumption $c_1 \neq c = c_1'$.

*Case* 3: *One and only one of the words* $jf$ *and* $jf'$, *say* $jf'$, *is reduced.* Arguing as in case 2, we may assume that $f = jf_1$, with $f_1$ reduced, and we set $G = (c', c_1, G_1)$. We must have $c_1 \neq c$; otherwise $(\mathrm{P}_c)$, applied to the galleries $(c, G_1)$ and $(c, G')$, would imply that $r_{f_1} = r_{jf'}$, and hence $r_f = r_{f'}$, in contradiction with the assumption made on $jf$ and $jf'$. Now, $(\mathrm{P}_c)$ applied to the galleries $(c, c_1, G_1)$ and $(c, G')$ shows that $r_{jf_1} = r_{jf'}$; hence $r_{f_1} = r_{f'}$, and, upon replacing $G'$ by a homotopic gallery, we may, by Proposition 1, assume that $f_1 = f'$. It then follows from $(\mathrm{P}_c)$ and (1) in Section 3.7.1 that $(c, c_1, G_1) = (c, G')$; hence $c' = c_1$, in contradiction with the simplicity of $G$. This completes the proof of (3), and hence of (2).

3.7.3. For $a, a' \in \mathcal{C}$, we denote by $w(a, a')$ the common value of $r_{\gamma(G)}$ for all simple galleries $G$ of reduced type joining $a$ and $a'$ (cf. Section 3.4 and (2) in Section 3.7.2). Clearly,

(4) $w(a', a) = w(a, a')^{-1}$.

Note also that

(5) *for* $a \in \mathcal{C}$, *the mapping* $\rho_a : x \mapsto w(a, x)$ *of* $\mathcal{C}$ *into* $W$ *is a morphism of chamber systems.*

In other words,

(5′) *if* $j \in I$ *and if* $x, x'$ *are two distinct* $j$*-adjacent chambers, then* $w(a, x') = w(a, x)$ *or* $w(a, x) \cdot r_j$.

Indeed, let $G$ be a simple gallery of reduced type $f$ joining $a$ and $x$. If $fj$ is reduced, the gallery $(G,x')$ provides the equalities $w(a,x') = r_{fj} = r_f r_j = w(a,x) \cdot r_j$. If $fj$ is not reduced and if $f_1$ denotes a reduced word such that $r_{fj} = r_{f_1}$, we may, upon replacing $G$ by a homotopic gallery and using Proposition 1, assume that $f = f_1 j$. Setting $G = (G_1, x)$ and considering the gallery $G_1$ or the gallery $(G_1, x')$ according as $x'$ is or is not the last term of $G_1$, we find that $w(a,x') = r_{f_1}$ or $r_f$, hence (5′).

Since morphisms of chamber systems diminish distances, (5) implies that

(6) *for* $x, x' \in \mathcal{C}$, *one has* $l(w(x,x')) \leqslant d(x,x')$.

3.7.4. We define a *strong isometry* of a subset $X$ of $W$ into $\mathcal{C}$, as a mapping $\alpha: X \mapsto C$ such that $w(\alpha(x), \alpha(x')) = x^{-1}x'$ for all $x, x' \in X$. We now prove that

(7) *every strong isometry* $\alpha: X \to \mathcal{C}$ *can be extended to a strong isometry of* $W$ *into* $\mathcal{C}$.

By Zorn's lemma, it suffices to show that

(8) *if* $X \neq W$, $\alpha$ *can be extended to a strong isometry of a strictly larger subset of* $W$ *into* $\mathcal{C}$.

We assume $X \neq \emptyset$ and $X \neq W$. Those inequalities imply the existence of $x_0 \in X$ and $j \in I$ such that $x_0 r_j \notin X$. Upon submitting $X$ to the translation $x \mapsto x_0^{-1}x$ (and modifying $\alpha$ accordingly), we may—and shall—assume that $x_0 = 1$. Thus, $1 \in X$ and $r_j \notin X$. If for all $x \in X$, one has $l(r_j x) > l(x)$, and if $a$ denotes any chamber of $\mathcal{C}$ distinct from $\alpha(1)$ and $j$-adjacent to it, it is readily seen that $\alpha$ extended by $r_j \mapsto a$ is a strong isometry of $X \cup \{r_j\}$ into $\mathcal{C}$, and (8) is proved. Suppose therefore that there exists $x_1 \in X$ such that $l(r_j x_1) < l(x_1)$, let $f$ be a reduced word such that $r_j x_1 = r_f$, so that $jf$ is also reduced, and let $a$ be the second term of a gallery of type $jf$ joining $\alpha(1)$ and $\alpha(x_1)$ (such a gallery exists, by Corollary 2, since $w(\alpha(1), \alpha(x_1)) = x_1 = r_{jf}$). For every $x \in X$, one has, by (5′), $w(\alpha(x), a) = w(\alpha(x), \alpha(1))$ or $w(\alpha(x), \alpha(1)) \cdot r_j$; hence, by (4), $w(a, \alpha(x)) \in \{x, r_j x\}$. Let $\beta: X \mapsto W$ be the mapping defined by $\beta(x) = r_j \cdot w(a, \alpha(x))$. We have $\beta(1) = 1$ and $\beta(x_1) = x_1$. Therefore, in view also of (6), $\beta$ satisfies the conditions of Lemma 2 (Section 3.5), and must be the inclusion mapping. Consequently, $\alpha$ extended by $r_j \mapsto a$ is a strong isometry of $X \cup \{r_j\}$ into $\mathcal{C}$, and (8)—hence also (7)—is proved.

As a special case of (7), we note that

(9) *given two chambers* $x, x'$ *in* $\mathcal{C}$, *there exists a strong isometry of* $W$ *into* $\mathcal{C}$ *whose image contains* $x$ *and* $x'$.

3.7.5. *End of the proof.* The set $\mathcal{I}$ of all strong isometries of $W$ in $\mathcal{C}$ and the mappings $\rho_a$ of (5) in Section 3.7.3 satisfy the conditions (i) and (ii) of Section

2.5: indeed, (ii) follows from the very definition of strong isometries, and (i) is nothing but (9). Now, Section 2.5 asserts that $\mathcal{C}$ is a building, and our theorem is proved.

## 4. Auxiliary Results on Graphs and Coxeter Systems

In this section, which can be omitted in a first reading, we collect some technical observations which will be used only to derive Corollary 3 from Theorem 3. Here, as before, we take the word "graph" in the sense of [1]; in particular, our graphs are undirected, they have no loops, and an edge is determined by its extremities.

### 4.1

Let $X$ be a connected graph, and let $A$ be a set of closed paths of $X$. By abuse of language, we say that "the fundamental group of $X$ is generated by $A$" if, given a vertex $v$ of $X$, the fundamental group of $(G,v)$ (that is, the group of homotopy classes of closed paths from $v$ to $v$) is generated by (classes of) paths of the form $faf^{-1}$, where $a \in A$, $f$ is a path from $v$ to the origin of $a$, and $f^{-1}$ is the "inverse" of $f$; it is clear that that property is independent of $v$. A mapping of the set of vertices of a graph $Y$ into the set of vertices of a graph $Y'$ is called a *morphism* (of graphs) if it maps the edges of $Y$ onto edges or vertices of $Y'$. (N.B. Connected graphs are also chamber complexes of rank 2, in the terminology of [14], but the notions of morphisms are different.)

### 4.2

**Lemma 3.** *Let $\alpha: X \to X'$ be a morphism of connected graphs such that the inverse image of every vertex of $X'$ is a connected subgraph of $X$, and let $A$ be a set of closed paths of $X$. Suppose that $A$ has the following properties:*

(i) *for every vertex $v$ of $X'$, the fundamental group of the graph $\alpha^{-1}(v)$ is generated by elements of $A$;*

(ii) *if $(a',b')$ is an edge of $X'$ and if $(a_1,b_1)$, $(a_2,b_2)$ are two edges of $X$ such that $\alpha(a_1)=\alpha(a_2)=a'$ and $\alpha(b_1)=\alpha(b_2)=b'$, then there exist a path $f$ from $a_1$ to $a_2$ in $\alpha^{-1}(a')$ and a path $g$ from $b_1$ to $b_2$ in $\alpha^{-1}(b')$ such that the path $f\cdot(a_2,b_2)\cdot g\cdot(b_1,a_1)$ belongs to $A$.*

*Then $A$ generates the fundamental group of $X$ if and only if $\alpha(A)$ generates the fundamental group of $X'$.*

The assertion is made intuitively clear by the following heuristic argument. "Modulo $A$," the "fibers" $\alpha^{-1}(v)$ of $\alpha$ (where $v$ denotes a vertex of $X'$) are homotopically trivial, and if we contract each such fiber to a point, any two edges of $X$ having the same image in $X'$ become "congruent mod $A$"; therefore, $X$ and $X'$ "coincide homotopically mod $A$."

The formalization of that argument is easy but tiresome. We shall only sketch it in somewhat loose terms.

We choose a vertex $v$ of $X$ and take $v$ and $\alpha(v)$ as origins to define the fundamental groups $\pi_1(X)$ and $\pi_1(X')$. The set of elements of $\pi_1(X)$ deduced from elements of $A$ by adding a head and a tail inverse to each other (as in Section 4.1) will also be called $A$. Let $A_1$ be the set of all elements of $\pi_1(X)$ represented by paths of the form $faf^{-1}$, where $a$ is a closed path contained in a fiber of $\alpha$ (inverse image of a vertex of $X'$), and let $A_2$ be the set of all elements of $\pi_1(X)$ represented by paths of the form $fege'^{-1}g'f^{-1}$, where $e$ and $e'$ are oriented edges of $X$ having the same image in $X'$, and $g$, $g'$ are contained in fibers of $\alpha$.

From (i), one readily deduces that $A_1$ "is generated by elements of $A$." Similarly, every element of $A_2$, multiplied on the right and left by suitable elements of $A_1$, is an element of $A$ of the form provided by (ii). Therefore, $A_2$ also "is generated by elements of $A$."

Let $K$ be the kernel of the homomorphism $\pi_1(X) \to \pi_1(X')$ induced by $\alpha$. The lemma will be proved if we show that every element $[k]$ of $K$ is a product of elements of $A_1 \cup A_2$. Let $k$ be a path in $X$ representing $[k]$, and let $k' = (v_0' = \alpha(v), v_1', \ldots, v_r' = \alpha(v))$ (where the $v_s'$ are vertices of $X'$) be its image by $\alpha$ in which repetitions are omitted (i.e. $v_s' \neq v_{s+1}'$ for all $s$). If $r = 0$, $[k] \in A_1$. Suppose therefore $r > 0$. Since $k'$ is null-homotopic, there is an integer $t \in \{1, \ldots, r-1\}$ such that $v_{t-1}' = v_{t+1}'$. But then, upon multiplying $[k]$ by a suitable element of $A_2$, one can modify it so that, for an appropriate choice of the representative $k$, $k'$ becomes equal to $(v_0', \ldots, v_{t-1}', v_{t+2}', \ldots, v_r')$. An induction on $r$ finishes the proof.

### 4.3. Generating Homotopies in Coxeter Chamber Systems

4.3.1. We recall that for $J \subset I$, $M_J$ denotes the restriction of the Coxeter diagram $M$ to $J \times J$, and we represent by $W_J$ the Coxeter subgroup of type $M_J$ of $W$ generated by all $r_j$, for $j \in J$. We call the set $J$ *spherical* if $W_J$ is finite.

4.3.2.
**Lemma 4.** *Let $w \in W$ and $J \subset I$ be such that $l(wr_j) < l(w)$ for all $j \in J$. Then $J$ is spherical, and if $w''$ denotes the "longest element" of $W_J$, one has $l(w) = l(ww''^{-1}) + l(w'')$, which means that there is a reduced word $f = f'f''$ such that $r_f = w$ and $r_{f''} = w''$.*

Set $w = w_1'w_1''$, where $w_1'$ is "$J$-reduced on the right"—i.e., $l(w_1'r_j) < l(w_1')$ for all $j \in J$—and $w_1'' \in W_J$; thus $l(w) = l(w_1') + l(w_1'')$ (cf. [14, Section 2.29] or [1, p. 33, Exercice 3]). For $j \in J$, we also have $l(wr_j) = l(w_1') + l(w_1''r_j)$ (same references); hence $l(w_1''r_j) < l(w_1'')$, by hypothesis. Therefore, $W_J$ is finite and $w_1'' = w''$ is its longest element (cf. [14, Section 2.36] or [1, p. 43, Exercice 22]). The lemma is proved.

4.3.3. Heuristically, the next proposition means that nontrivial self-homotopies of galleries in Coxeter complexes only occur in finite stars (cf. [14, Section 1.1]) of simplices of codimension 3, or, if one prefers, that $\pi_2$ (suitably defined) "lives" in such stars.

**Proposition 4.** *Let $w \in W$, and let $X$ be the graph whose vertices are the reduced words $f$ such that $r_f = w$ and whose edges are the pairs of such words forming an elementary homotopy (cf. Section 3.4.1). Then the fundamental group of $X$ is "generated" (in the sense of Section 4.1) by the closed paths of the following form:*

(a) $(f_0 \cdot p(i,j) \cdot f', f_1 \cdot p(i,j) \cdot f', \ldots, f_r \cdot p(i,j) \cdot f', f_r \cdot p(j,i) \cdot f', f_{r-1} \cdot p(j,i) \cdot f', \ldots, f_0 \cdot p(j,i) \cdot f', f_0, p(i,j) \cdot f')$, *where $i, j \in I$ are such that $M(i,j) \neq \infty$ (for the notation $p(i,j)$, see Section 3.1);*

(b) $(ff_0 f', ff_1 f', \ldots, ff_r f', ff_0 f')$, *where the words $f_0, f_1, \ldots, f_r$ involve only three elements of $I$ forming a spherical set.*

The proof will be by induction on $l(w)$, the assertion being clear for $l(w) = 0$. Let $j$ be the set of all $j \in I$ such that $l(wr_j) < l(w)$. By Lemma 4, $J$ is spherical.

Let $X'$ be the complete graph with set of vertices $J$, let $\alpha : X \to X'$ be the morphisms mapping every vertex $f$ of $X$ into its last factor (remember that a vertex of $X$ is a word in the elements of $I$), and let $A$ be the set of closed paths of $X$ of the form (a) or (b). To prove our assertion, it suffices to show that those data satisfy the hypotheses of Lemma 3 and that $\alpha(A)$ "generates" the fundamental group of the graph $X'$ (always in the sense of Section 4.1).

The connectedness of $X$ and of the inverse images by $\alpha$ of the vertices of $X'$ are direct consequences of Proposition 2(ii). Assumption (i) of Lemma 3 follows from the induction hypothesis. Assumption (ii) is readily implied by the fact that $A$ contains the paths of type (a) (using also Proposition 2(ii)).

There remains to be proved that $\alpha(A)$ "generates" the fundamental group of the complete graph $X'$. To that effect, it will suffice to show that any "triangle" $(i_0, i_1, i_2, i_0)$ in $J$ belongs to $\alpha(A)$. In what follows, the indices $0, 1, 2$ must be interpreted as elements of $\mathbb{Z}/3\mathbb{Z}$ (i.e., computations on them are performed mod 3). Set $J' = \{i_0, i_1, i_2\}$, and let $F'$ be the free monoid generated by $J'$. Being a subset of $J$, $J'$ is spherical. Let $w''$ be the longest element of $W_{J'}$, and set $w = w'w''$. By Lemma 4, $l(w) = l(w') + l(w'')$. Let $f'$ be a reduced word representing $w'$ (i.e. such that $w' = r_{f'}$). For $s \in \mathbb{Z}/3\mathbb{Z}$, let $w_s = r_{p(i_{s-1}, i_s)}$ be the longest word of the group $W_{\{i_{s-1}, i_s\}}$, and let $f_s \in F'$ be a reduced word representing $w''w_s$. Thus, for all $s$, the words $f_s \cdot p(i_{s-1}, i_s)$ and $f_s \cdot p(i_s, i_{s-1})$ represent $w''$. By Proposition 2(ii), there exists a homotopy $\eta_s$ from $f_s \cdot p(i_{s-1}, i_s)$ to $f_{s+1} \cdot p(i_{s+1}, i_s)$, all terms of which end with $i_s$. Let $\eta$ be the homotopy $(\eta_0, \eta_1, \eta_2, f_0 \cdot p(i_2, i_0))$. By putting $f'$ in front of all elements of $\eta$, one gets a closed path belonging to $A$ whose image by $\alpha$ is the triangle $(i_0, i_1, i_2, i_0)$. The proposition is proved.

## 5. Chamber Systems Covered by Buildings

### 5.1. Coverings

In this subsection, we drop the assumption that $(\mathcal{C}, (\mathcal{P}_i))$ is a chamber system of type $M$.

Let $m$ be a strictly positive integer $< \operatorname{Card} I$. A morphism $(\mathcal{C}', (\mathcal{P}'_i)_{i \in I}) \to (\mathcal{C}, (\mathcal{P}_i)_{i \in I})$ is called an *m-covering* if it is surjective and if for every subset $J$ of $I$ of cardinality $\leqslant m$, each element of $\mathcal{P}'_J$ (the join of all $\mathcal{P}'_j$ for $j \in J$) is mapped bijectively onto an element of $\mathcal{P}_J$. *Par abus de langage*, we also say that the chamber system $\mathcal{C}'$ is an $m$-covering of the system $\mathcal{C}$. If $\mathcal{C}$ is the chamber system of a complex $\Delta$ of finite rank such that $\Delta(\mathcal{C}) = \Delta$ (cf. Section 2.2), then the $(\operatorname{Card} I - 1)$-coverings of $\mathcal{C}$ correspond to the topological coverings of the geometric realization of $\Delta$; in general, $m$-coverings may be viewed as "ramified coverings" with ramification sets of codimension at least $m$.

Suppose the chamber system $\mathcal{C}$ connected, let $\alpha' : \mathcal{C}' \to \mathcal{C}$ and $\alpha'' : \mathcal{C}'' \to \mathcal{C}$ be two $m$-coverings, and let $c' \in \mathcal{C}'$, and $c'' \in \mathcal{C}''$ be such that $\alpha'(c') = \alpha''(c'')$. If $\mathcal{C}'$ is connected, there is at most one morphism $\varphi$: $\mathcal{C}' \to \mathcal{C}''$ such that $\alpha' = \alpha'' \circ \varphi$ (just note that for any gallery $G'$ in $\mathcal{C}'$ with origin $c$, there is a unique "lifting" of $\alpha'(G')$ in $\mathcal{C}''$ starting in $c''$). We say that $\alpha'$ (or $\mathcal{C}'$) is a *universal m-covering* of $\mathcal{C}$ if $\mathcal{C}'$ is connected and if $\varphi$ exists for an arbitrary choice of $\alpha''$, $c'$, and $c''$; for that, it suffices that $\alpha'$ factors *in some way* through every $m$-covering of $\mathcal{C}$.

*Every connected chamber system $\mathcal{C}$ possesses a universal m-covering*, which can be constructed, in the usual way, as follows. Denote by $\mathcal{C}'$ the set of equivalent classes of galleries in $\mathcal{C}$ with given (arbitrarily chosen) origin, for the finest equivalent relation satisfying the following condition: if two galleries $G, G'$ can be given the form $G = (G_1 G_2 G_3)$ and $G' = (G_1 G_2' G_3)$, where the subgalleries $G_2$ and $G_2'$ have the same origin and the same extremity and are contained in the same element of some $\mathcal{P}_J$ with $\operatorname{Card} J \leqslant m$, then $G$ and $G'$ are equivalent. Define the partition $\mathcal{P}_j'$ (for $j \in J$) of $\mathcal{C}'$ by the condition that the equivalence class of two galleries $G, G'$ belongs to the same element of $\mathcal{P}_j'$ if and only if $G'$ is equivalent to a gallery of the form $(G, x)$ where $x$ is $j$-adjacent to the last element of $G$. Finally, take for $\alpha'$ the morphism $\mathcal{C}' \to \mathcal{C}$ which maps every equivalence class of galleries onto the common last term of its members.

If $\alpha'$, $\alpha''$, $c'$, and $c''$ are as above, and if $\alpha'$ and $\alpha''$ are both universal, it follows from the uniqueness property that $\varphi$ is an isomorphism. Thus, all universal $m$-coverings of $\mathcal{C}$ are isomorphic, and we shall sometimes, *par abus de langage*, talk about *the* universal $m$-covering of $\mathcal{C}$ (although it is not canonically defined: see Remark (1) below).

For any $m$-covering $\alpha' : \mathcal{C}' \to \mathcal{C}$, the automorphisms $\psi$ of $\mathcal{C}'$ such that $\alpha' \circ \psi = \alpha'$ form a group, called the *group of deck transformations* of the covering. If $\alpha'$ is universal, that group is clearly regular (i.e. simply transitive) on the "fibers" $\alpha'^{-1}(x)$ (with $x \in \mathcal{C}$) of $\alpha'$.

If the universal $m$-covering of $\mathcal{C}$ is an isomorphism, we say that $\mathcal{C}$ is *m-connected*, in which case every $m$-covering of $\mathcal{C}$ is a disjoint union of copies of $\mathcal{C}$. Note that an $m$-covering is also an $m'$-covering for $1 \leqslant m' \leqslant m$; therefore, the $m'$-connectedness implies the $m$-connectedness if those inequalities hold.

*Remarks*

(1) To have canonically defined $m$-coverings, one must (as usual) consider "pointed" chamber systems, that is, systems with a privileged chamber.
(2) The "homotopy" of Section 3.4 is a refinement of the equivalence relation between galleries described above, for $m = 2$.
(3) The present notion of $m$-connectedness has nothing to do with the topological $m$-connectedness which would rather, if anything, be vaguely related to our $m'$-connectedness, for $m' = \operatorname{Card} I - m$.

### 5.2. A Second Characterization of Buildings

We now return to the convention of Section 3.2 and assume again that $(\mathcal{C}, (\mathcal{P}_i)_{i\in I})$ is a chamber system of type $M$.

**Theorem 3**

(i) *The buildings are 2-connected chamber systems.*

(ii) *The universal 2-covering of the chamber system $\mathcal{C}$ is a building if and only if the following condition* ($R_c$) *holds for some chamber $c \in \mathcal{C}$, in which case it holds for all $c$:*

**($R_c$)** *If two simple galleries with the same reduced type and common origin $c$ are homotopic, they coincide.*

Since a connected 2-covering of a chamber system of type $M$ is a chamber system of type $M$, assertion (i) is an immediate consequence of Theorem 2 and Corollary 1. Therefore, we only have to prove (ii). For that purpose, we choose a universal 2-covering $\kappa$: $(\tilde{\mathcal{C}}, (\tilde{\mathcal{P}}_i)_{i\in I}) \to (\mathcal{C}, (\mathcal{P}_i)_{i\in I})$ and a chamber $\tilde{c} \in \tilde{\mathcal{C}}$, and we set $c = \kappa(\tilde{c})$. For every gallery $G$ in $\mathcal{C}$ starting at $c$, there is a unique gallery in $\tilde{\mathcal{C}}$, starting at $\tilde{c}$ and projecting onto $G$; we denote it by $\lambda(G)$. Since $\kappa$ is a 2-covering, the image by $\kappa$ of a homotopy is a homotopy.

If $\tilde{\mathcal{C}}$ is a building, it satisfies the condition ($Q_{\tilde{c}}$) of Section 3.3 (see Theorem 2), and hence also the condition ($R_{\tilde{c}}$) (because homotopic galleries have the same extremities), and ($R_c$) follows, by projection.

To prove the converse, we suppose, from now on, that ($R_c$) holds. Let $\bar{\mathcal{C}}$ be the set of all homotopy classes of simple galleries in $\mathcal{C}$, of reduced type and starting at $c$. Let $\tilde{\kappa} : \bar{\mathcal{C}} \to \tilde{\mathcal{C}}$ (respectively, $\omega : \bar{\mathcal{C}} \to W$) be the mapping which, for every such gallery $G$, maps the homotopy class of $G$ onto the extremity of $\lambda(G)$ (respectively, onto $r_{\gamma(G)}$ : cf. Sections 3.1 and 3.2). Set $\bar{\kappa} = \kappa \circ \tilde{\kappa}$. For $\bar{x} \in \bar{\mathcal{C}}$ and $J \subset I$, we define an element $\rho_J(\bar{x})$ of $\bar{\mathcal{C}}$ as follows: consider the "canonical decomposition"

$$\omega(\bar{x}) = w' \cdot w'',$$

where $w'' \in W_J$ and $w'$ is "$J$-reduced on the right" (cf. the definition and references in Section 4.3.2), and choose a representative $(G', G'')$ of $\bar{x}$ such that $r_{\gamma(G')} = w'$ (cf. Sections 3.4.2 and 3.4.3); then $\rho_J(\bar{x})$ is the homotopy class of $G'$. By Proposition 1, Proposition 2 and ($R_c$), that class is independent of the choice of $(G', G'')$. We set $\bar{\mathcal{C}}_J = \rho_J(\bar{\mathcal{C}})$: that is, the set of all $\bar{x} \in \bar{\mathcal{C}}$ such that $\rho_J(\bar{x}) = \bar{x}$, and also the inverse image by $\omega$ of the set of all elements of $W$ which are $J$-reduced on the right. For $\bar{x} \in \bar{\mathcal{C}}_J$, set $\rho'_J(\bar{x}) = \rho_J^{-1}(\bar{x}) = \{\bar{y} \in \bar{\mathcal{C}} \mid \rho_J(\bar{y}) = \bar{x}\}$. The following two assertions are immediate consequences of those definitions and of the fact that $\mathcal{C}$ is a chamber system of type $M$ (we set $\rho_j = \rho_{\{j\}}$):

(1) If $\bar{x} \in \bar{\mathcal{C}}$ and $j \in J \subset I$, one has $\rho_J(\rho_j(\bar{x})) = \rho_J(\bar{x})$ and there is a sequence $j_1, \ldots, j_m$ of elements of $J$ such that $\rho_J(\bar{x}) = \rho_{j_m}(\ldots(\rho_{j_1}(\bar{x}))\ldots)$.

(2) If $J$ has cardinality at most 2 and if $\bar{x} \in \bar{\mathcal{C}}_J$, the mapping $\bar{\kappa}$ induces a bijection of $\rho'_J(\bar{x})$ onto the element of $\mathcal{P}_J$ containing $\bar{\kappa}(\bar{x})$.

For $i \in I$, the sets $\rho'_i(\bar{x})$ with $\bar{x} \in \bar{\mathcal{C}}_i$ form a partition $\bar{\mathcal{P}}_i$ of $\bar{\mathcal{C}}$. By (1) applied to the case $J = I$, the chamber system $(\bar{\mathcal{C}}, (\bar{\mathcal{P}}_i)_{i \in I})$ is connected. For $J \subset I$, it follows from (1) that the join $\bar{\mathcal{P}}_J$ of all $\bar{\mathcal{P}}_j$ $(j \in J)$ is the partition of $\bar{\mathcal{C}}$ in the sets of the form $\rho'_J(\bar{x})$, with $\bar{x} \in \bar{\mathcal{C}}_J$. Consequently, by (2), $\bar{\kappa} = \kappa \circ \tilde{\kappa}$ is a 2-covering, which implies that $\tilde{\kappa}$ is an isomorphism of chamber systems. In particular, $\bar{\mathcal{C}}$ is a chamber system of type $M$. Now, let $\bar{c}$ $(\in \bar{\mathcal{C}})$ denote the homotopy class of the gallery reduced to $(c)$. It is readily seen that if $\bar{G}$ is any simple gallery of reduced type $f$ in $\bar{\mathcal{C}}$ starting at $\bar{c}$ and with extremity $\bar{x}$, then $\bar{\kappa}(\bar{G})$ is a representative of the homotopy class $\bar{x}$ and one has $\omega(\bar{x}) = r_f$. From that it follows that the chamber system $\bar{\mathcal{C}}$ satisfies the condition $(\mathrm{P}_{\bar{c}})$ of Section 3.3, and hence is a building, by Theorem 2. The proof is complete.

### 5.3

**Corollary 3.** *The universal covering of $\mathcal{C}$ is a building if and only if, for every spherical subset $J$ of $I$ (cf. Section* 4.3.1) *of cardinality three, the universal covering of every element of $\mathcal{P}_J$ (considered as a chamber system over $J$) is a building of type $M_J$ (cf. Section* 1.1).

That is an immediate consequence of Theorem 3 and Proposition 4.

## 6. Application to Geometries

*From now on, the set $I$ is assumed to be finite* (of cardinality $\geqslant 2$: cf. Section 3). As before, $M$ denotes a Coxeter diagram over $I$.

### 6.1. Proof of Theorem 1

6.1.1. By a *chain* of elements of a geometry over $I$ (cf. Section 1.2), we understand a sequence of elements of the geometry such that any two consecutive elements of the sequence are incident.

6.1.2.
**Lemma 5.** *If $J$ is a subset of $I$ of cardinality at least two, any two elements $x, y$ of a residually connected geometry can be joined by a chain all of whose elements, except possibly $x$ and $y$, have their types in $J$.*

The proof by descending induction on Card $J$ is immediate.

6.1.3.
**Proposition 5.** *Let $I', I''$ be two complementary subsets of $I$ (i.e. $I = I' \cup I''$ and $I' \cap I'' = \emptyset$), and suppose that the diagram $M$ is the disjoint union of the diagrams $M_{I'}$ and $M_{I''}$ (i.e. $M(I', I'') = \{2\}$). Then every geometry of type $M$ is the "direct sum" of a geometry $\Gamma'$ of type $M_{I'}$ and a geometry $\Gamma''$ of type $M_{I''}$, each element of $\Gamma'$ being incident to each element of $\Gamma''$.*

We have to show that if the types of two elements $x$ and $y$ of such a geometry belong respectively to $I'$ and $I''$, then $x$ and $y$ are incident. Let us assume, without loss of generality, that $\operatorname{Card} I' \geqslant 2$, and let $x = x_0, x_1, \ldots, x_m$ be a chain of elements whose types belong to $I'$ and such that $x_m$ is incident to $y$ (cf. Section 6.1.2). Then it is easily seen, by descending induction on $s \leqslant m$, that $y$ is incident to $x_s$ for all $s$, and hence to $x$.

6.1.4.
**Corollary 4.** *For $m \in \mathbb{N} \cup \{\infty\}$, with $m \geqslant 2$, every geometry of type $\mathbf{A}_1 \times \mathbf{I}_m$ is a building. (Here, $\mathbf{I}_m$ stands for* $\overset{m}{\vdash\!\!\!-\!\!\!-\!\!\!\dashv}$.)

6.1.5.
**Proposition 6.** *For $n \in \mathbb{N}^*$, every geometry of type*

$$\mathbf{A}_n \quad \vdash\!\!\!-\!\!\!-\!\!\!+\!\!\!-\!\!\!-\!\!\!\dashv \ldots . \vdash\!\!\!-\!\!\!-\!\!\!\dashv \qquad (n \text{ vertices})$$

*is a building, i.e., is an $n$-dimensional projective geometry* (*cf. Section* 1.2).

Let us denote by $0, 1, \ldots, n-1$ the vertices of the diagram $\mathbf{A}_n$ in a natural order (say, from left to right). We consider a geometry $\Gamma$ of type $\mathbf{A}_n$ and represent by $\mathfrak{F}_i$ the set of its elements of type $i \in \{0, \ldots, n-1\}$. The elements of $\mathfrak{F}_0$ are also called *points*, and to designate the points incident to an element $x$ of $\Gamma$, we sometimes talk of the *points of* $x$. If $x, y$ are two incident elements of respective types $i, j$, with $i \leqslant j$, every point of $x$ is also a point of $y$ (by Sections 6.1.3 and 1.4 (Res)). The proposition is easily reduced to the following three assertions:

(i) given two elements $x, y$ of $\Gamma$ of types $i, j$ such that $i + j \leqslant n - 2$, there exists an element $z$ of type $i + j + 1$ incident to both of them;
(ii) if $x, y, z$ are as in (i) and if $x$ and $y$ have no point in common, then every element $z'$ of $\Gamma$ incident to both $x$ and $y$ is incident to $z$ and has type $k \geqslant i + j + 1$; in particular, $z$ is the only element of its type incident to $x$ and $y$;
(iii) two elements of $\Gamma$ having the same points coincide.

The proof of (i) is easy, by the method of [8, Section 7] (in fact, [8, $7.2_n$] essentially provides the proof of (i) for $j = 0$).

Let us show (ii). For that purpose, we suppose that $(\Gamma, x, y, z, z')$ is a counterexample to the assertion with minimal value of the sum $i + j + k$ and such that $i \geqslant j$ (which is no loss of generality, since $x$ and $y$ play a symmetric role). Let $p$ be a point of $x$, let $\Gamma'$ be the geometry of type $\mathbf{A}_{n-1}$, residue of $p$ in $\Gamma$, and let $y'$ be an element of $\mathfrak{F}_{j+1}$ incident to $p$, $y$, and $z'$ (the existence of such an element follows from (i) applied to the residue of $z'$). If $i \neq 0$ (hence $x \neq p$), the minimality of $(\Gamma, x, y, z, z')$ implies that every element of $\Gamma$ incident to $p$ and $y$, in particular $z$, is incident to $y'$; but then, $(\Gamma', x, y', z, z')$ contradicts the minimality assumption. Therefore, $i = j = 0$ and $x = p$. Let now $z''$ be an element of $\mathfrak{F}_2$ incident to $z$ and $y'$ (the existence of such an element follows from (i) applied to $\Gamma'$). Since the elements of $\mathfrak{F}_0$ and $\mathfrak{F}_1$ incident to $z''$ form a projective plane (by Sections 6.1.3 and 1.4), and since the "lines" $y'$ and $z$ of that plane have the two points $x$ and $y$ in common, we have $z = y'$. Therefore, $z$ is incident to $z'$, a contradiction.

Finally, (iii) is proved by induction on $n$: one shows, using (ii), that also in the residue of any one of their common points, the two elements in question have the same "points" (elements of $\mathfrak{F}_1$).

6.1.6. *Remarks*

(a) In the proof of Theorem 1, we shall only need the case $n=3$ of Proposition 6, a case where the above proof can be somewhat simplified. Once Theorem 1 is established, the general case of Proposition 6 follows right away; indeed, since every two points of a projective space of dimension $\geqslant 3$ are incident to a same flag of corank 2 (and even of corank 1), an automorphism group of such a space which satisfies (Q2′) and operates freely on the flags of corank 2 is necessarily reduced to the identity.

(b) It can be shown that there exist 2-connected chamber systems of types $\mathbf{A}_1 \times \mathbf{I}_m$ and $\mathbf{A}_n$ which are not buildings. The proof relies on a kind of "free construction" analogous to the one mentioned in Section 1.6.

6.1.7.
**Proposition 7.** *Let $\varphi : \Delta \to \Gamma$ be a morphism of geometries over $I$. Suppose that $\Delta$ is a building of type $M$ (cf. Section 1.6) and that one of the following holds:*

*$\varphi$ is a covering in the sense of Section 1.3 and* $\operatorname{Card} I \geqslant 3$;

*there exists an automorphism group $A$ of $\Delta$ with the properties of Theorem 1 such that $\varphi$ is the product of the canonical projection $\Delta \to \Delta/A$ and an isomorphism $\Delta/A \to \Gamma$.*

*Then the morphism of chamber complexes $\varphi_* : \mathcal{C}(\Delta) \to \mathcal{C}(\Gamma)$ induced by $\varphi$ is a universal 2-covering.*

It is clear that $\varphi_*$ is a 2-covering, which is universal by Theorem 3(i).

6.1.8.
**Corollary 5.** *Let $\Gamma$, $\Delta$, and $A$ be as in Theorem 1, let $\varphi : \Delta \to \Gamma$ be the product of the canonical projection $\Delta \to \Delta/A$ and an isomorphism $\Delta/A \to \Gamma$, and let $X$ be a flag of $\Delta$. Suppose that the residue $\Gamma_{\varphi(X)}$ of $\varphi(X)$ in $\Gamma$ is a building. Then $\varphi$ induces an isomorphism of the residue $\Delta_X$ of $X$ in $\Delta$ onto $\Gamma_{\varphi(X)}$.*

By the above proposition, the restriction of $\varphi$ to $\Delta_X$ is a universal covering of $\Gamma_{\varphi(X)}$. The assertion ensues, in view of Theorem 3(i).

6.1.9.
**Proposition 8.** *Buildings of finite rank are residually simply connected geometries.*

In view of the first assertion of Section 1.6, it suffices to show that a building $\Gamma$ of finite rank $\geqslant 3$ is simply connected. Let $\varphi : \tilde{\Gamma} - \Gamma$ be a covering of $\Gamma$ by a connected geometry $\tilde{\Gamma}$. The morphism of chamber complexes $\mathcal{C}(\tilde{\Gamma}) \to \mathcal{C}(\Gamma)$ induced by $\varphi$ is a 2-covering, and hence an isomorphism by Theorem 3(i), and the assertion follows.

6.1.10. *End of the proof of Theorem* 1. We use the notation of the statement of the theorem.

(i): Let $(\mathcal{C},(\mathcal{P}_i))$ be the chamber system of $\Gamma$, let $\tilde{\mathcal{C}}\to\mathcal{C}$ be a universal 2-covering of that system, let $A$ be the group of deck transformations of that covering (Section 5.1), and set $\Delta=\Gamma(\tilde{\mathcal{C}})$ (cf. Section 2.2). For every spherical set $J\subset I$ of cardinality 3, every element of $\mathcal{P}_J$ is 2-covered by a building: if $M_J$ is of type $\mathbf{A}_1\times\mathbf{I}_2(m)$ or $\mathbf{A}_3$, this follows from Section 6.1.4 or 6.1.5, and if $M_J$ is of type $\mathbf{C}_3$ or $\mathbf{H}_3$, it is part of the hypotheses of the theorem. By Section 5.3, it follows that $\tilde{\mathcal{C}}$, and hence $\Delta$, is a building. On the other hand, it is readily checked that $A$, considered as a group of automorphisms of $\Delta$, has the properties stated in the theorem, and that $\Gamma=\Delta/A$. Finally, the uniqueness of $(\Delta,A)$ is an immediate consequence of Proposition 7.

(ii) is easily deduced from (i) and Proposition 8, by induction on $\operatorname{Card} I$: the induction hypothesis implies that, if $\Delta$ and $A$ are as in (i) and if $\Delta/A$ is residually simply connected, then the canonical projection $\Delta\to\Delta/A$ is a covering.

## 6.2. Geometric Characterization of Some Buildings

6.2.1. In the spirit of [10, Section 4], or of F. Buekenhout's work [2, 3], one may wish to characterize the buildings among the geometries of type $M$ by "geometric" properties, more "concrete" than some sort of simple connectedness. Proposition 9 below gives an example of such characterization for each connected diagram $M$ of spherical type and rank $\geqslant 3$, with the exception of $\mathbf{A}_n$, $\mathbf{H}_3$, and $\mathbf{H}_4$, which are uninteresting for our purpose in view of Proposition 6 and [15].

6.2.2. For later reference, we first recall an important property of buildings, which is also—up to minor variation—used as an axiom in [10] and [3]. Let $\Gamma$ be a geometry over $I$. By definition, the *shadow* of a flag $X$ on a subset $Y$ of $\Gamma$ is the set of all elements of $Y$ which are incident to $X$. Suppose now that $\Gamma$ is a building; then the following assertion holds, as readily follows from [14, Sections 12.9 and 12.15]:

**(Int)** the intersection of the shadows of two elements $x$ and $y$ of $\Gamma$ on the set $\Gamma_i$ of all elements of a given type $i\in I$ is empty or is the shadow on $\Gamma_i$ of a flag incident to $x$ and $y$.

(In fact, the assertion remains true if one replaces $x$ and $y$ by arbitrary flags and $\Gamma_i$ by the set of all flags of a given type, but the weaker form (Int) better serves our purpose, as will be clear from Corollary 6 below.)

6.2.3. As a preliminary to the statement of Proposition 9, we now draw the diagrams to be considered and give names to some of their vertices:

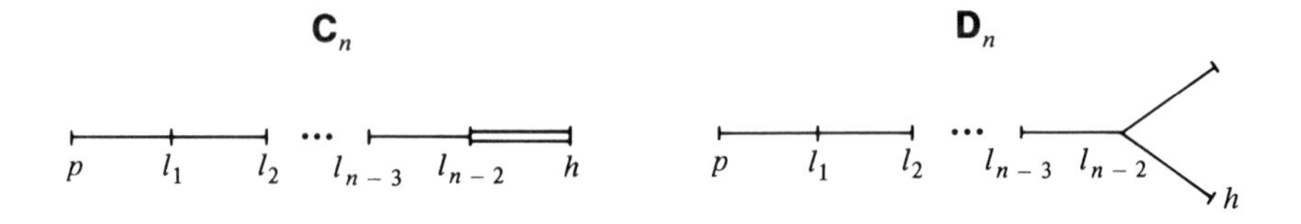

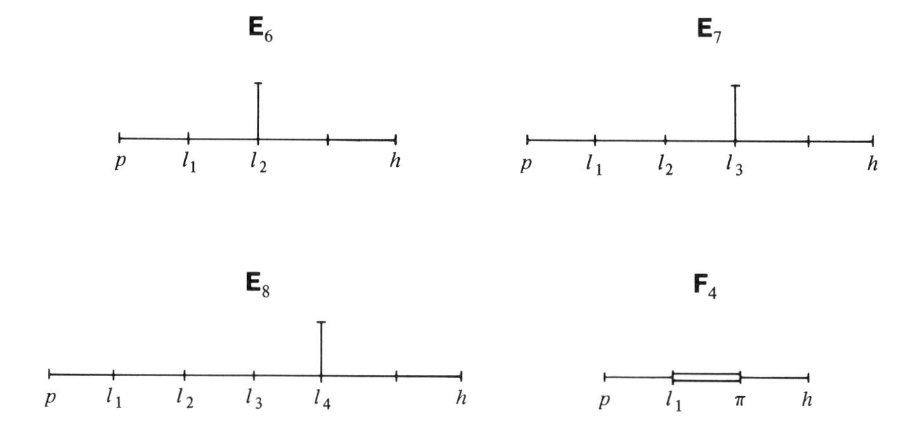


Let $M$ be one of the above diagrams and let $\Gamma$ be a geometry of type $M$. The elements of type $p$, $l_1$, and $h$ of $\Gamma$ will be called *points*, *lines*, and *hyperlines* respectively. From (Int) (in Section 6.2.2), (Res) (in Section 1.4), and Sections 6.1.5 and 6.1.4, it readily follows that, *if* $\Gamma$ *is a building*, then the following assertions hold:

**(LL)** If two lines are both incident to two distinct points, they coincide.

**(LH)** If a line and a hyperline are both incident to two distinct points, they are incident.

**(HH)** If two distinct hyperlines are both incident to two distinct points, the latter are incident to a line.

**(O)** If two elements of type $l_i$ (for some $i$) have the same shadow in the set of all points, they coincide.

Actually, it is equally true (in the case of a building) that *any two* elements of $\Gamma$ which have the same shadow in the set of all points coincide, but we shall not make use of that fact.

**Proposition 9.** *The geometry* $\Gamma$ *is a building if and only if it has the following properties*:

*if* $M = \mathbf{C}_n$, $\mathbf{D}_n$, *or* $\mathbf{E}_6$, *properties* (O) *and* (LL);
*if* $M = \mathbf{E}_7$, *properties* (O), (LL), *and* (LH);
*if* $M = \mathbf{E}_8$ *or* $\mathbf{F}_4$, *properties* (O), (LL), (LH), *and* (HH).

Since all those properties are consequences of (Int), we have

**Corollary 6.** *A geometry of irreducible spherical type different from* $\mathbf{H}_3$ *and* $\mathbf{H}_4$ *is a building if and only if it satisfies condition* (Int) *of Section* 6.2.2.

I do not know whether $\mathbf{H}_3$ and $\mathbf{H}_4$ are true exceptions.

6.2.4. *Proof of Proposition* 9. We suppose that $\Gamma$ has the properties in question, and we intend to show that $\Gamma$ is a building.

(1) *If* $M \neq \mathbf{F}_4$ *and* $\mathbf{E}_6$, *the residue* $\Gamma_x$ *of a point* $x$ *of* $\Gamma$ *also has the properties* (O) *and* (LL). For simplicity, we shall "identify" the elements of type $l_i$ (for any $i$) with their shadows in the set of all points, which are $i$-dimensional projective spaces. Let $y, z$ be two elements of $\Gamma_x$ of type $l_i$ for some $i$, which, *in* $\Gamma_x$, have the same shadow in the set of all elements of type $l_1$ ("points" of $\Gamma_x$). Every point $s$ of $y$ belongs to a line of $y$ containing $x$, and hence to a line of $z$; therefore, $s$ belongs to $z$ (by (Res) and Section 6.1.3). The converse being true for the same reason, $y$ and $z$ have the same set of points, and hence coincide by (O). Thus, $\Gamma_x$ satisfies (O). Let now $q$ and $r$ be two elements of type $l_2$ of $\Gamma_x$ incident to two distinct elements $u, v$ of type $l_1$. Every point $s$ of the projective plane $q$ which does not belong to the lines $u$ and $v$ belongs to a line $t$ joining a point $u'$ of $u$ and a point $v'$ of $v$. In the projective plane $r$, the points $u'$ and $v'$ are also joined by a line, which must coincide with $t$, by (LL). Therefore, $s$ also belongs to the plane $r$. The converse being also true, we have, as above, $q = r$, which establishes (LL) for $\Gamma_x$.

(2) *If* $M = \mathbf{C}_3$, $\Gamma$ *is a building*. Let $\mathbb{S}$ denote the set of all points of $\Gamma$. The elements of type $h$ of $\Gamma$, whose shadows on $\mathbb{S}$ are projective planes, by (Res), will be called *planes*. It is readily seen, using (LL) and (Res) (applied to the residue of a point), that distinct planes have different shadows on $\mathbb{S}$. As before, we can therefore "identify" the lines and the planes with their shadows on $\mathbb{S}$. Now, using (LL) and the fact that the lines and the planes through a given point form a generalized quadrangle (by (Res)), it is a simple exercise to show that $\mathbb{S}$ (with its lines and planes) is a polar space; hence the claim, by [14, Section 7.4].

(3) *If* $M = \mathbf{F}_4$, *the residue of a point in* $\Gamma$ *is a building*. We give the name of *planes* to the elements of type $\pi$ (cf. the diagram $\mathbf{F}_4$ in Section 6.2.3) of $\Gamma$. By (2), we only have to show that if two hyperlines $r$ and $s$ are incident to two distinct planes $t$ and $u$, then $r = s$. In view of (2) again, the shadow $\mathbb{S}$ of $r$ on the set of all points of $\Gamma$ is a polar space of rank 3, in which the shadows of $t$ and $u$ are two distinct planes. Consider in those planes two points which are not collinear in $\mathbb{S}$. By (HL), they are not incident to any line of $\Gamma$. Since they are incident to $s$, by (Res) and Section 6.1.3, it follows from (HH) that $r$ and $s$ coincide, q.e.d.

(4) *If* $M = \mathbf{C}_n$, *the residues of type* $\mathbf{C}_3$ *in* $\Gamma$ *are buildings*. The proof, by induction, is immediate, using (1) and (2).

In view of (3) and (4), Theorem 1 applies to $\Gamma$ in all cases. Therefore, we may —and shall—assume that $\Gamma = \Delta / A$, where $\Delta$ is a building of type $M$, and $A$ is a group of automorphisms of $\Delta$ satisfying the conditions of Theorem 1. We denote by $\varphi : \Delta \to \Gamma$ the canonical projection. From now on, we use induction on the rank of $M$; the induction starts because the proposition is true for $M = \mathbf{C}_3$ (by (2)) and for $M = \mathbf{D}_3 = \mathbf{A}_3$ (by Section 6.1.5).

(5) *If $x$ is a hyperline of $\Delta$, $\varphi$ induces an isomorphism of the residue $\Delta_x$ of $x$ in $\Delta$ onto the residue $\Gamma_{\varphi(x)}$ of $\varphi(x)$ in $\Gamma$.* By the induction hypothesis or by Section 6.1.5—as the case may be—$\Gamma_{\varphi(x)}$ is a building. Our assertion now follows from Section 6.1.8.

(6) *If $M = \mathbf{E}_6$, $\Gamma$ is a building.* The argument of [8, Section 7.6] shows that any two points are incident to a hyperline. Therefore, by (5), $\varphi$ is injective on the points. Our assertion readily follows.

(7) *The residue $\Gamma_x$ of a point $x$ in $\Gamma$ is a building.* In view of (1), (3), (6), and the induction hypothesis, it suffices to consider the case where $M = \mathbf{E}_8$ and to prove that $\Gamma_x$ satisfies the condition (LH). Let $r$ and $s$ be an element of type $l_2$ and a hyperline of $\Gamma_x$, both incident to two distinct elements $t$ and $u$ of type $l_1$, and let $t'$ and $u'$ be two points of $\Gamma$, respectively incident to $t$ and $u$ and distinct from $x$. By Section 6.1.3 applied to the residue of $r$ in $\Gamma$, there is a line $v$ of $\Gamma$ incident to $t'$ and $u'$. Since $t'$ and $u'$ are incident to $s$ (also by Section 6.1.3), it follows from (LH) that $v$ is incident to $s$. But the residue of $s$ in $\Gamma$ is a building of type $\mathbf{D}_7$ (by induction), and hence is the oriflamme geometry of a polar space (cf. [14, Section 7.12]). Considering the triangle $[x, t', u']$ in that space, we see that $t$ and $u$ are incident to a plane $r'$ incident to $s$. Since (LL) holds in $\Gamma_x$ by (1), we have $r = r'$, and $r$ is incident to $s$, q.e.d.

(8) *For any point $x$ of $\Delta$, $\varphi$ induces an isomorphism of $\Delta_x$ onto $\Gamma_{\varphi(x)}$.* By (7) and 6.1.8.

The end of the proof relies on the following assertion, which is easily proved by the method of [8] (in fact, for all types except $\mathbf{F}_4$, it is either well known or effectively proved in [8]):

(9) *In a building of type $\mathbf{C}_n$ or $\mathbf{D}_n$ (respectively: $\mathbf{E}_7$; $\mathbf{F}_4$ or $\mathbf{E}_8$), given a line (respectively: a line; a hyperline) $y$ and a point $x$, there exist a point $r$ incident to $y$ and a line (respectively: a hyperline; a hyperline) $s$ incident to both $x$ and $r$.*

(10) *End of the proof: if $M \neq \mathbf{E}_6$, one has $\Gamma = \Delta$.* It is clearly sufficient to show that $\varphi$ is injective on the set of all points. Assume the contrary, and let $x$ be a point of $\Delta$ and $a$ an element of $A$ with $x \neq ax$. Let $y$ be

a line of $\Delta$ incident to $ax$ if $M = \mathbf{C}_n$, $\mathbf{D}_n$, or $\mathbf{E}_7$,
a hyperline of $\Delta$ incident to $ax$ if $M = \mathbf{F}_4$ or $\mathbf{E}_8$,

and let $r$ and $s$ be as in (9). From (5) applied to a hyperline incident with $s$ (to $s$ itself if it is a hyperline), it follows that $ax$ is not incident to $s$ and that $\varphi(ax) \neq \varphi(r)$. There is no line $y'$ of $\Delta$ incident to both $ax$ and $r$: otherwise $\varphi(y')$ would be incident to $\varphi(s)$ (by (LL) or (LH)), therefore $y'$ would be incident to $s$ (by (8) applied to $r$) and so would be $ax$. Consequently, $M = \mathbf{F}_4$ or $\mathbf{E}_8$, the hyperlines $\varphi(y)$ and $\varphi(s)$ are distinct (again by (8)), and (HH) implies the existence of a line of $\Gamma$ incident with $\varphi(y)$, $\varphi(ax)$, and $\varphi(r)$. But then, by (5) applied to $y$, there also exists a line of $\Delta$ incident to $ax$ and $r$, in contradiction with what we have just seen. The proof is complete. □

6.2.5. *Remarks*

(a) The above proposition explains *a posteriori* the success of the method developed or used in [6], [7], [8], [9] for the geometric investigation of the exceptional Lie groups.

(b) The generalization of (Int) mentioned between parentheses in Section 6.2.2 readily implies the *a priori* weaker property

**(Int′)** For $i \in I$, the intersection of any set of shadows of elements of $\Gamma$ on the set $\Gamma_i$ of all elements of type $i$ is empty or is the shadow of a flag on $\Gamma_i$.

In the same spirit, (LL), (LH) and (HH) have "weaker set-theoretical versions" (LL′), (LH′), (HH′) which are consequences of (Int′). By way of example, we state:

**(LL′)** If the shadows of two lines in the set of all points have two distinct points in common, those shadows coincide.

In [10], it is suggested that, at least for the diagrams $M$ considered here, "good" geometries of type $M$ are those satisfying (Int′). One may wonder which geometries, besides the buildings, fall into that category and, more generally, what becomes of Proposition 9 if one drops property (O) and replaces (LL), (LH), and (HH) by (LL′), etc. Using Theorem 1 again, it should not be too difficult to answer those questions.

## References


[1] Bourbaki, N., *Groupes et Algèbres de Lie*, Chapters 4, 5, 6. Actu. Sci. Ind. No. 1337. Hermann, Paris 1968.

[2] Buekenhout, F., Diagrams for geometries and groups. *J. Comb. Theory (A)* **27** (1979), 121–151.

[3] Buekenhout, F., On the geometry of diagrams. *Geom. Dedicata* **8** (1979), 253–257.

[4] Chevalley, C., Sur certains groupes simples. *Tôhoku Math. J. (2)* **7** (1955), 14–66.

[5] Chevalley, C., Classification des groupes de Lie algébriques, I, II. Séminaire E. N. S., 1956–1958, mimeographed.

[6] Tits, J., Sur certaines classes d'espaces homogènes de groupes de Lie. *Mém. Acad. Roy. Belg.* **29** (3), 1955.

[7] Tits, J., Sur la géométrie des R-espaces. *J. Math. P. et Appl.* **36** (1957), 17–38.

[8] Tits, J., Les groupes de Lie exceptionnels et leur interprétation géométrique. *Bull. Soc. Math. Belg.* **8** (1956), 48–181.

[9] Tits, J., Les "formes réelles" des groupes de type $E_6$. Sém. Bourbaki, Exp. No. 162, févr. 1958.

[10] Tits, J., Groupes algébriques semi-simples et géométries associées. In *Proc. Coll. Algebraical and Topological Foundations of Geometry*, Utrecht 1959, Pergamon Press 1962, 175–192.

[11] Tits, J., Géométries polyédriques et groupes simples. In *Deuxième Réunion du Groupement de Mathématiciens d'Expression Latine*, Florence 1961, 66–88.

[12] Tits, J., Structures et groupes de Weyl. Sém. Bourbaki, Exp. No. 288, févr. 1965.

[13] Tits, J., Le problème des mots dans les groupes de Coxeter. *1st Naz. Alta Mat., Symposia Math.* **1** (1968), 175–185.

[14] Tits, J., *Buildings of Spherical Types and Finite BN-Pairs*. Lecture Notes in Math. No. 386, Springer, 1974.

[15] Tits, J., Endliche Spiegelungsgruppen, die als Weylgruppen auftreten. *Inventiones Math.* **43** (1977), 283–295.

[16] Tits, J., Buildings and Buekenhout geometries. In *Finite simple groups*, II, ed. M. J. Collins, Acad. Press 1980, 309–320.

# Representations and Coxeter Graphs[1]

David Ford*
John McKay*

## 1. Introduction

The properties described here were found while investigating relations between the Lie group $E_8$, Thompson's simple subgroup of the "Monster" which centralizes an element of order three therein, and the cube root of the modular function $j$.

What appears to happen is that the coefficients of the Fourier expansion of the cube root of $j$ are nonnegative integral combinations of the degrees of irreducible representations of both Thompson's group $T$ and the complex Lie group $E_8$, although $T \not\subset E_8$.

Since the automorphic functions describing the characters of the Monster may be the eigenvectors of some natural operator, we examined the eigenvectors of the Cartan matrix of the Killing form for the affine group $\overline{E}_8$. The subject matter of this note extends our findings to affine groups of type $\overline{A}_r$, $\overline{D}_r$, $\overline{E}_6$, $\overline{E}_7$, and $\overline{E}_8$, each of which has a Coxeter graph all of whose bonds are simple.

## 2. The Main Result

The binary polyhedral groups $\langle a,b,c \rangle$ are defined by generators and relations in Coxeter and Moser [6, Section 6.5] as

$$R^a = S^b = T^c = RST.$$

These groups have a center of order two whenever the corresponding polyhedral group $(a,b,c)$ is finite [4]. They are finite groups of quaternions and have a two-dimensional representation as subgroups of $SL(2,\mathbb{C})$.

[1] Partially supported by NSERC and FCAC research grants.
* Department of Computer Science, Concordia University, Montréal, Québec, H3G 1M8.

To each Coxeter graph of type $\bar{A}_r$, $\bar{D}_r$, $\bar{E}_6$, $\bar{E}_7$, and $\bar{E}_8$ there is a matrix, the Cartan matrix $C$, which is the matrix of a positive semidefinite integral quadratic form known as the Killing form. It is indexed by the nodes, and its nonzero entries are given by $c_{ii} = 2$ and $c_{ij} = -1$ if nodes $i$ and $j$ are adjacent. It is immediate that $C = 2I - A$, where $A$ is the adjacency matrix of the graph. $C$ therefore enjoys many properties of the adjacency matrix; in particular, $A$ and $C$ have the same eigenvectors, and their eigenvalues are simply related.

Our main result, which will later be stated in a different way, is:

**Proposition.** *The columns of the character tables of the cyclic group of order* $r+1$, *the dicyclic group (also known as the generalized quaternion group)* $\langle 2,2,r-2\rangle$ *of order* $4(r-2)$, *the binary tetrahedral group* $\langle 2,3,3\rangle$ *of order* 24, *the binary octahedral group* $\langle 2,3,4\rangle$ *of order* 48, *and the binary icosahedral group* $\langle 2,3,5\rangle$ *of order* 120 *are the (suitably normalized) eigenvectors of the Cartan matrices of type* $\bar{A}_r$, $\bar{D}_r$, $\bar{E}_6$, $\bar{E}_7$, *and* $\bar{E}_8$, *respectively.*

Table 1 gives the Coxeter graphs, the eigenvalues of the adjacency matrices, the eigenvectors of the Cartan matrices, and further information which will be explained in due course. The number of nodes in the graph is one more than the subscript in the group name.

**Table 1**

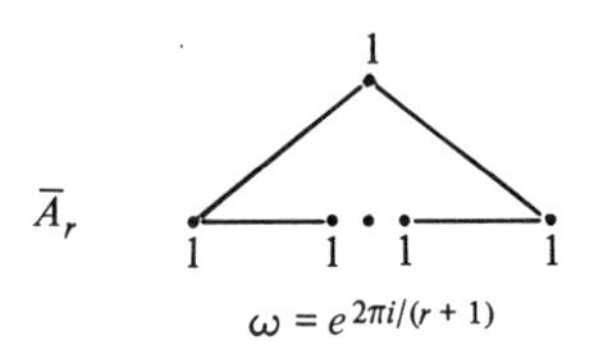


Eigenvalues: $\omega^j + \omega^{-j}, j = 0, \ldots, r.$

Eigenvectors: $v_{jk} = \omega^{jk}, j, k = 0, \ldots, r.$

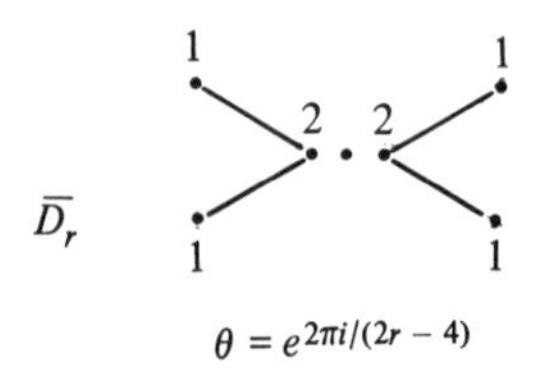


Eigenvalues: $0, 0, \theta^k + \theta^{-k}, k = 0, \ldots, r-2.$

Eigenvectors:

$$\begin{array}{ccc} 1 & 1 & 1 \\ -1 & -1 & 1 \\ 0 & 0 & \theta^{jk} + \theta^{-jk} \\ & & \\ -1 & 1 & (-1)^k \\ 1 & -1 & (-1)^k \end{array} \qquad \begin{array}{l} k = 0, \ldots, r-2. \\ j = 1, \ldots, r-3. \end{array}$$

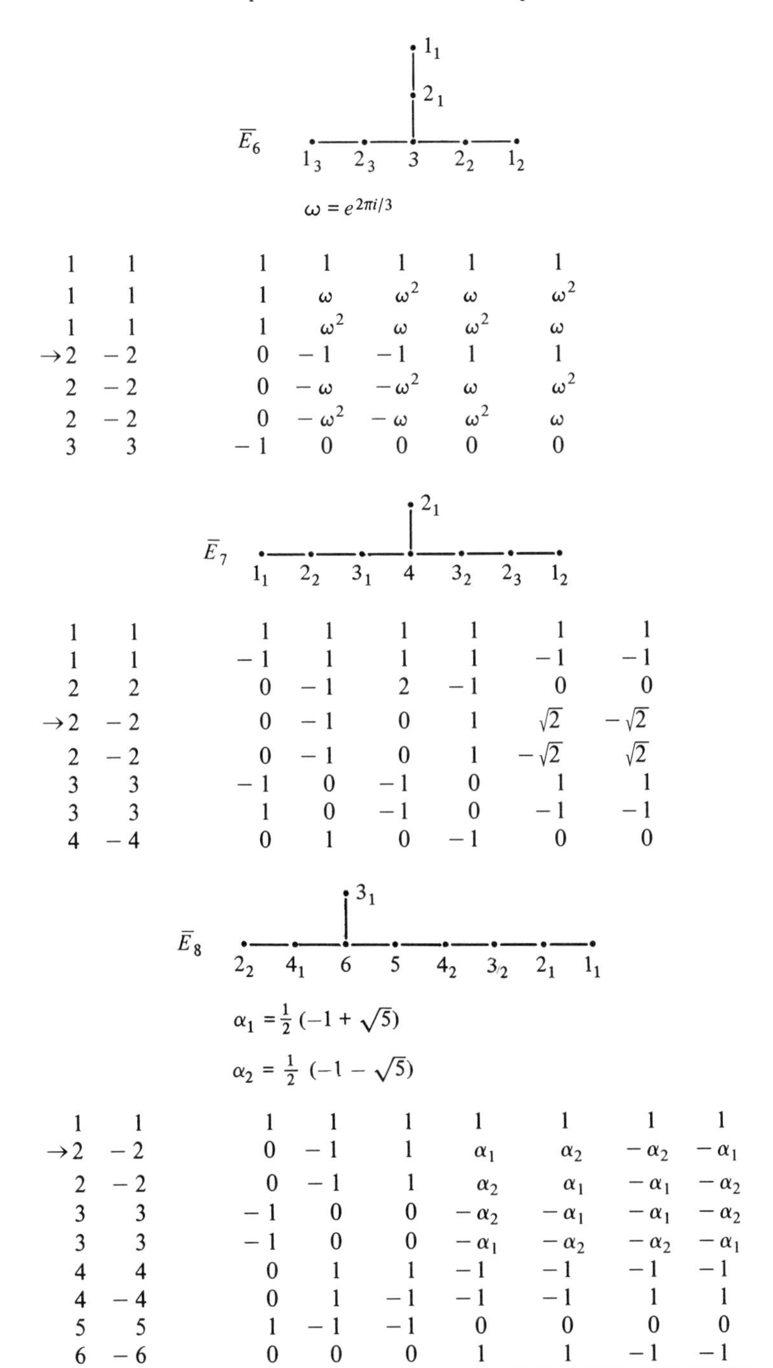


$\omega = e^{2\pi i/3}$

| | | | | | | |
|---|---|---|---|---|---|---|
| 1 | 1 | 1 | 1 | 1 | 1 | 1 |
| 1 | 1 | 1 | $\omega$ | $\omega^2$ | $\omega$ | $\omega^2$ |
| 1 | 1 | 1 | $\omega^2$ | $\omega$ | $\omega^2$ | $\omega$ |
| →2 | −2 | 0 | −1 | −1 | 1 | 1 |
| 2 | −2 | 0 | $-\omega$ | $-\omega^2$ | $\omega$ | $\omega^2$ |
| 2 | −2 | 0 | $-\omega^2$ | $-\omega$ | $\omega^2$ | $\omega$ |
| 3 | 3 | −1 | 0 | 0 | 0 | 0 |



| | | | | | | | |
|---|---|---|---|---|---|---|---|
| 1 | 1 | 1 | 1 | 1 | 1 | 1 | 1 |
| 1 | 1 | −1 | 1 | 1 | 1 | −1 | −1 |
| 2 | 2 | 0 | −1 | 2 | −1 | 0 | 0 |
| →2 | −2 | 0 | −1 | 0 | 1 | $\sqrt{2}$ | $-\sqrt{2}$ |
| 2 | −2 | 0 | −1 | 0 | 1 | $-\sqrt{2}$ | $\sqrt{2}$ |
| 3 | 3 | −1 | 0 | −1 | 0 | 1 | 1 |
| 3 | 3 | 1 | 0 | −1 | 0 | −1 | −1 |
| 4 | −4 | 0 | 1 | 0 | −1 | 0 | 0 |



$\alpha_1 = \frac{1}{2}(-1 + \sqrt{5})$

$\alpha_2 = \frac{1}{2}(-1 - \sqrt{5})$

| | | | | | | | | |
|---|---|---|---|---|---|---|---|---|
| 1 | 1 | 1 | 1 | 1 | 1 | 1 | 1 | 1 |
| →2 | −2 | 0 | −1 | 1 | $\alpha_1$ | $\alpha_2$ | $-\alpha_2$ | $-\alpha_1$ |
| 2 | −2 | 0 | −1 | 1 | $\alpha_2$ | $\alpha_1$ | $-\alpha_1$ | $-\alpha_2$ |
| 3 | 3 | −1 | 0 | 0 | $-\alpha_2$ | $-\alpha_1$ | $-\alpha_1$ | $-\alpha_2$ |
| 3 | 3 | −1 | 0 | 0 | $-\alpha_1$ | $-\alpha_2$ | $-\alpha_2$ | $-\alpha_1$ |
| 4 | 4 | 0 | 1 | 1 | −1 | −1 | −1 | −1 |
| 4 | −4 | 0 | 1 | −1 | −1 | −1 | 1 | 1 |
| 5 | 5 | 1 | −1 | −1 | 0 | 0 | 0 | 0 |
| 6 | −6 | 0 | 0 | 0 | 1 | 1 | −1 | −1 |

The eigenvectors of the Cartan matrix corresponding to a tree graph were computed by Frame in 1951 [9], but he did not notice the connection with character tables.

## 3. The Representation Ring

Let $\{R_s\}$ be a complete set of $t$ inequivalent irreducible representations of the group $G$ over $\mathbf{C}$ which satisfy

$$R_i \otimes R_j = \bigoplus_k a_{ijk} R_k, \qquad i, j, k = 1, \ldots, t.$$

Taking traces we find for $A^{(i)}_{jk} = a_{ijk}$, that

$$A^{(i)} \chi^j = \chi_i^j \chi^j, \qquad i, j = 1, \ldots, t,$$

where $\chi_i^j$ denotes the value of the character of $R_i$ on conjugacy class $j$ of $G$. In other words, the eigenvalues of $A^{(i)}$ are the character values afforded by $R_i$, and the columns of the character table of $G$ are eigenvectors. The dimension of this commutative associative representation ring is the number, $t$, of conjugacy classes of $G$, and *any matrix for which the columns of the character table of G are eigenvectors must commute with the* $A^{(i)}$ and so be a linear combination of these matrices.

## 4. The Representation Graph

From the ring above, and any representation $R$ of $G$, we may construct a directed graph (with multiple edges and loops) which we shall call the representation graph $\Gamma_R$ of $G$. The nodes are the irreducible representations, and an edge joins $R_j$ to $R_k$ with multiplicity $m_{jk}$ where $R \otimes R_j = \oplus_k m_{jk} R_k$. As usual, we convene that a pair of opposing directed edges with the same multiplicity between two nodes be replaced by an undirected edge of that multiplicity.

We can now restate our main result:

**Proposition.** *Each of the five types of finite group described above has a two-dimensional representation R such that* $\Gamma_R$ *is the Coxeter graph for the corresponding affine group.*

*Remarks*. For $G$ cyclic, we may choose $R = R_i \oplus R_i^*$, where $*$ denotes the dual and $R_i$ is any faithful irreducible representation. For the dicyclic case, any faithful two-dimensional irreducible representation is appropriate. In the other cases there is indicated by $\rightarrow$ a two-dimensional faithful representation of real character which may be chosen as $R$. It is easy to prove that these two representation-theoretic properties may be interpreted in the language of graphs:

(1) The representation graph $\Gamma_R$ is connected if and only if $R$ is faithful, and
(2) $\Gamma_R$ is undirected (possibly with directed loops) if and only if $R$ is self-dual.

In the graphs displayed earlier each node has been identified with an irreducible representation of $G$; thus $3_2$ denotes the second irreducible representation of degree 3 in the table. The number of edges (with multiplicities) which emanate from the node $1_1$ corresponding to the trivial representation is the number of (distinct) irreducible constituents of $R$; hence if $R$ is irreducible, the row $1_1$ of the adjacency matrix contains only one nonzero entry and so there is a row of the table of eigenvectors which consists of the eigenvalues of $A$ which are the character values of $R$.

Representation graphs may be superimposed according to the law

$$\Gamma_R = \sum_i a_i \Gamma_{R_i} \quad \text{for } R = \bigoplus_i a_i R_i.$$

Since $C = 2I - A$ and $R$ is faithful, we have that the null space of $C$ is one-dimensional and is spanned by the eigenvector whose components are the irreducible degrees of $G$.

*Remarks.* There appear to be representation-theoretic interpretations for the eigenvectors of some of the other extended diagrams (see [7] or [2]) obtained by "folding" the $A$, $D$, and $E$ diagrams.

What is the explanation underlying all these observations? We do not know, but there are several clues in the literature. On the basis of finiteness conditions Coxeter [5] has suggested that we can say that for each quaternion group $\langle a, b, c\rangle$ there is a Weyl group $[3^{a-1,b-1,c-1}]$. This recalls work of Du Val [8] in which he establishes a connection between certain finite groups and rational singularities in two dimensions. Artin [1] expands further to rational singularities in three dimensions. We find Brieskorn [3] also interested in the question. Steinberg [12, p. 156] gives the connection in terms of a ridge of singularities formed by the unipotent subregular subvariety. The singularities Steinberg mentions are constructed by Klein [10, Part I, Chapter 2, Sections 9–14] in his book on the icosahedron. Slodowy [11] in his 1978 Regensburg thesis gives an algebraic description of the theory of rational singularities. A final source on representations of graphs is Dlab and Ringel [7].

## References


[1] Artin, M., On isolated rational singularities of surfaces. *Amer. J. Math.* **88** (1966), 129–136.

[2] Berman, S., Moody, R., and Wonenburger, M., Certain matrices with null roots and finite Cartan matrices. *Indiana Univ. Math. J.* **21** (1971/72), 1091–1099.

[3] Brieskorn, E., Singular elements of semi-simple algebraic groups. In *Actes Congrès International Math.* 1970.

[4] Conway, J. H., Coxeter, H. S. M., and Shephard, G. C., The centre of a finitely generated group. *Tensor* **25** (1972), 405–418; **26**, 477.

[5] Coxeter, H. S. M., letter to author.

[6] Coxeter, H. S. M. and Moser, W. O. J., *Generators and Relations for Discrete Groups*, 2nd ed. Springer-Verlag 1965.

[7] Dlab, V. and Ringel, C., *Indecomposable Representations of Graphs*. Memoirs AMS #173, 1976.

[8] Du Val, P., On isolated singularities of surfaces which do not affect the conditions of adjunction. *Proc. Camb. Philos. Soc.* **30** (1934), 483–491.

[9] Frame, J. S., Characteristic vectors for a product of $n$ reflections. *Duke Math. J.* **18** (1951), 783–785.

[10] Klein, F., *Lectures on the Icosahedron*, 2nd ed. Reprint, Dover, New York 1956.

[11] Slodowy, P., *Simple Singularities and Simple Algebraic Groups*. Springer-Verlag Yellow Series #815, 1980.

[12] Steinberg, R., *Conjugacy Classes in Algebraic Groups*. Springer-Verlag Yellow Series #366, 1974.

# Coinvariant Theory of a Coxeter Group

Howard L. Hiller*

## 1. Introduction

Let $G$ be a finite group represented on a real vector space $V$. We can make $G$ act on the polynomial algebra $S(V)$ on $V$ by $g \cdot f(x) = f(g^{-1}x)$. Classical invariant theory studies the invariant subalgebra

$$S(V)^G = \bigoplus_{j=0}^{\infty} S_j(V)^G.$$

Alternatively, one has the graded, homogeneous ideal $I_G$, generated by the positive components of $S(V)^G$, and we can form the quotient algebra $S_G = S(V)/I_G$. For convenience, we call this the *coinvariant algebra* of $G$ and its elements *coinvariants* (though this terminology has been used for other purposes).

One new dimension to this coinvariant theory is that $S_G$ and its homogeneous components $S_{G,j}$ support potentially interesting $G$-module structures. Of course, by Chevalley [3] we know that if $G$ is a complex reflection group, $G$ acts on $S_G$ via the regular representation. More generally, R. Stanley has observed that for arbitrary finite $G$, $G$ acts by $t$ times the regular representation, where $t$ is the free rank of $S(V)^G$ as a Cohen–Macaulay module over the polynomial algebra on a homogeneous system of parameters. A great deal of information on the $G$-module structure of the pieces $S_{G,j}$ can be found in [10] (and the references there) for $G$ a Weyl group.

In this note we concern ourselves with the coinvariant theory of a Coxeter group. More specifically, let $(W, S)$ be a Coxeter system in the sense of Bourbaki [2]. It is well known that $W$ can be realized as the "Weyl group" of a (possibly noncrystallographic) root system $\Delta$ in a real Euclidean space $V$ of dimension $n = |S|$. This space possesses a basis $\Sigma$ of simple roots such that the reflection $s_\alpha$

*Mathematical Institute, Oxford University, 24-29 St Giles Street, Oxford OX1 3LB, U.K.

through the hyperplane perpendicular to $\alpha \in \Sigma$ precisely yields the generating set $S$. In this fashion, $W$ admits a natural representation on $V$, and we are in the situation described above.

Of course, Chevalley's theorem [3] tells us that $S(V)^W$ has $n$ algebraically independent generators whose degrees $d_1 \leqslant \cdots \leqslant d_n$ (the fundamental degrees) are useful in describing the gross structure of $S_W$. In particular, one can compute the Poincaré series of $S_W = \oplus_{j=0}^{\infty} S_{W,j}$:

$$PS(S_W, t) = \sum_{j=0}^{\infty} \dim_{\mathbb{R}}(S_{W,j}) t^j \prod_{i=1}^{n} (1 + t + t^2 + \cdots + t^{d_i - 1})$$

so that the real dimension is

$$PS(S_W, 1) = \prod_{i=1}^{n} d_i = |W|,$$

and $S_{W,j} = 0$, for $j > \deg(PS(S_W, t)) = \sum_{i=1}^{n}(d_i - 1)$. Recall that this last sum is equal to the number $N$ of reflections in $W$, by a formula of Solomon [11].

We are interested here in a finer analysis of the algebraic and $W$-module structure of $S_W$. Recall that when $W$ is a Weyl group, $S_W$ corresponds to the real cohomology of an appropriate flag manifold $G/T$. Mimicking algebraically the Bruhat decomposition of such varieties, we can hope to develop a Bruhat basis for $S_W$ and understand its structure with an appropriate Schubert calculus. We describe briefly how this can be done. As an application we mention an algebraic derivation of the Pieri formula of the classical Schubert calculus. The full statement and proofs of this work will appear elsewhere.

We note, in passing, that the sort of results described here have already been analyzed from a variety of viewpoints—for example, the Chow ring [5], Lie-algebra cohomology [9], and de Rham cohomology [12], to mention a few.

The advantage of our method, inspired by [4] and [1], is that once the algebra in question has been identified as the coinvariant algebra $S_W$, all of the Schubert machinery follows in a purely formal fashion.

It is hoped that an extension of this circle of ideas to affine Weyl groups will shed some light on the Bott decomposition of the space of loops on a Lie group [7].

## 2. Statement of Results

Roughly speaking, the basic ingredients of a "Schubert calculus" for a graded algebra are three: (1) a basis theorem that provides a certain homogeneous vector space basis, (2) a Pieri formula that describes the multiplication of these generators, and (3) a Giambelli formula that gives a procedure for writing the generators as polynomials in the original "variables" (see [8]).

Demazure [4] produced an integral basis for the coinvariant algebra $S_W$ of a Weyl group $W$ (relative to a choice of a $W$-invariant $\mathbb{Z}$-lattice). In particular, the group $\Sigma_n \int \mathbb{Z}_2$ supports two different Schubert calculi of type $B_n$ and $C_n$. For a general Coxeter group one can only expect an $\mathbb{R}$-basis result that will vary with the choice of the lengths of the simple roots.

**Theorem 1** (Basis theorem). *Let $W$ be an irreducible Coxeter group. There is a graded $\mathbb{R}$-algebra $H_W$ (relative to a choice of simple roots) with a homogeneous $\mathbb{R}$-basis $\{X_w\}_{w\in W}$ and a surjective map of graded $W$-algebras*

$$c : S(V) \to H_W$$

*such that* $\operatorname{Ker}(c) = I_W$.

We briefly describe the construction. Define an $S(V)^W$-endomorphism of $S(V)$ by

$$\Delta_\alpha(u) = \frac{u - s_\alpha(u)}{\alpha}.$$

Note that the division is legitimate, since $s_\alpha$ is the identity on the kernel of $\alpha$, i.e., $\alpha^\perp$. Let $\boldsymbol{\Delta}_W$ be the subalgebra of endomorphisms of $S(V)$ generated by the $\Delta_\alpha$, $\alpha \in \Delta^+$, and the multiplication operators $\omega^*$, $\omega \in S_1(V) = V^*$. Then we let $\overline{\boldsymbol{\Delta}}_W = \epsilon_* \boldsymbol{\Delta}_W$, where $\epsilon : S(V) \to S_0(V) \approx \mathbb{R}$ is the projection map. So we get a map

$$S(V) \to S(V)^{**} \to \overline{\boldsymbol{\Delta}}_W^*,$$

and we let $H_W = \overline{\boldsymbol{\Delta}}_W^*$, and this composition is $c$ (see [4]). The elements $X_w$ arise as duals to a basis $\{\epsilon \cdot \Delta_w\}_{w\in W}$ of $\overline{\boldsymbol{\Delta}}_W$ where $\Delta_w = \Delta_{\alpha_1}, \ldots, \Delta_{\alpha_n}$ for any reduced decomposition $w = s_{\alpha_1}, \ldots, s_{\alpha_n}$ of $w$.

Our desired Giambelli formula now amounts to finding a $c$-preimage for $X_w$. In its basic form, this is

**Theorem 2** (Giambelli formula). *If $w \in W$, then*

$$X_w = c\left( \frac{1}{|W|} \Delta_{w^{-1}w_0}(d) \right),$$

*where $d$ is the product of the positive roots.*

It is also possible to show that $X_{s_\alpha} = c(\omega_\alpha)$, where $\omega_\alpha$ is the fundamental weight given by the requirement $(\omega_\alpha, \beta^\vee) = \delta_{\alpha\beta}$, where $\alpha, \beta \in \Sigma$ and $\beta^\vee$ is the coroot $2\beta/(\beta, \beta)$. By "expanding" $\Delta_w(d)$ and using the Cartan matrix to replace roots by weights, one can write each $X_w$ as a polynomial in the $X_{s_\alpha}$'s, $\alpha \in \Sigma$. Hence, to understand the multiplication table of the $X_w$'s it suffices to prove

**Theorem 3** (Pieri formula). *If $\alpha \in \Sigma$, $w \in W$, then*

$$X_{s_\alpha} \cdot X_w = \sum_{\substack{\gamma \in \Delta^+ \\ l(ws_\gamma) = l(w)+1}} (\gamma^\vee, \omega_\alpha) X_{ws_\gamma}.$$

We mention that to prove this theorem requires writing the composition $\epsilon \cdot \Delta_w \cdot \omega^*$ in terms of the basis $\{\epsilon \cdot \Delta_w\}_{w\in W}$ and then simply dualizing (see [1]).

It is also possible to compute the action of $W$ on $H_W$ by giving a formula for $s_\alpha \cdot X_w$. One can exploit this computation to relativize the above Schubert machinery. Indeed, let $\theta \subseteq S$ and $(W_\theta, \theta)$ be the corresponding *parabolic* subsystem of $(W, S)$. It is well known (see, e.g., [6]) that

$$W^\theta = \{ w \in W : l(ws_\alpha) = l(w) + 1 \ \forall \alpha \in \theta \}$$

is a set of (minimal length) left coset representatives of $W_\theta$ in $W$. We have

**Theorem 4.** *The set $\{X_w\}_{w \in W^\theta}$ is an $\mathbb{R}$-basis for the algebra of invariants $H_W^{W_\theta}$. In addition, if $w, w' \in W^\theta$ and one computes the product $X_w \cdot X_{w'}$ in $H_W$ and then crosses out the basis elements not indexed by $W^\theta$, one obtains the correct product in $H_W^{W_\theta}$.*

In order to understand these formulae concretely, we analyze the Weyl group of type $A_{n+k-1}$, i.e., the symmetric group $\Sigma_{n+k}$ on $n+k$ letters. In addition, suppose $\theta = \{ s_{e_i - e_{i+1}} : 1 \leqslant i \leqslant n+k-1, i \neq k \}$, so that $W_\theta = \Sigma_k \times \Sigma_n$. Then it is not hard to show

$$W^\theta = \{ (d_1, \ldots, d_k) : 1 \leqslant d_1 < \cdots < d_k \leqslant k+n \},$$

where $(d_1, \ldots, d_k)$ denotes the permutation (in $\Sigma_{n+k}$)

$$(d_1, d_2, \ldots, d_k, d_1', \ldots, d_n')$$

where $d_1' < \cdots < d_n'$ is an ordered enumeration of $\{1, 2, \ldots, n+k\} - \{d_1, \ldots, d_k\}$. It is a result of Bernstein et al. [1] that $X_{(d_1, \ldots, d_k)}$ corresponds to the usual Schubert cocycle $\langle d_1, \ldots, d_k \rangle$ in the $2(\sum_{i=1}^k (d_i - i))$-cohomology group of the complex Grassmannian of $k$-planes in $\mathbb{C}^{n+k}$ (modulo reindexing, of course). By invoking Theorem 4 and checking that (up to sign) $X_{(1,2,\ldots,k-1,k+j)}$ is $c$ of the $j$th elementary symmetric in the last $n$ coordinates, one can eventually give an algebraic demonstration of the classical Pieri formula. (Recall that then the classical Giambelli formula follows by an easy induction.)

*Remark.* It should be possible to argue similarly for $W$ of type $C_n$ and get a Schubert calculus for the cohomology of $Sp(2n)$ modulo a maximal parabolic (i.e., the space of totally isotropic planes in a vector space equipped with a skew-symmetric form).

## References


[1] Bernstein, I. N., Gelfand, I. M., and Gelfand, S. I., Schubert cells and the cohomology of the spaces G/P. *Russian Math. Surveys*, **28** (1973), 1–26.

[2] Bourbaki, N., *Groupes et algèbres de Lie*, Chapitres IV, V, VI. Hermann, Paris 1968.

[3] Chevalley, C., Invariants of finite groups generated by reflections. *Amer. J. Math.* **77**, (1955), 778–782.

[4] Demazure, M., Invariants symétriques entiers des groupes de Weyl et torsion. *Inv. Math.* **21** (1973), 287–301.

[5] Demazure, M., Désingularisation des variétés de Schubert généralisées. *Ann. Scient. Éc. Norm. Sup., 4^e Serie* **7** (1974), 53–88.

[6] Deodhar, V., Some characterizations of Bruhat ordering on a Coxeter group and determination of the relative Möbius function, *Inv. Math.* **39** (1977), 187–198.

[7] Garland, H. and Raghunathan, M., A Bruhat decomposition for the loop space of a compact group: a new approach to results of Bott. *Proc. Nat. Acad. Sci. U.S.A.* **72** (1975), no. 12, 4716–4717.

[8] Kleiman, S., Problem 15. Rigorous foundations of Schubert's enumerative calculus. In *Proc. Symp. Pure Mat.*, Vol. 28. Amer. Math. Soc., Providence, R.I. 1976.

[9] Kostant, B., Lie algebra cohomology and generalized Schubert cells. *Ann. Math.* **77** (1963), 72–144.

[10] Lusztig, G. and Beynon, W., Some numerical results on the characters of exceptional Weyl groups. *Math. Proc. Camb. Phil. Soc.* **84** (1978), 417–426.

[11] Solomon, L., Invariants of finite reflection groups. *Nagoya Math. J.* **22** (1963), 57–64.

[12] Stoll, W., *Invariant Forms on Grassmann Manifolds*, Annals of Math. Studies. Princeton University Press, Princeton, N.J. 1977.

# Two-Generator Two-Relation Presentations for Special Linear Groups[1]

C. M. Campbell*
E. F. Robertson*

## 1. Introduction

A finite group defined by $n$ generators and $m$ relations must have $m \geqslant n$. A finite group is said to have *deficiency zero* if it has a presentation with $n$ generators and $n$ relations. In 1907 Schur [13] proved important results showing that certain finite groups could not have deficiency zero presentations. Let $SL(2, p)$ denote the group of $2 \times 2$ matrices of determinant 1 over the field $GF(p)$, $p$ an odd prime, and put $PSL(2, p) = SL(2, p)/\{\pm I\}$. Now $PSL(2, p)$ and $SL(2, p)$ can be generated by two elements, but Schur's result showed that $PSL(2, p)$ required at least three relations. However, the possibility of a 2-generator 2-relation presentation for $SL(2, p)$ was not excluded.

The number of relations necessary to define $PSL(2, p)$ and $SL(2, p)$ has slowly been reduced in the last hundred years. In 1882 Dyck [9] gave a 2-generator 4-relation presentation for $PSL(2,7)$, and in the 1890s E. H. Moore gave presentations for $PSL(2, p)$ and $SL(2, p)$ with two generators and approximately $2p^2$ relations. It may be worth noting that presentations of a somewhat similar type were studied in the context of algebraic $K$-theory by Steinberg and Milner [12]. A neater presentation for $PSL(2, p)$, but still with the number of relations increasing with $p$, was given by Bussey [3], and recently a similar but symmetric presentation was given by Beetham [1]. Details of Bussey's presentation may be found in Coxeter and Moser [7], which also contains the first presentations not to have the number of relations increasing with $p$, due independently to Frasch [10] and Todd [17]. During the 1930s Coxeter [5] and Sinkov

[1] The authors wish to thank the Carnegie Trust for the Universities of Scotland for a grant to assist the work of this paper.

* Mathematical Institute, University of St. Andrews, St. Andrews, KY16 9SS, Scotland.

[14] discovered 2-generator 4-relation presentations for some of the groups $PSL(2, p)$. In Section 3 we shall discuss some new results related to their work.

An important step was the discovery by Behr and Mennicke [2] of 2-generator 4-relation presentations for both $SL(2, p)$ and $PSL(2, p)$. Zassenhaus [18] and Sunday [16] reduced the number of relations for $PSL(2, p)$ to their theoretical minimum of three. Campbell and Robertson [4] reduced the number of relations for $SL(2, p)$ to their theoretical minimum of two.

The group

$$\langle a, b \mid a^5 = b^3 = (ab)^2 \rangle$$

appears in the work of Poincaré (1895) and was known to Dehn [8] to have order 120. It is in fact the group $SL(2,5)$ and, prior to [4], was the only finite perfect group known to have deficiency zero. (A perfect group $G$ is a group such that $G = G'$, where $G'$ is the derived group of $G$.) In Section 3 we give a finite perfect group of deficiency zero which is not a special linear group.

We would like to thank Peter D. Williams for some helpful ideas relating to this paper.

## 2. Schur Extensions

Let $G$ be a finite group, and suppose that $G$ has a presentation $F/R$, where $F$ is a free group of finite rank.

**Definition.** The *Schur multiplicator* $M(G)$ of $G$ is the subgroup $(F' \cap R)/[F, R]$.

$M(G)$ is independent of the presentation. It is a finite Abelian group.

**Definition.** A *representation group* $C$ of $G$ is a group $C$ with $A \leqslant C$ and

(i) $C/A \cong G$,
(ii) $A \leqslant C' \cap Z(C)$,
(iii) $|A| = |M(G)|$.

It can be shown that $A \simeq M(G)$.

**Definition.** We say that a group $H$ is a *Schur extension* of $G$ if $A \leqslant H$ with

(i) $H/A \cong G$,
(ii) $A \leqslant H' \cap Z(H)$.

In [4] the following result on Schur extensions was proved in the special case of $G$ a perfect group. However, for the applications in this paper we require a more general theorem.

**Theorem 2.1.** *A Schur extension $H$ of $G$ is a homomorphic image of some representation group of $G$. Moreover, if $(|G/G'|, |M(G)|) = 1$, then $G$ has a unique representation group.*

*Proof*. Since $H$ is a Schur extension, there is a subgroup $A$ satisfying (i), (ii) above. The argument in the proof of Theorem 23.5(e) of [11] shows that there exists an epimorphism $\sigma : F \to H$ with $R\sigma = A$ and $[F,R]\sigma = 1$. Hence $\sigma$ induces an epimorphism $\bar{\sigma} : F/[F,R] \to H$. Put $\bar{F} = F/[F,R]$, $\bar{R} = R/[F,R]$, so that $M = M(G) = \bar{F}' \cap \bar{R}$. Now

$$M\bar{\sigma} = (F' \cap R)\sigma = F'\sigma \cap R\sigma = H' \cap A = A.$$

Since $R\sigma = A$, we have $\bar{R}\bar{\sigma} = A$, so, given any $r \in \bar{R}$, there is an $m \in M$ with $m\bar{\sigma} = r\bar{\sigma}$. But then, letting $N = \ker \bar{\sigma}$, we have

$$r = (rm^{-1}) \cdot m \in NM,$$

so that $\bar{R} = NM$.

Now

$$\frac{N}{N \cap M} \cong \frac{NM}{M} \cong \frac{\bar{R}}{\bar{F}' \cap \bar{R}} \cong \frac{R}{F' \cap R} \cong \frac{RF'}{F'} \leqslant \frac{F}{F'}.$$

Hence $N/N \cap M$ is free Abelian, and so $N \cap M$ is a direct factor of $N$,

$$N = (N \cap M) \times E.$$

However, $\bar{R} = E \times M$; for $\bar{R} = NM = ((N \cap M) \times E)M$, showing that $\bar{R} = EM$. Also $E \cap M \subseteq N \cap M, E \cap M \subseteq E$, giving

$$E \cap M \subseteq (N \cap M) \cap E = 1.$$

But now $\bar{F}/E$ is a representation group of $G$.

(i) $$\frac{\bar{F}/E}{\bar{R}/E} \cong \frac{\bar{F}}{\bar{R}} \cong \frac{F}{R} \cong G.$$

(ii) First we show that $\bar{R}/E \leqslant (\bar{F}/E)'$. Now, since $\bar{R} = EM$, it is sufficient to prove that $M \leqslant \bar{F}'$. But this is clearly true, since $F' \cap R/[F,R] \leqslant F'/[F, R]$. Also, since $[\bar{R}, \bar{F}] = 1$, we have $\bar{R}/E \leqslant Z(\bar{F}/E)$.

(iii) $$\frac{\bar{R}}{E} = \frac{E \times M}{E} \cong M.$$

Finally, $H$ is a homomorphic image of the representation group $\bar{F}/E$. For, since $E \leqslant N = \ker \bar{\sigma}$, $\bar{\sigma} : \bar{F} \to H$ induces an epimorphism

$$\bar{\bar{\sigma}} : \frac{\bar{F}}{E} \to H.$$

The final result in Section 1 of [13] proves that $G$ has a unique representation group when $(|G/G'|, |M(G)|) = 1$. □

**Corollary 2.2.** *If $G$ is any finite group and $H$ is a Schur extension of $G$, then $H$ is finite and $|G| \leqslant |H| \leqslant |M(G)|\,|G|$.*

The next result is proved in [13]; see also Theorems 25.5 and 25.7 of [11].

**Theorem 2.3.** *If $p$ is an odd prime, $SL(2, p)$ has trivial Schur multiplicator. $PSL(2, p)$ has Schur multiplicator $C_2$, and $SL(2, p)$ is the unique representation group of $PSL(2, p)$.*

From Theorems 2.1 and 2.3 we deduce immediately the following result.

**Theorem 2.4.** *Let $H_1, H_2, \ldots, H_n$ be a sequence of groups such that $H_1 = PSL(2, p)$ and $H_{i+1}$ is a Schur extension of $H_i$, $1 \leqslant i \leqslant n-1$. Then $H_n$ is either $PSL(2, p)$ or $SL(2, p)$.*

## 3. The Groups $(l, m, n; k)$

The groups

$$(l,m,n;k) = \langle a,b \mid a^l = b^m = 1, (ab)^n = 1, (a^{-1}b^{-1}ab)^k = 1\rangle$$

are defined by Coxeter [5]. We define

$$[l,m,n;k] = \langle a,b \mid a^l = b^m = 1, (ab)^n = (a^{-1}b^{-1}ab)^k\rangle.$$

**Theorem 3.1.** *If $m$ is odd, $[2,m,n;k]$ is a Schur extension of $(2,m,n;k)$.*

*Proof.* Suppose $m$ is odd. Putting $G = (2,m,n;k)$, $H = [2,m,n;k]$, and $A = \langle (ab)^n \rangle$, we clearly have $H/A \cong G$. Now $A \leqslant H'$, since $(ab)^n$ is a power of a commutator. We must show that $(ab)^n$ is central. Now

$$(ab)^n = (ab)(ab)^{n-3}(ab)^2 = (ab^{-1}ab)(ab^{-1}ab)^{k-2}(ab^{-1}ab).$$

Hence

$$b(ab)^{n-3}ab^2 = (b^{-1}aba)^{k-1}. \tag{3.1}$$

Now, using (3.1) twice,

$$\begin{aligned}(ab)^n &= (ab)^{n-3}a \cdot b(ab)^2 \\ &= b^{-2}aba(b^{-1}aba)^{k-1}b \\ &= b^{-2}abab(ab)^{n-3}ab^2 \cdot b \\ &= b^{-2}(ab)^n b^2.\end{aligned}$$

Suppose $m = 2s+1$. Then $(ab)^n = b^{-2s}(ab)^n b^{2s}$. But $b^{2s+1} = 1$, so $(ab)^n = b(ab)^n b^{-1}$. Also $(ab)^n = a(ba)^n a^{-1} = a(ab)^n a^{-1}$, and so $(ab)^n$ is central as required. □

Two corollaries to Theorem 3.1 can be obtained using Theorems 2.1 and 2.4 respectively.

**Corollary 3.2.** *If $m$ is odd and $(2,m,n;k)$ is finite, then $[2,m,n;k]$ is finite and is a homomorphic image of a representation group of $(2,m,n;k)$.*

**Corollary 3.3.** *If $m$ is odd and $(2,m,n;k)$ is $PSL(2,p)$ for some prime $p$, then $[2,m,n;k]$ is either $PSL(2,p)$ or $SL(2,p)$.*

We can extend Corollary 3.3 to the following lemma.

**Lemma 3.4.** *If $m$ is odd, then $(ab)^{2n}=1$ in $[2,m,n;k]$. Hence either*

$$|[2,m,n;k]| = |(2,m,n;k)|$$

*or*

$$|[2,m,n;k]| = 2|(2,m,n;k)|.$$

*Proof.* From (3.1)

$$(ab)^{n-3}a = b^{-1}(b^{-1}aba)^{k-1}b^{-2}.$$

Hence, if $k$ is even,

$$(ab)^{n-3}a = b^{-1}(b^{-1}aba)^{(k-2)/2}(b^{-1}ab)(ab^{-1}ab)^{(k-2)/2}ab^{-2},$$

and if $k$ is odd,

$$(ab)^{n-3}a = b^{-1}(b^{-1}aba)^{(k-3)/2}(b^{-1}ab)a(b^{-1}ab)(ab^{-1}ab)^{(k-3)/2}ab^{-2}.$$

In either case $b(ab)^{n-3}ab^2a$ is a conjugate of $a$, and so $(b(ab)^{n-3}ab^2a)^2=1$. Hence $(b(ab)^{n-1})^2=1$, giving

$$b(ab)^{n-1}a\cdot ab(ab)^{n-1}=1,$$

which shows that $(ba)^n(ab)^n=1$. But, since m is odd, the proof of Theorem 3.1 shows that $(ba)^n=(ab)^n$. Hence $(ab)^{2n}=1$ as required. □

It is of interest to note that the condition $m$ odd is necessary. For example $(2,4,5;3)$ is the symmetric group $S_5$ (see p. 91 of [5]). But $[2,4,5;-3]$ is an extension of $C_{11}$ by $S_5$. However, $[2,4,5;3]$ is $S_5$.

Examples. In [5] it is shown that

$$(2,3,7;4) \cong PSL(2,7),$$
$$(2,3,7;6) \cong (2,3,7;7) \cong PSL(2,13),$$
$$(2,3,11;4) \cong PSL(2,23).$$

Using Corollary 3.3, it is not hard to show that

$$[2,3,7;4] \cong PSL(2,7),$$
$$[2,3,7;6] \cong [2,3,7;7] \cong PSL(2,13),$$
$$[2,3,11;4] \cong PSL(2,23).$$

We can also obtain results for finite groups $(l,m,n;k)$ which are not $PSL(2,p)$. For instance $(2,3,8;5) \cong [2,3,8;5]$ is of order 2160, while $[2,3,8;4] \cong [2,3,8;-4]$ is an extension of $C_2$ by $(2,3,8;4)$, $(2,3,8;4)$ being $PGL(2,7)$.

**Theorem 3.5.** *If* $(2,3,n;k)$ *is a finite perfect group, then*

$$H = \langle a,b \mid a^2b^3 = 1, (ab)^n = (a^{-1}b^{-1}ab)^k a^{(n\pm 1)/3}\rangle$$

*is a Schur extension of* $[2,3,n;k]$.

*Proof.* Since $(2,3,n;k)$ is perfect, $(n,6)=1$. Now putting $A = \langle a^2\rangle$, we have $H/A \cong [2,3,n;k]$. Clearly $A$ is central since $a^2 = b^{-3}$, and $A \leqslant H'$ since $H$ is perfect. □

**Corollary 3.6.** *If* $(2,3,n;k)$ *is* $PSL(2,p)$, *then* $SL(2,p)$ *has a 2-generator 2-relation presentation*

$$\langle a,b \mid a^2b^3 = 1, (ab)^n = (a^{-1}b^{-1}ab)^k a^{(n\pm 1)/3}\rangle.$$

Consequently we have deficiency-zero presentations for $SL(2,7)$, $SL(2,13)$, $SL(2,23)$. Sinkov [15] showed that $(2,3,7;8)$ has a homomorphic image of order 10752. In fact $|(2,3,7;8)| = 10752$. Putting $G = (2,3,7;8)$, we see that $G$ has a normal nilpotent subgroup $N$ with $|N| = 2^6$ and $G/N \cong PSL(2,7)$. By Lemma 3.4, $|[2,3,7;8]| = 10752$ or 21504. Then by Theorem 3.5

$$\overline{G} = \langle a,b \mid a^2b^3 = 1, (ab)^7 = (a^{-1}b^{-1}ab)^8 a^2\rangle$$

is a Schur extension of $[2,3,7;8]$, so by Corollary 2.2, $\overline{G}$ is a finite perfect group of deficiency zero.

## 4. The Groups $\langle l,m,n\rangle$

Our results on Schur extensions may be applied to obtain many of the "remarkable" results in [7]. As a typical example we examine the relation between the groups

$$(l,m,n) = \langle a,b,c \mid a^l = b^m = c^n = abc = 1\rangle$$

and

$$\langle l,m,n\rangle = \langle a,b,c \mid a^l = b^m = c^n = abc\rangle$$

proved in [6].

**Theorem 4.1.** *If* $n = 3$, 4, *or* 5, *then* $\langle 2,3,n\rangle$ *is a Schur extension of* $(2,3,n)$.

*Proof.* Let $H = \langle 2,3,n\rangle = \langle b,c \mid c^n = b^3 = (bc)^2\rangle$, and put $A = \langle c^n\rangle$. Clearly $H/A \cong (2,3,n)$ and $A \leqslant Z(H)$. However, $c^{n-6} \in H'$, and so for $n = 3$, 4, 5 we have $A \leqslant H'$ as required. □

**Corollary 4.2.** *For* $n = 3, 4, 5$, $|\langle 2, 3, n\rangle| = 2|(2, 3, n)|$. *Moreover* $\langle 2, 3, n\rangle$ *is* $SL(2, 3)$, $GL(2, 3)$, *or* $SL(2, 5)$ *respectively*.

*Proof*. $(2, 3, n)$ has Schur multiplicator $C_2$ for $n = 3, 4, 5$. The cases $n = 3$ and $n = 5$ follow from Theorem 2.3, since $(2, 3, 3) \cong PSL(2, 3)$ and $(2, 3, 5) \cong PSL(2, 5)$. However $(2, 3, 4) \cong S_4$, so the results of [11, p. 653] apply. Since $\langle 2, 3, n\rangle$ has deficiency zero, we see, using Theorem 2.2, that $\langle 2, 3, n\rangle$ is the unique representation group of $(2, 3, n)$ for $n = 3, 4, 5$. The result now follows again using Theorem 2.3 and [11, p. 653]. □

## References


[1] Beetham, M. J., A set of generators and relations for the group $PSL(2, q)$, $q$ odd. *J. London Math. Soc.* **3** (1971), 554–557.

[2] Behr, H. and Mennicke, J., A presentation for the groups $PSL(2, p)$. *Canad. J. Math.* **20** (1968), 1432–1438.

[3] Bussey, W. H., Generational relations for the abstract group simply isomorphic with the group $LF[2, p^n]$. *Proc. London Math. Soc.* **3** (1905), 296–315.

[4] Campbell, C. M. and Robertson, E. F., A deficiency zero presentation for $SL(2, p)$. *Bull. London Math. Soc.* **12** (1980), 17–20.

[5] Coxeter, H. S. M., The abstract groups $G^{m,n,p}$. *Trans. Amer. Math. Soc.* **45** (1939), 73–150.

[6] Coxeter, H. S. M., The binary polyhedral group and other generalizations of the quaternion group. *Duke Math. J.* **7** (1940), 367–379.

[7] Coxeter, H. S. M. and Moser, W. O. J., *Generators and Relations for Discrete Groups*, 3rd ed. Springer, Berlin 1972.

[8] Dehn, M., Über die Topologie des dreidimensionalen Raumes. *Math. Ann.* **69** (1910), 137–168.

[9] Dyck, W., Gruppentheoretische Studien. *Math. Ann.* **20** (1882), 1–45.

[10] Frasch, H., Die Erzeugenden der Hauptkongruenzengruppen für Primzahlstufen. *Math. Ann.* **108** (1933), 229–252.

[11] Huppert, B., *Endliche Gruppen I*. Springer, Berlin 1967.

[12] Milner, J., *Introduction to Algebraic K-theory*. Annals of Math. Studies 72, Princeton Univ. Press, Princeton 1971.

[13] Schur, I., Untersuchungen über die Darstellung der endlichen Gruppen durch gebrochene lineare Substitutionen. *J. Math.* **132** (1907), 85–137.

[14] Sinkov, A., A set of defining relations for the simple group of order 1092. *Bull. Amer. Math. Soc.* **41** (1935), 237–240.

[15] Sinkov, A., Necessary and sufficient conditions for generating certain simple groups by two operators of periods two and three. *Amer. J. Math.* **59** (1937), 67–76.

[16] Sunday, J. G., Presentations of the groups $SL(2, m)$ and $PSL(2, m)$. *Canad. J. Math.* **24** (1972), 1129–1131.

[17] Todd, J. A., A note on the linear fractional group. *J. London Math. Soc.* **7** (1932), 195–200.

[18] Zassenhaus, H., A presentation of the groups $PSL(2, p)$ with three defining relations. *Canad. J. Math.* **21** (1969), 310–311.

# Groups Related to $F^{a,b,c}$ Involving Fibonacci Numbers[1]

C. M. Campbell*
E. F. Robertson*

## 1. Introduction

The definition of the groups

$$F^{a,b,c} = \langle x, y \mid x^2 = 1, xy^axy^bxy^c = 1\rangle$$

was suggested by H. S. M. Coxeter because of their relevance to the search for trivalent 0-symmetric Cayley diagrams, and these groups are studied in [1]. The structure of these groups depends heavily on that of the groups

$$H^{a,b,c} = \langle x, y \mid x^2 = 1, y^{2n} = 1, xy^axy^bxy^c = 1\rangle,$$

where $n = a + b + c$. The groups $H^{a,b,c}$ and related groups are studied in [1] and [2]. Perhaps rather surprisingly, the orders of the groups $F^{a,b,c}$ and $H^{a,b,c}$ sometimes involve Fibonacci numbers; see for example Theorem 9.1 of [1]. Another connection with Fibonacci numbers emerges from a study of the apparently unrelated class of groups discussed in [3]. In the work of this last paper the groups $T(2m) = \langle x, t \mid xt^2xtx^2t = 1, xt^{2m+1} = t^2x^2\rangle$ are shown to have order $40mf_m^3$ if $m$ is odd and $8mg_m^3$ if $m$ is even, where $f_m$ and $g_m$ are Fibonacci and Lucas numbers respectively. (Definitions and useful properties of Fibonacci and Lucas numbers may be found in [3], [5], [8].) To see that the groups $T(2m)$ are in fact closely related to the groups $F^{a,b,c}$, put $a = xt$, $b = t$ to obtain

$$T(2m) = \langle a, b \mid b^{-1}a^2b = a^{-2}, abab^{-2}ab^{2m+1} = 1\rangle.$$

Then $\langle a^2\rangle$ is normal in $T(2m)$, and the factor $T(2m)/\langle a^2\rangle$ is isomorphic to $F^{1,-2,2m+1}$.

In an attempt to illuminate the involvement of the Fibonacci numbers in the two classes $F^{a,b,c}$ and $H^{a,b,c}$, we study in this paper the two related classes

$$M^{a,b,c} = \langle x, y \mid y^{-1}x^2y = x^{-2}, xy^axy^bxy^c = 1\rangle$$

[1] The authors wish to thank the Carnegie Trust for the Universities of Scotland for a grant to assist the work of this paper.

* Mathematical Institute, University of St. Andrews, KY16 9SS, Scotland.

and

$$N^{a,b,c} = \langle x, y \mid y^{-1}x^2y = x^{-2}, y^{2n} = 1, xy^axy^bxy^c = 1\rangle.$$

As part of the argument we find it useful to compute the Schur multiplicator of the groups $H^{a,b,c}$. This multiplicator, which is of independent interest, is calculated by techniques related to those used in the study of centropolyhedral groups in [4].

We shall use the following notation:

$Z(G)$ = the center of the group $G$,

$G'$ = the derived group of $G$,

$C_m$ = the cyclic group of order $m$,

$[a,b] = a^{-1}b^{-1}ab$,

$(m,n)$ = the highest common factor of the integers $m, n$,

$$\epsilon_i = \begin{cases} 1 & \text{if there is a } j \text{ with } 1 \leqslant j \leqslant n, j \equiv i \pmod{2n}, \\ -1 & \text{if there is a } j \text{ with } n+1 \leqslant j \leqslant 2n, j \equiv i \pmod{2n}. \end{cases}$$

## 2. The Groups $N^{a,b,c}$

The subgroup $\langle x^2\rangle$ of $N^{a,b,c}$ is clearly normal, and $N^{a,b,c}/\langle x^2\rangle$ is isomorphic to $H^{a,b,c}$. Denote the cyclic subgroup $\langle x^2\rangle$ by $K$. The crux of the problem is thus to determine the order of the subgroup $K$. We shall use the notation $n = a+b+c$.

The easily checked isomorphism $N^{a,b,c} \cong N^{-c,-b,-a}$ allows us to assume that $n \geqslant 0$. Since the case $n = 0$ is trivial, we shall consider only $n \geqslant 1$. First we consider the case when $n$ is even. Then either two of $a, b, c$ are odd or $a, b, c$ are all even. However it is clear that $N^{a,b,c} = N^{b,c,a} = N^{c,a,b}$, so the problem when $n$ is even is reduced to the following two cases:

*Case* (i). $n \geqslant 2$; $a, b$ odd, $c$ even,
*Case* (ii). $n \geqslant 2$; $a, b, c$ even.

Before we consider these cases separately, we find a presentation for $L^{a,b,c}$, the derived group of $N^{a,b,c}$, which is valid for all even $n$. Notice that $N^{a,b,c}/L^{a,b,c} \cong C_{2n}$. For any integer $i$ let $X_i = y^{i-1}xy^{n-i+1}$. Then if $i = 2n\lambda + j$, where $1 \leqslant j \leqslant 2n$, we have $X_i = X_j$. It is not hard to see that $L^{a,b,c}$ is generated by $X_i$, $1 \leqslant i \leqslant 2n$. Put $T = X_1X_{n+1}$. Then $T = x^2$ and, since $n$ is even,

$$[T, X_i] = 1, \qquad 1 \leqslant i \leqslant n. \tag{1}$$

Also $X_{n+1} = X_1^{-1}T$, $X_{n+2} = X_2^{-1}T^{-1}$ and, in general,

$$X_{n+i} = X_i^{-1}T^{(-1)^{i+1}}, \qquad 1 \leqslant i \leqslant n. \tag{2}$$

Now use the Reidemeister–Schreier algorithm to obtain a presentation for $L^{a,b,c}$ on the generators $T$, $X_i$, $1 \leqslant i \leqslant n$. The Reidemeister–Schreier algorithm is described in Section 2.3 of [10] and is elucidated geometrically in [4]. Using (1) and (2), the only relations for $L^{a,b,c}$ additional to (1) are

$$X_i^{\epsilon_i} X_{i+a}^{\epsilon_{n+i+a}} X_{i+a+b}^{\epsilon_{i+a+b}} = T^{\delta_i}, \tag{3}$$

where

$$\delta_i = (-1)^i \frac{1-\epsilon_i}{2} + (-1)^{n+i+a} \frac{1-\epsilon_{n+i+a}}{2} + (-1)^{i+a+b} \frac{1-\epsilon_{i+a+b}}{2}. \tag{4}$$

For simplicity of notation we put $Y_i = X_i^{\epsilon_i}$, $1 \leqslant i \leqslant 2n$, so $Y_i = Y_{i+n}^{-1}$, $1 \leqslant i \leqslant n$. We shall call the relation

$$Y_{i+a+b} Y_i = Y_{i+a} T^{\delta_i}$$

relation $(i)$ of $L^{a,b,c}$. We now have the following theorem.

**Theorem 2.1.** *If $n$ is a positive even integer $L^{a,b,c}$ has a presentation*

$$\langle Y_1, Y_2, \ldots, Y_n, T \mid Y_{i+a+b} Y_i = Y_{i+a} T^{\delta_i}, [T, Y_i] = 1, 1 \leqslant i \leqslant 2n \rangle.$$

First we consider case (i) and compute the period of $T$, which, of course, is the order of the subgroup $K$. The crucial step is given by the following theorem. Recall that, for any integer $m$, we are using the notation $f_m$ and $g_m$ for the Fibonacci and Lucas numbers respectively.

**Theorem 2.2.** *Suppose $n$ is a positive even integer and $a, b$ are odd. For any integers $s$ and $t$ the following commutator relations hold in $L^{a,b,c}$:*

$$[Y_i, Y_{i+sa+tb}] = T^{\mu(i,s,t)},$$

*where $\mu(i,s,t) = (-1)^{i+s} f_{s+t}$.*

*Proof.* First notice that, using the expression for $\delta_i$ given in (4), we have

$$\delta_i + \delta_{i+n} = (-1)^i. \tag{5}$$

From relations $(i)$ and $(i+n)$ we have

$$Y_i^{-1} Y_{i+a+b}^{-1} = Y_{i+a}^{-1} T^{-\delta_i},$$

$$Y_i Y_{i+a+b} = Y_{i+a} T^{-\delta_{i+n}}$$

respectively. Hence, using (5),

$$[Y_i, Y_{i+a+b}] = T^{(-1)^{i+1}}. \tag{6}$$

Relation $(i)$ together with (6) gives $[Y_i, Y_{i+a}] = T^{(-1)^{i+1}}$.

The proof is now completed by induction on $\max\{|s|, |t|\}$. The inductive step requires that four cases be considered. These cases are the same as in Theorem 3.3 of [2], and since the argument given there may be modified to prove the present inductive step, we omit further details. □

**Theorem 2.3.** *Suppose $n$ is a positive even integer and $a, b$ are odd. Let $\alpha = (a-b)/2$, $\beta = n/(2(n,a))$, and $\gamma = n/(2(n,b))$. Then the period of $T$ is*

$$\begin{array}{ll} (f_\alpha, f_\beta, f_\gamma) = f_{(\alpha,\beta,\gamma)} & \text{if } \alpha \text{ is odd and } n/2 \text{ is odd,} \\ (g_\alpha, f_\beta, f_\gamma) = (g_\alpha, f_{(\beta,\gamma)}) & \text{if } \alpha \text{ is even and } n/2 \text{ is odd,} \\ (f_\alpha, g_\beta, g_\gamma) & \text{if } \alpha \text{ is odd and } n/2 \text{ is even,} \\ (g_\alpha, g_\beta, g_\gamma) & \text{if } \alpha \text{ is even and } n/2 \text{ is even.} \end{array}$$

*Proof.* Obviously $[Y_i, Y_i] = 1$. But, from the commutator relations of Theorem 2.2, we have $[Y_i, Y_{i+ab-ba}] = T^{\mu(i,-b,a)}$. Since $\mu(i,-b,a) = (-1)^{i-b} f_{a-b}$, we have $T^{f_{a-b}} = 1$ and so

$$T^{f_{2\alpha}} = 1. \tag{7}$$

Again Theorem 2.2 gives $[Y_i, Y_{i+b}] = T^{\mu(i,0,1)}$ and $[Y_i, Y_{i+(a+1)b-ba}] = T^{\mu(i,-b,a+1)}$. Therefore, since $\mu(i,0,1) - \mu(i,-b,a+1) = (-1)^i(1+f_{a-b+1})$, we have $T^{1+f_{a-b+1}} = 1$ and so

$$T^{1+f_{2\alpha+1}} = 1. \tag{8}$$

Now

$$(f_{2\alpha}, 1+f_{2\alpha+1}) = \begin{cases} f_\alpha, & \alpha \text{ odd,} \\ g_\alpha, & \alpha \text{ even} \end{cases}$$

(see Theorem 2.3 of [3]). Hence by (7) and (8), $T^{f_\alpha} = 1$ if $\alpha$ is odd, while $T^{g_\alpha} = 1$ if $\alpha$ is even.

Obviously $[Y_i, Y_{i+n}] = [Y_i, Y_i^{-1}] = 1$. But, since $a$ is odd, $2\beta a$ is an odd multiple of $n$, so $[Y_i, Y_{i+n}] = [Y_i, Y_{i+2\beta a}] = T^{\mu(i,2\beta,0)}$. Hence, since $\mu(i,2\beta,0) = (-1)^{i+2\beta} f_{2\beta}$,

$$T^{f_{2\beta}} = 1. \tag{9}$$

Next $[Y_i, Y_{i+na}] = [Y_i, Y_{i+a}^{-1}] = T^{\mu(i,1,0)}$. But, since $a$ is odd, $2\beta a$ is an odd multiple of $n$, so $[Y_i, Y_{i+n+a}] = [Y_i, Y_{i+(2\beta+1)a}] = T^{\mu(i,2\beta+1,0)}$. Therefore, since $\mu(i,1,0) - \mu(i,2\beta+1,0) = (-1)^i(1+f_{2\beta+1})$,

$$T^{1+f_{2\beta+1}} = 1. \tag{10}$$

Hence by (9) and (10), $T^{f_\beta} = 1$ if $\beta$ is odd, while $T^{g_\beta} = 1$ if $\beta$ is even. Note that $\beta$ is odd if and only if $n/2$ is odd. Similarly we obtain $T^{f_\gamma} = 1$ if $n/2$ is odd, while $T^{g_\gamma} = 1$ if $n/2$ is even.

We now have the required result provided no further relations hold which restrict the period of $T$. Now all further relations arise in the following way. From $[Y_i, Y_{i+sa+tb}] = T^{\mu(i,s,t)}$ and $[Y_i, Y_{i+(s-b)a+(t+a)b}] = T^{\mu(i,s-b,t+a)}$ we obtain $T^{\mu(i,s,t)-\mu(i,s-b,t+a)} = 1$. But $\mu(i,s,t) - \mu(i,s-b,t+a) = (-1)^{i+s}(f_{s+t} + f_{s+t+a-b})$, and a straightforward induction argument shows that $(f_{2\alpha}, 1+f_{2\alpha+1})$ divides $(f_{2\alpha}, f_k + f_{2\alpha+k})$ for any integer $k$. □

The results of Theorem 2.3 show that $K$ can have order $f_{2m+1}$ or $g_{2m}$ for any $m$. For, taking $c = -2a$ gives $n = a - b$, so provided $a, b$ are coprime, $K$ has order $f_{(a-b)/2}$ if $(a-b)/2$ is odd and $g_{(a-b)/2}$ if $(a-b)/2$ is even. Other groups $N^{a,b,c}$ have the subgroup $K$ with order a Fibonacci or Lucas number; for example take $c = (\lambda - 1)a - (\lambda + 1)b$. Rather surprisingly, $K$ can only have order $f_{2m+1}$ or $g_{2m}$ for some $m$.

We now consider case (ii). Analogous to Theorem 2.2 we obtain, in this case, the following result.

**Theorem 2.4.** *Suppose $n$ is a positive even integer and $a, b$ are even. For any integers $s$ and $t$ the following commutator relations hold in $L^{a,b,c}$:*

$$[Y_i, Y_{i+sa+tb}] = T^{\nu(i,s,t)}$$

*where*

$$\nu(i,s,t) = \begin{cases} 0 & \text{if } s+t \equiv 0 \pmod 3, \\ 3(-1)^{i+1} & \text{if } s+t \equiv 1,2 \pmod 6, \\ 3(-1)^{i} & \text{if } s+t \equiv 4,5 \pmod 6. \end{cases}$$

*Proof.* An inductive argument as in Theorem 2.2 gives this result. □

The analogous result in case (ii) to Theorem 2.3 is considerably simpler in both statement and proof.

**Theorem 2.5.** *Suppose $n$ is a positive even integer and $a, b$ are even. Then the period of $T$ is infinite if $a \equiv b \equiv c \pmod 6$ and is 3 otherwise.*

*Proof.* If $a \equiv b \equiv c \pmod 6$ then $L^{a,b,c}$ has $L^{2,2,2}$ as a homomorphic image. It is straightforward to check that $T$ has infinite period in $L^{2,2,2}$. Otherwise we can assume $a \not\equiv b \pmod 6$. Then take $s = b$ and $t = -a$ in the commutator relations of Theorem 2.4 to obtain the required result. □

Notice that although Theorem 2.5 shows that $T$ has finite period unless $a \equiv b \equiv c \pmod 6$, most of the groups $N^{a,b,c}$ with $a, b, c$ even are infinite. For $H^{a,b,c}$ is infinite when $(a,b,c) = h \neq 1$ unless $H^{a/h,b/h,c/h}$ is Abelian. For example $H^{2,-6,8}$ is $C_8$, so $N^{2,-6,8}$ is a group of order 24. In fact $N^{2,-6,8}$ is isomorphic to $\langle -2,2,3 \rangle$ in the notation of [6, p. 70].

If $n$ is odd it is easy to prove that $K$ is central in $N^{a,b,c}$. For

$$x^2 \cdot xy^a xy^b xy^c = xy^a xy^b xy^c \cdot x^{-2},$$

since $x^2 y = yx^{-2}$. Thus $x^4 = 1$, so $\langle x^2 \rangle$ is a central subgroup. Also it is easy to see that $K$ is contained in the derived group of $N^{a,b,c}$. We have proved the following lemma.

**Lemma 2.6.** *If $n$ is odd, the subgroup $K \leqslant Z(N^{a,b,c}) \cap (N^{a,b,c})'$. Moreover $K$ is either trivial or $C_2$.*

The precise order of $K$ is obtained as a corollary to the main theorem of the next section.

## 3. The Multiplicator of $H^{a,b,c}$

Let $G$ be a finite group, and suppose $G$ has a presentation $F/R$ where $F$ is a free group of finite rank. Then let $D = F/[F,R]$. The Schur multiplicator $M(G)$ of $G$ is the subgroup $(F' \cap R)/[F,R]$ of $D$. A representation group $S$ of $G$ is a group such that $S$ has a subgroup $A$ with $S/A \cong G$, $A \leqslant Z(S) \cap S'$ and $|A| = |M(G)|$. We have the following theorem (for a proof see Section 9.9 of [7], in particular Proposition 8).

**Theorem 3.1.** *Let $G$ be a finite group. Suppose $G \cong H/B$, where $B \leqslant Z(H) \cap H'$. Then $H$ is a homomorphic image of some representation group of $G$.*

The group $H^{a,b,c}$ has two generators and three relations, and so $M(H^{a,b,c})$ is cyclic by Theorem 25.2 of [9].

Rather than calculate the orders of all representation groups of $H^{a,b,c}$, we consider the more general problem of the orders of the groups

$$H^{a,b,c}_{\alpha,\beta,\gamma} = \langle x, y, t \mid x^2 = t^\alpha, xy^axy^bxy^c = t^\beta, y^{2n} = t^\gamma, [t,x] = [t,y] = 1\rangle.$$

Consider the subgroup $\overline{H}$ of $H^{a,b,c}_{\alpha,\beta,\gamma}$ generated by $\{X_i, t : 1 \leqslant i \leqslant n\}$, where $X_i = y^{i-1}xy^{n-i+1}t^{-\gamma}$. Determining a presentation for $\overline{H}$ as in Theorem 2.1 gives

$$\overline{H} = \langle Y_1, Y_2, \ldots, Y_n, t \mid Y_{i+a}^{-1}Y_{i+a+b}Y_i = t^{\sigma(i)}, [t, Y_i] = 1, 1 \leqslant i \leqslant 2n\rangle$$

where $Y_i = X_i^{\epsilon_i}$ and $\sigma(i) = (2\beta - 3\alpha - \gamma + (1 - 2\alpha + 2\beta)(\epsilon_i - \epsilon_{i+a} + \epsilon_{i+a+b}))/2$. Notice that $\sigma(i) + \sigma(i+n) = 2\beta - 3\alpha - \gamma$. Similar arguments to those of Sections 3 and 4 of [2] are now applicable. The modified versions of Lemmas 4.1 and 4.2 of [2] give $t$ of period $|2\beta - 3\alpha - \gamma|$ when $(a-b, b-c, 3) = 1$, and $t$ of period $2|2\beta - 3\alpha - \gamma|$ when $(a-b, b-c, 6) = 3$. This proves the following theorem.

**Theorem 3.2.** *If $a \equiv b \equiv c$ (mod 6) or $2\beta - 3\alpha - \gamma = 0$, then $H^{a,b,c}_{\alpha,\beta,\gamma}$ is infinite; otherwise*

$$|H^{a,b,c}_{\alpha,\beta,\gamma}| = |H^{a,b,c}|\,|2\beta - 3\alpha - \gamma|\,\frac{(a-b, b-c, 3) + 1}{2}.$$

Notice that (4.14) of [4] solves the same type of problem for polyhedral groups.

The representation groups of $H^{a,b,c}$ are those groups $H^{a,b,c}_{\alpha,\beta,\gamma}$ with $t \in (H^{a,b,c}_{\alpha,\beta,\gamma})'$. However, $t \in (H^{a,b,c}_{\alpha,\beta,\gamma})'$ if and only if $|2\beta - 3\alpha - \gamma| = 1$. Thus, for fixed $a, b, c$, the representation groups of $H^{a,b,c}$ are the groups $H^{a,b,c}_{\alpha,\beta,1-3\alpha+2\beta}$, and for those groups,

$\overline{H}$ coincides with $(H^{a,b,c}_{\alpha,\beta,1-3\alpha+2\beta})'$. Using Theorem 3.1, we have the following result on the Schur multiplicator of $H^{a,b,c}$.

**Theorem 3.3.** *If* $a \equiv b \equiv c \pmod 6$, $H^{a,b,c}$ *is infinite. Otherwise* $H^{a,b,c}$ *is finite and its Schur multiplicator is*

$$\begin{array}{ll} 1 & \text{if } (a-b, b-c, 3) = 1, \\ C_2 & \text{if } (a-b, b-c, 3) = 3. \end{array}$$

This theorem, together with the proof of Theorem 9.1 of [1], allows us to verify the conjecture stated in [1] following Theorem 9.1, namely

$$|F^{a,b,-2a}| = 2|H^{a,b,-2a}| \quad \text{if } a \equiv b \pmod 3,$$
$$|F^{a,b,-2a}| = |H^{a,b,-2a}| \quad \text{if } a \not\equiv b \pmod 3.$$

**Corollary 3.4.** *If* $(a-b, b-c) = 1$, *then* $H^{a,b,c}$ *is isomorphic to* $F^{a,b,c}$.

*Proof.* By Theorem 4.1 of [1], when $(a-b, b-c) = 1$, $F^{a,b,c}$ contains a subgroup $B$ with $B \leqslant Z(F^{a,b,c}) \cap (F^{a,b,c})$, and $F^{a,b,c}/B \cong H^{a,b,c}$. Using Theorem 3.1 we have $F^{a,b,c} \cong H^{a,b,c}$. □

This corollary proves part of the conjecture given in Section 12 of [1].

**Corollary 3.5.** *If* $n$ *is odd, then* $K$ *is a trivial subgroup of* $N^{a,b,c}$ *when* $(a-b, b-c, 3) = 1$, *and* $K$ *is* $C_2$ *when* $(a-b, b-c, 6) = 3$.

*Proof.* The result follows from Lemma 2.6. □

We can now give the order of $M^{a,b,c}$ in the case $(a-b, b-c) = 1$. For when $(a-b, b-c) = 1$, then by Corollary 3.4 the relation $y^{2n} = 1$ is deducible from $x^2 = 1$ and $xy^axy^bxy^c = 1$. The same algebraic proof with the relations $y^{-1}x^2y = x^{-2}$ and $xy^axy^bxy^c = 1$ must yield $y^{2n} = x^{2\alpha}$ for some $\alpha$. Then $y^{4n} = 1$ and $M^{a,b,c}/\langle y^{2n}\rangle \cong N^{a,b,c}$.

## References


[1] Campbell, C. M., Coxeter, H. S. M., and Robertson, E. F., Some families of finite groups having two generators and two relations. *Proc. Roy. Soc. London A* **357** (1977), 423–438.

[2] Campbell, C. M. and Robertson, E. F., Classes of groups related to $F^{a,b,c}$. *Proc. Roy. Soc. Edinburgh* **78A** (1978), 209–218.

[3] Campbell, C. M. and Robertson, E. F., Deficiency zero groups involving Fibonacci and Lucas numbers. *Proc. Roy. Soc. Edinburgh* **81A** (1978), 273–286.

[4] Conway, J. H., Coxeter, H. S. M., and Shephard, G. C., The centre of a finitely generated group. *Tensor, N.S.* **25** (1972), 405–418.

[5] Coxeter, H. S. M., *An Introduction to Geometry*, 2nd ed. Wiley, New York 1969.

[6] Coxeter, H. S. M. and Moser, W. O. J., *Generators and Relations for Discrete Groups*, 3rd ed. Springer, Berlin 1972.

[7] Gruenberg, K. W., *Cohomological Topics in Group Theory*. Springer, Berlin 1970.

[8] Hardy, G. H. and Wright, E. M., *An Introduction to the Theory of Numbers*, 4th ed. Oxford University Press 1960.

[9] Huppert, B., *Endliche Gruppen I*. Springer, Berlin 1967.

[10] Magnus, W., Karrass, A., and Solitar, D., *Combinatorial Group Theory*. Interscience, New York 1966.

# Part V: The Combinatorial Side

# Convex Polyhedra

W. T. Tutte*

Contemplation of the convex polyhedra leads to some interesting enumerative problems. How many combinatorially distinct polyhedra are there with $n$ edges? Or, as Kirkman asked, how many with $p$ faces and $q$ vertices?

We count two convex polyhedra as *combinatorially equivalent* if there is a 1-1 mapping of vertices, edges, and faces of one onto vertices, edges, and faces respectively of the other which preserves incidence relations. If no such mapping exists then the polyhedra are *combinatorially distinct*. It should be noted that this definition makes a polyhedron combinatorially equivalent to its mirror image.

Steinitz's theorem allows us to replace our polyhedra by 3-connected maps on the sphere, and there are papers in the literature concerned with the enumeration of such maps. The expository paper of P. J. Federico [3] deserves special mention.

Enumeration of polyhedra or maps becomes much easier when we are allowed to "root" them. In a rooting we choose an edge $R$ to be called the *root edge*, one of its ends to be called the *root vertex*, and one of its incident faces to be called the *root face*. When $R$ is chosen we may expect there to be two choices for the root vertex and two for the root face, four choices in all. But some of these may be equivalent under the symmetry of the unrooted figure.

Combinatorial equivalence of rooted maps or polyhedra is defined in the same way as for unrooted ones. But the equivalence is now required to preserve the root edge, root vertex, and root face, as well as the incidence relations.

An *automorphism* of a polyhedron or map is a 1-1 mapping, or combinatorial equivalence, of that object onto itself which preserves incidence relations. The automorphisms of a polyhedron or map are the elements of a group called its *automorphism group*. It is easily seen that only the trivial or identical automorphism can preserve a rooting. We often express this result by saying that all

*C and O Department, Faculty of Mathematics, University of Waterloo, Waterloo, Ontario, Canada N2L 3G1.

rooted polyhedra are unsymmetrical. It is perhaps for this reason that rooted polyhedra are easier to enumerate than unrooted ones.

In [6] there appears an exact formula for the number $c_n$ of combinatorially distinct rooted polyhedra of $n$ edges. It makes use of the definite integral

$$J_n = \int_0^1 t^n(2t-1)^n \, dt, \tag{1}$$

defined for all nonnegative integers $n$. A number $R_n$ is defined as follows:

$$R_n = \frac{(2n+1)}{8/(n!)^2}\left\{(27n^2+9n-2)J_n - (9n-2)\right\}. \tag{2}$$

It is shown in the paper that

$$c_{n+1} = 2(-1)^{n+1} + R_n \tag{3}$$

if $n > 3$.

There is an associated recursion formula that is useful for numerical calculations. Let us write

$$S_n = 27n^2 + 9n - 2. \tag{4}$$

It is found in [6] that

$$S_n R_{n-1} + 2S_{n-1}R_n = \frac{2(2n)!}{(n!)^2} \tag{5}$$

if $n > 0$. Starting from the observation that $R_0 = 0$, we can use (5) to compute successively $R_1$, $R_2$, $R_3$, and so on. A table of $c_n$ constructed in this way is given in the paper. It runs from $c_4 = 0$ to $c_{25} = 1{,}932{,}856{,}590$. I understand that the entries in this table have been verified by actual counting up to $c_{22}$. This was done in the course of the computer search that culminated in the discovery by A. J. W. Duijvestijn of a perfect squared square of order 21 [2].

The problem of enumerating unrooted polyhedra is more difficult. We can begin by considering in how many ways such a polyhedron $P$ can be rooted. If there are $n$ edges, we can initially state the number of rootings as $4n$, remarking however that some of them may be equivalent under the symmetry of $P$. It is easy to see that if $h$ is the order of the automorphism group of $P$, then the number of combinatorially equivalent rooted forms of $P$ is $4n/h$ [4].

Let us write $d_n$ for the number of (combinatorially distinct) unrooted polyhedra of $n$ edges. By the above observations we have

$$c_n \geqslant d_n \geqslant c_n/4n. \tag{6}$$

For large $n$ we are tempted to assert the asymptotic approximation

$$d_n \sim \frac{c_n}{4n}. \tag{7}$$

This is based on the belief that almost all polyhedra of $n$ edges are unsymmetrical, i.e., that the proportion of unsymmetrical polyhedra (with $h = 1$) tends to unity as $n$ tends to infinity. No theoretical justification of this belief is given in [6]. I can still offer it only as a conjecture, not as a theorem.

If this conjecture of negligible symmetry is accepted, we can deduce a simple asymptotic formula for $d_n$. We get this by combining (7) with an asymptotic estimate of $c_n$ deduced from (5). It is found that

$$d_n \sim \frac{n^{-7/2} 4^n}{243\sqrt{\pi}} \tag{8}$$

[6, Section 9].

The need for a theorem of negligible symmetry occurs over and over again in the enumerative theory of planar maps. Often an exact formula for some class of rooted maps is found. Always there is the same difficulty in passing from this formula to an asymptotic formula for the unrooted case.

I have overcome this difficulty in only one nontrivial case, that of the convex polyhedra in which all faces are triangular. Equivalently the problem concerns the enumeration of 3-connected triangulations of the sphere. Success was made possible by the existence of an extensive enumerative theory of rooted "near-triangulations."

A *near-triangulation* is a spherical map in which at most one face is nontriangular. Conventionally the nontriangular face, if any, must be taken as the root face in any rooting. In the literature near-triangulations are usually classified in terms of two numbers $n$ and $m$. The valency of the root face is $m + 3$, and the number of internal vertices (that is, vertices not incident with the root face) is $n$.

There are several variants of the enumerative theory of rooted near-triangulations. In one form the near-triangulations are required to be 3-connected, that is, to correspond to convex polyhedra. An explicit formula for the number of rooted near-triangulations of this kind, with given values of $m$ and $n$, is given in [5]. In another form the near-triangulations are required to be *strict*, that is, to have no loops or multiple joins. It is also required that the boundary of the root face shall be a simple closed curve. In a strict near-triangulation there may be edges not incident with the root face but having both ends incident with the root face. A 3-connected near-triangulation can be characterized as a strict one without such "diagonal" edges.

A theory of strict near-triangulations is given by W. G. Brown in [1]. Let us write $A(n, m)$ for the number of rooted strict near-triangulations with given $n$ and $m$. It is found in [1] that

$$A(n,m) = \frac{2(2m+3)!\,(4n+2m+1)!}{(m+2)!\,m!\,n!\,(3n+2m+3)!}\,. \tag{9}$$

For $m = 0$ near-triangulations become true triangulations. Moreover, strict and 3-connected triangulations are the same. So the number of rooted convex polyhedra with all faces triangular is

$$A(n,0) = \frac{2(4n+1)!}{(3n+2)!\,(n+1)!}\,. \tag{10}$$

This formula appears also in [5].

Formula (9) can be used to establish crude upper bounds for numbers of rooted 3-connected triangulations whose unrooted forms have specified kinds of symmetry. Suppose for example that we are interested in those 3-connected triangulations with $E$ edges which have an automorphism combinatorially equivalent to reflection in the center of the sphere. Such a triangulation $T$ can be constructed from two isomorphic near-triangulations $N_1$ and $N_2$. Their root faces are deleted, and the root-face boundaries are suitably identified.

Let us consider the case in which $N_1$ is restricted to have a root face of valency $s + 3$ and a number $n$ of internal vertices, where $n$ depends on $s$ and $E$. Then the number of unrooted 3-connected triangulations derivable from $N_1$ by our construction, summed over all possible choices of $N_1$, does not exceed $A(n,s)$. We can expect this estimate to be excessive, partly because it enumerates rooted near-triangulations, not unrooted ones, and partly because some triangulations with double joins may be counted. Moreover some triangulations may be constructible in more than one way.

Our estimate $A(n,s)$ can be summed over all possible values of $s$ and then multiplied by $4E$. It then becomes an upper bound for the number of rooted 3-connected triangulations with $E$ edges whose unrooted forms have the stated kind of symmetry.

There are analogous arguments giving upper bounds for triangulations having a plane of symmetry, or an axis of symmetry of prime order $p$. Details are given in [7]. The plane of symmetry gives the most difficulty. It can bisect edges as well as contain them, and all the resulting possibilities have to be considered.

But the crude upper bounds can be found, and it is shown in [7] that they are all negligible in comparison with $A(n,0)$ when $n$ is large. So in this one case the theorem of negligible symmetry is established. The paper [7] concludes with the unqualified assertion that the number of combinatorially distinct convex polyhedra, with all faces triangular and with $2q$ faces in all, is asymptotically

$$\frac{1}{64(6\pi)^{1/2}}\, q^{-7/2}\left\{\frac{4^4}{3^3}\right\}^q .$$

## References


[1] Brown, W. G., Enumeration of triangulations of the disk. *Proc. London Math. Soc.* **14** (1964), 746–768.

[2] Duijvestijn, A. J. W., Simple perfect squared square of lowest order. *J. Combinatorial Theory B* **25** (1978), 240–243.

[3] Federico, P. J., The number of polyhedra. *Philips Research Reports* **30** (1975), 220–231.

[4] Harary, F. and Tutte, W. T., On the order of the group of a planar map. *J. Combinatorial Theory* **1** (1966), 394–395.

[5] Tutte, W. T., A census of planar triangulations. *Can. J. Math.* **14** (1962), 21–38.

[6] Tutte, W. T., A census of planar maps. *Can. J. Math.* **15** (1963), 249–271.

[7] Tutte, W. T., On the enumeration of convex polyhedra. *J. Combinatorial Theory* (B), **28** (1980), 105–126.

# Non-Hamilton Fundamental Cycle Graphs

Joseph Malkevitch*

The purpose of this note is to exhibit examples related to the work of Maciej Syslo [3, 5]. The examples concern the nonexistence of Hamilton circuits in some fundamental cycle graphs. Also, some problems concerning fundamental cycle graphs are posed.

We begin by reviewing some definitions. Terms not defined here can be found in Harary [1]. If $G$ is a connected graph and $T$ is a spanning tree of $G$, then as is well known, $T$ together with any edge $e$ of $G$ not in $T$ forms a unique cycle. The collection of such cycles as $e$ ranges over the edges in $G$ but not in $T$ is called a *fundamental cycle set* of $G$ with respect to $T$. For a survey of questions related to fundamental cycle sets, see [2, 3, 5].

Let a connected graph $G$ be given, and let $T$ be a spanning tree of $G$. Let $c$ be the set of fundamental cycles of $G$ with respect to $T$. Following Syslo [3], we define the *fundamental cycle graph* $G(c)$ of $c$ as follows: The vertices of $G(c)$ are in one-to-one correspondence with the cycles of $c$. Two vertices of $G(c)$ are joined by an edge if the cycles they represent have at least one edge in common. Thus, $G(c)$ can be thought of as the intersection graph of the fundamental cycles of $T$ with respect to $G$. Fundamental cycle graphs would appear to be of interest to persons interested in perfect graphs.

Figure 1 shows a plane, 3-valent, 3-connected graph $G$, and a spanning tree $T^*$ (hatched edges) such that the fundamental cycle graph (Figure 2) associated with the fundamental cycles $c^*$ of $T^*$ in $G$ is non-Hamilton. This can be seen by observing that $G(c^*)$ has 2-valent vertices at vertices 1, 3, 5. (The edges in $G$-$T^*$ have been labeled to correspond to the clrcuits in $c^*$, and thus to the vertices of $G(c^*)$.) Based on this example, one can easily construct an infinite family of non-Hamilton fundamental cycle graphs obtained from 3-valent, 3-connected graphs. Furthermore, by replacing each triangle in Figure 1 with a configuration

*Department of Mathematics, York College (CUNY), Jamaica, New York 11451, U.S.A.

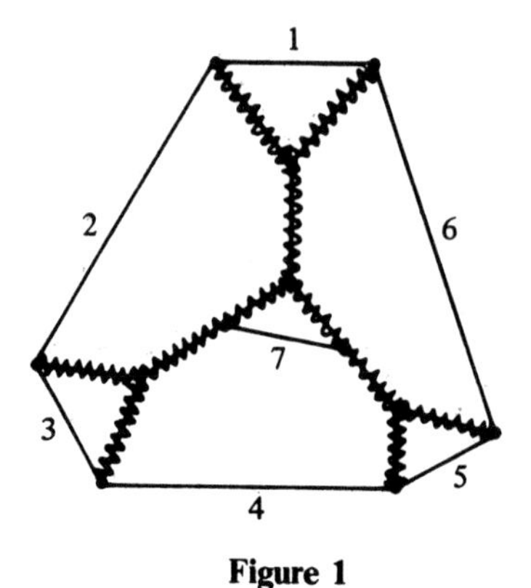


Figure 1

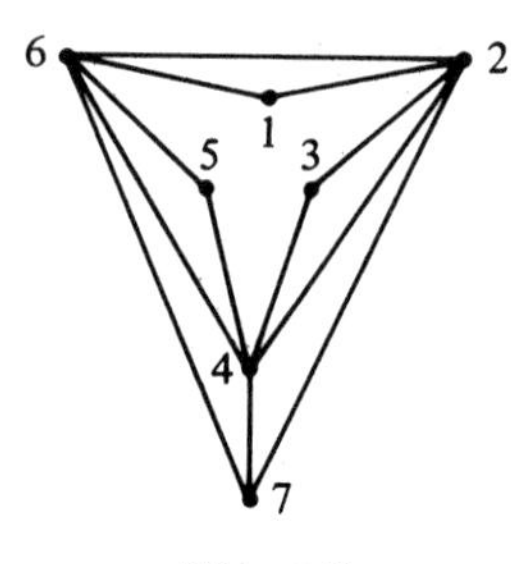


Figure 2

of three 4-gons at a vertex, one obtains a graph $G$ which is 3-valent, is 3-connected, and contains no triangle, and which has a tree for which $G(c)$ is non-Hamilton.

## Problems

(1) What conditions on $G$ and/or $T$ make $G(c)$ $k$-connected?
(2) What conditions on a plane $G$ and/or $T$ make $G(c)$ (a) $k$-connected, (b) plane and $k$-connected?
(3) What conditions on $G$ and/or $T$ are necessary and sufficient for $G(c)$ to be Hamilton?
(4) If $G$ is plane and 3-connected, does there exist a spanning tree $T$ of $G$ such that $G(c)$ is Hamilton? (Note: There exists such a tree for the graph in Figure 1.)
(5) Which graphs $G$ admit a tree $T$ so that $G(c)$ is a complete graph?

*Remark*. Partial solutions to (1) and (2) can be found in Syslo [4].

## References

[1] Harary, F., *Graph Theory*. Addison-Wesley, Reading, Mass. 1969.

[2] Syslo, M. M., On some problems related to fundamental cycle sets of a graph. Report No. 33, Institute of Computer Science, Wroclaw University, Poland. *Proceedings of the Stefan Banach International Mathematical Center, Warsaw* (to appear).

[3] Syslo, M. M., On characterizations of cycle graphs and other families of intersection graphs. Report No. 40, Institute of Computer Science, Wroclaw University, Poland, 1978 (submitted for publication).

[4] Syslo, M. M., On characterizations of outerplanar graphs. *Discrete Mathematics* **26** (1979), 47–53.

[5] Syslo, M. M., On characterizations of cycle graphs. In *Colloque CNRS. Problemes Combinatories et Theorie des Graphes, Orsay 1976*, Paris 1978.

# Some Combinatorial Identities

W. O. J. Moser*

For integers $n \geqslant 1$, $k \geqslant 0$, and $w \geqslant 0$, let $(n\,|\,k)_w$ denote the number of $k$-choices

$$1 \leqslant x_1 < x_2 < \cdots < x_k \leqslant n \tag{1}$$

from $\{1, 2, 3, \ldots, n\}$ satisfying the restrictions

$$x_i - x_{i-1} - 1 \geqslant w \quad \text{for } i = 2, 3, \ldots, k \tag{2}$$

and

$$n + x_1 - x_k - 1 \geqslant w.$$

Displaying $1, 2, 3, \ldots, n$ in a circle, these restrictions are seen as: every chosen integer is followed (clockwise) by (at least) $w$ nonchosen integers. It is well known and easy to prove [3, (17); 4, p. 222, Problem 2] that

$$(n\,|\,k)_w = \begin{cases} \dbinom{n-kw}{k}\dfrac{n}{n-kw} & \text{if } 0 \leqslant k \leqslant \dfrac{n}{w+1},\ (n,k) \neq (0,0), \\ 0 & \text{if } 0 \leqslant \dfrac{n}{w+1} < k. \end{cases}$$

Of course $(0\,|\,0)_w$ has no combinatorial meaning; by setting it equal to $w + 1$, the $(n\,|\,k)_w$ neatly satisfy (and are determined by) the recurrence

$$(n\,|\,0)_w = (n-1\,|\,0)_w \qquad \text{for } n \geqslant w+1, \tag{3}$$

$$(n\,|\,k)_w = (n-1\,|\,k)_w + (n-w-1\,|\,k-1)_w \qquad \text{for } n \geqslant w+1,\ k \geqslant 1, \tag{4}$$

the "initial values" being

$$(n\,|\,0) = 1 \quad \text{for } 1 \leqslant n \leqslant w, \qquad (n\,|\,k)_w = 0 \qquad \text{for } 0 \leqslant n \leqslant w,\ k \geqslant 1, \tag{5}$$

and

$$(0\,|\,0)_w = w + 1.$$

* Department of Mathematics, McGill University, 805 Sherbrooke Street West, Montreal, Quebec, Canada H3A 2K6.

An immediate consequence is that the generating function (polynomial of degree $[n/(w+1)]$)

$$f_n^{(w)}(z) = \sum_{k=0} (n \mid k)_w z^k$$

satisfies (and is determined by) the recurrence

$$f_n^{(w)}(z) = f_{n-1}^{(w)}(z) + z f_{n-w-1}^{(w)}(z) \quad \text{for } n \geqslant w+1, \tag{6}$$

with "initial values"

$$f_n^{(w)}(z) = 1 \quad \text{for } n = 1, 2, \ldots, w, \tag{7}$$

and

$$f_0^{(w)}(z) = w + 1. \tag{8}$$

Furthermore, it follows that

$$\begin{aligned} \sum_{n=0} f_n^{(w)}(z) x^n &= \frac{w+1-wz}{1-x-x^{w+1}z} \\ &= \frac{1}{1-\alpha_1 x} + \frac{1}{1-\alpha_2 x} + \cdots + \frac{1}{1-\alpha_{w+1} x} \\ &= \sum_{n=0} (\alpha_1^n + \alpha_2^n + \cdots + \alpha_{w+1}^n) x^n, \end{aligned} \tag{9}$$

where

$$1 - x - x^{w+1}z = (1-\alpha_1 x)(1-\alpha_2 x) \ldots (1-\alpha_{w+1} x),$$

i.e., $\alpha_1, \alpha_2, \ldots, \alpha_{w+1}$ are roots of the polynomial $x^{w+1} - x^w - z$ and are determined by

$$\begin{gathered} \alpha_1 + \alpha_2 + \cdots + \alpha_{w+1} = 1, \\ \sum_{i_1 < \cdots < i_j} \alpha_{i_1}\alpha_{i_2} \ldots \alpha_{i_j} = 0 \quad \text{for } j = 2, 3, \ldots, w, \\ \alpha_1 \alpha_2 \ldots \alpha_{w+1} = (-1)^w z, \end{gathered} \tag{10}$$

or equivalently (by the theory of symmetric functions; see e.g., [1, Chapter 11])

$$\begin{gathered} \alpha_1^i + \alpha_2^i + \cdots + \alpha_{w+1}^i = 1 \quad \text{for } i = 1, 2, \ldots, w, \\ \alpha_1^{w+1} + \alpha_2^{w+1} + \cdots + \alpha_{w+1}^{w+1} = 1 + (w+1)z. \end{gathered} \tag{11}$$

Equating coefficients of $x^n$ in (9), we have, for $n \geqslant 0$,

$$\sum_{k=0} (n \mid k)_w z^k = \alpha_1^n + \alpha_2^n + \cdots + \alpha_{w+1}^n,$$

or, for $n \geqslant 1$,

$$\sum_{k=0} \binom{n-kw}{k} \frac{n}{n-kw} z^k = \alpha_1^n + \alpha_2^n + \cdots + \alpha_{w+1}^n, \tag{12}$$

where $\alpha_1, \alpha_2, \ldots, \alpha_{w+1}$ satisfy (10) or (11).

When $w = 0$, (12) is just the binomial identity. When $w = 1$, (12) is the well-known identity, for $n \geqslant 1$,

$$\sum_{n=0} \binom{n-k}{k} \frac{n}{n-k} z^k = \alpha^n + \beta^n, \qquad \alpha + \beta = 1, \quad \alpha\beta = -z \tag{13}$$

[2, identity (1.64)], which has many interesting variations and special cases (e.g., identities (1.63), (1.64), (1.66), (1.68) in [2]). In particular, taking $\alpha = (1 + \sqrt{5})/2$ and $\beta = (1 - \sqrt{5})/2$, (13) becomes

$$\begin{aligned}\sum_{k=0} \binom{n-k}{k} \frac{n}{n-k} &= \frac{(1+\sqrt{5})^n + (1-\sqrt{5})^n}{2^n}, \qquad n \geqslant 1,\\ &= L_n,\end{aligned}$$

where $L_n$, the $n$th Lucas number, is usually defined by (6), (7), (8) with $w = 1$ and $z = 1$.

Taking $\alpha = t/(t+1)$ and $\beta = 1/(t+1)$, (13) takes the form

$$\sum_{k=0} (-1)^k \binom{n-k}{k} \frac{t^k}{(1+t)^{2k}} = \frac{t^n + 1}{(t+1)^n}, \qquad n \geqslant 1,$$

which for $t = 1$ and $t = -\frac{3}{2}$ are identities (1.65) and (1.69) in [2], while $t = -2$ yields

$$\sum_{k=0} \binom{n-k}{k} \frac{n}{n-k} 2^k = 2^n + (-1)^n, \qquad n > 1.$$

When $w = 2$, (12) is

$$\sum_{k=0} \binom{n-2k}{k} \frac{n}{n-2k} z^k = \alpha^n + \beta^n + \gamma^n, \qquad n \geqslant 1,$$

with

$$\alpha + \beta + \gamma = 1, \qquad \alpha^2 + \beta^2 + \gamma^2 = 1, \qquad \alpha^3 + \beta^3 + \gamma^3 = 1 + 3z. \tag{14}$$

In terms of the parameter $t = \alpha/\beta$ we find that

$$\alpha = \frac{t(t+1)}{t^2+t+1}, \qquad \beta = \frac{t+1}{t^2+t+1}, \qquad \gamma = -\frac{t}{t^2+t+1}, \tag{15}$$

and hence, for $n \geqslant 1$,

$$\sum_{k=0} (-1)^k \binom{n-2k}{k} \frac{n}{n-2k} \left[\frac{t^2(t+1)^2}{(t^2+t+1)^3}\right]^k = \frac{t^n(t+1)^n + (t+1)^n + (-t)^n}{(t^2+t+1)^n}.$$

For example, $t = 1$ yields

$$\sum_{k=0} (-1)^k \binom{n-2k}{k} \frac{n}{n-2k} \left(\frac{4}{27}\right)^k = \frac{2^{n+1} + (-1)^n}{3^n}, \qquad n \geqslant 1;$$

$t = 2$ yields

$$\sum_{k=0} (-1)^k \binom{n-2k}{k} \frac{n}{n-2k} \left( \frac{6^2}{7^3} \right)^k = \frac{6^n + 3^n + (-2)^n}{7^n}, \qquad n \geqslant 1;$$

and $t = 3$ yields

$$\sum_{k=0} (-1)^k \binom{n-2k}{k} \frac{n}{n-2k} \left( \frac{12^2}{13^3} \right)^k = \frac{12^n + 4^n + (-3)^n}{13^n}, \qquad n \geqslant 1.$$

This analysis, which we applied to the "circular" $k$-choices, can also be applied to linear $k$-choices. It is well known [3, (23)] that the number of $k$-choices (1) satisfying (2) is

$$\begin{array}{ll} \binom{n-(k-1)w}{k} & \text{if } 0 \leqslant k \leqslant \frac{n+w}{w+1}, (n,k) \neq (0,0), \\ 0 & \text{if } \frac{n+w}{w+1} < k. \end{array}$$

It is more convenient to work with the numbers

$$[n \mid k]_w = \begin{cases} \binom{n-kw}{k} & \text{if } 0 \leqslant k \leqslant \frac{n}{w+1}, \\ 0 & \text{if } 0 \leqslant \frac{n}{w+1} < k, \end{cases}$$

which satisfy (3), (4), (5) and

$$[0|0]_w = 1.$$

The generating function

$$h_n^{(w)}(z) = \sum_{k=0} [n \mid k]_w z^k$$

satisfies (6), (7), and $h_0^{(w)}(z) = 1$, and it follows that

$$\begin{aligned} \sum_{n=0} h^{(w)}(z) x^n &= \frac{1}{1 - x - x^{w+1} z} \\ &= \frac{A_1}{1-\alpha_1 x} + \frac{A_2}{1-\alpha_2 x} + \cdots + \frac{A_{w+1}}{1-\alpha_{w+1} x} \\ &= \sum_{n=0} (A_1 \alpha_1^n + A_2 \alpha_2^n + \cdots + A_{w+1} \alpha_{w+1}^n) x^n, \end{aligned} \tag{16}$$

where the $\alpha_i$'s and $A_j$'s are determined by (11) and

$$\alpha_1^i A_1 + \alpha_2^i A_2 + \cdots + \alpha_{w+1}^i A_{w+1} = 1 \quad \text{for } i = 0, 1, 2, \ldots, w. \tag{17}$$

Equating coefficients of $x^n$ in (16), we have

$$\sum_{k=0} \binom{n-kw}{k} z^k = A_1 \alpha_1^n + A_2 \alpha_2^n + \cdots + A_{w+1} \alpha_{w+1}^n \quad \text{for } n \geqslant 0. \tag{18}$$

When $w = 0$, (18) is the binomial identity. When $w = 1$, (18) is

$$\sum_{k=0} \binom{n-k}{k} z^k = A\alpha^n + B\beta^n,$$

with

$$A + B = 1, \quad \alpha A + \beta B = 1, \quad \alpha + \beta = 1, \quad \alpha\beta = -z,$$

so that

$$A = \frac{\alpha}{2\alpha - 1} = \frac{\alpha}{\alpha - \beta}, \qquad B = \frac{\beta}{2\beta - 1} = \frac{\beta}{\beta - \alpha},$$

or, for $n \geqslant 0$,

$$\sum_{k=0} \binom{n-k}{k} z^k = \frac{\alpha^{n+1} - \beta^{n+1}}{\alpha - \beta}, \qquad \alpha + \beta = 1, \quad \alpha\beta = -z,$$

a well-known identity [2, identity (1.61)] which has many interesting variations and special cases (see, for example, identities (1.60), (1.74), (1.75), (1.62), (1.71), (1.72), (1.73) in [2]).

When $w = 2$, (18) is

$$\sum_{k=0} \binom{n-2k}{k} z^k = A\alpha^n + B\beta^n + C\gamma^n,$$

with $\alpha$, $\beta, \gamma$ determined by (14), and $A, B, C$ by

$$A + B + C = 1, \qquad \alpha A + \beta B + \gamma C = 1, \qquad \alpha^2 A + \beta^2 B + \gamma^2 C = 1.$$

Solving for $A, B, C$ in terms of $\alpha$, $\beta, \gamma$, we find that

$$A = \frac{\alpha}{3\alpha - 2} = \frac{\alpha^2}{(\alpha - \beta)(\alpha - \gamma)}$$

$$B = \frac{\beta}{3\beta - 2} = \frac{\beta^2}{(\beta - \alpha)(\beta - \gamma)}$$

$$C = \frac{\gamma}{3\gamma - 2} = \frac{\gamma^2}{(\gamma - \alpha)(\gamma - \beta)},$$

and hence, for $n \geqslant 0$,

$$\sum_{k=0} \binom{n-2k}{k} z^k = \frac{\alpha^{n+1}}{3\alpha - 2} + \frac{\beta^{n+1}}{3\beta - 2} + \frac{\gamma^{n+1}}{3\gamma - 2}$$

$$= \frac{(\gamma - \beta)\alpha^2 + (\alpha - \gamma)\beta^2 + (\beta - \alpha)\gamma^2}{(\gamma - \alpha)(\gamma - \beta)(\beta - \alpha)}.$$

In terms of the parameter $t = \beta/\alpha$, $\alpha$, $\beta, \gamma$ are given by (15), and hence

$$\sum_{k=0} \binom{n-2k}{k} z^k = \frac{1}{(t^2 + t + 1)^n} \left\{ \frac{t^{n+1}(t+1)^{n+1}}{t^2 + t - 2} + \frac{(t+1)^{n+1}}{2t^2 - t - 1} \right.$$

$$\left. + (-1)^n \frac{t^{n+1}}{2t^2 + 5t + 2} \right\},$$

$$z = -\frac{t^2(t+1)^2}{(t^2 + t + 1)^3}.$$

Special instances are: with $t = 2$

$$\sum_{k=0} (-1)^k \binom{n-2k}{k} \left(\frac{6^2}{7^3}\right)^k$$
$$= \frac{1}{5 \times 7^n} \left\{5 \times 2^{n-1} \times 3^{n+1} - 3^{n+1} + (-1)^n 2^{n-1}\right\}, \qquad n \geqslant 0;$$

with $t = 3$

$$\sum_{k=0} (-1)^k \binom{n-2k}{k} \left(\frac{12^2}{13^3}\right)^k$$
$$= \frac{1}{35 \times 13^n} \left\{7 \times 2^{2n+1} \times 3^{n+1} - 5 \times 2^{2n+1} + (-1)^n 3^{n+1}\right\}.$$

It is not difficult to solve (17) for the $A_i$'s. Note that

$$\alpha_i^{w+1} - \alpha_i^w = z = (-1)^w \alpha_1 \alpha_2 \ldots \alpha_{w+1} \quad \text{for } i = 1, 2, \ldots, w+1, \tag{19}$$

so

$$\prod_{j \neq i} \alpha_j = (-1)^w \alpha_i^{w-1}(\alpha_i - 1) \quad \text{for } i = 1, 2, \ldots, w+1.$$

Also, differentiating

$$x^{w+1} - x^w - z = (x - \alpha_1)(x - \alpha_2) \ldots (x - \alpha_{w+1})$$

and then substituting $x = \alpha_i$, we have

$$(w+1)\alpha_i^w - w\alpha_i^{w-1} = \prod_{j \neq i} (\alpha_i - \alpha_j) \quad \text{for } i = 1, 2, \ldots, w+1. \tag{20}$$

Furthermore

$$\begin{vmatrix} 1 & 1 & \ldots & 1 \\ \alpha_1 & \alpha_2 & \ldots & \alpha_{w+1} \\ \vdots & \vdots & & \vdots \\ \alpha_1^w & \alpha_2^w & \ldots & \alpha_{w+1}^w \end{vmatrix} = \prod_{s<r} (\alpha_r - \alpha_s), \tag{21}$$

and (replacing the $i$th column by 1's)

$$\begin{vmatrix} 1 & \ldots & 1 & 1 & 1 & \ldots & 1 \\ \alpha_1 & \ldots & \alpha_{i-1} & 1 & \alpha_{i+1} & \ldots & \alpha_{w+1} \\ \vdots & & \vdots & \vdots & \vdots & & \vdots \\ \alpha_1^w & \ldots & \alpha_{i-1}^w & 1 & \alpha_{i+1}^w & \ldots & \alpha_{w+1}^w \end{vmatrix}$$
$$= \begin{vmatrix} 1 & \ldots & 1 & 1-w & 1 & \ldots & 1 \\ \alpha_1 & \ldots & \alpha_{i-1} & \alpha_i & \alpha_{i+1} & \ldots & \alpha_{w+1} \\ \vdots & & \vdots & \vdots & \vdots & & \vdots \\ \alpha_1^w & \ldots & \alpha_{i-1}^w & \alpha_i^w & \alpha_{i+1}^w & \ldots & \alpha_{w+1}^w \end{vmatrix} \tag{22}$$

(obtained by subtracting from the $i$th column all the other columns), and this is equal (for $i = 1, 2, \ldots, w+1$) to

$$\sum_{s<r} (\alpha_r - \alpha_s) - w(-1)^{i-1} \prod_{j \neq i} \alpha_j \prod_{\substack{s<r \\ r \neq i \\ s \neq i}} (\alpha_r - \alpha_s). \tag{23}$$

Of course $A_i$ is equal to the determinant (22) (its value is given by (23)) divided by determinant (21), yielding

$$A_i = 1 - w(-1)^{i-1} \prod \alpha_j / \prod (\alpha_i - \alpha_s) \prod (\alpha_r - \alpha_i)$$

or using (19) and (20),

$$A_i = 1 - \frac{w(-1)^{i-1}(-1)^w \alpha_i^{w-1}(\alpha_i - 1)}{(-1)^{w+1-i} \alpha_i^{w-1}((w+1)\alpha_i - w)}$$

$$= \frac{\alpha_i}{(w+1)\alpha_i - w}.$$

Thus, for $n \geqslant 0$,

$$\sum_{k=0} \binom{n-kw}{k} z^k = \frac{\alpha_1^{n+1}}{(w+1)\alpha_1 - w} + \cdots + \frac{\alpha_{w+1}^{n+1}}{(w+1)\alpha_{w+1} - w},$$

where

$$\alpha_1^i + \alpha_2^i + \cdots + \alpha_{w+1}^i = 1 \quad \text{for } i = 1, 2, \ldots, w, \qquad z = (-1)^w \alpha_1 \alpha_2 \ldots \alpha_{w+1}.$$

The recurrences for $(n \mid k)_w$ and $[n \mid k]_w$ are the same though the initial conditions are different. Thus we are led to consider any numbers $\{b_w(n,k)\}$, $n, k = 0, 1, 2, \ldots,$ determined by the recurrence

$$b_w(n,k) = b_w(n-1,k) + b_w(n-w-1,k-1), \qquad n \geqslant w+1, \quad k \geqslant 1,$$

$$b_w(n,0) = b_w(n-1,0), \qquad n \geqslant w+1,$$

with the initial conditions being the values assigned to $b_w(n,k)$ for $0 \leqslant n \leqslant w$, $k \geqslant 0$, subject to the condition that for each $n$ only finitely many $b_w(n,k)$, $k = 0, 1, 2, \ldots,$ are not 0. The generating function

$$g_n^{(w)}(z) = \sum_{k=0} b_w(n,k) z^k, \qquad n = 0, 1, 2, \ldots,$$

satisfies the recurrence

$$g_n^{(w)}(z) = g_{n-1}^{(w)}(z) + z g_{n-w-1}^{(w)}(z), \qquad n \geqslant w+1,$$

and it follows that

$$\sum_{n=0} g_n^{(w)}(z) x^n = \frac{g_0^{(w)}(z) + \left(g_1^{(w)}(z) - g_0^{(w)}(z)\right)x + \cdots + \left(g_w^{(w)}(z) - g_{w-1}^{(w)}(z)\right)x^w}{1 - x - x^{w+1}z}$$

$$= \frac{A_1}{1 - \alpha_1 x} + \frac{A_2}{1 - \alpha_2 x} + \cdots + \frac{A_{w+1}}{1 - \alpha_{w+1} x}$$

$$= \sum_{n=0} (A_1 \alpha_1^n + A_2 \alpha_2^n + \cdots + A_{w+1} \alpha_{w+1}^n) x^n,$$

i.e., with $\alpha_1, \alpha_2, \ldots, \alpha_{w+1}$ and $A_1, A_2, \ldots, A_{w+1}$ determined by

$$\begin{aligned}
\alpha_1 + \alpha_2 + \cdots + \alpha_{w+1} &= 1, & A_1 + A_2 + \cdots + A_{w+1} &= g_0^{(w)}(z),\\
\alpha_1^2 + \alpha_2^2 + \cdots + \alpha_{w+1}^2 &= 1, & \alpha_1 A_1 + \alpha_2 A_2 + \cdots + \alpha_{w+1} A_{w+1} &= g_1^{(w)}(z),\\
\alpha_1^3 + \alpha_2^3 + \cdots + \alpha_{w+1}^3 &= 1, & \alpha_1^2 A_1 + \alpha_2^2 A_2 + \cdots + \alpha_{w+1}^2 A_{w+1} &= g_2^{(w)}(z),\\
&\vdots & &\vdots\\
\alpha_1^w + \alpha_2^w + \cdots + \alpha_{w+1}^w &= 1, & \alpha_1^w A_1 + \alpha_2^w A_2 + \cdots + \alpha_{w+1}^w A_{w+1} &= g_w^{(w)}(z),
\end{aligned}$$

$$\alpha_1 \alpha_2 \ldots \alpha_{w+1} = (-1)^w z,$$

we have the identity

$$\sum_{k=0} b_w(n,k) z^k = A_1 \alpha_1^n + \cdots + A_{w+1} \alpha_{w+1}^n, \qquad n = 0, 1, 2, \ldots,$$

which contains identities (12) and (18) as special cases.

## References

[1] Archbold, J. W., *Algebra*. Pitman and Sons, London 1958.

[2] Gould, H. W., *Combinatorial Identities*. (Published and sold privately by the author.) Morgantown, W. Va. 1972.

[3] Moser, W. O. J. and Abramson, Morton, Enumeration of combinations with restricted differences and cospan, *J. Comb. Theory* **7** (1969), 162–170.

[4] Riordan, J., *An Introduction to Combinatorial Analysis*, Wiley, New York 1958.

# Binary Views of Ternary Codes[1]

Harold N. Ward*

## 1. Introduction

Recently Vera Pless, N. J. A. Sloane, and I completed the classification of ternary self-dual codes of length 20 [7]. We produced most of the codes by building on known codes of shorter length (the techniques are outlined and applied to codes of length 16 in a paper of Conway, Pless, and Sloane [1]). However, we used a different method for finding those codes having minimum weight 6. It is based on regarding the words of weight 6 as binary words and then examining the resulting set of binary vectors. The second section of this paper contains a summary of the results used in this approach; a detailed exposition will appear elsewhere [9]. Since the method links binary and ternary codes, the third section presents applications to the two most conspicuous of such linked codes, the ternary extended Golay code of length 12 and the binary extended Golay code of length 24.

## 2. Center Sets and Ternary Codes

A linear code of length $n$ over the finite field $GF(q)$ of $q$ elements is a subspace of $GF(q)^n$, the space of $n$-tuples with entries in $GF(q)$ (the recent book by MacWilliams and Sloane [4] is an encyclopedic reference). The members of $GF(q)^n$ are called words, and members of a code, codewords. The product $ab$ of two words $a$ and $b$ is their componentwise product, $a$ and $b$ being said to overlap on the set of coordinate positions where neither has a 0. The sum of the component products is the dot product $a \cdot b$. The weight $\mathrm{wt}(a)$ of a word $a$ is the

[1] This research was supported by National Science Foundation Grant MCS 78-01458.

* Department of Mathematics, University of Virginia, Charlottesville, Virginia 22903, U.S.A.

number of nonzero entries in $a$. This paper concerns the interplay between ternary ($q = 3$) and binary ($q = 2$) codes; for a ternary word $a$ the support $|a|$ of $a$ is the binary word of the same length obtained by changing the $-1$'s in $a$ to 1's and reading the resulting entries as though they were in $GF(2)$. It is also convenient to think of $|a|$ as the subset of coordinate positions where $a$ has nonzero entries.

Let $C$ be a linear self-orthogonal ternary code of length $n$ and minimum weight 6. That is, $a \cdot b = 0$ for $a$ and $b$ in $C$, and 6 is the smallest nonzero weight among codewords (all such weights will be multiples of 3). It is easy to verify that

$$\mathrm{wt}(a+b) + \mathrm{wt}(a-b) = 2(\mathrm{wt}(a) + \mathrm{wt}(b)) - 3\,\mathrm{wt}(ab),$$

so that if $a$ and $b$ have weight 6 in $C$,

$$\mathrm{wt}(a+b) + \mathrm{wt}(a-b) = 24 - 3\,\mathrm{wt}(ab).$$

If $|a| = |b|$, then $\mathrm{wt}(ab) = 6$ and $a = \pm b$. The weight of $ab$ cannot be 5, since one of $a \pm b$ would then have weight 3; nor can it be 1, since $a \cdot b$ is the sum of the components of $ab$. As

$$\begin{aligned}\mathrm{wt}(|a| + |b|) &= \mathrm{wt}(|a|) + \mathrm{wt}(|b|) - 2\,\mathrm{wt}(|a|\,|b|)\\ &= 12 - 2\,\mathrm{wt}(ab),\end{aligned}$$

$|a| + |b|$ has weight 6 exactly when $\mathrm{wt}(ab) = 3$, the only available odd weight for $ab$. In that case $|a| + |b|$ is the support of one of $a \pm b$ (the other has weight 9). Thus if $H$ is the set of *hexads*, or supports of words of weight 6, from $C$, then $H$ has the property that for $x$ and $y$ in $H$, $x + y$ will be in $H$ if and only if $x \cdot y = 1$.

$H$ is an example of a *center set*: if $V$ is a finite-dimensional vector space over $GF(2)$ carrying a symplectic form $\phi$ that may be degenerate, a center set $J$ in $V$ is a nonempty subset of nonzero members of $V$ with the property that if $x$ and $y$ are in $J$, $x + y$ is in $J$ exactly when $\phi(x, y) = 1$. For $H$, $V$ is the even subspace of $GF(2)^n$ consisting of the words of even weight, and $\phi$ is the dot product restricted to that space.

The members of a center set $J$ can be taken as the vertices of an undirected graph, $x$ and $y$ being connected when $\phi(x, y) = 1$ (so that $x$, $y$, and $x + y$ form a triangle). The group $G$ associated to $J$ is the group generated by the transvections

$$x \to x + \phi(x, y)\,y$$

for $y$ in $J$ (some of these may be trivial). $G$ preserves $J$, and the connected components of $J$ are the orbits of $G$ on $J$.

Suppose now $J$ is a connected center set; there is no great loss in assuming that $J$ spans its space $V$. $J$ then actually consists of all the centers of the transvections in $G$ (the reason for the name "center set"). On the quotient space $\bar{V} = V/\mathrm{rad}\,\phi$ of $V$ by the radical of $\phi$, let $\bar{G}$ be the group induced by $G$. $\bar{G}$ acts irreducibly on $\bar{V}$ and leaves invariant the form $\bar{\phi}$ induced by $\phi$ (as well as the image $\bar{J}$ of $J$). According to McLaughlin's classification of groups generated by transvections [5], $\bar{G}$ must be one of the following groups:

(1) the symplectic group of the form $\bar{\phi}$;
(2) the full orthogonal group of a quadratic form having $\bar{\phi}$ as the corresponding symplectic form;

(3) a symmetric group acting on $\bar{V}$ in one of the representations constructed by Dickson [2].

The symmetric case (3) actually includes some of the possibilities in the other two cases. Bearing that in mind, one can show that the connected center sets arising as components of the center set of hexads of a ternary self-orthogonal code of minimum weight 6 are all of symmetric type [9, Theorem 5.2].

In the symmetric case, $\bar{V}$ with the action of a symmetric group $S_m$ on $m$ letters (as $\bar{G}$) can be described this way: let $S_m$ act on $GF(2)^m$ by permuting coordinates (think of $GF(2)^m$ as the set of subsets of an $m$-set). $S_m$ preserves the dot product and the even subspace. The dot product yields a symplectic form on that space, as before, and $\bar{V}$ is the quotient of the even subspace by the radical of the form. This radical is 0 if $m$ is odd and the span of the all-one word if $m$ is even. $\bar{J}$ is the image of the set of pairs (words with two nonzero entries).

These results lead to the corresponding center sets (up to isometry; that is, a center set of symmetric type will be the image of one of those described by an isometry mapping the corresponding $V$ onto the vector space of the given center set). For any connected center set $J$, the set

$$T = \{t \in \operatorname{rad}\phi \mid x + t \in J\}$$

for a given $x$ in $J$ is a subspace of $\operatorname{rad}\phi$ that is actually independent of $x$. Let $E$ be the even subspace of $GF(2)^m$, and $P$ the set of pairs (as words). Take $e$ to be the all-one word, and for even $m$ let $E^*$ be $E/\langle e\rangle$, and $P^*$ the image of $P$ in $E^*$. $E$ and $E^*$ have the symplectic forms suggested above. Let $R$ be any finite-dimensional space over $GF(2)$, carrying the zero form as a degenerate symplectic form. Then these are the connected center sets of symmetric type having more than just one member [9, Theorem 4.1]:

(i) $V = E \perp R$ (orthogonal direct sum), $J = \{p + r \mid p \in P, r \in R\}$, $T = R$ ($m \geqslant 3$, $m \neq 4$).
(ii) $V = E \perp R$, $J = \{p + r, p + r + e \mid p \in P,\ r \in R\}$, $T = \langle e\rangle + R$ ($m$ even, $m \geqslant 8$).
(iii) $V = E^*$, $J = P^*$, $T = 0$ ($m$ even, $m \geqslant 8$).

The graph for $J$ will have $|J| = \binom{m}{2}|T|$ vertices, and each vertex will have valence $(2m-4)|T|$.

## 3. The Golay Codes

The extended ternary Golay code is a self-orthogonal linear code over $GF(3)$ of length 12 and dimension 6, having minimum weight 6. It is discussed, along with references, by MacWilliams and Sloane [4, Chapter 20]. The code is unique in the sense that any linear code with that same verbal description can be obtained from the Golay code by permuting and scaling components [6].

If $C$ is such a code, $C$ contains 264 words of weight 6 (as can be shown apart from the uniqueness), and the corresponding center set $H$ consists of 132 hexads. The orthogonal code $C^{\perp}$, the set of words orthogonal to all members of $C$, coincides with $C$ (and $C$ is called self-dual, somewhat inaccurately). From this it follows that if $a$ is a word of weight 6 in $C$, the subcode made up of words of $C$ having 0's on $|a|$ has dimension 1 (and thus is the span of a complementary word of weight 6). With $C_0$ denoting the projection of that subcode on the six positions outside $|a|$, $C_0^\perp$ is the projection of $C^\perp = C$. $C_0^\perp$ will have $2\binom{6}{3} = 40$ words of weight 3, each of which is the projection of two words in $C$ of weight 6 (and one of weight 9). Consequently $\mathrm{wt}(ab) = 3$ for 80 words $b$ of weight 6 in $C$. Thus the graph of $H$ is regular with valence 40.

For a valence of 40 the number of vertices in a connected center set of symmetric type can be 84, 132, or 231 ($|T|$ is a power of 2). It can only be, then, that $H$ is connected. Because $H$ is contained in $E$, the even space of $GF(2)^{12}$, an 11-dimensional space, $H$ must be equivalent to the center set $J$ described under (ii) at the end of Section 2, with $m = 12$ and $R = 0$. There is thus an isometry $x \to x'$ of $E$ with $J' = H$.

A similar discussion can be used for self-dual ternary linear codes of length 20 and minimum weight 6 to show that the graph of the center set of hexads is regular with valence 8. Although it is not connected, there are only two possible component sizes: 12 and 15. The resulting restrictions and descriptions for the connected components lead to the constructions of the various codes [7].

In the present situation, $H$ is a realization of $J$ by words of weight 6. There is actually only one such realization, up to coordinate permutations. To see that, let $e$ be the all-one word again, and abbreviate words by giving the coordinate positions where they have 1's (thus 12 stands for the word in $J$ with 1's in positions 1 and 2). For $i > 3$, the image $(123i)'$ of $123i$ is the sum of the two orthogonal hexads $(12)'$ and $(3i)'$ and therefore has weight 4 or 8. Let $f_i$ be the one of $(123i)'$ and $(123i)' + e$ having weight 4. As $f_i + f_j = (ij)'$ or $(ij)' + e$ is a hexad ($i \neq j$), $f_i f_j$ has weight 1, and $f_i$ and $f_j$ overlap in one position. Because of these numerical relations, the incidence structure obtained by calling the $f_i$ points ($4 \leqslant i \leqslant 12$), calling the 12 coordinate positions lines, and defining $f_i$ to be incident with a line if $f_i$ has a 1 in the corresponding position, is an affine plane with three points per line. Such a plane is unique, as is well known and easy to see [3, p. 6]. Any of the six hexads of the set

$$B = \{(12)', (13)', (23)', (12)' + e, (13)' + e, (23)' + e\},$$

when added to an $f_i$, gives another hexad; so each hexad of $B$ shares two 1's with $f_i$. That is, each point is on two of the six lines corresponding to the 1's in a hexad of $B$. This implies that the six lines are the lines of two parallel classes, and the six hexads of $B$ arise from all possible pairings of parallel classes. The hexads of $H$ are the members of $B$, the members of $B + f_i$, and the sums $f_i + f_j$, $f_i + f_j + e$, for $4 \leqslant i, j \leqslant 12$, $i \neq j$. Thus once the affine plane is labeled, $H$ is specified.

The center set $H$ in turn determines the Golay code (up to coordinate scaling): begin with the hexads $(12)'$, $(13)'$, and $(12)' + e$, arranged for convenience as

$$\begin{matrix} 1 & 1 & 1 & 1 & 1 & 1 & 0 & 0 & 0 & 0 & 0 & 0 \\ 1 & 1 & 1 & 0 & 0 & 0 & 1 & 1 & 1 & 0 & 0 & 0 \\ 0 & 0 & 0 & 0 & 0 & 0 & 1 & 1 & 1 & 1 & 1 & 1 \end{matrix}$$

These can be interpreted directly as the corresponding ternary words after scaling the ternary words and the coordinates. The words corresponding to the other hexads of $B$ come from the evident combinations. A hexad $(1i)'$ or $(1i)' + e$ $(i \geqslant 4)$ must have three 1's in common with each of the three displayed. For the corresponding ternary word to be orthogonal to the displayed ternary words, the three nonzero digits involved in the overlap in each case will have to be all 1's or all $-1$'s. Scaling the word so that the nonzero entries in the first six positions are 1's will force the entries in the second six to be 1's also (the exceptional possibilities all correspond to hexads in $B$). In other words, each hexad $(1i)'$ or $(1i)' + e$ can actually be taken to be a ternary word, too, with no $-1$'s. The remaining hexads now correspond to sums or differences of the ternary words created so far. Since more than a third of the Golay code has been obtained, the code is spanned by these words and thus forced.

If one didn't know the Golay code in advance, this discussion would provide both the code and its uniqueness. (As a matter of tactics, one should construct the ternary words corresponding to the hexads from $B$ and the hexads $(1i)'$ and $(1i)' + e$, $4 \leqslant i \leqslant 12$, as above, verify that they span a code of dimension 6, rule out words of weight 3 to get the proper minimum weight, and at the last claim that the existence of $H$ is thus established. This provides the code and $H$ together and avoids a direct check that $H$, as described by the affine plane, is really a center set.) Similar procedures were used in constructing the codes of length 20 mentioned before.

The center set $H$ can be used to obtain the binary extended Golay code of length 24 and dimension 12. That code is a self-dual code in which all words have weight divisible by 4 and the minimum weight is 8 (this code is also discussed by MacWilliams and Sloane [4, Chapters 2 and 20]; both Golay codes have an extensive literature). For the construction, use the isometry $x \to x'$ of $E$ given before, and form the subspace $C$ of $GF(2)^{24}$ made up of the words $(x, x')$ and $(x, x' + e)$, the pair symbol meaning that $x$ (from $E$) is put in the first 12 positions, and $x'$ or $x' + e$ in the second. The projection of $C$ on the first 12 positions is $E$, and with $(0, e)$ in $C$, $C$ has dimension 12. As $(x, x') \cdot (y, y') = x \cdot y + x' \cdot y' = 0$ and $(x, x') \cdot (0, e) = (0, e) \cdot (0, e) = 0$, $C$ is self-orthogonal. For $x$ in $J$, $(x, x')$ has weight 8. Because such words and $(0, e)$, of weight 12, span $C$, the self-orthogonality implies all words of $C$ have weights divisible by 4. Finally, a word of $C$ of weight 4 would have to be of the form $(a, b)$ with $a$ and $b$ both of weight 2. But $b$ is $a'$ or $a' + e$ and would actually have weight 6. So the minimum weight of $C$ is indeed 8.

If $f$ stands for the original isometry $x \to x'$ carrying $J$ onto $H$, and $g$ is another, the code constructed by using $g$ in place of $f$ will be equivalent to $C$. For if $*$ denotes the quotient mapping of $E$ onto $E/\langle e \rangle$, then $C$ can be described as the pairs $(x, y)$ for which $f(x)^* = y^*$. Now $g^{-1}f$ will be an isometry of $E$ preserving $J$, and $g^{-1}f$ induces one of $E/\langle e \rangle$ preserving $J^*$. By using the maximal totally connected subsets of $J^*$, which are just the sets $\{(ij)^* \mid j \neq i\}$, for each fixed $i$, one can show that such a map is induced by a member $s$ of the symmetric group (that is, a coordinate permutation). Then $f(x)^* = (gs(x))^*$; and $s$, acting on the first 12 coordinate positions, will carry $C$ onto the code constructed by using $g$ in place of $f$.

Now a monomial transformation preserving the ternary Golay code gives rise to a permutation of coordinates preserving $H$. Conjugation by $f$ yields an isometry of $E$ preserving $J$. If the corresponding member of the symmetric group is used on the first 12 positions and simultaneously the permutation preserving $H$ is used on the second 12, the combined permutation of the coordinates of $GF(2)^{24}$ preserves $C$. Thus the coordinate permutation group produced by the automorphism group of the ternary Golay code is injected into the automorphism group of the binary Golay code. These two permutation groups are, of course, the two Mathieu groups $M_{12}$ and $M_{24}$ [4]. In fact, $M_{12}$ now appears as the subgroup of $M_{24}$ stabilizing a word $(0, e)$ of weight 12 in the binary Golay code.

Finally, the uniqueness of the binary Golay code follows from what has been done, by the observation that if one takes a word of weight 12 in a binary code having the properties of the Golay code (such a word exists according to the weight distribution, which is again unique) and looks at the codewords meeting it in two positions, those words will meet the complementary word of weight 12 in six. These projections of weight 6 are readily shown to form a center set, which must be $H$ again, up to coordinate permutations. The code can then be assembled as was the Golay code.

The methods above can be used to obtain other properties of the codes and their groups; for example, the members of $H$ are just the blocks of the Steiner system associated with $M_{12}$. It should be mentioned that another approach to the passage from $M_{12}$ to $M_{24}$ by way of the Golay codes has been given by Rasala [8].

## References


[1] Conway, J. H., Pless, V., and Sloane, N. J. A., Self-dual codes over $GF(3)$ and $GF(4)$ of length not exceeding 16. *IEEE Trans. Information Theory* **IT-25** (1979), 312–322.

[2] Dickson, L. E., Representations of the general symmetric group as linear groups in finite and infinite fields. *Trans. Amer. Math. Soc.* **9** (1908), 121–148.

[3] Lüneburg, H., *Transitive Erweiterungen endlicher Permutationsgruppen*. Lecture Notes in Mathematics, Vol. 84. Springer-Verlag, Berlin 1969.

[4] MacWilliams, F. J. and Sloane, N. J. A., *The Theory of Error-Correcting Codes*. North-Holland, Amsterdam 1977.

[5] McLaughlin, J., Some subgroups of $SL_n(F_2)$. *Illinois J. Math.* **13** (1969), 108–115.

[6] Pless, V., On the uniqueness of the Golay codes. *J. Combinatorial Theory* **5** (1968), 215–228.

[7] Pless, V., Sloane, N. J. A., and Ward, H. N., Ternary codes of minimum weight 6 and the classification of the self-dual codes of length 20. *IEEE Trans. Information Theory* **IT-26** (1980), 305–316.

[8] Rasala, R., Split codes and the Mathieu groups. *J. Algebra* **42** (1976), 422–471.

[9] Ward, H. N., Center sets and ternary codes. *J. Algebra* **65** (1980), 206–224.